D0919089

Geography: Regions and Concepts

John Wiley & Sons

New York
Chichester
Brisbane
Toronto

Third Edition

Geography: Regions and Concepts

Harm J. deBlij
University of Miami

With a chapter by

Stephen S. Birdsall
University of North Carolina

Library of Congress Cataloging in Publication Data:

De Blij, Harm J
Geography, regions and concepts.

Includes indexes.
1. Geography—Text-books—1945-
I. Birdsall, Stephen S., joint author. II. Title.
G128.D42 1981 910 80-17961
ISBN 0-471-08015-2

Printed in the United States of America

10 9 8 7 6 5 4 3

Preface

This third edition of *Geography: Regions and Concepts* retains the structure of its predecessor, but contains substantial revisions of text and cartography. It is no longer necessary to elaborate on the rationale for this approach to the study of geographic regions and geographic concepts; the advantages of regional-conceptual linkages are well established.

Among major changes introduced are the following: (1) insertion of "ten major geographic qualities" of every region discussed; (2) the expansion of material on cultural geography in the Introduction; (3) the revision of text on physical geography, especially soils, also in the Introduction; (4) further amplification of the issue of national development in several chapters; (5) introduction of a more specific regional geography of the United Kingdom in the Europe chapter; (6) major rewriting of the North America chapter, giving greater stress to Canada and introducing new maps; (7) improved structuring of the chapter on North Africa and Southwest Asia including clearer regional definitions and discussions of Iran and Afghanistan; (8) revision of the essay on Southern Africa to place recent developments in geographic context; and (9) a general updating of text, tables, and maps where necessary.

This updating of the volume has taken several forms. A number of readers wrote to comment on the need for clear regional maps of North America similar to those accompanying other regional discussions, and these have been designed and inserted. All cartography was checked for current information (the energy map, for example, was completely redrawn). The Pin-yin form of spelling of all Chinese names was adopted for this edition, requiring a revision of maps not only of China but of other realms as well. The extensive table of data for individual countries in the Introduction was updated and new projections for the 1980s were added. Throughout the volume, incidental information in numerical form was checked for accuracy and currency. Every illustration was reconsidered and where it was possible to identify a superior or more effective photograph, a change was made.

As readers will perceive, numerous other items also received attention. The meaning of China's "Four Modernizations" is discussed in a new section. In the Middle America chapter, new material on tourism now alerts students to the cultural disadvantages as well as the material advantages of this "irritant" industry for developing countries. Advancing supranationalism in Europe, emerging regional-racial problems in Brazil, changing regionalisms in Canada, rising cultural tensions in Soviet border areas, and numerous other topics are newly introduced in geographic perspective. Since I changed the content (and to some extent the tone) of the North America chapter, the responsibility for this modification is mine, although my colleague Dr. S. S. Birdsall of the University of North Carolina, who wrote the original chapter in the first edition, approved the text.

Every chapter was subjected to an intensive and critical reading by regional specialists, and I am grateful for this assistance. Dr. Sen-dou Chang of the Department of Geography at the University of Hawaii commented in meticulous detail on the chapter on China and assisted me with the Pin-yin problem. Dr. David Kornhauser, also of the Univerity of Hawaii, gave me invaluable assistance with the essay on Japan. Dr. James W. King of the Department of Geography at the University of Utah made numerous useful suggestions on the African and Southwest Asian discussions. Dr. Roland Fuchs of the University of Hawaii annotated the chapter on the Soviet Union. Dr. Brian Murton made numerous helpful suggestions for the South Asia chapter. Dr. Donald Fryer gave me an enormous amount of his time in going through the material on Southeast Asia. Professor Eric Kemp Petiprin of the Department of Geography at the State University of New York at New Paltz wrote a lengthy commentary on the North America chapter and made additional suggestions as well.

This short list of helpful colleagues hardly does justice to the enormous amount of support I have again enjoyed since the appearance of the second edition. Sometimes a suggestion made in passing, a short note at the end of an examination, or some other informal comment can contribute importantly to a better book—but the revision file has no adequate record of it. So I thank again all those who have taken the trouble to comment on my previous efforts and trust that they will see their suggestions incorporated here.

I am also indebted to those on the staff of John Wiley and Sons who transformed 993 pages of my messy typing into this handsome volume. Editor Butch Cooper handled the logistics, coordinating my work with that of cartographers, reviewers, photo researchers, designers, and others. In the editor's office, Karen Grant facilitated the project in many ways. Eugene Patti was the manuscript editor; Malcolm Easterlin supervised the editing. The photographs were identified by Stella Kupferberg. The design was directed by Ed Burke. The production process was guided by Rosie Hirsch, and I am deeply grateful for her meticulous attention to detail and quality. Page make-up was handled by Blaise Zito Associates Inc.

Much of this book was written while I was teaching geography in a visiting capacity at the University of Hawaii. No brief acknowledgment could begin to measure my gratitude to my wife, Bonnie, who doubled the joy of every vivid rainbow and crimson sunset, and who halved the burden of every page.

Harm J. de Blij
Coral Gables, Florida
May 4, 1980

Contents

Geography: Regions and Concepts

Between desert and sown: a regional boundary demarcates the landscape.
Peru. (Loren McIntyre/Woodfin Camp)

Introduction

This is a book about the modern world's great culture realms, discussed in geographical perspective. Each of the major geographic realms of the human world (such as China, "Latin" America, "Black" Africa, or Southeast Asia) possesses a special combination of cultural, traditional, historical, and organizational qualities. These characteristic properties are imprinted on the landscape, giving each region its own atmosphere and social environment. Geographers take a particular interest in the way people have decided to arrange and order their living space. The street pattern of a traditional Arab town differs markedly from the layout of an Indian place of similar size. The fields and farms of China look quite unlike those of Africa. Thus the study of world realms provides an opportunity also to examine the concepts and ideas that form the basis of the field of geography. These are our twin objectives.

IDEAS AND CONCEPTS

- Regional concepts
- Scale
- Culture
- Cultural landscape
- Pleistocene cycles
- Natural environments
- Physiographic regions
- Development
- Population agglomeration
- World culture realms

Regional Geography of the World

Concepts of Regions

Modern concepts often are complicated constructions that require mathematical development, but we use others, almost without realizing it, in our everyday conversation. Among the fundamental concepts of geography are those involving the identification, classification, and analysis of regions. When we refer to some part of our country (the Midwest, for example), or to a distant area of the world (such as the Middle East), or even to a section of the town in which we may live, we employ a regional concept. We reveal our perception, our mental image of the region to which reference is made.

Everyone has some idea of what the word *region* means, and we use the regional concept frequently in its broadest sense as a frame of reference. But regional concepts are anything but simple. Take just one implication of a regional name just used: the Midwest. If the Midwest is indeed a region of the United States, then it must have limits. Those limits, however, are open to debate. In his book *North America* (1975), J. H. Paterson states that the Midwest includes the states of Ohio, Michigan, Indiana, Illinois, Wisconsin, Minnesota, Iowa, and Missouri, "but in the cultural sense it can also be said to include much of the area of heavy industry in Pennsylvania and West Virginia." Compare this definition to that of O. P. Starkey, J. L. Robinson, and C. S. Miller, who define the Midwest in their book *The Anglo-American Realm* (1975) as consisting of "most of the west North Central states (North and South Dakota, Nebraska, Kansas, Minnesota, Iowa, and Missouri) to which have been added the western parts of Wisconsin and Illinois." These two perceptions of the Midwest as an American region obviously differ. Does this invalidate the whole idea of an American Midwest? On the contrary: the apparent conflict arises from the use of different *criteria* to give specific meaning to a regional term that will always be a part of American cultural life. Your own personal impression of the Midwest as a region is based on certain properties you have reason to consider important. When you add to your information base, you may modify your definition. Regionalization is a geographer's means of classification, and regions, like classes, have their basis in established criteria. Classification schemes are open to change as new knowledge emerges, and so are regional definitions.

Regions obviously have *location.* Various means can be employed to identify a region's position on the globe, as the authors quoted above did when they enumerated the states that form part of their conception of the American Midwest. Often a region's name reveals much about its location. During the Vietnam War the name *Indochina* became familiar to us; it is a regional appellation identifying an area that has received cultural infusions from India and human migrations from China. Sometimes we have a particular landscape in mind when we designate a region—for example, the Amazon Basin or the Rocky Mountains. It would be possible, of course, to denote a region's location by reference to the earth's grid system and to record its latitude and longitude. That would give us its *absolute* location, but such a numerical index would not have much practical value. Location attains relevance only when it relates to other locations. Hence, many regional names give reference to other regions (*Middle* America, *Eastern* Europe, *Equatorial* Africa). This indicates a region's *relative* location, a much more meaningful and practical criterion.

Regions also have *area.* Again, this appears to be so obvious that it hardly requires emphasis, but some difficult problems are involved here. Some regions are identified as, for example, the (San Francisco) Bay Area, the Greater Chicago Area, the New York Metropolitan Area. Everyone probably would agree that such areas are focused on the urban concentration that lie at their heart, but what are the limits, the boundaries of these urban-centered regions? In quite another context, we use such terms as the *Corn Belt* (an agricultural region in the United States) and the *Sunbelt* (a zone across the southern United States that attracts a growing number of immigrants, including numerous persons of retirement age who seek to escape the rigors of northern winters). The areal or spatial extent of a region, whether Bay Area or Corn Belt, Midwest or Middle East, cannot be established and defined without reference to specific contents.

An overriding characteristic of a region's contents may be its *homogeneity* or sameness. Sometimes the landscape leaves no doubt where one region ends and another begins. In Egypt, the break between the green, irrigated, cultivated lands adjacent to the Nile River and the desert beyond is razor-sharp and all-pervading. On the map, the line representing that break is without question a regional boundary. Everything changes beyond that line—population density, vegetation, soil quality. But regional distinctions are not often so clear. The example of the United States Corn Belt provides a contrast. Traveling northward from Kentucky into Illinois or Indiana, you would undoubtedly be struck by the increasing number of fields under corn. Not all the farmland is under corn, even in the Corn Belt, so the difference between what you saw in Kentucky and Illinois is a matter of incidence. To define a Corn Belt and represent it on the map, it would be necessary to establish a criterion; for example, 50 percent or more of all the cultivated land must be under corn. The line so drawn would delimit an agricultural region, but in the landscape it would be far less evident than that delimiting Egypt's Nile Valley farmlands.

It is possible, of course, to increase the number of criteria, so that more than one condition must be satisfied before the region is delimited. To define a particular cultural region, such items as the use of a certain language, adherence to a specific religion, perhaps even the spatial patterns of architectural, artistic, and other traditions might be employed. Maps of the cultural geography of Canada, including those showing religious affiliation, dominant language used, land division, and settlement patterns, reveal the reality of Quebec as a discrete region within the greater Canadian framework.

A region can also be conceptualized as a *system.* Certain regions are not marked by internal sameness but by a particular activity, or perhaps a set of integrated activities, that connects its various parts. This is how it is possible to perceive of the Bay Area and similar urban areas as regions. A large city has a substantial surrounding area for which it supplies goods and services, from which it buys farm products, and with which it interacts in numerous ways. The city's manufacturers distribute their products wholesale to regional subsidiaries. Its newspapers sell in the nearby smaller towns. Maps showing the orientation of road traffic, the sources and destinations of telephone calls, the readership of newspapers, the audiences of radio stations, and other activities confirm the close relationship between the city and its tributary region or *hinterland.* Here again we have a region; this time it is not characterized by homogeneity but, instead, by the city-centered system of interaction that generates this, a *functional* region.

Regional Terminology

The internal sameness or homogeneity of regions can be expressed in human (cultural) as well as natural (physical) criteria. A country constitutes a political region, for within its boundaries certain conditions of nationality, law, and political tradition prevail. A natural region such as the Rocky Mountains or the Ozarks is expressed by the dominance of a particular physical landscape. Quebec is a cultural region; the Corn Belt is an agricultural region. Regions marked by internal homogeneity are grouped together as *formal* regions.

Regions conceptualized as spatial systems—for example, those centered on an urban core area, a node, a focus of regional interaction—are identified collectively as *functional* regions. Thus the formal region might be viewed as static, uniform, immobile; the functional region is dynamic, active, and the result of processes that continue to modify it.

This distinction between "formal" and "functional" regions is still debated among geographers. A formal region's sameness, some argue, is the result of processes just as the functional region is generated by processes. Perhaps formal regions are less affected by change, more durable, and therefore more visible, but they may not be fundamentally different from functional regions. Some easily accessible contributions to this discussion are cited in the bibliography at the end of this introduction.

Classifying Regions

Given the various qualities and properties of regions, their differences in dimensions and complexity, is it possible to establish a hierarchy, a ranking of regions based on some combination of their characteristics? Some geographers have proposed systems of classification, and although none of the results has gained general acceptance, it is interesting to see how they confronted this often frustrating problem. In a book entitled *Introduction to World Geography* (1977), R. H. Fuson suggests a comprehensive, seven-level regional hierarchy that would divide the earth into the European and non-European world and hence into realms, landscapes, superregions, regions, districts, and subregions. This system is complicated and quite difficult to apply consistently, but it underscores the elusive nature of the problem.

Cultural geographers often use a three-tier system. The culture *realm* (sometimes culture *world*) identifies the largest and most complex area that can be described as being unified by common cultural traditions. For example, North America (Canada and the United States) constitutes such a culture realm; Middle and South America together form another cultural realm. The great units on which this book is based (Europe, South Asia, Black Africa, and so on) all are, in this classification, culture realms.

Each of these culture realms, in turn, consists of an assemblage of culture *regions.* Within "Latin" America, Mexico is such a culture region, the Central American republics constitute a second region, and Brazil a third. In the European culture realm, Mediterranean Europe and Eastern Europe rank as culture regions.

The regions consist of *subregions.* Canada is a region within the North American realm; Quebec is a subregion within Canada. The Balkan states form a subregion within Eastern Europe.

Note that the regional and subregional areas tend to be

EFFECT OF SCALE

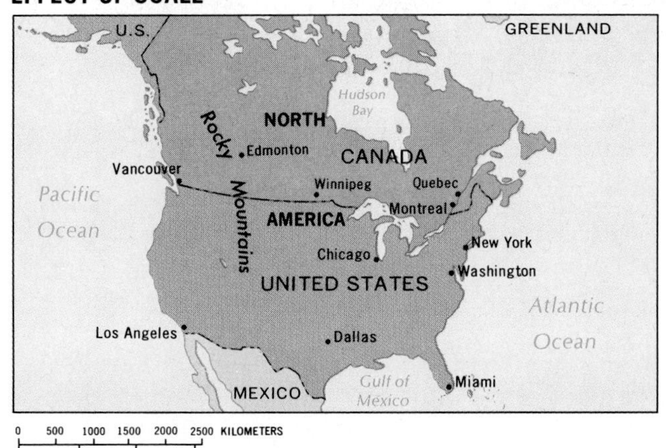

0 500 1000 1500 2000 2500 KILOMETERS
0 500 1000 1500 MILES
1:103,000,000

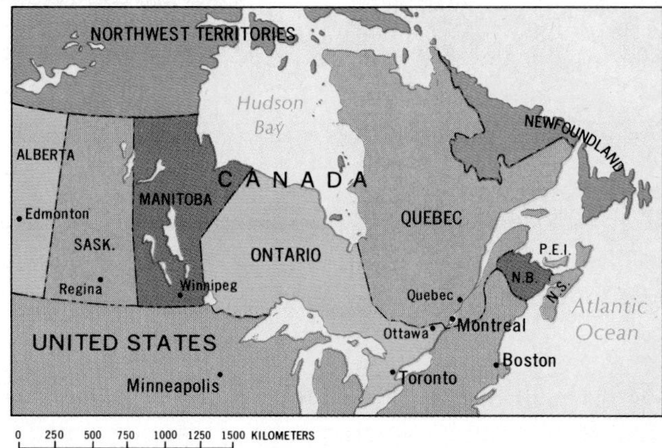

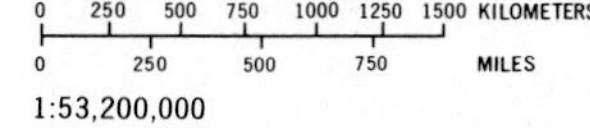

1:53,200,000

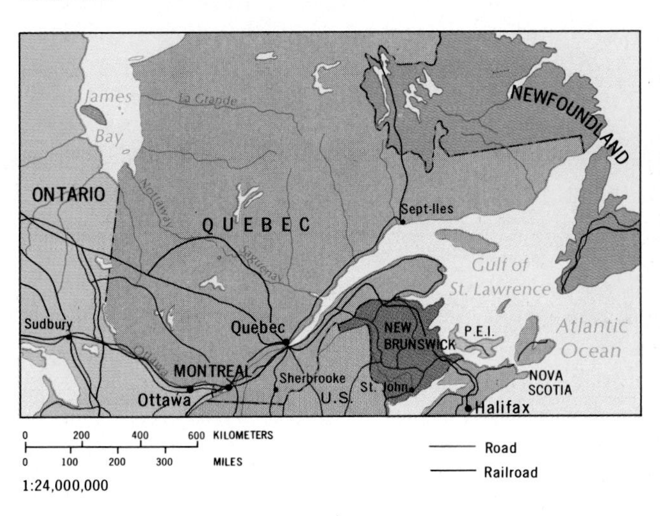

0 200 400 600 KILOMETERS
0 100 200 300 MILES
1:24,000,000

Road
Railroad

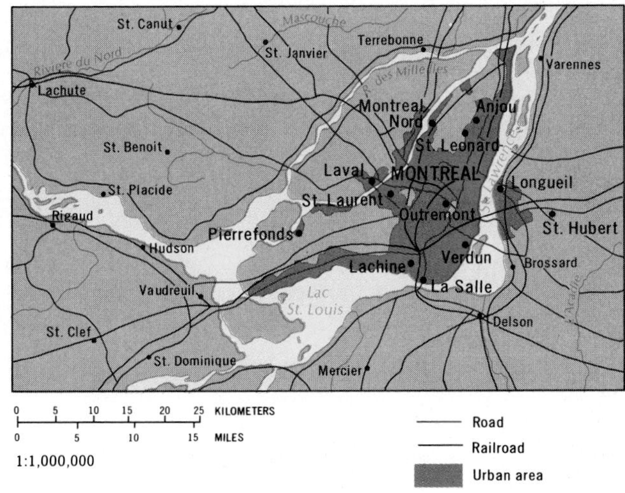

0 5 10 15 20 25 KILOMETERS
0 5 10 15 MILES
1:1,000,000

Road
Railroad
Urban area

Figure I-1

identified as countries or groups of countries, which can be misleading. The country names are used because of their convenient familiarity, but their boundaries do not necessarily coincide with those of the regions or subregions in question. Mexico's regional properties spill over into the North American realm. The Sahel—a distinct subregion in West Africa—lies across parts of several of that region's countries.

Although this hierarchy (realm, region, subregion) seems logical and convenient, the fact is that geographers have not adhered to it without exception. The term "region" still is substituted for "realm" and for "subregion" as well; and courses that deal with world realms still are called world *regional* geography courses. Indeed, the present book should be entitled *Geography: Realms and Concepts* if precision prevailed. Nevertheless, the three-level regional hierarchy has perhaps come closest to general adoption. It is worth keeping in mind, but it is important also to avoid confusion when inconsistent usage occurs in the literature.

Concepts of Scale

Regions can be conceptualized in various forms and at different levels of scale. Consider the four maps in Fig. I-1. On the first map (upper left) the North American realm is outlined, but very little spatial information can be provided, although the political boundary between Canada and the United States is shown. On the second map (upper right), East and Central Canada are depicted in sufficient detail to permit display of the provinces, several cities and towns, and some physical features (Manitoba's major lakes) not shown on the first map. The third map (lower left) shows the main surface communications of Quebec and immediate surroundings, the relative location of Montreal, and the St. Lawrence and James Bay drainage systems. The fourth map reveals the urban layout of Montreal and environs.

Each of the four maps has a scale designation, which can be shown as a bar graph (in kilometers and miles in this case) and as a fraction (1:103,000,000 on the first map). The fraction indicates that one unit of distance on the map (one inch or one centimeter) represents 103 million such units on the ground. The smaller the fraction, the smaller the scale of the map. Obviously the fraction on the first map (1:103,000,000) is the smallest of the four; that on the fourth map (1:1,000,000) is the largest. Comparing maps nos. 1 and 3, we find that on the *linear* scale, no. 3 has a fraction that is more than four times larger than no. 1. When it comes to *areal* representation, however, 1:24,000,000 is more than sixteen times larger than 1:103,000,000 because the linear difference prevails in both dimensions (the length *and* breadth of the map).

In this book it is obviously necessary to operate at relatively smaller scales. When studying regions or subregions in greater detail, our ability to specify criteria and to "filter" the factors we employ increases as we work at larger scales. On occasion that method will be used, for example, when urban centers are the topic of concern. But most of the time our view will be the more general—the small-scale view of the world's realms.

Concepts of Culture

The realms and regions to be discussed in the chapters that follow are defined by humanity's *cultures*. Geographers approach the study of culture from several vantage points, and one of these, the analysis of *cultural landscape,* is central to our regional interests. Therefore, we should look rather closely at the concept of culture. The word "culture" is not always used consistently in the English language, which can lead to some difficulties in establishing its scientific meaning. When we speak of a "cultured" individual we tend to mean someone with refined tastes in music and the arts, a highly educated, well-read person who knows and appreciates the "best" attributes of his or her society. But as a scientific term, culture refers not only to the music, literature, and arts of a society but also to all the other features of its way of life: prevailing modes of dress; routine living habits; food preferences; the architecture of houses as well as public buildings; the layout of fields and farms; and systems of education, government, and law. Thus culture is an all-encompassing term that identifies not only the whole life style of a people but also the prevailing values and beliefs.

This is not to suggest that anthropologists and other social scientists have not had problems with the concept of culture. If you read some of the basic literature in anthropology you will find that anthropologists have had as much difficulty with definitions of the culture concept as geographers have had with the regional concept. A culture may be the total way of life of a people—but is it their *actual* way of life ("the way the game is played") or the standards by which they give evidence of *wanting* to live, through their statements of beliefs and values ("the rules of the game")? There are strong differences of opinion on this, and as a result the various definitions have become quite complicated. Anthropologist E. Adamson Hoebel says in his *Anthropology: The Study of Man* (1972) that culture is "the integrated system of learned behavior patterns which are characteristic of the members of a society and which are not the result of biological inheritance . . . culture is not genetically predetermined; it is noninstinctive . . . [culture] is wholly the result of social invention and its transmitted and maintained solely through communication and learning." This definition raises still another question: how is culture carried over from one generation to the next? Is this entirely a matter of learning, as Hoebel insists, or are certain aspects of a culture indeed instinctive and in fact a matter of genetics? This larger question is the concern of sociobiologists and not cultural geographers, although some of its side issues, such as territoriality (an allegedly human instinct for territorial possessiveness) and proxemics (individual and collective preferences for nearness or distance in different societies) have spatial and therefore geographical dimensions.

But even without these theoretical additions, the culture concept remains difficult to define satisfactorily. In 1952 anthropologists A. L. Kroeber and C. Kluckhohn published a paper that identified no fewer than 160 definitions—all of them different—and from these they distilled their own: "Culture consists of patterns, explicit and implicit, of and for behavior and transmitted by symbols, constituting the distinctive achievements of human groups, including their embodiments in artifacts . . . the essential core of culture consists of traditional (that is, historically derived and selected) ideas and especially their attached values; culture systems may, on the one hand, be considered products of action, and on the other as conditioning elements of further action." Some of the definitions from which this one was distilled appear in the box on this page.

For our purposes it is enough to stipulate that culture consists of a people's beliefs (religious, political), institutions (legal, governmental), and technology (skills, equipment). This construction is a good deal broader than that adopted by many modern anthropologists, who now prefer to restrict the concept to the interpretation of human experience and behavior as products of symbolic meaning systems. It is important to remember that definitions of this kind are never final and absolute but, rather, arbitrary and designed for a particular theoretical purpose. The culture concept is defined to facilitate the explanation of human behavior. Anthropologists today tend to focus on what people know, on codes and values, on the "rules of the game." This was not always the case, as the

The Culture Concept

Below are several definitions of the concept of culture written by prominent anthropologists.

That complex whole which includes knowledge, belief, art, morals, law, custom, and any other capabilities and habits acquired by man as a member of society.
E. B. Tylor (1871)

The sum total of the knowledge, attitudes, and habitual behavior patterns shared and transmitted by the members of a particular society.
R. Linton (1940)

The mass of learned and transmitted motor reactions, habits, techniques, ideas, and values—and the behavior they induce.
A. L. Kroeber (1948)

The man-made part of the environment.
M. J. Herskovits (1955)

The learned patterns of thought and behavior characteristic of a population or society.
M. Harris (1971)

The acquired knowledge that people use to interpret experience and to generate social behavior.
J. P. Spradley and D. W. McCurdy (1975)

The sum of the morally forceful understandings acquired by learning and shared with the members of the group to which the learner belongs.
M. J. Swartz and D. K. Jordan (1976)

group of definitions shows. Sociologists, political scientists, psychologists, and ecologists have different requirements and would construct contrasting "operational" definitions. The same is true of cultural geographers. Geographers would be attracted to the Herskovits definition, because they have a particular interest in the way the members of a society perceive and exploit their resources, the way they maximize the opportunities and adapt to the limitations of their environment, and the way they organize that part of the earth that is theirs.

This last aspect, the way human societies organize that part of the earth that is theirs, goes to the heart of our study in this book. Human works carve long-lasting if not permanent imprints into the earth: Roman structures still mark some European countrysides, and Roman routes of travel today are among Europe's major highways. Over time, regions take on certain dominant qualities that together create a regional character, a personality, an atmosphere. This, in part, is the basis for our division of the human world into culture realms.

Cultural Landscape

Culture is expressed in many ways. Culture gives character to an area. Aesthetics play an important role in all cultures, and often a single scene in a photograph or a picture can reveal to us in general terms in what part of the world it was made. The architecture, the mode of dress of the people, the means of transportation, and perhaps the goods being carried reveal enough to permit a good guess. This is because the people of any culture, when they occupy their part of the earth's available space, transform the land by building structures on it, creating lines of contact and communication, parceling out the fields, and tilling the soil. There are but few exceptions; nomadic people may leave a minimum of permanent evidence, and some people living in desert margins (such as the Bushmen) and in tropical forests (the Pygmies, for example) alter their natural environment very little. But most of the time, there is change—asphalt roadways, irrigation canals, terraced hillslopes, fences and hedges, villages and towns.

This composite of human imprints on the surface of the earth is called the *cultural landscape,* a term that came into general use in geography in the 1920s. Carl O. Sauer, for several decades professor of geography at the University of California, Berkeley, developed a school of cultural geography that had the cultural landscape concept as its focus. In a paper written in 1927 entitled "Recent Developments in Cultural Geography," Sauer proposed his most straightforward definition of the cultural landscape. This constitutes, he said, "the forms superimposed on the physical landscape by the activities of man." He stressed that such forms result from cultural processes prevailing over a long time period, so that successive generations contribute to their development.

Sometimes these successive groups are not of the same culture. Farm settlements and villages built by European colonizers are occupied now by Africans. Minarets of Islam rise above the buildings of Eastern European cities, evincing an earlier period of hegemony of the Moslem Ottoman Empire. In 1929 D. Whittlesey introduced the term *sequent occupance* to categorize these successive contributions to the evolution of a region's cultural landscape. This concept is explored further in Chapter 7.

A cultural landscape consists of buildings and roads and fields and more, but it also has an intangible quality, an "atmosphere," that is often easy to perceive and yet difficult to define. The smells and sights and sounds of a traditional African market are unmistakable, but try recording those qualities on maps or in some other way for comparative study! Geographers have long grappled with this problem of recording the less tangible characteristics of the cultural landscape, which are often so significant in producing the regional personality.

Jean Gottmann, a European geographer, put it this way in his *Geography of Europe* (1969):

> . . . To be distinct form its surroundings, a region needs much more than a mountain or a valley, a given language or certain skills; it needs essentially a strong belief based on some religious creed, some social viewpoint, or some pattern of political memories, and often a combination of all three. Thus regionalism has what might be called *iconography* as its foundation: each community has found for itself or was given an icon, a symbol slightly different from those cherished by its neighbors. For centuries the icon was cared for, adorned with whatever riches and jewels the community could supply. In many cases such an amount of labor and capital was invested that what started as a belief, or as the cult of or even the memory of a military feat, grew into a considerable economic investment around which in interests of an economic region united . . . .

A mediterranean cultural landscape: red-tiled roofs, narrow streets, covered bridges, ornate arches, and decorative architecture evoke a sense of time as well as place. (Paul Slaughter/The Image Bank)

Gottmann is trying here to define some of the abstract, intangible qualities that go into the makeup of a cultural landscape. The more concrete properties are a bit easier to observe and record. Take, for example, the urban "townscape" (a prominent element of the overall cultural landscape), and compare a major United States city with, say, a leading Japanese city. Visual representations would reveal the differences quickly, of course, but so would maps of the two urban places. The United States city, with its rectangular layout of the central business district (CBD) and its far-flung sprawling suburbs contrasts sharply with the clustered, space-conserving Japanese city. Again, the subdivision and ownership of American farmland, represented on a map, looks unmistakably different from that of an African traditional rural area, with its irregular, often tiny patches of land surrounding a village. Still, the whole of a cultural landscape can never be represented on a map. The personality of a region involves not only its prevailing spatial organization but also its visual appearance, its noises and odors, even the pace of life.

Changing Natural Environments

Carl O. Sauer defined the cultural landscape as the forms superimposed on the physical landscape by human activity, and D. Whittlesey introduced the idea of cultural succession as sequent occupance. Both concepts contained the element of change—changing cultural traditions, changing societies. The earth itself, the physical landscape, was thought of as the static stage on which the cultural scenes were played out. Changes were brought to this natural landscape through humanity's works: building dams in river valleys, terracing hillsides, substituting cultivated crops for natural vegetation. The physical world was viewed as passive and stationary. The human world produced the dynamics of activity and transformation.

In recent decades this assumption (that the physical landscape is static and changeless except by human interference) has been challenged. Of course the earth always has been more receptive to human existence in certain areas than in others, a variability that is reflected by every map of world population distribution. However, the physical world varies not only in space, but also in time. Today, archeologists excavate ancient cities located in deserts, but when those cities were built, no desert prevailed there. In Roman times North Africa's farmlands produced shiploads of farm produce from fields amply watered by rain and irrigation, but today the Roman aqueducts lie in ruins and the fields are abandoned. Icy tundra conditions prevailed where the heart of the Soviet Union lies today. The earth is a variable stage, in time as well as in space.

This is true even of the surface itself—the plains, plateaus, hills, and mountains (Fig. I-2). The planet may be five billion years old, and a solid crust may have begun to form on it between three and four billion years ago. Eventually the earth differentiated into a number of shells, with the crust underlain by a thicker *mantle.* Physical geographers have known for more than a century that the surface of the crust has undergone momentous changes; mountain ranges have arisen only to be destroyed again, continental regions (such as the Great Plains, for example) were invaded by the ocean and lay inundated for millions of years as sediments accumulated. Then A. Wegener in 1915 published a book in which he presented evidence that the continents themselves are mobile and were once united in a gigantic landmass he called Pangaea. The breakup of this supercontinent, he reasoned, has taken place

Figure I-2

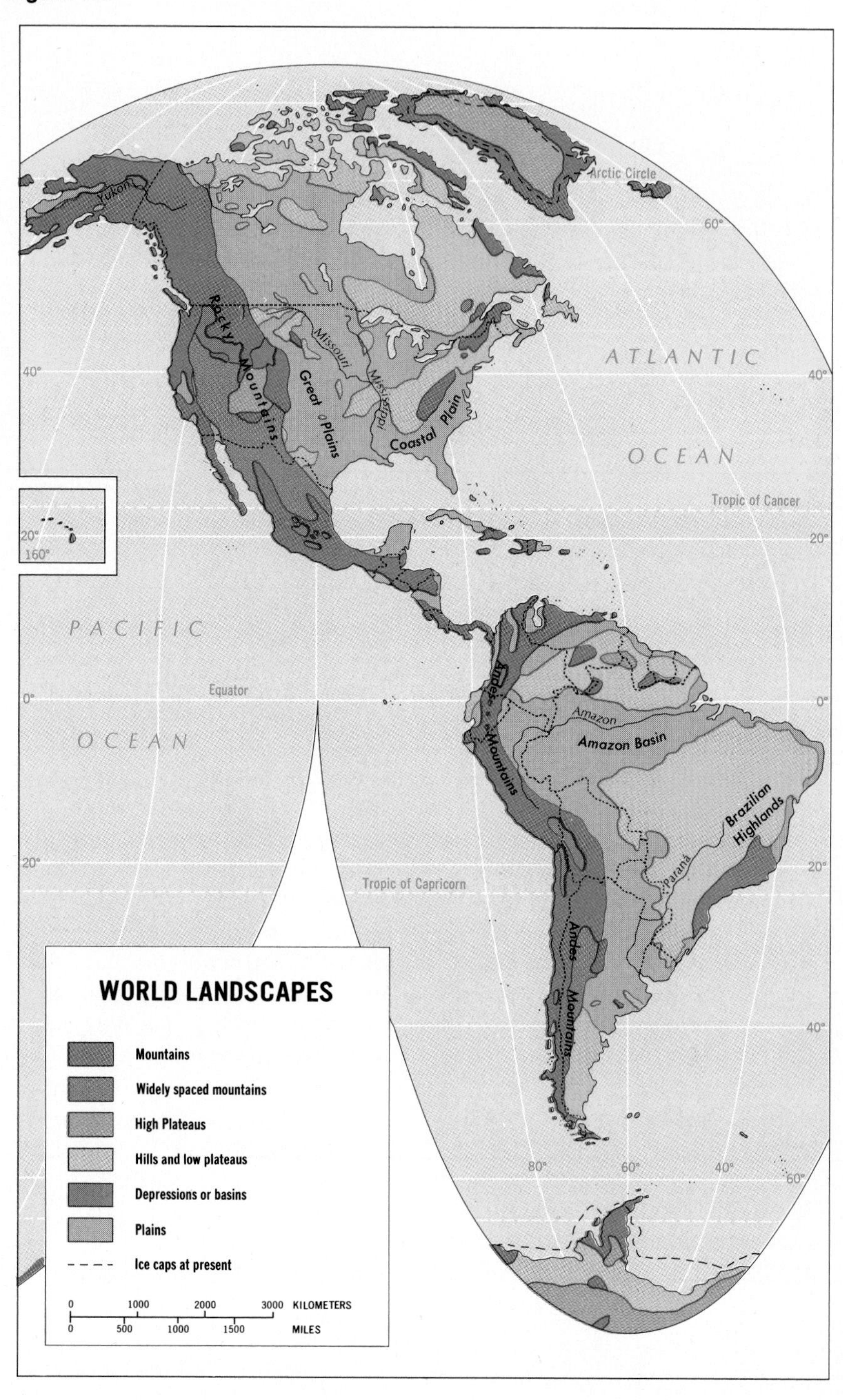

during the last 80 to 100 million years, and the continents are still moving.

Quite recently it was discovered that the earth's outer shell, the crust, and the upper layer of the mantle (together called the *lithosphere*) consist of a set of rigid, hard *plates.* The exact number and location of these tectonic plates are still not certain, but it is known that they average some 100 kilometers (60 miles) in thickness and that the largest plates are of continental dimensions (Fig. I-3). The plates are in motion, emerging along great fissure zones in the ocean basin and colliding elsewhere. Where they meet, one plate tends to descend under the other, and there is great deformation and crumpling of the crust. Earthquakes and volcanic

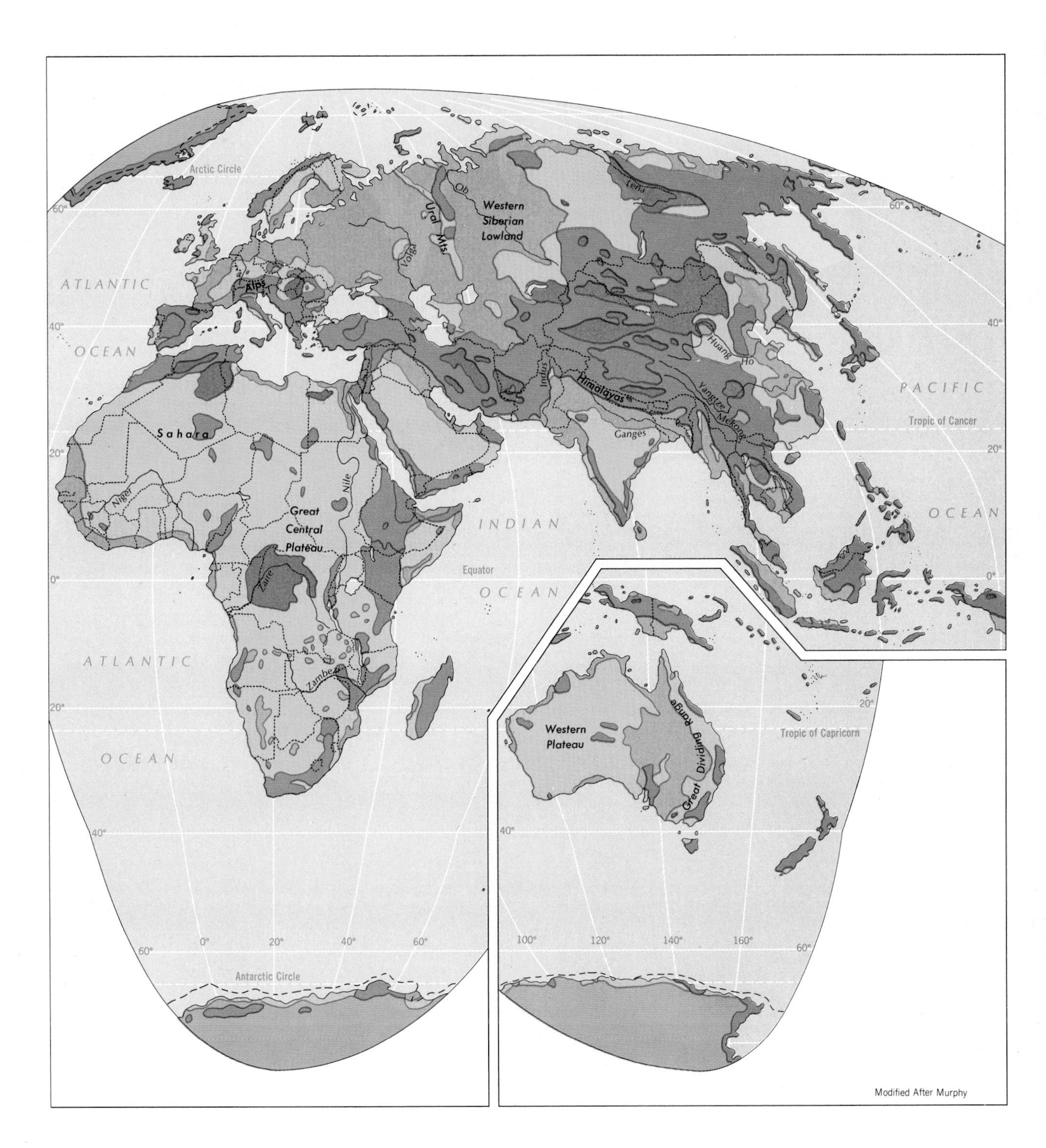

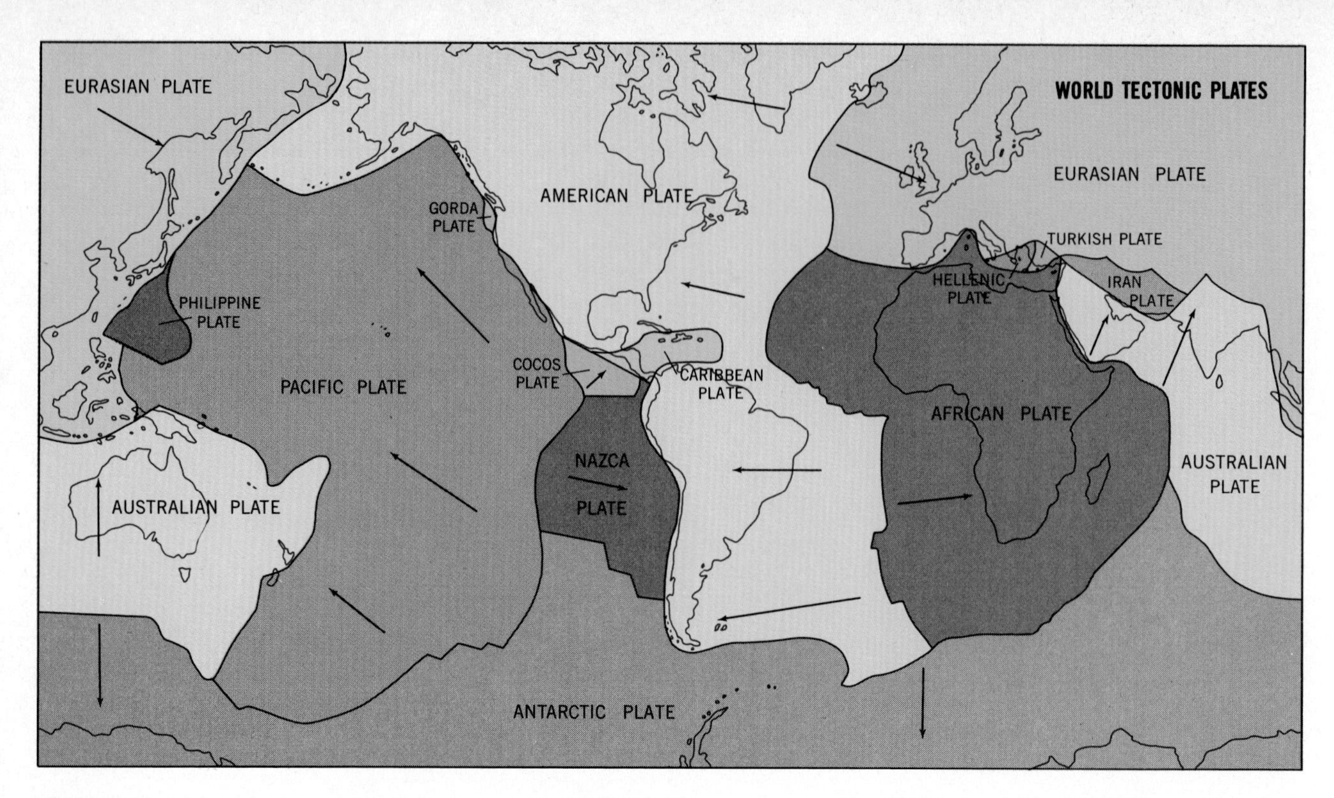

Figure I-3

eruptions attend this process. Along the earth's seismic zones, human communities are accustomed to sudden and sometimes violent changes in their physical environment. What thousands of years of quiet weathering and erosion cannot accomplish, an earthquake or volcanic eruption can achieve in seconds.

A comparison of Figs. I-2 and I-3 helps explain the spatial distribution of the earth's present landscapes. Great mountain ranges, complete with volcanic zones and earthquake-prone belts, extend from Western South America through Middle and North America to East Asia and the Pacific islands (including New Zealand) east of Australia. Here the Pacific Plate and its neighbors, the Philippine, Cocos, and Nazca Plates, collide against the Australian, Eurasian, and American Plates. Africa and Australia, on the other hand, lie at the heart of the African and Australian Plates, respectively, and are less subject to such deformation. Other relationships between Figs. I-2 and I-3 are readily apparent, but it is important to remember that the map of the world tectonic plates is being modified as additional evidence comes to light. For example, some physical geographers believe that the African Plate, shown here as an unbroken, cohesive unit of the lithosphere, in fact consists of several segments. About the major outlines of the world's tectonic plates, however, there is little doubt.

The mobile plates carry the continents along as they move, and Wegener's contention appears substantially correct. But it is important to view the process of continental drift in the perspective of human evolution on this earth. Current research in East Africa indicates that the human family began to emerge between 12 and 14 million years ago, and that the use of stone tools developed between 2 and 3 million years ago. A fossil site in Tanzania has yielded evidence of a stone structure, probably the foundation of a simple hut, dated as 1.8 million years old. But the emergence of larger human communities, made possible by the domestication of animals and plants and attended by the development of urban centers, is the product of the past 10,000 years—a mere fraction of time in the context of earth (and even evolutionary human) history. Recorded time has not been long enough to observe all the momentous changes we know the earth's landscapes have undergone.

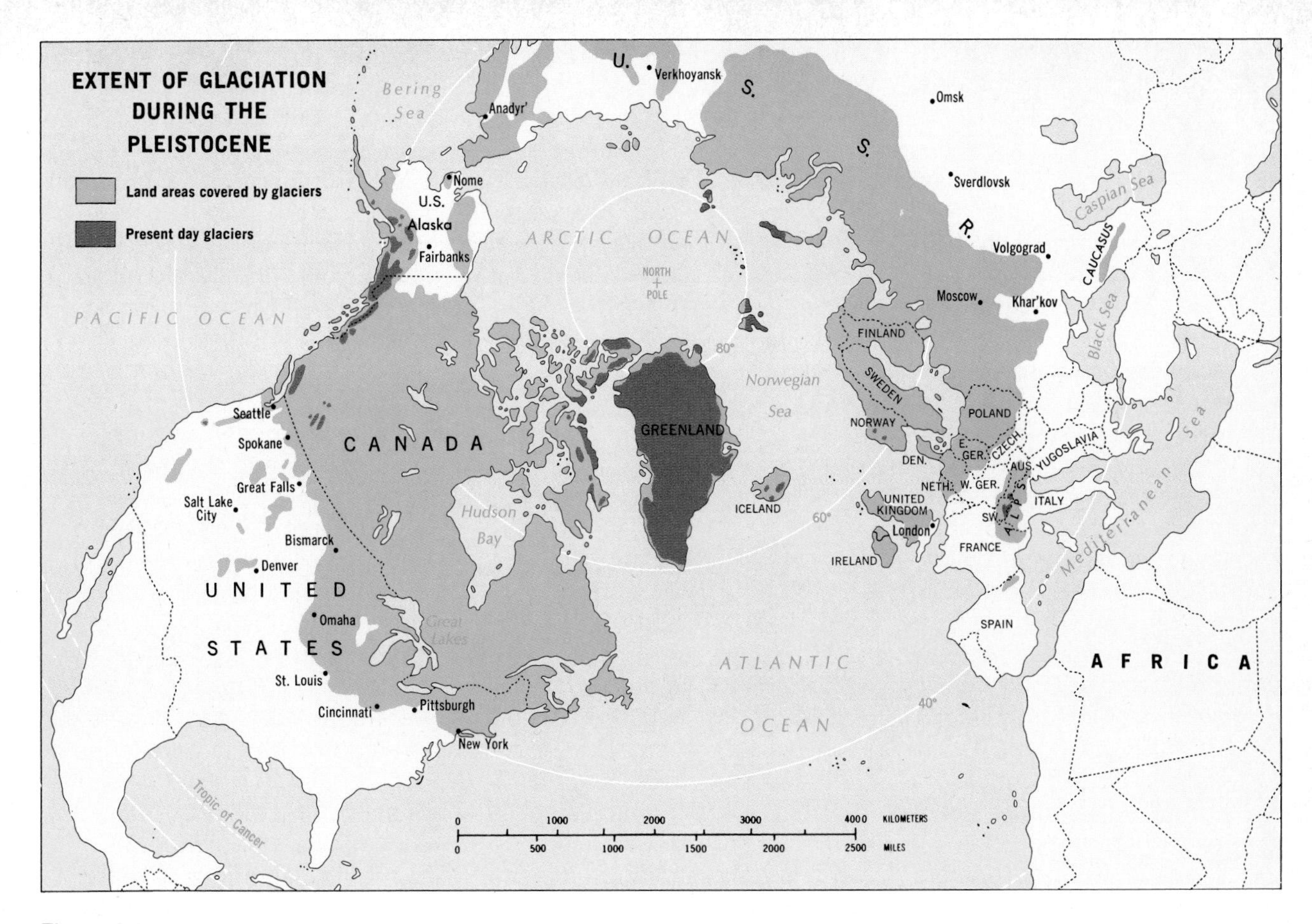

Figure I-4

Pleistocene Cycles

The earth's lifetime has been divided by geologists into four stages or *eras* on the basis of information derived from the study of rocks and fossils. Obviously, the least is known of the oldest of these eras, the Precambrian, covering the period from the planet's beginnings until about 600 million years ago. Next comes the Paleozoic era, from 600 to 225 million years ago, followed by the Mesozoic to about 70 million years before the present. The latest era, the one during which we live, is the Cenozoic. Each of these lengthy eras is subdivided into *epochs,* and the Cenozoic is marked by five of these. The latest Cenozoic epoch, the Pleistocene, witnessed the rise of human communities and civilizations.

Geologic eras and epochs are identified through the study of rock layering, fossil assemblages, mountain building, and other evidence yielded by the earth's crust. Many an epoch begins with the deposition of huge thicknesses of sedimentary rocks and ends when those rocks are bent and broken by tectonic activity (internal earth forces that deform the crust). But the Pleistocene is no ordinary epoch. Its beginning is marked by the development of large ice sheets that eventually covered all of Canada, most of the United States Midwest, and a substantial part of Europe (Fig. I-4). Human communities had first developed during the preceding Cenozoic epoch, the Pliocene. The earth's climatic and vegetative environments changed drastically when the glaciers of the Pleistocene appeared, probably between 2 and 3 million years ago. And the first outbreak of the ice sheets was not the last. Throughout the nearly 3 million years of the Pleistocene, the ice sheets have expanded, then contracted again, only to push outward once more. Many physical geographers believe that there were four major glacial advances and three periods of withdrawal (interglacials), but there is evidence to suggest that numerous additional advances and retreats have occurred, perhaps as many as 14. The Pleistocene epoch, therefore, is often called the Ice Age.

Some geologic time charts suggest that the Pleistocene is at an end and that we now exist during a new epoch, the "Recent." It is far more likely, however, that the human world of modern times has evolved during yet another Pleistocene interglacial, whose temperateness has enormously expanded the earth's living space—temporarily. The possibility is strong that the ice sheets will once again overspread large parts of the areas shown as affected by glaciation on Fig. I-4. If this occurs, the new ice age will have an impact far beyond those regions directly affected. World climates everywhere will change. Moist areas will dry up. Areas now dry will begin to receive ample moisture. Temperatures, even in tropical zones, will decline.

Whatever the future, there can be no doubt that the modern world's human cultures emerged and evolved in the wake of the last of the Pleistocene glaciers' withdrawal. Certainly there were villages and communities during earlier phases, but the great transformations—plant and animal domestication, urbanization, mass migration, agricultural and industrial revolutions—occurred during the past 10,000 years. We cannot know as yet why these momentous changes occurred when they did, but they have been attended, especially during the last century, by an unprecedented explosion of human numbers. This population explosion (about which more will be said in Chapter 8) is taking place now, while the earth's available living space is at a maximum. The implications of a return of the ice sheets and the impact on the earth's climates and habitable space stagger the imagination. Nevertheless, many scientists warn that such a sequence of events may well lie ahead.

Water—Essence of Life

The French geographer Jean Brunhes, writing in his book *Human Geography* (1952), remarked that "every state, and indeed, every human establishment, is an amalgam made up of a little humanity, a little soil, and a little water." He might have added that, without water, there would be neither humanity nor soil. When the United States in 1976 sent two space probes to Mars, scientists on earth waited for the crucial information to be relayed back from the landed vehicles: was any moisture present on the distant planet's surface? Moisture would be the key to life on Mars as here on earth. But the Mars surface proved to be as barren as the moon's. Alone among the planets of our solar system, the earth possesses a *hydrosphere*—a cover of water in the form of a vast ocean, frozen polar ice sheets, and a moisture-laden atmosphere. Technically all water on the earth, even lakes and streams, is part of the hydrosphere, but the great world ocean constitutes about 97 percent of it all by volume. The world ocean covers just over 70 percent of the earth's surface, but it would do little good if a mechanism did not exist whereby moisture from the ocean is brought to the land. This mechanism, the *hydrologic cycle,* functions as a circulation system. Moisture evaporates into the air from the ocean's surface, and the humid mass of air then moves over the land where, by various processes, condensation occurs and precipitation falls. Much of it returns to the ocean as runoff via streams, and the cycle continues.

The earth's landmasses do not share equally in this provision of moisture (Fig. I-5), and much of the historical geography of humanity involves the search and competition for well-watered areas. Again, the map of world precipitation distribution should be viewed as one frame from a piece of Pleistocene film, a still picture of changing conditions. It represents the earth's moisture conditions as they prevail today, but it differs from the map as it would have looked when the Middle East's Fertile Crescent witnessed the domestication of crops, or when North Africa, several thousand years later, was a granary of the Roman Empire—and it will look quite different a thousand years from now. Today the map reveals an equatorial zone of heavy rainfall, where annual totals exceed 200 centimeters (80 inches), extending from Middle America through the Amazon Basin, across smaller areas of West and Equatorial Africa, and into South and Southeast Asia. This equatorial zone of high precipitation gives way to dry conditions in both northward and southward directions. In equator-straddling Africa, for example, the Sahara lies to the north

Evapotranspiration

The map of world precipitation distribution (Fig. I-5) should be viewed in the context of temperature distribution. From the map it might be concluded that all the areas that receive over 100 centimeters (40 inches) of rainfall are thereby equally amply supplied with moisture. But there are equatorial areas where temperatures average over 25°C (77°F) that receive 100 centimeters and much cooler places—for example, parts of New Zealand and Western Europe—where 100 centimeters are also recorded. Obviously, evaporation from the ground goes on much more rapidly in the tropical areas than in the midlatitude zones.

Similarly, evaporation from vegetation also speeds up in equatorial regions. This evaporation from leaf surfaces is actually a three-stage process. Roots of plants absorb water in the soil. This water is then transmitted through the organism and reaches the leafy parts, which transpire in warm weather much as we perspire. From the surface of the leaves, the moisture evaporates. Thus, a plant acts like a pump, and the process of evaporation from vegetation is actually a process of transpiration plus evaporation, or *evapotranspiration.*

Thus 100 centimeters of rainfall in a tropical area may very well be inadequate, and if the amount lost by evaporation and evapotranspiration is calculated, it could exceed 100 centimeters—which means that the plants would use more moisture if it were available. Some of those "moist" tropical areas, even those with over 150 centimeters (60 inches) of rainfall annually, can be shown to be moisture deficient. In other areas, the seasonality of precipitation is so pronounced that there is deficiency during part of the year. On the other hand, in cooler parts of the world, just 75 centimeters (30 inches) of rainfall may be enough to keep the soil moist and the vegetation adequately supplied. A map such as Fig. I-5 is of necessity a generalization, and it is important to know what it fails to reveal.

of the low-latitude humid zone and the Kalahari Desert lies to the south. Interior Asia and central Australia also are very dry, as is the Southwest of North America.

The general pattern of equatorial moistness and adjacent dryness is broken along the coasts of all the continents, and it is possible to discern a certain consistency in the spatial distribution of precipitation. Thus the eastern coasts of continents and islands in tropical as well as midlatitude locations receive comparatively heavy rainfall, as in the southeastern United States, eastern Brazil, eastern Australia, and coastal China. Again, a narrow zone of higher precipitation exists at higher latitudes on the western margins of the continents, including the coasts of Oregon, Washington, and British Columbia, the coast of Chile, the southwestern tip of Africa, the southwestern corner of Australia, and, importantly, the western exposure of the great Eurasian landmass: Western Europe.

The distribution of world precipitation as reflected by Fig. I-5 results from an intricate combination of global systems of atmospheric and oceanic circulation, heat and moisture transfer. The analysis of these systems is the subject of physical geography, but we should remind ourselves that even a slight change in one of them can have a major impact on a region's habitability. Figure I-5 represents the *average* annual precipitation on the continents, but no place on earth has an absolute guarantee that it will receive, in any given year, precisely its average rainfall. In general the variability of precipitation increases as the recorded average total decreases, which means that rainfall is least dependable just where reliability is needed most—in the drier parts of the inhabited world. A prominent region of low rainfall that has suffered much from variable rainfall is West Africa's Sahel, where a seven-year drought intervened between moist years in the decade 1967–77, and hundreds of thousands of people died of starvation.

Even heavy, year-round precipitation is no guarantee that an area can sustain large and dense populations. In equatorial areas the heavy precipitation combined with high temperatures leads to the faster destruction of fallen leaves and branches by bacteria and fungi, so that little or no humus develops; the drenched soil is also leached of its best nutrients, so that oxides of iron and aluminum remain to give the tropical soil its characteristic reddish color. Such tropical soils (called latotols) support the rain forest, but they do not carry crops without massive fertilization. The rain forest thrives on its own decaying vegetative matter, but when the land is cleared, the soil proves to be quite infertile. The Amazon and Congo (Zaire) basins are not among the world's most populous regions.

Climatic Regions

It is not difficult to discern the significance of precipitation distribution in the map of world climates (Fig. I-6). The regionalization of climates has always presented problems for geographers. In the first place, climatic records are still scarce, short-term, or otherwise inadequate in many parts of the world. Second, weather and climate tend to change gradually from place to place, but the transitions must be represented as lines on the map. In addition, there is always room for argument concerning the criteria to be used, and how these criteria should be weighed. Vegetation, for ex-

Figure I-5

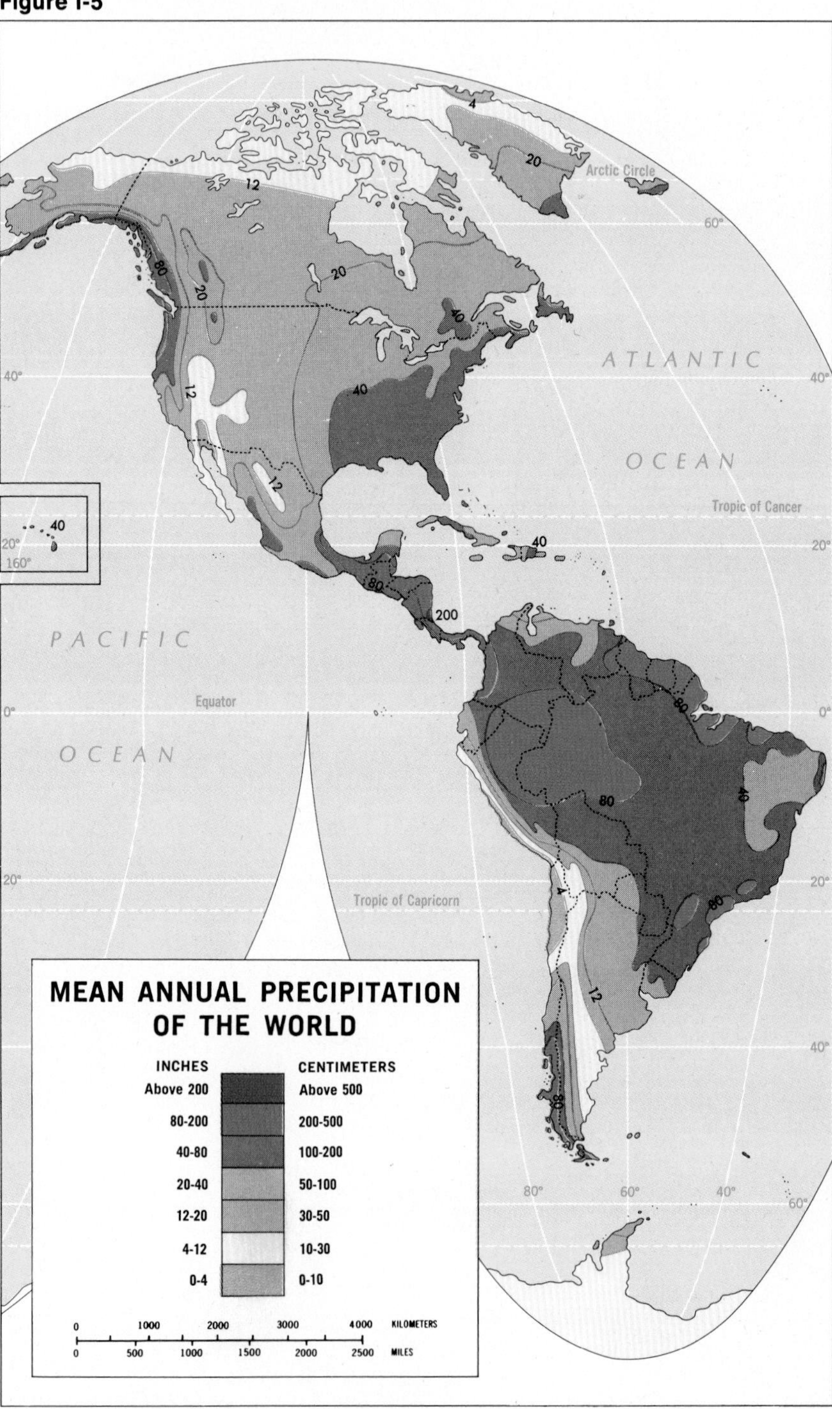

ample, is a response to prevailing climatic conditions. Should boundary lines between climatic regions therefore be based on vegetative changes observed in the landscape, no matter what precipitation and temperature records show? This debate still goes on.

Figure I-6 displays a regionalization system devised by Köppen and modified by Geiger. It has the advantage of comparative simplicity and is based on three-letter symbols. The first (capital) letter is the critical one; the *A* climates are humid and tropical, the *B* climates are dominated by dryness, the *C* climates are still humid and comparatively warm, the *D* climates reflect increasing coldness, and the *E* climates mark the frigid polar and near-polar areas.

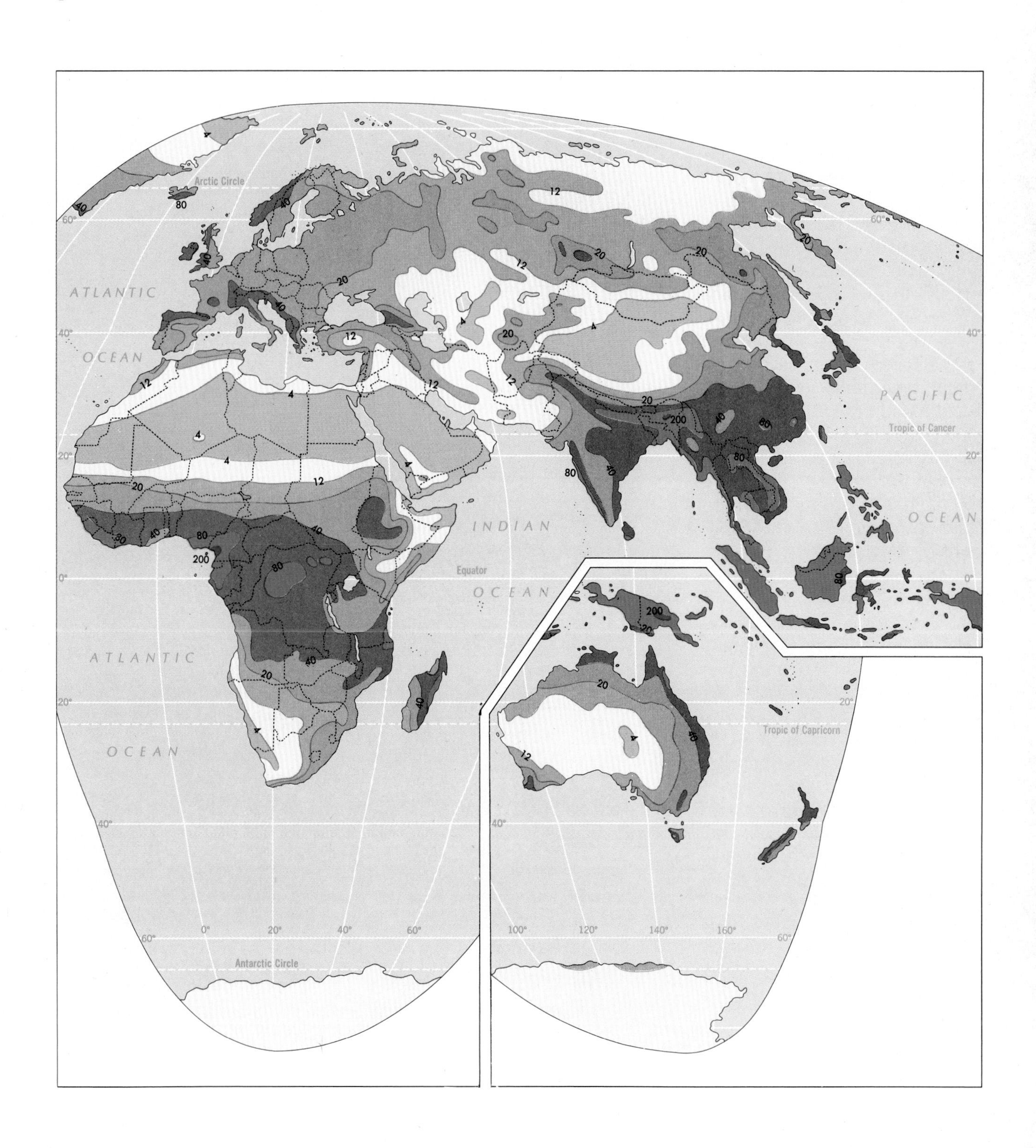

The *humid equatorial* climates *(A),* also referred to as the humid tropical climates, are marked by high temperatures all year and by heavy precipitation. In the *Af* subtype, the rainfall comes in substantial amounts every month, but in the *Am* areas there is a sudden, enormous increase due to the arrival of the annual monsoon. The *Af* subtype is named after the vegetation association that develops there—the rain forest. The *Am* subtype, prevailing in part of peninsular India, in a coastal area of West Africa, and in sections of Southeast Asia, is appropriately referred to as the monsoon climate. A third tropical climate, the savanna, has a wider daily and annual temperature range and a more strongly seasonal distribution of rainfall. As Fig. I-5 indicates, savanna rainfall totals tend to be lower than those in the rain-forest zone, and the associated seasonality often is expressed in a "double maximum." This means that each year produces two periods of increased rainfall separated by pronounced dry spells. In many savanna zones the people refer to the "long rains" and the "short rains" to identify these seasons, and a persistent problem in these regions is the unpredictability of the rain's arrival. Savanna soils are not among the most fertile, and when the rains fail the specter of hunger arises. Savanna regions are far more densely peopled than rain-forest areas, and millions of residents of the savanna subsist on what they manage to cultivate. Rainfall variability under the savanna regime is their principal environmental problem.

The *dry* climates *(B)* occur in low as well as higher latitudes. The difference between the *BW* (true desert) and *BS* (less arid steppe) varies, but may be taken to lie at about 25 centimeters (just over 10 inches). Parts of the central Sahara receive less than 10 centimeters (4 inches) of rainfall and, as Fig. I-5 shows, much of West Africa's Sahel is steppe country, the most tenuous of environments for farmers or herders. A pervasive characteristic of the world's dry areas is the enor-

Figure I-6

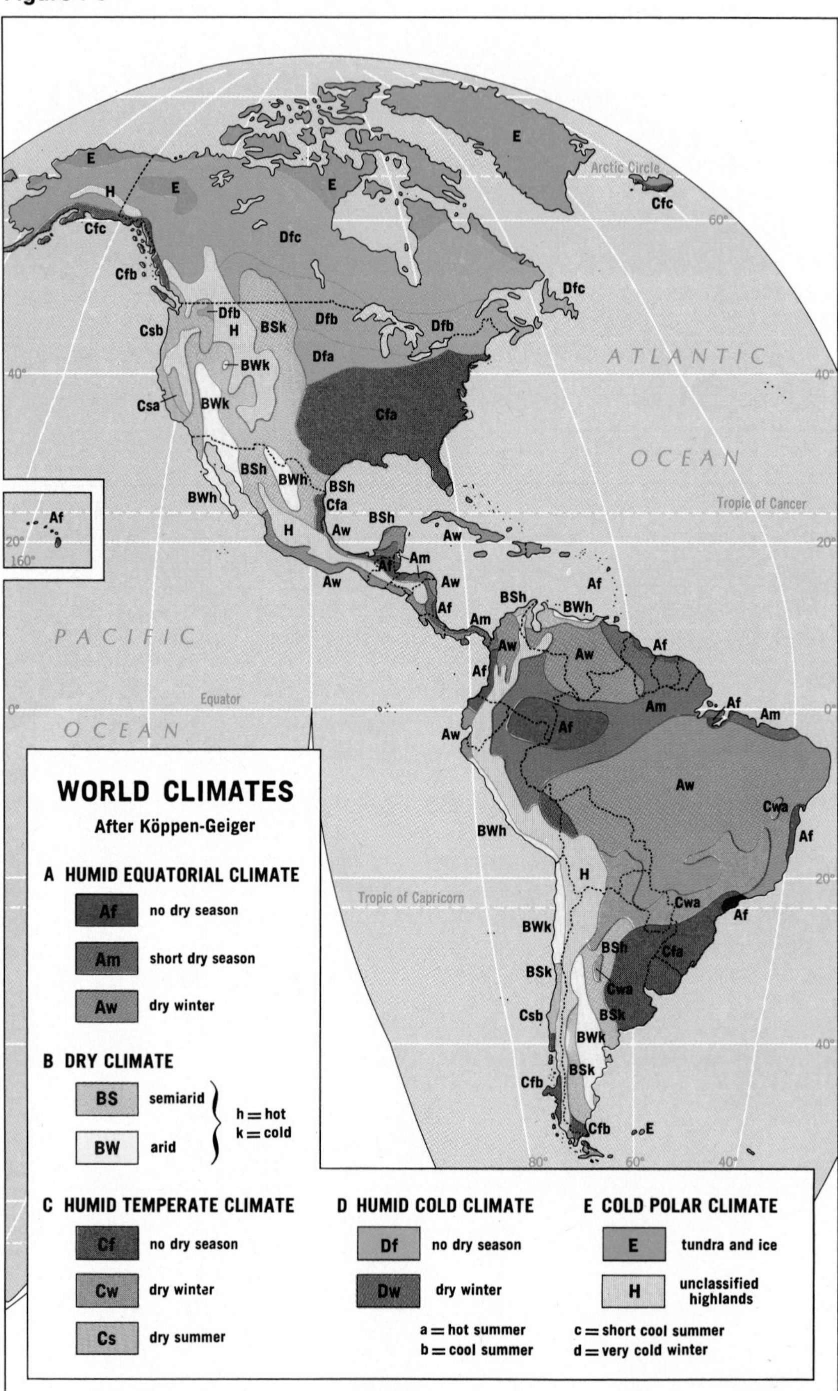

mous daily temperature range that may exceed 35°C (60°F) in low-latitude deserts. Recorded instances exist where the maximum daytime shade temperature was over 49°C (120°F) followed by a nighttime low of 9°C (48°F).

The *humid temperate* climates *(C)* are also referred to as the midlatitude climates; as the map shows, almost all the areas of *C* climate lie just beyond the Tropics of Cancer and Capricorn. This is the prevailing clime in the southeastern United States from Kentucky to Florida, in California and the coastal Northwest, and Western Europe and the Mediterranean, in southern Brazil and northern Argentina, in coastal South Africa and Australia, and in eastern China and southern Japan.

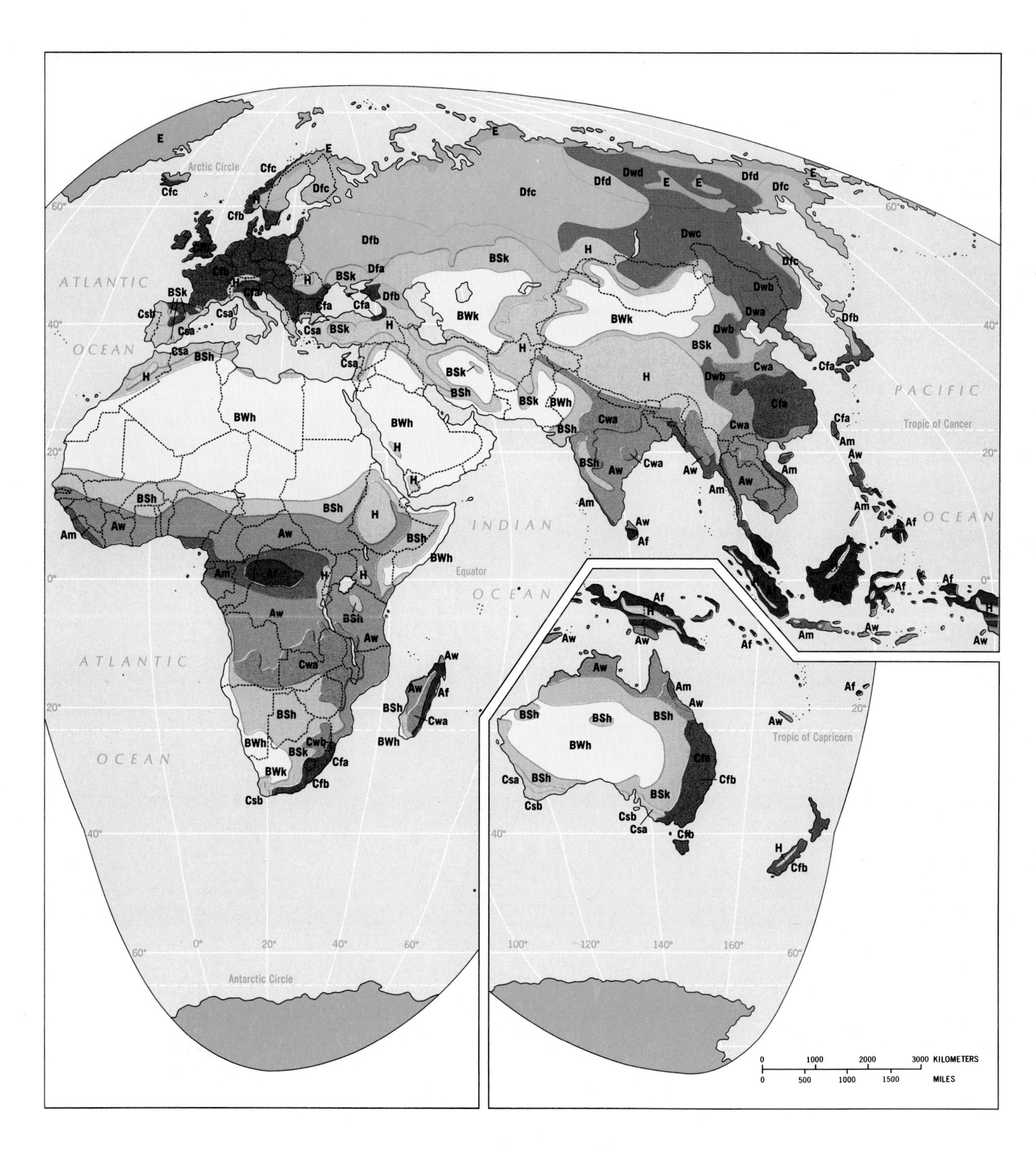

None of these areas is associated with climatic extremes or severity, but the winters can be fairly cold. These areas lie about midway between the winterless equatorial climates and the summerless polar zones.

The humid temperate climates range from quite moist, as along the densely forested coasts of Oregon, Washington, and British Columbia, to relatively dry, as in the so-called Mediterranean (dry-summer) areas that include not only coastal southern Europe and northwest Africa, but also the southwestern tips of Australia and Africa, middle Chile, and southern California. In these Mediterranean areas the scrubby, moisture-preserving vegetation creates a natural landscape very different from that of more lushly clothed Western Europe.

The *humid cold* climates *(D),* or snow climates as they are also known, may be called the continental climates, for they seem to develop where there is a lot of landmass, as in the heart of Eurasia and in interior North America. No equivalent land areas, at similar latitudes, exist in the Southern Hemisphere, and no *D* climate occurs there at all.

Great annual temperature ranges mark the humid continental climates, and very cold winters and relatively cool summers are the rule. In a *Dfa* climate, for example, the warmest summer month (July) may average as high as 21°C (70°F), but the coldest month (January) only −11°C (12°F). Total precipitation, a substantial part of which comes in the form of snow, is not very high, ranging from over 70 centimeters (30 inches) to a steppelike 25 centimeters (10 inches). Compensating for this paucity of precipitation are the cool temperatures, which inhibit loss of evaporation and evapotranspiration.

Some of the world's most productive soils lie in areas under humid cold climates, including the United States Midwest, parts of the Soviet Union's Ukraine, and North China. The period of winter dormancy, when all water is frozen, and the accumulation of plant debris during the fall combine to balance the soil-forming and enriching processes. The soil differentiates into well-defined,

Figure I-7

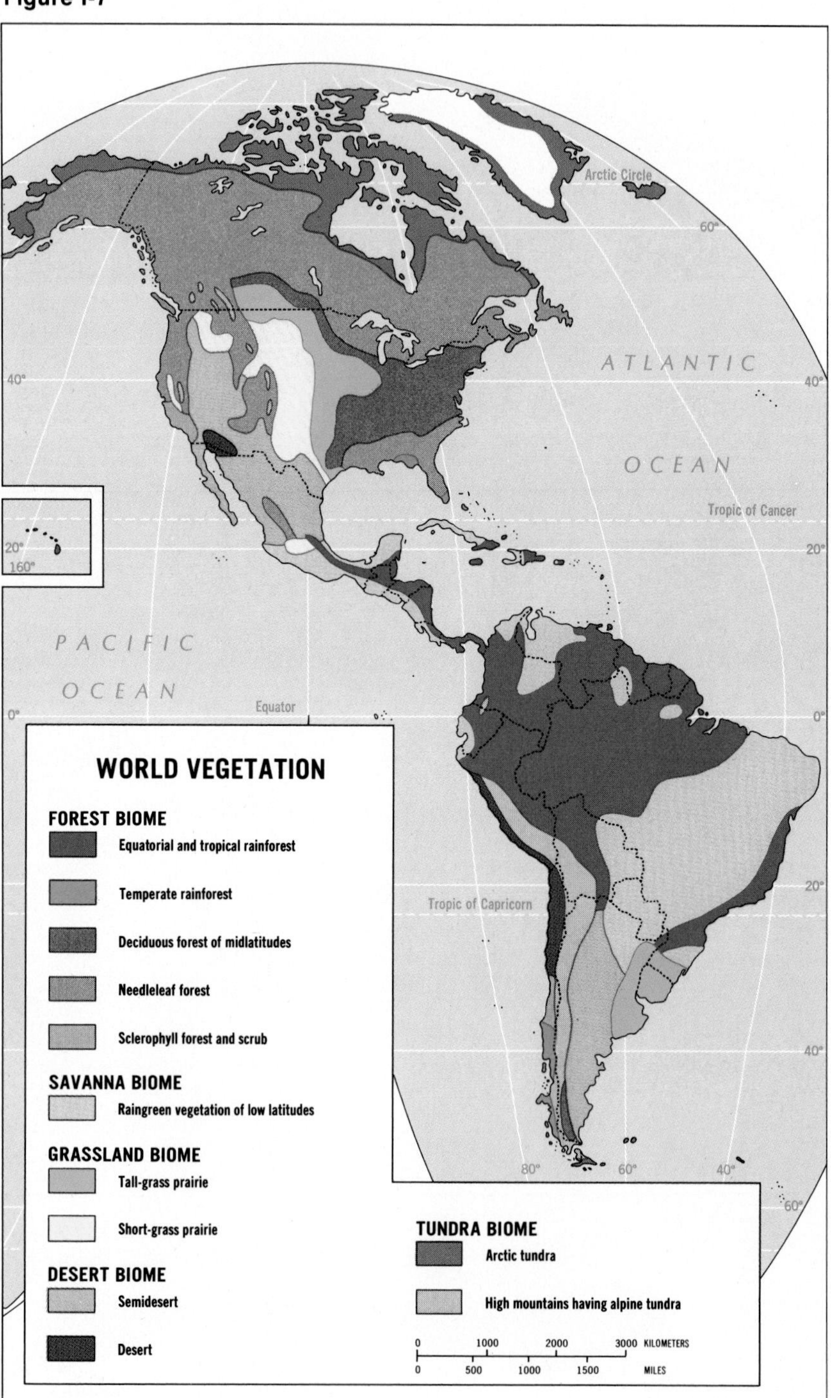

nutrient-rich layers, and a substantial store of organic humus accumulates. Even where the annual precipitation is light, this environment sustains extensive forests.

The *cold polar* climates *(E)* are differentiated into true ice-cap conditions, where permanent ice and snow exist and no vegetation can gain a foothold, and the "tundra," where up to four months of the year average temperatures may be above freezing. Like *rain forest* and *savanna,* the term *tundra* is a vegetative as well as a climatic appellation, and the boundary between *D* and *E* climates on Fig. I-6 can be seen to correspond quite closely to that between the needleleaf forests and arctic tundra on Fig. I-7.

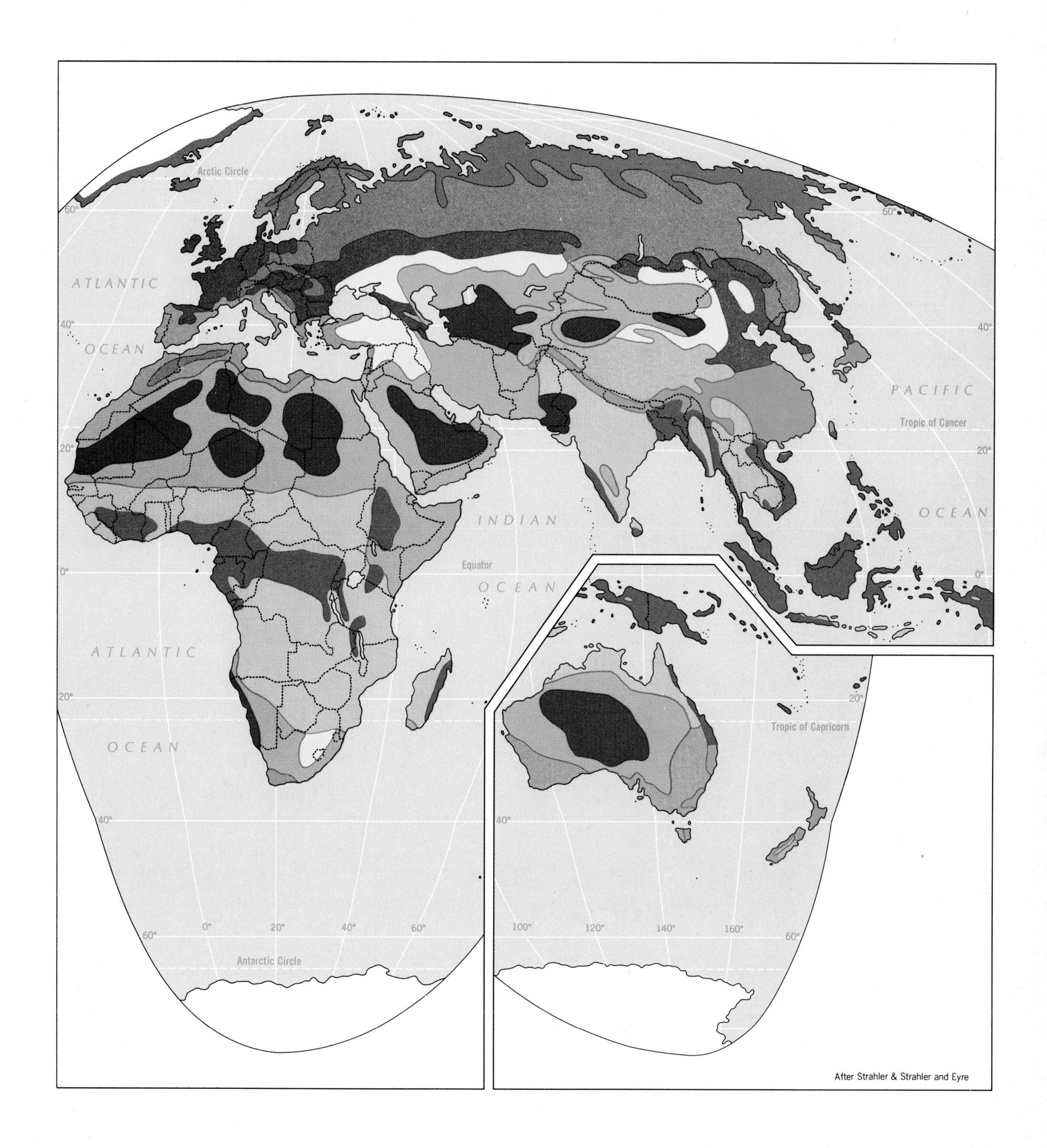

Vegetation Patterns

The map of world vegetation distribution (Fig. I-7) shows the closeness of the spatial relationship between climatic regions and plant associations. The map depicts the global distribution of natural vegetation, but much of this vegetation has been destroyed or modified by the human population. Hence the regions shown on Fig. I-7 represent the natural vegetation that exists or would exist as a result of long-term plant succession and adaptation to prevailing climatic conditions. When climate over an area remains the same for several thousand years, a plant association develops that is in equilibrium with this environment. Such a *climax* vegetation may be a rain forest, as in the Amazon Basin, or it may be a scrub-and-bush association that just manages to hold on in the driest steppe. We should not lose sight of the changeable character of Pleistocene environments, however; like the maps of precipitation and climates, this map of world vegetation is a still from a motion picture, not a permanent end product. Places exist where one association can be observed to be gaining on another, signaling change. In West Africa the steppe is encroaching southward on the savanna, and the Sahara in turn is gaining on the steppe. Inexorable climatic change may be involved in this, although human communities in the region have played a major role in modification of the natural environment. In any case, large parts of the regions shown on Fig. I-7 do not actually support the climax vegetation communities shown in the legend.

The vegetative cover of the earth consists of trees, shrubs, grasses, mosses, and an enormous variety of other plants. Plant geographers (phytogeographers) group this mass of vegetation into the five vegetative regions *(biomes)* shown in the map's legend: forest, savanna, grassland, desert, and tundra. Note that these biomes are not confined to particular latitudes; there is forest in the equatorial zone, in temperature regions astride the Tropics of Cancer and Capricorn, in middle latitudes such as in the eastern United States, and in high latitudes on the margins of the tundra. The adaptation of the forests' species is what differs. Equatorial and temperate forests have leafy evergreen trees, while cold-climate, high-latitude forests have trees with thin needles. In the middle latitudes the trees are deciduous and shed their leaves each autumn.

A prominent impression gained from Fig. I-7 relates to the vastness of the savanna lands. The bulk of Africa south of the Sahara is savanna country, as is most of eastern Brazil and India. The savanna also prevails in interior Southeast Asia and in northern Australia; in later chapters there will be frequent occasion to refer to the vagaries of the savanna environment, with which hundreds of millions of the world's farmers must cope.

Soil Distribution

We conclude this view of the Pleistocene stage for the human drama with an examination of another vital ingredient in the sustenance of life on this earth: the soil. The earth's soils have proved even more difficult to classify and regionalize than the climate and vegetation, and Fig. I-8 is only one possible alternative among several. Research concerning the processes of soil formation continues to produce new data, and as the new evidence becomes available the schemas change.

Because the parent material (the rocks beneath), the temperature, the moisture conditions, the vegetation, and the terrain (the degree of slope steepness) all vary from place to place over the globe, there is enormous diversity of soils as well. Some soils are infertile and do not sustain crops; others can carry two or even three crops per year, and do so year after year. And still today, liberating technologies notwithstanding, the great majority of the world's people depend directly on the soil for their food. Any map of global population distribution to a considerable degree reflects the productiveness of the soils of certain particular areas—and their infertility elsewhere.

Figure I-8 once again displays areas of correspondence with world distributions of climatic elements and vegetation, reminding us that the soil is a responsive element of the total environmental complex. It is important to remember that this map of the world distribution of soils represents a generalization of the true situation. After many years of experimentation, soil scientists produced an all-encompassing classification system called the *Comprehensive Soil Classification System* (CSCS) also called, because of the numerous attempts that had gone before it, the "Seventh Approximation." In this classification, the world's soils are grouped into 10 *Orders,* which in turn are divided into 47 Suborders, 185 Great Groups, about 1000 Subgroups, 5000 Families, and 10,000 Series. No small-scale map such as Fig. I-8 could possibly contain all this detail.

Some of the new soil names are self-explanatory and easy to understand. The *oxisols,* for example, are the excessively leached soils of the tropics, the familiar reddish-colored soils marked by high concentrations of oxides of iron and aluminum. Formerly these soils were called laterites and latosols, and these are the soils the subsistence farmers of the savannas and equatorial areas of Africa and South America must cultivate. The oxisols often are thick and deeply weathered, but the needed nutrients have to a large extent been washed downward. Tropical vegetation that grows in this soil sustains itself by absorbing nutrients directly from fallen leaf matter. When the farmer clears the land, the oxisols may carry a crop for a year or two, but then they are exhausted and fail.

Another soil order with a suitable name comprises the *aridisols.* These are salt-rich, infertile soils of rainfall-deficient areas and, as Fig. I-8 shows, they are widely distributed across the earth, occurring in the Southwestern United States, in Western South America, in large areas of Africa, and vast regions of Asia as well as Australia.

In high latitudes, where soil-forming conditions are inhibited by periods of extreme cold, limited warmth from the sun, and poor drainage, soils often are not well developed and called *inceptisols* (tundra vegetation prevails here, as Fig. I-7 shows). This is another name with an obvious derivation, but the neighboring *spodosols* are not. The spodosols, which extend across Northern Canada and much of high-latitude Eurasia, are better developed and support great stands of pine and spruce (the Needleleaf Forests on Fig. I-7). They owe their name to the peculiar, ashlike, bleached appearance that results from the constant and intense removal of matter from the upper layer; Russian farmers referred to this as *podzol* (ashy soil).

Between the oxisols (with the related ultisols and vertisols) and the aridisols of lower latitudes, and the inceptisols and spodosols of higher latitudes, lie the *mollisols* and the *alfisols.* As Fig. I-8 shows, these two soil orders extends across vast reaches of interior North America, Eurasia, and (to a lesser extent) South America. The mollisols (*mollis* is Latin for "soft") possess a dark, humus-rich upper layer, and they never become massive and hard as the aridisols do, even under comparatively dry conditions. As the map indicates, mollisols are located in intermediate positions between dry and moist climates (see Fig. I-5). These are the grassland soils that support large livestock herds or, when farmed, sustain vast expanses of grain crops. The alfisols (formerly called ped*alf*ers, hence the new name), occur in a broad zone in Canada, in the United States Midwest, and in Europe and Soviet Union. These soils evolve under a wide range of environmental conditions, and four major suborders are recognized. The alfisols are generally quite fertile, and can support intensive agriculture.

Figure I-8 shows limited areas of *entisols* (*recent*) in North America and Eurasia (but larger zones in Africa and Australia). Entisols, like inceptisols, have not existed long enough to develop maturely. They may lie on sand accumulations, on recently deposited river alluvium, or in areas subject to strong erosion where surface material is constantly removed. Except for intrinsically fertile alluvium, entisols are not usually good soils for farming. The *histosols* (from *histos,* Greek for "tissue") are the soils of bogs and moors; they consist primarily of organic rather than mineral matter. Often waterlogged, the histosols may develop as peat accumulations. The largest areas of histosol exposure lie in Canada, south of Hudson Bay and in the far northwest, just below the permafrost limit.

Finally, Fig. I-8 shows the distribution of the world's mountain soils, the soils of high-relief terrain. A comparison with Fig. I-2, however, proves that not all mountainous or high-elevation areas contain such thin, poorly developed, often stony soils. The highlands of Ethiopia, for example, form an area of comparatively fertile alfisols.

It is interesting to compare Fig. I-8 and Fig. I-9 (showing the world distribution of population). The valley and delta of the lower Nile River in North Africa, the basin of the Ganges River in South Asia, and the plain of the Huang River in China contain alluvial inceptisols, and by their almost legendary fertility they sustain many millions of people. In 1980, fully 95 percent of Egypt's 45 million people lived within 20 kilometers (12 miles) of the Nile River. Hundreds of millions of Indians and Chinese depend directly on the alluvial soils in the basins of the Ganges and Huang Rivers, where crops are grown that range from corn to cotton, wheat to jute, rice to soybeans. These are only the largest examples of such alluvium-based agglomerations. In Pakistan, Bangladesh, Vietnam, and many other countries, the alluvial soils in river valleys provide what the upland soils do not.

These riverine population clusters provide a link with humanity's past. In the fertile valleys of the Middle East's rivers, the art and science of irrigation may have been first learned, and two of the world's oldest continuous cultures, Egypt and China, still retain heartlands near their ancient foci of thousands of years ago.

As a result, the map of world population distribution (Fig. I-9) displays prominently the clustering of huge numbers of people in areas of fertile and productive soils. For thousands of years following the beginning of crop and animal domestication, communities remained dependent on soils and pastures in their immediate vicinity; the large-scale, worldwide food transportation of today is essentially a phenomenon of the past two hundred years, a product of industrial and technological rev-

Figure I-8

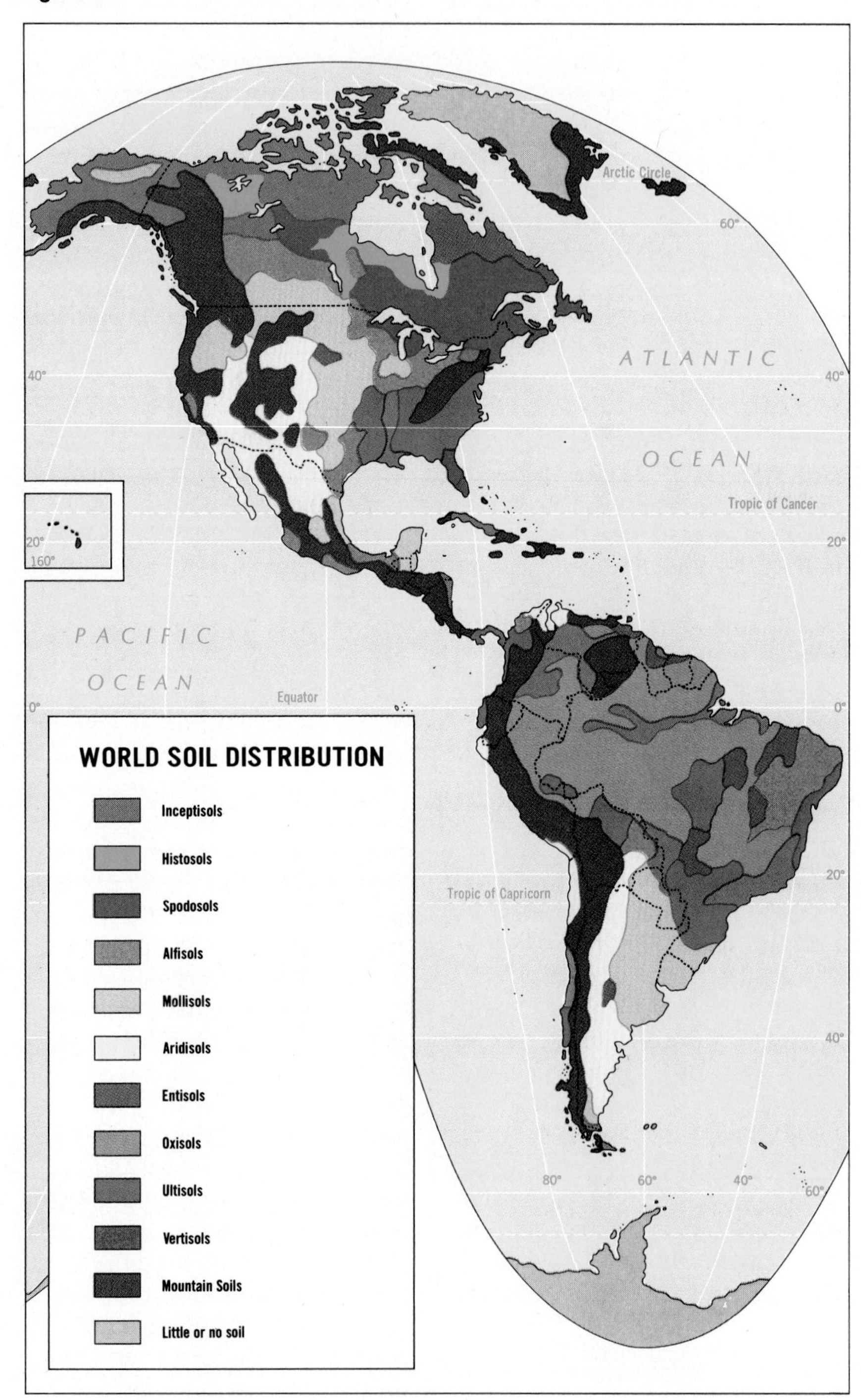

Space and Population

Less than 30 percent of the earth's surface is land area, and, as we observed in the course of a brief survey of global environments, large parts of that land area cannot support any substantial number of inhabitants. Dry deserts, rugged mountain ranges, and frigid tundras constitute just a few of the less habitable parts of the world, but together they cover nearly half the landmasses—and we have not even begun to account for places where unproductive tropical soils, persistent diseases, frequent droughts, and other more localized conditions keep population numbers comparatively low.

olutions. True, the ancient Romans imported grains from North Africa, and colonial powers even centuries ago brought shiploads of spices and sugar across the oceans. But the overwhelming majority of the world's peoples continued to subsist on what they could cultivate. When groups migrated, they did so in search of new lands to be opened up, new pastures to be exploited. If they succeeded, as the Chinese of North China did when they moved southward, they thrived. When they failed, they starved.

Today's map of world population is another one of those stills from that Pleistocene motion picture, for it shows the current stage of humanity's expansion and dispersal across the habitable space on the

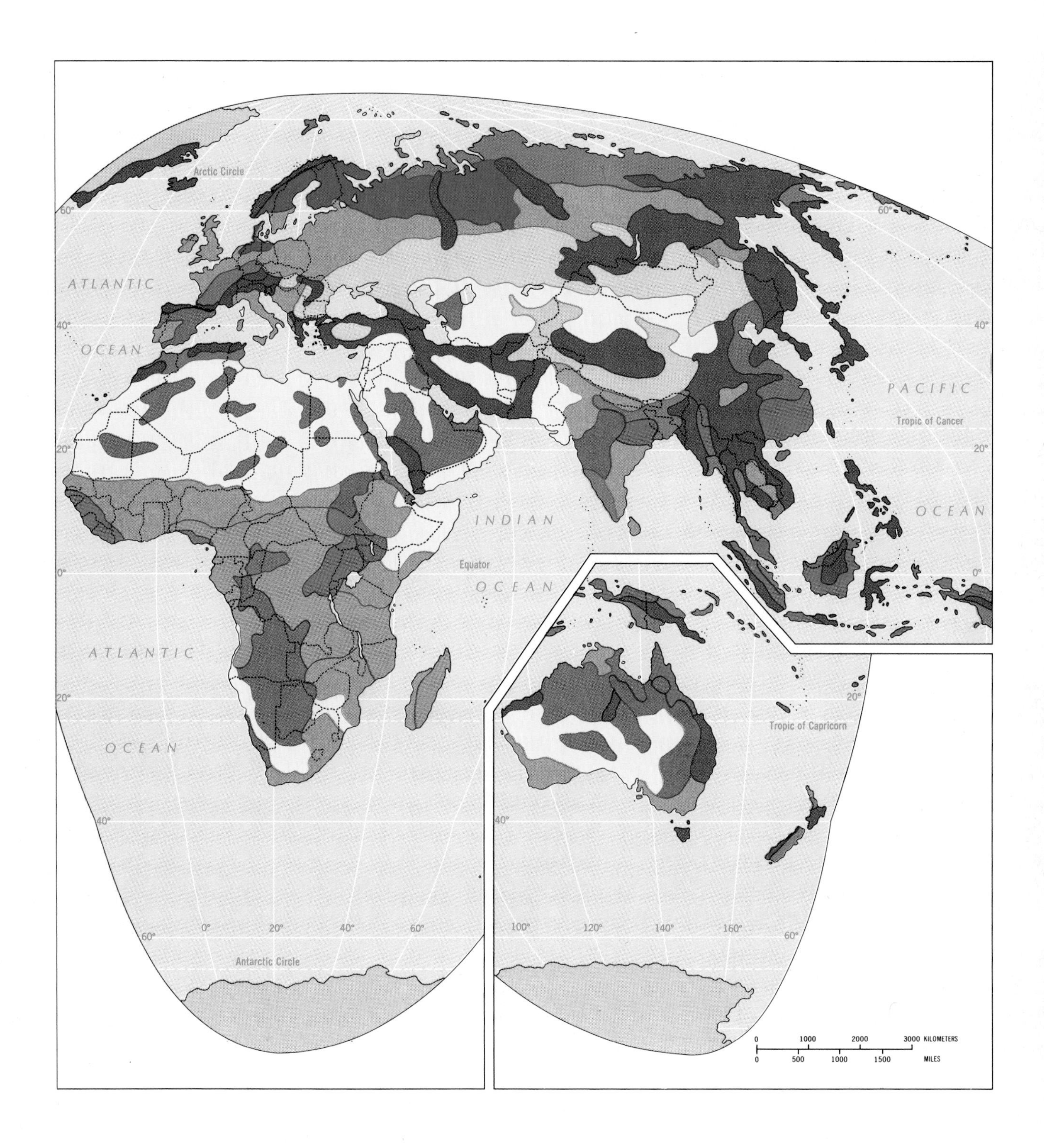

globe. And the picture is changing rapidly. After thousands of years of relatively slow growth, world population during the past two centuries has been expanding at an increasing rate. It took about 17 centuries from the time of the birth of Christ for the earth to add 250 million people. We will add the same number in less than the next three years. This is the subject of a discussion in Chapter 8, but it is important to realize that this explosive modern growth is *not* simply a matter of filling in the remaining available living space on the earth. Rather, it is the already-crowded areas that are becoming even more so, and the food produced in the rice fields of Asia (plentiful though it is) must be shared by ever more, hungrier people.

At present we confine ourselves to a view of the world population as it is distributed today. The 1980 world population was estimated to be some 4.7 billion. About one-fifth of this total resides in one country, China, and the next most populous single state, India, has more than 650 million inhabitants (see Table I-1). Yet both India and China still have extensive areas that are nearly devoid of permanent population, as Fig. I-9 reveals. China's habitable, agriculturally productive lands lie concentrated in that country's east, and the map leaves no doubt about the associated clustering of population.

Figure I-9 shows that the earth presently contains four large population agglomerations, the three largest of which lie on a single landmass: Eurasia. Greatest of all these is the East Asia cluster, adjoining the Pacific Ocean from Korea to Vietnam and centering, of course, on China itself. The map indicates that the number of people per unit area tends to decline from the coastal zone toward the interior, but several ribbonlike extensions can be seen to penetrate the deeper interior (*A* and *B* on Fig. I-9). Reference to the map of world landscapes (Fig. I-2) proves that these extensions represent populations concentrated in the valleys of China's major rivers. This serves to remind us that the great majority of the people of East Asia are farmers, not city dwellers. True, there are great cities in China, and some

Figure I-9

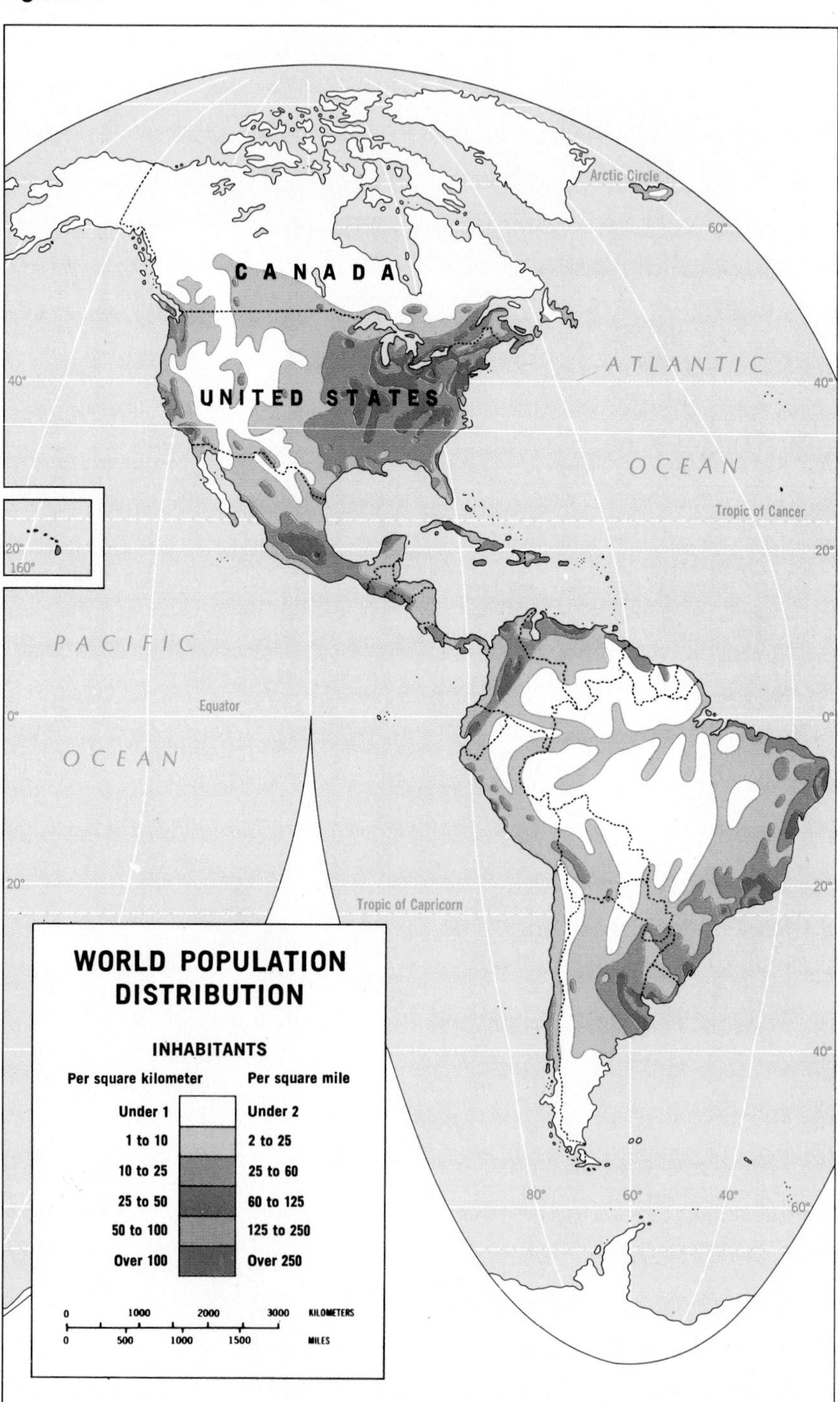

of them, such as Shanghai and Beijing (Peking), rank among the largest in the world. But the total population of these and the other cities is far outnumbered by the farmers—those who need the river valleys' soils, the life-giving rains, and the moderate temperatures to produce crops of wheat and rice to feed not only themselves but also those in the cities and towns.

The second major concentration of world population also lies in Asia, and it displays many similarities to that of East Asia. At the heart of this South Asia cluster lies India, but it extends also into Pakistan and Bangladesh and onto the island of Sri Lanka. Again, note the coastal orientation of the most densely inhabited zones, and the finger-

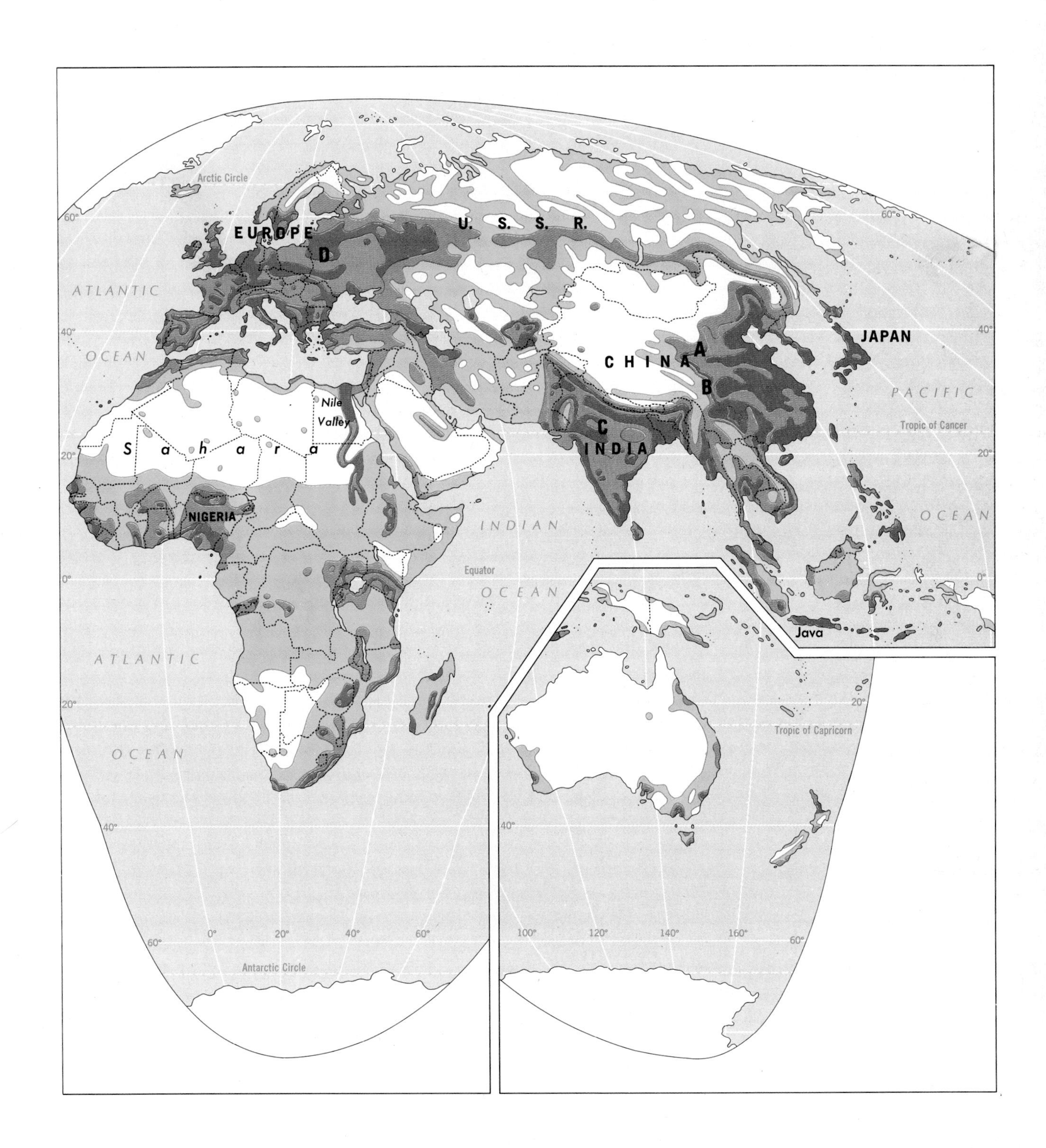

Table I-1 (Continued) Data for the World's States (Microstates Omitted)

Country	Area 1000 Km²	Area 1000 Miles²	Population (Millions) 1976	1979	1980	1985	Latest Census Date	Natural Rate of Increase (%) 1970–76	Population Density (No./Km²) 1976	1980
EUROPE	4,870.9	1,880.7	476.0	482.0	488.2	503.7			98	100
Albania	28.7	11.1	2.6	2.6	2.9	3.4	1975	3.0	88	101
Austria	83.8	32.4	7.5	7.5	7.6	7.6	1971	.2	90	90
Belgium	30.5	11.8	9.9	9.8	10.1	10.3	1970	.4	324	331
Bulgaria	110.9	42.8	8.8	8.9	9.0	9.2	1975	.5	79	81
Czechoslovakia	127.9	49.4	14.9	15.2	15.3	15.9	1970	.7	117	120
Denmark	43.1	16.6	5.1	5.1	5.2	5.3	1970	.5	118	121
Finland	337.0	130.0	4.7	4.8	4.8	4.9	1970	.4	14	14
France	547.0	211.0	53.0	53.4	54.5	56.4	1975	.7	97	100
West Germany	248.6	96.0	61.5	61.2	62.0	62.6	1970	.2	247	249
East Germany	108.2	41.8	16.8	16.7	16.6	16.4	1971	−.3	155	153
Greece	131.9	50.9	9.2	9.5	9.5	9.8	1971	.7	69	72
Hungary	93.0	35.9	10.6	10.7	10.8	11.0	1970	.4	114	116
Iceland	103.0	39.8	.2	.2	.2	.2	1970	1.3	2	2
Ireland	70.3	27.1	3.2	3.3	3.4	3.6	1971	1.2	45	48
Italy	301.2	116.3	56.2	56.9	58.0	60.4	1971	.8	186	193
Luxembourg	2.6	1.0	.4	.4	.4	.4	1970	.9	138	143
Netherlands	40.8	15.8	13.8	14.0	14.3	15.0	1971	.9	337	350
Norway	324.2	125.2	4.0	4.1	4.1	4.2	1975	.6	12	13
Poland	312.7	120.7	34.4	35.4	35.7	37.3	1970	.9	110	114
Portugal	92.1	35.6	(9.5)	10.0	(10.2)	(11.1)	1970	1.7	103	111
Romania	237.5	91.7	21.5	22.1	22.4	23.5	1977	1.0	90	94
Spain	504.8	194.9	36.0	37.6	37.6	39.7	1970	1.1	71	74
Sweden	450.0	174.0	8.2	8.3	8.3	8.5	1975	.4	18	18
Switzerland	41.3	15.9	6.4	6.3	6.5	6.6	1970	.4	154	157
United Kingdom	244.0	94.2	56.0	55.8	56.4	57.0	1971	.2	229	231
Yugoslavia	255.8	98.8	21.6	22.2	22.4	23.4	1971	.9	84	88
SOVIET UNION										
Soviet Union	22,402.2	8,649.5	256.7	264.0	266.1	278.3	1970	.9	11	12
AUSTRALIA	7,955.5	3,071.6	17.0	17.6	18.2	20.0			2	2
Australia	7,686.8	2,967.9	13.9	14.4	14.9	16.3	1976	1.8	2	2
New Zealand	268.7	103.7	3.1	3.2	3.3	3.3	1976	1.9	12	12
NORTH AMERICA	19,353.1	7,472.3	238.4	244.2	246.7	257.6			12	13
Bahamas	13.9	5.4	.2	.2	.2	.3	1970	3.6	15	17
Canada	9,976.1	3,851.8	23.1	23.7	24.4	26.2	1971	1.4	2	2
United States	9,363.1	3,615.1	215.1	220.3	222.1	231.1	1970	.8	23	24

like extension of dense population in northern India (*C* on Fig. I-9). This is one of the great concentrations of people on earth, in the valley of the Ganges River.

The South Asia population cluster numbers nearly 900 million people, and at present rates of growth it will exceed the one billion mark in the early 1980s. Our map shows how sharply this region is marked off by

Table I (Continued) Data for the World's States (Microstates Omitted)

Country	Area 1000 Km²	Area 1000 Miles²	Population (Millions) 1976	1979	1980	1985	Latest Census Date	Natural Rate of Increase (%) 1970–76	Population Density (No./Km²) 1976	1980
JAPAN										
Japan	372.3	143.7	112.8	115.9	118.7	126.7	1975	1.3	303	319
MIDDLE AMERICA	2,710.6	1,046.6	107.2	116.9	120.9	140.9			40	45
Barbados	.4	.2	.3	.3	.3	.3	1970	.7	574	589
Belize	23.0	8.9	.1	.2	.2	.2	1970	3.1	6	7
Costa Rica	50.7	19.6	2.0	2.2	2.2	2.5	1974	2.6	40	44
Cuba	114.5	44.2	9.5	9.9	10.1	11.0	1970	1.7	83	88
Dominican Republic	48.7	18.8	4.8	5.3	5.4	6.3	1970	3.0	99	112
El Salvador	21.4	8.3	4.1	4.5	4.7	5.6	1971	(3.4)	190	224
Guatemala	108.9	42.0	6.3	6.8	7.0	8.1	1973	2.9	57	64
Haiti	27.8	10.7	4.7	5.7	5.0	5.4	1971	1.6	168	179
Honduras	112.1	43.3	2.8	3.1	3.2	3.8	1975	(3.3)	25	29
Jamaica	11.0	4.3	2.1	2.2	2.2	2.4	1971	1.6	188	200
Mexico	1,972.5	761.6	62.3	67.7	71.5	84.9	1975	3.5	32	36
Nicaragua	130.0	50.2	2.2	2.5	2.5	2.9	1974	3.3	17	19
Panama	75.6	29.2	1.7	1.9	1.9	2.2	1970	3.1	23	25
Puerto Rico	8.9	3.3	3.2	3.5	3.6	4.1	1970	2.8	361	404
Trinidad and Tobago	5.1	2.0	1.1	1.1	1.1	1.2	1970	1.1	214	224
SOUTH AMERICA	17,715.4	6,840.0	218.0	233.6	241.9	275.4			12	14
Argentina	2,766.9	1,068.3	25.7	26.7	27.1	28.9	1970	1.3	9	10
Bolivia	1,098.6	424.2	5.8	5.2	6.5	7.4	1976	2.7	5	6
Brazil	8,512.0	3,286.0	109.2	118.7	122.0	140.0	1970	2.8	13	14
Chile	756.9	292.2	10.4	11.0	11.2	12.2	1970	1.8	14	15
Colombia	1,138.9	439.7	24.3	26.1	27.2	31.4	1973	2.9	21	24
Ecuador	283.6	109.5	7.3	8.0	8.3	9.9	1974	3.4	26	29
Guyana	215.0	83.0	.8	.8	.9	.9	1970	1.8	4	4
Paraguay	406.8	157.1	2.8	3.0	3.2	3.7	1974	3.2	7	8
Peru	1,285.2	496.2	16.1	17.3	18.1	21.0	1974	3.0	13	14
Surinam	163.3	63.1	.4	.4	.5	.6	1972	2.7	3	3
Uruguay	176.2	68.0	2.8	2.9	2.9	3.1	1975	(1.2)	16	16
Venezuela	912.0	352.0	12.4	13.5	14.0	16.3	1971	3.1	14	15

physical barriers: the Himalaya Mountains rise to the north above the Ganges lowland, the desert takes over west of the Indus River valley in Pakistan. This is a confined region, whose population is growing more rapidly than it is almost anywhere else on earth and whose capacity to support it has, by all estimates, already been exceeded. As in East Asia, the overwhelming majority of the people are farmers, but here in South Asia the pressure on the land is even greater. In Bangladesh, 90 million people are crowded into an area about the size of Iowa. Nearly all of these people are farmers, and even fertile Bangladesh has areas that cannot sustain as many people as others. Over large parts of Bangladesh the rural population is over 1000 per square kilometer. In comparison, the 1977 population of Iowa was about 2.9 million, of whom just 43 (by the 1970 census) lived on the land

Table I-1 (Continued) Data for the World's States (Microstates Omitted)

Country	Area 1000 Km^2	Area 1000 $Miles^2$	Population (Millions) 1976	1979	1980	1985	Latest Census Date	Natural Rate of Increase (%) 1970–76	Population Density (No./Km^2) 1976	1980
AFRICA	21,739.7	8,393.7	314.0	349.0	349.8	398.2			14	16
Angola	1,246.7	481.4	(6.4)	6.9	(7.0)	(7.7)	1970	(2.1)	5	6
Benin	112.6	43.5	3.2	3.5	3.6	4.1	1961	2.7	28	32
Botswana	600.4	231.8	.7	.7	.8	.9	1971	3.0	1	1
Burundi	27.8	10.7	3.9	4.4	4.2	4.8	1970	(2.4)	139	153
Cameroon	475.4	183.6	6.5	8.3	7.0	7.7	1974	1.9	14	15
Central African Republic	623.0	241.0	(2.3)	2.4	(2.5)	(2.9)	1973	(2.2)	4	4
Chad	1,284.0	496.0	4.1	4.4	4.5	5.0	1974	2.1	3	3
Comoro Islands	2.2	0.8	.3	.4	.3	.4	1975	2.6	145	158
Congo	342.0	132.0	1.4	1.5	1.5	1.8	1974	2.6	4	5
Equatorial Guinea	28.0	10.8	.3	.3	.3	.3	1974	1.7	11	12
Ethiopia	1,222.0	472.0	28.2	31.8	30.9	34.6	1970	2.3	23	25
Gabon	267.7	103.4	.5	.5	.6	.6	1970	1.0	2	2
Gambia	11.3	4.4	.5	.6	.6	.7	1973	2.6	48	53
Ghana	238.5	92.1	10.3	11.3	11.6	13.5	1970	3.0	43	49
Guinea	245.9	94.9	4.5	4.9	5.0	5.6	1972	2.4	18	20
Guinea-Bissau	36.1	13.9	.5	.6	.6	.6	1970	1.5	15	16
Ivory Coast	322.5	124.5	6.8	7.7	7.6	8.6	1975	2.6	21	24
Kenya	582.6	224.9	13.8	15.4	16.0	19.0	1975	3.6	24	27
Lesotho	30.4	11.7	1.2	1.3	1.3	1.4	1976	(2.0)	40	43
Liberia	111.4	43.0	1.8	1.8	2.0	2.2	1974	2.4	16	18
Madagascar	587.0	227.0	8.3	8.5	9.3	10.8	1975	3.0	14	16
Malawi	118.5	45.8	5.2	5.9	6.0	6.8	1977	2.6	44	51
Mali	1,240.0	479.0	6.0	6.5	6.7	7.5	1976	2.5	5	5
Mauritania	1,030.7	398.0	1.5	1.6	1.6	1.8	1976	(2.2)	1	2
Mauritius	2.0	0.8	.9	.9	.9	1.0	1972	1.3	438	471
Mozambique	783.0	302.0	9.4	10.2	10.3	11.6	1970	2.3	12	13
Namibia (S.W. Africa)	824.3	318.3	(.9)	1.0	(.9)	(1.0)	1970	(2.0)	(1)	(1)
Niger	1,267.0	489.0	4.7	5.1	5.3	6.0	1974	2.7	4	4
Nigeria	923.8	356.7	64.8	74.6	72.0	82.3	1973	2.7	70	78
Rwanda	26.3	10.2	4.3	4.9	4.8	4.4	1970	2.6	163	181
Senegal	196.2	75.8	5.1	5.5	5.6	6.3	1976	(2.4)	26	28
Sierra Leone	71.7	27.7	3.1	3.7	3.6	4.2	1974	3.4	43	50
Somalia	637.7	246.2	3.3	3.5	3.6	4.1	1975	2.6	5	6
South Africa	1,221.0	471.0	26.1	28.2	29.0	32.9	1970	2.6	21	24
Tanzania	945.1	364.9	15.6	17.0	17.4	19.8	1974	2.7	17	18
Togo	56.0	21.6	2.3	2.5	2.5	2.9	1970	2.6	41	45
Uganda	236.0	91.1	11.9	13.2	13.6	16.0	1975	3.3	51	58
Upper Volta	274.2	106.0	6.2	6.7	6.8	7.6	1975	2.3	23	25
Zaire	2,345.4	905.6	25.6	28.0	28.6	32.9	1970	2.8	11	12
Zambia	752.6	290.6	5.1	5.6	5.9	7.0	1974	3.5	7	8
Zimbabwe	390.6	150.8	6.5	7.2	7.5	8.9	1975	3.5	17	19

Table I-1 (Continued) Data for the World's States (Microstates Omitted)

Country	Area		Population (Millions)				Latest Census Date	Natural Rate of Increase (%)	Population Density (No./Km²)	
	1000 Km²	1000 Miles²	1976	1979	1980	1985		1970–76	1976	1980
NORTH AFRICA AND SOUTHWEST ASIA	15,096.1	5,828.6	240.1	256.3	267.8	307.0			16	18
Afghanistan	647.5	250.0	19.8	18.3	21.9	24.7	1975	2.5	31	34
Algeria	2,381.7	919.6	17.3	19.1	19.6	23.0	1974	3.2	7	8
Bahrain	.6	.2	.3	.3	.3	.3	1971	3.2	416	472
Cyprus	9.3	3.6	.6	.6	.7	.7	1973	.9	69	72
Egypt	1,001.4	386.6	38.0	40.6	41.5	46.3	1976	2.2	38	41
Iran	1,648.0	636.0	33.4	36.3	37.0	42.1	1976	2.6	20	22
Iraq	434.9	167.9	11.5	12.9	13.2	15.5	1975	3.4	26	30
Israel	20.8	8.0	3.5	3.8	3.9	4.5	1972	3.0	167	188
Jordan	97.7	37.0	2.8	3.0	3.2	3.7	1975	3.2	28	32
Kuwait	17.8	6.9	1.0	1.3	1.3	1.8	1975	6.0	60	75
Lebanon	10.4	4.0	3.0	3.1	3.3	3.9	1970	3.1	285	322
Libya	1,759.4	679.3	(2.5)	2.8	(2.8)	(3.2)	1973	(3.0)	1	2
Morocco	446.6	172.4	17.8	19.4	20.3	23.9	1971	(3.3)	40	45
Oman	212.5	82.0	.8	.9	.9	1.0	1974	3.1	4	4
Qatar	11.0	4.2	.1	.2	.1	.1	1972	3.1	9	9
Saudia Arabia	2,149.7	830.0	9.2	8.1	10.4	12.1	1974	3.0	4	5
Sudan	2,505.8	967.5	16.1	17.9	18.2	21.2	1973	(3.1)	6	7
Syria	185.2	71.5	7.6	8.4	8.6	10.2	1970	3.3	41	47
Tunisia	163.6	63.2	5.7	6.4	6.4	7.4	1975	(2.8)	35	39
Turkey	780.6	301.4	40.2	44.3	44.2	49.8	1975	2.4	51	57
United Arab Emirates	83.6	32.3	.2	.9	.3	.3	1974	3.2	3	3
Yemen (Aden)	333.0	128.6	1.8	1.9	2.0	2.3	1973	(3.0)	5	6
Yemen (San'a)	195.0	75.3	6.9	5.8	7.7	9.0	1975	(3.0)	35	40
INDIA AND PERIMETER	4,489.2	1,733.3	791.0	857.5	864.6	966.6			176	193
Bangladesh	144.0	55.6	80.6	87.1	90.0	103.3	1974	2.8	559	625
Bhutan	47.0	18.0	1.2	1.3	1.3	1.5	1974	2.2	26	28
India	3,287.6	1,269.3	610.1	660.9	663.0	735.6	1971	2.1	186	202
Maldives	.3	.1	.1	.1	.1	.2	1974	(2.3)	454	496
Nepal	140.8	54.5	12.9	13.7	14.1	15.8	1971	2.3	91	100
Pakistan	803.9	310.4	72.4	79.9	81.5	94.4	1972	3.0	90	101
Sri Lanka	65.6	25.3	13.7	14.5	14.6	15.8	1971	1.6	209	223

rather than in cities and towns. Rural density in Iowa is 8 per square kilometer.

Further inspection of Fig. I-9 reveals that the third-ranking population cluster also lies in Eurasia—and at the oppostie end from China. An axis of dense population extends from the British Isles into Soviet Russia; it includes large parts of West and East Germany, Poland, and the western Soviet Union. It also incorporates the Netherands and Belgium, parts of France, and northern Italy. This European cluster (including the contiguous U.S.S.R.) counts over 700 million inhabitants, which puts it in a class with the South Asia concentration—but there the similarity ends. A comparison of the population and physical maps indicates that in Europe, terrain and environment appear to have less to do with population distribution than in the two Asian cases. See, for example, that lengthy extension marked *D* on Fig. I-9, which protrudes far into the Soviet Union. Unlike the Asian extensions,

Table I-1 (Continued) Data for the World's States (Microstates Omitted)

Country	Area		Population (Millions)				Latest Census Date	Natural Rate of Increase (%) 1970–76	Population Density (No./Km²)	
	1000 Km²	1000 Miles²	1976	1979	1980	1985			1976	1980
CHINA AND ITS SPHERE	11,381.0	4,394.4	970.0	1,024.1	1,042.1	1,134.9			85	92
China	9,561.0	3,692.0	(900.0)	950.0	(966.2)	(1,051.1)	1953	(1.7)	94	101
North Korea	120.5	46.5	16.2	17.5	18.0	20.4	1974	2.6	135	149
South Korea	98.5	38.0	35.9	37.6	38.5	42.1	1975	1.8	364	391
Mongolia	1,565.0	604.0	1.5	1.6	1.7	1.9	1975	3.0	1	1
Taiwan	36.0	13.9	(16.4)	17.3	(17.7)	(19.4)	1970	(1.9)	456	492
SOUTHEAST ASIA	4,478.4	1,728.8	330.3	344.8	367.0	419.0			74	82
Brunei	5.8	2.2	.2	.2	.2	.3	1971	5.2	31	38
Burma	676.6	261.2	30.8	32.9	33.6	37.5	1973	2.2	46	50
Kampuchea	181.0	69.9	8.4	8.9	9.3	10.7	1974	2.8	46	52
Indonesia	1,904.3	735.3	139.6	140.9	154.7	175.9	1971	2.6	69	81
Laos	236.8	91.4	3.4	3.7	3.7	4.1	1975	2.2	14	16
Malaysia	329.7	127.3	12.3	13.3	13.8	15.9	1970	2.9	37	42
Philippines	300.0	116.0	43.8	46.2	49.1	56.7	1975	2.9	146	164
Singapore	.6	.2	2.3	2.4	2.4	2.6	1970	1.6	3,921	4,178
Thailand	514.0	198.0	43.0	46.2	48.0	55.1	1970	2.8	84	93
Vietnam	329.6	127.3	46.5	50.1	52.2	60.2	—	2.9	141	158
PACIFIC	480.0	185.4	3.4	3.7	3.7	4.1			7	8
Fiji	18.3	7.1	.6	.6	.6	.7	1974	1.8	32	34
Papua New Guinea	461.7	178.3	2.8	3.1	3.1	3.4	1971	2.2	6	7

which reflect fertile river valleys, the European population axis relates to the orientation of Europe's coalfields, the power resources that fired the industrial revolution. If you look more closely at the physical map, you will note that comparatively dense population occurs even in rather mountainous, rugged country—for example, along the boundary zone between Czechoslovakia and Poland. In Asia, there is much more correspondence between coastal and river lowlands and high population density than there generally is in Europe.

Another contrast lies in the number of Europeans who live in cities and towns. Far more than in Asia, the Europe population cluster is constituted by numerous cities and towns, many of them products of the industrial revolution. In the United Kingdom, about 80 percent of the people live in such urban places; in West Germany, nearly 80; in France about 70. With so many people concentrated in the cities, the rural countryside is more open and sparsely populated than in East and South Asia, where fewer than 20 percent of the people reside in cities and towns.

The three world population concentrations discussed (East Asia, South Asia, and Europe) account for over 3.1 of the world's more than 4.7 billion people. Nowhere else on the globe is there any population cluster with dimensions even half of any of these. Look at the dimensions of the landmasses on Fig. I-9 and consider that the populations of South America, Africa, and Australia combined total *less* than that of India alone. In fact, the next-ranking cluster, comprising the east-central United States and southeastern Canada, is only about one-quarter the size of the smallest of the Eurasian concentrations. As Fig. I-9 shows, this region does not have the large, contiguous high-density zones of Europe or East and South Asia.

Agricultural Practices

The majority of the earth's peoples, industrial and technological progress notwithstanding, still farm the soil for a living. But farming in the tropical rain forest of Africa is something very different from farming in the rice paddies of Asia—and Asia's rice farms look nothing like the vast wheat fields in the North American Great Plains.

Agricultural practices and systems, therefore, vary widely. In the tropics it is often necessary to cut down and burn the original forest vegetation to clear a patch of land that will support a crop (probably a root crop) for one year, perhaps two. It is all done by hand, and slash-and-burn agriculture is energy-efficient; machines mean very little in areas where cleared land must soon be abandoned in favor of a new patch. But neither can shifting agriculture support a very dense population, even on the forest margins where the rainfall becomes a little less and the soil somewhat less leached. On Fig. I-9, huge areas of equatorial Africa, South America, and Southeast Asia show a population density under 10 per square kilometer (25 per square mile). Extensive subsistence farming is the rule here, even where the rainfall declines to 100 centimeters (40 inches) annually (Fig. I-5), and a dry season permits the harvesting of corn and other hardy grains.

The paddies of South and East Asia sustain subsistence farmers too, but cultivation here is intensive and population densities are very high. Once again the Asian rice culture is highly efficient in terms of energy inputs; most of the paddies are still prepared by ox-drawn plow, and the rice is planted by hand—hundreds of millions of hands. But much of the South and East Asian rice land would be just another tropical zone of meager subsistence agriculture were it not for the alluvial soils in the great river basins. Highly fertile and replenished by rains that bring not only needed moisture but also new coatings of silt, these soils are sometimes capable of sustaining two, even three crops in a single year—one *after* the other. The great masses of South and East Asian population depend on these soils, and these great human clusters have grown on the strength of their productivity.

Commerical agriculture in the wheatlands of the Great Plains presents quite another picture. Vast, almost unbroken fields of grain clothe the countryside; the soil was fertilized by machine, the sowing was done by machine, and so was the harvesting. Modern commercial agriculture is labor-efficient, requiring few hands—but it is less efficient in terms of energy requirements.

Between these extremes lie other agricultural systems: the plantation, a specialized commercial enterprise in tropical and subtropical environments capable of supporting a population of medium density, and the complex, integrated exchange type in the hinterlands of the large urban areas of Western Europe, the eastern United States, and Japan, where space is at a premium, distance to markets is often crucial, and the soil is heavily fertilized and tended so that it will produce as much as possible.

The North American population cluster displays European characteristics, and it even outdoes Europe in some respects. Like the European region, much of the population is concentrated in several major cities, while the rural areas remain relatively sparsely populated. The major focus of the North American cluster lies in the urban complex along the Eastern Seaboard, from Boston to Washington, which includes New York, Philadelphia, and Baltimore. This great urban agglomeration is called *megalopolis* by urban geographers who predict that it is only a matter of time before the whole area coalesces into an enormous megacity. But there are other urban foci in this North American cluster; Chicago lies at the heart of one, and Detroit and Cleveland anchor a second. If you study Fig. I-9 carefully, you will note other prominent North American cities standing out as small areas of high-density population, including Pittsburgh, St. Louis, and Minneapolis-St. Paul.

Still further examination of Fig. I-9 leads us to recognize substantial population clusters in Southeast Asia. It is appropriate to describe these as discrete clusters, for the map confirms that they are actually a set of nuclei rather than a contiguous population concentration. Largest among these nuclei is the Indonesian island of Djawa (Java), with some 85 million inhabitants. Elsewhere in the region, populations cluster in the lowlands of major rivers, some of which came frequently to our attention during the Vietnam conflict—for example, the Mekong Delta. Neither these river valleys nor the rural surroundings of the cities have population concentrations comparable to those of either China to the north or India to the west, and under normal circumstances Southeast Asia is able to export rice to its more hungry neighbors. Decades of strife have, however, disrupted the region to such a degree that its productive potential has not been attained.

South America, Africa, and Australia do not sustain population concentrations comparable to those we have considered. Africa's 480 million inhabitants cluster in above-average densities in West Africa (where Nigeria has a population larger than its offical census, probably 90 million) and in a zone in the east extending from Ethiopia to South Africa. Only in North Africa is there an agglomeration comparable to the crowded riverine plains of Asia: the Nile Valley and Delta with 42 million residents. Importantly, it is the pattern—not the dimensions—that resembles Asia. As in East and South Asia, the Nile's Valley and Delta teem with farmers who cultivate every foot of the rich and fertile soil. But the Nile's gift is a miniature compared to its Asian equivalents. The Ganges, Yangtze, and Huange Rivers' lowlands contain many times the number of inhabitants who manage to eke out a living along the Nile.

The large light-shaded spaces in South America and Australia, and the peripheral distribution of the modest populations of these continents, suggest that here remains space for the world's huge numbers. And indeed, South America could probably sustain more than its present 240 million, as Australia can undoubtedly accommodate more than 15 million. But the population growth rate of South American countries is among the highest in the world, and Australia's environmental limitations hardly qualify that continent as a relief valve for Asia's farming millions.

Political Geography

We return in later chapters to additional issues involving world population: problems of density, growth patterns, migration, well-being. The discussion of world realms must, however, also be preceded by a brief introduction to the complexities of the politico-geographic world. The limited land area of the globe is presently divided among about 150 national states and nearly 60 dependent territories. The states range in population from China's 1000 million to Liechtenstein's 24,000; in terms of territory, the largest is the Soviet Union with 22,402,000 square kilometers (8,650,000 square miles), and the smallest is Monaco, with less than 2 square kilometers. Even when the so-called *microstates* are left out of such comparisons (as is done in Table I-1), the range remains enormous. Some full-fledged members of the United Nations such as Iceland and Bahrain have populations of about 250,000.

Inequalities among the world's national states are not confined to size and population. Large territorial size is no guarantee of resource wealth: some of the world's larger states are among the poorest. Zaire, in Equatorial Africa, Sudan, much

An ancient linear boundary in Europe: Hadrian's Wall on the Scottish-English border, built by the Romans to mark the limit of their empire in Britain. The concept of a boundary as a line (rather than a transition zone or frontier) developed early and independently in several regions of the world, including China and South America. (Brian Seed/Black Star)

of which is Sahara, India, with its huge population, and Brazil, better off but not rich despite its huge area, are among countries whose size is not matched by comparable shares of the earth's resources. Neither do large numbers of people ensure national strength. Indonesia has more than 150 million inhabitants; Nigeria contains more people than West Germany. Some of the states with small populations (Sweden, the Netherlands) are among those enjoying high standards of living, and many populous countries are poor.

The familiar world political map (Fig. I-10) is a recent product of national territorial competition and adjustment. Just one century ago large areas of Africa and Asia remained beyond the jurisdiction of encroaching imperialist powers. Finally, when colonial spheres of influence collided, boundaries often were drawn to assign dependencies and to facilitate European administration. The map of Africa was largely created or confirmed at the Berlin Conference held in the mid-1880s. At that time, most probably, no one realized that those hastily defined colonial boundaries would within a century constitute the borders of independent states.

The world's boundary framework always is under pressure in certain areas, and the past decade has witnessed several changes. The boundary between North and South Vietnam has been eliminated. Similarly, the border between Portuguese and Indonesian Timor has been erased. In 1980 the issue was still not finally settled, but officially the boundary between Morocco and former Spanish Sahara was erased and the territory divided between Morocco and Mauritania along a new border. Following a major conflict on the island of Cyprus, a *de facto* boundary between Greeks and Turks appeared and functioned in some respects as a recognized international boundary. In Lebanon, too, demarcated boundaries appeared between Christian and Moslem zones during the conflicts of the 1970s. The question of Israel's boundaries with its neighbors continues to dominate Middle East political affairs.

These actual and prospective boundary changes involve comparatively minor modifications of a boundary framework whose major outlines have proved quite durable. When African and Asian colonies attained independence, some observers predicted that the "boundaries of imperialism" would soon be abandoned in favor of more realistic dividing lines. But despite some significant steps in this direction (the India-Pakistan border, for example, created as the British colonial era came to an end), the chief elements of the preindependence boundary framework remained intact—as they are today.

The concept of the linear political boundary is quite old; Romans and Chinese built walls for this purpose 2000 years ago. In post-Roman Europe, rivers often served as trespass lines, and territorial competition and boundary delimitation were part of European political gestation. Europe's colonial expansion during the seventeenth and eighteenth centuries, and the power arising from the industrial revolution, hastened a process that had begun early but had progressed slowly. European concepts of territorial acquisition and delimitation, even at sea, now were imposed on much of the rest of the world. Within a century, what remained of open frontiers, even on ice-covered Antarctica, disappeared. We are witness today to the final stage of this process as states are about to claim and absorb the bulk (and perhaps all) of the seas and oceans and what lies beneath them. These and other politico-geographical realities will be discussed in several of the regional chapters that follow.

Figure I-10

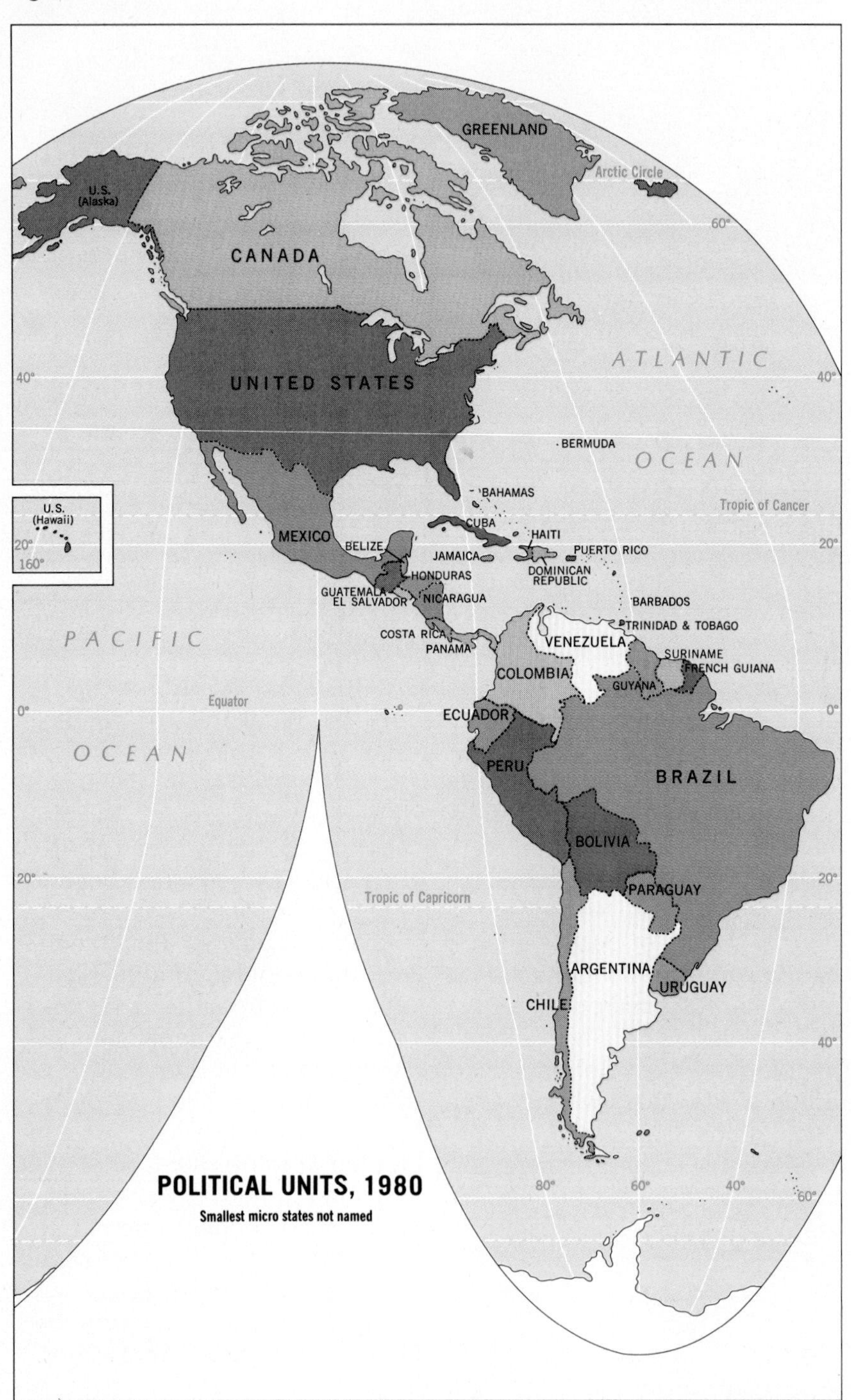

Developed and Underdeveloped Countries

The world geographic realms discussed in this book are

grouped under two headings: *developed* regions and *underdeveloped* (or *developing,* as some prefer) regions. These terms are used because they have become commonplace, as have such adjectives as rich and poor, "haves" and "have-nots," and the like. But perhaps the best appellation for the developed and underdeveloped countries would be *advantaged* and *disadvantaged.* In United Nations parlance, the disadvantaged countries are mainly the countries of the Third World—the capitalist and socialist systems dominating the first two. Third World countries find themselves in a disadvantageous economic position in relation to the developed, economically powerful states of the world—capitalist or socialist.

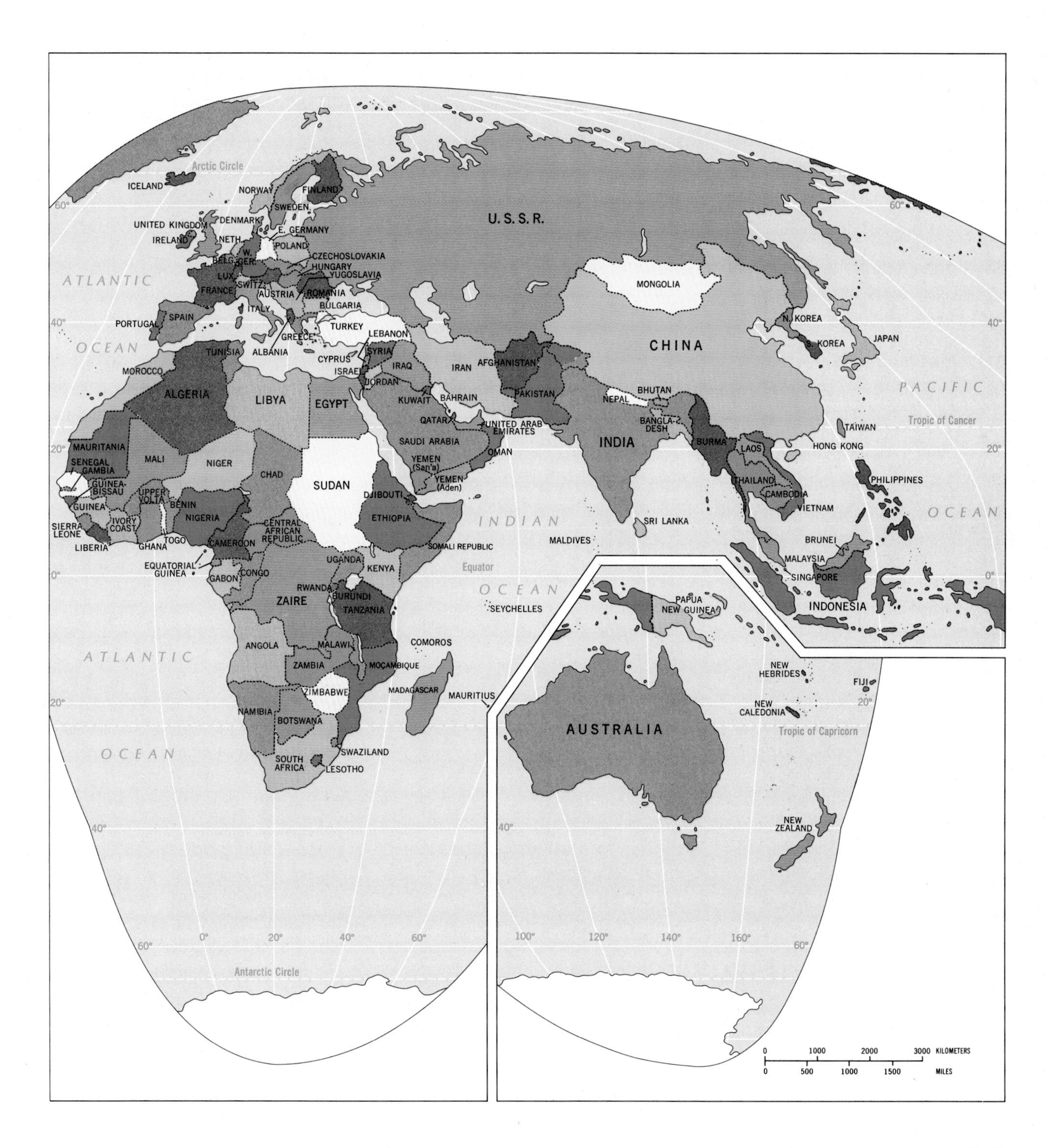

Grouping the advantaged countries together, they include most of the states of Europe, the Soviet Union, Canada and the United States, Australia and New Zealand, Japan, Singapore, Israel, and, by some definitions, South Africa. Against this comparatively short list (the combined population of these countries amounts to barely 30 percent of the world total) are the countries at various levels of underdevelopment: those of Middle and South America, North Africa and Southwest Asia, Black Africa, and South, Southeast, and East Asia. Not all these less developed countries are at the same level of underdevelopment (just as some developed countries are ahead of others). Few would argue that Haiti, Bolivia, Ethiopia, Bangladesh, and Indonesia are not underdeveloped countries (UDCs). But to place Argentina, Uruguay, Chile, Venezuela, Mexico, Brazil, Turkey, Taiwan, and South Korea in the same category creates an issue. Some economic geographers suggest that it is appropriate to recognize, between the developed countries (DCs) and the underdeveloped countries, a middle tier of countries which, while still possessing many characteristics of UDCs, are "emerging" or at a "takeoff" stage.

Although we routinely (and arbitrarily) divide our world into developed and underdeveloped areas, no universally applicable criteria exist to measure development. Leading indexes are listed under seven headings (box, p. 39), but they remain arbitrary and subject to debate. And what these numerical indexes fail to convey is the time factor. The developed countries are the advantaged countries—but why cannot the underdeveloped and "takeoff" countries catch up? It is not, as is sometimes suggested, simply a matter of environment, resource distribution, or cultural heritage (a resistance to innovation, for example). The sequence of events that led to the present division of our world began long before the industrial revolution occurred. Europe even by the middle of the eighteenth century had laid the foundations for its colonial expansion; the industrial revolution magnified Europe's demands for raw materials while its products increased the efficiency of its imperial control. While Western countries gained an enormous head start, colonial dependencies remained suppliers of resources and consumers of the products of the Western industries. Thus was born a system of international exchange and capital flow that really changed but little when the colonial period came to an end. Underdeveloped countries, well aware of their predicament, accused the developed world of perpetuating its advantage through *neo*colonialism—the entrenchment of the old system under a new guise.

Symptoms of Underdevelopment

The disadvantaged countries suffer from numerous demographic, economic, and social ills. Their *populations* tend to display high birth rates and moderate to high death rates; life expectancy at birth is comparatively low (see Chapter 8). A large percentage of the population (as much as half) is 15 years or younger. Infant mortality is high. Nutrition is inadequate, and diets are not well balanced. Protein deficiency is a common problem. The incidence of disease is high; health care facilities are inadequate. There is an excessively high number of persons per available doctor; hospital beds are too few in number. Sanitation is poor. Substantial numbers of school-age chidren do not go to school; illiteracy rates are high.

The *rural areas* are overcrowded and suffer from poor surface communications. Men and women do not share fairly in the work that must be done; women's workload is much heavier, and children are pressed into the labor force. Landholdings are often excessively fragmented, and the small plots are farmed with outdated, inefficient tools and equipment. The main crops tend to be cereals and roots; protein output is low because its demand on the available and is higher. There is little production for the local market, because distribution systems are poorly organized and the local market is weak and generates low demand. On the farms, yields per unit area are low, subsistence modes of life prevail, and the spectre of debt hangs constantly over the peasant family. Such circumstances preclude investment in such luxuries as fertilizers and soil conservation measures. As a result, soil erosion and land denudation scar the rural landscapes of many underdeveloped countries. Where areas of larger-scale, modernized agriculture have developed, these produce for foreign markets, and their impact on domestic conditions is slight.

In the *urban areas* overcrowding, poor housing, inadequate sanitation, and a general

Measures of Development

What distinguishes a developed economy from an underdeveloped one? Obviously it is necessary to compare countries on the basis of certain measures; the question cannot be answered simply by subjective judgment. No country is totally developed; no economy is completely underdeveloped. We are comparing *degrees* of development when we identify DCs and UDCs. Our division into developed and underdeveloped economies is arbitrary, and the dividing line is always a topic of debate. There is also the problem of data. Statistics for many countries are inadequate, unreliable, or unavailable.

The following list of measures is normally used to gauge levels of economic development:

1. National product per person. This is determined by dividing the sum of all incomes achieved by a country's citizens by the total population. Figures for all countries are then converted to a single currency for purposes of comparison. In DCs the level is as high as $2000 and over; in some UDCs it is as low as $100.
2. Occupational structure of the labor force. This is given as the percentages of workers in various sectors of the economy. A high percentage of laborers engaged in the production of food staples, for example, signals low development levels.
3. Productivity per worker. This is the sum of production over a period of a year by the total number of persons engaged in productive activity (that is, the labor force).
4. Consumption of energy supply per person. Apart from noncommercial sources, how much electric and other power is used in a country? High-quantity use reflects greater development, for example, but the data must be viewed to some extent in the context of climate.
5. Transport and communication facilities per individual. This measure reduces railroad, road, airlines, vehicle numbers, and so forth, to a per capita index.
6. Consumption of metals per person. A strong indicator of development levels is the quantity of iron and steel, copper, aluminum, and other metals used by a population during a given year.
7. Rates. A number of additional measures are employed, including literacy rates, caloric intakes per person, percentage of family income spent on food, and amount of savings per person.

lack of services prevail. Employment opportunities are insufficient, and unemployment is high. Per capita income is low, savings per person are minimal, and credit facilities are poor. Families spend a very large portion of their income on food and basic necessities. The middle class remains small; not infrequently a substantial segment of the middle class consists of foreign immigrants (see Chapter 7).

These are some of the criteria that signal underdevelopment, and the list is not complete. For example, one of the geographic properties that mark underdeveloped countries is the problem of *regional imbalance.* Even in UDCs, there are local exceptions to the general economic situation. The capital city may appear as a skyscrapered model of urban modernization, with thriving farms in the immediate surroundings, factories on the outskirts; road and rail may lead to a bustling port where luxury automobiles are unloaded for use by the privileged elite. Here in the country's core area the rush of "progress" is evident—but travel a few kilometers into the countryside and you may find that almost nothing has changed. And just as the rich countries become richer and leave the poorer countries ever farther behind, so the gap between progressing and stagnant regions *within* developing countries grows larger. It is a problem of global dimensions.

There can be no doubt that the world economic system works to the disadvantage of the underdeveloped countries, but sadly it is not the only obstacle the poorer countries face. Political instability, corruptible leaderships and elites, misdirected priorities, misuse of aid, and traditionalism are among circumstances that inhibit development. External interference by interests representing powerful developed countries have also had negative impact on the economic as well as the poltical progress of UDCs. Underdeveloped countries even get caught in the squeeze when other developing countries try to assert their limited strength; when the OPEC countries, mostly underdeveloped themselves, raise the price of oil, energy and fertilizers slip still further from the reach of the poorer underdeveloped countries not fortunate enough to belong to this favored group. As the DCs get stronger and wealthier, they leave the underdeveloped world ever farther behind: the gap is widening, and the prospects for the UDCs are not bright.

Geographic Realms of the World

Thirteen world realms form the structural basis for our study; five of them consist mainly of developed countries (Europe, Australia, the Soviet Union, North America, Japan), and the remaining eight are constituted by underdeveloped and emerging countries (Fig. I-11). We begin with a series of regional outlines that should be read in conjunction with the map, and that form the intro-

Figure I-11

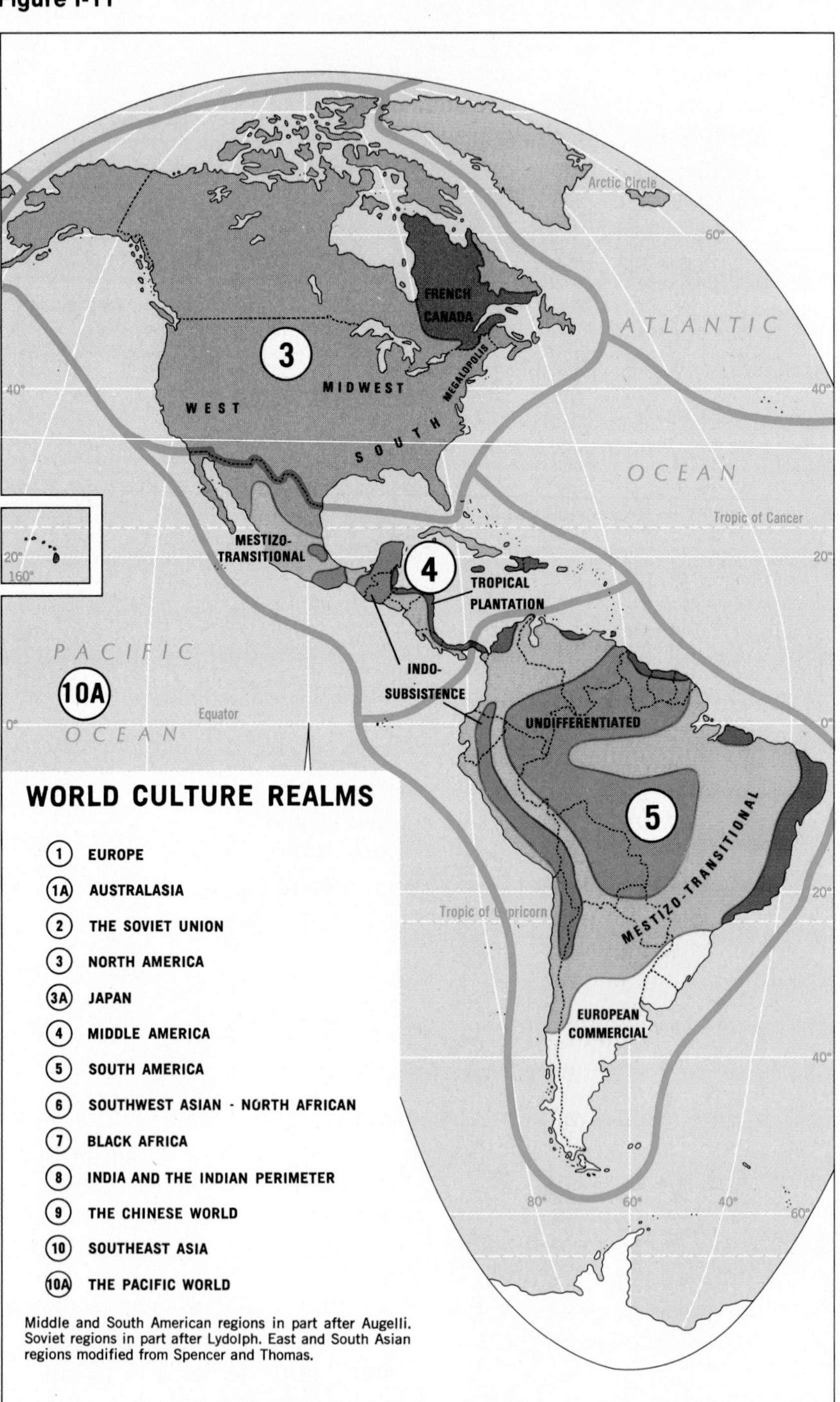

ductions to the coming chapters.

Europe (1)

Europe merits identification as a culture realm despite the fact that it occupies a mere fraction of the total area of the Eurasian landmass—a fraction that, moreover, is largely made up of that continent's western peninsular extremities. Certainly Europe's size is no measure of its world significance; probably no other part of the world is or ever has been so packed full of the products of human achievement. Innovations and revolutions that transformed the world originated in Europe. Over centuries of modern times the evolution of world interaction focused on European states and European capitals. Time and

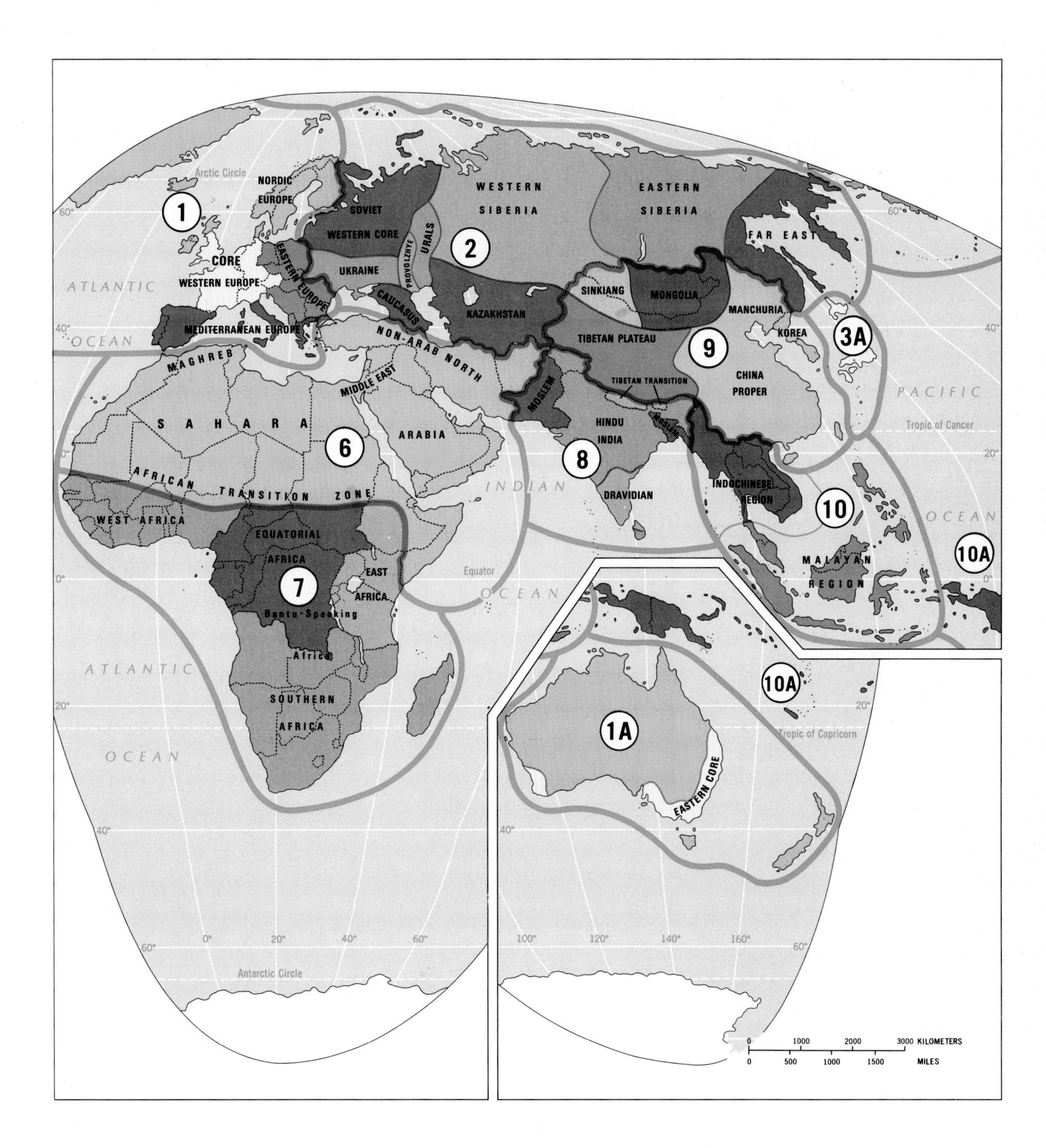

again, despite internal wars, despite the loss of colonial empires, despite the impact of external competition, Europe proved to contain the human and natural resources needed for rebounding and renewed progress.

Among Europe's greatest assets is its internal natural and human diversity. From the warm shores of the Mediterranean to the frigid Scandinavian Arctic and from the flat coastlands of the North Sea to the grandeur of the Alps, Europe presents an almost infinite range of natural environments. An insular and peninsular west contrasts against a more continental east. A resource-laden backbone extends across Europe from England eastward. Excellent soils produce harvests of enormous quantity and variety. And the population includes people of many different stocks, peoples grouped under such familiar names as Latin, Germanic, and Slavic. Europe has its minorities as well—for example, the Hungarians and the Finns. Immigrants continue to stream into Europe, contributing further to a diversity that has been an advantage to Europe in uncountable ways. Today Europe is a realm dominated, especially in the west, by great cities, intensive transport networks and mobility, enormous productivity, dynamic growth, a large and in many areas very dense population, and an extremely complex technology.

Australia (1A)

Just as Europe merits recognition as a continental realm although it is merely a peninsula of Eurasia, so Australia has achieved identity as the island realm of Australasia. Australia and New Zealand are European outposts in an Asian-Pacific world, as unlike Indonesia as Britain is unlike India.

Although Australia was spawned by Europe and its people and economy are Western in every way, Australia as a continental realm is a far cry from the crowded, productive, populous European world. The image of Australia is one of impressive, large, open cities, wide expanses, huge herds of livestock (perhaps also the pests, rabbits and kangaroos), deserts, and beautiful coastal scenery. There is much truth in such a picture: 10 of Australia's nearly 15 million people are concentrated in the country's seven largest cities! In this as well as other respects, Australia is more like the United States than other realms. With its British heritage, its homogeneous population, single language (excepting only the small indigenous minority), and type of economy, Australia's identification as a realm rests on its remoteness and spatial isolation.

The Soviet Union (2)

The Soviet Union—the world's largest country territorially—constitutes a culture realm not just because of its size. To its south lie the Japanese, Chinese, Indian, and Southwest Asian realms, all clearly different culturally from the realm that Russia forged. The Soviet realm's western boundary is always subject to debate, but neither Finland nor Eastern Europe have been areas of permanent Soviet or Russian domination.

The events of the twentieth century have greatly strengthened the bases for the recognition of the Soviet Union as a culture realm. Russian expansionism was halted, the revolution came, and a whole new order was created, transforming the old Russia and its empire into a strongly centralized state whose hallmark was economic planning. The new political system awarded the status of Soviet Socialist Republic to areas incorporated into the Russian empire, including those inhabited by Moslems and other minorities. In the economic sphere, the state took control of all industry and production, and agriculture was collectivized.

Notwithstanding the disastrous dislocation of World War II, the past half century has seen much progress in the Soviet Union. The state has risen from a backward, divided, near-feudal country to a position of world power with a strong, individualistic (and also centrally directed) culture, a highly advanced technology, and a set of economic and social policies that have attracted world attention and, in some instances, emulation.

North America (3)

The North American culture realm consists of two of the most strongly urbanized and industrialized countries in the world. In the United States there are more than 60 cities with populations in excess of half a million, within which live fully 50 percent of all the people in the country. The vast majority of the remainder, moreover, live in urban places larger than 10,000.

The North American realm is characterized by its large-

A North American cultural landscape: skyscrapers, suburbs, and expressways of Los Angeles, California. (Tom McHugh/Photo Researchers)

scale, massive technology, its enormous consumption of the world's resources and commodities, and its unprecedented mobility and fast-paced life-styles. Suburbs grow toward each other as cities coalesce, surface and air transport networks intensify, skylines change. Skyscrapers, traffic jams, waiting lines, noise, pollution of air and water—these are some of the attributes of North American technocracy.

Both the United States and Canada are plural societies, Canada's cultural sources lying in Britain and France and those of the United States in Europe and Africa. In both Canada and the United States, minorities remain separate from the dominant culture: Quebec is Canada's French province, and in the United States, patterns of racial segregation persist, with black Americans concentrated strongly in particular urban areas. The problems associated with cultural pluralism are prominent modifiers of this culture realm.

Japan (3A)

On Fig. I-11 Japan hardly seems to qualify as a culture realm. Can a group of comparatively small islands qualify as a culture realm? Indeed it can—when 116 million people inhabit those islands and when they produce there an industrial giant, an urban society, a political power, and a vigorous nation. Japan is unlike any other non-Western country, and it exceeds many Western-developed countries in many ways. Like its European rivals, Japan built and later lost a colonial empire in Asia. During its tenure as a colonial power Japan learned to import raw materials and export finished products. Even the calamity of World War II, when Japan became the only country ever to suffer atomic attack, failed to extinguish Japan's forward push. In the postwar period Japan's overall economic growth rate has been the highest in the world. Raw materials no longer come from colonies, but they are bought all around the world. Products are sold practically everywhere. Almost no city in the world is without Japanese cars in its streets, few photography stores are without Japanese cameras, many laboratories use Japanese optical equipment. From tape recorders to television sets, from ocean-going ships to children's toys,

Japanese manufacturers flood the world's markets. The Japanese seem to combine the precision skills of the Swiss with the massive industrial power of prewar Germany, the forward-looking designs of the Swedes with the innovativeness of the Americans.

Japan constitutes a culture realm by virtue of its role in the world economy today, the transformation, in one century, of its national life, and its industrial power. Yet Japan has not turned its back on its old culture; modernization is not all that matters to the Japanese. Ancient customs still are adhered to and honored. Japan is still the land of *Kabuki* drama and *sumo* wrestling, Buddhist temples and Shinto shrines, tea ceremonies and fortune-tellers, arranged marriages, and traditional song and dance. There is still a reverence for older people, and despite all the factories, huge cities, fast trains, and jet airplanes, Japan is still a land of shopkeepers, handcraft industries, home workshops, carefully tended fields, and space-conserving villages. Only by holding on to the culture's old traditions could the Japanese have tolerated the onrush of their new age.

Middle America (4)

Middle and South America combined are often called "Latin" America because of the Iberian imprint placed on them during the European expansion, when the ancient Indian civilizations present in the region were submerged and destroyed. The Latin imprint is far more pervasive in South America than in Middle America, however, for while South America contains only three small non-Spanish/Portuguese countries, Middle America has a much larger number of territories with British, Dutch, and French legacies.

Middle America consists of Mexico, the Central American republics from Guatemala to Panama, and the numerous islands of the Caribbean. In pre-European times this was a significant hearth of human development, where great cities arose a thousand years ago, crops were domesticated, progress was made in the sciences and arts, and empires were forged. This region, too, was the scene of the first European arrivals in the Americas, and for some time it remained the focus from where the white invaders' influences radiated outward. The early importance of the slave trade is reflected still in the high percentages of black people among Middle American island (and some mainland) populations.

South America (5)

The continent of South America also is a culture realm—a region shared by Portuguese-influenced Brazil and a Spanish colonial domain now divided among nine separate countries. As in mainland Middle America, the Catholic religion dominates life and culture; systems of land ownership, tenancy, taxation, and tribute were transferred from the Iberian Old World to the New. Today, South America still carries the impress of its source region in the architecture of its cities, in the visual arts, in the music.

South America remains one of the more sparsely peopled realms of the world, and much of the continent's interior is virtually empty terrain (Fig. I-9). Population is not only coast-oriented, it is also strongly clustered, and to a considerable extent the clusters exist in isolation from each other. Internal communication and interaction have not been strong qualities in South America, whose individual countries often have stronger ties with Europe than they do with their neighbors.

While South America does not yet suffer from the population pressure that bedevils other parts of the underdeveloped and emerging world (including several countries in neighboring Middle America), the region nevertheless confronts some serious liabilities. These include the nature of land ownership and control of the means of production, and the resistance of those in authority against the pressures of change. Under the circumstances, South American experiments in democratic government have failed and authoritarian rule meets violent response.

Southwest Asia-North Africa (6)

The Southwest Asian-North African culture realm, as we noted earlier, is known by several names, none of them satisfactory: the Islamic realm, the Arab world, the dry world. Undoubtedly the all-prevading influence of the Islamic religion is this realm's overriding cultural quality, for Islam is more than a faith; it is a way of life. The con-

An image of the world of Islam: market day during the Bazaar at Zari, Afghanistan. (Courtesy American University Field Staff)

trasts within the realm are underscored when we note its contents, extending as it does from Morocco and Mauritania in the west to Iran and Afghanistan in the east, and from Turkey in the north or Ethiopia in the south. Huge desert areas separate strongly clustered populations whose isolation perpetuates cultural discreteness, and we can distinguish regional contrasts within the realm quite clearly. The term *Middle East* refers to one of these regions, countries of the eastern Mediterranean (Fig. I-11); the *Maghreb* is a region constituted by the population clusters in the countries of northwest Africa and centered on Algeria. Still another region extends along the transition zone to Black Africa, where Islamic traditions give way to African life styles. In the realm's northern region, Turkey and Iran dominate as non-Arab countries.

Rural poverty, strongly conservative traditionalism, political instability, and conflict have marked the realm in recent times, but this is also the source area of several of the world's great religions and the site of ancient culture hearths and early urban societies. Had we drawn Fig. I-11 several centuries ago, we would have shown Arab-Turkish penetrations of Iberian and Balkan Europe and streams of trade and contact reaching to Southeast Africa and East Asia. So vigorously was Islam propagated beyond the realm that, to this day, there are more adherents to the faith outside the Arab world than within it.

Black Africa (7)

Between the southern margins of the Sahara Desert and South Africa's Cape Province lies the Black Africa culture realm. As we noted, its boundary with the realm of Islam is generally a zone of transition, and this is a

Africa is a realm of farmers. A rural landscape on the margins of the Great Eastern Rift Valley, Kenya. (Harm J. de Blij)

case where coincidence with political boundaries cannot be employed. In Sudan, for example, the north is part of the Arab world, but the south is quite distinctly African. So strong are the regional contrasts within Sudan that it would be inappropriate to include the entire country in either the Islamic realm or the Black Africa realm. Therefore, the Sudan (as well as Ethiopia, Chad, Niger, Mali) is bisected.

The African realm is defined by its mosaic of hundreds of languages belonging to specific African language families and by its huge variety of traditional, local religions that, unlike world religions such as Islam or Christianity, remain essentially "community" religions.

With some exceptions, Africa is a realm of farmers. A few people remain dependent on hunting and gathering, and some communities depend primarily on fishing. But the principal mode of life is farming. The subsistence way of life was changed very little by the European colonialists. Tens of thousands of villages all across Black Africa were never brought into the economic orbits of the European invaders. Root crops and grains are grown in ancient, time-honored ways with the hand hoe and the digging stick. It is a backbreaking low-yield proposition, but the African farmer does not have much choice.

India and the Indian Perimeter (8)

The familiar triangle of India is a subcontinent in itself—a clearly discernible physical region bounded by mountain range and ocean, inhabited by a population that constitutes one of the greatest human concentrations on earth. The scene of one of the oldest civilizations, it became the cornerstone of the vast British colonial empire. Out of the colonial period and its aftermath emerged four states: India, Bangladesh, Pakistan, and Sri Lanka.

Europe was a recipient of Middle Eastern achievements in many spheres; so was India. From Arabia and the Persian Gulf came traders across the sea. The wave of Islam came over land, across the Indus, through the Punjab, and into the Ganges Valley. Along this western route had also come the realm's modern inhabitants, who drove the older Dravidians southward into the tip of the peninsula.

Long before Islam reached India, major faiths had already arisen here, religions that still shape lives and attitudes. Hinduism and Buddhism emerged before Christianity and Islam, and the postulates of the Hindu religion have dominated life in India for several thousands of years. They include beliefs in reincarnation and the *karma* (see p. 435), in the goodness of holy men and their rejection of material things, and in the inevitability of a hierarchical structure in life and afterlife. This last quality of the Hindu faith had expression in India's caste system, the castes being social strata and steps of the universal ladder. It had the effect of locking people into social classes, the lowest of which—the untouchables—suffered a miserable existence from which there was no escape. Buddhism was partly a reaction against this system, and the invasion of Islam was facilitated by the alternatives it provided. Eventually the conflicts between Hinduism (primarily India's religion today) and Islam led to the partition of the realm into India and Pakistan, an Islamic republic.

If the realm is one of contrasts and diversity, there are overriding qualities nevertheless; it is a region of intense adherence to the various faiths, a realm of thousands of villages and several teeming, overpopulated cities, an area of poverty and frequent hunger, where the difficulties of life are viewed with an acquiescence that comes from the belief in a better new life, later.

The Chinese World (9)

China is a nation-state as well as a culture realm. The Chinese world may be the oldest of continuous civilizations. It was born in the upper basin of the Huang Ho (Yellow River) and now extends over an area of over 9 million square kilometers and has about 1 billion inhabitants. Alone among the ancient culture hearths we discussed earlier, China's spawned a major, modern state of world stature, with the strands to the source still intact. Chinese people still refer to themselves as the "People of Han," the great dynasty (202 B.C. to 220 A.D.) that was a formative period in the country's evolution. But the cultural individuality and continuity of China were already established 2000 years before that time, and perhaps even earlier.

In the lengthy process of its evolution as a regionally distinctive culture and a great nation-state, China has had an ally in its isolation. Mountains, deserts, and sheer distance protected China's "Middle Kingdom" and afforded the luxury of stability and comparative homogeneity. Not surprisingly, Chinese self-images were those of superiority and security; the growing Chinese realm was not about to be overrun. There were invasions from the steppes of inner Asia, but invaders were repulsed or absorbed. There would always be China. The European impact of the nineteenth century finally ended China's era of invincibility, but it was held off far longer than in India, it lasted a much shorter time, and it had less permanent effect.

Unlike India, China's belief system was always concerned more with the here and now, the state and authority, than with the hereafter and reincarnations. The century of convulsion that ended with the communist victory witnessed a breakdown in Chinese life and traditions, but there is much in the present system of government that resembles China under its dynastic rulers. China is being transformed into a truly communist society, but Chinese attitudes toward authority, the primacy of the state, and the demands of regimentation and organization are not new.

China is moving toward world-power status once again, and there are imposing new industries and growing cities. But for all its modernization, China still remains a realm of crowded farmlands, carefully diked floodplains, intricately terraced hillsides, cluttered small villages. Crops of rice and wheat are still meticulously cultivated and harvested; the majority of the people still bend to the soil.

Southeast Asia (10)

Southeast Asia is nowhere nearly as well-defined a culture realm as either India or China. It is a mosaic of ethnic and linguistic groups, and the region has been the scene of countless contests for power and primacy. Even spatially the realm's discontinuity is obvious; it consists of a peninsular mainland, where population tends to be

A natural landscape transformed: a hillslope terraced into paddies in Sri Lanka. (Harm J. de Blij)

clustered in river basins, and thousands of islands forming the archipelagoes of the Philippines and Indonesia.

During the colonial period the term *Indochina* came into use to denote part of mainland Southeast Asia. The term is a good one, for it reflects the major sources of cultural influence that have affected the realm. The great majority of Southeast Asia's inhabitants have ethnic affinities with the people of China, but it was from India that the realm received its first strong, widely disseminated cultural imprints: Hindu and Buddhist faiths, architecture, arts, and aspects of social structure. The Moslem faith also arrived via India. From China came not only ethnic ties but also culture elements: Chinese modes of dress, plastic arts, boat types, and other qualities were adopted widely in Southeast Asia. In recent times a strong immigration from China to the cities of Southeast Asia further strengthened the impact of Chinese culture on this realm.

Pacific Regions (10A)

Between Australia and the Americas lies the vast Pacific Ocean, larger than all the land areas of the world combined. In this great ocean lie tens of thousands of islands, large (such as New Guinea) and small (some even uninhabited). This fragmented, culturally complex realm defies effective generalization. Population contrasts are reflected to some extent in the regional diversification of the realm. Thus the islands from New Guinea eastward to the Fiji group are called *Melanesia* (*mela* means black). The people here are black or very dark brown and have black hair and dark eyes. North of Melanesia—that is, east of the Philippines—lies the island region known as *Micronesia* (*micro* means small), and the people here evince a mixture between Melanesians and Southeast Asians. In the vast central Pacific, east of Melanesia and Micronesia and extending from the Hawaiian islands to the latitude of New Zealand, is *Polynesia* (*poly* means many). Polynesians are widely known for their good physique; they have somewhat lighter skin than other Pacific peoples, wavier hair, and rather high noses. Their ancestry is complex, including Indian, Melanesian, and other elements. Anthropologists recognize a second Polynesian group, the Neo-Hawaiians, a blend of Polynesian, European, and Asian ancestries.

Culture in the Pacific realm is similarly complex. In their songs and dances, their philosophies regarding the nature of the world, their religious concepts and practices, their distinctive building styles, their work in stone and cloth, and in numerous other ways the Polynesians built a culture of strong identity and distinction.

We have defined the 13 realms depicted on Fig. I-11 on the basis of a set of criteria that include not only cultural elements but also political and economic circumstances, relative location, and modern developments that appear to have lasting qualities. Undoubtedly our scheme has areas where it is open to debate, and such a debate can itself be quite instructive. We have indicated some locations where doubts may exist; there are others, as further reading and comparisons of our framework to others will prove.

Additional Reading

The volume on *North America* by J. H. Paterson was published in 1975 in a fifth edition by Oxford University Press, New York; the quote is from page 181. Also see G. Hutton, *The Midwest at Noon,* London, George Harrap, 1946, and J. W. Brownell, "The Cultural Midwest," *Journal of Geography,* Vol. 59 (1960), pp. 81–85. The quote from O. P. Starkey, J. L. Robinson, and C. S. Miller, *The Anglo-American Realm,* published in 1975 by McGraw-Hill, New York, is from page 138; see the map on page 139.

On regions and the regional concept, consult J. R. McDonald, *A Geography of Regions,* published in 1972 by William C. Brown, Dubuque. The article by R. Symanski and J. L. Newman entitled "Formal, Functional, and Nodal Regions: Three Fallacies," was published in the *Professional Geographer,* Vol. XXV (1973), pp. 350–352. Discussions of the regional concept occur throughout the geographic literature. For a summary of its development, see P. E. James, *All Possible Worlds,* Odyssey, New York, 1972. Another interesting volume is R. E. Dickinson, *Regional Concept,* published in London by Routledge, Kegan, and Paul in 1976 and concerned principally with the scholars in geography who forged the regional concept. J. Gottman's *Geography of Europe* was published in 1969 by Holt, New York. The quotation on page 8 is from page 76. R. H. Fuson's *Introduction to World Geography: Regions and Cultures* was published by Kendall-Hunt, Dubuque, in 1977. The regional ranking appears on pages 2 to 5. On culture, the definition by Kroeber and Kluckhohn is from the article "Culture: A Critical Review of Concepts and Definition," *Papers of the Peabody Museum of American Archeology and Ethnology,* Vol. 47, 1952, page 181.

Jean Brunhes' remark on the role of water in human life occurs on p. 40 in *Human Geography,* published (in translation by E. F. Row from the 1947 French edition) by George Harrap, London, in 1952. For an idea of the scope and contents of physical geography see A. N. Strahler and A. H. Strahler, *Elements of Physical Geography,* a John Wiley, New York, publication of 1976, the source of two of our maps.

Nothing evokes the image of a population explosion as powerfully as the figures on numbers and growth rates published annually in United Nations Yearbooks and in the publications of the Population Reference Bureau, whose *Bulletins* contains statistics, maps, and commentaries on population problems. Among the more depressing conclusions drawn from these data are those of P. and A. Ehrlich in their *Population, Resources, Environment,* published by Freeman, San Francisco, in 1975. If no-growth world population is to be achieved, it may occur along one of the routes described by T. Frejka in *The Future of Population Growth,* a Wiley-Interscience publication of 1973. Still a classic is the volume by D. H. Meadows, et al., *The Limits to Growth,* published by Universe Books, in New York in 1972.

Numerous commentaries exist on the question of development, especially in a periodical entitled the *Journal of Developing Areas,* but also in books on economic geography. The final chapter in a volume by E. C. Conkling and M. Yeates, *Man's Economic Environment,* published by McGraw-Hill in New York in 1976, constitutes a useful overview of the phenomenon. Other volumes on basic economic geography also contain discussions of the manifestations and problems of national development. Probably the best available source on development problems is *The Underdevelopment and Modernization of the Third World* by A. R. de Souza and P. W. Porter, published by the Association of American Geographers in Washington in 1974. Similarly, it is instructive to compare the framework of 13 world realms proposed in this book with alternative schemes contained in, for example, *Culture Worlds* by R. J. Russell, F. B. Kniffen, and E. L. Pruitt, published by Macmillan in New York in 1969; J. O. M. Broek and J. W. Webb, *A Geography of Mankind,* a McGraw-Hill, New York publication of 1977; and a volume by J. H. Wheeler, Jr., J. T. Kostbade, and R. S. Thoman, *Regional Geography of the World,* published by Holt, Rinehart, and Winston, New York, in 1974.

Part One

Developed Regions

Bustling waterways: key lifelines of prosperous Europe. The port of Rotterdam. (©Charles E. Rotkin/Photography for Industry)

One

For centuries Europe has been the heart of the world. European empires spanned the globe and transformed societies far and near. European capitals were the foci of trade networks that controlled distant resources. Millions of Europeans migrated from their homelands to the New World as well as the Old, to create new societies from Canada to Australia. While Europe's colonial powers competed abroad they fought each other at home. In agriculture, in industry, and in politics Europe went through revolutions—revolutions that were promptly exported throughout the world and served to consolidate European advantage. Europe's decline (it may be temporary) began during the second of two disastrous twentieth-century wars. In the aftermath of World War II (1939–1945), Europe's weakened powers lost the colonial possessions that had so long provided wealth and influence. But it is a measure of the region's fundamental strength that European states, even without their empires, continued to thrive.

IDEAS AND CONCEPTS

- Location theory
- Complementarity
- Functional specialization
- Models
- Isolated state
- Organic theory
- Mercantilism
- Nation-State
- Devolution
- Balkanization
- Irredentism
- Primate city
- Site and situation
- Supranationalism

The Mosaic of Europe

As noted in the introductory chapter, Europe constitutes a discrete culture realm despite its comparatively small size on the peninsular margins of Western Eurasia. For more than two thousand years Europe has been a focus of human achievement, a hearth of innovation and invention. And Europe's human resources were matched by its enormous and varied raw material base. When the opportunity or the need arose, Europe's physical geography proved to contain what was required. During the 1960s and 1970s the sea floor adjacent to Europe's North Sea coasts yielded enough oil to make several European countries *exporters* of energy.

For so limited an area, Europe's internal natural diversity is probably unmatched. From the warm shores of the Mediterranean to the frigid Scandinavian Arctic and from the flat coastlands of the North Sea to the grandeur of the Alps, Europe presents an almost infinite range of natural environments. The insular and peninsular west contrasts strongly against a more interior, continental east. A resource-laden backbone extends across Europe from England eastward to Poland, yielding coal, iron ore, and other minerals. The North Sea contains oil and natural gas. And this diversity is not confined to the physical makeup of the continent. The European realm includes peoples of many different stocks. Not only does Europe have Latins, Germans, and Slavs, but it also includes numerous minorities, including Finns and Hungarians.

This diversity of physical and human content alone, of course, does not constitute such a great asset to Europe. If differences in habitat and race or culture automatically led to rapid human progress, Europe would have had many more competitors for its world position than it did. But in Europe there were virtually unmatched advantages of scale and proximity. Globally, Europe's relative location is one of maximum efficiency for contact with the rest of the world (Fig. 1-1). Regionally, Europe is not just a western extremity of the Eurasian landmass—almost nowhere is Europe far from that essential ingredient of European growth: the sea. Here in Europe the water interdigitates with the land as it does nowhere else. Southern and Western Europe are practically made of peninsulas and islands, from Greece, Italy, France, and the Iberian Peninsula (Spain and Portugal), to the British Isles, Denmark, Norway, and Sweden. Southern Europe faces the Mediterranean, and Western Europe virtually surrounds the North Sea and looks out over the Atlantic Ocean. Beyond the Mediterranean lies Africa, and across the Atlantic are the Americas. Europe has long been a place of contact between peoples and cultures and of circulation of goods and ideas. The hundreds of kilometers of navigable waterways, the bays, straits, and channels between numerous islands and peninsulas and the mainland, the Mediterranean and North Seas, and later the oceans provided the avenues for these exchanges.

Figure 1-1
RELATIVE LOCATION: EUROPE IN THE LAND HEMISPHERE

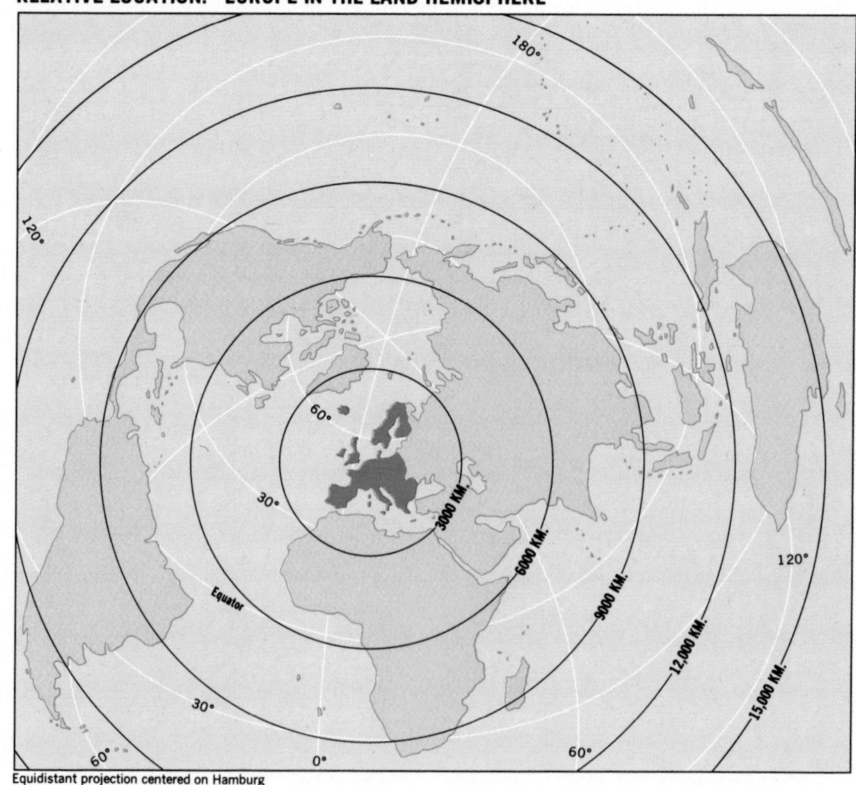

Equidistant projection centered on Hamburg

Ten Major Geographic Qualities of Europe

1. The European realm is constituted by the western extremity of the Eurasian landmass, a location of maximum efficiency for contact with the rest of the developed world.
2. The European natural environment displays a wide range of topographic, climatic, vegetative, and soil conditions.
3. Europe's population is generally healthy, well fed, has low birth and low death rates, enjoys long life expectancies, and constitutes one of the three great population clusters on earth.
4. Europe's population is highly urbanized, highly skilled, and well educated.
5. European agriculture achieves high levels of productivity and is mainly market-oriented.
6. Europe is marked by strong regional differentiation (cultural as well as physical), has a high degree of functional specialization, and provides multiple internal exchange opportunities.
7. European economies are dominantly industrial and levels of productivity are high. Levels of development, in general, decline from west to east.
8. Europe is served by efficient transport and communications networks.
9. Europe's nation-states emerged from durable power cores that formed the headquarters of world colonial empires.
10. Europe's lingering and resurgent world influence results largely from advantages accrued over centuries of global political and economic domination.

Small wonder, then, that navigation skills were quickly adopted, learned, and developed in Europe. Again, the advantage of scale presents itself—the advantage of moderate distances. With the exception of Iceland, the islands of Europe are all quite near the mainland. There is a remarkably low variance in the widths of such water bodies as the Aegean Sea, the Adriatic, the English Channel, the Skagerrak-Kattegat, the Baltic Sea, and the Gulf of Bothnia. Again, the coasts of the North Sea are about as far from each other as are those of the western Mediterranean, the Tyrrhenian, and Ionian Seas. Europe's offshore waters are not the almost endless expanses that are those of the Americas, Africa, South Asia, and Australia. They invited exploration and utilization; what lay on the other side was not some dark unknown but, more often than not, a visible coast of land of known qualities.

This applies on land as well. Europe's Alps seem to form a transcontinental divider, but what they divide still lies in close juxtaposition (and in any case, the Alps provide several corridors for contact). But consider Rome and Paris: the distance between these long-time headquarters of Mediterranean and northwestern Europe is less than that between New York and Chicago or Miami and Atlanta. No place in Europe is very far from anyplace else on the continent, and nearby places are often sharply different from each other in terms of economy and outlook. Short distances and large differences make for much interaction—and that is what Europe has had for many centuries.

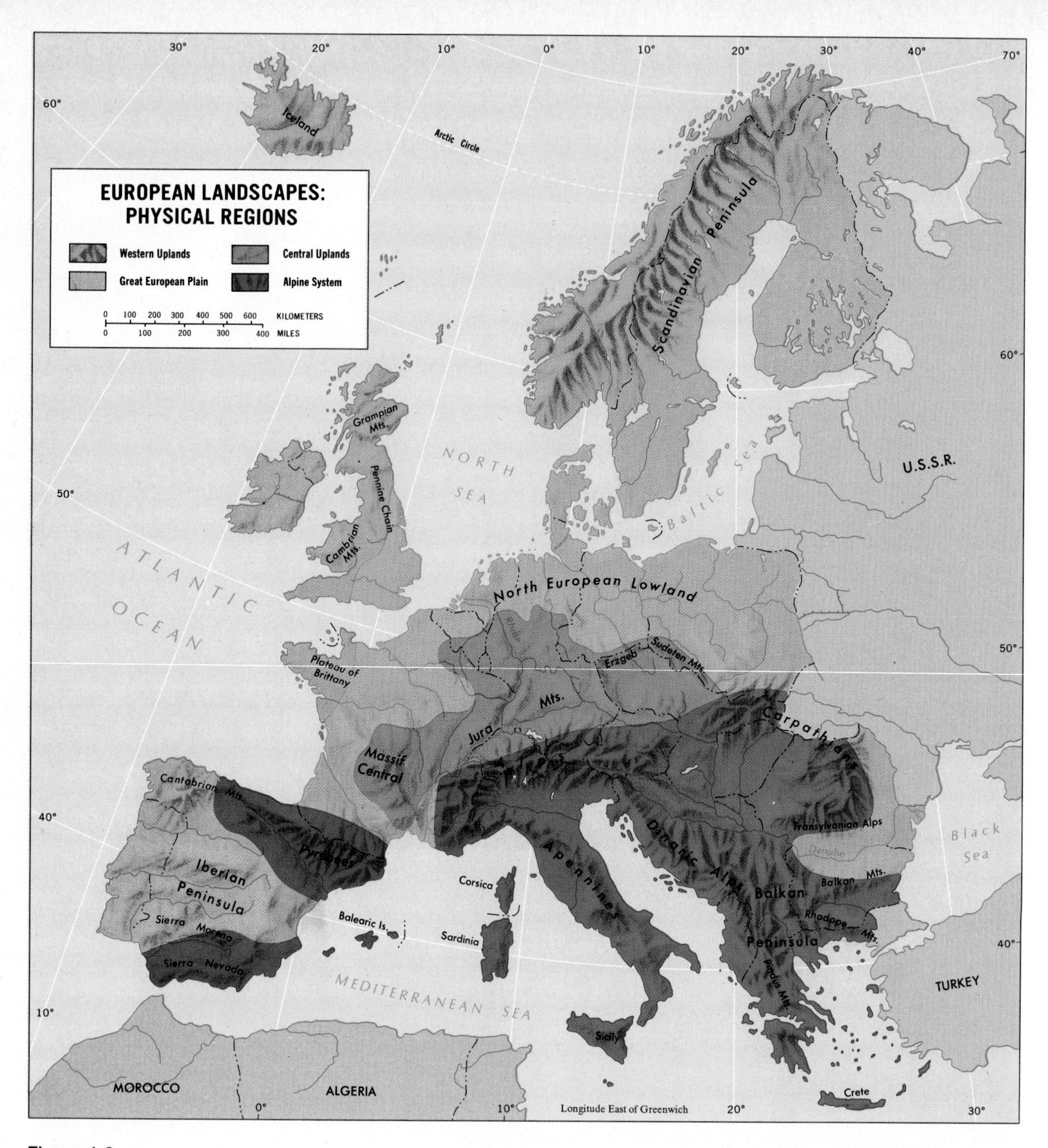

Figure 1-2

Landscapes and Rivalries

Europe may be small (it is among the three smallest world regions discussed in this book), but its landscapes are varied and complex. It would be possible to establish a large number of physiographic regions, but in doing so we might lose sight of the overall regional pattern—a pattern that has much to do with the way the human drama developed.

Europe's landscapes can be grouped regionally into four units (Fig. 1-2): the Central Uplands, the Alpine Mountains in the south, the Western Uplands, and the great North European Plain. The very heart of Europe is occupied by an area of hills and small plateaus, with forest-clad slopes and fertile valleys. These *Central Uplands* also contain the majority of Europe's productive coalfields, and when the region emerged from its long medieval quiescence and stirred with the stimuli of the industrial revolution, towns on the uplands' flanks grew into cities, and farms gave way to mines and factories.

The Central Uplands are flanked to the south by the much higher *Alpine Mountains,* and to the west and north by the North European Lowland. The Alpine Mountains include not only the famous Alps themselves, but also other ranges that belong to this great mountain system. The Pyrenees between Spain and France (one of Europe's few true barriers), Italy's Apennines, Yugoslavia's Dinaric Ranges, and the Carpathians of Eastern Europe all are part of the system that extends even into North Africa (the Atlas Ranges) and eastward into Turkey and beyond. Although these Alpine mountains are rugged and imposing, they have not been a serious obstacle to contact and communication. Europe's highest mountain, Mont Blanc (4813 meters, 15,781 feet) towers over valleys that have for centuries served as passes for traders and warriors.

Europe's western margins also are quite rugged, but the *Western Uplands* of Scandinavia, Scotland, Ireland, Brittany (France), Portugal, and Spain are not part of the Alpine system. Maximum elevations are lower than those in the Alps, ranging to over 2500 meters (8200 feet) in several locations in Spain and approaching such heights in a few locales in Scandinavia. This western arc of highlands represents an older segment of Europe's geomorphology compared to the relatively young, still active, earthquake-prone Alpine mountains. Scandinavia's uplands form part of an ancient shield zone, underlain by old crystalline rocks now bearing the marks of the Pleistocene glaciation. Spain's plateau, or *meseta,* is also supported by comparatively old rocks, worn down now to a tableland.

The last of Europe's landscape regions is also its most densely populated. The *North European Lowland* (other names have also been given to this region) extends from southwestern France through the Paris Basin and the Low Countries across northern Germany and eastward into Poland. Southeast England and the southern tip of Sweden also are part of this region, which forms a continuous belt on the mainland from southern France to the Gulf of Finland (Fig. 1-2). Most of the North European Lowland lies below 150 meters (500 feet) in elevation, and local relief rarely exceeds 30 meters (100 feet). But make no mistake; this region may topographically be low-lying and flat or gently rolling, but there its uniformity ends. Beyond this single topographic factor there is much to differentiate it internally. In France it includes the basins of three major rivers, the Garonne, Loire, and Seine. In the Netherlands, for a good part, it is made up of land reclaimed from the sea, enclosed by dikes and lying below sea level. In southeast England, the higher areas of the Netherlands, northern Germany, southern Sweden, and farther eastward, it bears the marks of the great glaciation that withdrew only a few thousand years ago.

Each of these particular areas affords its own opportunities for human endeavor, as soils and climates vary. Along with wheat and corn, there are areas of viticulture in the Garonne lowland. There are fine pastures in the Netherlands and England, where dairying is an important industry. Fields of rye cover the north German countryside. Toward the northeast the glaciated country is clothed by stands of mixed forest (including areas of conifers familiar to anyone who has seen the woodlands of Canada and northern Michigan), with pastures and cropland in intervening belts.

The North European Lowland has been one of Europe's major avenues of human contact. Entire peoples migrated across it; armies marched through it. At an early stage there were trade ties with southern parts of the continent. As stock raising gave way to agriculture, the forests that once covered most of this region were cut down. As time went on, new techniques of cultivation and new crops led to ever-greater diversification. It is hardly possible to speak of a "wheat belt" or "ranching area" in Europe, such as exist in North America. Even where one particular activity or crop predominates in the farming scene, some other pursuit takes place only a few hundred yards away. A short train ride almost anywhere will lead past fields of carefully demarcated pasture, then suddenly along fields of well-tended potatoes, back to pasture, perhaps through fields of bulb flowers, a stretch of rye, and then—we must be nearing a city—neat rows of vegetable gardens. Trimmed hedges, narrow canals, or fences limit the

Europe: The Eastern Boundary

The European culture realm is bounded to the west, north, and south by Atlantic, Arctic, and Mediterranean waters, respectively. But Europe's eastern boundary has always been a matter for debate. Some scholars place this boundary at the Ural Mountains, thus recognizing a "European" Soviet Union, and presumably, a non-European one as well. Others argue that, because there is transition from west to east (a transition that continues into the Soviet Union), there is no point in trying to define a boundary.

Still, the boundary used in this chapter (marking Europe's eastern boundary as the border with the Soviet Union) has geographic justifications. Here, diversity changes into uniformity, variety into sameness. Eastern Europe shares with Western Europe its fragmentation into several states with distinct cultural geographies—a condition that sets it apart from the giant to the east. Historical and cultural contrasts mark large segments of this boundary, notably in Finland, Hungary, and Romania (see Fig. 1-7). The Soviet Union's recent ascendancy in Eastern Europe has not wiped out basic ideological and iconographic differences between the two regions. The Soviet Union remains the great monolith; Eastern Europe retains its discrete nationalisms, latent conflicts, pressures, and vicissitudes.

various fields; every bit of available land seems somehow in productive use. Small wonder that trade in agricultural produce long was Europe's mainstay.

It is not just the low relief that has favored contact and communication in Europe's lowland. The region has another crucial advantage: its multitude of navigable rivers, emerging from higher, adjacent areas, and wending their way to the sea. In addition to the rivers of France already mentioned, the Rhine-Maas (Meuse) river system serves one of the world's most productive industrial and agricultural areas and reaches the sea in the Netherlands; the Weser, Elbe, and Oder penetrate northern Germany; the Vistula traverses Poland. In Eastern Europe the Danube rivals the Rhine in regularity of flow and navigability. Thus Europe's major rivers create a radial pattern outward from the region's higher central areas, although the Mediterranean south is less well endowed in this respect. France's Rhône-Saône system flows from Alpine flanks into the Mediterranean Sea; Italy's Po River, also fed from Alpine slopes, enters the Adriatic.

In this way the natural waterways as well as the land surface of the North European Lowland favor traffic and trade; over the centuries the Europeans improved the situation still more. Navigable stretches of rivers were connected by artificial canals, and through these and a system of locks a network of water transport routes evolved. Waterways, roads, and later railroads combined to bring tens of thousands of localities in contact with each other. New techniques and innovations spread rapidly; trade connections and activities intensified continuously. No region in Europe provided greater opportunities for all this than the Lowland.

Heritage of Order

Modern Europe was peopled in the wake of the Pleistocene's most recent glacial retreat—a lingering withdrawal that saw cold tundra turn into deciduous forest and ice-filled valleys into grassy vales. It was on Mediterranean shores that Europe witnessed the rise of its first great civilizations—on the islands and peninsulas of Greece and later in Italy. Greece lay exposed to the influences radiating from the advanced civilizations of the Middle East (see Chapter 6), and the eastern Mediterranean was crisscrossed by maritime trade routes.

As the ancient Greeks forged their city states and intercity leagues, they made intellectual achievements as well. Their political philosophy and political science became important products of Greek culture—and, indeed, here was a culture of world significance. Plato (428–347 B.C.) and Aristotle (384–322 B.C.) are among the most famous of the Greek philosophers of this period, but they stand out among a host of other contributors to the lasting greatness of ancient Greece. The writings they left behind have influenced politics and government ever since, and constitutional concepts that emerged at that time are still being applied in modified form today. But there was more to Greece than politics. In such fields as architecture, sculpture, literature, and education, the Greeks displayed their unequaled abilities as well. Because of the fragmentation of their habitat, there was local experimentation and success—followed by exchanges of ideas and innovations.

Individualism and localism were elements the Greeks turned to great advantage, but internal discord was always present, actually or potentially. Eventually it got the better of them, especially because of the endless struggle between the two major cities, Athens and Sparta. After the fourth century B.C., there was a period of ups and downs, and finally the Romans defeated the last sovereign Greek intercity league, the Achaean, in 147 B.C. But what the Greeks had accomplished was not undone. True, they had borrowed from their Middle Eastern neighbors in such fields as astronomy and mathematics, but they had by their own energy transformed the eastern Mediterranean into one of the culture cores of the world.

Greek culture was a major component of Roman civilization; but the Romans made their own essential contributions. The Greeks never achieved politico-territorial organization on a scale accomplished by the Romans. In such fields as land communications, military organization, law, and public administration, the Romans made unprecedented progress. Comparative stability and peace marked the vast realm under their domination, and for centuries these conditions promoted social and economic advances. The Roman Empire during its greatest expansion (which came during the second century A.D.) extended from Britain to the Persian Gulf and from the Black Sea to Egypt (Fig. 1-3). In North Africa the desert and the mountains formed its boundaries; to the west it was the sea, and to the north tribal peoples lived beyond the Roman domain. Only in Asia was there contact with a state of any significance. Thus the empire could organize internally without interference. But its internal diversity was such that isolation of this kind did not lead to stagnation. The Roman Empire was the first truly interregional political unit in Europe. Apart from the variety of cultures that had been brought under its control, and the consequent diffusion of ideas and innovations, there were a multitude of possibilities for economic exchange.

This process of economic development (for such it really was) made a profound impact on the whole structure of southern and western Europe. Areas that had hitherto supported only subsistence modes of life were drawn into the greater economic framework of the state, and suddenly there were distant markets for products that had never found even local markets before. In turn these areas received the farming knowhow of the heart of the Roman state, so that they could increase their yields and benefit even further. Foodstuffs came to Rome from across the Mediterranean as well as from southeastern and southwestern Europe; with its ever more complex population of perhaps a quarter of a million, the city was the greatest individual market of the empire and the first real metropolitan urban center of Europe. In addition to these agricultural advances, the Romans undertook a sociopolitical reorganization of their dominions that served economic growth—and was to have a lasting impact on Europe in general. This was their decision to support the development of a number of secondary Roman capital cities throughout the empire, cities that would have authority over surrounding rural areas and that would be the chief organizing centers for such areas. It is not difficult to see the Greek city-state concept in this, modified to serve Roman needs. But unlike the Greek cities, the Roman towns were connected by that unparalleled network of highways referred to earlier. European integration was proceeding at an unprecedented pace. On current maps Roman towns (and, here and there, Roman routes as well) are

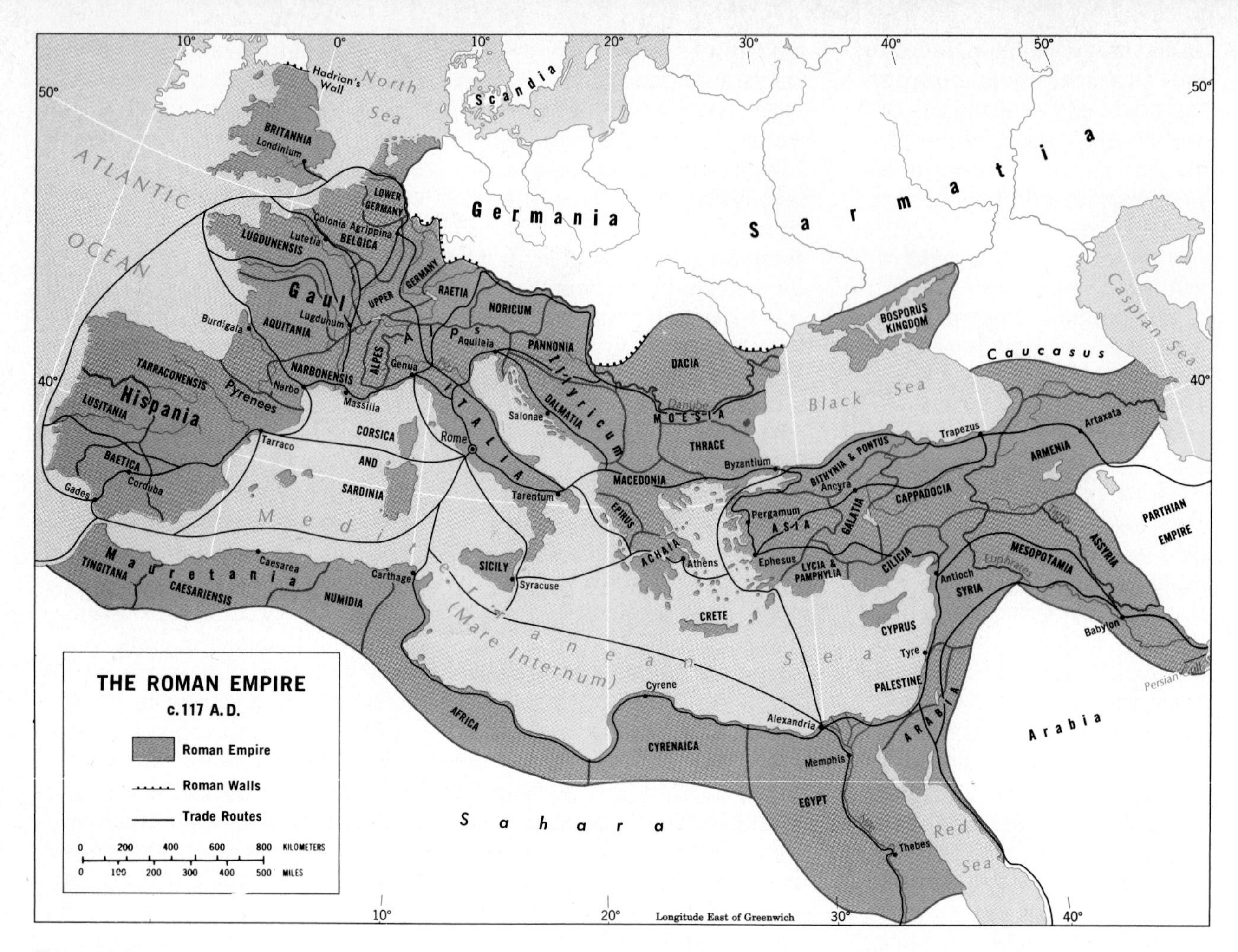

Figure 1-3

still represented: Lugdunum (Lyons in France), Singidunum (Belgrade), Eburacum (York), and Colonia Agrippensis (Cologne), among others.

Roman culture was, above all else, an urban culture, and many urban centers founded or developed by the Romans continue to function and grow today. While the economic base of many of these places has changed, others, notably the ports, still perform as they did in Roman times. And while the original surface of the old Roman roads has mostly disappeared, some modern highways in Europe follow exactly the routes laid out by Roman engineers. But more than anything else, the Roman Empire left Europe a legacy of ideas, of concepts that long lay dormant but eventually played their part when Europe again found the path of progress. In terms of political and military organization, effective administration, and long-term stability, the Roman Empire was centuries ahead of its time. Concepts of federal and confederal relationships as applied by the Romans had great currency more than a thousand years later. At no time was a larger part of Europe unified by acquiescence than was the case under the Romans. Never did Europe come closer to obtaining a *lingua franca* than it did during Roman times.

Europe's transformation under Roman hegemony also involved a geographic principle of *functional specialization.* Before the Romans brought order and connectivity to their vast empire, much of Europe was inhabited by tribal peoples whose livelihood was on a subsistence level; many of these groups lived in virtual isolation, traded little, and fought over territory when encroachment occurred. Peoples under Rome's sway, however, were brought into Roman economic as well as political spheres, and farmlands, irrigation systems, mines, and workshops appeared. Thus Roman areas of Europe began to take on a characteristic that has marked Europe ever since: particular peoples and particular places produced specific goods. Parts of North Africa became granaries for urbanizing (European) Rome. Elba, a Mediterranean island, produced iron ore that was shipped to Puteoli, near present-day Naples. Inland from today's Cartagena in Spain lay important silver and lead mines whose yields were exported through Cartagena's Roman predecessor, the port of Nova Carthago. Many other locales in the Roman Empire specialized in the production of certain farm products, manufactured goods, or minerals. The Romans knew how to exploit their natural resources; they also used the talents of their subjects.

Heirs to the Empire

Today, 2000 years later, the contributions of ancient Rome still are imprinted in the cultural landscape of Europe. The eventual breakdown and collapse of the empire could not undo what the Romans had forged in the diffusion of their language, in the dissemination of Christianity (in some ways the sole strand of permanence through the Dark Ages), in education, in the arts, and in countless other spheres. But Rome's collapse was attended by a momentous stirring of European peoples as Germanic and Slavic populations moved to their present positions on the European stage. The Anglo-Saxons invaded Britain from Danish shores, the Franks moved into France, the Allemanni traversed the North European Lowland and settled in Germany. Capitalizing on the wane of Roman power, numerous kings, dukes, barons, and counts established themselves as local rulers. Europe was in turmoil, and in its weakness it was invaded from Africa and the Middle East. In Spain and Portugal the Arab-Berber Moors conquered a large region; in Eastern Europe the Ottoman Turks created a large Islamic empire.

The townscapes of Iberia and the Balkans still carry the cultural imprint of the Moslem invasions; Europe's politicogeographic map still shows the marks of feudal fragmentation. Counties, duchies, and marches *(marks),* although now incorporated into modern nation-states, still evince by their boundaries the domains of the old feudal rulers. And it was not until the second half of the eleventh century that reform began. In England the Norman invasion (1066) destroyed the Anglo-Saxon nobility and replaced it with one drawn from the new immigrants. A great Germanic state evolved as an eastern fragment of Charlemagne's empire (it was named the Holy Roman Empire to confirm its political, religious, and cultural inheritance derived from the conquest and absorption of Rome itself). By the end of the twelfth century, elements of the modern map of Europe were already evident. (Fig. 1-4).

The Rebirth of Europe

The emergence of modern Europe dates from the second half of the fifteenth century—some put this rebirth at 1492, the year of Columbus' first arrival in the New World. At home, the growing strength of monarchies forged the beginnings of nation-states; feudal lords and landed aristocracies began to lose their power and privilege. Abroad, Western Europe's emerging states were on the threshold of discovery—the discovery of continents and riches across the oceans. Europe's developing nations were fired by a new national consciousness and pride, and there was renewed interest in Greek and Roman achievements in science and government. Appropriately, the period is referred to as Europe's *Renaissance.*

The new age of progress and rising prosperity was centered in Western Europe, and this was not a matter of chance. Western Europe's countries lay open to the new pathways to wealth—the oceans. While Columbus ventured to the Americas, Eastern Europe was under attack from the Ottoman Turks. Having captured Constantinople, capital of the eastern remnant of the Roman Empire, the Ottomans pushed into the Balkans. They took Hungary, and penetrated as far as present-day Austria. Had the Ottomans not been preoccupied with conflicts in Asia and the Near East as well, they might have succeeded, as the Romans had done, in making the Mediterranean Sea an interior lake to their empire. In that case Western Europe would have been concerned with self-defense instead of with a scramble for the wealth of the distant lands of which it had become newly aware.

Thus protected, the competing monarchies of Western Europe could afford to engage in economic rivalry without interference from the east. Indeed,

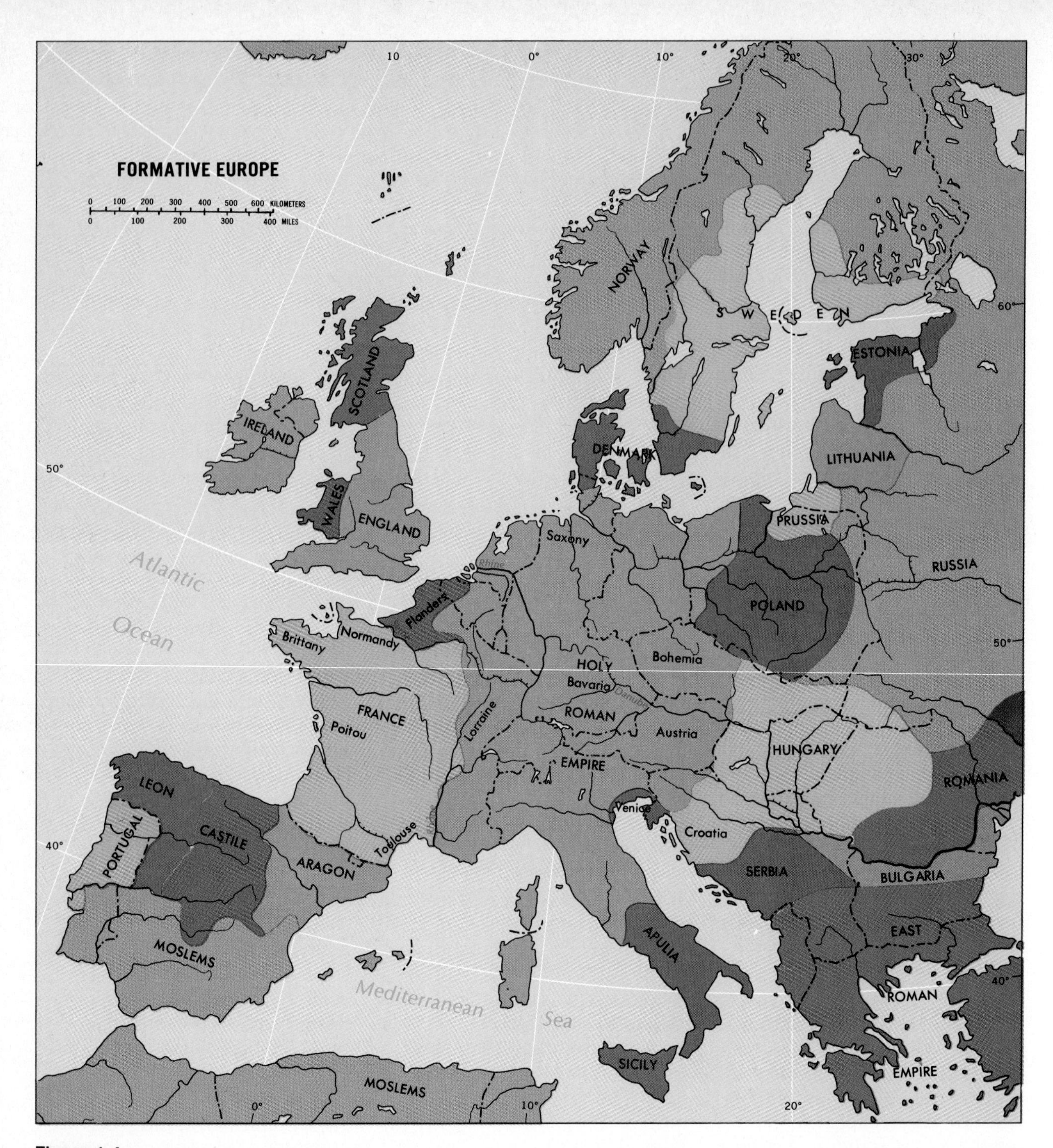

Figure 1-4

to the political nationalism that focused on the monarchies we must now add economic nationalism in the form of *mercantilism.* The objectives of the policy of mercantilism were the accumulation of as large a quantity of gold and silver as possible and the use of foreign trade and colonial acquisition to achieve this. Mercantilism was a policy promoted and sustained by the state; it was recognized to be in the interests of the people of the state in general. Wealth lay in precious metals, and precious metals could be obtained either directly by the conquest of peoples in possession of them (as Spain did in Middle and South America) or indirectly by achieving a favorable balance of international trade. Thus, there was stimulus not

Nation-State as Concept

As Europe went through its periods of rebirth and revolutionary change the politico-geographic map was transformed. Smaller entities were absorbed into larger units, conflicts resolved (by force as well as negotiation), boundaries defined, and internal divisions reorganized. European *nation-states* were in the making.

But what is a nation-state and what is not? The question centers in part on the definition of the term *nation*. The definition usually involves measures of homogeneity: a nation should speak a single language, have a common long-range history, share the same racial-cultural background, and be united by common political institutions. Accepted definitions of the term suggest that many states are not nation-states, because their populations are divided in one or more important ways.

But cultural homogeneity may not be as important as a less tangible national spirit or emotional commitment to the state and what it stands for. One of Europe's oldest states, Switzerland, has a population that is divided along linguistic, religious, and historical lines, but Switzerland has proved a durable nation-state nevertheless. A nation-state may therefore be defined as a political unit comprising a clearly defined territory and inhabited by a substantial population, sufficiently well organized to possess a certain measure of power, the people considering themselves to be a nation with certain emotional and other ties that are expressed in their most tangible form in the state's legal institutions, political system, and ideological strength.

This definition essentially identifies the European model that emerged in the course of the region's long period of evolution and revolutionary change. France is often cited as the best example among Europe's nation-states, but Italy, the United Kingdom (Great Britain), Germany (before World War I), Spain, Poland, Hungary, and Sweden are also among countries that satisfy the terms of the definition to a great extent. European states that cannot at present be designated as nation-states include Yugoslavia and Belgium.

only to seek new lands where such metals might lie, but also to produce goods at home that could be sold profitably abroad. The government of the state sought to protect home industries by imposing high duties on imported goods and by entering into favorable commercial treaties. The network of nation-states in Western Europe was taking shape more and more rapidly; Europe had entered the spiral that was to lead to great empires and temporary world domination.

The Revolutions

Europe's march to world domination did not proceed without strife and dislocation. Much of what was achieved during the Renaissance was destroyed again as monarchies struggled for primacy, religious conflicts dealt death and misery, and the beginnings of parliamentary government fell under new despotisms. Once again the landowning nobility rose to power and privileged status. France's King Louis XIV (1638–1715) personified the absolutism that had returned to Western Europe. Builder of Versailles and patron of the arts, Louis XIV also mercilessly persecuted the Huguenots and financed wars to achieve his personal ambitions.

Revolutions, however, were in the making—and in several spheres. Economic developments in Western Europe ultimately proved to be the undoing of monarchical absolutism and its systems of patronage. The urban-based merchant gained wealth and prestige, and the traditional measure of affluence—land—began to lose its relevance in the changing situation. The merchants and businessmen of Europe demanded political recognition on grounds the noblemen could not match. Urban industries were thriving; Europe's population, more or less stable at about 100 million since the mid-sixteenth century, was on the increase.

Already, an *agricultural revolution* was in progress. In our changing world we sometimes tend to look on the industrial revolution as the beginning of a new era, thereby losing sight of a less dramatic but nevertheless significant and revolutionary transformation in European farming. This began even earlier than the industrial revolution, and it helped make possible the sustained population increase in Europe during the late seventeenth and eighteenth centuries.

The agricultural revolution was focused in the Netherlands and northern Belgium as well as northern Italy. These areas experienced population growth and increased urbanization because of their success in commerce and manufacture, and the stimulus provided by the expanding markets led to improved organization of land ownership and cultivation. Some of the new practices spread to England and France, where traditional communal land ownership began to give way to individual landholding by small farmers. Land parcels were marked off by fences and hedges, and the owners readily adopted new innovations to improve yields and increase profits. Methods of soil preparation, crop rotation and care, and harvesting improved. For the first time in centuries, innovations in the form of more effective farm equipment were adopted on European farms. There was better planning and experimentation; storage and distribution systems became more efficient. In the growing cities and towns, farm products fetched higher prices. The pastoral industry also benefited. New breeding techniques and the provision of winter fodder for cattle, sheep, and other livestock enlarged the herds and increased the animals' weight. New crops were introduced, especially from the Americas; the potato became a European staple. More and more European farmers were drawn from subsistence into marketing economies. Later, the products of the industrial revolution further stimulated the transformation of agriculture in Europe. But on the farms, a revolution had already begun.

Europe was also not without industries before the *industrial revolution* began. In Flanders and England, specialization had been achieved in the manufacture of woolen and linen textiles. In several parts of present-day Germany (especially in Thüringen, now in southwestern East Germany), iron was mined, and smelters refined the ores. European manufacturers produced a wide range of goods for local markets. But the quality of those products could not match the superior textiles and other wares from India and China. So good were the Indian textiles, for example, that British textile makers staged a riot in 1721, demanding legislative protection against these imports. Thus the European manufacturers had much incentive to refine and mass produce their products. Raw materials could be shipped home in virtually unlimited quantity. If they could find ways to mass produce these raw materials into finished products, they could bury the Indian and Chinese industries under growing volume and declining prices. The search for improved machines was on, especially improved spinning and weaving equipment. The first steps in the industrial revolution were not especially revolutionary, for the larger spinning and weaving machines that were built were still driven by the old source of power: water running downslope. But then Watt and others who were trying to devise a steam-driven engine succeeded during the period 1765–1788, and soon this new invention was adapted for various uses. At about the same time, it was realized that coal could be converted into coke and that coke was a superior substitute for charcoal in the smelting of iron.

These momentous innovations had rapid effect. The power loom revolutionized the weaving industry. Iron smelters, long dependent on Europe's dwindling forests for fuel, could now be concentrated near coalfields. Engines could move locomotives as well as power looms. Ocean shipping entered a new age. England had an enormous advantage, for the industrial revolution occurred when British influence was worldwide and the significant innovations were made in Britain itself. The British controlled the flow of raw materials, they held a monopoly over products that were in world demand, and they alone possessed the skills necessary to make the machines that manufactured the products. Soon the fruits of the industrial revolution were being exported, and British influence around the world reached a peak.

Meanwhile, the spatial pattern of modern, industrial Europe began to take shape. In Britain, industrial regions, densely populated and heavily urbanized, developed in the "Black Belt"—near the coalfields. The largest complex, called the Midlands, is positioned in north-central England. Secondary coal-industrial areas developed near Newcastle in northeast England, in southern Wales, and along the Clyde River in Scotland.

In mainland Europe, a belt of major coalfields extends from west to east, roughly along the southern margins of the North European Lowland—due eastward from southern England across northern France and southern Belgium, the Netherlands, Germany (the Ruhr), western Bohemia in Czechoslovakia, and Silesia in Poland. Iron ore is found in a broadly similar belt, and the industrial map of Europe reflects the resulting concentrations of economic activity (Fig. 1-5). But nowhere are the coasts, the coalfields, and iron ores located in such close proximity as they are in Britain.

Another set of industrial regions developed in and near the growing urban centers of Europe, as Fig. 1-5 demonstates. London already was Europe's greatest urban focus and Britain's largest and richest domestic market. Many local industries had been established here, taking advantage of the large supply of labor, the ready availability of capital, and the proximity of so great a number of potential buyers. Although the industrial revolution thrust other places into prominence, London's primacy was sustained, and industries in and around London multiplied. Similar developments occurred in the Paris region and near other, major urban concentrations in Europe. As the new methods of manufacturing were adopted in mainland Europe, old industrial areas were rejuvenated, and regional growth was stimulated in France and Italy (Fig. 1-5).

Europe had long experience with experiments in democratic government, but the *political revolution* that swept the realm from the 1780s brought transformation on an unprecedented scale. Overshadowing these events was the French Revolution (1789–1795), but Europe's political catharsis lasted into the twentieth century and affected every monarchy on the continent.

In France the popular revolution plunged the country into years of chaos and destruction. There had been dissatisfaction everywhere: among the nobility and clergy, critical of the rule of Louis XVI; among the middle classes, where a constitutional, parliamentary monarchy along English lines was wanted; and among the peasants, who faced rising prices, declining living standards, and harsh exploitation by landowners. Ultimately the most extreme elements came to control the movement, and in 1792 a republic was proclaimed amid terrible excesses. The king was executed in 1793, and revolutionary armies began to advance into neighboring countries, calling on all peoples to dispose of their rulers and to join in a war of liberation. Europe's growing nationalism proved to be the downfall of the dynastic monarchies.

Even while the revolutionary forces swept into Europe, disorder continued in France itself. Only when Napoleon took control (1799) was stability restored. Napoleon personified the new French republic; he reorganized France so completely that he laid the foundations of the modern nation-state. His empire extended from Prussia and Austria to Spain and from the Netherlands to Italy. Although his armies were repelled by the British in Portugal and more seriously by the Russians two years later (1812) Napoleon had forever changed the politico-geographic map of Europe. His French forces had been joined by revolutionaries from all over the continent; one monarchy after another had been destroyed. Even after his defeat at Waterloo, there were popular uprisings in Spain, Portugal, Italy, and Greece. In France itself, the monarchy was temporarily reestablished through outside intervention—but France was to be a republic, as it still is today. If the monarchy survived elsewhere, its powers were limited or shortlived. Europe had its taste of democracy and nationalist power and it would not revert to its old ways.

Geographic Dimensions

Modern Europe has emerged from an age of revolutionary change—in livelihoods and commerce, industrialization and technology, revolution and war. During the nineteenth century, some geographers tried to interpret the forces and processes that were shaping the new Europe, and some of their work still holds interest today.

The agricultural revolution, for example, was observed attentively by a geographer-farmer named J. H. Von Thünen (1783–1850). Von Thünen owned a large farming estate not far from the German town of Rostock. For four decades he kept meticulous records of his estate's transactions, and he became interested in a subject that still excites economic geographers today—the effects of distance and transportation costs on the location of productive activity. Using all the data he gathered, Von Thünen began to publish books on the spatial structure of agriculture. His series, under the title *Der Isolierte Staat (The Isolated State)* in some ways constitutes the foundations of spatial theory. Von

Figure 1-5

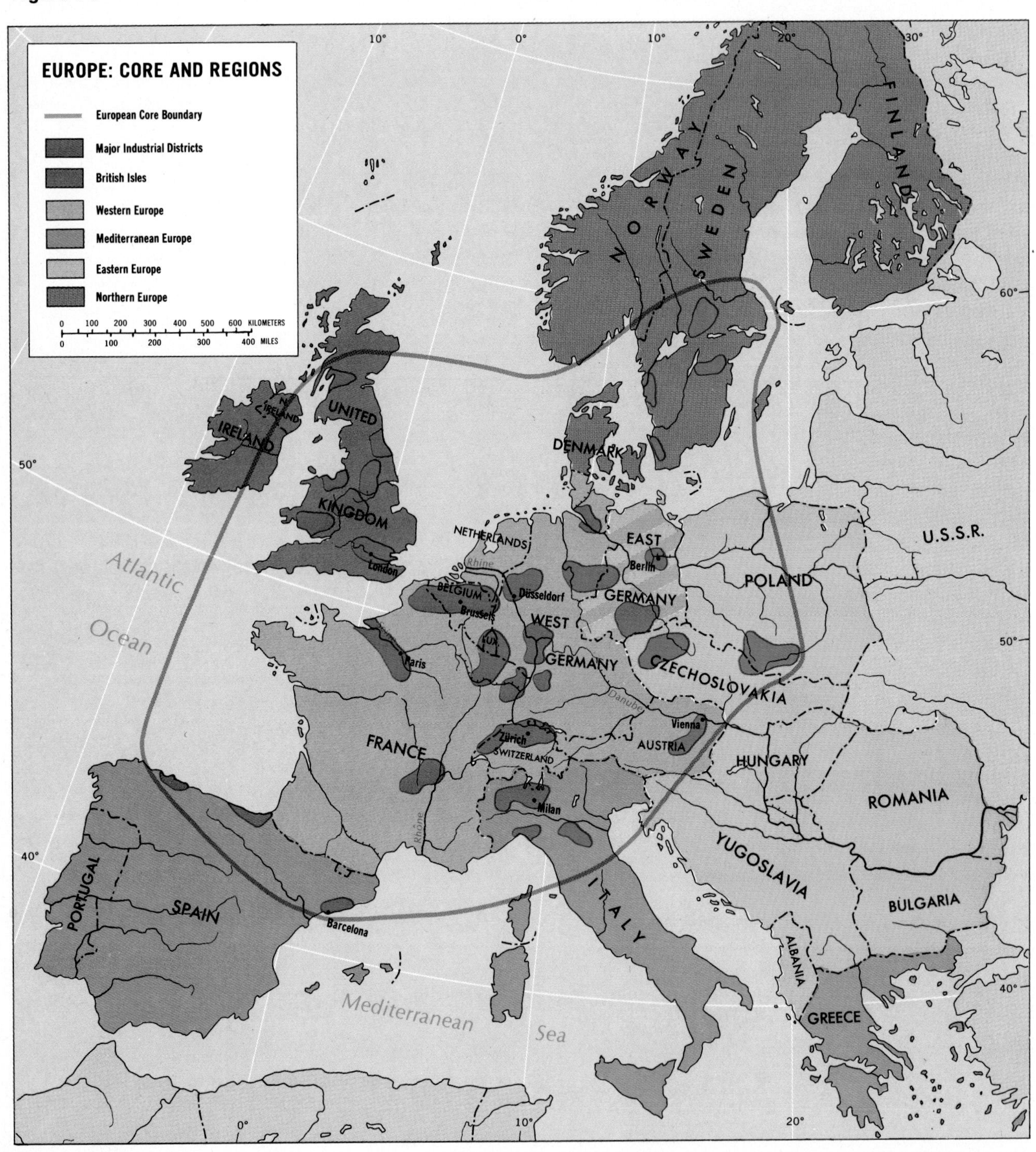

Thünen's conclusions are still being discussed and debated in the modern literature.

Von Thünen called his work *The Isolated State* because he wanted to establish, for purposes of analysis, a self-contained country, devoid of outside influences that would disturb the internal workings of the economy. Thus he created a sort of regional laboratory within which he could identify the factors that influence the location of farms around a central urban area. In order to do this, he made a number of assumptions. First, he stipulated that the soil and climate would be uniform throughout the region. Second, there would be no river valleys or mountains to interrupt a completely flat land surface. Third, there would be a single, centrally positioned city in the "isolated state," which was surrounded by an empty, unoccupied wasteland. Fourth, Von Thünen postulated that the farmers in the "isolated state" would transport their own products to market (no transport companies here), and that they would do so by oxcart, directly overland, straight to the central city. This, as you can imagine, is the same as assuming a system of radially converging roads of equal and constant quality; with such a system, transport costs would be directly proportional to distance.

Von Thünen combined these assumptions with what he had learned from the actual data collected while running his estate, and he now asked himself: what would the spatial arrangement of agricultural activities be in his "isolated state?" He concluded that farm products would be grown in a series of concentric zones outward from the market center (i.e., the central city). Nearest to the city would be grown those crops that perished easily and/or yielded the highest returns—for example, vegetables; dairying would also be carried on in this innermost zone. Farther away would be potatoes and grains. And eventually, since transport costs to the city increased with distance, there would come a line beyond which it would be uneconomical to produce crops. There the wasteland would begin.

Von Thünen's model, then, incorporated five zones surrounding the market center (Fig. 1-6). The first and innermost belt would be a zone of intensive agriculture and dairying. The second zone, Von Thünen said, would be an area of forest, used for timber and firewood. Next, there would be a third belt of increasingly extensive field crops. A fourth zone would be occupied by ranching and animal products. The fifth and outermost zone would be the unproductive wasteland. If the location of the second zone, the forest, seems inappropriate, it should be noted that the forest was still of great importance during Von Thünen's time as a source of building materials and fuel. All that was about to change with the onslaught of the industrial revolution; but there are lots of towns and cities left in the world that are still essentially preindustrial. When R. J. Horvath made a study of Addis Ababa, the capital of Ethiopia, in this context, he found a wide and continuous belt of eucalyptus forest surrounding that city, positioned more or less where Von Thünen would have predicted it to be and serving functions similar to those attributed to the forest belt of the "isolated state."

Figure 1-6

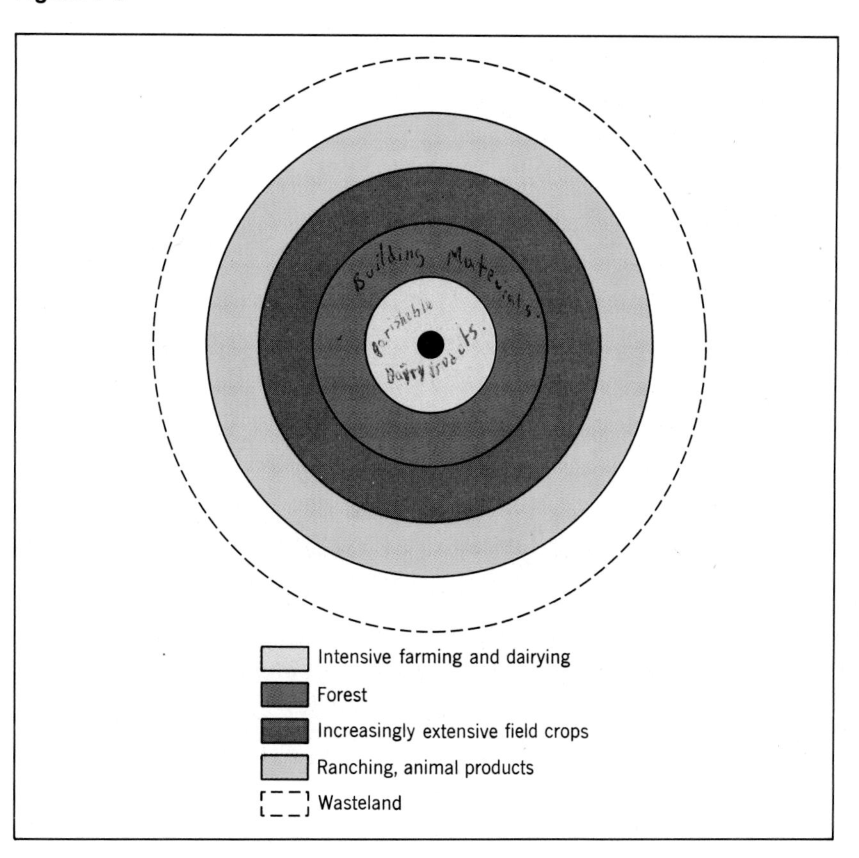

Von Thünen knew, of course, that the real Europe (or the world) did not present situations exactly as he postulated them. Transport routes serve certain areas more efficiently than others. Physical barriers can impede the most modern of surface communications. External economic influences invade every area. But Von Thünen wanted to eliminate these disruptive conditions in order to discern the fundamental processes at work in shaping the spatial layout of the agricultural economy. Later, the distorting factors could be introduced one by one and their influence measured. But first he developed his model in theoretical isolation and based it on total regional uniformity.

It is a great tribute to Von Thünen that his work still retains the attention of geographers. The economic-geographic landscape of Europe has changed enormously since his time, but geographers still compare present-day patterns of economic activity to the Thünian model. The role of distance (to markets, from raw materials, and so on) in the development of the spatial pattern of the economy remains the subject of many modern studies, and geographers acknowledge that it was Von Thünen who first formulated the crucial questions.

Industrial Intensification

It is not surprising that the industrialization of Europe (or, rather, its industrial *intensification* following the industrial revolution) also became the topic of geographic research. What influences affect the location of industries? How was Europe's industrialization channeled? Again, the first important studies were done by German geographers and economists, but well after Von Thünen's time—during the second half of the nineteenth century. Much of this work was incorporated in a volume by A. Weber, published in 1909, entitled *Uber den Standort der Industrien (Concerning the Location of Industries).* Like Von Thünen, Weber began with a set of assumptions in order to minimize the complexities of the real Europe. But unlike Von Thünen, Weber dealt with activities that take place at particular points rather than over large areas. Manufacturing plants, mines, and markets are located at specific places, and so Weber created a model region marked by sets of points where these activities would occur. He eliminated labor mobility and varying wage rates, and so he could calculate the "pulls" exerted on each point in his theoretical region.

In the process Weber discerned various factors that affect industrial location. He defined these in various ways. For example, he recognized what he called "general" factors that would affect all industries, such as transport costs for raw materials and finished products, and "special" factors such as perishability of foods. He also differentiated between "regional" factors (transport and labor costs) and "local" factors. Local factors, Weber argued, involve *agglomerative* and *deglomerative* forces. Take the case of London we discussed previously; industries located there in large part because of the advantages of locating together. The availability of specialized equipment, a technologically sophisticated labor force, and a large-scale market made London (and Paris, and other cities not positioned on rich resources) an attractive site for many manufacturing plants that could benefit from agglomeration. On the other hand, such agglomeration may eventually create strong disadvantages, chiefly in terms of competition for space and rising land rents. Eventually an industry might move away, and deglomeration sets in. This subject arises again in the context of North American industry, in Chapter 3.

Weber singled out transport costs as the critical determinant of regional industrial location and suggested that the least-transport-cost location is the place where it would be least expensive to bring raw materials to the point of production and finished products to the consumers. But economic geographers following Weber have concluded that some of Weber's assumptions seriously weakened the usefulness of his conclusions, especially in the area of markets and consumer demand. Consumption does not take place at a single location, but over a wide (in some cases a worldwide) area. But practically all modern study of industrial location has a direct relationship to Weber's work, and he was a pioneer in a class with Von Thünen.

Urban Patterns

Europe's industrialization speeded the growth of many of its cities and towns. In Britain in the year 1800, only about 9 percent of the population lived in urban areas, but by 1900 some 62 percent lived in cities and towns (today the figure approaches 90 percent); all this was happening as the total population skyrocketed as well. As industrial modernization came to Belgium and Germany, and to France and the Netherlands and other areas of Western Europe, the whole urban pattern changed. The nature of this process—the growth and strength of towns and cities—and related questions also became topics of geographic study, just as agriculture and industrialization had. The German geographer W. Christaller addressed himself to these topics, and today he is recognized as one of the founders of a whole field of geographic research. But it took almost a generation for Christaller's work to come to general attention. His major book was published in 1933, just as the Nazi regime took power; in Germany it was political geography, not urban geography that held center stage. Not until 1954 did an English translation of *The Central Places of Southern Germany* appear in the United States.

Christaller was interested in the spacing of towns and the forces that govern their distribution. As we shall note in greater detail in Chapter 3, among Christaller's major contributions was the notion of *centrality*—the idea that cities and towns have surrounding areas with which they are in functional contact (classic functional regions, therefore). Not all settlements are central places; a mining town or border village may not function as a central place to any area at all. And some places, Christaller emphasized, are more "central" and have greater centrality than others. It is not merely city size that determines the strength of this centrality, but also the town's capacity to attract consumers through its provision of goods and services. And so Christaller proceeded to develop a central-place theory (using, again, a set of simplifying assumptions) that proposed that cities and towns, large and small, are not positioned where they are by accident, but as a result of their competitive centrality. His model (see page 208) involves a set of interlocking hexagons, with larger cities lying farther apart than towns or villages, because cities have larger regions in which their economic strength dominates.

A look at the map of Europe does not immediately produce images of hexagonal regularity as envisaged by Christaller, but he argued that disruptions of the model were normal consequences of barriers imposed by terrain, transport routes, and other impediments. He participated enthusiastically in the debate that followed (it is still going on today); in 1950 he published *The Foundations of Spatial Organization in Europe,* in which he endorsed his position once more: "When we connect the metropolitan areas with each other through lines, and draw such a network of systems on the map of Europe, it indeed becomes eminently clear how the metropolitan areas everywhere lie in hexagonal arrangements." We return to this topic in Chapter 3.

Politico-Geographical Order

Europe's convulsive political revolution also was the object of geographic study, and in the process the field of modern political geography was born. Still another German geographer, F. Ratzel (1844–1904), observed the fortunes and failings of European states and sought to identify underlying principles and processes. Ratzel suggested that political states are like biological organisms, passing through stages of growth and decay. Just as an organism requires food, Ratzel reasoned, so the state needs space. Any state, to retain its vigor and to continue to thrive, must have access to more space, more territory—the essential, life-giving force. Only by acquiring territory and through the infusion of newly absorbed cultures could a state sustain its strength. The acquisition of frontier areas with their resources and populations would renew the state's energies, Ratzel argued, but boundaries were dangerous straightjackets.

Ratzel believed that a nation, being constituted by an aggregate of organisms (human beings) would itself behave as an organism capable of birth, growth, maturity, and death. From his study of European historical geography, he proposed an *organic theory* of state evolution, and he likened the struggles among European nation-states to biological competition. The German nation-state was forged during Ratzel's lifetime (1871), and in the 1880s the German *Reich* acquired a colonial empire overseas. In his book *Political Geography* (1897) Ratzel detailed the rise of empires and the fortunes of alliances, and the subjugation of weaker countries. Invariably, states in ascendancy were gaining territory; those in decline were losing it. But what might determine whether a state would rise or fall? Like Von Thünen and other geographers, Ratzel sought to identify the fundamental processes that affected the course of events he could observe. In an article entitled "Laws of the Spatial Growth of States," published in 1896, he attempted to discern laws that govern the growth of states. Essentially, each of the seven laws forms an element of the organic theory: growth is vitality.

Certainly Europe appeared to provide ample confirmation of Ratzel's theory. The British Empire spanned the world, and British power was nearing its zenith. France, having lost its European empire, expanded its spheres of influence in Africa and Asia. Other European states thrived on their overseas domains. Ratzel's writings, of course, appeared in learned journals and were couched in theoretical language. But some of his readers were less theoretical or scholarly than he, and ultimately there arose in Germany a school of geopolitics. The practitioners of geopolitics gave practical expression to the determinist character of Ratzel's writings, and they laid the intellectual foundations for German expansionism—not in distant colonies but in Europe itself. Twice during the twentieth century Germany plunged Europe and the world into war; the recovery from the second of these conflicts (1939–1945) has been the formative period of modern Europe.

The European Realm Today

Europe has emerged from the devastation of World War II to enter a new era of reconstruction, realignment, and resurgence. Reconstruction was aided by $12 billion made available through the Marshall Plan by the United States. Realignment came in the form of the "Iron Curtain" separating the Soviet-dominated East from Western Europe, and also involved the emergence of several international "blocs" consisting of states seeking to promote multinational cooperation. The resurgence of Europe continued despite the loss of colonial empires, recurrent political crises, energy shortages, labor problems, and other obstacles. Europe's momentum has carried the day again.

Although Europe is quite clearly a culture realm, it has little of the homogeneity that such regional identity might imply. It is sometimes postulated that Europe may be viewed as a regional unit because its peoples share Indo-European languages (Fig. 1-7), Christian religious traditions (Fig. 6-1), and common European racial ancestry (Fig. 7-4). But these culture traits extend well beyond Europe, and in any case they are not strong unifying elements in the European mosaic. As Fig. 1-7 shows, Hungarians and Finns are among Europeans who do not speak Indo-European languages, and for so small a region Europe is a veritable tower of Babel. As for common religious traditions, Europe has a history of intense and destructive conflict over religious issues. Shared Christian principles have done little to bring unity to Northern Ireland, and the recurrent conflict in that area is only the most recent manifestation of the depth of latent religious division in parts of Europe. And again, Europe's common racial ancestry (the term *European race* has replaced *Caucasian*) masks strong differences between Spaniard and Swede, Frisian and Sicilian.

Among the most important geographic properties of Europe are the following:

Opportunities for Exchange

Perhaps to a greater degree than any world area of similar dimensions, Europe's environments and resources present opportunities for exchange and interaction. European economic-geographic development has always been stimulated by internal *complementarities.* When two areas or places are in a position of complementarity, it means that area A requires products supplied by area B, and at the same time area B needs products available from area A. A specific example involves Italy. Among Mediterranean countries other than France, Italy is foremost in terms of economic development. But Italy is a coal-poor country, and for a long time its industries depended on coal supplies from the rich resources of Western Europe. On the other hand, Italian farmers grow crops that cannot be grown in the north: citrus fruits, olives, grapes, early vegetables. Such products are wanted on Western Europe's markets. Hence, Italy needs Northern Europe's coal; Northern Europe imports Italian fruits and wines. There are thousands of examples of this kind in Europe alone, and if we possessed all trade information for the past 2000 years and drew a series of maps to represent these transactions, we would see a growing web of ever-increasing complexity, involving more and more places, eventually overflowing the continent and reaching out across the oceans to other parts of the world.

Advantages of Durability

European nation-states are among the world's oldest, and the colonial empires of European powers have been among the most durable. Wars and revolutions notwithstanding, European nations survived, and out of their long-term permanence and stability they forged a confidence that is a hallmark of European culture. Centuries of exploitation of overseas domains amassed fortunes at home and established a foreign influence that continued when the colo-

Figure 1-7

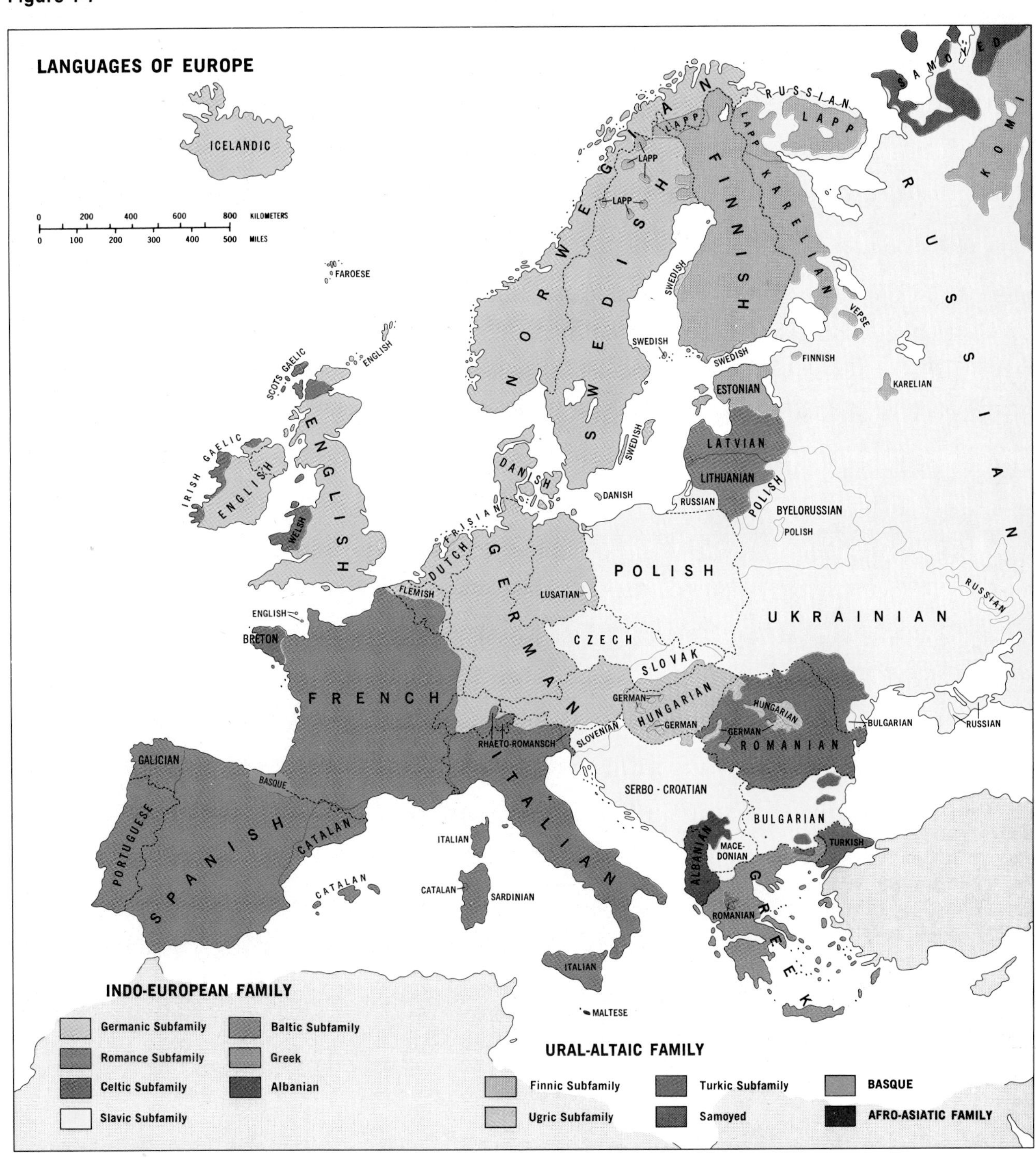

nial era ended. The impressive merchants' homes that line the canals of Amsterdam, the architectural splendor of Paris, London, Lisbon, those castles and country villas—much of all this was built by the slave trade, by the domination of colonial economies, by policies that no longer prevail in the modern world. But the advantages of the accumulated wealth continue.

Marshall Plan

On June 5, 1947, the United States Secretary of State George C. Marshall proposed, during a lecture at Harvard University, that the United States finance a European Recovery Program. Postwar Europe was on the brink of chaos, and the survival of free institutions was imperiled. A committee representing 16 Western European states (and West Germany) presented the U. S. Congress with a program for economic rehabilitation, and Congress approved. From 1948 to 1952 about $12 billion was given to Europe under the "Marshall Plan." This investment not only revived European national economies, it also constituted a major push toward international cooperation among European states. Out of the original committee of 16, the Organization for European Economic Cooperation (OEEC) was born; and the OEEC in turn gave rise to further progress in interstate association. Today's European Economic Community (the *Common Market*), a major step toward European integration and unification, is a product of the European Recovery Program—one of the proudest achievements of the Truman administration.

Problems of Political Fragmentation

Unlike the United States or the Soviet Union, Europe's stronger competitors in terms of world power and influence, the European realm is fragmented politically and ideologically—and economically as well—although efforts are being made to overcome this problem in Western Europe. Ever since the unifying period of the Roman Empire, Europe has been divided into numerous political entities; not counting the smallest of these (Monaco, Vatican City, Liechtenstein etc.) there are 26 states in 1980, as Table I-1 (page 28) indicates. Many of these states are true nation-states, with strong identities and traditions.

World War II left Eastern Europe under Soviet domination, Germany fractured, Western Europe in chaos. Even before the end of the war, three countries (the Netherlands, Belgium, and Luxembourg) had laid plans to create a cooperative organization to be called *Benelux.* Later the terms of the Marshall Plan required greater international cooperation, and 16 countries of Western Europe formed the Organization for European Economic Cooperation (OEEC). Ever since the OEEC began operations (1948), the states of Western Europe have pursued the prospect of European unification. By 1980 the greatest success was the nine- (soon to be twelve-) member European Economic Community, incorporating most of Europe's core area. But Germany and Berlin still remained divided, and Eastern and Western Europe continued to be separated along an externally imposed ideological barrier.

Assets of Urban Tradition

Europe is among the most highly urbanized regions in the world (though more so in the west than in the east). In the United Kingdom, more than 85 percent of the people live in cities and towns; in West Germany, nearly 80; in France, 70. Even in Eastern Europe, where the percentage of urbanization is lower (45 to 55 percent average), the towns and cities contain a much larger share of the population than in Asian or African countries.

Cities are the crucibles of culture. As M. Jefferson wrote (see box, right) Europe's largest cities are expressions of national cultures and traditions. But Europe's great urban population clusters also contain concentrations of one of the world's most technologically advanced labor forces. Specializations and skills that have been learned over many years now are available to the urban-based industries. And Europe's cities constitute large and wealthy markets, because personal incomes in Europe (while not quite as high as in the United States and Canada) are among the world's highest.

Law of the Primate City

Commenting on the central role of great cities in the development of national cultures, M. Jefferson in 1939 published an article entitled, "The Law of the Primate City." The "law" states that "a country's leading city is always disproportionately large and exceptionally expressive of national capacity and feeling." It is a rather imprecise "law," of course, but Europe does provide numerous examples. Certainly Paris personifies France in countless ways, and there is nothing in England to rival London, where the culture and history of a nation and an empire are etched in the urban landscape. Again, the architecture, museums, public art works, public squares and parks, and other attributes make Amsterdam the focus of culture in the Netherlands. Prague is the unrivaled heart of Czechoslovakia, its atmosphere of centuries is still there despite the new conditions of life. In many ways Vienna is Austria, Athens is Greece, Lisbon is Portugal. And perhaps no city reflects the nation as faithfully as divided Berlin represents fragmented Germany.

Achievements in Communication Systems

Circulation and interaction have characterized Europe from Roman times onward, and Europeans have always been among the most mobile peoples in the world. Today, the barge traffic on rivers and canals still is substantial, but Europe is connected even more efficiently by excellent railroad and road networks, air routes, and pipelines. Again, these systems are best developed in Western Europe, but even in Eastern Europe few areas remain remote from some means of modern communication.

The evolution of Europe's transport networks was not a coordinated process. Individual countries built systems designed to serve their internal needs; regional integration of the various networks became a concern of the postwar planners of a unified Europe. Especially in the Common Market countries of the European core area, considerable progress has been made not only in the integration of the transport routes, but also, as will be noted later, in the reduction of the barrier effect of international boundaries.

Benefits of Well-being

The European population, in general, is well housed, well educated, and well fed. European cities have their slums, but nothing to compare to the miserable shantytowns that surround so many cities of the underdeveloped world. Levels of education are high, and educational achievement is held in esteem throughout Europe. Measures of such attainment include literacy rates (over 95 percent everywhere except Spain and Portugal, Italy, and the Balkan countries), newspapers and magazines printed (among the largest number per capita in the world), book titles published (over 30,000 in 1979 in Britain alone) and, less tangibly, the premium placed on high school and university graduation.

Europeans also eat well-balanced and adequate meals and enjoy good health. Several countries rank ahead of the United States in such areas as longevity and infant survival, and medical services are generally very good. European countries no longer confront the explosive population growth that marked the turbulent nineteenth century (and still beleaguers underdeveloped countries today). Under such circumstances, progress and improved living standards are attainable.

Industrial Leadership

Europe no longer enjoys the industrial monopolies held during the early stages of the industrial revolution, but in aggregate European industries still constitute one of the four great manufacturing complexes in the world (the other three are the United States, the Soviet Union, and Japan). Many European specialized products still command world markets, and the growing prosperity of the European market itself has sustained the national industries.

But European industry faces serious problems as well. These are perhaps most evident at the very source of the industrial revolution—in the United Kingdom. British industry, once the world's leader, now is comparatively inefficient. Output per worker is low and declining, and overall productivity lags far behind other industrial countries, including West Germany, Japan, and the United States. British manufacturing plants, once among the world's most modern, are now outmoded; the investments needed to keep them current and competitive have not been made. As the British economy has suffered, government representatives and labor leaders have struggled for control over the course of events. Will the British experience diffuse into mainland Europe as the industrial revolution itself did? Are the political problems of Italy and the labor unrest in France signs of a similar situation in the making? The tenuous fabric of Europe's Common Market may not survive such an economic crisis.

Commercial Agriculture

When Von Thünen calculated his farming estate's transactions, most Europeans were still farmers, and enough food was produced in Europe itself to feed city dwellers and farmers alike. But in the aftermath of industrialization and urbanization, only one in six Europeans still farms, and foodstuffs must be imported to supplement what the local farmers can produce. With very few exceptions (and those mostly in southeastern Europe), European farmers grow their crops for the market and for profit.

European agriculture is not as large-scale nor as mechanized as it is in the United States, but European farmers are very productive, especially in the Netherlands and Britain. In Eastern Europe, agriculture is being reorganized on the Soviet pattern, but despite the advantages (greater availability of mechanized equipment on the collectives) production per person and per hectare still remain far below those of the West.

Regions of Europe

Earlier in this chapter a European core area was defined (see Fig. 1-5, page 68). This is Europe's heartland, a functional region of dynamic growth, dense population, high degree of urbanization, enormous mobiltiy, and great productivity. We now turn to a more traditional view of Europe's regions as we attempt to group sets of countries together so that the regional groups reflect their proximity, historical-geographic association, cultural similarities, common habitats, and shared economic pursuits. On these bases it is possible to identify five regions of Europe: (1) the British Isles, (2) Western Europe, (3) Nordic or Northern Europe, (4) Mediterranean Europe, and (5) Eastern Europe.

The British Isles

Off the coast of mainland Western Europe lie two islands called the British Isles (Fig. 1-8).

The largest of the two, also lying nearest to the mainland (a mere 34 kilometers or 21 miles at the closest point) is the island of Britain, and the smaller island is known as Ireland. Surrounding Britain and Ireland are several thousand small islands and islets, none of great significance.

Although Britain and Ireland continue to be known as the British Isles, the British government no longer rules over the state that occupies most of Ireland, the Republic of Ireland or *Eire.* After a very unhappy period of domination, the London government gave up its control over that part of the island, which was overwhelmingly Catholic; in 1921 it became the Irish Free State. The northeastern corner of Ireland, where English and Scottish Protestants had settled, was retained by the British and called *Northern Ireland.*

Figure 1-8

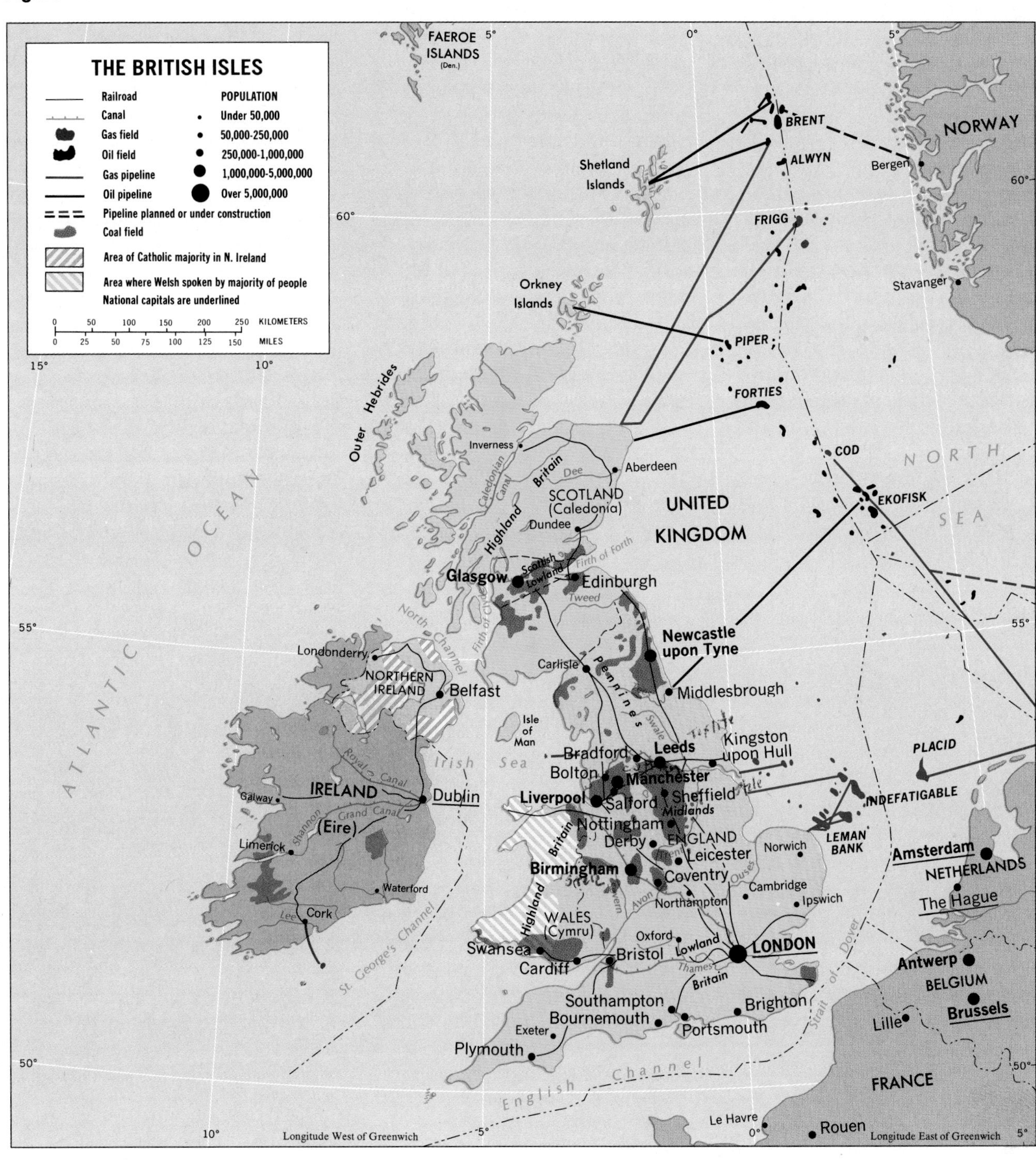

Although all of Britain is part of the United Kingdom, political divisions exist here, too. *England* is the largest of these units, the center of power from where the rest of the Isles were originally brought under unified control. The English conquered *Wales* in the Middle Ages, and *Scotland* was first tied to England in the seventeenth century, when a Scottish king ascended the English throne. Thus England, Wales, Scotland, and Northern Ireland became the *United Kingdom* (U.K.). In recent years there has been a revival of nationalism in both Wales and Scotland, and demands for greater local autonomy are being heard.

Highland and Lowland Britain. Britain's *physiography* (that is, the island's topography and relief, climate and drainage, soils and natural vegetation—the total physical landscape) is varied, and the relative location of the different areas matters much. The most simple, and at the same time the most meaningful, division of Britain is into a *lowland* region and a *highland* region. Most of England, with the exception of the Pennine Mountains and the peninsular southwest, consists of lowland Britain. Highland Britain includes Scotland and Wales. Lowland Britain, therefore, lies opposite the western periphery of the North European Lowland on the mainland, and it is essentially a continuation of it. In this part of England, the relief is low, soils are generally good, rainfall is ample, and agricultural productivity is high.

Throughout the evolution of British society, the Thames River (on which London lies) and its basin were avenues of penetration into England. Several good natural harbors and places of easy access also rendered Britain open to foreign influences. From the east, northeast, and southeast they came: the builders of the great Stonehenge structures; the Celtic-speaking peoples, whose linguistic imprints still remain on the map as Irish and Scottish Gaelic (Fig. 1-7); the Romans, who made Dover their port of entry and who established routes that remain to this day among England's major highways; the Angles and the Saxons, Germanic peoples who settled on England's eastern bulge and in the Thames Basin, respectively; Danish and Norwegian Vikings who penetrated the interior and settled on many surrounding islands; and the Normans, those Scandinavian Vikings who had been acculturated through generations of residence in French Normandy and whose invasion under William the Conqueror in 1066 closed one era in British historical geography and started another.

Through all this, lowland Britain held center stage in the British Isles. True, Christianity—which had reached Britain during Roman times—proved more durable in remote Ireland than in exposed, frequently strife-torn England. But the British Isles' destiny was determined in the English countryside, where the invaders struggled for power and where, after the eleventh century, the British nation and empire were born. In its political development, Britain benefited from its insular separation from Europe, and England became the center of regional power and political as well as economic integration and organization. Centered on London, England steadily evolved into the core area of the British Isles. Ultimately the British achieved a system of parliamentary government that had no peer in the Western world, and England rose from a national core area into the headquarters of a global empire.

The economic developments that to a considerable extent precipitated this political progress came in stages that could be accommodated without wrecking the whole society. Britain's original subsistence economy gave way to the commercial era of mercantilism, and the industrial revolution was foreshadowed by the development of manufacturing based on the energy provided by streams flowing off the Pennines, lowland Britain's mountain backbone. Then, when the industrial revolution came, coalfields in the southern part of the Pennines provided the needed power source. The eastern deposits supported the industries of Sheffield, Derby, and Nottingham; those to the west sustained Manchester, Salford, and Bolton. Smaller fields were south of the Pennines, near Birmingham, and to the northeast, near Newcastle.

The impact of the industrial revolution was felt most strongly in the industrial regions of England near significant coal deposits in the Midlands and the northeast—London was not directly involved. Working conditions in the burgeoning areas were bad, living conditions were often worse, and population totals soared. Labor organized and became a potent force in British politics. Meanwhile, Britain prospered. Nowhere is the principle of functional specialization better illustrated than here. Woolen-producing cities lie east of the Pennines, centered on Leeds and Bradford.

Focus of a former world empire: the Houses of Parliament, the Thames River, and the urban landscape of central London, England. (Brian Seed/Black Star)

The cotton textile industries of Manchester are on the other side of the range. Birmingham, now a city of 2.5 million people, and its nearby cities concentrate on the production of steel and metal products, including automobiles, motorcycles and bicycles, and airplanes. The Nottingham area specializes in hosiery, as does Leicester; boots and shoes are made in Leicester and Northampton. In northeast England, shipbuilding and the manufacture of chemicals are the major large-scale industries, along with iron and steel production and, of course, coal mining. Coal became an important British export, and with its seaside locations in northeast England, Wales, and Scotland, it could be shipped directly to almost any part of the world—more cheaply than it could be produced from domestic sources in those faraway countries.

Today, the industrial specialization of British cities continues, but the resource situation has changed drastically. A one-time exporter of large quantities of coal, England has in recent years imported even this commodity. Iron ores have been depleted, and two-thirds of the ore required by British industries is bought elsewhere. Of course, Britain's industries always did use raw materials from other countries: wool from Australia, New Zealand, and South Africa; cotton from the United States, India, Egypt, and the Sudan. But with the breakdown of the empire and greater competition from rising industrial powers, Britain is now hard-pressed to maintain its position as an industrial power. West Germany far outdistances Britain in the production of steel, and Britain has lost the lead in the manufacture of other products as well. These developments have induced some industrial enterprises to flee the declining cities of the Midlands, to relocate near the old primate city, London. This reflects not only the Midlands' decline, but also the industries' declining dependence on coal (the use of nuclear energy is on the rise), the decision to start afresh with new, modern machinery, and the attraction of London not only as a huge and still wealthy domestic market, but also as a convenient port and a source of specialized labor.

The United Kingdom's energy situation, on the other hand, has improved markedly during the 1970s, and in 1980 Britain was self-sufficient in oil and natural gas (see box, page 83). In the late 1970s the British holdings on the North Sea floor continued to yield new discoveries of petroleum and natural gas reserves, and it was estimated that self-sufficiency would last well into the 1990s and perhaps much longer.

Lowland Britain, in addition to its industrial development, also has the vast majority of Britain's good agricultural land. Actually, the land is not nearly enough to feed the British people (56 million of them) at present standards of calorie intake. The good arable land is concentrated largely in the eastern part of England, although there is a sizable belt of good soils behind the port of Liverpool, in the west. Where the soils are adequate there is very intensive cultivation, and grain crops (wheat, oats, barley) are grown on tightly packed, irregularly shaped fields covering virtually the entire countryside. Potatoes and sugar beets are also grown on a large scale. But much of lowland Britain is suitable only for pasture and cannot, for reasons of soil quality, coolness, excessive precipitation, or other factors, sustain field crops. Hence Britain requires large food imports every year; the country cannot adequately feed even half its people, and perhaps not even one-third. There are various ways to estimate this, but in all probability Britain imports nearly two-thirds of the food consumed by its population. This makes Britain the greatest importer of food in the world, and, for those countries that have food to sell, the world's top market. Meat, grains, dairy products, sugar, cocoa, tea, fruits—Britain is either the leading or one of the world's leading importers.

In a country where some 90 percent of the population live in urban centers of one kind or another, and where only 1 of every 40 gainfully employed persons makes a living out of farming, none of this is surprising. Most Britishers work in commerce, manufacturing, transportation, or other urban-oriented economic activities; this is the most highly urbanized society in the world. The dependence on food imports just described developed during the nineteenth century, with Britain's population explosion and the competition from outside (especially Australia) of cheap farm produce—cheaper wheat came from America and Australia, for instance, than British farmers could profitably grow at home. Hence they turned in ever larger numbers to raising cattle and sheep, the main domestic livestock, and turned more acreage over to pasture and meadow. With their milk and meat, they would have a better chance to beat foreign competitors to the local market than with grain crops. In recent decades, the British government has attempted to turn this tide and has subsidized farmers to encourage them to return to cultivating wheat and potatoes, among other crops. But it has been an expensive and not very fruitful exercise. Only a crisis involving a cutoff of overseas supply lines is likely to stimulate British agriculture to full calorie-producing capacity.

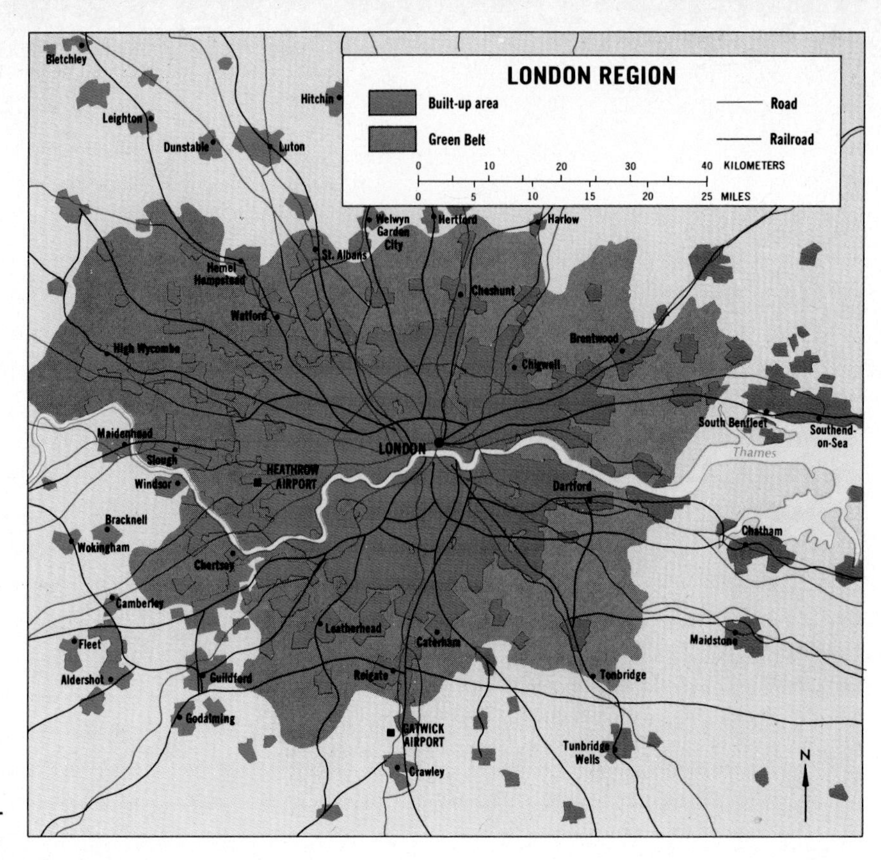

Figure 1-9

Regions of Britain

The British Isles constitute a region of the European realm, and both Britain and Ireland can be divided into subregions. In the case of Britain, nine such units can be identified: (1) the Southeast, including London, (2) the Southwest, (3) the Midlands, industrial heartland, (4) East Anglia, historic settlement zone, (5) the Northeast, between the Pennines and the North Sea, (6) the Northwest, or Lancashire, (7) Northern England with the Pennines and the famous Lake Country, (8) Scotland, and (9) Wales.

England's *Southeast* is essentially London and its immediate hinterland, a region that contains nearly 20 million of the United Kingdom's 56 million people. London itself is a city of more than 8 million people, one of the world's great conurbations (Fig. 1-9). It is the country's historic focus, the seat of government, the headquarters of numerous industrial and commercial enterprises, the major port, and most concentrated and richest market. London lies at the center of Britain's transport networks,

London's "Green Belt" and "Green Wedges" have preserved open spaces and breathing room for a vast city greatly in need of such relief. Land set aside for recreational and other nonresidential uses separates crowded suburbs. (Harm J. de Blij)

and the city is served by two major international airports (Heathrow and Gatwick), while a third airport is under construction at Foulness near the mouth of the Thames River.

Figure 1-9 reveals the dimensions of the London metropolitan area. To stem and channel the city's vast urban sprawl, a "Green Belt" was designated in a development plan for Greater London published in 1944. This plan recommended that a zone surrounding built-up London would be permanently set aside for nonresidential uses including recreation and farming. The original proposal was modified under the pressures of London's continued growth, so that today there are segments of a Green Belt but also "green wedges" between residential areas. Notwithstanding these modifications, the Green Belt proposal had the effect of preserving open space in and around a city badly in need of breathing room.

The *Southwest* is a world apart from the crowded, populous Southeast. This is a portion of Highland Britain, including Cornwall, a declining stronghold of Celtic Britain. This area has long been in economic stagnation, but its equable climate, spectacular coastlines, and impressive national parks (Dartmoor and Exmoor) have generated a growing tourist industry. Eastward the counties of Hampshire, Wiltshire, and Gloucestershire display, through new industries and revived economies, their proximity to the burgeoning London area.

Kaolin bone china

The *Midlands* are sometimes divided into the East Midlands, centered on Nottingham and Leicester, and the West Midlands, whose focus is Birmingham. The Midlands are appropriately named, for they do occupy the center of England and many of the region's cities have thrived because of their fortuitous location. Industry thrived in the Midlands from the time of the industrial revolution and, as noted earlier, enormous functional specialization occurred here. The cities and towns of the Midlands are not England's most attractive (those of the *Black Country* are indeed afflicted by grime and soot from factories and mines), but the labor force is large, the unions strong, the wages high.

East Anglia is England's historic landfall, a region of early settlement by the Angles, invaders from Germany. Long before the industrial revolution this area exported wool to the textile makers of Flanders, and its outward orientation continued afterward, when fishing grew in importance. Without coalfields, East Anglia could not compete with other regions in Britain, and in addition to fishing, mechanized agriculture and tourism have provided alternatives. During the 1970s East Anglia looked outward once more when its location, opposite the newly productive North Sea oil and gas reserves, brought pipelines to its coast from the gas fields at Leman Bank (Fig. 1-8).

The *Northeast* is the region of Sheffield and its world-famous steel, Leeds and Bradford and their textiles—but also the farms of Yorkshire, the deep-sea fishing fleets of Hull, the tourist hotels along a scenic North Sea coast, the impressive mansions of a verdant countryside and the magnificent churches and cathedrals in towns and cities. The Northeast, too, has shared in the development of the North Sea's energy reserves; an oil pipeline connects Middlesbrough to the Ekofisk oil field. Diversification is the hallmark of the productive and varied Northeast.

The *Northwest* is centered on Manchester and Liverpool. This is the region of Lancashire, one of England's leading areas when the industrial revolution gained momentum. With water power and available coal, soft water (needed in the dyeing process of cotton textiles), and moist air (which facilitated spinning before air conditioning became possible), this region had all the advantages—including its port of Liverpool, where raw cotton from America was unloaded. Liverpool grew from a fishing village into England's second port; Manchester rose to become the country's textile capital. But today the Northwest shows the effects of modern competition, the obsolescence of unreplaced equipment, overspecialization, and inadequate planning. The grimy factory towns of Lancashire, set against gray sandstone ridges defoliated by two centuries of smoke and pollution, symbolize the current problems of a region that once was the pulse of industrializing Britain.

Northern England consists of the island's mountain backbone, the Pennines, as well as the scenically beautiful Lake District. This is a segment of what we previously identified as Highland Britain, and there is much rugged topography with pastureland and high-relief coastlines studded with fishing villages. Fewer opportunities for industrialization presented themselves here, and the major city in the region is Barrow with some steel mills and shipyards based, in part, on high-grade iron ores scattered through the mountains and valleys of the Lake District.

Scotland and Wales are distinct regions of Britain because they have political as well as economic and physiographic identity. Their relationships with long-dominant England, furthermore, are changing. By European standards of scale, these are by no means small provinces; Scotland covers nearly 80,000 square kilometers (30,000 square miles) and has just under 5.5 million inhabitants, and Wales, with 2.8 million people, extends over 21,000 square kilometers (8,000 square miles). These are Britain's most rugged, remote, highland territories, where ancient Celtic peoples found refuge from encroaching, alien invaders. Celtic languages still survive here (Fig. 1-7), after centuries of domination by the English and their strong culture. The same hills and mountains that provided sanctuary and long repelled the power of London also tended to keep the Scots and Welsh divided into relatively isolated groups or clans. But communications are better now, and Scottish and Welsh nationalisms are reviving.

Wales and Scotland shared in Britain's industrial revolution, but with mixed results. Initially the coalfields of southern Wales generated valuable export revenues and attracted industries, but when the most accessible coal seams were exhausted and production costs went up, competition from fields in England proved too strong. A creeping depression struck Cardiff, Swansea, and the other cities that had attracted Welsh labor from the hills. The industrial revolution left Wales' countryside ravaged by strip mines, its people dislocated, its cities slumridden. Many Welsh citizens left their country for other parts of the world and new opportunities.

Scotland was more fortunate. Along the narrow "waist" of the Clyde and Firth of Forth, Scotland possessed an extensive coalfield, a nearby iron ore deposit, and (near Glasgow) an excellent port. This narrow lowland corridor between Scotland's mountains has become the region's core area, with industrial Glasgow at one end and the cultural focus of Edinburgh, the primate city, at the other. The Scottish coalfields were not simply a supply area for other parts of the world; they formed the basis for major manufacturing development, notably in shipbuilding and textiles. A variety of healthy and competitive industries managed to stave off the fate that had befallen Wales, and today the Scottish economy still is in better condition than that of the Welsh, although it shares with England the decline caused by obsolescence. Scotland now finds itself fronting a North Sea that produces not only the old commodity, fish, but also the new, oil and natural gas. A new age may be dawning, and not only in politics.

Wales and Scotland share a resurgent nationalism that has modern economic-geographic as well as historical roots. The government in London (where Scottish and Welsh representatives are far outnumbered by their English counterparts) has always pursued economic policies that favor England, often at the clear disadvantage of Scotland and Wales. The 1970s brought a recession that left unemployment high; few major industrial or financial firms are in Scottish or Welsh hands, and no political or economic decisions affecting the two territories are made by the Scots or Welsh themselves. The centralization of authority in London and the disadvantageous position of Wales and Scotland sometimes lead local newspapers to refer to their "colonial" status.

During the 1970s stronger demands by Scottish and Welsh nationalists were heard for regional autonomy. These demands had a varied background. For several decades local leaders had been encouraging the revival of Celtic languages not only in Scotland and Wales, but even in the Southwest, where the *Mebyon Kernow* (Sons of Cornwall) led this effort. Furthermore, Scotland's eastern cities were experiencing an economic resurgence as a result of the success of the new energy industries of the North Sea. And the crisis in Northern Ireland seemed to be a warning against any failure to adjust at an early stage to regional pressures. Thus the British Parliament in 1976 began formal discussions under the theme of *devolution*— the redistribution of authority in the United Kingdom and the restructuring of the country's political framework. In the 1979 general election, separatist proponents had only limited success in Scotland and Wales, but the issue is likely to remain current in British politics for decades to come.

The United Kingdom and North Sea Petroleum

Britain's dwindling coal reserves have recently been supplemented by a new source of energy: oil and natural gas discovered under the North Sea. European nation-states facing the North Sea have allocated sectors of the seafloor to themselves (see Fig. 1-8), and the British sector has proved to contain rich fuel reserves. In addition, the United Kingdom has purchased concessions in Norway's productive Ekofisk field.

London anticipates that the country will become self-sufficient in petroleum during the early 1980s. In 1975 Britain's own Forties field was linked by undersea pipeline to Scotland (a pipeline from the Ekofisk field also was completed), and throughout the 1975 to 1979 period new discoveries buoyed British hopes for early self-sufficiency. Because Scotland's eastern coasts (and ports) lie nearer to several major production areas in the North Sea than England does, the oil boom raised Scottish economic expectations and helped revive demands for a greater share in economic decision-making. In 1980, oil was flowing to the United Kingdom from four British-owned North Sea reserves as well as the Ekofisk field.

Fractured Ireland

The smallest of the United Kingdom's four units has also been London's most serious domestic problem of the 1970s. Northern Ireland (14,000 square kilometers, 5450 square miles) occupies the northeastern one-sixth of the island of Ireland, a legacy of England's colonial control. With a population of just over 1.5 million, Northern Ireland also is the United Kingdom's poorest area in terms of resources for primary industries. Still, a diversified economy has emerged here, as a result of infusions of British capital and the import of raw materials, mainly from Britain. Belfast's shipbuilding industry grew to significant proportions, as did Londonderry's textile (mostly linen) manufacture. But unemployment has been an erosive reality, and Northern Ireland did not fare well during the 1970s—economically or socially.

Northern Ireland was severed from the Republic of Ireland when the latter became independent from London in 1921. The political entity incorporates six of the nine counties of the historic Irish province of Ulster. What makes Northern Ireland different from the rest of Ireland is its substantial Protestant population. About two-thirds of the people trace their ancestry to Scotland or England, from where their predecessors emigrated to settle across the Irish Sea. Only 35 percent of Northern Ireland's people are Catholics, in contrast to Ireland itself, which is overwhelmingly Catholic. No spatial separation marks the religious geography of Northern Ireland: Protestants and Catholics live in clusters throughout the area. The two eastern counties are most strongly (but by no means exclusively) Protestant (see Fig. 1-8).

During the 1970s Northern Ireland was on the brink of civil war as Protestants and Catholics combated each other in terrorist campaigns. Catholics have long protested what they perceive as discrimination by the Protestant-dominated local administration, especially in housing and employment; Protestants accuse Catholics of seeking union with Ireland. The deterioration of the economy during the early 1970s plunged the region into a deepening crisis, and in 1972 the British Parliament suspended local administration and began to rule the territory directly from London. Neither British administration nor a British-armed force could stop the violence, which was even carried to London itself in the form of random bombings. In the early 1980s there was hope that the discussions on devolution might produce a solution for Northern Ireland as well as Scotland and Wales, but the destruction continued.

Ireland. Ireland (Eire) is one of Europe's youngest independent states, and the country from which it fought itself free is its neighbor, Britain. In fact, it is more correct to say that the struggle was one between Irish and English, for certainly in Wales and Scotland there were pro-Irish sympathies. But to those who think that colonialism and its oppressive policies and practices, especially where land ownership is concerned, is something that can only be perpetrated by European peoples upon non-European peoples, Ireland's history is a good object lesson.

Without protective mountains and without the resources that might have swept their country up in the great forward rush of the industrial revolution, the Irish—a nation of peasants—faced their English adversaries across a narrow sea; and a sea was slim protection against England, whatever its width. Wales had given up the struggle against English authority centuries earlier (shortly before 1300); Scotland had been incorporated soon after 1600. But the Irish continued to resist, and their resistance led to devastating retaliation. During the seventeenth century the English conquered the Irish, alienated the good agricultural land, and turned it over to (mostly absentee) landlords; those peasants who remained faced exorbitant rents and excessive taxes. Many moved westward, to the lesser farmlands, there to seek subsistence as far away from the British sphere as possible. The English placed restrictions on every facet of life—especially in Catholic southern Ireland, though rather less in the Protestant north. Irish opportunity and initiative were put down.

The island on which Eire is located is shaped like a saucer with a wide rim, a rim that is broken toward the east, where lowlands open to the sea. Although less prominent or rugged than their Scottish and Welsh counterparts, the hills that form the margins of Ireland are made of the same kinds of rocks, and in places elevations exceed 900 meters (3000 feet). A large lowland area is thus enclosed, and certainly Ireland would seem to have better agricultural opportunities than Scotland and Wales. But in fact the situation is not much better at

Devolution—Global Symptom?

The possibility that regional self-government may come to Wales and Scotland is unthinkable to many Britons; numerous Canadians view the prospect of an independent Quebec as inconceivable and Nigeria went to war to prevent the secession of Biafra. Has the evolution of the state system come to an end, and will a reverse process—devolution—become commonplace?

The indications are that many states will confront the specter of fragmentation. Yugoslavia's fragile federal structure is under pressure from Croatian separatists. Spain faces a secessionist effort in the Basque country and strengthening regionalism in the northeast. Corsican nationalists want independence from France. Eritrean partisans are fighting for separation from Ethiopia. Pakistan has for years combated a separatist movement in Pathanistan, its northwestern zone. Kurds in Iran, Moslems in the Philippines, Saharans in Morocco, Palestinians in Lebanon, Turks in Cyprus—the number of regional secessionist movements appears to be growing, and the list of potential problem areas expands as well. Such movements are vulnerable to involvement in the greater ideological struggle in the modern world as the major powers exploit them to their own advantage.

all. As Fig. 1-5 shows, Ireland is even wetter than England, and excessive moisture is the great inhibiting element in agriculture here. Practically all of Ireland gets between 100 and 200 centimeters (40–80 inches) of rain annually. Hence pastoralism once again is the dominant agricultural pursuit, and a good deal of potential cropland is turned over to fodder. The country's major export is livestock and dairy products, and most of it goes to Britain.

How can 150 centimeters of rainfall be too much for cultivation, when there are parts of the world where 150 centimeters is barely enough? Here, the *evapotranspiration* factor comes into play, and this is a good illustration of the deceptive nature of such climatic maps as Fig. 1-5—deceptive unless more is known than the map tells us. Ireland's rain comes in almost endless, soft showers that drench the ground and keep the air cool and damp—in Scotland they are called "Scottish mists." Rarely is the atmosphere really dry enough to permit the ripening of grains in the fields. Without such a dry period the crops are damaged or destroyed, and so there are severe limitations on what a farmer can plant without too great a risk. In Ireland, the potato quickly took the place of other crops as a staple when it was introduced from America in the 1600s; it does well in such an environment. But even the potato could not withstand the effects of such wetness as prevailed in several successive years during the 1840s. Having long been the nutritive basis for Ireland's population growth, which by 1830 had reached 8 million, the potato crop, ravaged by blight and soaking, failed repeatedly; the potatoes simply rotted in the ground. Now the Irish faced famine in addition to British rule, and while over a million people died, nearly twice that number left the country within 10 years of this new calamity. It set a pattern that has only recently been broken, as emigration offset and even exceeded natural increase. Ireland and Northern Ireland combined had a 1980 population of 4.8 million (3.24 million in Ireland proper), barely 60 percent of what it was when the famine of the 1840s struck. Not many places in the world can point to a decline in population since the days of the industrial revolution.

The Irish have not forgotten their colonial experiences at the hands of the British. Independence in 1921 was followed by neutrality during World War II and withdrawal from the Commonwealth in 1949. The continued partition of Ireland is a source of friction, although Dublin has officially remained quite aloof from Northern Ireland's civil strife. Nevertheless, other signs point toward improving relations. Britain is Ireland's leading trade partner, and more Irish emigrants leave for the United Kingdom than any other destination. Still, Ireland seeks to foster domestic nationalism by encouraging the use of Gaelic and through vigorous support of Irish cultural institutions.

A determined effort is being made to diversify Ireland's economy and to speed the country's development. Dublin, the capital, has become the center for a growing number of light industries. For many years the great majority of Ireland's exports were meats, live animals, dairy items, and other farm-derived products, but the balance is swinging toward manufactures, now accounting for nearly half the total sales annually. Ireland's decision to join the European Economic Community (see box, page 119) is further evidence of the country's desire to modernize. But in the face of its resource and environmental limitations, the task is difficult. In many ways Ireland still is more reminiscent of agrarian Eastern Europe than the European heartland it has recently joined.

Western Europe

The essential criteria on the basis of which a Western European region can be recognized are those that give Western Europe the characteristics of a regional core: this is the Europe of industry and commerce, of great cities and enormous mobility, of functional specialization and areal interdependence. This is dynamic Europe, whose countries founded empires while forging parliamentary, democratic governments.

But if it is possible to recognize some semblance of historical and cultural unity in British, Scandinavian, and Mediterranean Europe, that quality is absent here in the melting pot of Western Europe. Europe's western coreland may be a contiguous area, but in every other way it is as divided as any part of the continent is or ever was. Unlike the British Isles, there is no *lingua franca* here. Unlike Scandinavia, there is considerable religious division, even within individual countries. Unlike Mediterranean Europe, there is not a common cultural heritage. And indeed, the core region we think we can recognize today based on economic realities could be shattered at almost any time by political developments. It has happened before.

France and West Germany

The leading states of Western Europe, of course, are West Germany and France. In addition, there are the Low Countries of Belgium, the Netherlands, and Luxembourg (also known collectively as Benelux), and the Alpine states of Switzerland and, rather peripherally, Austria. It will be recalled that northern Italy, by virtue of its industrialization and its interconnections with Western Europe, was identified as part of the European regional core; for the purposes of the present regionalization of Europe, however, all of Italy will be considered as part of Mediterranean Europe. Thus the region comprises nearly 1 million square kilometers (414,000 square miles), not including East Germany's 108,000 square kilometers (41,800 square miles).

The partition of Germany into West and East poses a regional problem, for East Germany has stronger cultural-geographic and historical ties with Western than Eastern Europe. But East Germany, following World War II, was incorporated into the Soviet sphere of influence and has been reorganized along socialist lines, in common with Eastern European countries. East Germany will therefore be mentioned both in the overall German context in the present section and again under Eastern Europe. Reference to Germany is intended to involve all of the divided German state, West and East.

Territorially, Western Europe is dominated by France; economically, West Germany is paramount. In many other ways, the two mainland powers present interesting and sometimes enlightening contrasts. France is an old state, by most measures the oldest in Western Europe; Paris has for centuries been its primate city and cultural focus. Germany is a young country, born in 1871 after a loose association of German-speaking states had fought a successful war against . . . France! Berlin became the national focus only during the nineteenth century (although it was a prominent Prussian center earlier), having reached 100,000 inhabitants about 1750. Modern France bears the imprint of Napoleon, who died just a few years after the political architect of Germany, Bismarck, was born.

From the map of Europe it would seem that Germany, smaller than France in area, also has a disadvantageous position on the continent in comparison with its rival (Fig. 1-10). France has a window on the Mediterranean, where Marseilles, at the end of the cross-continental route of which the Rhône-Saône valley is the southernmost link, is its major port. In addition, France has coasts on the North Atlantic Ocean, the English Channel, and, at Calais, even a corner on the North Sea. Germany, on the other hand, has short coastlines only on the North Sea and the Baltic, and for the rest it seems heavily landlocked by the Netherlands and Belgium to the west, by the Alps to the south, and by Poland, to which Germany lost important territory after World War II, in the east. But such appearances can be deceptive. In effect, France is at

Figure 1-10

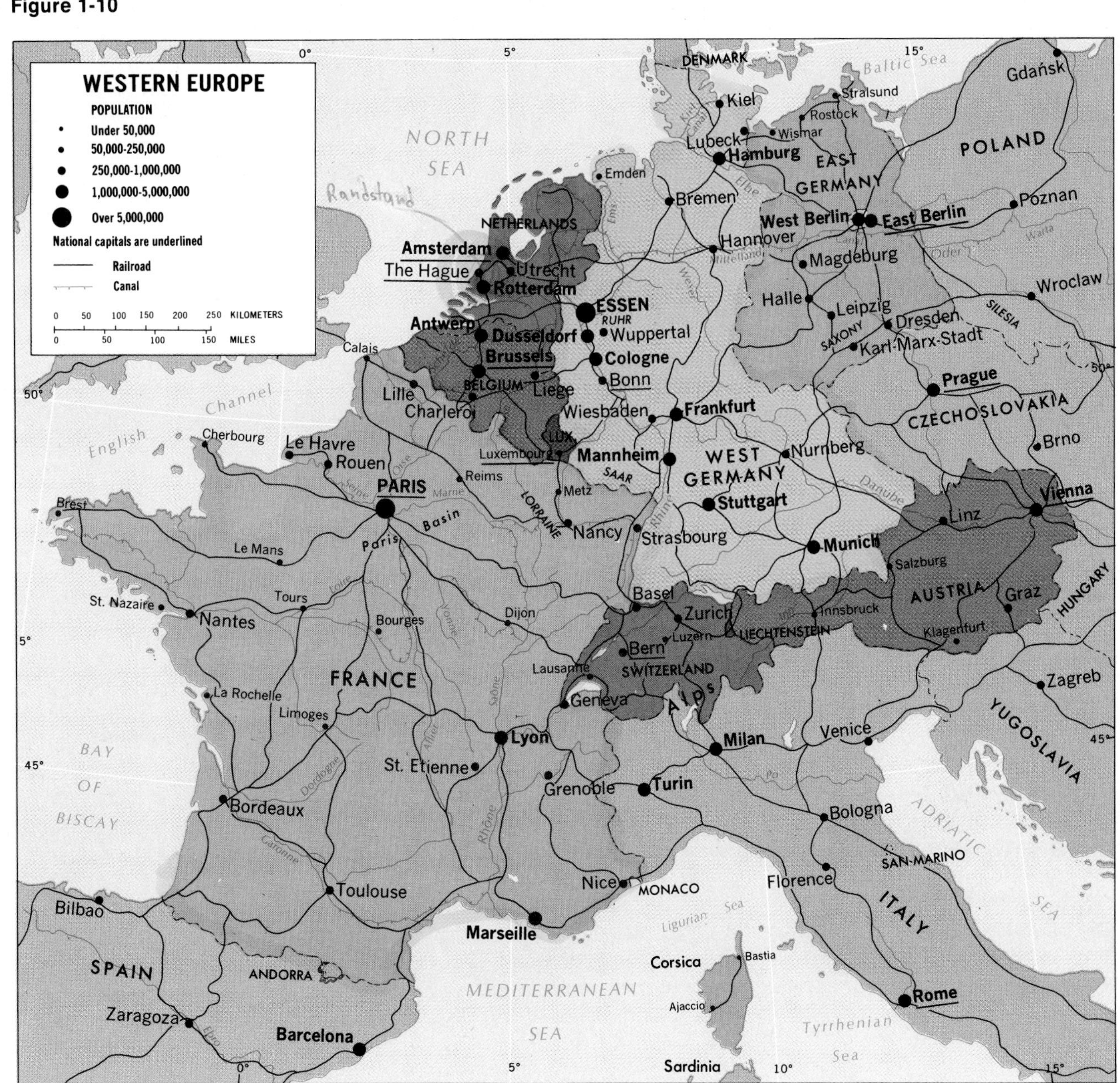

a disadvantage when it comes to foreign trade, not Germany. None of France's harbors, including that of Marseilles, is particularly good. Very few of France's many miles of rivers and waterways are navigable to modern, large, oceangoing vessels. Indeed, France in an earlier day was better served by water transport than it is at present; larger ships and deeper drafts have caused a decline in the usefulness of France's waterways. Among the Atlantic ports that France does have, several have major disadvantages. Le Havre and Rouen serve as outlets for Paris and the Paris Basin; large ships can navigate up the Seine only as far as Rouen, now France's second port. Brest and Cherbourg lie at the end of peninsulas, far from the major centers of production and population. Nantes, La Rochelle, Saint Nazaire, and Bordeaux have only local significance. A good deal of trade destined for France goes not through French, but through other European ports, such as London, Rotterdam, and Antwerp.

The great Rhine waterway, critical artery for Europe's leading industrial complex. More than 20 vessels can be seen on this short stretch of the river alone. (German Information Center)

Germany is much better off. Although the mouth of the great Rhine waterway lies in the Netherlands, Germany has most of its course—a course that runs through Europe's leading industrial complex, the Ruhr. For the western part of Germany, the Rhine is almost as effective a connection with the North Sea and the Atlantic as a domestic coast and harbor would be; Rotterdam, the world's largest port in terms of tonnage handled, is a more effective outlet for Germany than any French port is for France. But Germany has its own ports as well: Hamburg, 100 kilometers (60 miles) inland on the Elbe River, is the major break-of-bulk point where cargoes from oceangoing vessels are transferred to barges, and vice versa. Hamburg has suffered from the consequences of World War II; through a series of canals its hinterland extended as far across the North European Lowland as Czechoslovakia, and the partition of Germany has reduced its trade volume with most of this area. A North-South Canal has been constructed to connect Hamburg with the Mittelland Canal within West Germany. Not far to the west, on the Weser River, is the port of Bremen, a smaller city and located in a less productive area, but of historic and more recently of military importance, as it became the major American military port after World War II. Prior to the war,

Hamburg and Bremen had benefited from the completion of the Kiel Canal across the German "neck" of Denmark; they thereby attained new significance as gateways to the Baltic. And on the Baltic side, Germany (including East Germany) has several ports that are easily in a class with the French entries of Nantes, La Rochelle, and Bordeaux: such places as Kiel, Lubeck, and Wismar, each with well over a quarter of a million population, and each with some industrial as well as port development.

In aggregate, then, Germany's outlets are far superior to those of France. The same is true for the inland waterways that interconnect each country's productive areas. In France, although canals link the Garonne and Rhône Rivers and the Loire and Seine, the important waterways that carry the heavy traffic lie in the northeast and north, where the productive areas of the Lorraine, the Paris Basin, and the Belgian border areas are opened to the Meuse (Maas) and Rhine, and thus to Rotterdam and Antwerp. In Germany, the Mittelland Canal, with an east-west orientation, literally connects one end of the country to the other. In turn it intersects each of the northward-flowing rivers: the Oder-Neisse, the Elbe (to the port of Hamburg), the Weser (to Bremen), and the Dortmund-Ems Canal that links with the Ems River and with the port of Emden. The Dortmund-Ems Canal provides the Ruhr with a German North Sea outlet (and inlet—most of Sweden's Kiruna iron ores arrive through the port of Emden); the whole system constitutes a countrywide network of great effectiveness.

A part of the urban landscape of Paris as seen from the Eiffel Tower. Vertical development in the old section of Paris is limited to preserve its unique atmosphere, but modern high-rise development is taking place nearby. (Harm J. de Blij)

Another strong contrast between France and Germany lies in the nature and degree of urbanization in the two countries. True, Paris is without rival in France and in mainland Europe, but there is a huge gap in France between it (population: over 7 million in the city proper, over 9 million with adjacent urbanized areas) and the next ranking city, Lille (near the coalfields on the Belgian border: population about 1 million). Two questions arise, one no less interesting than the other: why should Paris, without major raw materials in its immediate vicinity, be so large; and why should Lille, Lyons, and Marseilles (all near 1 million in population) be no larger than they are?

Paris. Paris owes its origins to advantages of *site,* and its later development to a fortuitous *situation* (Fig. 1-11). Whenever an urban center is considered, these are two very important aspects to take into account. A city's *site* refers to the actual, physical attributes of the place it occupies: whether the land is flat or hilly, whether it lies on a river or lake, whether the port (if any) has shallow or deep water, whether there are any obstacles to future expansion such as ridges or marshes. By *situation* is meant the position of the city with reference to surrounding or nearby areas of productive capacity, the size of its hinterland, the location of competing towns—in other words, the greater regional framework within which the city finds itself. Paris was founded on an island in the Seine River, a place of easy defense and in all probability also a place where the Seine was often crossed. Exactly when settlement on this, the Ile de la Cité, actually began is not known, but probably it was in pre-Roman times, and it functioned as a Roman outpost some 2000 years ago. For many centuries this defense aspect of the site continued to be of great importance, for the authority that existed here was by no means always strong. Of course, the island soon proved to be too small, and the city began to sprawl along the banks of the river, but that did not diminish the importance of the security it continued to provide to the government.

The advantages of the situation of Paris were soon revealed. The city lies near the center of a large and prosperous agricultural area, and as a growing market its focality increased continuously. In addition, the Seine River, itself navigable to river traffic, is joined near Paris by several tributaries, all of them navigable as well,

Figure 1-11

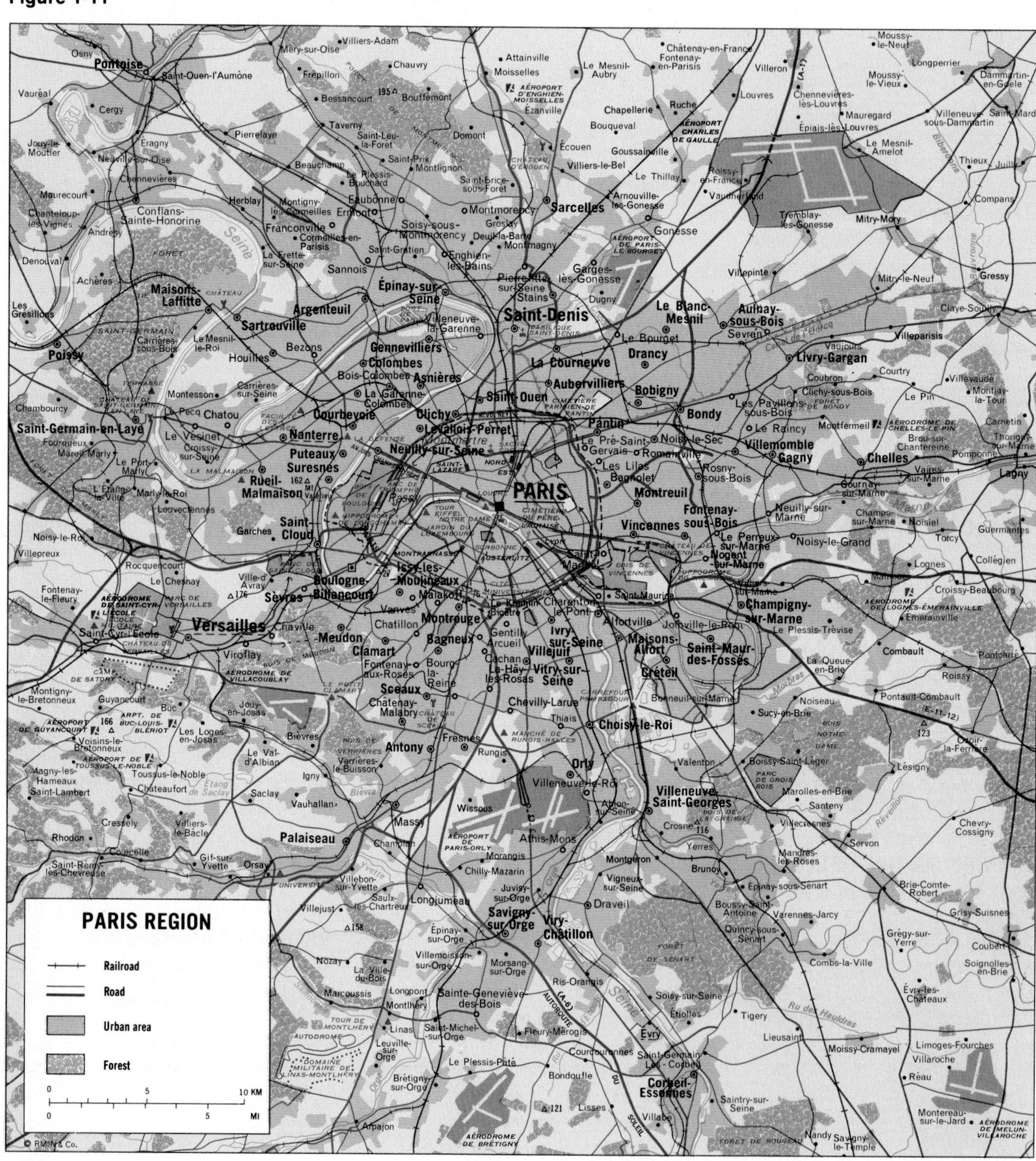

leading to various sections of the Paris Basin. Via the rivers that join the Seine (the Marne, Oise, Yonne) from the northeast, east, and southeast, and the canals that extend them even farther, Paris can be reached from the Loire, the Rhône-Saône, the Lorraine industrial area, and from the Franco-Belgian area of coal-based manufacturing. And, of course, there are land communications as well. The political reorganization brought to France by Napoleon involved not only a reconstruction of the internal administrative divisions of the country, but a radial system of roads that, like Roman roads focused on Rome, all led directly to Paris.

If it is not difficult to account for the greatness of Paris, it is quite another matter to account for the relatively limited development of French industrial centers elsewhere. In France there is no Birmingham, no Glasgow—and certainly no Ruhr. And yet there is coal, as we have seen earlier, and there is good iron ore. There is also a long history of manufacturing: the linen industry of Lille existed in the Middle Ages, and the silk industry of Lyons also is no newcomer to the industrial scene. What is lacking? For one thing, large amounts of readily accessible high-quality coal; for another, the juxtaposition of such coal with cheap transport facilities and large existing population concentrations. So French manufacturers have done what Europeans have done almost everywhere—they have specialized. French industry produces precision equipment of many kinds, high-quality china, luxury textiles, automobiles, and, of course, wines and cheese. Unable to compete in volume, they compete in quality and specialty.

France achieved greater strength in agriculture with a much larger area of arable land than Britain or Germany, and with the benefits of the temperate climate without excesses of moisture or temperature. Apart from the Paris Basin and adjacent parts of the North European Lowland in France, there are several other major agricultural areas, including the valleys of the Rhône-Saône, Loire, and Garonne. Elsewhere, as in the sandy and marshy parts of the Aquitaine, in the higher parts of the Massif Central, and in the Jura and Vosges Mountains, the soil is not suitable or the slopes are too steep for farming. Between these extremes there is a wide variety of conditions, each with its special opportunities and limitations; wheat is grown on the best soils, and it is France's leading crop. Oats will grow under slightly less favorable conditions, and barley will tolerate even greater disadvantage. Southern and Mediterranean France (the *Midi*), of course, produce grapes and the usual association of fruits. French agriculture is marked by an enormous diversity of production, and apart from the predominance of wheat it is hardly possible to isolate regions of exclusive land use—except in the vineyards of the Rhône-Saône, Garonne, and Loire Valleys. Much crop rotation is practiced; animal manure and commercial fertilizer are widely used (France is Europe's top beef and milk producer), and despite a rather low level of farming efficiency, especially in the south, France's annual agricultural output is second to none in Europe.

Germany—West or East—has no Paris, and West Germany is far from self-sufficient in farm produce. But despite the absence of a city with the primacy and centrality of Paris, West Germany today is a far more highly urbanized country than France (by official definitions the comparison is over 81 against under 70 percent; for East Germany, 74 percent). West Germany has many more urban centers with over a half million population: 19 in 1980 (France has 6). And as might be expected, West Germany also is much more strongly industrialized than its southwestern neighbor.

The majority of Germany's cities lie in or near the zone of contact between the two major physiographic regions, the plain of the north (part of the North European Lowland) and the uplands of the south (Fig. 1-2). This, as we know, is also the chief zone of coal occurrences, and of course the development of several of these cities is bound up with the availability of this power base. Prewar Germany counted three major areas of industrial growth in this zone extending from west to east: the Ruhr, based on the Westphalia coalfield, the Saxony area near the Czechoslovakian boundary (now in East Germany), and Silesia (now in Poland). In the south, near the French border, there are coal deposits in the Saar, and minor coal deposits lie scattered in many other parts of Germany. The best fields, however, are those serving the Ruhr and Silesia; Germany has lost Silesia, but the Ruhr has become the greatest industrial complex of Europe.

While the Ruhr specializes in heavy industries, Germany's second industrial complex, that of Saxony, is oriented to skill and quality. Today this region,

Divided Berlin

Berlin was the centrally positioned capital of the German state when Germany still included East Prussia and other areas of what is today Poland. As the German headquarters the city grew rapidly; it was endowed with magnificent, wide avenues and impressive public architecture. In the 1930s its population exceeded 4 million, and Berlin was a major industrial center in Germany, although (like Paris and London) the city does not lie on major sources of raw materials.

Berlin was devastated during World War II. The Allied forces in 1944, even before the war's end, divided defeated Germany into occupation zones and carved Berlin (although it lay in the Soviet zone) into four sectors: a Soviet East Berlin and a West Berlin divided among the United States, the United Kingdom, and France. In effect, the city was split in half, with 403 square kilometers (and about 1 million people) in the Soviet sector and 480 square kilometers (and just over 2 million) in the combined West. Soon West Germany's recovery and free economy began to contrast sharply with East Germany's more austere existence, a contrast that was especially vivid in the divided city. When the Allies unified their occupation zones in 1948 and included West Berlin in a new currency structure, the Soviets blockaded the Western sectors, but the blockade was broken by a massive airlift.

In August 1961 the East Germans, with Soviet approval, built a wall nearly 3 meters high, topped with barbed wire, to stop the flow of refugees from communist-dominated East Germany who used the West Berlin "window" as an escape route. Numerous people have been shot attempting to scale the wall, and Berlin today is two cities reflecting the two systems of modern Germany, interacting only very slightly. West Berlin is an exclave of the Germany Federal Republic, accessible only by East German and Soviet acquiescence; East Berlin is the headquarters of the German Democratic Republic.

which includes such famous cities as Leipzig (printing and publishing), Dresden (ceramics), and Karl-Marx-Stadt (textiles), lies in East Germany. Again, this was a region of considerable urban and manufacturing development long before the industrial revolution occurred, and nearby coal supplies, although of lower quality than those of the Ruhr, stimulated new industries and accelerated growth.

West and East Germany together have a population of nearly 80 million, and a great deal of good agricultural land would be required to feed so large a population. Farming is limited by extensive areas of hilly country in the south, and there are problems with sandy and otherwise difficult soils in the north. Nevertheless, with staples including the ever-present potato, rye, and wheat, Germany manages to produce about three-quarters of the annual calorie intake of the population. This is not quite as good as France, but it is a great deal better than Britain manages to do; Germany also has greater industrial productivity than France. In view of the size of the population (even West Germany alone has substantially more people than the United Kingdom or France), the German achievement in agriculture is remarkable.

Benelux

Three political entities are crowded into the northwestern corner of Western Europe: Belgium, the Netherlands, and tiny Luxembourg, together referred to by the first letters BeNeLux. In combination these countries also are often called the Low Countries, which is a very appropriate name; most of the land is extremely flat and lies near (and, in the Netherlands, even below) sea level. Only toward the southeast, in Luxembourg and Belgium's Ardennes, is there a hill-and-plateau landscape with elevations in excess of 1000 feet.

As between France and Germany, there are strong contrasts between Belgium and the Netherlands. Indeed, there are such contrasts that the two countries find themselves in a position of complementarity. Belgium, with its coal base along an axis through Charleroi and Liège, where there are heavy industries, and with its considerable manufacturing of lighter and varied kinds along a zone extending from Charleroi through Brussels to Antwerp, produces a surplus of industrial products including metals, textiles, chemicals, and items such as pianos, soaps, cutlery, etc. The Netherlands, on the other hand, has a large agricultural base (along with its vitally important transport functions); it can export dairy products, meats, vegetables, and other foods. Hence, there was mutual advantage to the Benelux union, for it facilitated the flow of needed imports to both countries and it doubled the domestic market.

The Benelux countries are among the most densely populated in the world, and space is at a premium. Some 25 million people inhabit an area about the size of West Virginia. For centuries the Dutch have been expanding their living space—not at the expense of their neighbors, but by wresting it from the sea. The greatest project so far drained almost the whole Zuider Zee; in the southwest the islands of Zeeland are being connected by dikes and the water pumped out, creating additional land for the Delta Plan. A future project involves the islands that curve around the northern Netherlands. Again, they may be connected and the intervening Wadden Sea laid dry.

Fertile polders lying below sea level constitute a substantial portion of the Netherlands' productive farm land. Shown here are part of the country's extensive bulb fields, meticulously laid out. (Russ Kinne/Photo Researchers)

The three cities of the Netherlands' core triangle each have about 1 million inhabitants. Amsterdam, the constitutional capital, has canal connections to the North Sea as well as the Rhine, but it obviously does not have the locational advantages of Rotterdam. Nevertheless, Amsterdam remains very much the focus of the Netherlands, with a busy port, a bustling commercial center, and a variety of light manufactures. Rotterdam, a gateway to Western Europe, commands the entries to the Rhine and Maas (Meuse) Rivers, and its development, especially during the past century, mirrors that of the German Ruhr-Rhineland. With its major shipbuilding industry, Rotterdam has importance also in fields other than transportation. The third city in the triangle, The Hague, lies near the North Sea coast but is without a port. It is the seat of the Dutch government, and in addition to its administrative functions, it benefits from the tourist trade of the beaches of suburban Scheveningen. In combination the three cities of the core triangle are called the *Randstad,* because their nearly coalescing outskirts are creating a ring-shaped urban complex that surrounds a still rural center: thus, ring-city (the literal translation of *rand* is edge or margin, but the intent as used is to convey the urban layout).

These cities of the core area stand out among a dozen other urban centers with populations in excess of 100,000, but the Netherlands is still somewhat less strongly urbanized than Belgium. Hence, the population density in the rural area is very high, and practically every square foot of available soil is in some form of productive use, even the medians of highways and the embankments of railroad lines. In contrast to Belgium, the local resource base has always been largely agricultural. Only in the southeast, the province of Limburg, wedged between Belgium and Germany, shares the belt of coal deposits that extends from France through Belgium. But other than ensuring self-sufficiency in coal for heat and fuel—and some minor industrial development in the province itself—this coal has not been as important to the Netherlands as have adjacent deposits to its neighbors. Not long ago, newly discovered natural gas reserves were opened up in the northeast; these reserves, incidentally, extend across the border into West Germany. But all in all, what the Dutch need most is farmland—and this they are adding by their own hands.

With the existing limitations of space and raw materials, the Netherlands especially, but Belgium as well, have turned to handling international trade as a major and profitable activity. It is hard to think of any countries that could be better positioned for this. Not only do the Rhine and Maas form highways that begin in the distant interior and end in the Low Countries, but Benelux itself is also surrounded by the three leading productive countries in Europe—and in the world.

This is not to suggest that the Low Countries share equally in this great locational advantage. Holland has the better position, and through centuries of competition between Rotterdam and Antwerp, the latter has fared badly. Antwerp cannot be reached except through the Schelde River, which cuts through the Dutch province of Zeeland. After the Eighty-Year War, Holland closed the Schelde, cut off Antwerp, and did not permit the city full access to the oceans until nearly a century and a half later. The Netherlands subsequently turned down Belgian requests for a canal to link Antwerp with the Maas (in part this canal was to lie through Dutch territory). And repeatedly the maintenance of the Zeeland portion of the Schelde has been an issue of contention. Thus Antwerp, after an early period of greatness and promise, became mostly a Belgian outlet; Rotterdam forged ahead to a lead it has never yielded. In this respect Antwerp resembles Amsterdam; Amsterdam, too, has canal and rail connections with the interior, but they are simply not efficient enough to affect Rotterdam's primacy. Hence Amsterdam, after its early period of greatness as the chief base for Dutch fleets and the center of a growing colonial empire, also gradually yielded its position to Rotterdam. Today Amsterdam's port handles mainly Netherlands trade, while Rotterdam specializes in the transit of international trade.

Belgium's capital city, Brussels, lies in Antwerp's hinterland, connected by river and canal. But Brussels, an agglomeration of 19 municipalities with a combined population of over 1 million, is not a port city of consequence. Instead, this historic royal headquarters, positioned awkwardly just north of the Flemish-French (Walloon) dividing line across Belgium but with dominantly French-speaking residents, has become an international administrative center. Hundreds of international companies with European interests have their main offices here; in addition, Brussels has become the administrative headquarters for international economic and political associations such as the European Economic Community (the Common Market) and its many subsidiaries and the North Atlantic Treaty Organization (NATO). It is, in addition, a financial center and commercial-industrial complex.

Switzerland and Austria

Switzerland and Austria share a landlocked location and the mountainous topography of the Alps—and little else. On the face of it, Austria would seem to have the advantage over its western neighbor; it is twice as large in area and has a substantially larger population than Switzerland. A sizable section of the Danube Valley lies in Austria, and the Danube is to Eastern Europe what the Rhine is to Western Europe and the Volga is to the Soviet Union. In addition, no city in Switzerland can boast of a population even half as large as that of the famous capital of Austria, Vienna (nearly 2 million). Austria also has considerably more land that is relatively flat and cultivable—a prize possession in these parts of Europe—and, as Fig. I-7 (page 20) suggests, more of the remainder of it is under forest, a valuable resource. That is not all. From what is known of the raw materials buried under the Alpine topography, it appears that Austria is again the winner; in the east, where the Alps drop to the lower elevations, iron ores have been found, and elsewhere there are deposits of coal, bauxite, graphite, magnesite, and even some petroleum. Switzerland, for all practical purposes, does not have any usable mineral deposit.

The Alpine topography of Switzerland presents obstacles to overland communications; Switzerland's domestic natural resources are limited as well. Still the Swiss have achieved a standard of living that is among the world's highest. Skills and talents have overcome a seemingly restrictive environment. (Craig Aurness/Woodfin Camp)

From all this we might conclude that Austria should be the leading country in this part of Europe, and that impression would probably be strengthened by a look at some cultural maps. Take, for example, that of European languages (Fig. 1-7). Austria is a unilingual state—that is, only one language is spoken throughout the country. In Switzerland, on the other hand, no less than four languages are in use. German is spoken in the largest part, French over the western quarter, and in the south there is an area where Italian is dominantly used. In the mountains of the southeast lies a small remnant of Romansh usage. This is hardly a picture of unity, and when we examine a map of religious preferences, we note that somewhat more than half of Switzerland's people are of Protestant orientation and most of the remainder Catholic, a division similar to that prevailing in the Netherlands, while Austria is 90 percent Roman Catholic.

Yet in the final analysis, it is the Swiss people who have forged for themselves a superior standard of living, and it is the Swiss state, not Austria, that has achieved greater stability, security, and progress. Although poor in raw materials, Switzerland is an industrial country, and it exports on an average about four times as high a value of manufactured products annually as Austria. Its resources amount to the hydroelectric power supplied by mountain streams and the specialized skills of its population, and yet over half of all Swiss employment is in industry. While both Austria and Switzerland are clearly Western European countries, a good case can be made for the position that Switzerland is part and parcel of the European core, but Austria is not.

The world map sometimes seems to suggest that mountain countries share a set of limitations on development that preclude them from joining the "developed" nations. It is very tempting to generalize about the impact of mountainous terrain (and the frequent corollary, landlocked location) as preventing productive agriculture, obstructing the flow of raw materials and the discovery of resources, hampering the dissemination of new ideas and the diffusion of innovations. Tibet, Afghanistan, Ethiopia, Lesotho, and the Andean portions of South American states seem to prove the point. That is why Switzerland is such an important lesson in human geography: all the tangible evidence suggests that here, at last, is a European area of stagnation and lack of internal cohesion—but the real situation is exactly the opposite. The Swiss, through their skills and abilities, have overcome a seemingly restrictive environment; they have made it into an asset that has permitted them to keep pace with industrializing Europe. First they used the passes to act as "middlemen" in interregional trade, then they used the water cascading from the mountains to reduce their dependence on power from imported coal by hydroelectric means, and finally they learned to accommodate those who came to visit the mountain country—the tourists—with professional excellence.

Farmers as well as manufacturers in Switzerland are skilled at getting the most (in terms of value) out of their efforts. The majority of the country's population is concentrated in the central plateau, where the land is at lower elevations. Here lie three of the major cities, Bern (the capital), Zurich (the largest city), and Geneva (the city of international conference headquarters). Here, too, despite the fact that the land on this "plateau" is far from flat, lie most of the farms. The specialization is dairying, for several reasons: (1) the industry produces items that can command a high price, (2) little of the central plateau is suitable for the cultivation of grain crops, and (3) the industry affords an opportunity to use a mountain resource—namely, the Alpine pastures that spring up at high elevations when the winter snows melt. In the summer, much of the herd is driven up the slopes to these high pastures. Ther herders—and sometimes the whole farm family—take up living in cottages built specifically for this purpose near the snow line. With the arrival of autumn the cattle and goats and their keepers abandon pasture and cottages and descend to the plateau or to the intermontane valleys. Swiss farming is famous for this practice, known as *transhumance,* but it is not by any means unique to Switzerland.

Thus Swiss farmers are engaged in substantially the same activity as their British and Dutch counterparts. Dairy products, cheeses, and chocolate are the chief products, and the last two are exported to a wider market. In manufacturing, also, the situation in Switzerland is quite like that in other countries of the European core—except that no other country has to import virtually *all* its raw materials. But again, specialization is the rule and high-value exports the result. Swiss industries attempt to import as few bulky raw materials as possible, while manufacturing items whose price is determined more by the skills that have gone into them than the materials used in them. Precision machinery, instruments, tools, fine watches, and luxury textiles are among the major exports, and the reputation of Swiss manufactures guarantees them a place on the world market.

Nevertheless, Switzerland must import over one-third of its food requirements in addition to its industrial raw materials, and without further sources of income the unfavorable trade balance would have the country in trouble. Instead, there are no problems—thanks to the thriving tourist industry and the country's role as an international banker and insurer. The tourist industry, of course, makes use of Switzerland's magnificent scenery, but the Swiss have made tourism another field of specialization. In their hotels, lodges, rest homes, and other temporary abodes for visitors they have set a standard of excellence that is enough to draw thousands of visitors each year. And the banking-insurance industry is founded on centuries of confidence inspired by Switzerland's stability, sovereignty, and neutrality.

Austria is a much younger state than Switzerland and its history is far less stable. Austria at one time lay at the center of the Austro-Hungarian Empire, one of the great focuses of power in Europe and one of the casualties of World War I. Modern Austria, remnant of the Empire, suffered through convulsions reminiscent of Eastern Europe rather than Switzerland's Alpine isolation. Thus while Switzerland could keep abreast of economic changes in industrializing Europe, Austria was constantly involved in costly power struggles. When Austria emerged as a separate entity in 1919, it faced not a time of plenty, but a period of reconstruction and reorganization of the national economy. And before the country had the opportunity to recover from the economic disaster of the early 1930s, it lost its independence to Nazi Germany, which forced incorporation upon it in 1938. Most recently Austria regained a measure of independence in stages, first in 1945 with the end of World War II (but under allied occupation), and then in 1955, when foreign forces withdrew under condition of continued Austrian neutrality.

Since 1945 Austria's most difficult problem has been its reorientation to Western Europe. Even Austria's physical geography seems to demand that the country look eastward; it is at its widest, lowest, and most productive in the east, the Danube flows eastward, and even Vienna lies near the eastern perimeter. But the days of domination over Eastern European countries are gone, and the markets there are no longer available. And with the interruptions and setbacks of the twentieth century, Austria is poorly equipped to catch up with competitive Western Europe.

Nordic Europe (Norden)

In three directions from the core area, Europe changes quite drastically. To the south, Mediterranean Europe is dominated by Greek and Latin influences and by the special habitat produced by a combination of Alpine topography and a "Mediterranean" climatic regime; to the east lies "continental" Europe, the Balkans, less industrialized, urbanized, or mobile than the core; and to the north lies Nordic Europe, almost all of it separated by water from what we have here defined as the core of Europe. Despite its peripheral location, Nordic Europe is not "underdeveloped" Europe. Quite the contrary; in general, a great deal has been achieved in many places without much help from nature. But in terms of resources, and in a European context, the northern areas of Europe are not particularly rich, although the discovery and exploitation of the North Sea oil fields have added an important dimension to the Norwegian economy.

Scandinavia and Finland (as well as Iceland) have the most difficult environments. Cold climates, poorly developed soils, steep slopes, and sparse mineral resources mark much of this region, although small Denmark presents better opportunities with its more fertile soils and lower relief. Northern Europe's conditions are reflected by its comparatively small population: its five countries have a total population (1980) of just 23 million, which is less than that of Benelux. The total area, on the other hand, is some 1,257,000 square kilometers (486,000 square miles)—larger than the entire European core, including Britain, France, Germany, Benelux, Switzerland, and northern Italy. People go where there is a living to be made; the living in most of Scandinavia is not easy.

Several aspects of Nordic Europe's location have much to do with this. First, this is the world's northernmost group of states; while the Soviet Union, Canada, and the United States possess lands in similar latitudes, each of these much larger countries has a national core area in a more southerly position. The North Europeans themselves call their region *Norden,* an appropriate term indeed. Second, Norden, from Western Europe, is on the way to nowhere. How different would the relative location of the Norwegian coast be if important world steamship routes rounded the North Cape and paralleled the shoreline on their way to and from the European core. Norway's ports of Trondheim and Bergen, and the capital of Oslo as well, would be different places today. Third, all of Norden except Denmark is separated by water from the rest of Europe. As we know, water has often been an ally rather than an enemy in the development of Europe—but mostly where it could be used for the *interchange* of goods. Norden lies separated from Europe *and* relatively isolated in the northwest corner. Denmark and southern Sweden are really extensions of the North European Lowland and an exception to the Scandinavian rule. Denmark, the most populous Scandinavian country after Sweden, is also the smallest state in the region; its average population density is over six times that of Sweden. On Denmark's flat and partially reclaimed country an intensive dairy industry has developed—so intensive in fact, that cattle feed has to be imported similar to the way an industrial country imports iron ore. The produce is sold to Europe's greatest food importer, Britain, and across the border to Germany; these two countries annually buy about half of Denmark's export production. For these transactions,

Denmark is very favorably situated (Fig. 1-12).

Nordic Europe's relative isolation did have some positive consequences as well. The countries have a great deal in common; they were not repeatedly overrun or invaded by different European groups, as was so much of the rest of Europe. The three major languages, Danish, Swedish, and Norwegian, are mutually intelligible, and people can converse with each other without the need of an interpreter. Icelandic belongs to the same language family; only Finnish is of totally different origins—but the long period of contact with Sweden has left a sizable Swedish-speaking resident population in Finland, where this language has official recognition. Furthermore, in each of the Scandinavian countries there is overwhelming adherence to the same Lutheran

Figure 1-12

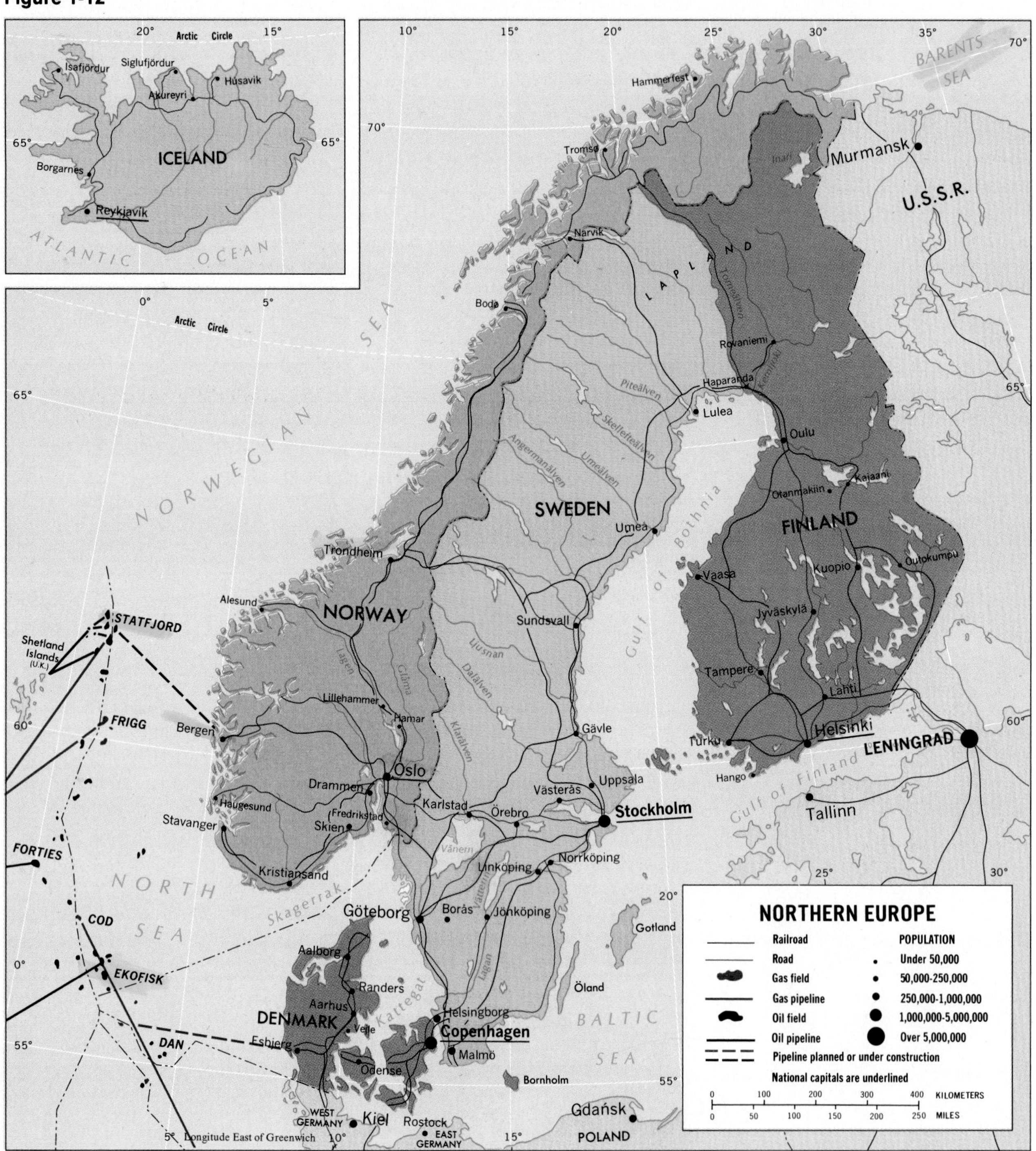

church, and in each it is the recognized state religion. Finally, there is considerable similarity in the political evolution of the Scandinavian states and in their socioeconomic policies. Democratic, representative parliaments emerged early; individual rights have long been carefully protected. This progress was possible, as it was in England, because of the lack of immediate outside threat to Scandinavia through most of its modern history. The manner in which the common cultural heritage that marks Scandinavia evolved in many ways parallels that of the British Isles.

The five Nordic countries share more than cultural ties. In their northerly location they also—within a certain range of variation, of course—face common conditions of climate and habitat. The three largest countries, Sweden, Finland, and Norway, all have major concentrations of population in the southern part of their land area. Although the waters of the North Atlantic Ocean help temper the Arctic cold and keep Norway's Atlantic ports open, they rapidly become less effective both northward and landward. Northward, the frigid polar conditions reduce the temperature of the water and thus its effect on overlying air masses. But most important is Scandinavia's high mountain backbone, which everywhere stands in the way of air moving eastward across Norway and into Sweden. Not only does this high upland limit the maritime belt to a narrow strip along the Norwegian coast, but by its elevation it also brings Arctic conditions southward into the heart of the Scandinavian Peninsula. This is well illustrated on Fig. 1-6, as is the "shadow" effect of the highland upon Sweden. Note that Denmark and southern Sweden lie in the temperate, marine-influenced *C* climatic region, and that the remainder of Sweden, despite its peninsular position, has a climate of continental character. One cannot but speculate on what Western Europe's environment might have been if that Scandinavian backbone had continued through the Netherlands, Belgium, France, and into Spain!

Only Denmark combines the advantages of temperate climate with land sufficiently level and soils good enough to sustain intensive agriculture. Norway, with its long Atlantic coast, is almost entirely mountainous; only in the southeast (around Oslo), in the southwest (south of Bergen), and on the west coast (near Trondheim) are there areas of agricultural land and reasonably good soils (apart from tiny patches of bottomland in fjorded valleys near the coast). Less than 4 percent of Norway's area can be cultivated, and even in the small areas that

Rough, often inhospitable terrain and severe climates prevail over much of Scandinavia. This rather typical scene shows an upland and glacial (U-shaped) valley near Stalheim, Norway. (Fritz Henle/Photo Researchers)

make up this 4 percent, conditions are far from ideal. Norway is by far the wettest part of Norden and, as we saw in Ireland, with cool temperatures the moisture is soon excessive. So it is in Norway, where potatoes and barley generally replace the crops the farmers would rather grow—namely, wheat and rye. Thus much of the farmland lies under fodder, and pastoralism is the chief agricultural activity. Certainly Norway came off second best in the division of the Scandinavian Peninsula, and it turned to the sea to make up for this. And in the seas the Norwegians have found considerable profit; the Norwegian fishing industry is one of the world's largest, and fishing fleets from Norwegian coasts ply all the oceans of the world. One of the most productive fishing grounds of all happens to lie very close to Norway, in part in its own territorial waters. Here, in the North Atlantic, Norwegian fishermen take large catches of herring, cod, mackerel, and haddock. But the oceans have served the Norwegians in yet another manner. Over the years Norway has developed a merchant marine that, in tonnage, is the fourth largest in the world, competing with such giants as the British, American, and Japanese fleets. This fleet carries little in the way of Norwegian products; it handles those of other countries, performing transfer functions. Norway, through its location, is otherwise denied. And now Norway's share of the North Sea is yielding undreamed-of wealth in petroleum (see box), so that the seas have favored Norway once again.

Sweden, Norway's eastern neighbor on the Scandinavian Peninsula, is more favored in almost every respect; it may not have Norway's access to the Atlantic, but it needs the Atlantic less. Sweden has two agricultural zones, of which the southernmost and leading one lies at the very southern end of the country, just across from Denmark. In fact, this area resembles Denmark in many ways, including its agricultural development, except that somewhat more grain crops, especially wheat, are grown here. Malmö, the area's chief urban center, is Sweden's third largest city with a quarter of a million people; its main function is that of an agricultural service center. Sweden's other agricultural zone lies astride a line drawn from the capital, Stockholm, to Göteborg. Here dairying is the main activity, but agriculture is overshadowed by manufacturing. Swedish manufacturing, unlike that of some of the Western European countries we have considered, is scattered through dozens of small and medium-sized towns; unlike Denmark and Norway, there are resources in Sweden to sustain such industries. For a long time Sweden served in large part as an exporter of raw or semifinished materials to industrializing countries, but increasingly the Swedes are making finished products themselves, specializing much in the way the Swiss have done. Already the list of famous products is quite long; it includes Swedish safety matches, furniture, stainless steel and ball bearings (based on steel produced through hydroelectric refining processes), and automobiles. And there is a great deal more, much of it based on relatively small local ores; apart from iron, of which there is a great deal, there is also copper, lead, zinc, manganese, and even some silver and gold. There are small metallurgical industries at one end of the scale—and the huge shipbuilding works at Göteborg at the other. in electronics and engineering, glassware, and textiles, Sweden shows once again that the skill and expertise of the people is as important as the resource base itself.

All three of the larger Nordic countries possess extensive forest resources, and all three have exported large quantities of pulp and paper. At first the mills used direct water power, but with the advent of steam power, they relocated at river mouths and used, in part, their own waste as fuel. No Nordic country depends to a greater extent on its forest resources than does Finland, whose wood and wood product exports normally account for between two-thirds and three-quarters of all annual export revenues. In this respect the region long resembled the colonial dependencies of Western European powers, underscoring their "underdeveloped" position compared to raw-material-consuming Western Europe. But even logs and timber can be transformed into something specialized before export: the Norwegians make much high-quality paper; the Swedes, matches and furniture; and the Finns, plywood and multiplex, veneers, and even prefabricated cottages. Not all the forest resources are thus transformed, of course, and Finland, like Sweden and Norway, continues to export much pulp and low-grade paper. In Finland, however, the alternatives are fewer. Most of the country is too cold to sustain permanent agriculture, and where farming is possible, mostly along the warmer coasts of the south and southwest, the objective is self-sufficiency rather than export. Known mineral deposits are few; some copper lies at Outokumpu and iron ore at Otanmakiin. Nevertheless, the Finns have succeeded in translating their limited opportunities into a healthy economic situation, in which, once again, the skills of the population play a major role. The country is nearly self-sufficient in its farm products, and the domestic market sustains a textile industry (145 kilometers inland at Tampere) and metal industries for locally needed machinery and implements, at Turku and the capital, Helsinki. For these, and for the shipyards at Helsinki and Turku, the raw materials must, of course, be imported.

The westernmost of Norden's countries—and the most westerly state of Europe—is Iceland, a hunk of basaltic rock that emerges above the surface of the frigid waters of the North Atlantic at the Arctic Circle. Three-quarters of Iceland's 103,000 square kilometers (40,000 square miles) are barren and treeless; one eighth of the island's rough and mountainous terrain is covered by imposing glaciers. Below, the earth still rumbles with earthquakes and hot springs are numerous. Volcanic eruptions have left their mark on the country's past, and have claimed thousands of lives.

Iceland's population has the dimensions of a microstate (230,000), and nearly half this population is concentrated in the capital, Reykjavik. Iceland shares with Scandinavia its difficulties of terrain and climate, only in even greater degree, and it also has ethnic affinities with continental Norden. The majority of the earliest settlers came from Norway during the ninth century, and the Icelanders claim that theirs was Europe's first constitutional democracy; in 930 A.D. a parliament was elected. Later, Iceland fell under the sway of Norway and then Denmark, and it was not until 1918 that it regained its autonomy. In 1944 all remaining ties with the Danish crown were renounced and the republic was finally resurrected.

Ekofisk! Statfjord!

The electrifying news came in 1970: exploration by the U. S. Phillips Petroleum Company had proved the existence of a large oil and natural gas field in the continental shelf under the North Sea—and in the sector allocated to Norway. In the late 1960s Norway had been consuming some 8 million tons of oil annually, but the new Ekofisk source promised to produce at least 15 million tons per year. Norway could become Europe's first petroleum exporting country.

The following years proved that initial estimates had been low, although technical difficulties and exceptionally bad weather in the North Sea delayed full production. An oil pipeline from Ekofisk to Teeside, England, was completed in 1974. A gas pipeline to Emden in West Germany was laid in 1975, and in 1976 Norway became, for the first time in its history, an exporter of energy resources. Norway's own petrochemical industry thrived. An then, late in 1976, the full dimensions of the Statfjord field became clear; the reserves here are larger even than those at Ekofisk. In the Barents Sea, north of the Scandinavian Peninsula, Norway has a maritime boundary with the Soviet Union—and here, it now appears, lies still another major reserve. But the Norway-Soviet boundary in the Barents Sea is under dispute, and the political difficulties must be resolved before exploitation can begin. Europe's historic marginal seas are now a lifeline of a different sort.

The possibilities for agriculture are severely restricted in Iceland, and so the country has turned to the sea for its survival. More than 80 percent of Iceland's exports in the late 1970s consisted of fish products, but Iceland has had its problems protecting its nearby fishing grounds. To avoid overfishing, Iceland in 1972 announced that it would claim exclusive rights in a zone 50 miles (80 kilometers) wide around the island. This led to confrontations at sea between Icelandic fishing boats and British boats protected by naval vessels, but eventually a treaty was signed that permitted British trawlers inside the 50-mile zone in certain areas. Then in 1975 Iceland extended its claim to 200 miles (320 kilometers) from its coasts, and the agreement with the United Kingdom expired. Again the North Atlantic was the scene of angry challenges. The Icelanders point to Britain's diversified economy and argue that they have no choice: the preservation of their fishing grounds is a matter of national survival. To Britain and Germany, those fish catches are merely another minor element in a much wider economy.

Denmark is by far the smallest of the countries of Northern Europe, but its population of well over 5 million is the second largest (Table I-1). Consisting of the Jutland Peninsula and numerous adjacent islands between Scandinavia and Western Europe, Denmark has a comparatively mild, moist climate; more than 70 percent of its area is devoted to agriculture. Denmark exports dairy products, meats, chickens, and eggs, mainly to its chief trading partners, the United Kingdom and West Germany. It has been estimated that Denmark's farm production could feed some 17 million people annually.

Figure 1-13

Denmark's capital, Copenhagen (1.5 million) is also Northern Europe's largest urban center. It has long been the place where large quantities of goods are collected, stored, transshipped, and dispatched, because it lies at the point where many large oceangoing vessels must stop and unload; they cannot enter the shallow Baltic Sea. Conversely, ships with smaller tonnages ply the Baltic and take their cargoes to Copenhagen's collecting stations. Thus Copenhagen is an *entrepôt*.

Like Copenhagen, Stockholm (1.4 million) and Oslo (0.5 million) are ancient cities, centers of old national states, and the primate cities of their countries today. It is a measure of Copenhagen's primacy that Denmark's second city, Aarhus, has fewer than 250,000 inhabitants; Stockholm, which has lost its position as Sweden's first port to Göteborg, still is three times as large as its competitor.

Nordic Europe has overcome its resource limitations through ingenuity and the pursuit of diversification. Long before oil was found in the North Sea, Scandinavia had secured energy from its hydroelectric power sources. Iceland made its mid-Atlantic location an asset and turned strategic position into revenues. The sea became Norway's avenue to prosperity. But problems persist; Denmark's productive economy is unbalanced, Finland's dependence on its forest resource is excessive. Still, a great deal of

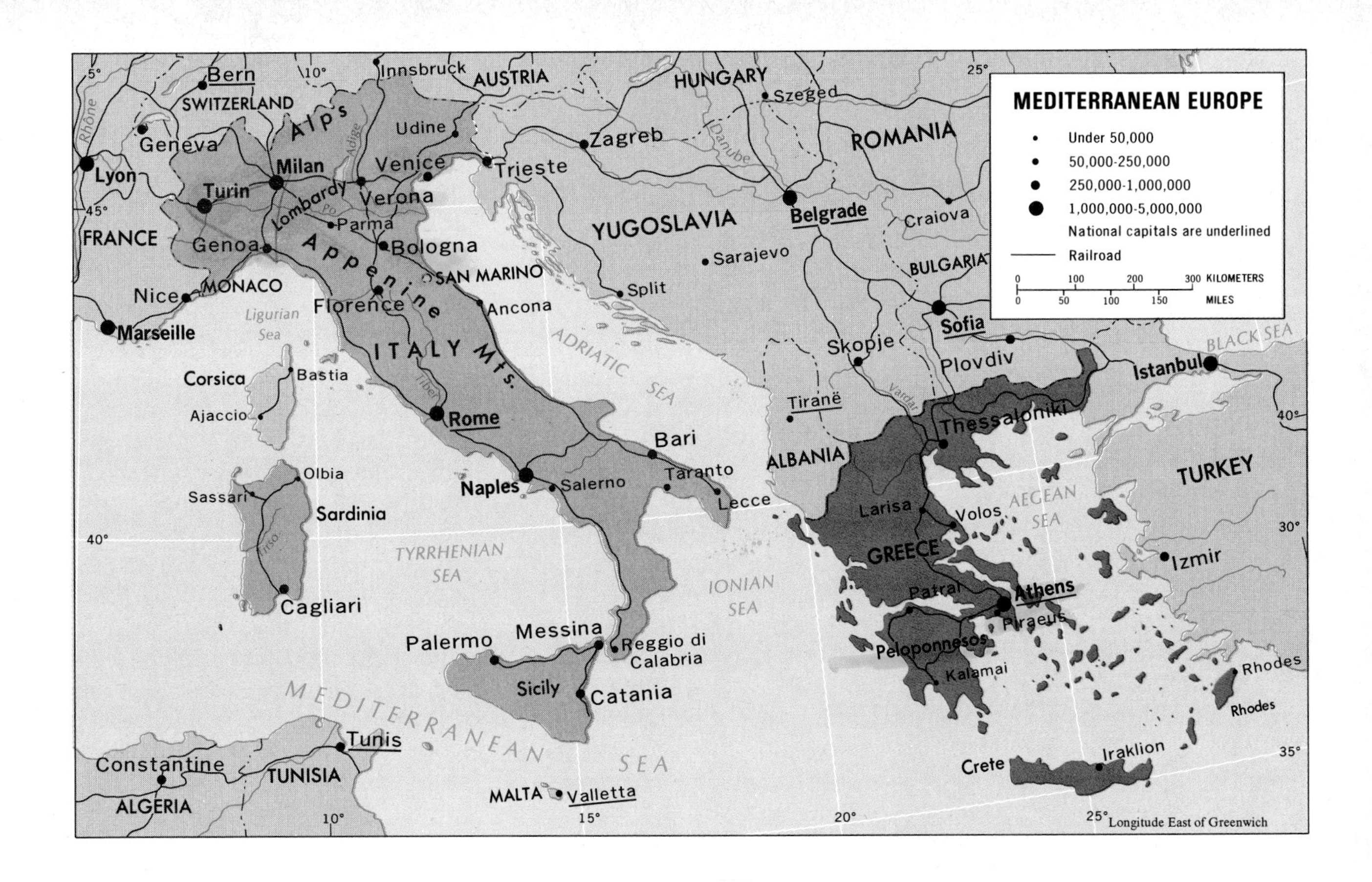

actual and potential complementarity and interdependence exists among the Nordic countries. Norway, with its merchant marine, buys tankers built in the shipyards of Göteborg. Sweden and Norway import Danish meat and dairy products. Denmark, most treeless of the Nordic countries except Iceland, needs timber, paper, pulp, and associated products and can import from Sweden, Finland, and Norway. Norway, best suited of all for the cheap production of hydroelectric power, could supply this vital commodity to Denmark, which needs it most, and Sweden. Norway, in turn, can use equipment and machinery built in Denmark, Sweden, and Finland. The opportunities are many; regional associations are already strong, and they will doubtlessly grow stronger.

Mediterranean Europe

From northern Europe we turn now to the four countries of the south: Greece, Italy, Spain, and Portugal (Fig. 1-13). From near-polar Europe we now look into near-tropical Europe; it is reasonable to expect strong contrasts, and there are many. But there are also some similarities, and some very telling ones. Once again we are dealing with peninsulas—three of them; this time, two are occupied singly by Greece and Italy and one jointly by Spain and Portugal (the Iberian Peninsula). Once again there is effective separation from the Western European core; Greece lies at the southern end of Eastern Europe and has the sea and the Balkans between it and the west; Iberia lies separated from France by the Pyrenees, which through history has proved to be quite a barrier; and then there is Italy. Southern Italy lies far removed from Western Europe, but the north is situated very close to it and presses against France, Switzerland, and Austria. For many centuries northern Italy was in close contact with Western Europe, and developed less as a Mediterranean area than as a part of the core area of Western Europe. In a very general way northern Italy is as much an exception to Mediterranean Europe as Denmark is to Scandinavian Europe, and these two areas happen to lie opposite one another across Germany and Switzerland, in physical as well as functional contact with the European core.

The Scandinavian countries share a common cultural heritage; so do the countries of Mediterranean Europe. Firm interconnections were established by the Greeks and Romans; under the Greeks, ports as far away as Marseilles and Alexandria formed part of an integrated trade system, and under the Romans virtually the whole Mediterranean region was endowed with similar cultural attributes. This unity, we know, did not last. New political arrangements replaced the old, and languages differentiated into Portuguese, Spanish, and Italian; in Greece, the Roman tide was to a considerable extent withstood. But the underlying, shared legacy remains strong to this day.

Like Nordic Europe, Mediterranean Europe lies largely within a single climatic region, and from one end to the other the opportunities and problems created by this feature of the environment are similar. The opportunities lie in the warmth of the near-tropical location, and the problems are related largely to the moisture supply: its quantity and the seasonality of its arrival. And in the topography and relief, too, Mediterranean countries share similar conditions—conditions that would look quite familiar to a Norwegian or a Swede. Much of Mediterranean Europe is mountain or upland country with excessively steep slopes and poor, thin, rocky soils; for its agricultural productivity the region largely depends on river basins and valleys and coastal lowlands. As with practically every rule, there are exceptions, as in northern Italy and northern and interior Spain, but generally the typical Mediterranean environment prevails.

Neither is Mediterranean Europe much better endowed with mineral resources than Scandinavia. Both Greece and Italy are deficient in coal and iron ore; for many years Italy, industrializing despite this shortcoming, has been one of the world's leading coal importers. Recently Italy's fuel position has improved somewhat through the exploitation of minor oil fields in Sicily and by the use of natural gas found beneath the Po Valley surface, but these are far from enough to satisfy domestic needs. Only in Spain are there really sizable fields of coal and iron ore positioned close to each other, in the Cantabrian Mountains of the north. But Spain, best endowed with raw materials of all Mediterranean countries, has not responded with manufacturing industries on the scale of Western Europe. On the contrary, a good deal of the highest-grade iron ore has been exported, as has the good coking coal, as though Spain were destined to be an underdeveloped, raw-material-supplying country rather than an industrial power. In this respect Spain only mirrors the whole Mediterranean region; there are scattered mineral deposits, some of them valuable and capable of sustaining local, skill-dependent metallurgical industries, chemical industries, and other enterprises—but instead, almost all these raw materials are exported to the core region. Scandinavia and Mediterranean Europe may share a resource poverty, but Scandinavia gets the most out of what it has—and Mediterranean Europe often the least.

This is to a lesser extent the case with both Scandinavian and Mediterranean Europe's answer to the fuel shortage: hydroelectric power. Like Norway, Italy is one of the world's leading states in the development of its hydroelectric power potential; in the 40 years after 1939, the country has quadrupled its output, further reducing its dependence on imported sources of power. But Italy is most favorably endowed in this respect, much more so than either Greece or Iberia: its hydroelectric production is about double that of its three Mediterranean partners combined. The Mediterranean environment, for all its mountains and uplands, is not all that advantageous for hydroelectric power development; with its seasonal and rather low precipitation there are often water shortages, when streams fall dry and water levels in dams go down. In northern Italy, where the largest market is located, conditions are better; rainfall distribution is much less seasonal, and water supply from the Alpine ranges is much more dependable.

Some centuries ago, Scandinavians and southern Europeans would have recognized another area of similarity between their environments: the prevalence of an extensive forest cover, a valuable resource in preindustrial as well as industrial times. The Scandinavians still have it, and their export revenues include a sizable percentage derived from these forests—but Mediterranean Europe stands largely bare and denuded. Trees were cut down for housebuilding, for centuries of shipbuilding, for fuel, and to make way for agriculture. In the period of Spain's greatness the forests of the Meseta constituted one of Iberia's major assets. Today this area stands windswept and treeless.

Mediterranean Europe's four countries in 1980 had a combined population of 115 million, notwithstanding enormous emigration, especially from Spain and Portugal, during the colonial period. From what we know of Mediterranean dependence on agriculture, the historical geography of trade routes and urban growth, and the general topography of Mediterranean Europe, we can fairly well predict the distribution of population. In Italy, nearly 25 million of the country's 58 million people are concentrated in and near the basin of the Po River, and elsewhere heavy concentrations also exist in coastal lowlands and riverine basins all the way from Genoa to Sicily. Thus, the least heavily peopled part.of the peninsula is the Appenine mountain chain (and the rim of Alpine areas in the far north), but the eastern and western flanks of the Appenines are very crowded. In Greece, today as in ancient times, heavily settled coastal lowlands are separated from each other by relatively empty and sometimes barren highlands. Especially dense concentrations occur in the lowland dominated by Athens and on the western side of the Peloponnesus. Both Spain and Portugal also have heavy settlement on coastal lowlands, although the Meseta is somewhat more hospitable than Greece's rocky uplands. The Mediterranean shorelands around Barcelona and Valencia have attracted a high population total; Barcelona stands at the center of Spain's leading industrial area (Catalonia). The Basque provinces from Bilbao to the Pyrenees are a major manufacturing area, especially for metal and machine industries. Spain's other population agglomerations have developed in the northwest, near the mining areas of the Cantabrian Mountains, in the south, where the broad lowland of the Rio Guadalquivir opens into the Atlantic, and where the Huelva-Cadiz-Cordoba triangle incorporates a sizable urban and rural population, and in the center of the country, in and near—especially southwest of—the capital, Madrid. In Portugal, which has Atlantic but no Mediterranean coasts, the majority of the population is nevertheless located on the coastal lowlands rather than on the Iberian plateau; Lisbon and Porto form centers for these coastal concentrations.

Thus Mediterranean population distribution is marked by a dominant peripheral location, by a heavy clustering of high concentrations and great densities in productive areas, usually coastal and riverine lowlands, and by a varying degree of isolation on the part of these clusters; contact between central and southern Greece, between the east and west coasts of Italy, and between Atlantic and Mediterranean Spain is not always effective. Terrain and distance are the obstacles. And although it is difficult to say exactly what constitutes overpopulation, there obviously is excessive population pressure on land and resources in many parts of Mediterranean Europe. Other than Ireland, no country in Europe has sent more of its people to overseas realms than has Portugal; standards of living here, in Greece, and to a lesser extent in Spain and Italy, are quite low.

Perhaps the most stunning contrast between Scandinavian and Mediterranean Eruope lies in the living standards of the people. While economic specialization and limited population growth, along with generally enlightened government policies and attempts to ensure a fairly equitable distribution of wealth, have produced standards of living in most of Scandinavia that are comparable and even superior to those prevailing in the European core, much of Mediterranean Europe seems hopelessly backward. Greece is perhaps the least favorably endowed of the four countries; less than a third of its area is presently capable of supporting some form of cultivation, and on this land the average density of population is around 300 per square kilometers (800 persons per square mile)—less than an acre per person. Thus many of Greece's farmers are engaged in sheer subsistence, their income is low, and their ability to buy improved farm equipment or fertilizer is minimal. Water supply is an ever-present problem, and the capital required to remedy it is very scarce. Yet agriculture is Greece's mainstay, for industrial opportunities are few. Greek workers by the thousands leave to seek work in the industries of Western Europe. Greek farmers, where conditions permit, turn to crops such as wheat and corn for the home market and tobacco, cotton, and typical Mediterranean produce such as olives, grapes, citrus fruits, and figs for exports. Greece ranks third among the world's leading exporters of olive oil (after Spain and Italy), and in some years nearly half its export revenues come from tobacco.

If this list of Greek agricultural products has an unusual ring to it, it should be remembered that the Peloponnesus is really Greece's Mediterranean zone, from which most Mediterranean crops are derived, while northward Greece takes on more continental characteristics—including more field and less garden agriculture. Eastward, the two areas meet in a point at Athens, whose urban area (including the port of Piraeus) counts some 2 million inhabitants. Athens is the administrative, commercial, financial, cultural, and indeed the historic focus of Greece, and despite its limited industry it has grown to a size far beyond what would seem reasonable for such a relatively poor and agrarian country. With Piraeus it stands at the head of the Aegean Sea; Athens also has a large and busy airport. With its heritage of ancient structures still a direct reminder of past glory, Athens has become one of the Mediterranean's major tourist attractions, and thus a source of much-needed revenues. Another source of revenues is one resembling that developed by Norway—a large, worldwide merchant marine competing for cargoes wherever and whenever they need to be hauled.

At the other end of the Mediterranean, the Iberian Peninsula is less restrictive in the opportunities it presents for development. Iberia is much larger than Greece, and proportionately less of it is quite as barren and rocky as the Greek land. Also, raw materials are in far more plentiful supply. This is not to suggest that these opportunities have been put to maximum use. The rural areas are overpopulated; one price Spain paid for its slow industrial development was that its population "explosion" had to be accommodated largely in the rural areas, where pressure was already high. Land was divided and subdivided, farms grew smaller and smaller, and less and less efficient; poorer soils were turned over to farming even though they were marginal and their productivity was bound to be low. Most of northern Spain is fragmented into these tiny parcels; the situation is most serious in Galicia (the northwest corner, north of Portugal), where farms have been subdivided beyond the level of viability.

These, then, are some of the reasons why Southern Europe's per-acre yields are always shown to be so much lower than those of Western Europe—60 percent lower, on the average—and why so many farmers are caught up in a cycle of poverty. Another reason lies in the division of land, something that is less of a problem in Greece, where most of the land is already held in small, private holdings. But in both Spain and Portugal problems of land ownership have impeded agricultural development, because huge estates remain in the hands of (often absentee) landowners, to be farmed by tenants. Land reform has been resisted by the conservative estate owners, although a program of reallocation was begun in Portugal following the political crises of the mid-1970s. In Spain, almost nothing changed under the long Franco regime, but political reform has been accomplished, and agrarian reform may follow.

Spain's major industrial area, we know, is located in Catalonia, and not on the coast of Biscay along the mineral-rich Cantabrian ranges. Thus Cantabrian coal would have to be shipped all the way around the Iberian Peninsula to provide power for Catalonian industries; that being the case, it might as well be imported from outside Spain. And indeed, much of it is imported. No, it is not the favorable location or the rich local resource base that has stimulated industrialization in Catalonia; it seems in the first instance to be the different attitude and outlook of the Catalans that has produced this development. Vigorous and progressive, the Catalans have forged ahead of the rest of Spain, aided by a strong regional identity in the form of a distinct language. Catalan differs from Castilian Spanish, the language used in most of Spain, a major urban-cultural focus (Barcelona, always a competitor of Madrid), and a certain local, Catalonian nationalism.

Catalonia does what many industrializing although resource-poor areas do—it imports most of its raw materials and depends on a few local assets for success. Among the assets are the hydroelectric power available from the streams coming off the Pyrenees, the local labor force and its skills, and the local market, comparatively poor as it may be. But although it has had considerable success, it cannot be counted among Europe's leading industrialized areas. Unlike the Po Valley, Catalonia never stimulated effective trade contact across the mountains to the north. Compared to the Midlands of England, there is less diversification here, although the last decade has seen considerable expansion; most of the industrial establishments still produce either textiles (mostly cotton goods) or chemicals. In large measure, of course, this reflects the very limited capacity of the Spanish market, with its millions of poor families and its underpaid labor. But in going against the trend in reluctant Spain, the Catalonian achievement is a major one, and it reminds us how people, by their determination and skills, can transform the economic map.

Very slowly, Spain has recently been altering course—all the while falling farther behind the accelerating development of the European core. In the northwest, local iron and steel production is increasing, as is coal production. The hydroelectric power output has multiplied. But the general situation has changed little; this is a characteristically underdeveloped economy—exporting a number of untreated raw materials and importing a wide variety of foods and consumer goods. The exports reveal Spain's varied resources: in addition to iron ore and coal they include copper, zinc, lead, mercury (Spain is the world's leading producer of this element), and potash; the agricultural exports sound more familiar, including olive oil from Andalusia in the south, citrus fruits from the coastal zone around Valencia, wines from the Ebro Valley. At the center of it all, at the foot of the Guadarama Range, stands Madrid, capital since the sixteenth century and still by far the dominant city of the whole Iberian plateau. Chosen because of its position of centrality on the Iberian Peninsula, Madrid (2.5 million) mirrors the problems of Spain; a facade of splendor hides large areas of severe urban blight, just as the tourist-admired beauty of Iberia conceals a great need for social reform.

Much of what has been said concerning Mediterranean habitats and economies applies to southern Italy and the islands of Sicily and Sardinia. But in a way Italy is not one country; it is two. While the north has had the opportunities and advantages (including those of proximity to the European core) to sustain development on the Western European model, the south has for centuries been stagnant and backward. While the north has developed a real urban complex counting several cities with over a half million people and many with over a hundred thousand, the south counts one major city—Naples (2 million), undoubtedly the poorest of Italy's large cities with staggering urban blight. Together, north and south count over 56 million inhabitants (more than Spain, Portugal, and Greece combined), bound by Rome, situated, fortuitously from this point of view, in the transition zone between the two contrasting regions. In every way Italy is Mediterranean Europe's leading state—in the permanence of its contributions to Western culture, in the productivity of its agriculture and industries, in the percentage of its people engaged in manufacturing, in living standards. But the focus of Italy has shifted from where it was during Roman times; Latium and Rome no longer form the peninsula's center of gravity. True, government and church are still headquartered in the historic capital and its adjunct, Vatican City; in terms of population totals, Milan, the northern industrial rival, has established no clear lead. And neither is Rome (3 million) ever likely to lose its special position in Italy and the world; it was chosen for psychological reasons to be the new Italy's capital, and no doubt would be chosen today if the choice had to be made again. But Italy's core area, certainly in economic terms, has moved into the area called Lombardy, centering on the valley of the Po River.

Northern Italy has a number of advantages. The Appenines, which form the backbone of peninsular Italy, bend westward, leaving the largest contiguous low-lying area in the Mediterranean between it and the Alps. This area, narrow in the west (where Alps and Appenines meet; Turin is located here), opens eastward to a wide and poorly drained coastal plain on the Adriatic Sea. Here lies one of the great centers of medieval Europe: Venice, Italy's third port, still carrying the imprint of the splendor brought by that early age. As the climatic map (Fig. I-6) shows, this area has almost wholly a non-Mediterranean regime, with more even rainfall distribution throughout the year. Certainly the Po Valley has great agricultural advantages, but what marks the region today is the greatest development of manufacturing in Mediterranean Europe. It is all a legacy of the early period of contact with Flanders and the development of transalpine routes; when the stimulus of the industrial revolution came, the old exchange was vigorously renewed. As we have seen, hydroelectric power from Alpine and Appenine slopes is the only local resource, other than a large, skilled labor force. But northern Italy imports large quantities of iron ore and coal and today ranks as Europe's fourth largest steel producer, after France—although its production is less than half that of France. The iron and steel are put to a variety of uses. Competing for the first position among the industries are the metal industries and the textile industries, the latter enjoying the benefit of a much longer history, dating back, indeed, to the days of glory during the Middle Ages. The metal industries are led by the manufacture of automobiles, for which Turin is the chief center. Italy is famous for high-quality automobiles; other metal products, such as typewriters, sewing machines, and bicycles, are also produced. The Italian industry seeks to create precision equipment, in which a minimum of metal and a maximum of skill produce the desired revenues. Italy also has an impressive shipbuilding industry at Genoa on the Ligurian coast; with 1 million people, Genoa is Italy's leading port, located as it is on the Atlantic side of the peninsula at the head of the Gulf of Genoa.

The principal city in the region is Milan (3 million), leading industrial center in Mediterranean Europe. This is Italy's financial and manufacturing headquarters, and although it has seen unprecedented growth in recent years it has its roots in an earlier age of greatness. This period is still visible in the urban landscape with its impressive public buildings, palaces, and churches, but towering above these are the modern multistoried office buildings that

Italy enjoys a wide range of natural environments and produces a large variety of agricultural crops. Farmlands extend to nearly every available hectare. Shown here are farms and fields on the Tuscan Plain in Tuscany, between Florence and Pisa. (Charles E. Rotkin/Photography for Industry)

house the offices of Italian industrial concerns. No city in Italy rivals the range of industries based here—from farm equipment to television sets, from fine silk (Milan competes with Lyons in this field) to medicines, from chinaware to shoes. The Milan-Turin-Genoa triangle is Italy's industrial heart, and it is the center of a larger, integrated region that forms part of the greater Western European core area.

Eastern Europe

Between the might of the Soviet Union and the wealth of industrialized Western Europe lies a region of transition and fragmentation: Eastern Europe. Its position with reference to the major cultural influences in this part of the world at once explains a great deal: to the west lie Germanic and Latin cultures, politically represented by Germany and Italy, and to the east looms the culture realm that is Russia's. And to the south lie Greece and Turkey, whose impact also has been felt strongly in Eastern Europe; the Byzantine and Ottoman Empires took in sizable portions of this ever-unstable region.

Balkanization

The southern half of what has been defined here as Eastern Europe is sometimes referred to as the Balkans, or the Balkan Peninsula. This refers to the triangular landmass whose points are at the tip of the Greek Peloponnesus, the head of the Adriatic Sea, and the northwestern end of the Black Sea. The name comes from a mountain range in the southeasterly part of this triangle, but it has become more than merely a name for this area—it has become a concept, born of the reputation for division and fragmentation of this whole Eastern European region. Any good dictionary of the English language will carry a definition for the terms *Balkanize* and *Balkanization*—for example, "to break up (as a region) into smaller and often hostile units," to quote Webster. Certainly division and hostility have been outstanding characteristics of Eastern Europe. Its peoples—Poles, Czechs, Slovaks, Magyars, Bulgars, Romanians, Slovenes, Croats, Serbs, and Albanians—have often fought each other; at other times they have been in conflict with forces from outside the region. Various empires have tried to incorporate all or parts of Eastern Europe in their domains; none was completely successful, and the most recent effort, by the Soviet Union, is not likely to succeed either. Countless boundary shifts have taken place; time and again local minorities rose in revolt against their rulers. Culturally Eastern Europe is divided by strong regional differences, expressed in language and religion and sustained by intense nationalisms. Among these, none has emerged with sufficient strength to impose unification on the region; the nearest thing to an "indigenous" empire was that of the Austro-Hungarians.

Balkanized Eastern Europe is a region of endless contrast and division, with Slavic and non-Slavic peoples, Roman Catholic, Eastern Orthodox, and even Moslem religions, and mutually unintelligible languages with Slavic, Romanic, and Asian roots. History has seen peoples with common ties separated by political boundaries, and peoples with few similarities thrown together in a single state. In a world in which boundary disputes and struggles for territory have long been commonplace, the case of Eastern Europe has been sufficiently unusual to have given its name to this phenomenon in political geography: *Balkanization.*

As here defined, then, Eastern Europe consists of seven states: Poland, the largest in every way; Czechoslovakia and Hungary, both landlocked; Romania and Bulgaria, facing the Black Sea; and Yugoslavia and Albania, political mavericks that have Adriatic coastlines. These countries form the easternmost and fifth regional unit of Europe; their eastern boundaries also form the eastern limit of Europe itself. This is Eruope at its most continental, most agrarian, most static. In total it was not as well endowed as Western Europe or the Soviet Union with the essentials for industrialization, and it was the most remote of all parts of Europe from the sources of those innovations that brought about the industrial revolution. Since World War II, Eastern Europe has looked eastward rather than westward for directions in its political and economic development. The Soviet Union gained control, and with the cooperation of local communist parties, communist forms of resource utilization and political organization were introduced. Although the countries of Eastern Europe were not remade into Soviet Socialist Republics (as Estonia, Latvia, and Lithuania were), they did become virtual satellites of the Soviet Union and were drawn completely into the Soviet economic and military sphere.

But nothing has ever succeeded in unifying Eastern Europe, and it is doubtful that even Soviet power can do it. As early as 1948, one of the satellites, Yugoslavia, began to move away from the Soviet course; an uprising was suppressed by Soviet armed forces in Hungary in 1956. Polish workers in the country's industrial regions wrested political reforms from the authorities, and restive Czechoslovakia secured a new deal with the Soviets in 1968. Albania, Eastern Europe's smallest state, for a time aligned itself with Maoist China. The pendulum of power has swung across Eastern Europe many times, and it is likely to do so again.

The boundary framework of Eastern Europe today has evolved from the 1919 Paris Peace Conference, where a new Eastern European map was drawn following World War I. But the boundaries with which Eastern Europe was endowed in 1919 did not eliminate the internal ethnic problems of the region. In fact, it really was impossible to arrive at any set of boundaries that would totally satisfy all the peoples involved; so intricate is the ethnic patchwork that different people simply had to be joined together. Hence people who had affinities with each other were also separated by the new international borders, and every country in Eastern Europe as constituted in 1919 found itself with minorities to govern. Frequently these minorities were located near the boundaries, and adjacent states began to call for a transfer of their authority, on ethnic, historical, or some other grounds. For example, Transylvania, a part of the Hungarian Basin, was severed from Hungary and attached to Romania—but Hungary openly laid claim to it. Macedonia had been divided between Yugoslavia and Greece; Bulgaria and Albania wanted parts of it. Between Yugoslavia and Italy the area known as the Julian March became the scene of territorial competition. These are just a few instances, and there were several more—not to mention German claims on Poland and Czechoslovakia and Soviet claims on Romania.

When a certain state, through appeals to a regionally concentrated minority in an adjacent state, seeks to acquire the people and territory on the other side of its boundary, its action is termed *irredentism.* We discuss irredentist policies in greater detail in Chapter 7, but the term comes from the name of a political pressure group in Italy whose principal objective was the incorporation of *Italia Irredenta* (Unredeemed Italy), a neighboring part of Austria.

Eastern Europe has been beset by irredentist problems. Only after the end of World War II were some of them eliminated by further boundary revision (Fig. 1-14). Poland was literally moved westward; the Soviet Union took almost half of that country in the east, but Poland gained a very large area from Germany in the west. The Soviet Union also took eastern portions of Czechoslovakia and Romania, and Bulgaria and Romania settled their dispute over the southern Dobruja, which was returned to Bulgaria. The problem of the Julian March was also settled, though not immediately: through United Nations' mediation the port of Trieste came under Italian administration, while most of the disputed territory involved went to Yugoslavia.

Another major factor in the simplification of the ethnic situation in Eastern Europe has been the dominance of the Soviet Union, not only over this region, but over East Germany as well. East Germany in a sense has become an Eastern European state, certainly as far as economic and political reorganization are concerned; but the new order has also involved a migration, voluntary as well as enforced, of Germans from several East European countries—where they formed significant minorities—to this part of their homeland. Thus the German minorities of Poland and Czechoslovakia have been drastically reduced (and in the former, practically eliminated), and long-troublesome East Prussia no longer exists. Elsewhere, migrations have carried Hungarians from Czechoslovakia and Romania to Hungary, Ukrainians and Belorussians from Poland to the Soviet Union, and Bulgars from Romania to Bulgaria.

This is not to suggest that Eastern Europe's minority problems are solved. Many Hungarians still live in Romania, and Hungary has not forgotten its interests in Transylvania. Yugoslavia still has a large Albanian

Figure 1-14

minority in the south, a Hungarian minority in the north and some Romanians in the east. When the umbrella of Soviet overlordship is removed, Eastern Europe may still not be free of its irredentist problems.

Poland and Czechoslovakia

Several geographic bases exist to separate Eastern Europe's two northernmost countries from the others. Poland, on the Baltic Sea, is the region's largest state territorially (313,000 square kilometers, 121,000 square miles) and demographically (36 million in 1980). It ranks sixth in Europe in both categories. Czechoslovakia, mountainous and surrounded, is less than half as large—but it is ahead in economic development. Poland's strongly agrarian economy began to yield to industrialization only after World War II, when the innovation of national planning made possible the exploitation of the country's industrial opportunities. Czechoslovakia, on the other hand, has been more directly exposed to influences from Western Europe. It lay more directly in the paths of commercial exchange and industrial development along Europe's east-west axis, and the Czechs had long-standing ties with the west. One of these ties is the Elbe River, which originates in the basin that is Bohemia, cuts through the Erzegebirge (Ore Mountains), and flows to the North Sea port of Hamburg. Before the iron curtain was lowered, this waterway was a major factor in Czechoslovakia's westward orientation.

Poland lies largely in the Northern European Lowland, so that there are few barriers to effective communication. In the south, along the Czech border, Poland shares the foothills of the Sudetes and Carpathians, where the raw materials necessary for heavy industry exist. Warsaw, the capital (1.9 million) lies near the center of the country, in a productive agricultural area, near the head of navigation on the Vistula River, and at the focus of a radiating network of transport lines that reaches all parts of the country. Warsaw is Poland's primate city; it is the historical, cultural, and political center of Polish life. A major industrial complex is developing in southern and southwestern Poland, and the Krakow–Czestochowa–Wroclaw triangle has become the country's industrial core area. Polish planners also want to stimulate development in the country's heartland, which includes Warsaw and the textile center of Lodz (with a population of 1 million, Poland's second city), the "Polish Manchester." But in almost every way, Poland's best opportunities lie in the southern half of the country, including good farmland. Southern Poland has a black soil belt—it broadens eastward into the Ukraine—that sustains intensive farming, with wheat as the major crop. To the north the poorer glacial soils support rye and potato cultivation, and farther north still, pastureland and moors predominate.

Poland has not accepted Soviet economic innovations and political dominance without reservations, which have at times been expressed in violence. It cannot, however, be doubted that Soviet "cooperation" has boosted Poland's industrial and urban progress. On the other hand, a measure of the failure of the new order can be gained from the attempts that were made to collectivize agriculture. In Poland, as in some other countries of Eastern Europe, this program had to be slowed down or temporarily shelved because of peasant resistance, and today less than a quarter of Poland's farmland is under collective management. The communist pattern of great investment and support for industry, even at the cost of agricultural progress, is evident in Eastern Europe as well.

Czechoslovakia shares with Poland the industrial region of which the Krakow–Czestochowa–Wroclaw triangle is a part. The source of the raw materials lies astride the gap between the Ore Mountains and the Carpathians, and in Czechoslovakia the area is referred to as Moravia. Lying midway between Bohemia in the west and Slovakia in the east, this growing manufacturing area is of increasing importance to the country, with its coal supply and its heavy industry Ostrava (300,000) lies at the focus of a manufacturing complex that includes metallurgical and chemical industries.

The western sector of Czechoslovakia, mountain-enclosed Bohemia, has always been an important core area in Eastern Europe, most cosmopolitan in character and Western in its exposure and development. With its Elbe River outlet, the westward orientation of this part of the country was maintained for centuries. But in 1919 the Slovaks were attached to Bohemia, and the Slovak eastern part of Czechoslovakia, mountainous and rugged, lies in the drainage basin of Eastern Europe's greatest river, the Danube. And the Danube flows not

Yugoslavia and Carinthia

In southern Austria, on the border with Yugoslavia and Italy, lies the province of Carinthia, home of an estimated 40,000 Slovenes. Slovenia, the Slovenes' ancient homeland, is Yugoslavia's northernmost "socialist republic."

A minority census proposed for 1976 by the Austrian government produced opposition by the Slovenes (and also by the Croats, another Yugoslavian minority living in Burgenland, Austria's easternmost area). The Slovenes expressed fear of intimidation by Austrian nationalist groups; their leaders militated against the census count. In the Yugoslavian capital, Belgrade, the newspapers began to support the Austrian Slovenes' position, and when the Slovenes' leaders visited President Tito, the Yugoslav government proposed to take the issue to the United Nations. As the quarrel intensified there were calls for the "incorporation" of areas of Austrian Carinthia where Slovenes are in the majority. It was, for Eastern Europe, a familiar sequence of events; in 1980 the issue remained unsolved as another of the region's numerous latent territorial disputes.

westward, but eastward to the Black Sea. Slovakia is much more representative of Eastern Europe than is the Czech part of the state. It is mostly rural and neither as industrialized nor as urbanized as the Bohemian-Moravian west. In the days of the Austro-Hungarian Empire, this was a peripheral frontier of comparatively little importance, while Bohemia was significant as a manufacturing area even then; during the early decades of the Czechoslovakian state, the region and its inhabitants took second place to the more advanced West. This situation has continued lately: a steady stream of manpower leaves Slovakia every year, in search of work in the factories of Bohemia-Moravia. Labor shortages occur on the farms at every harvest time.

Czechoslovakia's center of gravity, then, lies in the west, in Bohemia-Moravia. Unlike Warsaw, Prague is not a centrally positioned capital; it is located in the west, and it is very clearly a Bohemian city. It is more than four times as large as the country's next urban center (the Moravian headquarters of Brno with 350,000 people). Slovakia's capital, Bratislava (on the Danube), is somewhat smaller still. In every way Prague is Czechoslovakia's primate city. It is the political and cultural focus of the country and constitutes its major industrial region. Founded at a place where the Vltava River can be easily crossed, the city lies near the Elbe River and in the middle of the country's greatest concentration of wealth. The mountains that surround Bohemia contain a variety of ores, and in many of the valleys stand small manufacturing towns that specialize, Swiss-style, in certain kinds of manufacture—for example, pencils are produced from local graphite at Budejovice, glass and crystal at Teplice-Sanoy and Yablonec, and so on. In Eastern Europe the Czechs have always led in the fields of technology and engineering skills. Prague became a major manufacturing center, and Pilsen is the site of the famous Skoda steelworks and (perhaps even more famous) Pilsen breweries. Czech products, from automobiles to textiles, find their way to capitalist as well as communist markets.

The South

The five countries that make up the southern tier of Eastern Europe (excluding Greece) seem almost to have been laid out at random, with little if any regard for the potential unifying features of this part of Eastern Europe. This apparent randomness is the result of centuries of territorial give and take, of migrations and invasions, of consolidation and shattering of states and empires. The comparative orderliness of Poland and Czechoslovakia, one a land of uniform plains and the other a rather well-defined mountain state, is lost here, especially in the Balkans, where mountain ranges and mountain masses abound, large and small basins are sharply differentiated and separated, and where the ethnic situation is even more confused than elsewhere in Eastern Europe. At least the Slovaks had this in common with the Czechs: they were neither Poles nor Hungarians, and their attachment to the Czech state seemed a reasonable solution. But in the Balkans such solutions have been hard to come by.

Prague, Czechoslovakia's capital and primate city, personifies the country's western sector, Bohemia. Its westward orientation is reflected by the townscape of the old city. (Tibor Hirsch/Photo Researchers)

The great unifier that might have been in this area is Danube River, which comes from southern Germany, traverses northern Austria, and then crosses Eastern Europe forming first the Czech-Hungarian boundary, then the Yugoslavia-Romania boundary, and finally the Romania-Bulgaria border. It is indeed anomalous that a great transport route such as this, which could form the focus for a large region, is instead a dividing line. After it emerges from the Austrian Alps, the Danube crosses the Hungarian Basin, which, although largely occupied by Hungary, is shared also by Yugoslavia, Romania, and Czechoslovakia. Then it flows through the Iron Gate (near Orsova) and into the basin that forms its lower course; this basin is shared by Romania and Bulgaria. No other river in the world touches so many countries, but the Danube has not been a regional bond. Only two Eastern European capitals—Budapest (Hungary) and Belgrade (Yugoslavia)—lie directly on the river. Only Hungary is truly a Danubian state, as the river turns southward just north of Budapest and crosses the entire country, along with its tributary, the Tisza.

Hungary. In a very general way it can be argued that progress and development in Eastern Europe decline from west to east and also from north to south. In both Poland and Czechoslovakia the western parts of the country are the most productive. Hungary has the largest and most productive share of the Hungarian Basin; it is better off than Yugoslavia immediately to the south. In Yugoslavia itself, the core of the country lies in the Danubian lowland, and the southern mountainous areas lag by comparison. As a whole, however, Yugoslavia is far ahead of its small southern neighbor, Albania. On the other side of the peninsula, Romania has the advantage over Bulgaria to the south. Thus Hungary is southeastern Europe's leading state in several respects; this, with its pivotal position in Eastern Europe and its role in the Austro-Hungarian Empire, is not surprising.

The Hungarians (or Magyars) themselves form a minority in the Balkans, since they are neither of Slavic nor of Germanic stock. They are a people of Asian origins, distantly related to the Finns, who arrived in Hungary in the ninth century A.D. Ever since, they have held on to their fertile lowland, retaining their cultural identity (including their distinctive language), although at times losing their political sovereignty. The capital, Budapest, was a Turkish stronghold for more than a century during the heyday of the Ottoman Empire. Its recent growth (to over 2 million) was achieved during the period of the Austro-Hungarian Empire and the creation of the all-Magyar state of Hungary after World War I. Today Budapest is about 10 times as large as the next-ranking Hungarian city, a reflection of the rural character of the country. With its Danube port and its extensive industrial development, its cultural distinctiveness and its nodal location within the state, Budapest epitomizes the general situation in Eastern Europe, where urbanization has been slow and where the capital city is normally the only urban center of any size.

Hungary's rural economy has not been without problems. When the country was delimited in 1919, there was a great need for agrarian reform. Large estates were carved into small holdings, and productivity rose. Then World War II and its attendant destruction, especially of livestock, set the rural economy back; after the war, the Soviets saw fertile Hungary as a potential breadbasket for Eastern Europe. When farm production failed to rise, collectivization was encouraged, but the peasants who had become small landholders resisted the effort. But Hungary had been on the German side during World War II, and the Soviets were conquerers here, not liberators; hence they pushed their reform program vigorously (as they did in East Germany). Today three-quarters of Hungary's land is in collective operation, but farm yields are not expanding nearly as fast as the country's planners would like. Certainly the country is agriculturally self-sufficient, with its harvests of wheat, corn, barley, oats, and rye, but there is less surplus for export than desired.

Industrially, also, Hungary has not yet been able to take full advantage of its potential. There is coal, notably near Pecs, not far from the Danube in the southern part of the country, and iron ore can be brought in via the Danube for iron and steel production. This has been done in quantity only recently; for a long time Hungary, like so many countries in the developing world, has been exporting millions of tons of raw materials to other producing areas. This is especially true of the one mineral Hungary has in major quantities, bauxite, mined near Gant but refined for manufacture in Czechoslovakia and the Soviet Union rather than in Hungary itself.

Yugoslavia. Yugoslavia, south of Hungary, shares a part of the Danube Basin, and this region has become Yugoslavia's heartland. Yugoslavia was created after World War I, when Serbs, Croats, Slovenes, Macedonians, Montenegrons, and other ethnic groups were combined in an uneasy union under the royal house of Serbia. The monarchy collapsed in the chaos of World War II, but the country was resurrected as a federal republic under the leadership of Marshal Tito in 1945. Tito, who had led the fight against the Germans, took Yugoslavia on a communist course. But when the Soviets tried to impose their directives on him, he asserted his country's independence from Moscow. Since 1948 the relationship between the Soviet Union and Yugoslavia has been difficult. Seeking a balanced approach to its economic problems, Yugoslavia went slow in its collectivization efforts (less than one-sixth of the farmland is under collectivized production today), while industry was guided by national planning. Before the war, Yugoslavia had been an exporter of some agricultural products, some livestock, and a few ores. After the war, with effective political control and economic centralization, its government sought to change this—not at the expense of agriculture, as was the case in so many communist-influenced countries, but in addition to it.

Yugoslavia today is a federal state consisting of the socialist republics of Slovenia, Croatia, Bosnia-Hercegovina, Montenegro, Macedonia, and Serbia (Serbia is divided into two "autonomous regions," Vojvodina and Kosovo). This complex federal structure dates from 1963, when a new constitution was approved. But the republics, while created along ethnic lines, contain many minorities: Yugoslavia has Hungarians, Albanians, Slovaks, Bulgarians, Romanians, Italians, and even Turks within its borders. About 43 percent of Yugoslavia's 23 million people (1980) are Serbian, and perhaps 24 percent are Croat (or Croatian). Less than 10 percent of the population is Slovene. Yugoslavia's politicogeographic problems can be discerned from the map—the complicated mixture of peoples, their unequal access (through location) to the country's productive opportunities, regional inequalities in progress and development. Add these to residual nationalisms and latent hostilities and it is something of a miracle that Yugoslavia has so long survived with stability. Internal dissent, however, has been put down quite ruthlessly, and nationalist elements representing several Yugoslavian groups are in exile. It is a situation that could be exploited now that Tito's unifying force no longer prevails.

If our generalization concerning declining development and southerly location in the Balkans is to hold true, then Albania, Yugoslavia's Adriatic neighbor, should be less developed even than mountain and plateau Yugoslavia. And so it is. With less than 3 million people and under 30,000 square kilometers (11,000 square miles), Albania ranks last in Europe (excluding Iceland) in both territory and population. Its percentage of urbanized population (about 3 percent) is also Europe's lowest. Most Albanians eke a subsistence out of livestock herding and farming the one-seventh or so of this mountainous country that can be cultivated at all; the largest town is the capital, Tirane, which has about 100,000 inhabitants and a few factories.

Perhaps because of its abject poverty and the limited opportunities for progress (consisting of some petroleum exports, tobacco cultivation, and chrome ore extraction), Albania turned to Maoist China for ideological as well as material support. Unlike Yugoslavia, which moved away from the Soviet orbit in the direction of moderation, Albania committed itself to the Maoist version, thus choosing the opposite way. Certainly Albania was worth more to China in this context than it was to the Soviet Union, which already dominated much of Eastern Europe; Albania clearly preferred a prominent place in China's priorities than a lowly one in the Soviet Union's. This country has little to bargain with except a somewhat strategic position in Eastern Europe, on the Mediterranean, and at the entry to the Adriatic Sea. In recent years China's new ideological directions have deprived Albania of its alternative.

The region's two Black Sea states, Romania and Bulgaria, also confirm the southward lag of progress and development. Romania is both richer and larger—twice as large as its southern neighbor in terms of territory as well as population. Both countries' boundaries epitomize those of the Balkans in general: areas of considerable homogeneity are divided (for example, the Danube lowland, shared by these two countries); areas that would seem to fit better with other countries are incorporated (such as the Transylvanian part of the Hungarian Basin, now part of Romania); national aspirations are denied (such as Bulgaria's long-time desire for a window on the Aegean Sea). But on the whole, Romania, despite some major territorial losses to the Soviet Union as a result of World War II, remains quite well endowed with raw materials.

The potential advantages of Romania's compact shape are to some extent negated by the giant arch of the Carpathian and Transylvanian Mountains. To the south and east of this arch lie the Danubian plains and the hills and valleys of the Siretul River (Moldavia), respectively, and to the west lies the Romanian share of the Hungarian Basin—with a sizable Hungarian population forming one of Eastern Europe's ubiquitous minorities here. As we have noted on previous occasions, a mountain-plain association often provides mineral wealth, and Romania is no exception. In the foothills of the Carpathians, near the westward turn of the Transylvanian Alps, lies Europe's major oil field. From Ploesti, the urban center of this area, pipelines lead to the Black

Sea port of Constanta, to Odessa in the Soviet Union, to Bucharest and to Giurgiu on the Danube. While the Soviet Union is of course Romania's chief customer, both oil and natural gas are sold in Eastern Europe as well; the gas is piped not only to Bucharest, but to Budapest also. Pipelines for oil are being laid into several Eastern European countries.

After World War II, the Romanian state found itself with some boundary revisions costing it over 50,000 square kilometers, and with 3 million fewer people. It also found itself under Soviet domination, and thus began the familiar sequence of national planning, with emphasis on industrialization, further agrarian reform involving collectivization, and accelerated urbanization. But progress has not been spectacular—not as impressive, even, as in some of the other Soviet-dominated Eastern European countries. Agricultural mechanization and the provision of adequate fertilizers have been delayed, and farm yields never have reached the levels they should. Despite Romania's considerable domestic resources, industrial development has largely been geared to the local market. Bucharest at 1.5 million people is still several times as large as the next-ranking town, the interior industrial center of Cluj with chemical, leather, and wood manufactures (260,000). The causes of all this are difficult to pinpoint, but there seems to have been a lack of national commitment to the cause of planned development, which may be a short-term factor; in the longer sense Romania has suffered from a chronic ailment of Eastern Europe—political instability.

The Romanians are a people of complex roots, whose distant Roman heritage is embodied in a distinctive language, but whose subsequent adventures have brought strong Slavic influence, especially to the people in the rural areas, as well as admixture with Hungarian, German, and Turkish elements.

The Bulgars, too, have been affected by Slavic as well as other influences. These people, numbering 9 million today, also have a history of Turkish domination, and a Bulgarian state did not appear until 1878. Bulgaria in a way was the westernmost buffer state in the long zone that emerged in Eurasia between Russian and English spheres of influence; the Russians helped push the Turks from Bulgaria, and British fears of a Russian penetration to the Aegean led to the Treaty of Berlin, at which time Bulgaria's boundaries were delimited.

Bulgaria is a mountain state, except in the Danube lowland, which in this country is narrower than in Romania, and in the plains of the Maritsa River, which to the south forms the boundary between Greece and Turkey. Even the capital, Sofia, is located in the far interior, away from the plains, the rivers, and the Black Sea; the mountains were used as protection for the headquarters of the weak embryo state, and it has remained here. In any case, Bulgaria is not a country of towns—it would be more appropriate to see it as a country of peasants. Nearly 60 percent of the population is still classed as rural, and after Sofia with nearly one million people, the next-ranking town has 200,000 (Plovdiv, the center of the Maritsa agricultural area). The east-west mountain range that forms the country's backbone and separates the Danubian lowland area from the southern valleys carries the region's name—the Balkan Mountains. Much of the rest of Bulgaria, especially the west and southwest, is mountainous as well, and people live in clusters in basins and valleys and are separated by rough terrain. The Turkish period destroyed the aristocracy and eliminated the wealthy landowners, and for many decades Bulgarian farms have been small gardens, carefully tended and productive. These garden plots survive today, although collectivization has been carried further here than in any other Eastern European country except perhaps Albania. On the big *kolkhozy* the domestic staples of wheat, corn, barley, and rye are grown, but the smaller gardens produce vegetables and fruits; plums, grapes, and olives remind us that we are approaching the Mediterranean. From this point of view Bulgaria is a valuable economic partner to the Soviet Union, where fruits, vegetables, and tobacco are always needed and where the conditions favoring the cultivation of this particular assemblage are very limited.

At the time of transition to Soviet domination, Bulgaria was in a different position from both of its Eastern European neighbors, Romania and Yugoslavia. It was poorer than both, and still had less in terms of industrial and agricultural resources. But there was a lengthy history of association of one kind or another with the Russians and Soviets, and the Bulgars are more Slavicized than the Romanians.

Bulgaria is among Europe's least-developed countries, having lain remote from mainstreams of change on the continent. This is one of Europe's least urbanized states. Over half the population farms the land, often by hand as shown in this photograph. (Marilyn Silverstone/Magnum)

Thus Eastern Europe is going through a phase of eastward orientation and communist political and economic organization—an organization that is in most cases interrelated with that of the Soviet Union itself. But, viewing the history and historical geography of Eastern Europe, it appears likely that this phase will also lead to something else. Stability has never been a quality of Eastern European life, and the superimposed stability of the postwar period has already been disturbed in some instances. And despite the repatriation of large numbers of people who formed minorities in Eastern Europe's countries, old national goals have not been forgotten. Bulgaria still considered the Macedonian question to be unsettled—and through Macedonia, of course, this country could obtain a way to the open sea. Hungary is very well aware of the Magyar minority in Romania's Transylvania. And there are signs of dissatisfaction with the status quo as it relates to the Soviet Union—in Czechoslovakia, in Romania, in Hungary, and in Poland. Eastern European nationalism is a potent force, and national pride and Soviet planning suggestions are not always compatible.

European Unification

Individually, no European country constitutes a power of world stature comparable to the United States or the Soviet Union. But in combination Europe west of the iron curtain would indeed have world power status. Europeans have long recognized the potential advantages of unification; as long ago as the 1500s the philosopher Erasmus called for it, and many after him have also done so, including Rousseau, Kant, and Victor Hugo. But until the twentieth century, only the Romans had succeeded in unifying Italy and Iberia, Greece and Britain.

Supranationalism in Europe

1944, September	Benelux Agreement signed
1947, June	Marshall Plan proposed
1948, January	Organization for European Economic Cooperation established
1949, January	Council for Mutual Economic Assistance *(Comecon)* formed
1949, May	Council of Europe created
1951, April	European Coal and Steel Community Agreement signed (effective July 1952)
1957, March	European Economic Community Treaty signed (effective January 1958)
	European Atomic Energy Commission Treaty signed (effective January 1958)
1959, November	European Free Trade Association Treaty signed (effective May 1960)
1965, April	EEC-ECSC–Euratom Merger Treaty signed (effective July 1967)
1972, January	Accession Treaty for entry to EEC by United Kingdom, Denmark, Ireland, and Norway signed
1972, September	Norway's voters defeat EEC membership proposal
1973, January	United Kingdom, Denmark, Ireland beome members of EEC, creating "The Nine"
1976, November	Greece makes application for EEC membership
1977, March	Portugal makes application for EEC membership
1977, December	Spain makes application for EEC membership
1978, April	The nine member states of the European Community (EEC has been shortened to EC) commit themselves to admit Greece, Portugal, and Spain when formalities are concluded
1978, November	Liechtenstein is admitted to the Council of Europe as its 21st member
1979, March	European Monetary System linking the currencies of EC members goes into effect
1979, June	First general elections to a European Parliament are held. First session of the 410-member legislature is held the following month and Simone Veil presides
1981, January	Greece formally enters membership in the European Community

In the aftermath of World War I, the nations of Europe held a congress in Vienna to consider the possibility of Pan-European unification, but their aspirations soon were dashed by the events of the 1930s and World War II. Since 1945, however, Europe has made unprecedented progress toward economic integration and political unification. There is, as yet, no United States of Europe—but critical first steps have been taken, and the push toward still stronger coordination continues.

It all began with Benelux, as we noted earlier, and this first move was followed by the infusion of Marshall Plan aid and the creation of the Organization for European Economic Cooperation (OEEC) (see page 74). An important principle of *supranationalism* (international cooperation involving the voluntary participation of three or more nations in an economic, political, or cultural association) is that political integration, if it is ever to occur, follows rather than precedes economic cooperation. In this way the infusion of Marshall Plan funds, by generating the need for an international European economic-administrative structure, also promoted coordination in the political sphere. As early as 1949, the year after the Marshall Plan began, the *Council of Europe* convened for the first time in a city that was proposed as a future capital for a united Europe, Strasbourg. For 30 years the Council was little more than a forum for the exchange of ideas and opinions; the Council discussed such issues as cultural cooperation, legal coordination, human rights, crime problems, and migration and resettlement. The representatives of the Council's 17 member states had no executive authority, but the Council's views did have an impact on national governments.

The year 1979 brought a momentous development when, for the first time, representatives to a European Parliament were elected rather than appointed, European political parties played a role in the formation of this regional legislature, and its composition began to reflect more accurately the political mainstreams of Europe west of the Soviet sphere. It was another step in the direction of a united Europe, and the Council of Europe may yet come to be recognized as the beginning of a true European government.

Achievements in economic cooperation have been more difficult to come by—but so far they have been more consequential. Soon after the OEEC was established, the (then) foreign minister of France, Robert Schuman, proposed the creation of a *European Coal and Steel Community* (ECSC), with the principal objective of lifting the restrictions and obstacles in the way of the flow of coal, iron ore, and steel among the mainland's six prime producers: France, West Germany, Italy, and the three Benelux countries. The mutual advantages of this arrangement are obvious, even from a map showing simply the distribution of coal and iron ore and the position of the political boundaries of Western Europe; but the six participants did not stop here. Gradually, through negotiation and agreement, they enlarged their sphere of cooperation to include reductions and eliminations of tariffs, and a freer flow of labor, capital, and nonsteel commodities, and ultimately, in 1958, they joined in the "Common Market," the *European Economic Community* (EEC). This organization incorporated virtually all of the European core area of the mainland, and its total assets in terms of resources, skilled labor, and market were enormous. Its jurisdiction was strengthened by various commissions and by legislative and judicial authorities.

One very significant development related to the creation of the Common Market was the decision of the United Kingdom not to join it. This was itself a move based on supranational considerations; there was fear in Britain that participation would damage evolving relationships with Commonwealth countries; for many of these countries Britain was the chief trade partner, and there were pressures on the United Kingdom not to endanger these ties—pressures both within the country and from the far-flung Commonwealth. Thus Britain stayed out of the Common Market, but it made its own effort to create closer economic bonds in Europe; in 1959 it took the lead in establishing the so-called *European Free Trade Association,* comprising, in addition to the United Kingdom, three Scandinavian countries (Sweden, Norway, and Denmark), the two Alpine states (Switzerland and Austria), and Portugal. This scattered group of countries, with their relatively small populations, generally limited resources, and restricted purchasing power, added up to something much less than the Common Market; they became known as the "Outer Seven," while the contiguous states of the core came to be known as the "Inner Six."

EUROPEAN SUPRANATIONALISM

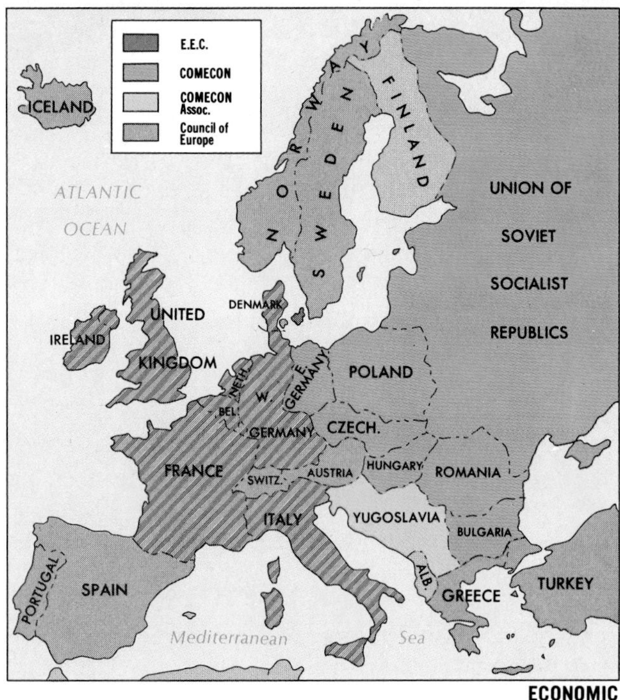

Figure 1-15

Within a few years of the creation of the Outer Seven, the United Kingdom changed its position on EEC membership and decided to seek entry. Now, however, the political attitude on the continent had changed; France took an inflexible position under De Gaulle and obstructed British participation, despite a desire on the part of other Common Market members to ratify admission. Early in the 1970s, the path was cleared for Britain's entry, and the British officially became members of the Common Market on January 1, 1973. At the same time, Denmark and Ireland also joined the EEC so that the Common Market is now known as "The Nine" (Fig. 1-15). In 1978 the nine EEC member states voted in principle to view favorably the future admission of Greece, Spain, and Portugal. By the early 1980s, the EEC will be "The Twelve."

Even before this expansion of the EEC occurred, still another step had been taken toward greater unity on the OEEC model, but with the United States and Canada as full members in a nonmilitary Atlantic association. The objective was to reduce the divisive nature of the Common Market–Outer Seven split, and at the same time to broaden the basis for Western economic cooperation. Ratified in 1961, the *Organization for Economic Cooperation and Development* counted 20 members, including all those who signed the OEEC papers in 1948. The OECD, unlike the Common Market, has little power and no binding authority, and some participants argued that it represents nothing very constructive in the European drive toward unification. But it may yet play its role in finding a way to overcome some of the obstacles in the path to this objective.

Behind the highly visible Common Market, a large number of other European associations function to further European unity. The European Atomic Energy Community *(Euratom)* aims at the development of peaceful uses of atomic energy, the sharing of research and information, and the formation of European Space Research Organization (ESRO) promotes coordinated space research. The Western European Union coordinates defense in Western Europe. And, of course, many of the states of Europe west of the iron curtain are members of the North Atlantic Treaty Organization (NATO), adding military cooperation to coordination in other spheres.

The existence of the iron curtain and the regional ideological division of Europe prevents still wider European integration. In 1949, the Soviet Union took the lead in establishing the Council for Mutual Economic Assistance (CMEA, also known as *Comecon*). This union was designed to promote industrial specialization in Eastern European countries, to finance investment projects planned by two countries or more, and generally to promote economic integration.

World economic problems during the 1970s somewhat slowed Europe's drive toward unification. Nor were all countries determined to participate: in 1972 Norway voted against membership in the Common Market. But the new Europe gives evidence that its old divisions are fading. Mobility is greater than ever before, Italians work in Amsterdam, French workers labor in German factories, Belgian capital invests in Italy. International boundaries function less to divide. License plates bearing the exhortation "Europa!" can be seen on automobiles from Rotterdam to Rome. Europe has transformed the world—now Europe restructures itself.

Additional Reading

General works on the geography of Europe include *A Geography of Europe* by J. Gottmann, published by Holt, Rinehart and Winston in New York (fourth edition, 1969) and T. G. Jordan, *The European Culture Area,* published by Harper and Row, New York, in 1973. Also see V. H. Malmström's *Geography of Europe,* published by Prentice-Hall, Englewood Cliffs, in 1971. For a view of the past, a book by W. G. East, *An Historical Geography of Europe* published in London by Methuen (1962) is still an excellent source.

Geographies on the regions of Europe abound. *The British Isles* are described by D. Stamp and S. H. Beaver in a volume so titled and published by St. Martin's Press, New York, in a sixth edition, 1971. F. J. Monkhouse wrote *A Regional Geography of Western Europe,* published by Longman, London, in 1974, the fourth edition of a standard work since it first appeared in 1959. On Nordic Europe see V. H. Malmström, *Norden: Crossroads of Destiny,* a Van Nostrand publication (1965) in the Searchlight Series. Also consider *Scandinavia: an Introductory Geography* by B. Fullerton and Al F. Williams, published in 1972 by Praeger in New York. Books in English on Mediterranean Europe are not plentiful, but D. S. Walker's *Mediterranean Lands,* published by Wiley in New York in 1962, is still readable, as is a book by J. M. Houston, *Western Mediterranean World: an Introduction to its Landscapes,* published by Longman, London, in 1964. On Eastern Europe, G. W. Hoffman has edited a volume entitled *Eastern Europe: Essays in Geographical Problems,* published in New York by Praeger, 1970.

Individual countries also have been the subject of many geographies (or groups of countries, such as the Low Countries and the Alpine states). A good beginning is a series of sketches published in *Focus,* available from the American Geographical Society in New York. The Van Nostrand Searchlight Series also includes discussions of European countries and areas, as does Aldine's World's Landscapes Series. A. R. Orme's volume on *Ireland* in the Aldine Series (Chicago, 1970) is noteworthy. Also try the Praeger *Country Profiles* Series, which includes volumes on Luxembourg by W. A. Fletcher, Norway by V. H. Malmström, and France by G. E. Pearcy. In any research for a term paper, the Methuen *Advanced Geographies* Series also constitutes a useful start, including volumes on Italy by D. S. Walker, Ireland by T. W. Freeman, and countries and regions beyond Europe as well. On France, see a book by I. B. Thompson, *Modern France: a Social and Economic Geography,* published by Rowman and Littlefield, Totowa, in 1970.

In the present chapter, the reference to R. Horvath's article is "Von Thünen's Isolated State and the Area Around Addis Ababa, Ethiopia," *Annals of the Association of American Geographers,* Vol. 59, No. 2, June, 1969. A reference to other literature (for example, the Jordan volume mentioned previously) will prove that Von Thünen's ideas still arouse debate. A. Weber's *Über den Standort der Industrien,* published in 1909, was published in translation in 1929 as *Theory of the Location of Industries* by the University of Chicago Press. The 1933 book by W. Christaller, *Central Places of Southern Germany,* was translated by C. Baskin and published in English by the University of Virginia in 1954. The quote (page 71) is from pages 17–18 of Christaller's *Foundations of Spatial Organization in Europe* (1950).

References to F. Ratzel can be found in most histories of geography, and one of his more famous articles was reprinted by R. E. Kasperson and J. V. Minghi in *The Structure of Political Geography,* published by Aldine in Chicago, 1969. Ratzel's book *Political Geography* appeared in 1897, combining much of his previous work published in the journal literature. His "Laws of the Spatial Growth of States" appeared in *Petermann's Mitteilungen,* Vol. XLII, 1896. M. Jefferson's article entitled "The Law of the Primate City" was published in the *Geographical Review,* Vol. 29, No. 2, April, 1939. And the dictionary referred to on page 109 (on the subject of Balkanization) is Webster's *Seventh New Collegiate Dictionary,* G. and C. Merriam Company, Springfield, 1963.

Australia: A European Satellite

IDEAS AND CONCEPTS

Migration
Intervening opportunity
Federalism
Remoteness

Sydney, largest city and major port of Australia. The first European settlers landed on these shores in 1788, and from their settlement the great city evolved. The Harbour Bridge (opened in 1932) connects south (across the bay) to north. Note the famous Opera House beyond the bridge. The Pacific Ocean lies in the distance. (G.R. Roberts/Photo Researchers)

Australia evokes images of far-away isolation, enormous expanses and uninhabited spaces, modern cities, vast livestock ranches, beautiful coastlines. Australia is a continent and a nation-state—a discrete realm of the world because its population and culture are European. If Australia were populated by peoples of Malayan stock with ways and standards of living resembling those of Indonesia and mainland Southeast Asia, and had a comparable history, then it is quite possible that it would be viewed today only as an exceptionally large island sector of the Asian continent. But just as Europe merits recognition as a continental realm despite the fact that it is merely a peninsula of "Eurasia," so Australia has achieved identity as the island realm of "Australasia." Australia and New Zealand are European outposts in an Asian Pacific world, as unlike Indonesia as Britain and America are unlike India.

Although Australia was spawned by Europe and its people and economy are Western in every way, Australia as a culture realm is a far cry from the crowded, productive, complex European world. Australia's entire population (nearly 15 million in 1980) is smaller than that of the Netherlands, but Australia is nearly 200 times as large territorially. This gives an average population density of under 2 persons per square kilometer (4.6 per square mile) and suggests that Australia is a virtual population vacuum on the very edge of overpopulated Asia. But so much of Australia is arid or semiarid (Fig. I-6) that only about 8 percent of its total area is agriculturally productive, and much of this moister part of the continent is too rugged for farming. By some calculations, only 1 percent of Australia's total area of more than 7.5 million square kilometers (3 million square miles) is prime land for intensive cultivation. Certainly Australia could support many more people than it does today, but is no feasible outlet for Asia's millions. Australia's *total* population is less than China's annual *increase.*

Ten Major Geographic Qualities of Australia

1. Australia lies remote from the places with which it has the strongest cultural and economic ties.
2. Australia and New Zealand constitute a culture realm by virtue of territorial dimensions, relative location, and cultural distinctiveness, not population size.
3. Australia has the lowest average elevation and the lowest overall relief of all the continental landmasses.
4. Australia is marked by a vast arid and semi-arid interior, extensive, open plainlands, and marginal moister zones.
5. Australia's population has a very low arithmetic and a low physiologic density.
6. A very high percentage of Australia's population is concentrated in a small number of major urban centers.
7. Australia's population distribution is strongly peripheral as well as highly clustered.
8. Australia's indigenous (black) population was almost completely submerged under the European invasion, remains numerically small, and participates only slightly in the modern society.
9. Australian agriculture is highly mechanized and produces large surpluses for sale on foreign markets. Huge livestock herds feed on vast pasturelands.
10. Australia has a large and diverse natural resource base and contains substantial untapped potential.

Migration and Transfer

Australia's indigenous peoples probably arrived on the island continent by way of New Guinea and across a land bridge that existed when enlarged Pleistocene glaciers lowered global sea level. The original Australians were black peoples with ancestral roots in East Asia, and they never numbered more than 300,000 to 350,000; they were hunters and gatherers who lived in small groups. Cultural anthropologists have identified over 500 tribes and clans using different and mutually unintelligible languages, and cultural geographers have determined that they were more numerous in eastern Australia than in the west. The first Australians skillfully adapted themselves to a difficult natural environment and developed a strong sense of territoriality. Almost every narrative about Australia's indigenous peoples remarks on their detailed knowledge of the terrain and its opportunities and limitations.

Australia's original inhabitants once numbered over 200,000, but only about 50,000 survive today (in addition to some 100,000 aboriginal Australians with mixed European ancestry). Some indigenous Australians live in camps in the northern interior; This group was photographed in the Northern Territory. (Herbertus Kanus/Rapho-Photo Researchers)

Today, the first Australians are a nearly forgotten people. They could not withstand the impact of the European invasion, which was delayed because of Australia's remoteness and its apparent unattractiveness. Portuguese, Spanish, and Dutch explorers saw and landed on Australian coasts during the sixteenth, seventeenth, and eighteenth centuries (Tasman, the Dutch seaman, reported on Tasmania's coasts in 1642), but it was not until the journeys of James Cook, the British sea captain, that Australia finally entered the European orbit. When Cook visited the east coast in 1770, he was the first European to see this part of Australia. At the behest of the British government he returned in 1772 and again in 1776, and by that time the British were determined that *terra australis incognita* should become a British settlement. In 1778—just two centuries ago—white settlement in Australia began, and with it the demise of the aboriginal societies.

The Europeanization of Australia involved the movement of hundreds of thousands of emigrants, most of them British, from Western Europe to the shores of a little-known landmass in another hemisphere. This process of *migration* has affected human communities from the earliest times, but it reached an unprecedented level during the nineteenth century. Before the 1830s, fewer than 3 million Europeans had left their homelands to settle in the colonies. But between 1835 and 1935, perhaps as many as 75 million departed for other lands—for the Americas, for Africa, and for Australia (Fig. A-1).

If so many people had not left Europe during the great population expansion that accompanied the industrial revolution, Europe's population problems would have been far more serious than they were. Even so, Europe during the nineteenth century was, for many people, an unpleasant place to be. Famine swept Ireland, war and oppression overtook parts of the mainland, and the cities of industrializing Britain, Belgium, and Germany contained some of the worst living conditions ever. Many Europeans abandoned their homes to seek a better life across the ocean, even though they were uncertain of their fate in their new environment.

Figure A-1

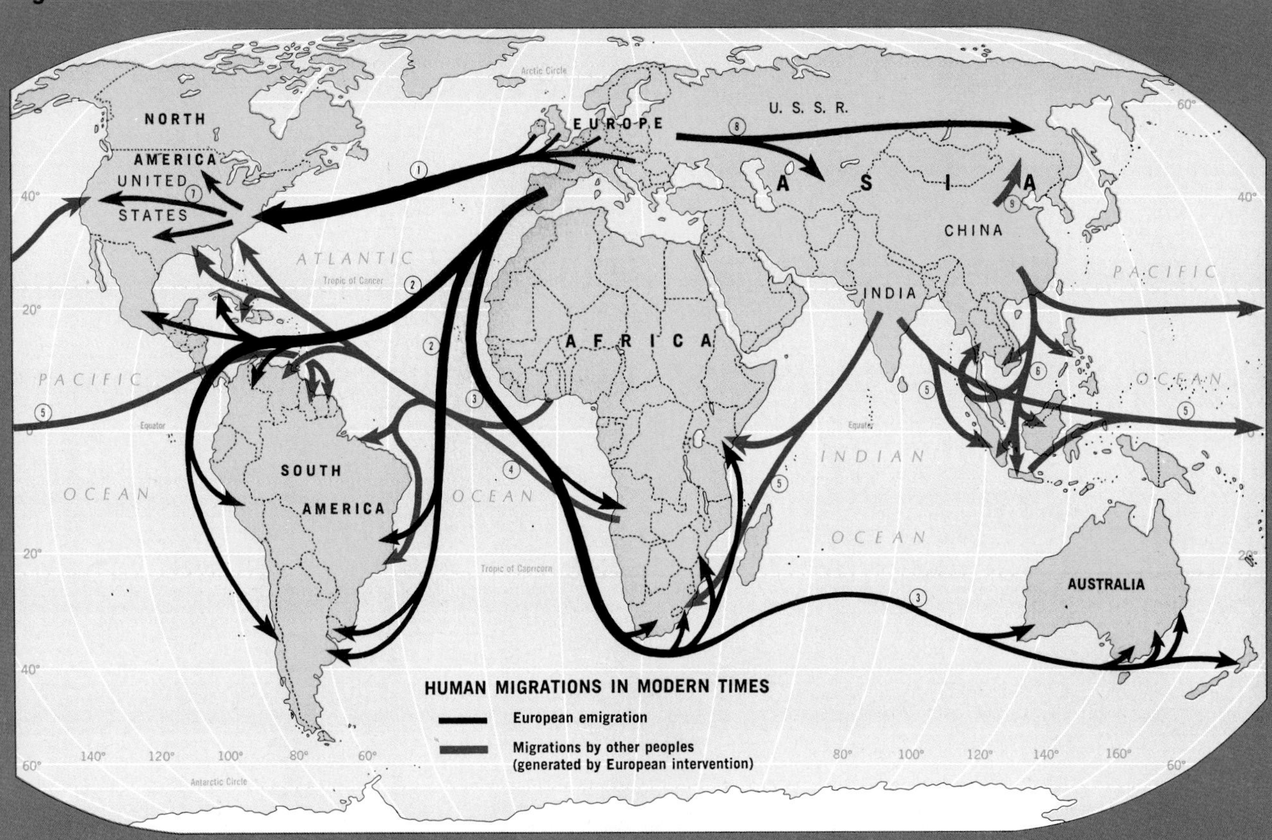

Many more Europeans came to the Americas than went to Africa or Australia. What led to their decision to select the New World? Studies focusing on the migration-decision indicate that the intensity of a migration flow varies with such factors as (1) the perceived degree of difference between one's home and the planned destination, (2) the effectiveness of the information flow—the news sent back by those who migrated to those who stayed behind, awaiting details, and (3) distance. A century ago a British social scientist, E. G. Ravenstein, studied migration in England and concluded that an inverse relationship exists between the incidence of migration and the distance between source and destination. Subsequent studies have modified Ravenstein's conclusions. The concept of *intervening opportunity,* for example, holds that people's perception of a faraway destination's comparative advantages is changed when there are closer opportunities. In the days when Britishers emigrated to lands other than North America, many chose South Africa rather than cross still another ocean to Australia.

Studies of migration also refer to *push* factors (conditions that tend to motivate people to move away) and *pull* factors (circumstances that attract them to a new destination). Australia was a new frontier, a place where one might acquire a piece of land, a herd of livestock, or where a piece of property might prove to hold valuable minerals; the skies, it was said, were clear, the air fresh, the climate much better than that of England.

Many of the first Europeans to arrive in Australia, however, were not free, but in bondage. The process we have just described involves *voluntary* migration, and the migrants made their decisions to relocate based on their minds. This is not the way thousands of the first European Australians reached their new homeland. European countries used their overseas domains as dumping grounds for people who had been convicted of some offense and sentenced to deportation—or "transportation," as the practice was called. This is a form of *forced* migration, in which push-pull factors play no role. "Transportation" had long been bringing British convicts to American shores, but in the late 1770s this traffic was impeded by the Revolutionary War. British jails soon were overcrowded, and judges continued to sentence violators to be deported; this was a time when Britain was undergoing its own economic and social revolution, and the offenders were numerous. An alternative to America for purposes of "transportation" simply had to be found, and it was not long before Australia was suggested. From the British viewpoint, the far side of Australia, the east coast, was an ideal place for a penal colony. It lay several thousand miles away from the nearest British colony and would hardly be a threat; its environment appeared to be such that the convicted deportees might be able to farm and hunt. In 1786 an order was signed making the corner of Australia known as New South Wales such a penal colony. Within two years, the first party of convicts arrived at what is today the harbor of Sidney and began to try to make a living from scratch.

For those of us who have learned of the horrible treatment of black slaves transported to America, it is revealing to see that European prisoners sentenced to deportation were not any better off, despite the fact that the offense of many of them was simply their inability to pay their debts. The story of the second group of deportees to Australia gives an idea: more than a thousand prisoners were crammed aboard a small boat, and 270 died on the way and were thrown overboard. Of those who arrived at Sydney alive, nearly 500 were sick, and another 50 died within a few days. What the colony needed was equipment and healthy workers; what it got was hardly any tools and ill and weakened people. The first white Australians hardly seemed the vanguard of a strong and prosperous nation. But by the middle of the nineteenth century, as many as 165,000 deported offenders had reached Australian shores, and London was spending a great deal of money on the penal stations. Meanwhile, the numbers of free colonists were increasing, convicts were entering free society after serving their sentences, and new settlements were founded. Australia's image in England began to change. As its penal image faded, the continent's spacious beauty and its economic opportunities beckoned. The last convict ship arrived in 1849. The flow of voluntary immigrants grew, and along Australia's coastline several thriving colonies emerged whose tentacles soon reached into the hinterland.

Regions

Australia's physiographic regions are identified on Fig. A-2, but another regionalization of Australia is more meaningful. Australia is a coastal rimland with cities, towns, farms, and forested slopes giving way to a dry, often desertlike interior that Australians themselves call the "outback" or the "inland." The habitable coastal rimland is not continuous, for both in the south and in the northwest the desert reaches the coast (Fig. I-6). In the east lies the Great Dividing Range, extending from Cape York in the north to the island of Tasmania in the south; between these highlands and the sea lie the fertile, well-watered foothills where Australia had its start. Across the Great Dividing

Figure A-2

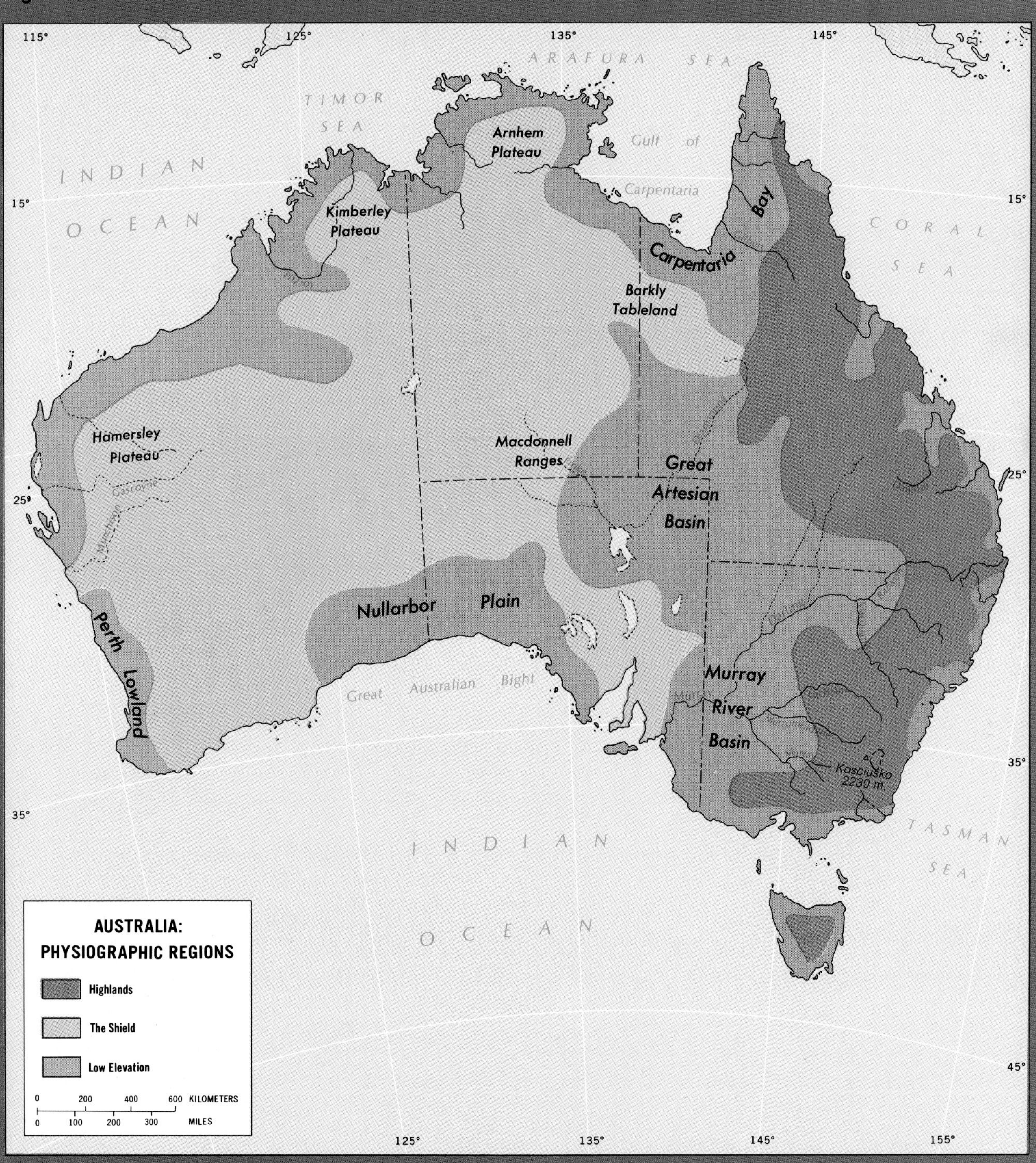

Range lie the extensive grassland pastures that catapulted Australia into its first commercial age, and on which still range the greatest sheep herds in the world (nearly 200 million sheep in 1976 produced one-third of all the wool sold in the world that year).

But Australia is not just a nation of stock-breeders. The steppe and near-desert country where sheep are raised, and the moister northern and eastern regions where cattle by the millions are ranched—these are Australia's sparsely inhabited areas (Fig. I-9), where sheep stations lie separated by dozens of kilometers of parched countryside, where towns are small and dusty, where a true frontier still exists. The great majority of modern Australians—more than 85 percent by the latest census—live in the cities and towns on or near the coast. And in this respect Australia's situation is quite similar to Europe itself, for in Britain, too, nearly nine out of ten people live in urban areas.

Australia's cities began as penal colonies (as was the case at Sydney Cove and Brisbane), as strategic settlements to protect trade and confirm spheres of influence (Perth in the southwest), or as centers of local commerce and regional markets (Adelaide, Melbourne). They became the foci of self-governing colonies, and for a long time Australia and Tasmania were seven ocean ports with separate hinterlands. These political regions were delimited by Australia's now-familiar pattern of straightline boundaries, completed by 1861 (Fig. A-3). Notwithstanding their shared cultural heritage, the Australian colonies found themselves at odds not only with London over colonial policies, but also with each other over economic and political issues. National integration was a slow and difficult process, even on the ground. Each of the Australian colonies laid down its own railroad lines, not with a view toward continent-wide transport, but with local objectives; in the process, three different railroad gauges came into use. The reorientation and unification of these separate systems was costly and difficult; not until 1970 was it possible to go by railroad from Sydney to Perth without changing trains.

Australians share with Europeans their great mobility. Australia counts more automobiles per 100 persons even than the United States, and Australians, like North Americans, are quite accustomed to long-distance travel. Australia is nearly as large as the contiguous United States, but its population of 15 million is what the United States white population was in 1825. In Australia, neighbors are often far removed. Perth in Western Australia is as far from Sydney (New South Wales) as Los Angeles is from Atlanta, and Brisbane and Adelaide are as far apart as Washington, D. C. and Miami. Not surprisingly, there are more road kilometers per person in Australia than in the United States.

Australia's cities retain some of the competitive atmosphere of the colonial period, although they are as modern, spacious, generally well planned, and suburbanized, as many cities in the Western world are. Sydney (3.1 million), capital of New South Wales and Australia's largest city and leading port, has a long history of rivalry with Melbourne which at one time overtook Sydney in size and was Australia's temporary capital. Sydney lies on a magnificent site, its skyscrapered central city overlooking the famed Sydney Harbour Bridge; but apart from the new Opera House its architecture is no match for less spectacular but historically more interesting Melbourne, still capital of Victoria. Between them, Sydney and Melbourne contain nearly 6 of Australia's 15 million inhabitants, and Brisbane (Queensland) and Adelaide (South Australia) each have a population of approximately one million. Perth, Western Australia's capital, has more than 800,000 residents, while Hobart, Tasmania, has under 250,000. By far the smallest of the regional capitals is Darwin (Northern Territory), devastated by the Christmas Day cyclone of 1974 when it had a population of about 37,000. Thousands have never returned to the scene of Australia's worst natural disaster. Australia's federal capital, the planned city of Canberra, still has a population of less than 250,000.

Farming

Sheep raising thrust Australia into the commercial age, and the technology of refrigeration brought European markets within reach of Australian beef producers—but there is more to Australian rural land use than herding livestock. Commercial crop farming concentrates on the production of wheat; Australia is one of the world's leading exporters of wheat, which brings in nearly half the amount of income derived from wool.

Wheat production is concentrated in a broad belt that extends from the vicinity of Adelaide into Victoria and New South Wales (along the rim of the Murray Basin) and even into Queensland, where a separate area is shown on Fig. A-4 in the hinterland of Brisbane. It also exists in an area behind Perth in Western Australia. Although these zones are mapped as "commercial grain farming,"

Figure A-3

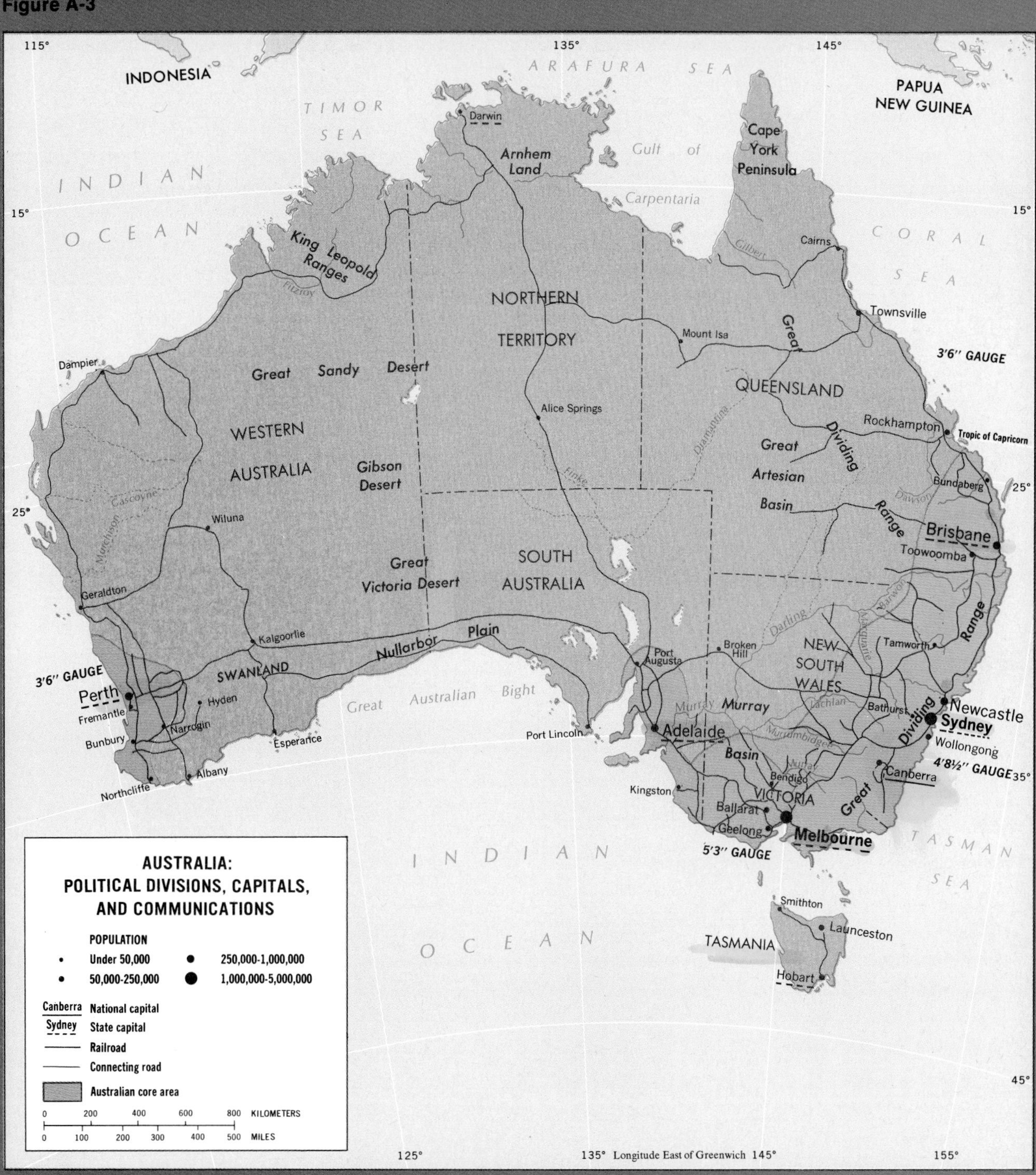

a unique rotation system exists here, whereby sheep and wheat share the land. Under this mixed crop and livestock farming, the sheep use the cultivated pasture for several years, and it is then plowed for wheat sowing; after the wheat harvest the soil is rested again. This system, plus the innovations in mechanized equipment the Australians themselves have made, has created a highly lucrative industry. Australian wheat yields per acre are still low by American standards, but the output per worker is about twice that of the United States.

As Fig. A-4 shows, dairying, with cultivated pastures, takes place in what Fig. I-5 shows to be areas of high precipitation. The

Figure A-4

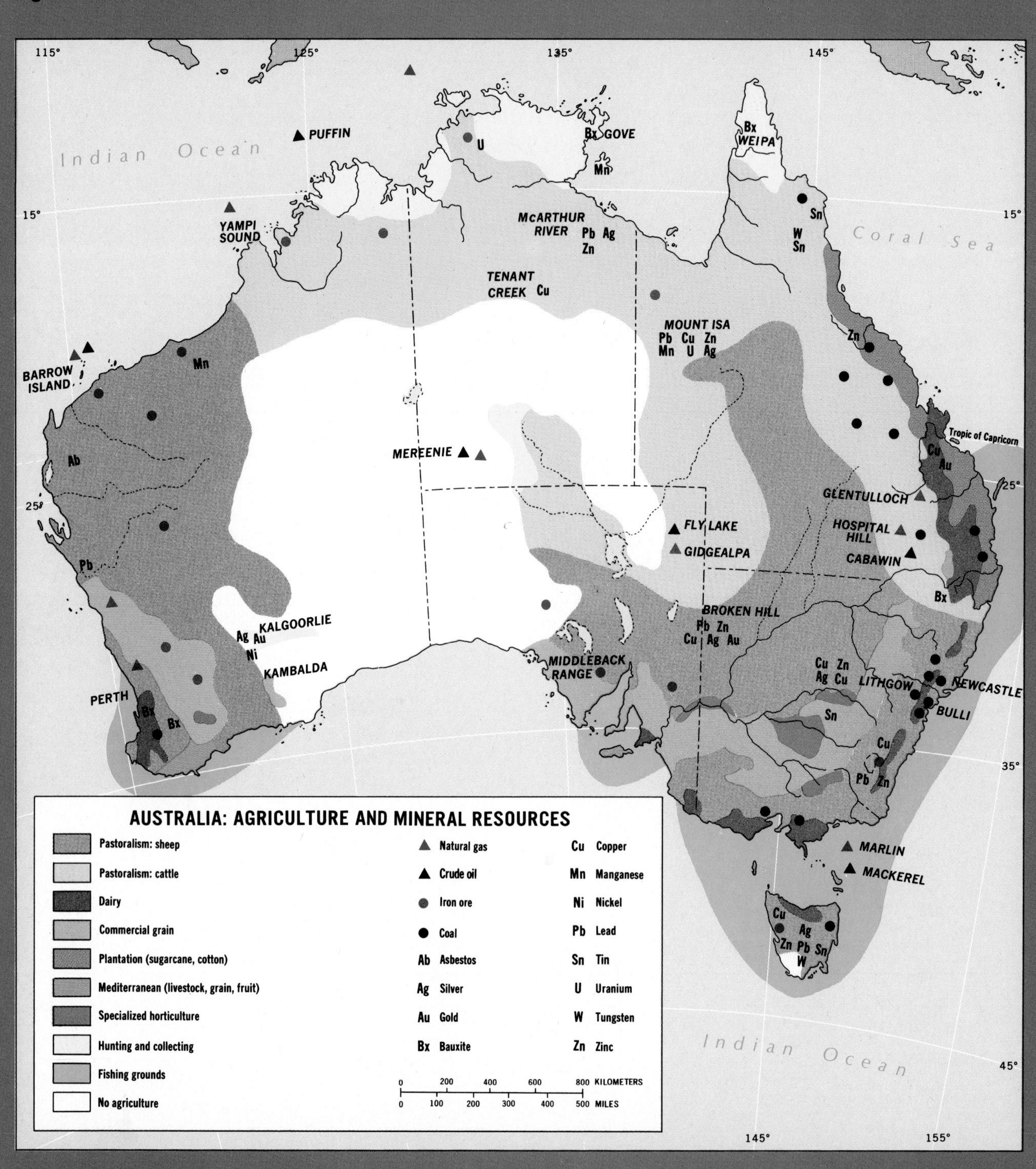

location of the dairy zones suggests also their development in response to Australia's comparatively large urban markets. Even these humid areas of Australia are occasionally afflicted by damaging droughts, but the industry is normally capable of providing more than the local market demands. Dairying is, in fact, the most important rural industry measured by the number of people who find work in it.

It is natural that irrigation should be attempted in a country as dry as Australia. Unfortunately, the opportunities for irrigation-assisted agriculture are quite limited: the major potential lies in the basin of the Murray River, shared by the states of Victoria (which leads in irrigated acreage) and New South Wales. Rice (produced at one of the highest per-hectare yields in the world) in the Wakool and Murrumbidgee valleys, grapes, and citrus fruits are among the irrigated crops in the Murray River Basin. On the east coast, in Queensland and northern New South Wales, the cultivation of sugar cane is partly under irrigation.

Mineral Resources

Nothing shook the Australian economy as much as the discovery of gold in 1851, in the territories of New South Wales and Victoria. In no way—politically, socially, or materially—was Australia prepared for what followed the news of the gold finds. True, gold had been known to exist prior to 1851, but only when some Australian diggers returned from California's fields to prove the lucrative character of the Australian ores did the rush begin. During the decade from 1851 to 1861, the population of the colony of Victoria increased more than sevenfold, from just over 75,000 to well over a half million; yet Melbourne, the colony's leading town, was just 16 years old when the rush began and had none of the amenities needed to serve so large a number of people. New South Wales, which in those days still included all of what is now Queensland and the Northern Territory, saw its population rise to over 350,000, and its pastoral economy was rudely disturbed by the rush of diggers to Bathurst. Overall, Australia's population nearly tripled in the 1850s; gone was the need for the subsidized immigration in all areas but the west and northeast (Queensland).

The new wealth of Australia brought problems of accommodation for the new settlers, and inevitably it brought on some of the disorder and political conflict that also marked the gold rushes in North America and Africa. But it also brought great advantages. A new prosperity came to every Australian colony; in the 1850s the continent produced 40 percent of the world's gold. It brought successful searches for other minerals (Tasmania, with its tin and copper, was an early beneficiary). Discoveries are still being made. Long believed poor in petroleum and natural gas (oil always ranks high on the import list for this urbanized, industrialized society), Australia seems to have sizable supplies of both fuels after all, if recent finds on the margins of the Great Artesian Basin are large enough to warrant exploitation. Additional reserves have been located on the continental shelf between the mainland and Tasmania, and these, in the Bass Strait, now produce half of Australia's crude oil requirements. Recently, a deposit of nickel was found near the center of Western Australia's gold mining, Kalgoorie. For nearly 80 years people have been mining and searching in this area, and yet the nickel discovered recently at Kambalda may constitute the world's largest known reserve of this important alloy. The Australian frontier still holds its secrets and surprises.

As Fig. A-4 suggests, Australia's mineral deposits are scattered and very varied. The country is fortunate in being well endowed with coal deposits when its water power prospects are so minimal (except in Tasmania and the southeast) and its petroleum discoveries so long delayed (the Bass Strait fields were not found until 1964). The chief coalfields, as the map indicates, lie in the east, notably around Sydney—north at Newcastle, west at Lithgow, and south at Bulli. In the hinterland of Brisbane and about 200 kilometers inland from Rockhampton (Queensland), bituminous coal is mined. From the New South Wales field, coal is sent by coastal shipping to areas that are deficient in coal. But coal is widely distributed—

even Tasmania and Western Australia have some production and can thus keep their import necessities down. There is no doubt that coal is Australia's more important mineral asset: it is used for the production of 90 percent of the electricity the country consumes, for the railroads, factories, manufactured gas production, and a host of other purposes.

The most famous Australian mining district undoubtedly is Broken Hill, which has neither gold, coal, nor iron. Discovered in the aftermath of the gold rush, Broken Hill became one of the world's leading lead and zinc-producing areas (in the beginning it was important also for silver), and the enormous income derived from the export sales of these minerals provided much of the capital for Australia's industrialization. Japan has become one of the principal buyers, and Australia's "Japanese connection" has proved very lucrative. In the late 1970s, Japan bought nearly 30 percent of Australian exports. In Queensland, the Mount Isa ore body yields a similar mineral association, and in both areas uranium also has been mined. But these products are only a small part of the total Australian inventory. Tasmania's copper, South Australia's large, recently discovered bauxite (aluminum ore) deposit in the York Peninsula, Queensland's tungsten, and Western Australia's asbestos only represent the range of the continent's minerals; there are but a few nonmetallic resources in which the country is really deficient. And the search has only just begun. In terms of mineral resources, Australia is indeed in a fortunate position.

Manufacturing

For a long time Australia was to Britain what colonies always were, a supplier of needed resources and a ready market for British-manufactured products. But World War I, at a crucial stage, cut these trade connections—and Australia for the first time was largely on its own. Now it had to find ways to make some of the consumer goods it had been getting from Britain. By the time the war was over, Australian industries had made a great deal of progress. Wartime also pushed the development of the food-processing industries, and today Australian manufacturing is varied, producing not only machinery and equipment made by locally manufactured steel, but also textiles and clothing, chemicals, foods, tobacco, wines, and paper—among many other items.

Australia's industries, naturally, are situated where the facilities for manufacturing are best (and where markets exist)—the large cities. Of a labor force numbering slightly under 6.5 million in 1980, nearly one-quarter is employed in manufacturing, although agriculture (which employs only 6 percent of the labor force) and mining remain the mainstays of the economy. Australia is not, as yet, a major exporter of finished products; manufacturing of steel products, automobiles, textiles, chemicals, and electrical equipment are oriented to the domestic market. Although this market is small numerically, it is not a minor factor: per capita income in Australia in 1980 was approaching $8000. Although Australia in 1980 had only 15 million inhabitants, it had over 6.5 million vehicles on its roads, nearly 5.5 million of them passenger cars. (The United Kingdom, with a population of 56 million, had fewer than three times as many.)

The success of domestic industries is a matter of prime concern for the Australian government. Australian manufactures have not taken their place on world markets, and foreign products pose a challenge. During the late 1970s Australian industries experienced a serious downturn accompanied by unemployment problems; at the same time Japan pressed Australia to lower tariff barriers against Japanese imports, notably automobiles. Thus, plentiful raw materials notwithstanding, Australian manufacturing faces high production costs and long distances to foreign markets, and cannot expect to compete overseas. Indeed, Australia's list of imports still includes machinery and equipment, chemicals, foods, and beverages that could, in part at least, be produced at home. In terms of its manufacturing industries, Australia still has symptoms of a developing country.

Population Policies

Not surprisingly, the state capitals of Australia became the country's major industrial centers. These urban clusters, at the foci of the railroad networks of the individual states, were also the main ports for overseas as well as coastwide shipping; they had the best amenities for industry—concentrated labor force, adequate power supply, water, access to government—and they were at the same time the major markets. The process of agglomerative self-perpetuation to which reference was made in the context of some of Europe's larger cities is well illustrated here. Additionally, the cities and manufacturing industries have absorbed the vast majority of Australia's recent European immigrants.

The Australian government pursues a vigorous policy of encouraging selective immigration; the majority of the new Australians come from Britain, but numerous Italians, Hollanders, Germans, and Greeks also seek new homes in Australia. About half of Australia's population today is of English origin; about 20 percent has Irish roots; 10 percent is of Scottish ancestry. Nearly 20 percent has non-British, mostly mainland European, roots. In recent years about 180,000 immigrants have annually come to Australia, and most of them have settled in the cities and found work in industry. The advantages of immigration to Australia are obvious: a strengthening of the country's still-small population, an expansion of the skills and numbers of the labor force, an increase in the size of the local market, and a growing consolidation of the Australian presence on the large island-continent, from a strategic viewpoint.

Australia has long been known for its selective immigration policies, virtually excluding persons who are not white from the opportunity to settle there. Since 1901 successive Australian governments adhered to an unofficial "White Australia" policy, but after World War II international criticism mounted, and in the 1950s the practice began to soften. First the government approved the admission of family members of Australians who were not white, and in 1966 it became possible for Asians to apply for Australian residence; but only those with needed skills are admitted even today. The rate of immigration by people who are not white is still quite small, but has approached 10,000 annually in recent years.

In the meantime, the indigenous Australian population that numbered as many as 350,000 two centuries ago has been reduced to a mere remnant. It is the familiar story of contact between indigenous peoples living a life of subsistence in small groups and the onslaught of European power: the black Australian communities were driven away, broken up, often destroyed. Today, fewer than 50,000 Australians still have a total aboriginal ancestry; another 100,000 have mixed origins. Only a few communities still live the original life of hunting and gathering, mainly in the Northern Territory, Western Australia, and northern Queensland. The remainder are scattered over the continent, working on the cattle stations, living near Christian missions, or subsisting on reserves set aside for them by the government; many do menial jobs in the towns. In recent years the Australian conscience has been aroused, and an effort has begun to undo centuries of persecution and neglect; these attempts are met, understandably, with suspicion and distrust. In affluent Australia the original Australians remain victims of poverty, disease, insufficient education, and even malnutrition.

In recent years Australia has faced a new immigration problem: the arrival of boatloads of Asians, most from Vietnam and Cambodia, whose desperation drove them out to sea in the hope of a landfall in a hospitable place. The phenomenon of the "boat people" is not solely a Southeast Asian one. Overcrowded boats with Haitians and Cubans aboard have been arriving on the shores of Caribbean islands and the United States for years; their accommodation has presented governments in the Americas with problems as well.

In Australia, immigration policies and procedures were modified significantly in 1979, raising immigration quotas and liberalizing admissions requirements. These changes were not adopted without opposition; labor leaders charged that the immigration of a large number of workers without guaranteed jobs would create a docile work force and sustain unemployment, which would be advantageous to employers but not to workers and their unions. Certainly the 1970s had been difficult for Australia, as the energy crisis of 1973 and its aftermath reduced the economy's ability to absorb newly arrived workers. In 1978, unemployment among new immigrants was reported to be nearly 20 percent.

Against this background came the issue of the Vietnamese (and other) refugees. They came by way of Malaysia and Indonesia and arrived, mostly, at or near the northern port of Darwin. Soon the inevitable problems of adjustment mounted, and a national debate arose. The Australian labor movement always was the focus for the "White Australia" immigration policy and traditionally opposed Asian immigration; furthermore, the Australian Labor Party (the opposition party) had opposed the Vietnam War. Refugees from the new order in Vietnam were not welcome, thus, on several grounds—in addition to their circumvention of established immigration policy. Australian government leaders were dispatched to Thailand and Malaysia to seek assistance in restraining would-be "boat people," but they reported that many in Southeast Asia, officials and refugees alike, believed Australia to be a vast, open land of promise and opportunity.

Quite probably Vietnam's "boat people" are but a vanguard of what is likely to be a growing stream of irregular immigrants. Just as United States shores beckon the oppressed and the poor in Middle America, so Australia is perceived as a desirable objective by Asian emigrants, and not just the Vietnamese "boat people." Political and economic crises in South, Southeast, and East Asia as well as the Pacific will generate new waves of emigrants. Indonesia's stepping stones, its corridor of islands stretching toward Australia, lead toward a threshold unlike any other in the entire Asian-Pacific hemisphere. Australia's relative location may, during the last decades of the twentieth century, become its greatest challenge.

Australian Federalism

Australia may be a European offshoot, but in at least one respect it differs significantly from most European countries. Australia is a federal state; all but a few European countries are unitary states.

The term *unitary* derives from the Latin *unitas* (unity), which in turn comes from *unus* (one). A unitary state, then, is unified, centralized. It is not surprising that many European states, long ruled with absolute authority from a royal headquarters, developed into unitary states. The distant colonies had no such tradition, whether in Canada, the United States of America, or Australia. In Australia, six colonies for most of the nineteenth century went their own way, competing for hinterlands and arguing over trade policies. Eventually, however, the idea that they might combine in some greater national framework began to take hold. But this would be no unitary state; each colony wanted to retain certain rights. Thus, Australia became a *federal* state.

The federal concept also has ancient Greek and Roman roots, and the term originates from the Latin *foederis*, meaning league or association. In practice the federal system is one of alliance and coexistence, and the recognition of difference and diversity. Of course there must be an underlying foundation for a federal union, and in Australia this was, in the first instance, the colonies' common cultural heritage. But there were other bases as well. The colonists worried about a foreign invasion from Asia or some mainland European power; there was a particular fear of Germany, for in the 1880s the Germans embarked on a vigorous colonization campaign in Africa and the Pacific. In 1883 the colonial government of Queensland took control of the eastern part of New Guinea because it feared a German takeover there. Also, the colonies shared the objectives of selective immigration. A coordinated White-Australia policy would be more easily administered on a national basis. The colonies realized that their internal trade competition was damaging and precluded a coordinated commercial policy on world markets.

With such positive prospects, it would seem that federation would have been accomplished quickly and without difficulties, but this was not the case. Victoria's manufacturers wanted protective tariffs; public opinion in New South Wales favored freer trade. There was wrangling over the site of the future federal capital, and no colony was willing to yield to another capital city the advantage of becoming the federal headquarters. But in 1900 lengthy conferences succeeded in creating a federal constitution approved by the voters in the colonies and confirmed by London. From January 1, 1901, the colonies became the Commonwealth of Australia consisting of six states and the federal territories of the capital (the site of Canberra was chosen in 1908) and Northern Territory, the sparsely peopled, rectangular area that extends from Arnhem Land to the heart of the Australian Desert (Fig. A-3).

The Federal Capital

More than one federal government has found the selection of a federal capital a difficult task requiring compromise and consent. Australia experienced this problem, arising mainly from the rivalry between the adjacent southeastern states of Victoria and

Melbourne, Australia's second city, is the capital of the state of Victoria. Modern skyscrapers envincing the city's financial wealth tower above historic buildings that grace wide, tree-lined avenues and parks. (Harm J. de Blij)

Table A-1 The Commonwealth of Australia

	Area		Population
	sq. km.	sq. m	(1981 est.)
States			
New South Wales	801,400	309,400	5,320,000
Victoria	227,600	87,900	4,010,000
Queensland	1,727,500	667,000	2,140,000
South Australia	984,400	380,100	1,390,000
Western Australia	2,527,600	975,900	1,400,000
Tasmania	68,300	26,400	430,000
Territories			
Northern Territory	1,347,500	520,300	130,000
Australian Capital Territory	2,430	940	240,000

New South Wales. Victoria's capital, Melbourne, was the chief competitor for Sydney, headquarters of New South Wales. Shortly before the implementation of federation, the premier of New South Wales managed to secure agreement that a federal capital would *not* be Sydney, but a totally new city to be built at a site to be jointly selected. Second, Melbourne would function as the temporary capital until the federal headquarters were ready for occupation.

The site of Canberra was chosen in 1908, and the buildings necessary for the government functions were ready by 1927. Canberra, unlike other major Australian cities, lies away from the coast, at the foot of Mount Ainslie in the Great Dividing Range (Fig. A-3). Once the site was selected, a worldwide architectural design competition was held in order to acquire the best city plan, for this would be a totally new city. The United States architect Walter Burley Griffin was chosen for the task, and today Canberra is a spacious, modern, widely dispersed urban area with some 240,000 residents. Its nearly 2500 square kilometers (940 square miles) form the Australian Capital Territory, a federal area separated from the state of New South Wales.

Canberra shares with other comparatively new federal capital cities a sense of remoteness and isolation. In 1980 the highway to Sydney, just under 200 miles (300 kilometers) long, still was interrupted by lengthy stretches of overcrowded, two-land road. The city itself, endowed with impressive public architecture, ceremonial places, and parks, still retains an atmosphere of coldness and distance. It lacks Sydney's drive and bustle, Melbourne's style and comfort, Adelaide's sense of history. The presence of the government and the Australian National University notwithstanding, Canberra is a far cry from the pulse of Australia.

A Youthful State

If states do indeed experience life cycles, Australia exhibits the energy and vigor of youth. Many observers have likened Australia today to the United States of a century and a half ago: still discovering major resources, still in the early stages of the penetration of national territory, still essentially peripheral, the interior still a frontier—albeit a modern frontier with airstrips rather than stagecoaches.

Australia's cities, where the great majority of Australians live, reflect the country's youthfulness in several ways. The largest, Sydney, contains well over 20 percent of the entire Australian population. Sydney has an urban area of over 12,000 square kilometers, larger than New York or London. Internal circulation in fast-growing Sydney is hampered by the absence of adequate highways; available facilities have been overtaken by the mushrooming population and its automobiles. The city is severely congested; traffic jams on the approaches to picturesque but inadequate Harbour Bridge linking the city's north and south (opened in 1932) are among the world's worst. Major arteries optimistically called the Great Western Highway and the Pacific Highway begin, at best, as wide city streets whose traffic is slowed by numerous stop lights, lane inconsistencies, and inadequate posted directions.

Australia never has had a space problem, and the cities reflect this; suburbs sprawl far and wide. Downtown Sydney lies at the heart of a conurbation consisting of five individual cities and more than 30 municipalities, a built-up region that extends over nearly 50 miles (80 kilometers) from north to south. Melbourne's dimensions are only slightly less. Few remnants in the urban landscape reveal the cities' age: modern, towering skyscrapers dominate the central city. Melbourne, capital of Victoria, retains more architectural interest and is a more cultured city than Sydney; it is also better served by major traffic arteries. Perth, the capital of enormous Western Australia, is the youngest of the state capitals, a thoroughly modern urban center with a population now approaching 1 million, still the most frontier-like among Australia's large cities.

The Australian economy reflects also the country's youth. Notwithstanding its urbanized and industrialized character, Australia still imports large quantities of consumers goods—many of which could be manufactured at home. Certainly Australians can afford to pay for these imports; they enjoy one of the world's highest living standards. But the situation reveals the limitations of a small domestic market and the liabilities of distance to world markets as well as the absence of any urgent need to compete on faraway marketplaces. Unlike the United Kingdom, where manufactures must be sold so that food may be bought, Australia has a small population and secure world markets for its wool and beef and wheat. Explanations for the condition of Australia's industries sometimes refer to the Australian preference for the leisurely life, and indeed Australians are a great sports people, excelling in competition from tennis to cricket and rugby to lawn bowling. But it more likely to be a question of opportunity rather than preference: Australia does not (yet) face the need to maximize efficiency. In their remote "down under," Australians have advantages of which many Europeans can only dream.

New Zealand

Like Australia, New Zealand is a product of European expansion. In another era New Zealand, located about 1700 kilometers (over 1000 miles) southeast of Australia, would have been included in the Pacific realm, for before the Europeans' arrival it was occupied by the Maori, a people with Polynesian roots. But today New Zealand's population of just under 3.5 million is more than 90 percent European, and the Maori form a minority of about 250,000—many of mixed ancestry.

New Zealand consists of two large mountainous islands and numerous, scattered smaller islands. The two large islands, with South Island somewhat larger than North Island, together cover just over 260,-000 square kilometers (100,000 square miles), and while they look dimunitive in the great Pacific Ocean, they are nevertheless larger than Britian (Fig. A-5). In sharp contrast to Australia, the islands are mainly mountainous or hilly, with several mountains rising far higher than any on the Australian landmass. South Island has a spectacular, snowcapped range appropriately called the Southern Alps, with peaks such as Mts. Cook and Tasman reaching beyond 3500 meters (11,700 feet). Smaller North Island has proportionately more land under low relief, but it also has an area of central ranges and high mountains, including Mts. Raupehu (nearly 2800 meters) and Egmont, on whose lower slopes lie the pastures for New Zealand's chief dairying district. Hence whereas Australia's land lies relatively low in elevation and has much low relief, New Zealand's is on the average quite high and its relief is mostly rugged. Thus the most promising areas must be the lower-lying slopes and lowland fringes on both islands. On North Island, the largest urban area, Auckland (nearly 800,000) lies on a comparatively low-lying peninsula, and on South Island the largest lowland is the Canterbury Plain, centered on Christchurch (300,000). What makes these lower areas so attractive, apart from their availability as cropland, is their magnificent pastures. Such is the range of soils and pasture plants that both summer and winter grazing can be carried on, and a wide variety of vegetables, cereals, and fruits can be produced. About half of all New Zealand is

Figure A-5

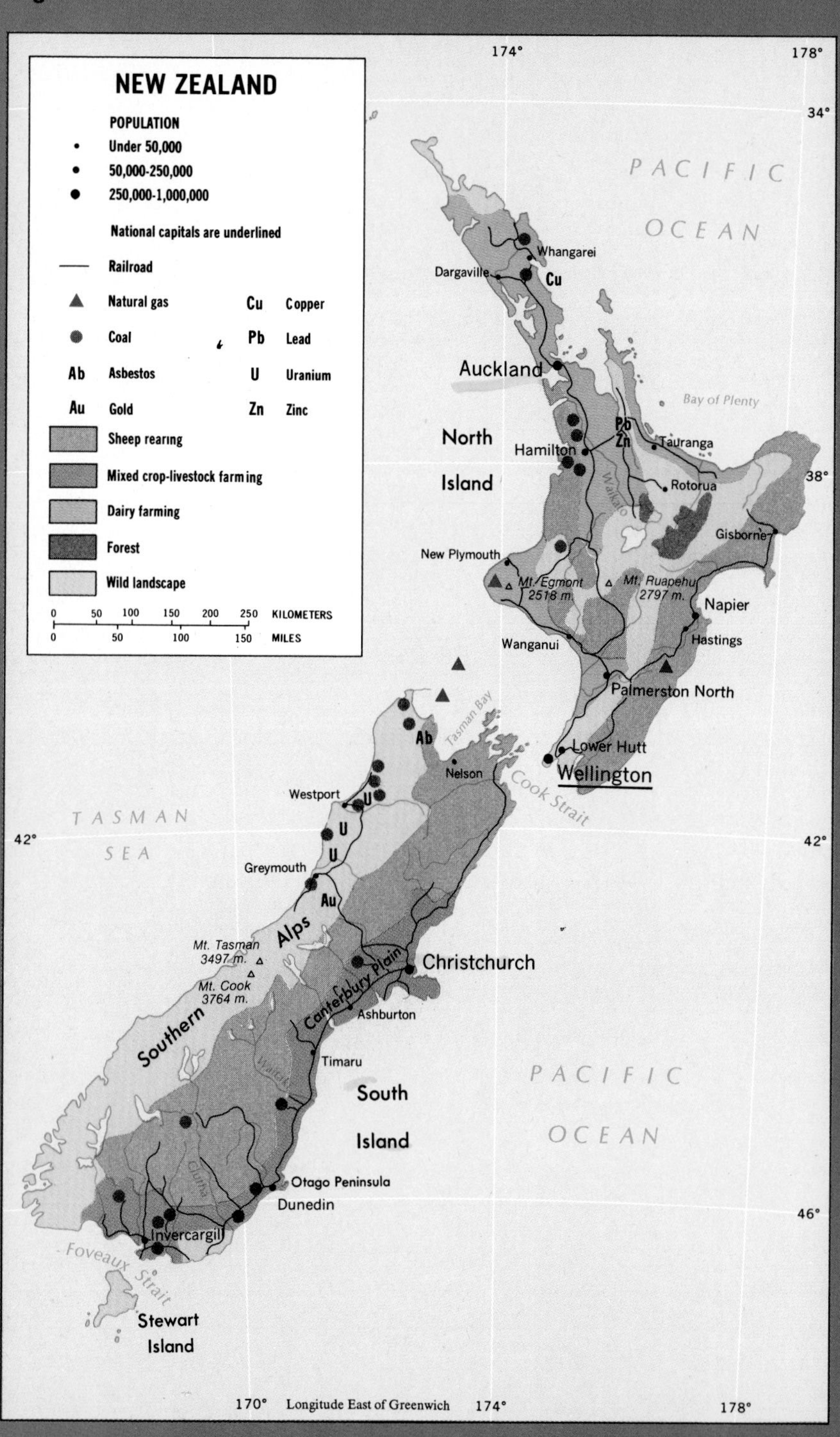

pastureland, and much of the farming is done to supplement the pastoral industry, to provide fodder when needed. But in the drier areas of South Island wheat is grown, truck farming exists around the cities, and fruit orchards of apples, pears, and grapes are widely distributed from Otago in the south to Nelson in the north.

Thus New Zealand's pastoral economy is based on sheep and cattle—wool, meat, and dairy products. As in Australia, sheep greatly outnumber cattle (over 60 million against about 9 million in 1977), and in modern times New Zealand has been among the world's leaders in terms of per capita trade; its standard of living is very high. Meat (mutton and beef) combine with wool to provide nearly 75 percent of the islands' export revenues.

Despite their contrasts in size, shape, physiography, and history, New Zealand and Australia have a great deal in common. Apart from their joint British heritage, they share a substantially pastoral economy, a small local market, the problem of great distances to world markets, and a desire to stimulate and develop domestic manufacturing.

The fairly high degree of urbanization in New Zealand indicates another similarity to Australia: the high employment in the city-based industries. About 80 percent of all New Zealanders live in cities and towns, where the industries are mainly still those that treat and package the products of the pastoral-agricultural industries. But although New Zealand is behind Australia in its industrial development, the country's manufactures are becoming more diverse all the time. Textiles and clothing, wood products, fertilizers, cement and building materials, metal products, and some machinery are now made in New Zealand. Among important developments in the industrial sphere are the opening of the first steel mill near Auckland and the construction of an aluminum smelter (converting Australian bauxite) at the port of Bluff on South Island.

The Canterbury Plain, New Zealand's chief farming region centered on Christchurch, lies between the Southern Alps (background) and the coast of South Island. (Harm J. de Blij)

Wool, meat, and dairy products form New Zealand's principal exports. Magnificent pastures such as this near Mount Egmont sustain the pastoral industry. (New Zealand Consulate General, N.Y.)

Spatially, New Zealand shares with Australia its pattern of peripheral development, imposed not by deserts but by high, rugged mountains. The country's major cities, Auckland and the capital of Wellington (with its satellite of Hutt, 330,000) on North Island and Christchurch and Dunedin (150,000) on South Island, are all located on the coast, and the whole railroad and road system (Fig. A-5) is peripheral in its orientation. This is true to a greater extent on South Island than in the north, for the Southern Alps form New Zealand's most formidable barrier to surface contact.

Compared to brash, progressive, modernizing Australia, New Zealand seems quiet. A slight regional contrast might be discerned between perhaps more forward-looking North Island and conservative South Island, but the distinction fades in the light of Australia's historic urban and regional rivalries. Nothing in New Zealand compares to Australia's urban architecture; a sameness pervades New Zealand that is quite unlike Australia. This may be partly why, in the late 1970s, New Zealand's population was actually dwindling slightly, its natural increase and small immigration exceeded by a high emigration rate, mainly to Australia.

If New Zealand fails to satisfy many of its young citizens, it is not because of a lack of personal security. The government has developed an elaborate cradle-to-grave system of welfare programs, that are affordable because of the country's high incomes and standards of living (taxes also are high). Whether this has suppressed entrepreneurship and initiative is a matter for debate, but New Zealand is a country with few excesses and much stability.

Nevertheless, significant change can be discerned. Auckland not only is New Zealand's largest city; it also may be the largest Polynesian city of all, with nearly 80,000 Maoris, Samoans, Cook Islanders, Tongans, and others making up 10 percent of the total population. The twentieth century has witnessed a revival of Maori culture and a still-slow but quickening pace of Maori integration into New Zealand society (nearly all Maori reside on North Island, where nearly three-quarters of all New Zealanders live). This has enlivened New Zealand's visual arts and music. But Australia and New Zealand prove a point: there is a difference between remoteness and isolation.

Additional Reading

The book by O. H. K. Spate, *Australia,* published by Praeger in New York in 1968, still remains an excellent source. Another useful introduction is K. W. Robinson, *Australia, New Zealand, and the Southwest Pacific,* published in London by the University of London Press (third edition, 1974). Also see T. L. McKnight's *Australia'a Corner of the World,* published by Prentice-Hall, Englewood Cliffs, in 1970, and, for something unusual, a book by the same author entitled *The Camel in Australia* (Melbourne University Press, 1969). A well-illustrated and fascinating account of Australia's colorful past is R. Cameron, *Australia: History and Horizons,* published by Columbia University Press in New York in 1971. A much older work that still retains its flavor and interest is T. Griffith Taylor's *Australia, A Study of Warm Environments and their Effect on British Settlement,* a volume that has gone through numerous Methuen editions in London—and one that should be read critically. On New Zealand, consider starting with a book in the World's Landscapes Series written by K. B. Cumberland and J. S. Whitelaw, entitled *New Zealand* and published in Chicago by Aldine in 1970. G. J. R. Linge and R. M. Frazer have published an *Atlas of New Zealand Geography* (Wellington, Reed, 1966) that is becoming dated but still contains much that retains interest.

Moscow: winter morning (George Holton/Ocelot)

Two

The Soviet Union: Region and Realm

From the Baltic Sea in Europe to the shores of the Pacific Ocean and from the Arctic to the borders of Iran, lies the Soviet Union—political region, cultural realm, ideological hearth, world power, and historic empire. Its 22.4 million square kilometers (8.65 million square miles) constitute fully one-sixth of the world's land surface and half the vast landmass of Eurasia; the Soviet Union is two and a half times as large as the United States or China. A country of continental proportions, the Soviet Union stretches across 11 time zones. When people awake in the morning in Vladivostok on the Sea of Japan, it is still early the previous evening in Leningrad on the Gulf of Finland. The ground is permanently frozen not far from Norilsk in the northern tundra, but subtropical crops grow near Yalta on the shores of the Black Sea.

The Soviet Union does not, however, have a population to match its huge territory. Its 265 million people (1980) are far outnumbered by 1 billion Chinese and 650 million Indians, neighbors on the Asian continent. And, as Fig. I-9 reveals, the Soviet population is still strongly concentrated in what is sometimes called "European" Russia—the Soviet Union west of the Ural Mountains. Most of Siberia and the Soviet Far East still remain virtually empty country, an unchanged, only partially known frontier. In those vast expanses the Soviets may discover additional mineral resources—but the cost of building transport systems to carry them to the industrial centers of the west may be prohibitive for a long time to come. As the map of Soviet railroads (page 163) shows, there is nothing east of the Urals to rival the network that has developed in the west. Surface contact between the Soviet Far East and the populous west is still tenuous.

IDEAS AND CONCEPTS

- State planning
- Imperialism
- Migration (2)
- Analogue area analysis
- Core area
- Acculturation
- Heartland theory
- Frontier
- Buffer state
- Boundary genesis

Ten Major Geographic Qualities of the Soviet Union

1. The Soviet Union is by far the largest territorial state in the world; its area is more than twice as large as the next ranking country (Canada).
2. The Soviet Union's enormous area lies mainly at high latitudes; much of it is very cold or very dry. Large, rugged mountain zones accentuate harsh environments.
3. The Soviet Union has more neighbors than any other state in the world. Boundary conflicts and territorial disputes involve several of these neighboring states, ranging from capitalist Japan and revisionist China to theocratic Iran, satellite Romania, and nonaligned Finland.
4. For so large an area, the Soviet Union's population of 265 million is comparatively small; it is, moreover, strongly concentrated in the westernmost one-quarter of the country.
5. Development in the Soviet Union remains focused in "European" Russia west of the Ural Mountains; here lie the major cities, industrial regions, densest transport networks, and most productive farming areas.
6. The Soviet Union is the heart of the socialist world (as opposed to the capitalist world and the "Third World").
7. The Soviet Union constitutes the world's largest-scale experiment in national economic planning.
8. National integration and economic development east of the Ural Mountains (in Soviet Asia) extend mainly along a ribbon that connects Omsk, Novosibirsk, Irkutsk, Khabarovsk, and Vladivostok.
9. Notwithstanding its large territorial size, the Soviet Union suffers from land encirclement in Eurasia; it has few good and suitably located ports.
10. The Soviet Union is a contiguous, multinational empire dominated by Russia. It is the product of Russian colonialism and imperial subjugation of nations ranging from Estonia, Latvia, and Lithuania in the north to Moslem Khanates in the south.

A World Power

The Soviet Union today is a state whose military power and political influence in the world rival that of the United States of America. Some observers suggest that the Soviet Union is in the process of surpassing the United States in terms of military strength, endangering the approximate balance between the two giants that has so far forestalled a calamitous, atomic World War III. But the competition between the two countries is not only a matter of military buildup. The world has become an arena of ideological and political as well as economic rivalry and conflict between the United States and its scattered allies and the Soviet Union with its Eurasian and third-world associates. At issue is the social system that will prevail in the world of the future: will capitalist enterprise survive or will communism ultimately triumph? The cost to the world of this twentieth-century power struggle has been incalculable. Apart from the billions of rubles and dollars that have been poured into Soviet and American military-industrial complexes, the greater conflict has magnified minor discords among nations, damaged smaller economies, intensified national politics, destroyed democracies, and fueled terrorism. When states quarrel, the great powers are quick to exploit the dispute, and soon leaders who might otherwise have been identified as representing national viewpoints are reported to be "pro-Western" or "left-leaning." Then the flow of money and weapons starts. The civil strife in Vietnam, the political convulsions of Chile, the decolonization of Angola—the list of countries caught up in the cold-war struggle is lengthening.

Although a champion of peoples trying to throw off repressive colonial regimes, the Soviet Union may be seen as a colonial empire itself. Just as the British and French built colonial realms in nearby (Ireland, Algeria) and distant (India, Vietnam) lands, the Soviet Union is the ultimate product of Russian imperialism—an imperialism whose acquisitions were not liberated when the revolution changed the social order in 1917. Nor has Moscow been especially tolerant of national self-determination in its Eastern European realm, acquired and consolidated when Western European colonial powers were beginning to lose their dependencies. The Soviets used tanks and other heavy arms to suppress assertions of independence in Hungary (1956) and Czechoslovakia (1968). Ideological dissent at home also has been harshly put down.

A European Heritage

For many centuries the main area of human interaction and spatial organization was in what is today western Russia. Although our knowledge of Russia before the Middle Ages is but fragmentary, it is clear that peoples moved in great migratory waves across the plains on which the modern state eventually was to emerge. The dominant direction seems to have been from east to west; many peoples came from central Asia and left their marks in the makeup of the population. Scythians, Sarmathians, Goths, and Huns—they came, settled, fought, and were absorbed or driven off. Eventually the Slavs emerged as the dominant people in what is today the Ukraine; they were peasants, farming the good soils of the plain north of the Black Sea. Their first leadership came not from their own midst, but from a Scandinavian people known as the Varangians, who had for some time played an important role in the fortified trading towns (gorods) of the area.

Soviet Union

The Soviet Union is the product of the Revolution of 1917, when more than a decade of rebellion against the rule of Tsar Nicholas II led to the tsar's abdication. Russian revolutionary groups were called *soviets* or "councils," and these soviets had been engaged in revolutionary activity since the first Russian workers' revolution of 1905. In that year thousands of workers marched on the palace in protest, and the tsar's soldiers opened fire on the marchers, killing and wounding hundreds. Throughout the years that followed, Russia was engaged in costly armed conflicts, government became more corrupt and less effective, and social disarray prevailed. In 1917 a coalition of military and professional men forced the tsar to abdicate, and Russia was ruled by a Provisional Government until November (by the Gregorian Calendar), when the country experienced its first and only truly democratic election. In the mean time, however, the Provisional Government had allowed the return to Russia of exiled Bolshevik activists: Lenin from Switzerland, Trotsky from New York, and Stalin from Siberia. In the political struggle that ensued, the Bolsheviks gained control over the revolutionary soviets, and this ushered in the communist era. From 1924 the country has been officially known as the Union of Soviet Socialist Republics (U.S.S.R.), for which Soviet Union is shorthand. Sometimes the country still is referred to by its prerevolutionary name, *Russia.* But Russia is only one (though by far the largest and dominant) of the 15 republics in the union.

The first Slavic state (ninth century A.D.) came about for reasons that will be familiar to anyone who remembers the importance of the transalpine routes across Western Europe. The objective was to render stable and secure an eastern crossing of the continent, from the Baltic Sea and Scandinavia in the northwest to Byzantine Europe and Constantinople in the southeast. This route followed the Volkhov River to Novgorod (positioned on Lake Ilmen), then led to Kiev and southward to the Black Sea along the Dnieper. Novgorod, so near Scandinavia and close to the shores of the Baltic, was the "European" center, with its cosmopolitan population and its Hanseatic trade connections. Kiev, on the other hand, lay in the heart of the land of the Slavs, near an important confluence on the Dnieper, and not far from the zone of contact between the forests of central Russia and the grassland steppes of the south. Kiev had centrality; it served as a meeting place of Scandinavian and Mediterranean Europe, and for a time during the eleventh and twelfth centuries it was a truly European urban center as well as the capital of Kievan Russia. Kiev and Milan had something in common: both were positioned near major breaks in the landscape, Milan at the entry to the difficult Alpine crossing, and Kiev at the edge of the dense and dangerous forests that covered middle Russia.

About the middle of the thirteenth century Kievan Russia fell to yet another invasion from the east. Now the Mongol empire, which had been building under Genghis Khan, sent its Tatar hordes into Kiev (led, incidentally, by the grandson of Genghis, Batu), and the city as well as the state fell. Many Russians sought protection by fleeing into the forests that lay north of Kiev; the horsemen of the steppes were not very effective in that terrain. What remained of Russia really lay in the forests between the Baltic and the steppes, and there, for a time, a number of weak feudal states arose, many of them ruled by princes who paid tribute to the Tatars in order to retain their position. From among these feudal states, the one that centered on Moscow emerged as supreme. In the fifteenth century its ruler conquered Novgorod; then in the sixteenth, Ivan IV (the Terrible) took over authority and assumed the title of tsar (another Mediterranean contribution: czar or *caesar*).

Ivan the Terrible made an empire out of Muscovy. He established control over the entire basin of the Volga River, succeeding in his campaigns against the Tatars, and extended Moscow's authority into western Siberia.

This eastward expansion of Russia was carried out by a relatively small group of seminomadic peoples who came to be known as Cossacks. Opportunists and pioneers, they sought the riches of the east, chiefly fur-bearing animals, as early as the sixteenth century. By the middle of the seventeenth century they reached the Pacific Ocean, defeating Tatars in their path, and consolidating their gains by constructing *ostrogs,* strategic way stations along river courses. Before the eastward expansion halted in 1812, the Russians had moved across the Bering Strait to Alaska and down the western coast of North America into what is now northern California (Fig. 2-1).

By the time Peter the Great took over the leadership of Russia (he reigned from 1682 to 1725), Moscow lay at the center of a great empire—great, at least, in terms of the territories under its hegemony. As such, Russia had many enemies. The Mongols were finished as a threat, but the Swedes, no longer allies in trade, threatened from the north; to the northwest there were the Lithuanians. To the west there was a continuing conflict with the Germans and the Poles. And to the south, the Ottoman Turks were heirs to the Byzantine Empire and they, too,

Figure 2-1

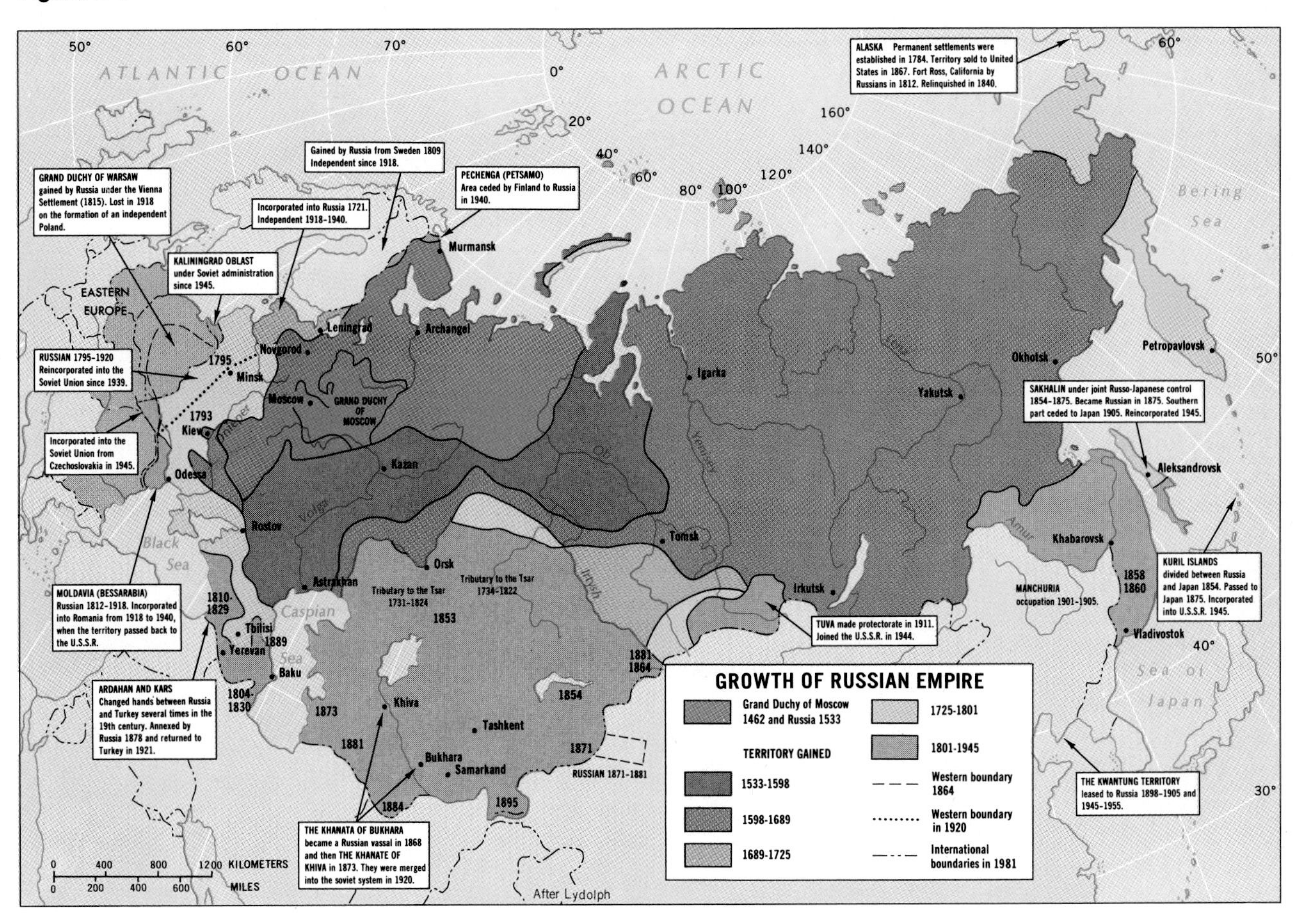

Leningrad has been called the Venice of the Baltic. The old city was designed, at Peter the Great's behest, by Italian architects who used the area's numerous waterways to great advantage. Shown here is one of Leningrad's canals and some of its ornate architecture, including the Church of Spilled Blood. (George Holton/Ocelot)

posed a threat. Peter consolidated Russia's gains and did much to make a modern, European type of state out of the loosely knit country. He wanted to reorient the empire to the Baltic, to give it a window on the sea and to make it a maritime as well as a land power. In 1703, following his orders, the building of Petersburg (later St. Petersburg) began. This city, known today as Leningrad, is positioned at the head of the Gulf of Finland, which opens into the Baltic Sea. Not only did it provide Russia with an important maritime outlet, but St. Petersburg also was designed to function as a forward capital: it lay on the doorstep of Finland, which at that time was a Swedish possession. In 1709 Peter's armies defeated the Swedes, confirming Russian power on the coast; in 1713 Peter took the momentous step of moving the Russian capital from Moscow to the new Baltic headquarters, where it remained until 1918.

Peter was an extraordinary leader, in many ways the founder of modern Russia. In his desire to remake Russia, to pull it from the forests of the interior to the western coast, to open it to outside influences, to end its comparative isolation, he left no stone unturned. Not only did he move the capital; he himself, aware that the future of Russia as a major force lay in strength at sea as well as power on land, went to Holland to work as a laborer in the famed Dutch shipyards, to learn how ships were most efficiently built. Peter wanted a European Russia, a maritime Russia, a cosmopolitan Russia. He developed Petersburg into one of the most magnificent cities in the world and employed Italian architects to make it so; with its wide avenues and elaborate palaces, it remains to this day a city apart in all of the Soviet Union, reminiscent of Paris, London, or Milan. The new capital became and long remained one of Europe's principal cities, a seat of power with which the whole continent had to reckon.

During the eighteenth century the tsarina Catherine the Great, who ruled from 1762 to 1796, continued to build Russian power, but on another coast and in another area: the Black Sea in the south. Here the Russians confronted the Turks, who had taken the initiative from the Greeks; the Byzantine Empire had been succeeded by the Turkish Ottoman Empire. But the Turks were no match for the Russians. The Crimea soon fell, as did the old and important trading city of Odessa, and before long the whole northern coast of the Black Sea was in Russian hands. Soon afterward the Russians penetrated the area of the Caucasus, and early in the nineteenth century they took Tbilisi, Baku, and Yerevan. But as they pushed farther into the corridor between the Black and the Caspian Seas, they faced growing opposition from the British (who held sway in Persia, now Iran) as well as the Turks, and their advance was halted short of its probable ultimate goal: a coast on the Indian Ocean.

But Russian expansionism was not yet satisfied. While extending the empire southward, the Russians also took on the Poles, old enemies to the west, and succeeded in taking most of what is today the Polish state including the capital of Warsaw; to the north, Russia took over Finland from the Swedes (1809). During most of the nineteenth century, however, Russian preoccupation was in Asia, where Tashkent and Samarkand came under St. Petersburg's control. The Russians were bothered still by raids of nomadic Mongol horsemen, and they therefore sought to establish their authority over the steppe country to the edges of the high mountains that lay to the south. Thus Russia gained a considerable number of Moslem subjects, for this was Moslem Asia they were penetrating, but under tsarist rule these people acquired a sort of ill-defined protectorate status and retained some autonomy. Farther to the east, a combination of Japanese expansionism and a decline of Chinese influence led Russia to annex from China several provinces along the Amur River. Soon afterward (1860), the port of Vladivostok was founded.

Now began the course of events that was to lead, after five centuries of almost uninterrupted expansion and consolidation, to the first setback to the Russian drive for territory. In 1892 the Russians began building the Trans-Siberian Railway, in an effort to connect the distant frontier more effectively to the western core. As the map shows, the most direct route to Vladivostok was across Chinese Manchuria. The Russians wanted China to permit the construction of the last link of the railway across Manchurian territory, but the Chinese resisted. Russia responded (taking advantage of the Boxer Rebellion in China) by annexing Manchuria and occupying it. This brought on the RussianJapanese war of 1905, in which the Russians were disastrously defeated; Japan took possession of southern Sakhalin. For the first time in nearly five centuries, Russia sustained a setback that resulted in a territorial loss.

Thus Russia—recipient of British and European innovations as were Germany, France, and Italy—was just as much a colonizer too. But where other European powers traveled by sea, Russian influence traveled overland, into South Asia toward India, into China, and to Pacific coasts. What emerged was not the greatest empire, but the greatest *contiguous* empire in the world; it is tempting to speculate what would have happened to this sprawling realm had European Russia (for such it still was) developed politically and economically in the manner of other European power cores. At the time of the Japanese war, the Russian tsar ruled over more than 22 million square kilometers, just a fraction less than the area of the Soviet Union today. The modern Communist empire, to a very large extent, is a legacy of St. Petersburg and European Russia—not the product of Moscow and the Revolution.

Russians in North America

The first white settlers in Alaska were Russians, not Western Europeans, and they came across Siberia and the Bering Strait, not across the Atlantic and overland. Russian hunters of the sea otter (valued for its high-priced pelt) established their first Alaskan settlement at Kodiak Island in 1784. Moving along the coast, the Russians founded additional villages and forts to protect their tenuous holdings, until they reached as far as the area just north of San Francisco Bay, where they built Fort Ross in 1812.

But the Russian settlements were isolated and vulnerable. European fur traders began to put pressure on their Russian competitors, and Moscow found the distant settlements a burden and a risk. In any case, American, British, and Canadian hunters were decimating the sea otter population, so that profits declined. When United States Secretary of State William Henry Seward in 1867 offered to purchase Russia's holdings, Moscow quickly agreed—for $7.2 million. Thus Alaska, including its lengthy coastal extension southward, became a United States sphere of influence, but Seward was ridiculed for his decision. Alaska was "Seward's Folly" and "Seward's Icebox" until gold was discovered in the 1890s. The twentieth century has proved Seward's wisdom many times over, strategically as well as economically. At Prudhoe Bay, off Alaska's northern Arctic slope, large oil reserves have been discovered; like Siberia, Alaska probably contains other, yet unknown riches.

Physiographic Regions

Before we return to the modern Soviet Union's human geography, we briefly examine the physiographic regions of this vast realm. As Figure I-6 (page 19) indicates, the Soviet Union's natural environments are harsh, even bleak. More than 80 percent of the total area of the country lies farther to the north than the American Great Lakes. Thus it lies open to the frigid air masses of the Arctic, but shielded (by Europe) from the moderating influences of Atlantic waters and (by mountains) from the warmer south. The prevalence of D climates (Fig. I-6) underscores the prevailing coldness, whose severity increases eastward; the map of precipitation (Fig. I-5, p. 17) emphasizes that low and variable precipitation is another environmental problem in the Soviet Union. And Fig. I-8 (page 25) reveals that the country's best soils lie in a belt extending from Eastern Europe to the heart of the Soviet Union—but a large part of this belt lies in a zone of low and variable rainfall. For an insight, locate the comparable soil zones in North America and the Soviet Union and relate them to the precipitation map (Fig. I-5) in both regions.

Notwithstanding its great size, the Soviet Union is almost a landlocked country, whose continentality is accentuated by coldness, drought, short growing seasons, high winds, and other environmental difficulties that make the productive areas of the southwest and south mere islands in a forbidding, severe expanse of tundra and forest, mountain and desert (Fig. I-7, p. 21). Neither will you discern physiographic regions as distinct as those of North America; the Soviet land surface, for all its size, divides much less clearly.

A map of Soviet terrain shows that the east and south tend to be high and rugged, while the north and west are lower and more level. Only the Ural Mountains, trending north-south right through the middle of the western lowlands, constitute a major exception to this broadest division of the Soviet Union's physiography.

The simplest subdivision of the Soviet Union's physical geography produces nine regions (Fig. 2-2). The *Russian Plain* (1) is the eastward continuation of the North European Plain—the theater of the rise of the Russian state. At its heart lies the Moscow Basin; to the north the countryside is covered by needleleaf forests similar to those of Canada, while to the south lie the grain fields of the Ukraine. The Russian Plain is bounded to the east by the *Ural Mountains* (2), not a high range but prominent because it separates two extensive plains. The Ural range forms no barrier to transportation, and its southern end is quite densely populated; the region has yielded a variety of minerals including oil.

South of the Urals lies the *Caspian-Aral Basin* (3), a region of windswept plains of steppe and desert. Its northern section is the Kirgiz Steppe, and in the south the Soviets are trying to conserve what little water there is here through irrigation schemes on rivers draining into the Aral Sea. East of the Urals lies Siberia. The *West Siberian Plain* (4) has been described as the world's largest unbroken, true lowland; it is the basin of the Ob River. Over the last 1600 kilometers (1000 miles) of its course, the Ob falls only 90 meters (less than 300 feet). The northern area of the West Siberian Plain is permafrost-ridden, the central areas are marshy, but the south carries significant settlement, including the cities of Omsk and Novosibirsk on the Trans-Siberian Railroad.

East of the West Siberian Plain the country begins to rise, first into the *Central Siberian Plateau* (5), a sparsely settled,

Figure 2-2

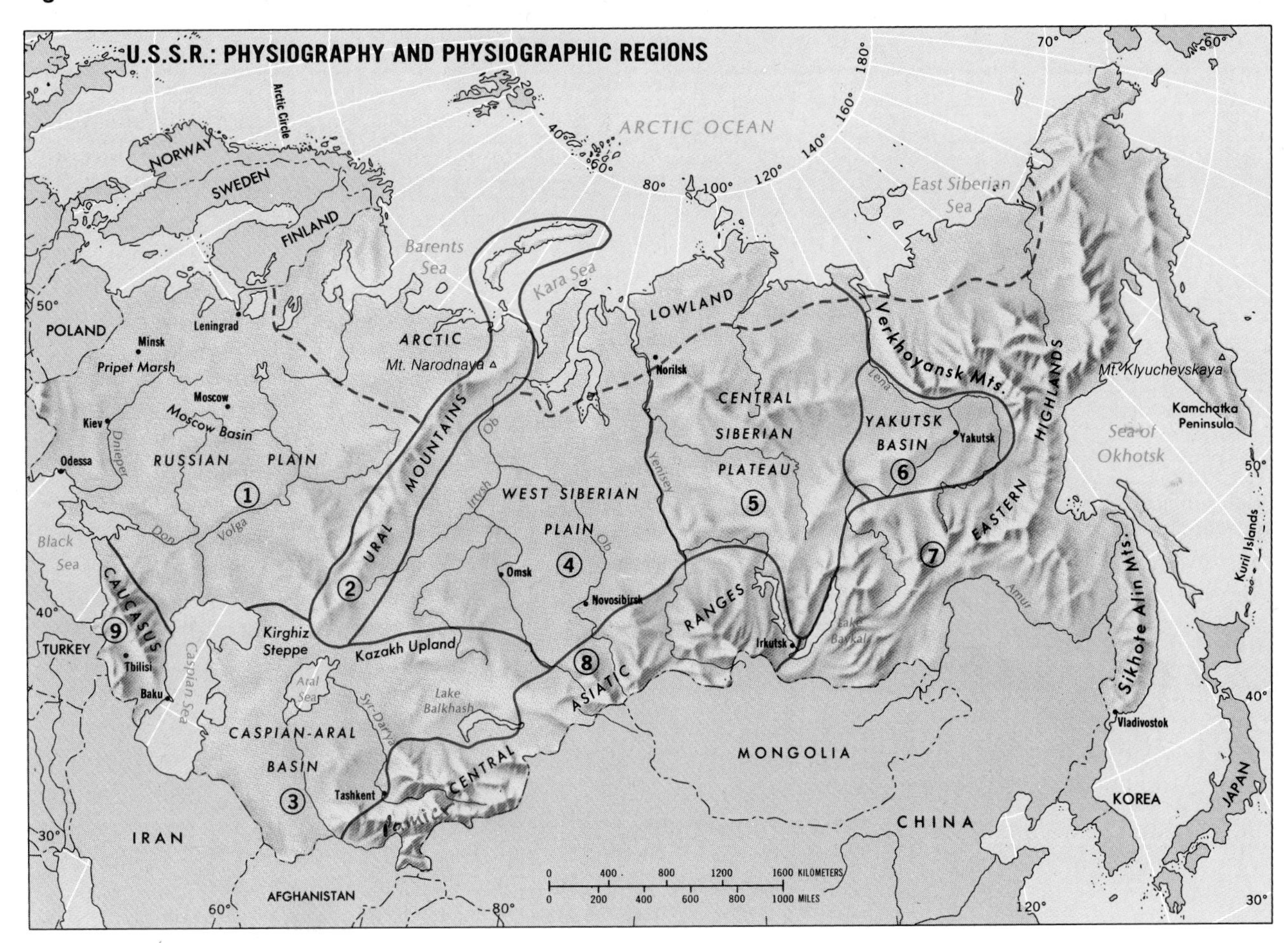

Surface communications always have been difficult in the vast Soviet land area. Shown here is a segment of the Trans-Siberian Railroad in the area of Lake Baykal. (Sovfoto)

remote, permafrost-afflicted region. Here winters are long and extremely cold, the summers short. Some mineral finds have been made (including nickel, copper, and platinum around Norilsk in the northwest), but the area remains barely touched by human activity. Beyond the *Yakutsk Basin* (6), the terrain becomes mountainous and the relief high. The *Eastern Highlands* (7) are a jumbled mass of ranges and ridges, precipitous valleys, and volcanic mountains. Lake Baykal lies in a trough that is over 1500 meters (5000 feet) deep; on the Kamchatka Peninsula, volcanic Mount Klyuchevskaya reaches nearly 4750 meters (15,600 feet). The northern area is the Soviet Union's most inhospitable zone, but southward, along the Pacific coast, the climate is less severe. Still, this is a true frontier region. The forests provide opportunities for lumbering, a fur trade exists, and gold and diamonds are found by hardy miners.

The southern margins of the Soviet Union are also marked by mountains: the *Central Asiatic Ranges* (8) from the vicinity of Tashkent in the west to Lake Baykal in the east, and the *Caucasus* (9) between the Black and Caspian Seas. The Central Asiatic Ranges rise above the snow line and carry extensive mountain glaciers. The annual melting brings alluvium to enrich the soils on the lower slopes and water to irrigate them. The Caucasus form an extension of Europe's Alpine Mountains and produce a very similar topography, but they do not provide convenient passes. The trip by road from Tbilisi to Ordzhonikidze (via the Georgian Military Highway) is still an adventure in surface travel.

Soviet Union in the Twentieth Century

Russia's enormous territorial expansion produced an empire, but nineteenth-century Russia was infamous for the abject serfdom of its landless, hopeless peasants and the exploitation of its workers, the excesses of its nobility, and the palaces and riches of its rulers. The arrival of the industrial revolution had brought a new age of misery for those laboring in factories. There were ugly strikes in the cities; from time to time the peasants rebelled against the nobility. But retribution always was severe, and the tsars remained in control. And then, in 1904, Russian forces fighting in the Far East against the Japanese faced defeat. It was a milestone in the destruction of the old order, because this loss was followed by ever-increasing opposition throughout Russia against the tsars' authority. Local revolutions took place, upheaval was the order

Moscow, the capital of the U.S.S.R. is a city of historic buildings, modern government and apartment structures, and wide, windswept squares. This wintry view was photographed in the Kremlin area. (Inge Morath/Magnum)

of the day. Certainly Russia was not ready for its own defense when World War I erupted, and again Russian forces were disastrously defeated.

Finally the 1917 Revolution succeeded where earlier rebellions had failed. The revolution was no more a unified uprising than its predecessors had been; while the Bolsheviks' "red" armies battled the "whites," both fought against the forces of the tsar. Ultimately the Bolsheviks prevailed and the capital, Petrograd (as St. Petersburg had been renamed in 1914 to rid it of its German nomenclature), witnessed the end of tsarist rule. Even before the Bolsheviks' victory was complete, the signs of things to come were already evident. In 1918 the headquarters of the government were moved from Petrograd to Moscow, the focus of the old state of Muscovy, deep in the interior, not even on a major navigable waterway, amid the remnants of the same forests that centuries earlier had afforded the Russians protection from their enemies. It was a symbolic move, symbolic of a new period in Russian history, an expression of distrust toward a Europe that had contributed war and misery to Russia, but whose promises—of maritime power, of industrial modernization, of political advancement—had never been fulfilled. The new Russia looked inward; it sought a means whereby it could achieve with its own resources and its own labor the goals that had for so long eluded it. The chief political and economic architect in this effort was also a revolutionary leader, V. I. Lenin. Among his solutions to the country's problems are the present-day political framework of the Soviet Union and the planned economy for which it is famous.

The Political Framework

Russia's great expansion had brought a large number of nationalities under the tsars' control, and now it was the turn of the revolutionary government to seek the best method of organizing all this variety into a smoothly functioning state. The tsars had conquered, but they had done little to bring Russian culture to the peoples they ruled; the Georgians, Armenians, Tatars, the Islamic Khanates of middle Asia (among dozens of individual cultural, linguistic, and religious groups) had not been "Russified." The Russians themselves, however, in 1917 constituted only about one-half of the population of the entire country (the proportion today remains about the same), and thus it was impossible simply to create in an instant a Russian state over the whole of the political region. Account had to be taken of the nationalities.

This question of the nationalities became a major issue in the Soviet state after 1917. Lenin brought the philosophies of Marx to Russia; in his earliest proclamations he talked about the "right of self-determination for the nationalities." The first response on the part of a number of Russia's subject peoples

Figure 2-3

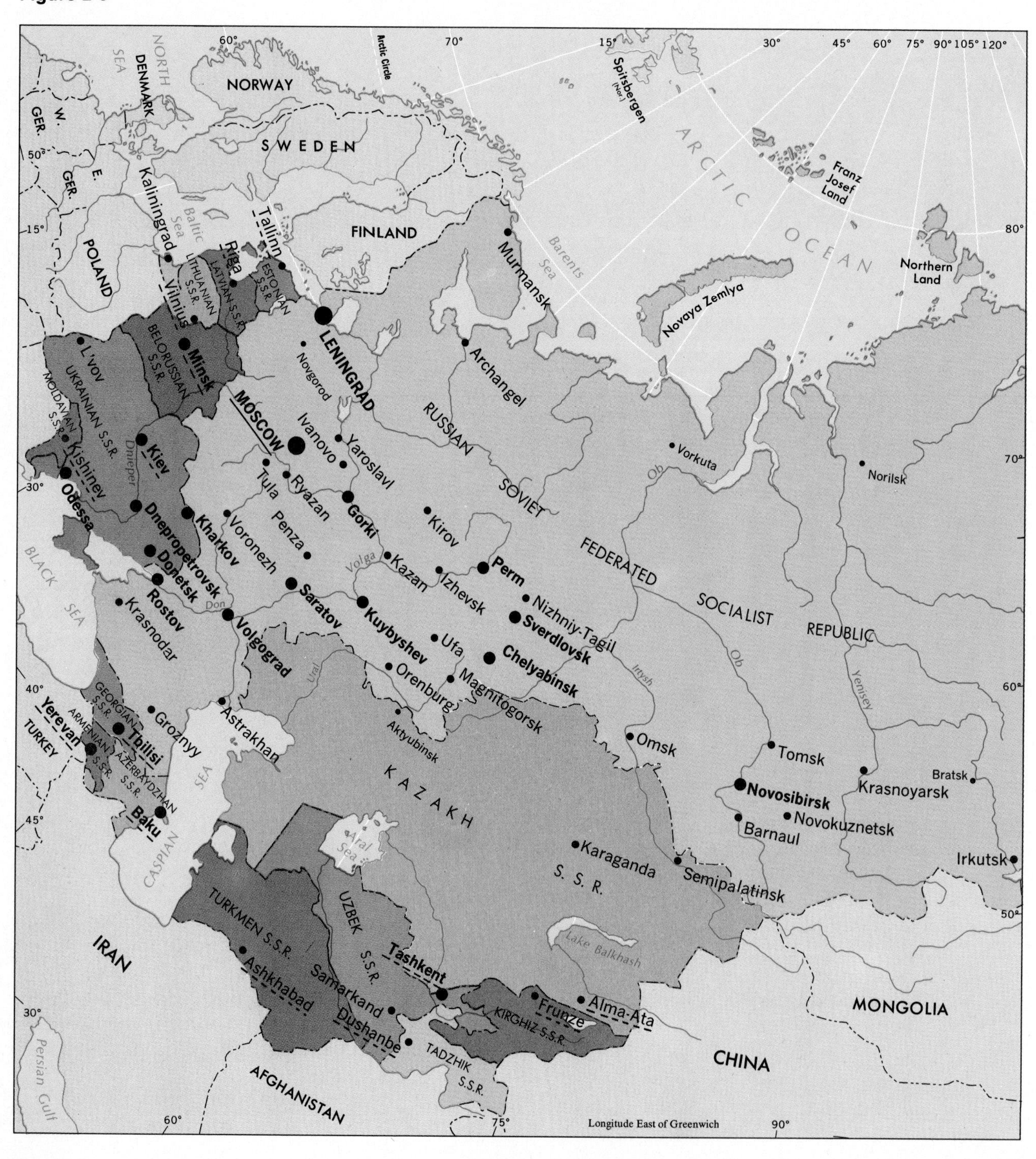

was to proclaim independent republics, as was done in the Ukraine, in Georgia and Armenia, in Azerbaydzhan, and even in middle Asia. But Lenin had no intention of permitting the actual breakup of the state, and in 1923, when his blueprint for the new Soviet Union went into effect, the last of these briefly independent units was absorbed into the sphere of Moscow. Nevertheless, the whole framework of the Soviet Union remains essentially based on the cultural identities of the peoples it incorporates. Since the days of Lenin there has been modification, but the major elements of the original system are still there. The country is divided into 15 Soviet Socialist Republics (S.S.R.s) (Fig. 2-3), each of which broadly corresponds to one of the major nationalities in the state. As Fig. 2-3 shows, by far the largest of these S.S.R.s is the Russian Soviet Federated Socialist Republic, which extends from west of Moscow to the Pacific Ocean in the east. The remaining 14 republics lie in a belt that extends from the Baltic Sea (where Estonia, Latvia, and Lithuania retain their identities as S.S.R.s) to the Black Sea (north of which lies the Ukrainian S.S.R.), and through Georgia and Armenia, both Union Republics, to the Caspian Sea, beyond which lie several Asian Soviet Socialist Republics.

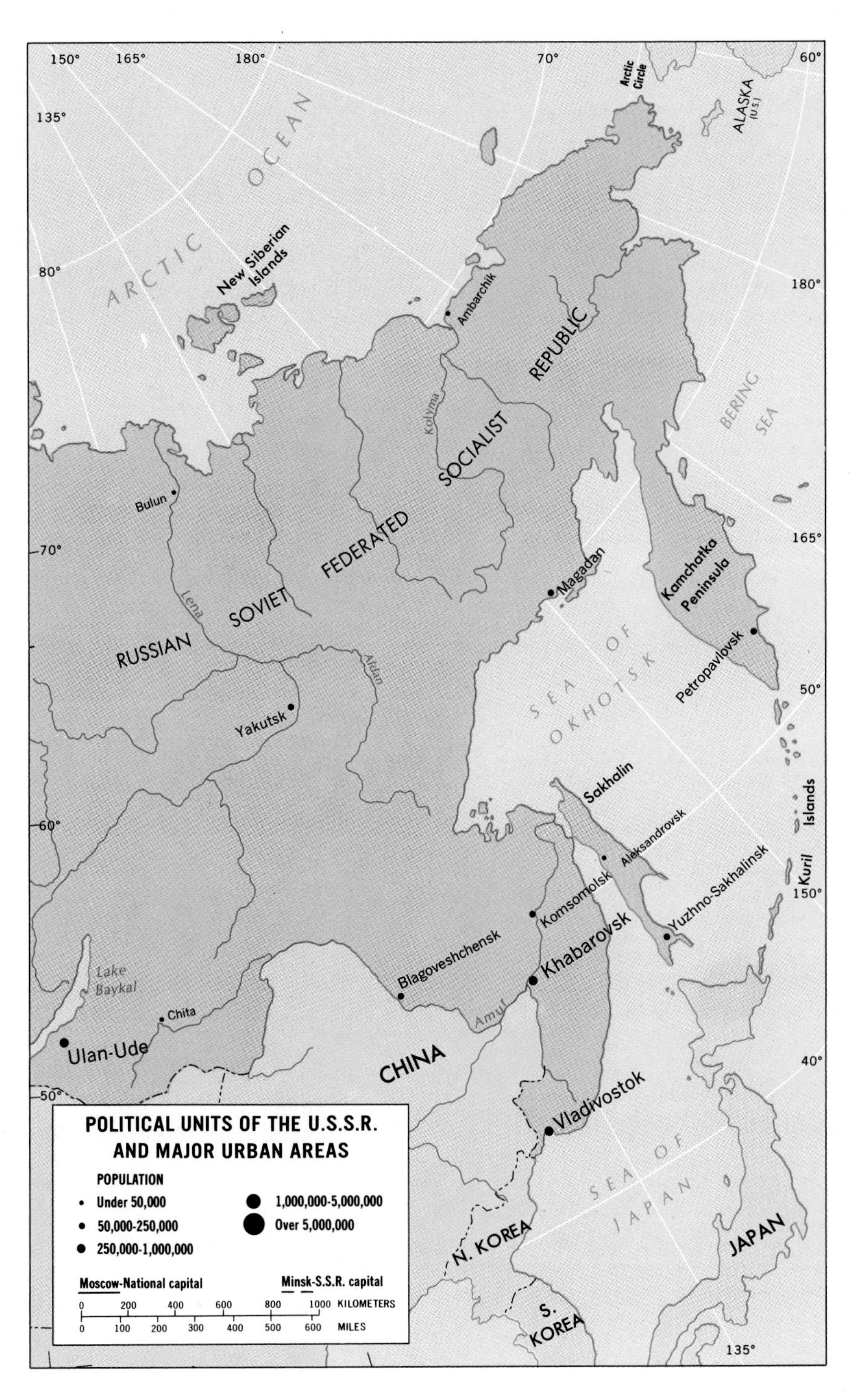

Theoretically these 15 republics have equal standing in the Union, but in practice the Russian Republic has the leadership and the others look upward. With half the country's population, the capital, most of its major cities, and over three-quarters of its territory, Russia remains the nucleus of the Soviet Union. The Russian language is taught in the schools of the other republics; the languages of these other areas are not taught compulsorily in Russia. Although the country's constitution, again theoretically, would permit each republic to carry on its own foreign policy, to issue its own money, and even to secede from the union, none of this has occurred in practice. On the other hand, even though it can be argued that the Soviet Union is highly centralized and a federal state only in theory, no republic has made use of wartime conditions

The Soviet Union incorporates numerous non-Russian nationalities. Any other Western country controlling a city whose cultural landscape looks like this (Moslem Khiva) might be accused of colonialism—but in the case of the U.S.S.R. this scene merely represents another Soviet Republic, Uzbekistan. (George Holton/Ocelot)

to break away. The Germans, during their invasion of the Soviet Union in World War II, hoped that an anti-Russian nationalism would provide them with support in the Belorussian S.S.R. and the Ukrainian S.S.R., but they found only isolated instances of cooperation and no attempts were made to proclaim sovereign states outside Moscow's sphere. (Persistent postwar guerilla movements did arise in the Baltic Republics and in the Ukraine.) Undoubtedly there is a great deal of acquiescence to the *status quo* among Soviet peoples, and, in view of the racial, cultural, linguistic, and religious complexity of the population (Fig. 2-4) this is no small achievement.

Figure 2-4

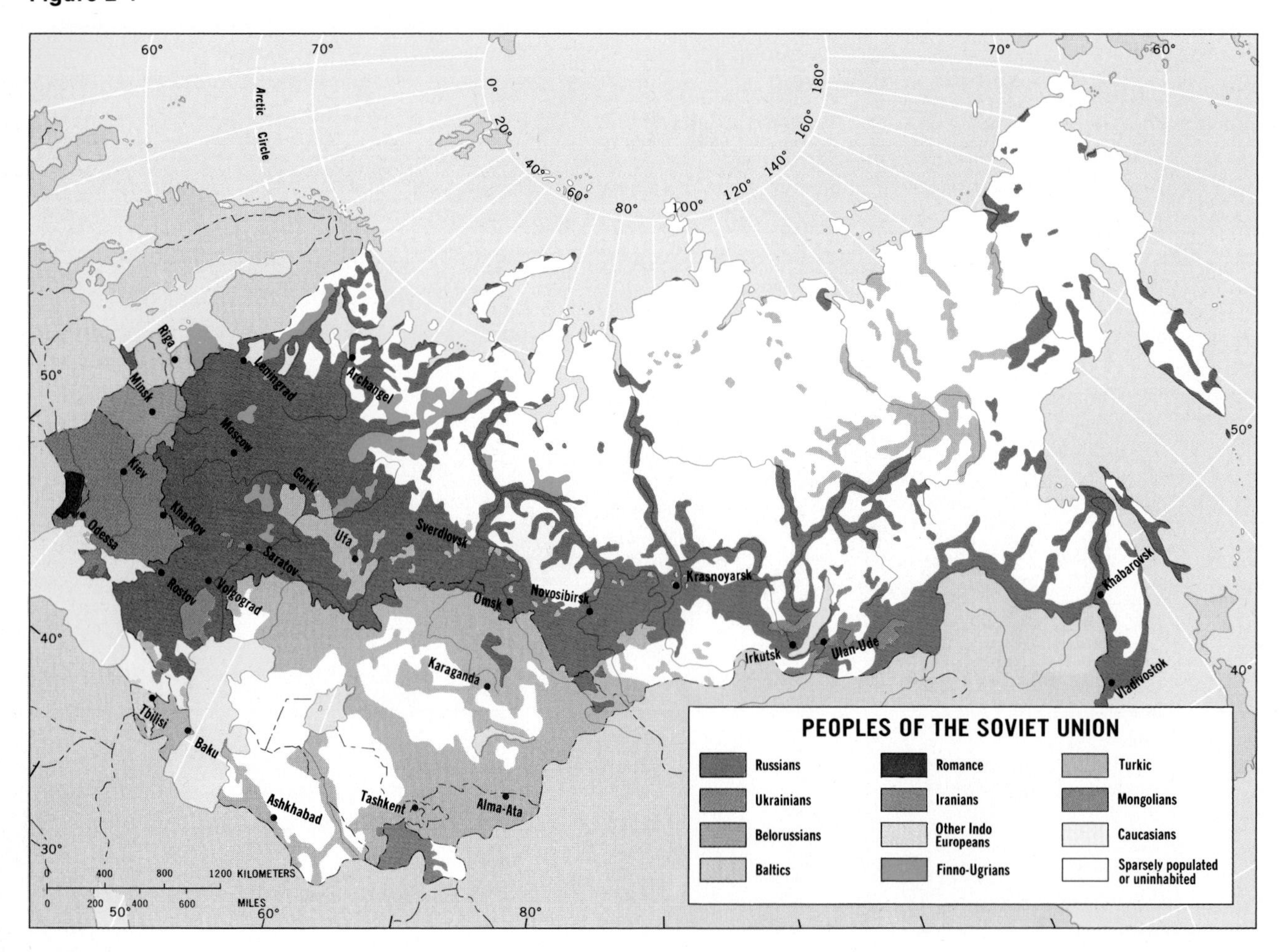

But if the system developed after 1917 proved to have some of the liabilities of the pre-1917 era, it also had some hitherto unknown assets. For the first time the entire country was brought under effective national control. For the first time the Russians, recipients of European innovations, now began to transmit these innovations to non-Russian and non-Slavic parts of the realm (Fig. 2-5). For the first time there were elements of compromise in the relationships between Russians, other Slavic peoples, and the non-Slavic population of the country. Certainly there were conflicts, and individual groups at various times have held separatist feelings. But in general there has been an awareness of progress brought by the Soviet administration; after the terror of the Russian conquest itself, the Soviet contributions in such areas as Armenia and Georgia are proof positive that membership in the Soviet Union does have its advantages. Standards of living have been raised, educational opportunities have improved, housing projects have taken the place of city slums, and the economy has accelerated. If these accomplishments resemble those boasted of by colonial powers defending the benevolence of their rule, we are reminded that the Soviet Union is a Russian empire. Local culture has been sustained (but organized religion discouraged); still, Russification is a form of *acculturation,* a process we will encounter in stronger perspective in Chapter 4. The Soviets, of course, like to claim that their six decades of control have brought unanimous approval of the system throughout the country. While this is not so, their treatment of minority nationalities does appear to have achieved greater domestic approval than that of some colonial powers and, indeed, of some governments of sovereign states.

Figure 2-5

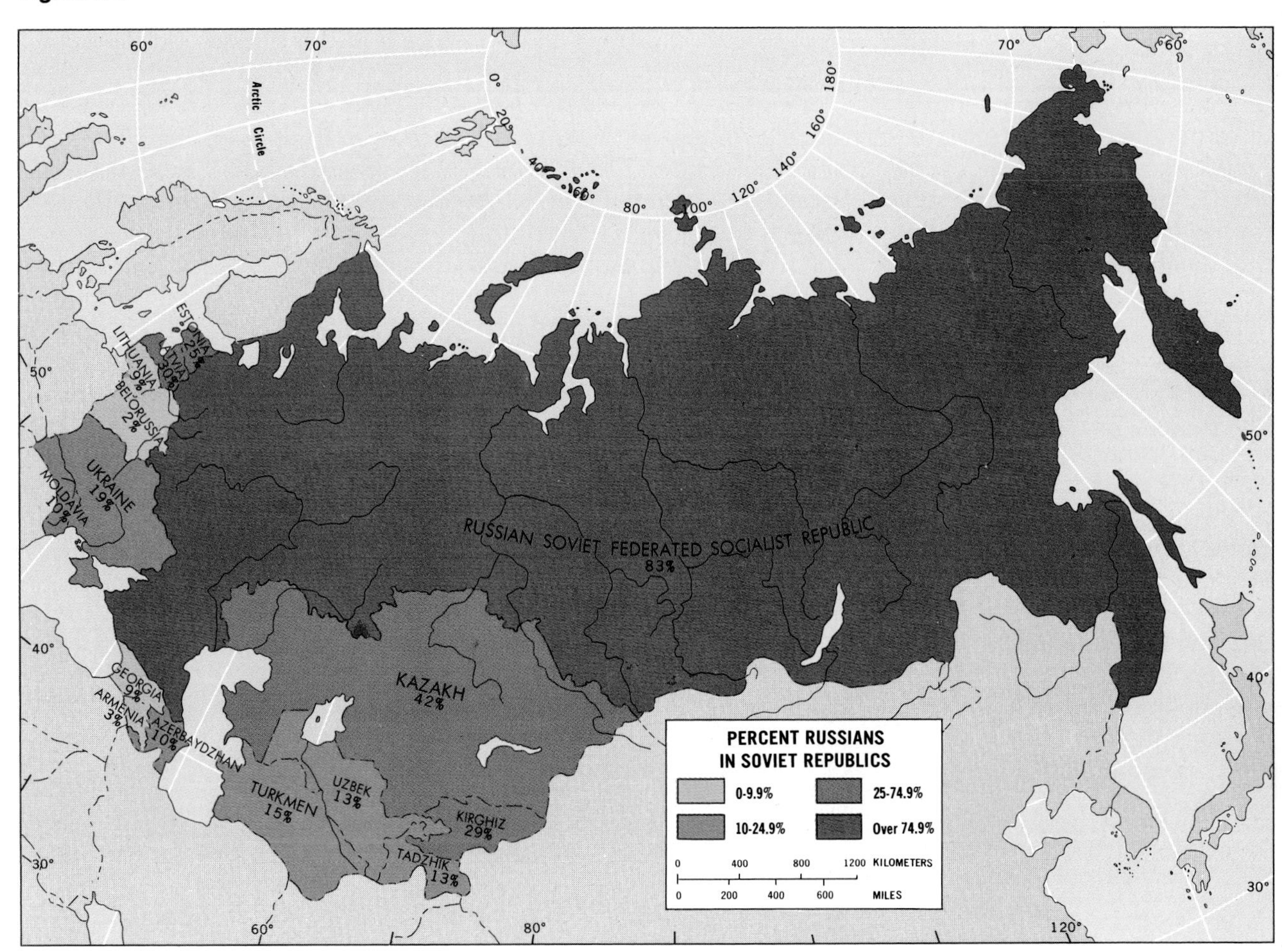

Economic Planning

The sweeping changes that came to Russian politics after the success of the 1917 Revolution are more than matched by the great Soviet economic experiment. After an early transitional period that saw some private enterprise continue while Soviet economists charted the socialist course, the revolutionary leaders' economic aspirations were embodied in the first of a series of *Five-Year Plans.* The objectives of the Soviet economic planners were (1) to speed the industrialization of the country and (2) to collectivize Soviet agriculture. In order to accomplish this, the entire country was mobilized, with a State Planning Commission *(Gosplan)* at the helm. For the first time ever on such a scale, and in accordance with Marxist principles, a whole state and its peoples worked in concert toward national goals that were prescribed by a central government. The Five-Year Plans serve first as guidelines and aims for the economy during the particular period (although the guidelines are often changed and the aims almost never exactly achieved), and second as exhortations for the Russian workers to labor as hard as they can toward the goals put before them by the state, goals that are invariably tied to promises of better days ahead.

The first Five-Year Plan was designed to accomplish the elusive collectivization of agriculture, delayed, by then, for fully a decade after the revolution. A major reason for this program was to secure more grain for the urban-industrial population through the extension of government control over the farmers. The farmers, dissatisfied with the terms of trade under the new order, had been withholding grain from the markets. The collectivization program during the period from 1928 to 1933 led to the incorporation of nearly 70 percent of all individual peasant holdings into larger units operated jointly and under government supervision. There was much opposition to the program, from the slaughter of livestock to avoid their being surrendered to the larger collective herds to outright revolt. On the other hand, the poorest peasants, those without land or livestock, voluntarily and gladly participated. But the initial impact on grain supplies was negligible, and in 1929 it was agreed that the process would have to be speeded up. The communist ideal was the huge state farm or *sovkhoz,* literally a grain and meat factory in which agricultural efficiency through mechanization and minimum labor requirements would be at its peak. But such a huge enterprise is not easily established and it can often result in overmechanization and inefficiency; so, by way of an intermediate step, the smaller, local collective farm or *kolkhoz* was recognized as a more easily attained replacement. By the end of the second Five-Year Plan, in the late 1930s, well over 90 percent of all peasant farms were collectivized.

Soviet Agriculture. A generalized map of agricultural regions in the Soviet Union (Fig. 2-6) emphasizes the limitations of the Soviet environment. The leading agricultural zone lies south of Moscow, extending from the shores of the Black Sea to the valley of the Irtysh River in the east. This region is named Large-Scale Diversified (1) because here lie the country's most extensive, highly mechanized collectives where the bulk of the wheat is harvested and crop variety is enormous. From newspaper reports of Soviet crop failures we can conclude that this is also a region of unreliable rainfall, and from time to time disastrous droughts affect certain areas, notably toward the drier east. In addition to winter and spring wheat, this region can produce a wide range of food and industrial crops including sugar beets and potatoes, fruits and vegetables (including grapes), while also supporting large herds of cattle, sheep, and pigs. But although the landscape will frequently remind you of the North American Great Plains, you will miss the American house-and-barn units that mark our countryside. Rather, the old Russian villages of the days of the empire have been converted into Soviet collectives, at whose focus lie the newer administration buildings—including the medical clinic, school, recreation facilities, meeting hall. Large, patriotic billboards still exhort the Soviet worker to labor hard for the country. But walk to some of the small, often old, and outmoded wooden houses of the collective's inhabitants, and you may find that each has a small plot of land where the farmer still can grow some crops for his own use (even sale). In those meticulously cultivated gardens the Soviet farmer still retains some individuality.

Analogue Area Analysis

The small private plots that Soviet farmers still are permitted to farm produce a huge harvest—over 40 percent of all the eggs and more than 30 percent of all the vegetables and milk consumed in 1975. If Soviet economic planners had not collectivized the farmlands, would production of wheat and other crops be higher than it is on the *kolkhozes* and *sovkhozes?*

In 1968 S. S. Birdsall used the method of analogue area analysis to compare Soviet and United States agriculture. He chose two areas with essentially the same physical base (climate, soils, relief, drainage): the Minskaya Oblast in the central part of Belorussia—region (2), Fig. 2-6—and six farming counties in central Wisconsin. Reports Birdsall: the two areas "appear nearly identical in terms of their respective physical bases for agriculture . . . areas so widely separated on the earth's surface could hardly be more similar physically." Having established the areas as approximately analogous, Birdsall next compares their productivity: in grain yields, the Wisconsin area produces 1.5 to 2.0 times as much as Minskaya. In pastoral output, too, the Wisconsin counties are far ahead: in milk yield per cow (1.8 to 3.6) and egg production (2.0 to 3.0). Only in the production of root crops do Minskaya's farmers match United States yields, exceeding in one—potatoes. But, overall, the Soviet production lags far behind. Birdsall leaves no doubt about his conclusion: "The lower agricultural productivity of the Minskaya Oblast cannot be said to be caused by factors beyond the control of those ultimately in charge of farming operations . . . the Minskaya Oblast could ultimately have agricultural productivity at least 150 percent of its present level. This could be achieved by adopting the managerial methods (including decisions on technology utilization) used in the selected counties in Wisconsin."

In order of importance, the region identified as *Mixed* (2) follows. Actually, three regions fall into this category: the large, triangular area within which Moscow lies, the mountains and foothills of the Caucasus (2A), and the small but significant zone in the Soviet Far East, fronting the Chinese border and the Pacific Ocean (2B). Farming here is on a rather smaller scale than in region (1), but the variety of crops and livestock is great. Rye and barley, potatoes, fodder crops, and dairy cattle form part of the farm scene here; in many ways the area looks like parts of the region adjacent to the North American Great Lakes. The zone centered on the Caucasus has a different mixture of crops, because the lower areas here have subtropical conditions (citrus groves and tea plantations mark the lower slopes). On the other side of the Caspian Sea, the climate is also warm and the growing season long—but rainfall is so low that crops must be *Irrigated* (3). Sheep and hardy cattle support themselves on the scattered pastures, but the crops—cotton, tobacco, fruits and vegetables, some grains including rice—must be artificially watered.

These three regions are clustered in the southwestern part of the Soviet Union, and the two remaining areas cover the rest of this vast realm. Region (4) is characterized by *Localized Farming*. In the difficult terrain that would be familiar to Canadians, farming tends to concentrate near the scattered, often isolated settlements that serve as local markets for such

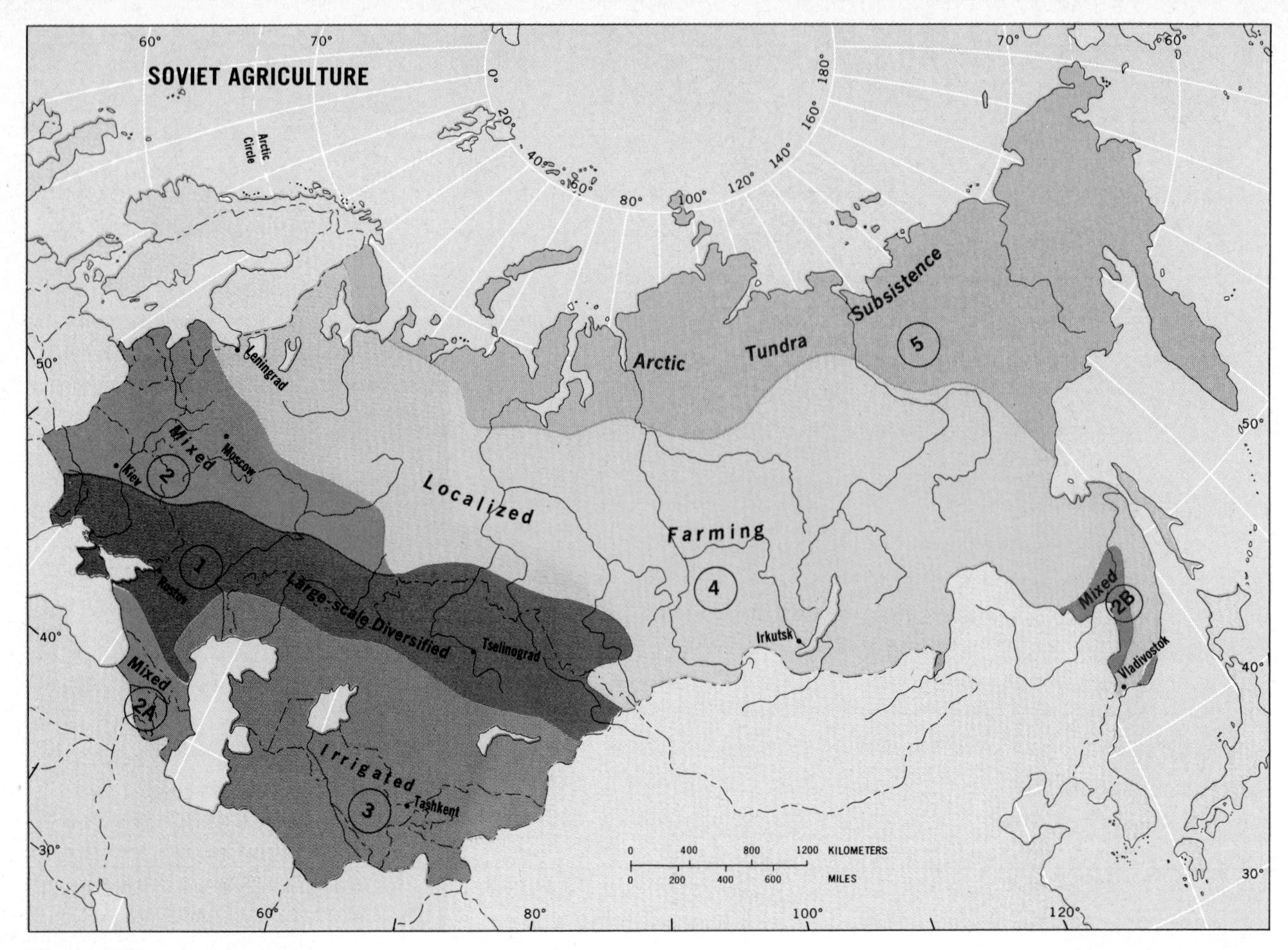

Figure 2-6

products as potatoes, rye, barley, oats, some hardy vegetables, and milk, butter, cheese, eggs, and meats. Although growing seasons are short in this region and farming difficult, things get even worse in region (5), where *Arctic-Tundra Subsistence* prevails. A few root crops manage to mature in the brief summer, but this is hunting country. As in other arctic areas, reindeer help sustain human communities.

Soviet Industry. Since the 1917 Revolution, the Soviet Union has been recognized less for its achievements in agriculture than for its massive industrialization; hence, the goals of the first Five-Year Plan, even in the field of agriculture, were in considerable measure related to other objectives in industry. The collectivization and mechanization of agriculture would free many thousands of peasants for industrial work in the cities, and with a larger and more dependable volume of farm production, the urban populations would be better fed. Major investments were made in transportation, electrification, and other heavy-industry-oriented facilities, and certainly the results were spectacular (Fig. 2-7). During the first Five-Year Plan, while the rest of the world experienced the effects of the Great Depression, Soviet industrial production climbed continuously and sharply. In terms of steel production, for example, the Soviet Union in 1932 was ahead of all countries but the United States; needless to say, it heralded this achievement as proof of the effectiveness of the plan, and it would be difficult to argue otherwise. From then on, Soviet manufacturing has been in the forefront of the world.

In a way, then, the Soviet Union operates as a giant corporation, with the government and its agencies acting as a board of directors that determines the levels of productivity, the resources that will be exploited or opened up, the kinds of goods that will be produced and made available, the wages

Figure 2-7

paid, and so forth. The government's economic planners set the country on a certain course, and although there are year-to-year modifications in all such programs, the major features of the plans normally remain visible. But it would be a mistake to see the Soviet Union's economy (and that of other communist states) as the only economy in the world that involves "planning." Economic changes in the United States, too, are the result of planning activities, but here the plans are laid at different levels, from the giant corporations to small industries. Certainly the government takes a hand in shaping the progress of the United States economy, but it does not control as the Soviet government does. On the other hand, it is easy to exaggerate the effectiveness of Soviet national planning in bringing about the desired goals. The measures used to gauge the degree of success in meeting production schedules have varied, and the psychological impact that is so important a part of each plan has not always had the desired result. Always agriculture seems to have been the problem area; the long haul toward adequate farm production was probably unnecessarily long because of the low prices paid, the lack of incentives, and the general secondary position of agriculture compared to that of industry. But the Soviets sought power, and the road to power, they realized, lay in heavy industry. In this field there really has been revolutionary development, so much so that two generations after the 1917 Revolution, the Soviet Union was in a position to make a challenge for world power. Agriculture still presents problems, its production insufficient to release the country from foreign imports, an enormous labor force still tied to the land, farming at comparatively low efficiency while industrial labor shortages persist. Still, the situation has improved as investment in agriculture has increased and yields have risen.

Consumer goods, of which the Soviet citizen has long been virtually deprived, are increasing in volume on the Soviet market. Accompanying this are signs of political and economic relaxation; the profit motive is helping to solve agricultural problems on collectives where farmers may be rewarded for individual output as well as time.

It is tempting, naturally, to speculate whether the Soviet economy could have developed so spectacularly without state control over all productive capacity, and without the economic planning for which it has become known. Many Western economists, pointing to the rapid industrialization of Russia during the 1880s and 1890s, argue that if the trends set at that time had been continued, the country would be further ahead industrially than it is today. But the revolution and ensuing civil wars largely obliterated this early progress, and what the Soviets remember is the rise of their country out of a divided, agrarian, poor, unproductive, inefficient, and stagnant past. Hundreds of thousands of the country's citizens have seen this emergence of the Soviet Union from poverty and obscurity to greatness; millions have seen promises made by economic planners come true. Certainly most Soviet citizens would have little doubt.

Regions of the Soviet Realm

The Soviet Union today is a country of large, spacious, and impressive cities, efficient communications, modern industries, and mechanized farms. It is also a country of vast empty spaces, remote settlements, inadequate transport facilities, and out-moded rural dwellings. Against the confidence and accomplishments of the Soviet planners, the treatment of many Jewish citizens and dissident authors, scientists, and artists stands in stark contrast. The Soviet Union has powerful armed forces, great universities, magnificent orchestras, spectacular space programs—but still it sends nonconformists to "mental hospitals"; the specter of Siberia still looms. Soviet world influence works to liberate faraway colonies from foreign domination, but Moscow finds itself unable to accommodate expressions of nationalism in Czechoslovakia or Hungary. In short, the modern Soviet Union is a complex of contradictions.

Some of those contradictions can be identified in the country's regional geography. Earlier in this chapter we examined its physical geography, but we turn now to the human regions that give the Soviet Union its varied character. Geographers who study the properties of states call the heartland of a state its *core* or core area. In the core area, the state is likely to display its national iconographies most strongly; here lie its largest cities (normally including the capital), its most productive farmlands, its greatest population cluster. In the core area, surface communications are more efficient, networks denser than anywhere else. States that have grown from ancient core areas have their lengthiest history there, and cultural imprints remain strongest.

1. The Soviet Core

The Soviet Union's core area centers on Moscow and includes the major cities of the west. Moscow was heir to Kiev as the early center of Russian consolidation and expansion. Although located in the southern forests and in rather meager country, Moscow had centrality—a situation that proved very favorable during the decades of conquest and growth. On the map it is the river system west of the Urals that really emphasizes Moscow's favored position, with the Volga and its Oka tributary draining toward the southeast, the Dvina north to the Arctic, the West Dvina west to the Baltic, and Dnieper south to the Black Sea. These were the routes of expansion; in the lands between the rivers the tsars' control was secured through land grants to loyal citizens. Southward, the frontier of expansion penetrated the better soils of the black earth belt, and a Russian breadbasket developed there. *Ostrogs* on the Volga developed into towns, and Moscow began to feel some urban competition: as early as the seventeenth century these Volga sites were manufacturing products from locally grown flax and hemp, from wood, and from the furs trapped in the forest margins.

Moscow, then, lies at the heart of what is commonly called the *Central Industrial Region.* The definition of this region varies, as all regional definitions are subject to debate. Some geographers prefer to use the name *Moscow Region,* thereby emphasizing the point that for over 400 kilometers (250

Building projects such as the Kalinin apartments and shops have transformed Moscow's townscape and have lessened the Soviet Union's urban housing shortage. Cities in Russia are crowded and waiting lists for housing are long; personal living space is limited. (George Holton/Ocelot)

miles) in all directions from the capital, regional orientations are toward the historic focus of the state. One advantage in this last name is that it avoids the suggestion that this region is the only industrial center of the Soviet Union, or that it leads in all sectors of the industrial economy. The fact is that the Moscow region has several competitors; the focus of heavy industry, for example, has long been to the south, in the Ukraine, and certainly Leningrad, to the north, has been an important center—there was a time when it seemed destined to take over the lead permanently from Moscow itself. Thus other geographers prefer a wider definition of the Central Industrial Region, to include not only the Moscow–Gorky industrial coreland but also most of southern Russia to the Ukrainian border, including such cities as Kursk, Voronezh, and Borisoglebsk.

The Moscow–Gorky area is the leading urban-industrial cluster within the Central Industrial Region (Fig. 2-8). Moscow (8.6 million, 13 million in the whole conurbation) still remains exceptionally large and exceptionally expressive of national culture and ideology, traits that helped Moscow maintain its strength in the Russian state during the rise of St. Petersburg. With the present distribution of population, Moscow has great centrality; roads and railroads radiate in all directions to the Ukraine in the south, to Minsk, Belorussia, and Europe in the west, to Leningrad and the Baltic coast in the northwest, to Gorky and the Ural area in the east, and to the cities and waterways of the Volga in the southeast; a canal links the city to this, the Soviet Union's most important navigable river. Moscow is the focus of an area that includes some 50 million inhabitants (about one-fifth of the country's total population), many of them concentrated in such major cities as Gorky (1.4 million), the automobile-producing "Soviet Detroit"; Yaroslavl (0.6 million), the tire-producing center; Ivanovo, the "Soviet Manchester" with its textile industries; and, to the south of Moscow, the mining and metallurgical center of Tula, where lignite deposits are worked.

Leningrad, as the Soviets renamed Petrograd in 1924 after Lenin's death, remains the Soviet Union's second city, with 4.7 million people. Leningrad has none of Moscow's locational advantages, at least not with reference to the domestic market; it lies well outside the Central Industrial Region and, in effect, in a northwestern corner of the country, 650 kilometers (400 miles) from Moscow. Neither is it better off than Moscow in terms of resources. Fuels, metals, and foodstuffs all must be brought in, mostly from far away; the Soviet emphasis on self-sufficiency has even reduced Leningrad's asset of coastal location—raw materials might be imported much more cheaply across the Baltic than across Middle Asia. Only bauxite deposits lie nearby, at Tikhvin. But Leningrad was at the vanguard of the industrial revolution in Russia, and its specialization and skills have remained. In the late 1970s the city and its immediate environs contributed about 5 percent of the country's manufacturing, much of it through high-quality machine building. In addition to the usual association of industries (metals, chemicals, textiles, and food processing), Leningrad has major shipbuilding plants and, of course, its port and the nearby naval station of Kronstadt. The aggregate, although not enough to maintain its temporary advantage over Moscow, has kept the city in the forefront of modern Soviet development.

2. Povolzhye: The Volga Region

Povolzhye is the Russian name for an area that extends along the middle and lower Volga River. It would be appropriate to call this the Volga region, for the great river is its lifeline, and most of the cities that lie in the Povolzhye are situated on its banks. In the early 1950s a canal was completed to link the lower Volga with the lower Don River, extending this region's waterway system still further.

The Volga River was an important historic route in old Russia, but for a long time neighboring regions overshadowed it. The Moscow area and the Ukraine were ahead in industry and agriculture; the industrial progress that came late in the nineteenth century to the Moscow area left the *Povolzhye*

Figure 2-8

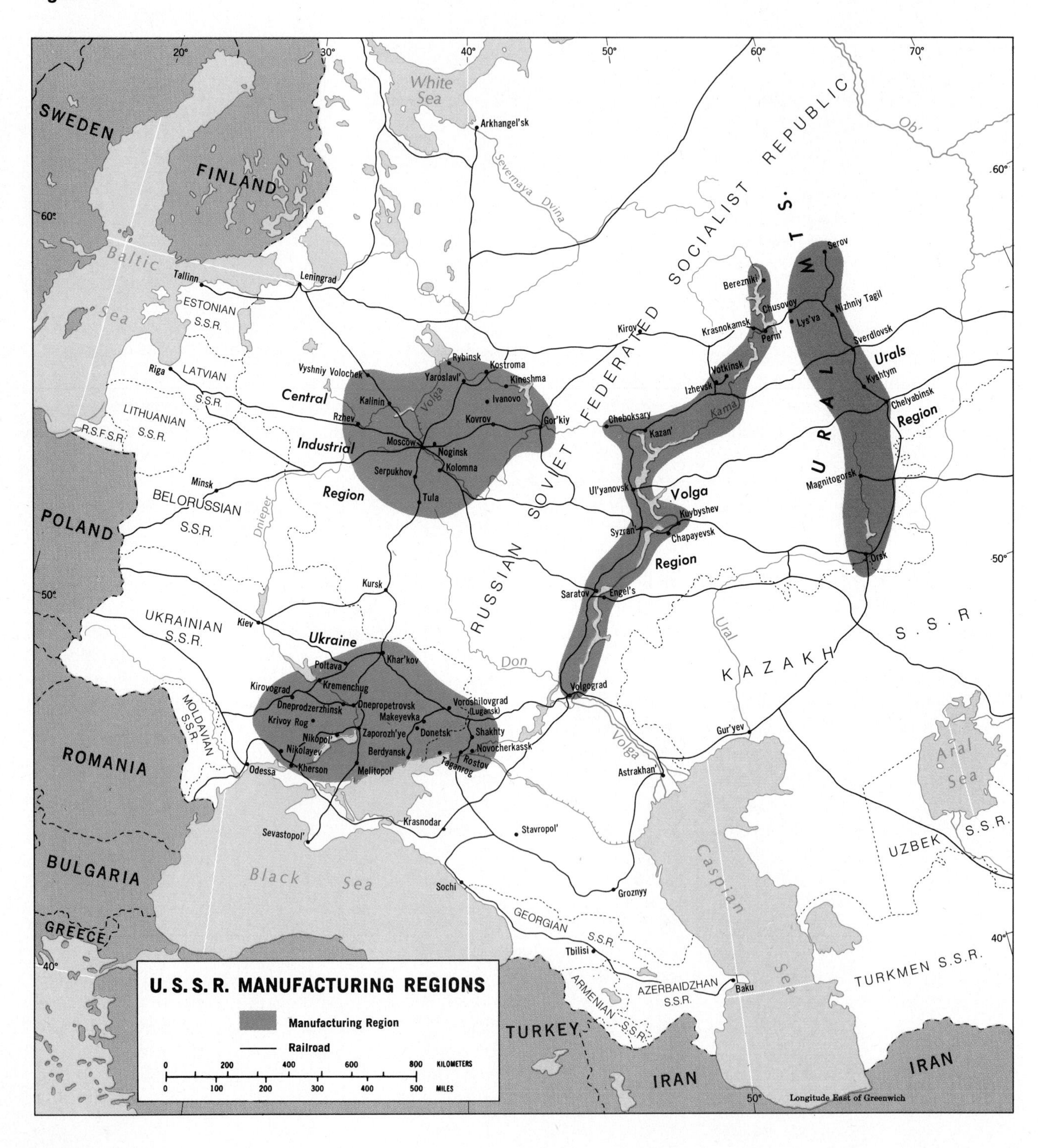

little affected. Its major function remained the transit of foodstuffs and raw materials to and from other regions.

This transport function is still important, but things have changed in the *Povolzhye.* First, World War II brought a time of furious development because the Volga region, located east of the Ukraine, was protected by distance from the invading German armies. Second, the Volga region proved to be the greatest source of petroleum and natural gas in the entire Soviet Union. From near Volgograd in the southwest to Perm on the Urals' flank in the northeast, an enormous oil field exists. Once, this was believed to be the Soviet Union's largest reserve, but recent discoveries of oil and gas in Western Siberia indicate that even more extensive fields lie east of the Urals. Still, because of their location and size, the Volga region's reserves have much importance. And third, the transport system has been greatly expanded. The Volga-Don Canal links the Volga waterway to the Black Sea; the Moscow Canal extends the northern navigability of the system into the very heart of the Central Industrial Region, and via the Mariinsk Canals the Baltic Sea can be reached. Today, the Volga region's population approaches 18 million, and the cities of Kuybyshev, Volgograd, Kazan, and Saratov are in the 1-million range.

3. The Ukraine

Soviet strength has been built on a mineral wealth of great variety and volume. With its immense area, the country possesses an almost limitless range of raw materials for industry, although some deposits are located in remote areas and require heavy investments in transportation. In the Ukraine the Soviet Union has one of those regions where major deposits of industrial resources lie in relatively close proximity. The Ukraine began to emerge as a major region of heavy industry toward the end of the nineteenth century, and one major reason for this was the Donets Basin, one of the world's greatest coalfields. This area, known as *Donbas* for short, lies north of the city of Rostov; in the early decades this Donbas field produced over 90 percent of all the coal mined in the country. Most of the *Donbas* coal is high grade. Today, the Donets Basin alone still accounts for between one-quarter and one-third of the total Soviet output, which is still about double that of the second-ranking producing area.

What makes the Ukraine unique in the Soviet Union is the location, less than 320 kilometers (200 miles) from all this Donets coal, of the Krivoy Rog iron ores. Again, the quality of the deposits is high, although the better ores are being worked out, and the industry is now turning to the concentration of poorer-grade deposits. Major metallurgical industries arose on both the Donets coal and the Krivoy Rog iron: Donetsk and its satellite Makeyevka dominate the Donets group (these constitute what might be called a Soviet Pittsburgh), while Dnepopetrovsk is the chief center of the Krivoy Rog cluster (Krivoy Rog remains the largest iron and steel producer). One way or another all the major cities located nearby have benefited from the fortuitous juxtaposition of minerals in the southern Ukraine: Rostov, Volgograd, and Kharkov near the cluster of Donbas cities and Odessa, and even Kiev not far from the Krivoy Rog agglomeration. And like the Ruhr, the Ukraine industrial region lies in an area of dense population

Winter wheat is harvested mechanically on the fields of the Iskra Collective Farm. The U.S.S.R. continues to depend on external sources to supplement its inadequate domestic harvests. (E. Shulepov/Sovfoto)

(and hence available labor), good agricultural productivity, and adequate transportation systems; it lies near large markets as well. In addition, it has provided alternatives when exhaustion of the better ores began to threaten; not only are large lower-grade deposits capable of sustaining production in the foreseeable future, but iron ores also exist near Kerch, on the eastern point of the Crimean Peninsula, quite close enough for use in the established plants of the Ukraine. The biggest development in recent decades is the opening up of the Kursk Magnetic Anomaly south of Moscow; this area may technically lie outside the Ukraine, but its ores are of critical value in keeping the heavy industrial complexes of the Ukraine going.

And this is not all. The Ukraine and areas immediately adjacent to it have several other essential ingredients for heavy industry. Chief among these is manganese, which is needed in the manufacture of steel. About 12 pounds of manganese ore, on the average, go into the making of a ton of steel. The Soviet Union has ample supplies of this vital commodity. Deposits just south of Krivoy Rog at Nikopol and between the greater and lesser ranges of the Caucasus at Chiatura are the two leading deposits in the world.

Still there is more. To the southeast, along the margins of the Caucasus in the Russian Soviet Federated Socialist Republic (S.F.S.R.) and on the shores of the Caspian Sea in the Azerbaydzhan S.S.R., there are significant oil deposits—certainly not far from the Ukrainian industrial hubs by Soviet standards of distance. A pipeline connection runs from the Caspian coast along the Caucasus foothills to Rostov and Donetsk. The Western Siberia and Volga-Urals fields have overtaken these southern oil fields, which center on the old city of Baku, but production continues here.

The Ukraine provided the Soviets with their opportunity to gain rapidly in strength and power, and when World War II broke out, this was the center of heavy industry in the country. But fortuitous as its concentration of raw materials and manufacturing was, it also contributed to Soviet vulnerability. The industries of the Ukraine and the oil fields of the Caucasian area were prime objectives of Germany's invading armies. As the Soviets were forced to withdraw, they dismantled and even destroyed more than a thousand manufacturing plants so that they might not fall into the hands of the aggressors. And thus began a series of developments that has pulled the Soviet center of gravity steadily eastward; Soviet planners were impressed with the need for greater regional dispersal of industrial production, not just because the Ukrainian and other western mines might eventually be worked out, but also for strategic reasons.

The Ukraine (loosely translated, the name means "frontier"), for all the Soviets' encouragement of the eastward march of population and economic development, remains one of the cornerstones of the country. With nearly 60 million people, it has between one-quarter and one-fifth of the entire Soviet population (though less than 3 percent of the land), and its industrial and agricultural production is enormous. With all that spectacular industrial growth, it is easy to forget that part of the Ukraine was pioneer country because of its agricultural possibilities, not its known minerals. Even today half of the people live on the land rather than in those industrial cities; in the central and western parts of the republic, rural densities may be as high as 155 per square kilometer (400 per square mile). Wheat and sugar beets cover the landscape where the soils are best—in the heart of the Ukraine—and where the moisture is optimum.

4. The Urals

The Ural Mountains, seen by some as the eastern boundary of Europe, form a north–south break in the vast Russian–Siberian plainland. In the north the Urals are rather narrow, but to the south the range broadens considerably; nowhere are the Urals particularly high, nor do they form any real obstacle to east–west transportation. Roads, railroads, and pipelines cross the range in many places. The range of metals located in and near the Urals is enormous.

This area, more or less on a par with the Ukraine in terms of total production of manufactures, rose to prominence during World War II and in its aftermath. The whole Urals industrial region is so well developed, so specialized, and the product of such heavy investment that it is likely that Soviet planners will see to it that the flow of raw materials into the Urals is not interrupted as the indigenous materials are utilized.

It would perhaps be reasonable to speak of a Urals–Volga industrial region, for the problem of energy supply in the coal-poor Urals area has been considerably relieved by the oil fields found to lie west of the Urals; this fuel is pipelined into various parts of the Soviet core area, to Eastern Europe, and even as far into Soviet Asia as Irkutsk near Lake Baykal. Natural gas, too, is piped cheaply from distant sources to the Urals, for instance, from Bukhara in the Uzbek S.S.R. Thus, the Urals industrial area, the Volga area (centered on Kuybyshev), and the Moscow–Gorky area are growing toward each other, and the core region of the Soviet Union is being consolidated. But eastward, too, there is growth; the core region by any measurement has long ago crossed the Urals and is expanding into the Asian interior.

5. The Karaganda Area

More than 1000 kilometers (600 miles) east-southeast of Magnitogorsk lies one of those "islands" of mining and industrial development that form part of the ribbonlike eastward extension of the Soviet heartland. This is the Karaganda–Tselinograd area, positioned in the northeastern part of the Khazakh S.S.R. Although it is much smaller than the Ukraine and the Urals, this is no longer merely a mining outpost, as it was before World War II when it sent coal to the Urals. Today it possesses a variety of industries, including iron and steel, chemicals, and related manufactures, and Karaganda, no longer a frontier town, has a population approaching three-quarters of a million. Rather than a supplier of raw materials, the Karaganda area now exchanges commodities with other productive areas; in exchange for the coal it continues to send to the Urals, it receives iron ores from the Urals' own source area at Kustanay.

Among Karaganda's advantages is its position in the path of the Soviet economy's eastward march. There are rail connections and cost-reducing exchanges with the Urals, but opportunities for such arrangements also exist to the east. In addition, further mineral discoveries have been made in the vicinity, including a sizable deposit of medium-grade iron ore. Less than 320 kilometers (200 miles) to the northwest and on the railroad to the Urals, Tselinograd has emerged as the urban focus for the great Virgin Lands agricultural project, launched in the 1950s. And the problem of water supply, a serious one here in the dry Kazakh steppe, has been relieved by the construction of a 500-kilometer canal from the Irtysh River. Nevertheless, Karaganda and its environs cannot compare to the Ukraine or the Urals, with their clusters of resources, urban centers, transport networks, agricultural productivity, and dense populations. Karaganda remains far from the present-day hub of the Soviet Union; it is still a marginally located, subsidiary developing area. Distances are great, markets are far, costs of production are high. But the future is likely to see the ever-greater integration of the Karaganda area into the spreading Soviet core. Soviet planners, by their attempts to bring vast eastern agricultural areas into production and by their willingness to have the state bear the costs involved in the industrial growth of the Karaganda region (again in large measure for strategic reasons born of World War II), have shown their hand to favor such developments.

6. The Kuznetsk Basin (Kuzbas)

Fully 2000 kilometers (1200 miles) east of the Urals lies the third-ranking area of heavy manufacturing in the Soviet Union—the Kuznetsk Basin, or *Kuzbas.* In the 1930s this area was opened up as a supplier of raw materials, especially coal, to the Urals, but this function has steadily diminished in importance while local industrialization has accelerated. The original plan was to move coal from the Kuzbas to the Urals and to let the returning trains carry iron ore to the coalfields, but subsequently good iron ores were located in the area of the Kuznetsk Basin itself. As the resource-based Kuzbas industries grew, so did the urban centers: Novosibirsk (with over a million inhabitants) stands on the Siberian Railroad as the symbol of Soviet enterprise in the vast eastern interior. To the northeast lies Tomsk, one of the oldest Russian Siberian towns in the whole region, founded three centuries before the Bolshevik takeover and caught up in the modern development of the Kuznetsk area. Southeast of Novosibirsk lies Novokuznetsk, a city of a half million people specializing in the manufacture of heavy engineering products such as rolling stock for the railroads; aluminum products (of Urals bauxite) are also made here.

Impressive as the proximity of coal, iron, and other resources may be in the Kuznetsk Basin, the industrial and urban development that has taken place here must in large measurement be attributed, once again, to the ability of the state and its planners to promote this kind of expansion, notwithstanding what capitalists would see as excessive investments. In return they were able to push the country vigorously ahead on the road toward industrialization, with the hope that certain areas would successfully reach "takeoff" levels, after which they would require less and less direct investments and would to an ever-greater extent be self-perpetuating. The Kuzbas, for example, was expected to grow into one of the Soviet Union's major industrial regions, with its own important market, and with a location, if not favorable to the Urals and points west, then at least fortuitous with reference to the developing markets of the Soviet Far East.

Khabarovsk, on the Amur River in the Soviet Union's Pacific Area, is rapidly developing into this region's leading urban center. Large investments have been made to improve the housing situation, and Khabarovsk has well over a half million residents. (Yuri Muravin/Sovfoto)

Between the Kuznetsk Basin and Lake Baykal, east of the Kuzbas, lies one of those areas that has not yet reached the takeoff point. This area has impressive resources, including coal, but Soviet planners have been uncertain exactly where to make the necessary investments to make this a significant industrial district. In addition to hydroelectric power plants already operating at Bratsk and elsewhere along the Angara River, there are further possibilities of hydroelectric power from the Yenisey River; timber is in plentiful supply; brown coal is used for electricity production; and oil is piped all the way from the Volga–Urals fields to be refined at Irkutsk. Thus, Krasnoyarsk and Irkutsk, both cities in the half-million class, lie several hundred miles farther again from the established core of the country, and while the distant future may look bright (especially if Soviet plans for the Pacific areas become reality), the present situation still presents difficulties.

7. The Pacific Area

The Soviet Union has about 8000 kilometers (5000 miles) of Pacific coastline—more than the United States including Alaska. But most of this coastline lies north of the latitude of the state of Washington, and, from the point of view of coldness, the Soviet coasts lie on the "wrong" side of the Pacific. The port of Vladivostok (500,000), the endpoint of the Trans-Siberian Railway, lies at a latitude about midway between San Francisco and Seattle, but must be kept open throughout the winter by icebreakers—something unheard of in these American ports. The climate, to say the least, is severe. Winters are long and bitterly cold, and summers are cool. Nevertheless, the Soviets are determined to develop their Far Eastern Region as intensively as possible, and their resolve has recently been spurred by the growing ideological conflict between the Soviet Union and China. Farther to the west, high mountains and empty deserts mark the Chinese–Soviet boundary; south of the Kuznetsk Basin the Lake Baykal, the Republic of Mongolia functions as a sort of buffer between Chinese and Soviet interests. But in the Pacific area, the Soviet Far Eastern Region and Chinese Manchuria confront each other across an uneasy river boundary (along the Amur and its tributary, the Ussuri). Hence, the Soviets want to consolidate their distant outpost and are offering inducements to western residents to "go east."

Although it has a rigorous climate and difficult terrain, the Far Eastern Region is not totally without assets. While most of it still consists of very sparsely populated wilderness, the endless expanses of forest (see Fig. I-7) have slowly begun to be exploited by a timber industry whose major problems are the small size of the local market, the bulkiness of the product, and—as always in Soviet Asia—the enormous distances to major markets. Lumber stations and fishing villages break the emptiness of the countryside; fish is still the region's major product, and from Kamchatka to Vladivostok Soviet fishing fleets sail the Pacific in search of salmon, herring, cod, and mackerel, to be frozen, canned, and railed to the western markets.

In terms of minerals, too, the Far Eastern Region has possibilities. A deposit of coking-quality coal lies in the Bureya River valley (a northern tributary of the Amur), and a lignite field near Vladivostok is being exploited; on Sakhalin there are minor quantities of bituminous coal and oil. Not far from Komsomolsk (250,000) lies a body of poor-grade iron ore, and this city has become the first steel producer in the region. In addition to lead and zinc, the tin deposits north of Komsomolsk are important because they constitute the country's major source.

The Soviet invasion of Afghanistan produced a wider and costlier conflict than Soviet strategists seem to have predicted. Rebels in this village called for government help; when the Soviet helicopter arrived they ambushed the crew and forced the navigator to report engine trouble. A second helicopter arriving to assist suffered a similar fate. Afghan as well as Soviet losses have been heavy. (Alain DeJean/Sygma)

Thus, the major axis of industrial and urban development in the Pacific area has emerged along the Ussuri–lower Amur River system, from Vladivostok in the south, with its naval installations, shipbuilding, and fish processing plants, to Komsomolsk, the iron and steel center, in the north. Near the confluence of the Amur and Ussuri rivers lies the city with the greatest advantages of centrality, Khabarovsk. Here metal manufacturing plants process the iron and steel of Komsomolsk, chemical industries benefit from Sakhalin oil, and timber is used in sizable furniture industries. Khabarovsk, a city of somewhat under a half million people, has become the region's leading urban center. Railroads lead south toward Vladivostok, north to Komsomolsk and beyond, and west to the Bureya coalfields, to Svobodny, and eventually to the Kuzbas and on to Moscow.

The farther eastward the Soviet Union's developing areas lie, the greater is their isolation and the larger their need for self-sufficiency, in view of the ever-increasing cost of transportation. Thus, the Pacific area exchanges far fewer commodities with points westward than does the Kuzbas, although some raw materials from the Pacific area do travel to the west in return for relatively small tonnages of Kuzbas resources. In terms of food supply, the Pacific area cannot come close to feeding itself, despite its rather small population (the 1980 estimate is 6.4 million) and the farmlands of the Ussuri and lower Amur River valleys. But Soviet planners have not been sure just how to direct industrialization here. Large investments are needed in energy supply (and there are magnificent opportunities for hydroelectric development in the Amur and tributary valleys), but the choice is difficult: to spend here—in an area of long-term isolation, modest resources, labor shortage, and harsh environment—or to spend money where conditions are less bleak. Future relationships with China may have much to do with the pace of Soviet development in the Far East.

Soviet Heartlands

In the late nineteenth century there was little indication that Russia would within a half century emerge as a world power. Still an agrarian country, remote from the major sources of industrial innovation, vast but poorly organized, subject to unrest and resistant to change, Russia seemed to most scholars to have little prospect for rapid development. When Japan's forces defeated the Russians (1904) and the country plunged into more than a decade of chaos, its fundamental weakness appeared to be confirmed.

But some remarkable scholarship was done by those nineteenth-century geographers we identified in Chapter 1 (Von Thünen, Ratzel, and their colleagues), and one of them, Sir Halford Mackinder (1861–1947), did foresee the rise of Russian power. Mackinder studied Britain's imperial expansion along the world's maritime routes as well as Russia's overland growth. He noted that areas subject to penetration along river routes were vulnerable to a strong maritime power, for much of Europe's colonial acquisition was first accomplished along the corridors provided by rivers. Thus, he was led to speculate on the comparative strengths of land and sea power. Scanning the world map, Mackinder concluded that a large area in Eurasia, not penetrated by major navigable rivers, would be safe as a fortress and thus able to develop strength in its secure isolation. This area, he argued in an article published in 1904, was located in western Russia, from the Moscow Basin and across the Urals into Western Siberia (Fig. 2-9). Mackinder predicted that this region (called the *pivot area*) would be of crucial significance in the political geography of the twentieth century; he asserted that any power based in the pivot area could gain such strength as to dominate the world eventually.

Mackinder's article aroused a heated debate, and he was an enthusiastic participant, convinced to the end that he had been correct. He proved willing to adjust his concept of the pivot area when it was pointed out that much of adjacent Eastern Europe had similar qualities but greater known resources. Long before Russians became masters of Eastern Europe, Mackinder proposed his heartland theory:

> Who rules East Europe commands the Heartland
>
> Who rules the Heartland commands the World Island
>
> Who rules the World Island commands the World

EURASIAN HEARTLANDS

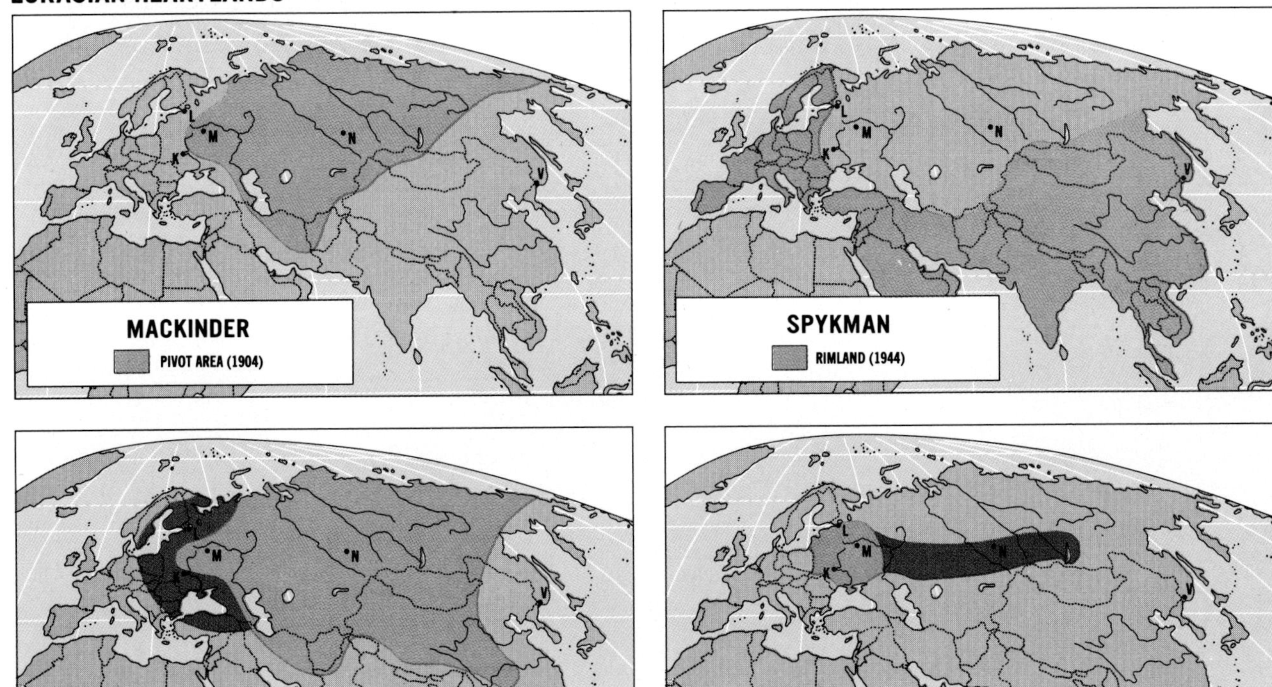

Figure 2-9

In Mackinder's hypothesis, the *heartland* was the new name for the pivot area with its Eastern European appendage, and *world island* meant the entire land cluster of Eurasia and Africa. Soon other geographers sought to evaluate the resource content and strategic assets of the Soviet–East European heartland, and there were those who concluded that Mackinder had overestimated the capacities of the region. N. J. Spykman, concluding that the countries on the outer edges of Eurasia contained greater power potential than the heartland, responded to Mackinder in a parody of his "theory":

Who controls the Rimland rules Eurasia

Who rules Eurasia controls the destinies of the world

But the rimland, of course, always consisted of a large number of countries, and at no time were all these countries under a single authority. The heartland, on the other hand, fell to the Soviet Union when, after World War II, Eastern Europe came under Moscow's domination. When Mackinder wrote his famous essay, Germany was dominant in Central Europe and parts of Eastern Europe, and Russia hardly seemed likely to challenge German power there. Mackinder could not foresee the political consolidation of the heartland—but the fact is that the heartland today sustains one of the two most powerful states in the world.

The Soviet Union has made major investments in its drive to develop its Siberian and Far Eastern regions. Among the reasons for this eastward push is the awareness that Soviet productive capacity and strategic strength are heavily concentrated in a core area that, by virtue of its clustering, is vulnerable to attack by modern weapons of massive destruction. In addition, Soviet planners remember the shattering impact of the German invasion during World War II, when cities, factories, and farms were laid waste in Western Russia, but the east lay protected by distance. Hence the urge to develop the eastern interior is not only a matter of economic policy, it is also a question of strategic advantage. As D. M. Hooson has

Mongolia: Frontier to Buffer

Although unclaimed frontier areas have virtually disappeared from the world map, territories still exist whose origins lie in the period of great-power competition for land. Landlocked Afghanistan, Nepal, Sikkim, and Bhutan have such a past, interposed between the British colonial sphere in South Asia on one hand and Russian and Chinese interests on the other. Mongolia is also a *buffer* state, as such countries are appropriately called, but between Russian and Chinese spheres.

Mongolia, a vast country of more than 1.5 million square kilometers but with barely over 1.5 million people, was under Chinese control from the end of the seventeenth century until 1911, when it declared independence amid the chaos of the Chinese revolution against the Manchu rulers. Bypassed by the major thrust of Russian imperialism, Mongolia nevertheless was a pawn in the Russian–Chinese settlement and treaty of 1915, when it became virtually a Chinese colony. Chinese attempts to confirm their hegemony were aborted by the 1917 Russian Revolution, and fighting between Red and White Russian armies spilled over into Mongolia. Mongolians saw the Bolshevik victory as a chance for help against the covetous Chinese, and in 1921 Mongolia became a People's Republic, although outside the Soviet Union.

Today Mongolia, with its vast deserts, grassy plains, forest-clad mountains, and fish-filled lakes, lies uneasily between the Soviet Union and China, neither of whom can permit its incorporation into the other. Notwithstanding its historic associations, ethnic affinities, and cultural involvement with China, Mongolia's development in the past 60 years has been guided by the Soviets. Today, Soviet army divisions are based in Mongolia, and the country is a member of Soviet-dominated COMECON. The capital, Ulan Bator (Ulaanbaatar), with the largest population concentration, lies near the northern Soviet boundary; China lies far from here, across the nearly endless Gobi Desert and Altai Mountains. The Soviet presence has transformed Mongolia in many ways (apartment buildings now mark the urban scene in the capital; a railroad traverses the country; trucks often replace camels on roads), but the cultural landscape nevertheless remains East Asian. Mongolia survives as a true buffer state by the acquiescence of powerful adversaries. Will it escape Afghanistan's fate?

pointed out in a further contribution to the "heartland" discussion, this process is creating a new Central Siberian Soviet heartland, an eastward extension of the original Soviet core.

Eastern Boundaries

A stronger confirmation of the Soviet presence in its Far Eastern Region is important to Moscow for other reasons as well. The Soviet Union has boundaries with as many as twelve countries, from Norway and Finland in the west to Mongolia, China, and North Korea in the east. The European boundaries have undergone several shifts (Poland and Romania have lost territory to the Soviet Union in the aftermath of World War II), but the central Asian boundaries lie in uncontested,

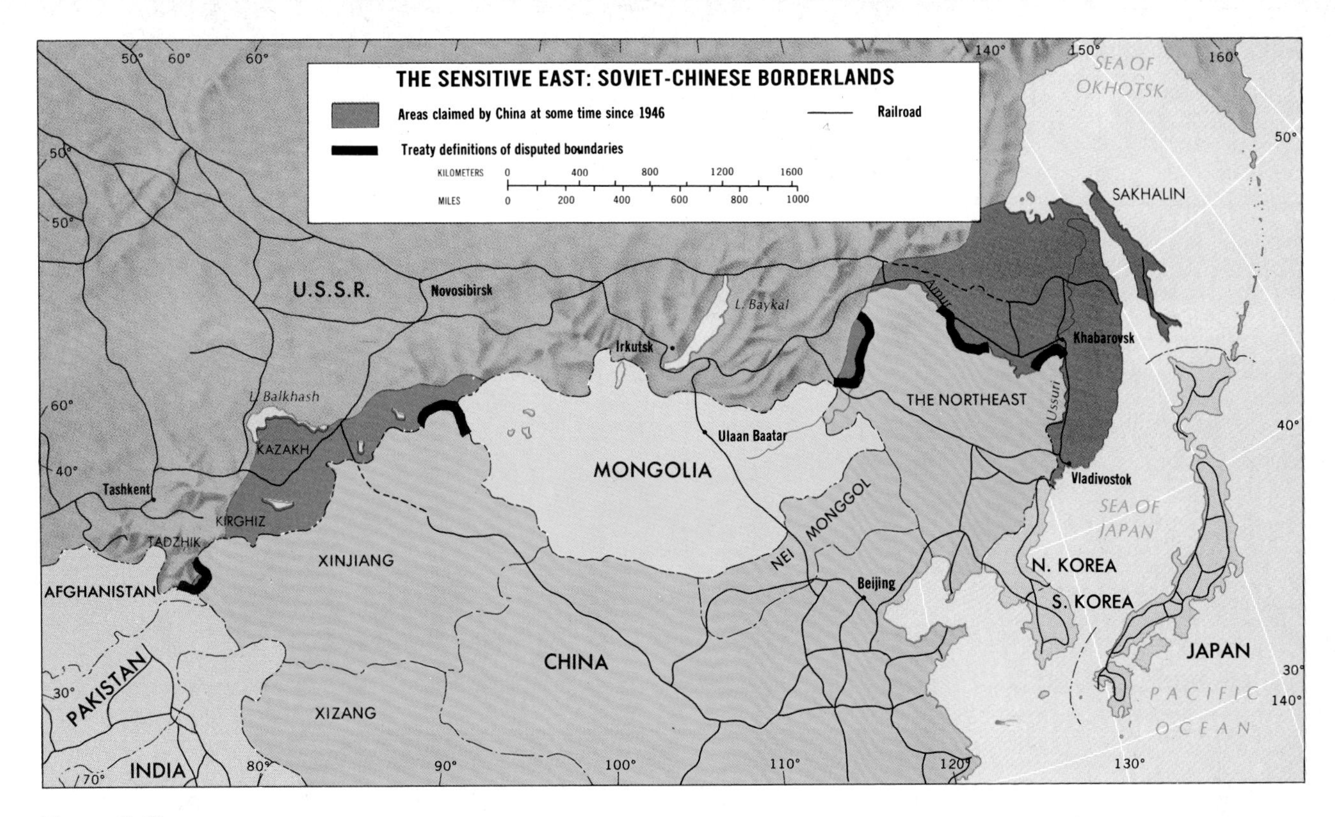

Figure 2-10

remote, high mountains. In the Far Eastern Region, however, the Soviet Union faces China across a boundary that may yet be contested (Fig. 2-10).

While states were still expanding into unclaimed areas, as Russia did during its era of territorial growth under the tsars, boundaries often were temporary trespass lines, obliterated with the next forward push. In those times the core areas of national states were separated as much by *frontiers* as by boundaries. A frontier was a zone of separation, perhaps one of contention; it might be an area of dense forest, high mountains, a swamp, a desert zone. Or a frontier might exist simply by virtue of distance. But eventually the effects of Europe's technological and political revolutions, the colonial scramble, explosive population growth, and the rush for resources and empires combined to produce competition for even the most remote frontier areas. Before long the last frontiers disappeared, divided and parceled out among states. Boundaries replaced them—boundaries that were sometimes established without much planning or concern for the peoples whose lives they would affect. With the creation of certain boundaries in Asia, Africa, and America were sown the seeds of conflict.

So it was in East Asia. To the Chinese and Japanese, the Russians had as much business on Pacific shores as the British had in East Africa, but for a long time the Russians had the power, China was weak, and Japan was yet to become a major force in Asian affairs. Russia was able to acquire huge areas of East Asia and to secure its territorial position through treaties imposed on the Chinese between 1858 and 1864. But the terms of those treaties are under dispute, and at various times China has announced its intentions to reclaim lands lost during the Russian imperial expansion (even though Chinese control in most of those areas was marginal and the Chinese presence minimal.)

Although the boundaries between the Soviet Union and China extend over some 6000 kilometers (nearly 4000 miles), problems have so far arisen in two areas (Fig. 2-10). One of these lies between Chinese Sinkiang and the Soviet Kazakh, Kirghiz, and Tadzhik Republics. At various times people under Soviet rule have sought refuge in China (for example, during Stalin's excesses); at other times Chinese minorities living in Sinkiang have crossed the border to seek safety from Peking's harsh rule. Inevitably, the border region became a zone of stress and frequent violence, worsened by the growing ideological split between the two countries. And in recent years serious conflicts have involved the lengthy Sino–Soviet boundary along the Amur and Ussuri Rivers in the Far East. In 1969 there was fighting between Soviet and Chinese troops at Chen Pao Island (Damansky Island) in the Ussuri River, and a dispute over navigation rights on the border rivers had to be settled anew. Then the Central Asian boundary conflict flared up again, with major armed forces on both sides and a substantial number of casualties. The Soviet Union may eventually find itself in difficulties quite familiar to colonial powers trying to hold remote, vulnerable territories.

Population Movements

In discussing Australia, we took note of the migration process, specifically in the context of voluntary and forced migration. The migration streams that carried people from Europe to Australia (and to the Americas and Africa) can be classified as *intercontinental* migrations. But it is also possible to identify *regional* migrations, involving migration within a culture realm but across political boundaries (for example, the movement of people from Mediterranean European countries into the states of the European core area in search of work), and *internal* migrations, when a substantial population shift occurs within a country.

The United States of America has witnessed large-scale internal migration flows, notably from the country's eastern heartland toward the West (see Chapter 3). Another internal population movement of large dimensions involves the Soviet eastward drive (Fig. A-1, page 127, the route numbered 8). As Fig. I-9 (page 27) reveals, this movement has proceeded along a fairly narrow, well-defined corridor to the Soviet Far East, but it has not even begun to bring to the eastern side of the Urals a population comparable to that of the western Soviet Union. For so vast a country, the Soviet population remains quite small.

Little more than a century ago, Russia had twice as many inhabitants as the United States. Even as late as 1900 the Russian population was about 125 million, while that of the United States was not much more than half this figure. Yet today the Soviet population is not much larger than that of the United States (nearly 270 against 220 million). What suppressed Soviet population growth?

The unhappy answer is that the twentieth century brought repeated disaster to the Soviet people, notwithstanding the fact that their country rose to a position of world power during that period. By comparison, the population of the United States, though suffering casualties in war as well, sustained far fewer losses. Part of the Soviet Union's losses were self-inflicted; for example, during the collectivization period of the early 1930s, famines and killings probably cost almost 5 million lives. But the major destruction, staggering in its magnitude, was caused by the two world wars. World War I, its aftermath of civil war, and the attendant famines resulted in some 17 million deaths and about 8 million deficit births (that is, births that would have taken place had the population not been reduced in such numbers). World War II cost the Soviet Union approximately 27 million deaths and 13 million deficit births. Thus, during the twentieth century, the Soviet population lost over 70 million in destroyed lives and unborn children, while United States losses in all its twentieth-century wars stand well under 1 million. In addition, the Soviet Union was marked by fairly heavy emigration, while the United States received millions of immigrants. Hence, the United States population grew rapidly while the Soviet increase was severely limited: in the half century from the beginning of World War I to 1964 the United States gained some 90 million people (from a base of 100 million) while the Soviet Union gained about 65 million. Thus, United States population nearly doubled in that period, while the Soviet population in a half century grew by only one-third.

Naturally the results of these recent calamities have concerned Soviet governments and planners. The deficit of births is today reflected in a shortage of younger people in the labor force; at higher ages females still outnumber males to a greater extent perhaps than in any other country in the world. For a period after World War II the government pursued policies whereby families were encouraged to have a large number of children, and mothers who had 10 children received the Order of Mother Heroine from the Presidium of the Supreme Soviet. This vigorous pursuit of expansionist population policies has now been slackened, and Soviet planners have other objectives: to accommodate the strong movement from rural areas to the cities, and to lure labor to population centers in the more remote areas of the country. Still, concern over low birth rates has emerged again in recent years, and incentives for larger families may be reinstituted.

Just 50 years ago, less than one-fifth of the Soviet population was urbanized, a reflection of the country's state of development shortly after the takeover of the communist administration. In 1940 it reached one-third of the population, and by 1960, just under half of the population (then 210 million) resided in cities and towns. Today about three-fifths live in urban areas, a figure that is still low by Western European standards. The continuing movement to the cities has produced housing problems and serious overcrowding in the many miles of medium-sized apartment buildings that line the streets of Soviet cities from Leningrad to Vladivostok. The desire to open the more remote areas of the country for development has led Soviet governments to offer incentives to families willing to move to places such as Komsomolsk and Khabarovsk. These incentives include better living space and higher wages.

Although Soviet planners ever since the 1920s have encouraged (and forced) the eastward movement of the population, to strengthen the transasian ribbon that now represents the Soviet presence east of the Urals, no government policy could have had the impact of World War II. The war ravaged western Russia, Belorussia, and the Ukraine, but it left the east comparatively unscathed. There, in the Povolzhye and the Urals and eastward, the Soviet Union had the means to survive the German attack—and the resources to strike back. Hundreds of thousands of citizens fled eastward before the invading German armies and many remained there after the war. Before World War II, the population of the Urals and regions farther east constituted just over 25 percent of the country's total, but today it is about 40 percent. Still, the eastward push is slower than Soviet planners would like, available apartments and wage differentials notwithstanding. If the eastern expansion of the Soviet Heartland is indeed slowing down again, it may be doing so in the face of distance and difficult environments—two old enemies of this giant country.

Additional Reading

Paul E. Lydolph's *Geography of the U.S.S.R.* was published in a new edition by Wiley, New York, in 1977. Another useful source is a volume of readings translated from the Russian and edited by G. J. Demko and R. J. Fuchs, *Geographical Perspectives in the Soviet Union,* published by the Ohio State University Press, Columbus, in 1974. Also see the new edition (1971) of J. P. Cole and F. C. German, *A Geography of the U.S.S.R.: Background to a Planned Economy,* a Rowman and Littlefield, Totowa publication. In the Aldine World's Landscapes Series is W. H. Parker's *The Soviet Union,* published in Chicago in 1969. Other books about the general geography of the Soviet Union include J. S. Gregory, *Russian Land, Soviet People: A Geographical Approach to the U.S.S.R.,* published in London by Harrap in 1968; R. E. H. Mellor, *Geography of the U.S.S.R.,* St. Martin's Press, New York, 1965; D. J. M. Hooson, *The Soviet Union: Peoples and Regions,* Wadsworth, Belmont, 1966; G. M. Howe, *The Soviet Union,* Macdonald and Evans, London, 1968; and J. C. Dewdney, *A Geography of the Soviet Union,* Pergamon, New York, 1965. A standard work, although old, is T. Shabad's *Geography of the U.S.S.R.: A Regional Survey* (Columbia University Press, New York, 1951). An enormously useful source is Shabad's journal *Soviet Geography: Review and Translation.* Among more specialized works, consult a volume edited by I. P. Gerasimov and colleagues, translated by J. I. Romanowski and further edited by W. A. D. Jackson, *Natural Resources of the Soviet Union: Their Use and Renewal,* published in San Francisco by Freeman in 1971. A good way to review Russian history is by looking at one of several historical atlases of the country, for example A. F. Chew's *Atlas of Russian History: Eleven Centuries of Changing Borders,* published by Yale University Press in 1967. The Russian adventure in North America is described by J. R. Gibson in *Imperial Russia in Frontier America,* published by Oxford University Press in New York in 1976. On the planned economy, see the translation by I. Nove of the book by P. J. Bernard *Planning in the Soviet Union,* published in 1966 by Pergamon Press in New York. On Soviet agriculture and the impact of Soviet practices in the East European satellites, J. F. Karcz edited a volume entitled *Soviet and East European Agriculture,* published in 1967 by the University of California Press in Los Angeles. Also see, on this subject, the book edited by R. D. Laird, *Soviet Agriculture: The Permanent Crisis,* a Praeger publication, New York, 1965. With reference to the physical environment, it is probably best to consult the relevant chapters in the regional volumes mentioned; one book not yet listed is N. T. Mirov's *Geography of Russia,* a 1951 Wiley publication that is strongly oriented toward the physical features of the country. Many details about permafrost, soils, vegetation, and so forth, appear in a book by S. P. Suslov, *Physical Geography of Asiatic Russia,* a translation from the Russian published by W. H. Freeman in San Francisco in 1961. On the population question, see a publication by the Milbank Memorial Fund, *Population Trends in Eastern Europe, the U.S.S.R. and Mainland China,* published in New York in 1960, and an older work by F. Lorimer, *The Population of the Soviet Union; History and Prospects,* published by Cambridge University Press in 1946. On population policies and problems see the work edited by E. Goldhagen, *Ethnic Minorities in the Soviet Union,* published by Praeger, New York, in 1968. Another revealing book in the Oxford University Press paperback by G. Wheeler, *Racial Problems in Soviet Muslim Asia,* published in London in 1962.

Descriptions of individual Soviet regions are also available, Columbia University Press in 1967 published *Central Asia: A Century of Russian Rule,* edited by E. Allworth. A volume by T. Armstrong, *Russian Settlement in the North,* was published by Cambridge University Press in 1965. The eastern realm of the Soviet Union is covered by E. Thiel in *The Soviet Far East: A Survey of Its Physical and Economic Geography,* published by Praeger, New York, in 1957. Also in this connection, see W. Kolarcz, *The Peoples of the Soviet Far East,* a Praeger publication of 1954. Urbanization in the Soviet Union is detailed by C. D. Harris in *Cities of the Soviet Union,* a volume in the Association of American Geographers Monograph Series and published in 1970.

Sir H. J. Mackinder's 1904 article entitled "The Geographical Pivot of History" can be found in *The Geographical Journal,* Vol. XXIII; but it has been reprinted numerous times and a more accessible source may be a textbook on political geography. The "heartland theory" still arouse debate, and a summary of the lengthy discussion and a reproduction of the article itself are in H. J de Blij, *Systematic Political Geography,* a Wiley production (second edition) of 1973. Also see Mackinder's *Democratic Ideals and Reality,* a 1919 publication by Holt, New York, but reprinted in paperback, and an article he published in *Foreign Affairs,* Vol. 21, July 1943, entitled "The Round World and the Winning of the Peace." N. J. Spykman postulates his "rimland" idea in *The Geography of the Peace,* Harcourt, New York, 1944.

Rocky Mountain backbone: North America's great divide. (Harm J. de Blij)

Three

North America

The North American realm consists of two countries that share numerous characteristics. In the United States as well as Canada, European cultural imprints dominate. English is the official language in the United States; French shares this status with English in Canada. The overwhelming majority of churchgoers adhere to Christian faiths. Most (but not all) of the people trace their ancestries to various European countries. In the arts, architecture, and other spheres of cultural expression, European norms prevail. North American society is highly urbanized, and nothing symbolizes the New World quite as strongly as a skyscrapered panorama of New York, Toronto, Chicago, or Vancouver. Like Europe—but even more so—North American society is mobile. Networks of superhighways, railroads, and air routes connect the far-flung cities and regions of North America with great efficiency. Commuters stream into and out of American cities by the millions each working day. A typical family relocates, on the average, once every five years.

Notwithstanding the high degree of urbanization in North America, agriculture remains an important element of the economy. In both Canada and the United States, vast expanses of the landscape are clothed by fields of grain. Great herds of livestock are sustained by pastures and fodder crops—because North America can afford the luxury of feeding animals from its farmlands. North Americans consume more meat per person than the people of any other world realm. No overpopulation threatens here, and each year Canada and the United States export food to less fortunate countries.

IDEAS AND CONCEPTS

- Plural society
- Site and situation (2)
- Hinterland
- Agglomeration
- Central place theory
- Urban hierarchies
- Urban structure
- Megalopolis
- Time–space convergence

Ten Major Geographic Qualities of North America

1. The North American realm is marked by unmatched physiographic regionalization.
2. North America comprises two of the world's largest states territorially (Canada, second in the world; the United States, fourth).
3. Both Canada and the United States are plural societies. Canada's pluralism is most strongly expressed in regional bilingualism; in the United States the major division is along ethnic lines.
4. North America's population, not large by world standards, is strongly urbanized and highly mobile.
5. The North American realm is the world's largest and most productive industrial complex. The region's industrialization generated its unparalleled urban growth.
6. By world standards, North America is a rich realm where high incomes and high rates of consumption prevail. Raw materials are consumed at prodigious rates.
7. Agriculture in North America employs less than 5 percent of the work force, is highly mechanized, and produces a huge surplus for sale on overseas markets.
8. Both Canada and the United States are federal states, but the systems differ; Canada's is adapted from the British parliamentary system and divides into ten Provinces and two Territories; the United States separates executive and legislative branches and consists of 50 states, the Commonwealth of Puerto Rico, and a number of island territories under U.S. jurisdiction in the Caribbean Sea and the Pacific Ocean.
9. North America possesses a varied resource base, but rates of depletion are high and energy prospects remain uncertain.
10. Notwithstanding North America's linguistic and ethnic regionalisms and its regional economic inequalities, the realm is unified by the prevalence of European cultural norms set within North America's broader physical landscapes.

Living standards in North America are, on the average, the highest in the world, although even so rich and developed a region as this nevertheless has its poor and underprivileged people. Surpassed in per capita income by some small, oil-rich states of the Persian Gulf and by several Western European countries, the North American population as a whole still is the most prosperous on earth. The amenities available to Americans in the form of living accommodations, transport and communications, health and education, food and nutrition, clothing, and leisure options are unequalled, at this level of magnitude, in the world today. And while it is true that these material comforts and conveniences are not equally available to all North Americans, the assets of life in America accrue even to the disadvantaged. The poverty

North American cultural landscape: the image of mobility. Crowded expressway on the approach to the Bay Bridge. Oakland-San Francisco, California. (Joe Munroe/Photo Researchers)

and malnutrition that exist in this realm are serious and very real for the individuals involved, but they pale in comparison to those found in most other parts of the world. The most blighted areas of North American cities still have facilities unavailable in the slums of cities in, say, South Asia or West Africa. Malnutrition in North America is the exception rather than the rule, as it is in entire regions of the underdeveloped world.

North America's evident good fortune raises numerous questions to be answered in this chapter. Two prominent ones are: How did North America achieve world leadership in so many spheres? And why do all Americans not share equally in this realm's unparalleled advantages? The answer to the first question lies in a combination or history and geography. Just as the empire-builders of Europe secured an early advantage in the accumulation of colonial wealth and domestic capital, so the early American entrepreneurs turned the productive opportunities of this continent (and the slave labor from another) into affluence and influence. North America proved to contain an enormous and varied inventory of natural resources, and when the impact of the industrial revolution came the essentials for primacy were there—raw materials, capital, labor, and markets. The further advantage of modern methods of production and the datedness and decline of European competition thrust the United States into world industrial leadership. Coupled with the region's growing agricultural output, it held a position of unchallenged primacy, which made it possible to control resources the world over. Today, North America—with about 6 percent of the world's population—consumes as much as 33 percent of the world's resources annually. The United States alone used nearly one-third of all the oil consumed in the world in 1979.

No population, not even in theoretically egalitarian communist countries, shares equally in the assets of a state. In the case of Canada and the United States, it is tempting to ask why such great wealth as exists here still leaves some people comparatively disadvantaged. The causes are many and range from the long-term outgrowth of racial discrimination to the inevitable inequalities inherent in our system of free-enterprise economics. Also involved are the varying fortunes of particular regions, whose resources may be in great demand in one decade and overproduced, overexploited, or undervalued the next. In recent years such North American regions as Appalachia, the Ozarks, and parts of the Southeast have contained populations with comparatively low incomes; in earlier decades other regions were so afflicted.

But it is important to remember that such *pockets of poverty,* as low-income population clusters and their home areas came to be known, are the exception rather than the rule in North America.

Two Nation-states?

Although it is true that Canada and the United States share a number of historical, cultural, and economic qualities, the two countries differ in important ways. The differences, in fact, tend to have geographic dimensions. Although the United States is somewhat smaller territorially than Canada, the U.S. occupies the heart of the North American continent and, as a result, encompasses a

Figure 3-1

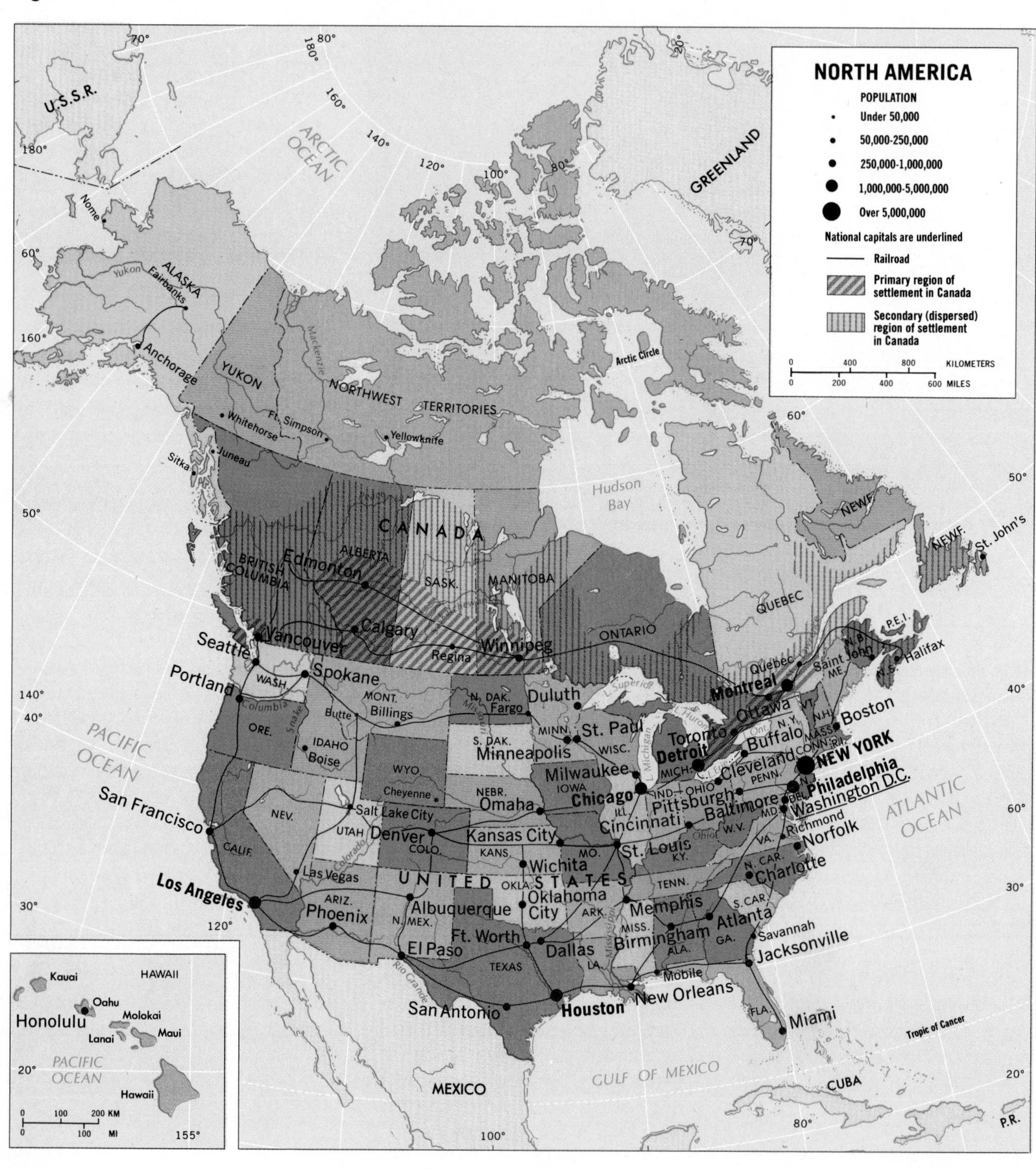

greater environmental range. Whereas, in the United States, we tend to think of an "East" and "West" as a major geographical breakdown, in Canada it is appropriate to identify a *Southern* Canada and a Canadian *North* (Fig. 3-1). The overwhelming majority of Canadians reside in Southern Canada, the zone that adjoins the northern boundary of the United States from Maine to Washington. Furthermore, the United States includes North America's northwestern extension, Alaska, so that—unlike Canada—the U.S. is in effect a *fragmented* state, a country whose territory consists of more than one major unit (see Chapter 10).

Differences also become apparent when population numbers and composition are examined. In 1980 the population of the United States approached 220 million; that of Canada was about 24 million, just over one-tenth as large. Although comparatively small, Canada's population is divided by culture and tradition, and the division has strong regional expression. According to the most recent census, 45 percent of Canada's citizens trace their ancestries to Britain, 29 percent have French origins, and about 23 percent come from other European stock. Today, English is the home language of 61 percent of Canadians, French is spoken by 26 percent, and other languages are used by 13 percent of the population (Indians and Eskimos make up less than 1.5 percent of the total). Certainly such a multilingual situation is not unique in the world; peoples of countries from Guatemala to Nigeria to Switzerland to India speak several different languages. Canada's problem is compounded by the strong spatial concentration of French speakers in one of the country's ten provinces, Quebec. With 6.5 million people, Quebec ranks second in population (after Ontario), and the province contains one of Canada's two largest cities, Montreal (3 million). Eighty percent of Quebec's population is French Canadian, and Quebec is the historic, traditional, and emotional focus of French culture in Canada. Quebec straddles the St. Lawrence River, one of the access routes into North America for the Early French settlers.

During the 1960s and especially the 1970s the Canadian state was shaken by a rising storm of separatism in Quebec. Dissatisfied with the position of French institutions in the Canadian federation, with the level of development of Quebec, and with many other aspects of life in bicultural Canada, Quebec nationalists proposed the "devolution" of Canada (see page 85)—the creation of an independent, French-speaking state.

The geographic implications of this proposal are far-reaching, as Fig. 3-1 indicates. Not only is Quebec one of Canada's cornerstones, but its separation would have the effect of fragmenting Canada into a group of Atlantic provinces in the east and five western provinces anchored by Ontario. Quebec, as the map shows, is Canada's largest province territorially, and it borders northern Maine, New Hampshire, Vermont, and New York State. The loss of Quebec would render the future of the Canadian state doubtful. Even today, however, the issue overshadows all else in the political geography of Canada, and it has certainly set the state back in its search for true nationhood. What is happening in Canada underscores the relevance of the last segment of the nation-state definition on page 65, the need for an emotional attachment to the concept of state and nation.

Nor is Canadian regionalism confined to Quebec and its Francophone issues. The four Atlantic provinces (New Brunswick, Nova Scotia, Prince Edward Island, and Newfoundland) have a strong regional personality; about 37 percent of New Brunswick's more than 700,000 inhabitants are French-speaking. These provinces, with their severe climate and mainly extractive industries, would be the most seriously affected by a Quebecan succession. The *Maritimes,* as they are informally called, would stand isolated, economically underdeveloped, but unable to go it alone under the new circumstances. The possibility that they might join the United States has been broached.

Regionalism also affects Canada west of Quebec. The country's core area lies mainly in Ontario, Canada's most populous province, centered on Toronto (3 million). Ontario, facing the United States across the Great Lakes, had a population of approximately 9 million in 1980, with a French-speaking minority of 10 percent mostly clustered in counties near the Quebec border. To the west lie the interior provinces of Manitoba, Saskatchewan, and Alberta, where other minorities contribute to Canada's complex cultural mosaic: Germans (19 percent in Saskatchewan, 14 percent in Alberta, 12 percent in Manitoba), Ukrainians (12 percent in Manitoba, 9 percent in

Canadian biculturalism: the French imprint in Quebec City. (©Photo E. Otto from Miller Services - Toronto)

Saskatchewan and Alberta), and others, including French. These western people of central and eastern European stock have been unwilling to accede to the demands of Francophone Canadians, given their own minority status and language preferences, and in 1977 this attitude led to formal opposition against any future economic association with a future independent Quebec.

Canada's westernmost province, British Columbia, is third in population (approaching 3 million in 1980) and centers on Vancouver (nearly 2 million). Here nearly 60 percent of the population has British ancestries, 9 percent are of German stock; only 4 percent of British Columbians are French-speaking. Like Ontario and the plains provinces, British Columbia has been unwilling to support to French Canadian aspirations in Quebec. The fortunes of this province have been closely tied to changing accessibility and relative location: from pack-animal, canoe, and porterage to wagon roads, then to transcontinental railroads, followed by the impact of the opening of the Panama Canal and the development of trans-Pacific trade routes. The secession of Quebec would, for the first time in a century, bring a negative change to British Columbia's relative location, and the province has been restive at the prospect.

No problem comparable to the Quebec question affects the North American realm's other federation—but pluralism of another kind prevails in the United States. As we note in greater detail later, the population of the United States also is divided, but not by language. Although "ethnic" regions continue to exist in large cities and bilingualism appears in places as a result of immigration flows (as, for example, in the Miami area and in the Spanish-influenced Southwest), the "melting pot" has quite consistently absorbed such transitory events.

More persistent in the U.S. cultural mosaic is the division between peoples of European descent (87 percent of the 220 million population) and those of African origins (just under 12 percent). It is important to keep the dimensions in mind, because there are more black Americans than there are Canadians, and during the 1980s the black population will exceed 30 million. Another contrast lies in this minority's geographical distribution. From its regional origins in the American Southeast, the black population of the United States has spread throughout the country, and especially to its major urban areas. Vancouver in Western Canada is only 4 percent French, but Los Angeles in the Western U.S. has a black minority in excess of 10 percent—close to the national average. Thus, although local racial segregation has characterized society in the United States almost from the beginning, regional separation on the Quebecan model did not prevail. Remnants of the original pattern of black concentration in the Southeast remain visible on the map, but almost half of all black Americans now reside in urban areas from New York to Detroit, Chicago, and the west.

Although it may be argued that, by several measures, there still are two "nations" rather than one in the United States today, there are also good reasons to take the opposite view.

Recurrent frustrations and setbacks notwithstanding, enormous progress has been made in the integration of American society. No barriers of language, religion, or fundamental political orientation are superimposed in the ethnic division. In certain respects the two plural societies of the North American realm have been moving in opposite directions.

Continent and Culture Realm

North America's *culture realm,* as here delimited, is constituted by Canada and the United States. But the North American *continent* usually is defined as containing Mexico and the smaller republics on the land bridge to South America as well. An atlas map titled "North America" will show the entire mainland from the boundary between Colombia and Panama to the Canadian Arctic. To avoid confusion, it is best to designate all mainland and island countries between the United States and South America as *Middle* America.

North American Physiography

Before we proceed to examine North America's human geography more closely, it is useful to consider briefly the physical stage on which this culture realm is based. Although the North American continent extends from Alaska to Panama, we confine ourselves here to the region north of Mexico—a region that still extends from the near-tropical latitudes of Florida and Texas to the Arctic lands of Alaska and Canada's Northwest Territories.

North America's physiography is characterized by its clear and well-defined division into physically homogeneous regions called *physiographic provinces.* A physiographic province is a region marked by a certain degree of uniformity in relief, climate, vegetation, soils, and other physical criteria, resulting in a scenic sameness that often comes readily to mind. For example, we identify such regions when we refer to the Rocky *Mountains,* the Great *Plains,* the Colorado *Plateau.* Not all the physiographic provinces of North America are so easily delineated, however. To obtain a picture of the continent's physiography, we need Fig. 3-2.

The most obvious aspect of the map of North America's physiographic provinces is the approximate longitudinal alignment of the continent's great mountain backbone, the Rocky Mountains, whose rugged, often snow-covered topography dominates the western sector of the continent from Alaska to New Mexico. The eastern part of North America is dominated by another, much lower set of mountain ranges called the Appalachian Highlands. These eastern mountains also trend approximately north–south, and extend from Canada's Atlantic Provinces to Georgia. The orientation of the Rockies and Appalachians is important, because they do not form a topographic barrier to polar or tropical air masses flowing southward or northward across the continent's interior. In Europe, the generally east–west trending Alpine Mountains do form such a barrier, protecting the Mediterranean region from polar cold.

Between the Rocky Mountains and the Appalachian Highlands lie North America's vast interior plains that extend from the shores of Hudson Bay to the coast of the Gulf of Mexico. These plainlands can be divided into several provinces. The major regions are the great Canadian Shield (see page 188), which is the geologic core area of North America; the Interior Lowlands, consisting mainly of glacial debris laid down during the Pleistocene glaciation; and

Great Plains, the extensive sedimentary surface that adjoins the Rocky Mountains. Along the southern margin these interior plainlands merge into the Gulf–Atlantic Coastal Plain, which extends along the southern and eastern edge of the Appalachian Highlands and the neighboring Piedmont until it disappears in New Jersey.

On the western (Pacific) side of the Rocky Mountains lies a zone of intermontane basins and plateaus (Fig. 3-2). This zone includes such well-known physiographic provinces as the Colorado Plateau with its thick sediments and spectacular Grand Canyon, the lava-supported Columbia Plateau, which forms the watershed of the Columbia River, and the Great Basin of Nevada and Utah in which lie several extinct lakes from the

Figure 3-2

glacial period, as well as survivors such as Great Salt Lake. Although the zone of intermontane topography continues into Canada and includes much of Yukon and central Alaska, the landscape is much more complex, and distinct regions such as those just described cannot be identified.

The reason this zone is called *intermontane* has to do with its position between the Rocky Mountains to the east and the Pacific Coast Ranges to the west. From Alaska to California the west coast of North America is formed by a series of major mountain ranges whose origins are connected to the contact between the American and Pacific plates (Fig. 1-3). The mountains of the Alaska Peninsula, the fiorded coastline of British Columbia, the Olympic Mountains of Washington, and the Klamath Mountains of Oregon and Northern California all form part of this coastal mountain belt.

Two major river systems have played a crucial role in the development of the European settlement pattern in North America. The St. Lawrence River provided the avenue for the first penetration, by French missionaries and traders in the late 1630s. Spanish explorers reached the mouth of the Mississippi River as early as the 1530s, but it was not until 1682 that this area was claimed for King Louis XIV of France. Both the St. Lawrence and the Mississippi valleys became the scenes of vigorous colonial competition.

Climate

The physiography just described should be viewed in the context of North America's varied climates and soils. The realm is quite clearly depicted in Fig. 1-5 to 1-8, and the maps show the comparative advantages of the United States with its more moderate climates, more extensive zones of productive soils, and (except for the west coast) higher precipitation totals. As Fig. 1-5 shows, the dry zone (below 20 inches or 50 centimeters of precipitation) that extends across interior western North America widens markedly in Canada. Where Canada's west coast receives higher amounts of precipitation, rugged topography limits agricultural possibilities. The same Rocky Mountain ranges that permit the northward and southward flow of tropical and polar air also interfere with the eastward flow of moist Pacific airmasses.

Figure 1-6 reveals the absence of humid temperate climates from Canada (except along the narrow Pacific coast zone), and the prevalence of coldness in Canadian natural environments. East of the Rocky Mountains, Canada's most *moderate* climes correspond to the United States' *coldest*. Nevertheless, Southern Canada does share the environmental conditions that mark the United States' upper Midwest and the Great Lakes area, so that agricultural productivity in the plains provinces and in Ontario is substantial. Canada is a food exporting country as is the United States (chiefly wheat), in spite of its comparatively short growing seasons.

Patterns in Colonial America

For fully 100 years following Columbus' initial voyage, almost no penetration of North America by Europeans occurred. Spain's American empire was the most extensive by the end of the sixteenth century, but even this was still limited to the continent's southern fringe, immediately north of Mexico and the Gulf of Mexico. In the same year that Vasco da Gama rounded he southern tip of Africa to be the first European to reach India directly by ship within the modern era, John Cabot was pursuing the first of two exploratory trips from England to North America. It was on the basis of these early voyages that later English claims were made. France sent out explorers in the early sixteenth century as did England again later in that century and the Netherlands early in the seventeenth century.

The French were able to claim, and initially control, the two major water routes into and out of the continent. French interest in "New France," however, was largely as pecuniary as Spanish interest in "New Spain." Furs were the primary economic reason for control of the northerly portion of New France; this meant that the population of this segment of the empire remained sparse. The Mississippi River and its tributaries were seen as an approach to the fur-trapping regions, and excluding other competing European interests was of primary concern. Major efforts were needed only at the mouths of the rivers and at selected, strategic locations along their routes. Montreal and Quebec on the St. Lawrence and New Orleans on the Mississippi became the French strong points, both militarily and culturally. Thus, French influence in North America, like Spanish influence earlier, was localized and basically peripheral.

In contrast to the French penetration, the general thrust of English settlement in North America was coastal throughout the seventeenth century. Granted, initial movement was up the major rivers emptying into the Atlantic Ocean; nevertheless, enough separate towns and associated farming settlements were established to provide an extremely broad front for inland penetration by the first half of the eighteenth century. The French holdings were greatly overextended for the manpower available, while the English developed a broad coastal base that could not be stifled by the loss of only one or two sites.

Sharp differences also evolved between the individual English colonies. Following several abortive attempts in North Carolina, the first formally established and permanent colony was begun on the James River in Virginia in 1607. Initial settlement was along the short navigable stretches of the many rivers crossing the flat coastal plain both north and south of Jamestown. By learning from the Indians the techniques of growing tobacco (once it became apparent that gold and gems were not locally available), the plantation economy became quickly established in Virginia and the newly formed colonies of Maryland and the two Carolinas. As in Virginia, tobacco was the basis of the economy in early Maryland; to the south, commercial rice, indigo, and sea-island cotton provided the base for the plantations of North and South Carolina and coastal Georgia.

Several significant results followed the persistence of this form of economic development in the Southern colonies. One was the sharp distinction that developed between Northern and Southern colonial economies. This distinction remained after the American Revolution and in many ways laid the bases for nonecomonic differences that have continued to the present. Another is related to the character of a plantation economy. The plantation system is basically a feudal agricultural economy with a few owners of large properties each employing a considerable amount of labor. Each of the major cash crops grown in the Southern colonies required great amounts of hand labor. Slaves from Africa came to be a major source of this labor. The Africans were supplemented by bonded or "indentured" servants from Europe. These latter were virtually short-term contract slaves in spite of the careful distinction (carried down even to the present) between white "servants" and black "slaves." A third result was the settlement pattern in the Southern colonies, where the population was relatively dispersed with only a few large coastal ports. The dominating plantation economy, again, did not lead to dense settlement or require numerous small urban centers. Any reasonably prosperous plantation was nearly self-sufficient with respect to the goods and services ordinarily obtained in such urban centers.

The middle colonies were different in many respects from those to the south. Taken from the Netherlands in mid-seventeenth century, Delaware and much larger Pennsylvania were granted to William Penn. New Jersey and New York, where the primary Dutch influence had been felt, were granted to the Duke of York. In New York, the Dutch "patroons," or large landholders, were allowed to remain in control of their holdings, and similar-sized grants were given to English friends of the colony's governors. Thus, there were similarities to land ownership in the Southern colonies along much of the Hudson River valley, but a sharp contrast to the landlord pattern developed immediately to the east in New England. Pennsylvania, on the other hand, was opened by the Quaker governor to small farmers from many countries regardless of their religion. Therefore, in economic organization and cultural makeup, it tended to have more in common with the New England colonies than any of its major colonial neighbors.

As in the South, entry to the interior was gained primarily along the major rivers. North of Maryland, however, the coastal plain narrows sharply so that penetration was not as easy as southward. Even an outline map of the east coast of North America illustrates this narrowing of the flat coastal plain; there are no major river estuaries north of the Delaware River comparable to those to the south, such as Chesapeake Bay (Susquehanna and Potomac Rivers) and the James River estuary. Penetration inland by large, deep-draft vessels is therefore possible in the Southern coastal region (as long as sandbars do not block the rivers mouths, as they do in

North Carolina), whereas movement inland by these ships is sharply limited in New England by rapids.

The New England colonies were still different in their settlement patterns, political foundations, and economic organization. Three of the four New England colonies (New Hampshire was the exception) were founded on dissension and separatism. New England's political fragmentation was reflected in its economic organization. From the beginning, the settlers who migrated to New England came for the express purpose of living according to their own, or their group's, standards. The primary goal, therefore, was not commercial gain but subsistence. One result was that during the early years, many more small farmers entered New England than did initially in the Southern colonies. This approach to farming also was sustained to a great extent by the environment, relatively inhospitable to large-scale commercial farming. Cold, wet winters and cool, often wet summers in a region with virtually no coastal plain, hilly topography, and thin, rocky soils meant that the individual surviving in agriculture must be accustomed to self-sustaining efforts rather than operations with hired or slave labor. These conditions resulted in a much higher average population density in the northern colonies. Equally important, the population lived in many small, nucleated settlements that became interconnected and interdependent urban centers.

During the decades immediately preceding the American Revolution, the character of the colonies became more similar in detail. While the coastal Southern colonies remained a region of plantation agriculture, the Piedmont Region attracted small-scale farmers, in many respects similar to the New Englanders. The Piedmont lies between the flat coastal plain and the Appalachian Mountains. It is a hilly region with rocky soil, much of which is better left in forest than cultivated. The farmers settling the Virginia and Carolina Piedmont were immigrants avoiding wage labor on the coastal plantations or were indentured servants who had served their bond period and were free to leave the coast. Needless to say, except for the low percent freed, the black slaves remained with the plantation economy until they died and were replaced. With this major regional exception of Southern coastal plantations, the dominant form of agricultural pursuit became increasingly small-scale, one-family farms throughout most of British America.

Revolution and Expansion

The colonists' success in the American RevolutionaryWar had several significant results. Of course, this success heralded the beginning of political and economic expansion across the breadth of the continent, but the expansion required several generations to complete. Within the context of the immediate aftermath of the war, a number of implications may be drawn from the geographic character of the population and settlement patterns as well as the 1790 economic and political organization of the continent.

First, there was little likelihood of restraining the westward movement of the farming population. Before the revolution, in 1763, the Indian peoples had been guaranteed by proclamation that European settlement would not expand beyond a line approximately following the front range of the Appalachians through Maryland and the North and Northeast irregularly, allowing for the European farmers, at most, several hundred miles between the coast and the "Indian Reserve." Although this mild restraint was not accepted by all colonies as binding even before the war, it was legally removed after independence from Britain was achieved. Furthermore, the Indians were generally too loosely organized to resist the westward expansion. The only peoples having sufficient organization to have managed such resistance for some time, the Iroquois, had fought for and with the defeated British. (This is not to imply that allegiance with the colonial revolutionaries had significant benefit for those Indians who fought on the winning side; "human rights," it was clear, were relevant only for "civilized" people, and the Indians, like their black counterparts within the system, were hardly civilized in the eyes of most European-Americans.)

Second, the economic activities of the successful revolutionaries perpetuated the immediate prewar patterns of development. The small-farm operator continued to move onto the Piedmont, between coastal plain and mountain barrier. The plantation economic organization, however, was no longer restricted by political considerations to the plain along the eastern margin of the continent. It could now be carried westward onto the lower Piedmont, and south of the mountains onto the Gulf of Mexico plain.

Third, the very diversity of the separate colonies in economy, in outlook, in population composition, in religion, and in other ways required a great deal of compromise and conciliation in order to form a single country. The form used—indeed, the only form liable to succeed—was a federation. Federalism in America bound the diverse and different states together; it made unification possible but it also set the stage for the events of the mid-nineteenth century. The federal form of government had much to do with the evolution of the social spatial structure of present-day North America, both in the United States and Canada.

And fourth, the results of the American Revolution carried much significance for the development of the remainder of the continent's lands. The war was not popular with all settlers in the British colonies. Three colonies, Quebec, Nova Scotia, and Newfoundland, did not revolt, and many citizens in the other new lands remained loyal to the established order. Many such "loyalists" were expelled from the United States during and after the war. Those who stayed in North America migrated to the British portions of the continent—that is, to Canada. This, of course, provided a sudden influx of English population into an area that for some time had been predominantly French. This influx, in turn, affected the political and economic organization of Canada.

Beginnings of pluralism: the African connection

Blacks were among the first immigrants to early British America. Africans had been brought to Spanish colonial holdings in the Caribbean Sea and in Florida for nearly a century before successful English settlement in North America began in 1607 along the James River in Virgina. Within five years, it was clear that gold and gems were not readily available, while Indian-domesticated tobacco could be grown and sold to England. For profitable cultivation, tobacco demands many worker-hours of labor, a long growing season, fertile sandy soil, and moderate amounts of rainfall well distributed over the growing season. The environment in coastal Virginia met these requirements, but the sparsely populated colony was in severe need of labor. This labor had to be cheap, but not necessarily skilled, if the production of tobacco was to meet demands of rapidly expanding markets in Europe. By the middle of the seventeenth century, it was clear that blacks, held in hereditary lifetime slavery, were looked upon as the optimum source of the needed labor.

The number of black slaves in North America expanded slowly from the first twenty in 1619 to about three hundred in Virginia by 1649. During the next 140 years, however, as European immigration accelerated, as agricultural production of plantation crops such as indigo, rice, and some cotton in addition to tobacco expanded rapidly southward along the coastal plain, and as it became even clearer that the temporary bondage of indentured whites was not as satisfactory as the permanent bondage of blacks, the African population in America grew rapidly.

Trade routes became established as the demand for slaves increased. The pattern of these routes illustrated the growing complementarity between sections of colonial America, and also between the new colonies and other parts of the world. The notorious "triangular trade" became established between North American colonies, Caribbean colonies, and Africa (Fig. 3-3).

The diversity of the North American colonies was clear even at this early period. In terms of the pattern of black population, differences in colonial economies resulted in an overwhelming preponderance of slaves in the southern four colonies (Maryland, Virginia, and North and South Carolina). The middle and New England colonies had little need for the kind of labor supplied by slaves. Small farms, individually owned and worked, dominated the northern rural economy. While extra hands could nearly always be used, crops grown in the North—chiefly grains, supplemented by cattle and fur sales—were not sufficiently remunerative to recover the payment necessary to purchase a man or woman. In addition, religious scruples, beginning with the Quakers, raised doubts in many minds concerning the moral issue of slavery. An increasing divergence of social and political attitudes as well as economies followed. The changing attitudes in the North culminated in the state-by-state abolition of slavery in the middle and northern states, beginning with Vermont in 1777 and ending with

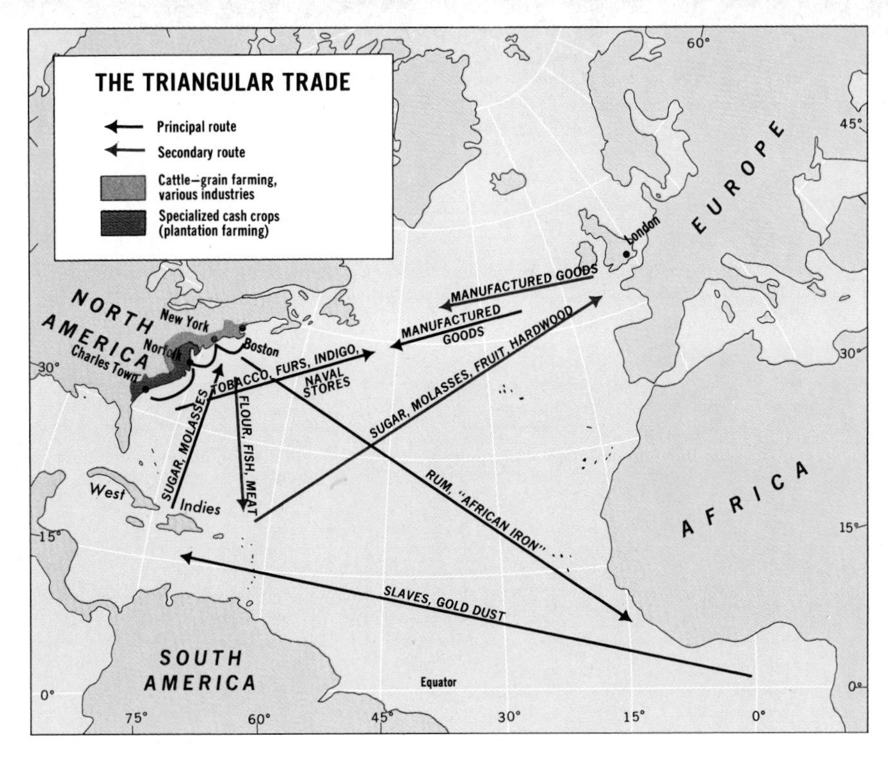

Figure 3-3

New York and New Jersey in 1799 and 1805, respectively. With this, the seeds were planted for the American Civil War and the regional separatism that smoldered for a century after the war's conclusion.

The Emergence of National Patterns: the United States

The end of the Revolutionary War heralded the beginning of a century of settlement expansion in the United States. Although the Appalachian Highlands had been breached even before the war, interior settlement in large numbers did not begin in earnest until after 1790 (Fig. 3-4). Primary expansion during the first 20 years of nationhood was in settlement of the Ohio River valleys and its major tributaries east of the Tennessee River. There was also a major two-pronged thrust toward the Great Lakes, one northward from Fort Pitt (Pittsburgh) to Lake Erie and the other westward through the great Mohawk Valley in central New York State to Lake Ontario

It is easy to note the importance of the river system to the settlers migrating into this region. As earlier with the first colonists, rivers provided the only easy transportation; they were the only means of ready access to and from the developed portion of the country. A farm and homesite located on a navigable river during the early nineteenth century is closely analogous to a commercial farm or merchant located a short distance from a railroad depot in the middle of the century.

The use of, and dependence on, rivers as means of extending the settlement frontier during the period 1810–1830 was not lessened. In addition to filling in the earlier areas of sparse settlement, there was considerable population movement north from the Ohio River, up the Scioto and Miami rivers into Ohio, up the Wabash into Indiana, and north from the Ohio along the Mississippi River. Settlement north and west from St. Louis had begun by 1830—west along the Missouri River and north along the Illinois and Mississippi. In the South, too, settlement was densest along the major rivers, primary expansion occurring along the southern Mississippi, up the Arkansas, and along the Alabama River and its tributaries. By 1850, little in the East could be called "frontier" settlement. And by 1870, the forest margins had finally been left behind for the frighteningly open grass plains, with settlement expanding toward central Texas, Kansas, and Nebraska. The striking explosion of settlement, 1870–1890, along the west coast, along the front of the Rocky Mountains, and throughout the dry West was the final rapid movement of the frontier, with settlement expansion in the twentieth century primarily in urban clusters, especially in the major developing core areas of the nation.

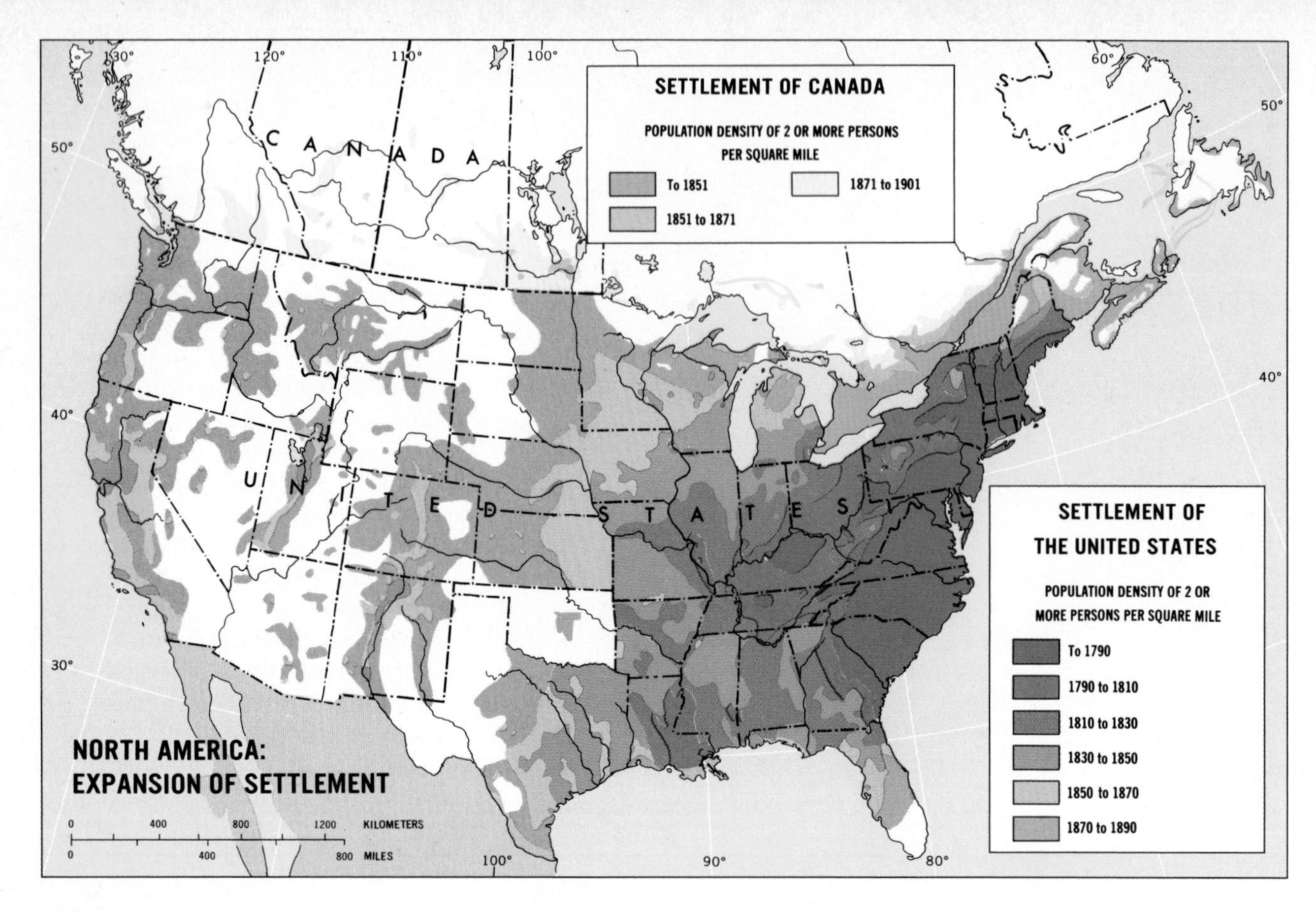

Figure 3-4

Canadian Patterns

The initial cultural heritage of the United States was overwhelmingly English. Although millions of immigrants from other European cultures were accepted into the country and the national character changed slightly with each wave, the process was primarily one of assimilation into a modified English culture rather than a mutual change. Even the African cultures carried to North America were altered by the dominant English culture just as they generated alterations in it. Generally, these contributions were neither desired nor recognized by the Europeans. Canada, on the other hand, was distinctly bicultural virtually from its beginning.

The pattern of Canadian biculturalism began with the British military victory over the French, finally settled in 1763. The French had been firmly in control of the St. Lawrence River valley since Champlain's explorations in the early seventeenth century. French settlement along the St. Lawrence, mostly by farmers from northern and central France, had become sufficiently well established for these farmers to stay with their land even after Canada became a British possession.

In spite of the fact that New France had never been settled as densely as the British colonies, the Francophone community in Canada managed to remain distinct in the face of increasing political and numerical dominance by the Anglophone community. It is important to remember that the French in Canada had several distinguishing characteristics, only one of these being linguistic. They were also Catholics, exclusively. The Protestant French Huguenots had been prohibited from settling in French Canada. Their migration to British America had contributed to the low population of New France. Religious and linguistic differences and the complete range of cultural associations that distinguish one such group from another permitted and encouraged the French Canadians to maintain their group solidarity in the face of increasing formal and informal pressures for assimilation by the British Canadians.

By the end of the eighteenth century (1791), there were three distinct regions in Canada: Upper Canada (Ontario—English), Lower Canada (Quebec—French), and the Maritimes (Nova Scotia, New Brunswick, and Newfoundland—mixed, but mostly English). Although the French settlers had numerical superiority during the years immediately following the Treaty of Paris (1763), this changed rapidly during and after the American Revolution. Loyal British subjects fled the rebelling colonies in large numbers, many migrating to Nova Scotia, there to be given their own, separate colony (New Brunswick), while many others moved to settle upriver from Quebec in Ontario.

For almost a century, population growth in Canada occurred within these three regions. Except for the St. Lawrence River valley and the lowlands of the Southern Ontario Peninsula, the landscape of eastern Canada is dominated by the *Canadian* (or Laurentian) *Shield* (Fig. 3-2). Extending monotonously to the Arctic, this physiographic region of gently rolling relief spawned the continental glaciers of North America during the recent Ice Age. As a consequence, the bedrock was scoured clean during the glaciers' advances, and temperatures were too low and effective precipitation too slight to have allowed much replacement of soil in the last 6000–8000 years. The soil is therefore thin and unproductive. The paucity of population north of Lakes Huron and Superior reflects the poor agricultural base even today.

This poor environmental base retarded the westward expansion of Canadian population relative to that in the United States, although the expansion south of the border was an indirect stimulus to expansion to the north. For over 100 years after United States independence, relations between these neighbors were marked by suspicion and occasional "jingoism" on the part of the United States, which in turn appeared to justify Canadian distrust. As settlement spread westward south of the border, the Canadian population grew only in Ontario and Quebec. Even by the 1880s settlement had not moved in significant quantities onto the grass plains of Canada—the "Prairie Provinces"—or into the mountain and Pacific regions. Only Winnipeg on the Red River in Manitoba and Victoria on Vancouver Island showed any indication of future urban development. Although there were still fewer than a thousand settlers on the prairies, Canada's transcontinental railroad was constructed during the decade of 1880s, following by more than ten years the completion of the United States' first east–west rail connection. In addition to the important function of connecting for the first time both coastal extremities of this very broad country, the railroad also stimulated settlement by making possible the development of a wheat economy. Also, in what may have been one of the primary factors of the day, it was believed that the railroad and the accompanying population increase prevented a much-feared infiltration of the western territories by the energetically expansionist American population.

The increased accessibility of the Prairie Provinces and the West Coast generated population growth in both regions. By 1911, Winnipeg in the east and the city of Vancouver on the mainland in the west had grown to more than 100,000 population; Regina, Edmonton, and Calgary exceeded 25,000, and the rural population distribution patterns of present-day Canada were established.

Economic Patterns and Regional Formation

Contained within the evolving patterns of rural settlement in the United States and Canada were the seeds of the economic, cultural, and political variations that define the present human geographies of these two countries. In the United States, farmers left the eastern Coastal Plain and Piedmont for the less crowded, often richer farmlands of the Interior Plains, and residents of the long-established New England towns and farming regions also migrated westward. In New England, a commercial-industrial economy had been established during the colonial period based on the timber (shipbuilding) and fishing resources of the region. Rum distilleries, small-scale iron works, and similar activities focused on the ports of the Northeast. This in turn developed the region into the import-export and financial core for the colonies. During the early years of independence, the textile industry became increasingly important because this region contained the needed capital for industrial establishment as well as the technical ability and abundant labor in the rapidly growing cities. Urbanization and industrialization grew apace as immigrants from crowded European cities and towns brought with them the skills and knowledge (or at least the attitudes) that were driving the industrial revolution in the Old World.

The immigrants settled in nearly every town and city in the East, but the overwhelming majority selected the largest cities: Boston, New York, Philadelphia, and Baltimore along the Atlantic coastline and later such interior cities as Cleveland, Detroit, Chicago, St. Louis. The four Atlantic ports established themselves early in the colonial and postcolonial period as the dominant urban centers among the many harbor towns along the coast north of Chesapeake Bay. The growth pattern of each city was primarily a function of the city's locational factors. As in the case of the older European cities (see p. 89), situation and site gave certain places competitive advantage over others.

The present four major port cities in the northeastern United States all possess important advantages of site and situation. New York, for example, grew astride the convoluted coastline where Long Island almost touches Manhattan, where the East River and the Hudson River enter the Atlantic. The many miles of harbor also lie at the mouth of the Hudson, the single best route (with the Mohawk Valley) through the Appalachian Mountains to the Great Lakes and beyond.

The site of New York City, North America's largest urban agglomeration, from two perspectives. The vertical view shows the intensity of land use on Manhattan (flanked by the Hudson and East Rivers); the oblique photograph displays Manhattan's twin cores of great vertical development in the Wall Street area (foreground) on the island's southern tip and midtown (center of photo). Millions of immigrants to North America saw these shores first, as the port of New York was a leading gateway for European settlers. (left, U.S. Geological survey; right, Skyviews Survey, Inc.)

All four cities also benefited from their relatively easy accessibility to Northern Europe and from productive local *hinterlands.* The hinterland of a town or city is the area that participates significantly in the exchange of goods and services with that city. As one travels outward from the urban center, the total costs of transport gradually increase as distance increases, barriers are encountered, or transportation facilities decline in quality. The edge of the hinterland is therefore reached when it is generally too costly to send products to that city. The strong early growth in New England supported equivalent growth in the region's primary focus, Boston. As the rich lands west of the Appalachians began to be exploited, however, Baltimore, Philadelphia, and especially New York took advantage of their superior locations and competitively extended their hinterlands rapidly toward the continental interior while Boston's growth rate began to lag.

Characteristics of site, situation, and hinterlands are also clear in Canada's early urban development. In French Canada, both Quebec and Montreal were originally located at sites of

navigable limits on the St. Lawrence River—Quebec at the limit of oceangoing sailing ships and Montreal immediately below the Lachine Rapids, the first encountered by vessels moving upriver on the St. Lawrence itself. Although Quebec had the added advantage of an excellent defensive position overlooking the river, the future lay with Montreal, which had superior access to the growth around the southern Great Lakes.

Montreal's primary urban competition came from Toronto, the major focus of Southern Ontario. The site advantages of Toronto are not outstanding. Its small harbor is not better than Montreal's riverfront access, but Toronto is much better located with respect to the continental interior, and the agricultural productivity of Southern Ontario is high. Development of transportation connecting the southern Great Lakes with these cities and those along the United States Atlantic coast established an urban-economic growth pattern by the mid-nineteenth century that dominated North American development for the next century.

In the Southeast, the plantation economy was well established as a productive form of commercial enterprise. Cotton production expanded both in area under cultivation and in revenue earned, moving south along the Piedmont and Coastal Plain of the Carolinas into Georgia. As of 1810, however, most of the land comprising the present states of Georgia, Alabama, and Mississippi was occupied by numerous Indian peoples. As in the Northwest Territory, land was ceded piecemeal by the Indians until 1830, when the Indian Removal Bill was passed by Congress. This led to a series of forced mass migrations during the 1830s from the Indians' traditional homes to the relatively barren Indian Territory (Oklahoma) on the margin of the "Great American Desert." The removal of the Indians, accompanied by great hardship (4000 Cherokee, for example, died on the trip west from northern Georgia), was completed by the middle of the decade. Millions of acres of new land were thus made available to the Southern cotton planters.

In an irregular, patchwork pattern of expansion, plantation cotton production gradually shifted westward across the Gulf Coastal Plain. By 1850, large-scale cotton cultivation had spread onto the comparatively fertile chernozem plains of Texas, and during the next decade, cotton consolidated its position of importance in the national economy as the primary export product of the United States. So important was cotton to the country as a source of foreign exchange ($192 million out of a total $316 million worth of goods exported in 1860) that the Southern states believed the North would have difficulty surviving without them.

Accompanying this southward and westward movement of the plantation system was a marked change in the population pattern in the country. The development of the cotton plantations was attended by a large-scale transfer of slaves from the East Coast to the cotton-producing Deep South, modifying these regions' cultural character. As cotton production moved westward across the lower South, the threat of black dominance in the older colonies was reduced (or so it was believed) by the transfer of slaves into these developing areas. The slave trade, legally, had become an internal affair in 1808 when a federal ban on the importation of new slaves went into effect.

By 1860, the major transformation in the pattern of black population in the South was virtually complete. Although the actual numbers moved are not known, it was undoubtedly many hundreds of thousands. Most trade was overland with chained slaves marched in gangs over hundreds of miles from Kentucky and Virginia into Alabama, Mississipi, and beyond. The change in the pattern over two generations, 70 years, is startling. While in 1790 slaves were concentrated in coastal Virginia through northern Georgia, by 1860 a linear region through Georgia and Alabama, the Mississippi lowland from the Tennessee border to the Gulf, and portions of Texas and Florida, contained counties with over 50 percent of the total county population black. Also striking is the great similarity between this 1860 distribution of black population and its counterpart for rural areas in 1980.

The presence of high densities of blacks before 1860 meant, almost without exception, a predominance of economic activities for which slave labor was suitable—that is, activities demanding large amounts of cheap, largely unskilled hand labor. Considering that the slaves themselves were naturally resentful of their servitude, the activities for which they were used also had to be such that sabotage and work slowdowns could be guarded against by an overseer. In sum, these requirements limited most slavery to plantation agriculture of labor-intensive, high-value-per-acre crops. (Output per worker-hour was obviously less important although still relevant). The continuation of slavery in the South reflected the continued importance of specialized agricultural activities in this region rather than manufacturing, as it developed in the Northeast, and commercial food crop production, as it increased in the Middle West.

Emergence of Modern Spatial Patterns

The half century following the end of the Civil War in the United States was one of economic growth and transformation; in Canada it was a time for national integration and territorial consolidation. In the emerging northern core area of the U.S., centers of mercantilist activity, basing their raison d'être on commerce, increasingly became manufacturing instead of trade centers. In the South, however, this was a period of bitter and stubbornly slow economic, political, and social reorganization. This was also the period during which the West was populated and the basic patterns of continent-wide human organization were laid. Northern development was powered by mineral resource use and absorbed great numbers of immigrants from Europe. It was further stimulated by increasingly efficient means of interaction and the growing markets, East and West, that were reached by the expanding network of railroads. After a relatively brief period of attention from the rest of the country, during which both exploitative and egalitarian forces were present, the South was allowed by a preoccupied North to revert to many of the antebellum forms of economic, social, and political exploitation.

In Canada, the half century preceding the outbreak of World War I (in 1914) was a period of growing national consciousness. Territorial consolidation, the expansion of the transport network, and political compromise all proceeded as the Canadian state took shape. The historic British–French conflict was overcome politically in 1867 when the British North America Act established the Dominion of Canada, a confederation of Nova Scotia, New Brunswick, and Middle and Upper Canada (Quebec and Ontario). The other provinces joined the federal union later: Manitoba in 1870, British Columbia in 1871 on condition that the government construct a railroad across the continent to the west coast, Prince Edward Island in 1873, Alberta and Saskatchewan in 1905, and Newfoundland as recently as 1949. The transcontinental railroad was completed in 1885, a milestone in the unification of Canada, and larger numbers of European immigrants entered the country as less "empty" land became available in the more populous United States.

Canada also urbanized rapidly during the last half of the nineteenth century. While the country's total population doubled (from under 2.5 million in 1851 to over 5 million in 1901), urban centers mushroomed: Montreal during the same period grew from under 60,000 to nearly 270,000 residents, Toronto went from just over 30,000 to nearly 210,000. Other cities were founded during the period, including Ottawa, Winnipeg, and Vancouver, major cities just a century later.

The United States to 1980

In the United States, the more than sixty years following World War I have been a time of national economic and social transformation. At the root of the changes that affected spatial patterns during the twentieth century are (1) the relative decline and reorganization of agriculture, (2) the increased importance of service activities complementing the previously developed manufacturing-industrial sector, (3) a series of migration flows: South to North, East to West, rural to urban, and recently North to South, (4) the changed methods and rates of movement for both people and goods by air and highway carriers, and (5) during the last two decades, the revolution in social attitudes required of the white population majority by a large, hitherto excluded, black minority. Each of these changes has spatial manifestations. Existing geographic patterns must be viewed in light of the social and economic trends that generated them.

Forms of industrial activity generally are divided into three categories: *primary,* or extractive industries (e.g., mining, lumbering, agriculture), *secondary,* or manufacturing industries (e.g., steel, chemicals, automobiles, furniture), and *tertiary,* or service industries (such as insurance, entertainment, wholesaling and retailing, and transportation). In the underdeveloped countries, the primary industries tend to employ a larger proportion of the labor force than do secondary and tertiary industries (see page 39). As the economy develops and becomes more complex—and the per capita standard of living rises—the secondary and tertiary industries become proportionately larger and begin to dominate the economy. In 1880, when the U.S. was still a developing country, 50 percent of the labor force was engaged in primary industries, 24 percent in secondary industries, and 26 percent in tertiary industries. In the 1970s, only 6 percent still worked in the primary industries, 34 percent were in the secondary industries, and 60 percent were engaged in tertiary activities. A similar—though not quite so striking—change occurred in Canada.

These economic changes had dramatic spatial effects. Secondary and tertiary industries are concentrated in and near urban areas, and as the economic transformation progressed, a relocation of population accompanied it. These changes were even more comprehensive in the United States, with its larger population base, than they were in urbanizing Canada. The cities of the United States expanded and coalesced into megalopolitan areas, not only in the East but also in the Midwest and West. As we note in detail later, the rural areas were transformed as well. The North American realm's core area took shape, encompassing a great rectangle from Boston to Washington to Kansas City to Minneapolis-St. Paul (Fig. 3-5), and including southernmost Ontario.

None of these developments could have occurred were it not for North America's exceptional resource base, the quality, quantity, and relative location of its raw materials. We turn next to this aspect of the North American realm.

Mineral Resources of North America

The recurrent energy crises of the 1970s made clear to North Americans the extent to which they had become dependent upon foreign supplies of an expendable, nonrenewable commodity found in substantial quantities on their own continent as well: petroleum. The energy crisis is not, however, a simple question of inadequate domestic supply and excessive local demand. It also is a crisis of convenience: when Americans could afford it, it was easier to buy foreign oil to supplement supplies from domestic reserves than to prepare for shortages by developing ways to convert other raw materials (such as coal or oil shale) into needed fuels.

North America may contain the world's largest reserves of coal, and the United States may be the single richest country in these terms. Large areas of the Eastern, Central, and Western United States are underlain by coalfields (Fig. 3-6). The extensive Appalachian fields are the easternmost, and they include a small field of hard, anthracite coal in eastern Pennsylvania. They are easy to work, lying in seams between one and nine feet thick, in many areas close to the surface, and nearly continuous from Kentucky northward. The mines in this field, primarily eastern Kentucky, West Virginia, and western Pennsylvania, account for over 75 percent of the nation's coal production. Although mined for decades, the reserves remaining are still large. Even now they comprise

Figure 3-5

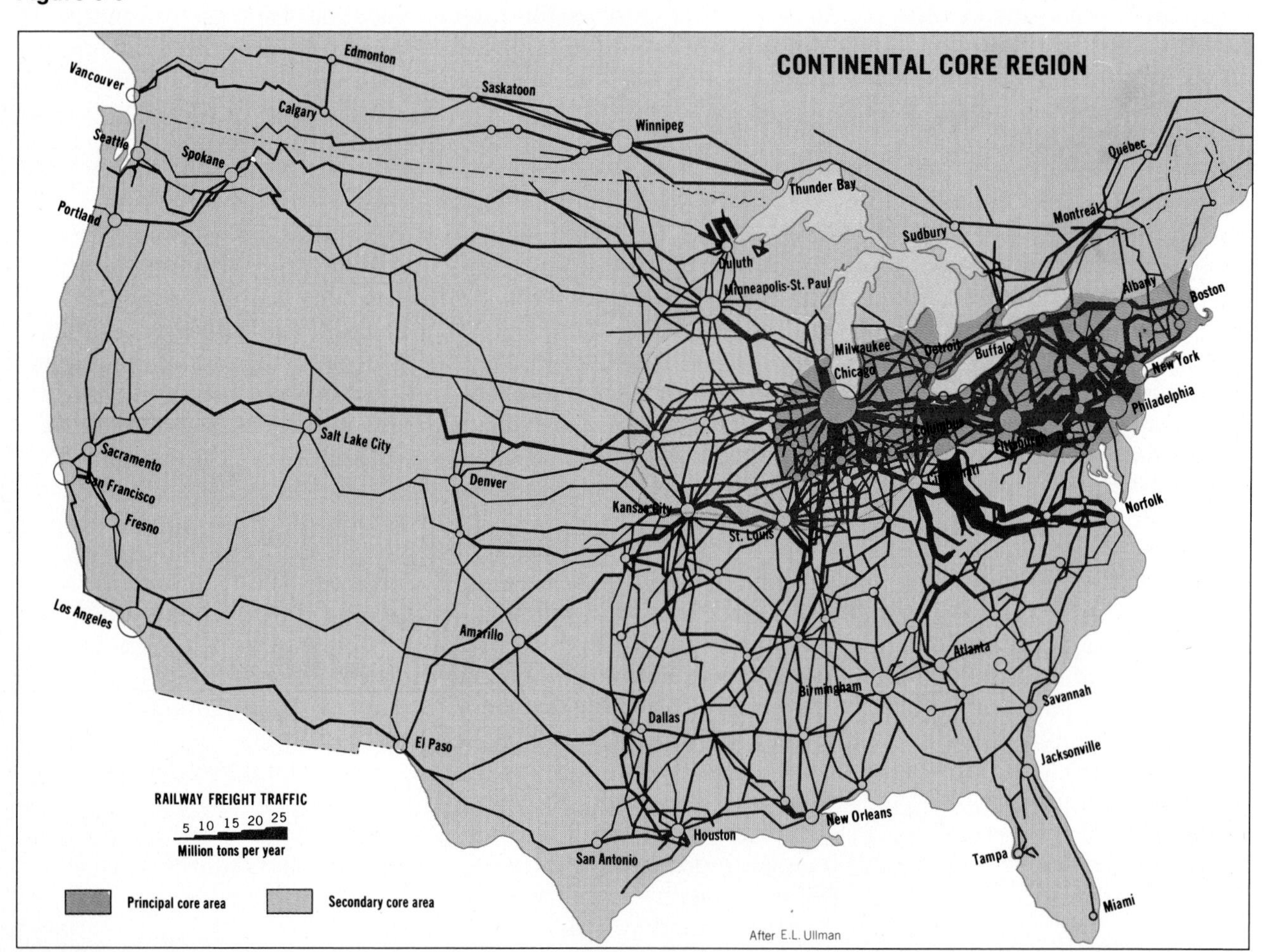

about 20 percent of the nation's total supply of bituminous coal. Most of the remaining bituminous coal presently mined in the United States is obtained from the Eastern Interior field underlying most of Illinois and extending into western Kentucky. Of slightly poorer quality than the excellent coal found in the two Eastern fields, the Western Interior field is also extensive and, as yet, little mined. Many small, somewhat lower-quality coal deposits are found throughout the Rocky Mountains. Because of the abundance of good, easily obtainable coal in the East, the tilted and fragmented reserves in the Rockies have not been much used until recent years. The tremendous growth of population along the Pacific Coast and its concomitant growth in demand

Figure 3-6

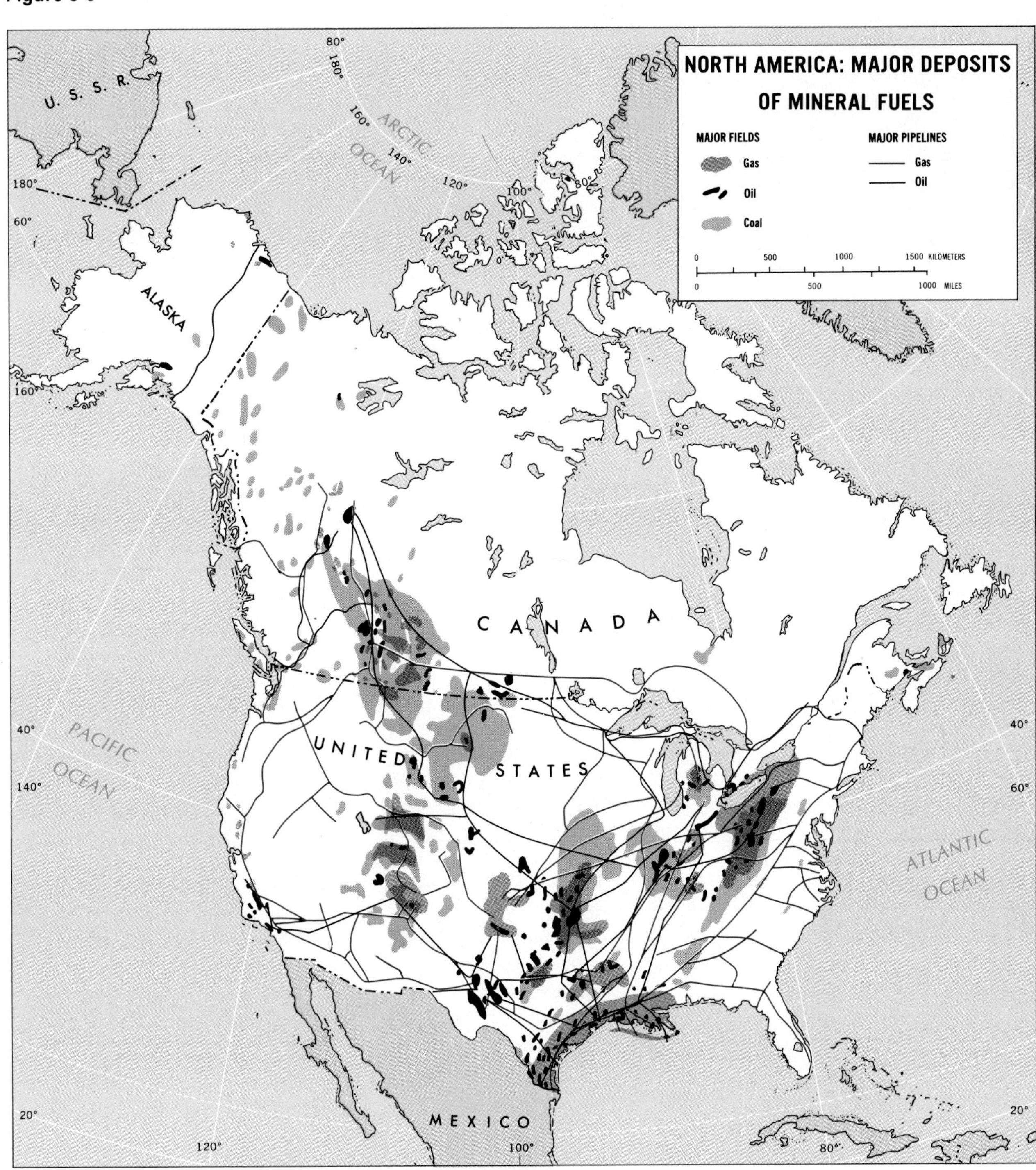

The energy crisis of the 1970s has intensified the exploitation of the United States' plentiful coal reserves. Much of the coal is found in comparatively shallow layers that can be mined by removing the overlying soil and rock in a process called *strip mining*. As this photograph, taken in Missouri, shows quite dramatically, the impact of strip mining on the landscape can be devastating. Mining companies sometimes commit themselves to redistribute rock and soil over the area where coal has been removed, so that damage will not be permanent. But strip mining has already destroyed many square kilometers of the natural landscape. (Grant Heilman)

for electricity have led to construction of very large coal-based generating plants in the "Four Corners" region of Arizona–New Mexico–Utah–Colorado and more extensive mining in the Green River basin of Wyoming. These changes have not always been welcomed by local inhabitants and environmentalists.

As Fig. 3-6 shows, North America's coalfields extend into Canada, especially west of the great Canadian Shield. Coal was mined early in Canadian history in Nova Scotia, which was a coal exporting province when the Confederation was established. But the larger reserves lie in British Columbia and neighboring Alberta. Coal has been exported to Japan from these provinces for decades.

The distribution of petroleum, often found in association with natural gas, also favors the United States. Major petroleum reserves occur along the Gulf of Mexico coastline, both on land and on the continental shelf, in Southern California, and several areas of the interior in Texas, Oklahoma, and Kansas. Extensive deposits also lie along the coastal plain of Northern Alaska, well above the Arctic Circle, and in Northern Canada.

Although Canada's oil reserves may not be comparable to those of the United States, Canada's overall position as far as energy is concerned is better than that of the U.S., mainly because of discoveries and developments in the province of Alberta. Oil first flowed from an Alberta field as long ago as 1947, but the price increases on international markets during the 1970s spurred the oil and gas boom Alberta has recently experienced. Alberta's energy resources take three forms: oil reserves in several scattered fields, natural gas reserves that extend in a diagonal zone from British Columbia (itself a producer) across the southern half of the province, and vast tar sands in the north and northeast (the Athabasca reserve alone may contain more oil than the entire Middle East).

The fortunes of Alberta's energy industries provide an illustration of the role of relative location in economic geography. Alberta lies 3000 kilometers (nearly 2000 miles) from Canada's eastern heartland, where most of the domestic consumers are. But a lucrative U.S. market is just across the southern boundary of the province. Thus, although Canada could be self-sufficient in oil, it was economically more profitable to sell Albertan oil on U.S. markets and to buy cheaper Venezuelan oil for use in Ontario and Quebec. As international prices of oil continued to rise during the 1970s, a dispute arose between the province and the Canadian federal government. The federal government wanted to keep the price of domestic oil below world levels, and taxed exported oil to match world prices. After reluctantly agreeing to this policy, Alberta became involved in a new conflict with Ottawa in the late 1970s over the sale of natural gas. Alberta wanted to sell excess natural gas to the United States at comparatively high prices; the Canadian government preferred to create internal Canadian markets by lowering the price for domestic consumers. Alberta's energy boom, which is likely to grow as improved technologies bring the tar sands into production, is another factor contributing to Canadian regionalism.

Metallic Minerals

North America's metallic mineral deposits are located in three zones: the Laurentian Shield, sections of the Appalachian Highlands, and scattered zones throughout the western mountain ranges (Fig. 3-7). The

Figure 3-7

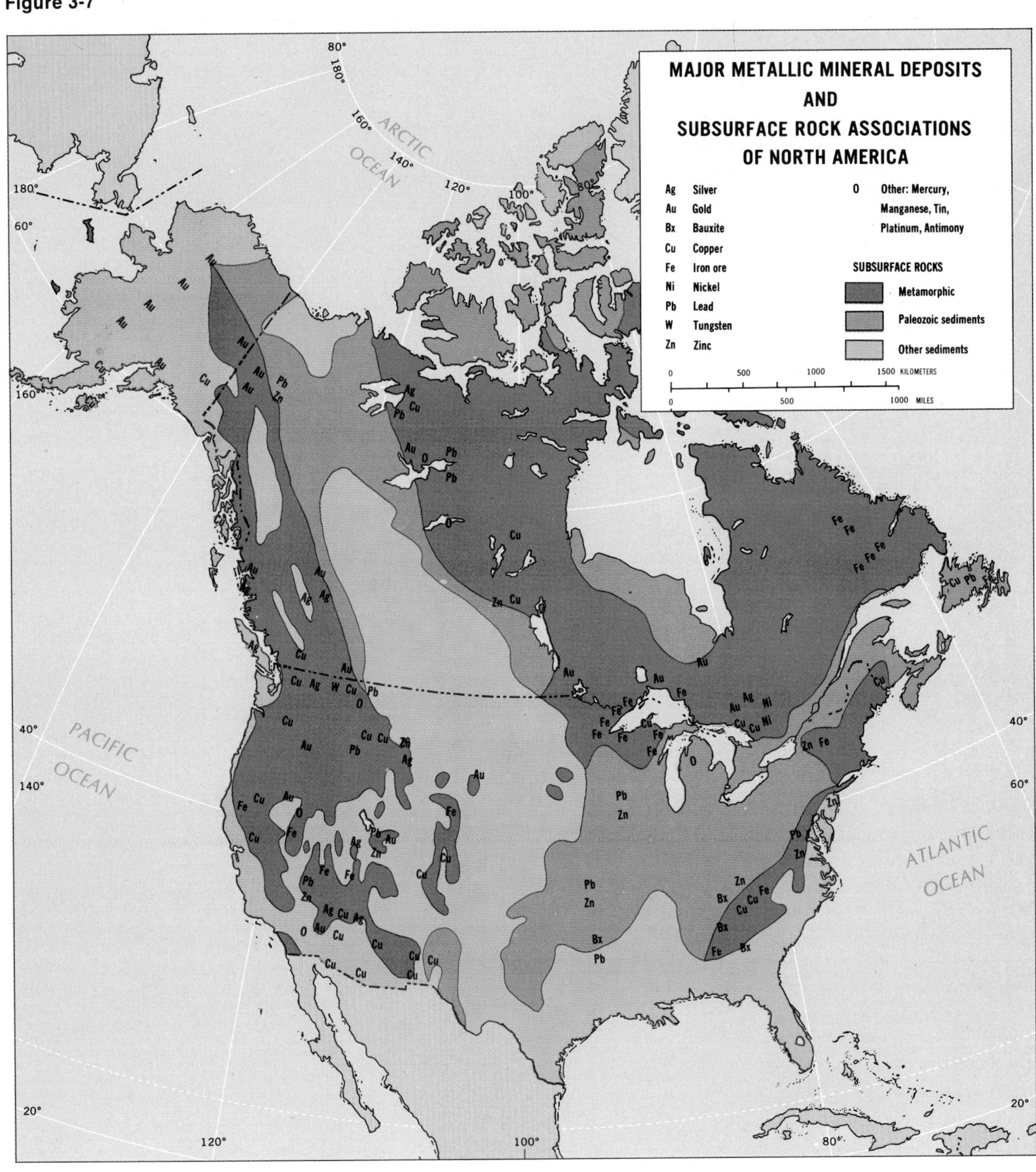

The Trans-Alaska Pipeline

The first significant oil strike on Alaska's North Slope on March 1968 promised a major alteration in Alaskan life-styles and the Alaskan environment. These changes have been initiated by construction of the Trans-Alaska pipeline. With an estimated final cost of almost $8 billion, it is described as the most costly privately financed construction project in history. The rewards, the environmental problems to be overcome, and the long-term impact on Alaska's economy and culture are also very great.

The pipeline was built to transport oil from the largest oil field yet discovered in North America. With estimated reserves of 9.6 billion barrels and continually rising world demand for petroleum products, direct and indirect economic impacts within the state are expected to exceed $50 billion before 1988. The 1275-kilometer (800-mile) pipeline crosses three mountain ranges, 34 major rivers and streams, and several zones of intense earthquake activity. Much of the land is underlain by permafrost (p. 153). To prevent this ground from thawing due to the high oil temperature generated in the pumping, almost half of the pipeline had to be built on vertical supports elevating the pipeline four or more feet above ground.

Thousands of migrants have arrived in the state since 1974, many from such oil-producing states as Texas and Oklahoma, seeking high-paying jobs. This has lent a boom atmosphere to the state. Many recent in-migrants, or "boomers" as they are called, are unable to find employment, however, and yet they remain in the state, swelling its unemployment rate.

Further changes can be expected as "Seward's Folly" continues to be explored and its vast resources exploited. The Prudhoe Bay field tapped by the Trans-Alaska pipeline is expected to last 25 to 30 years at planned extraction rates. In addition, estimates of undiscovered recoverable oil range from 12 billion to more than 60 billion barrels. The changes created by these developments have been resisted by longer-term residents of Alaska, who resent the disruptions to their life-style, and by environmentalists, who fear permanent damage to the fragile Arctic ecology and the end of frontier isolation in the state's interior.

shield's minerals extend in a broad arc from Quebec and Labrador deposits of iron ore at Schefferville, Wabush, and Gagnon, through Sudbury (Ontario) nickel deposits, copper and iron deposits in northern Michigan and Minnesota, iron at Steep Rock in western Ontario, and copper and zinc at Flin Flon (Manitoba) to the major gold, uranium, and copper deposits from extreme northern Saskatchewan to the Arctic waters.

The second major zone of metallic minerals is located throughout the Appalachian Highlands, from the recently closed iron ore mines at Wabana, Newfoundland, to the Alabama iron ore used in the industrial development at Birmingham. Although this mineralized district is often forgotten by American students, it was very important during the early rise of United States industry and remains a significant contributor of selected minerals in the East.

The third region containing extensive concentrations of metallic minerals is the western cordillera, or mountain region. No student familiar with North American frontier lore can be unaware of the gold rushes to California and later to Alaska, the similar silver "boom" in Nevada, and the many frontier tales associated with mineral prospecting in the Western United States. Also of great economic importance are the less romantic but substantial deposits of copper, zinc, lead, silver, molybdenum, uranium, and other minerals scattered throughout this region from Alaska to Mexico.

Much of the mineral wealth of North America can be attributed to the intensity with which it has been sought during the last century and more. It cannot be disputed, however, that the United States and Canada contain fuel and metallic minerals in an abundance and quality found in few other regions. Few minerals required by modern industry are not found here; only high-grade bauxite (for aluminum), tin, manganese, and several less important minerals are absent or in small quantity. This does not mean that the United States is self-sufficient in virtually all its minerals. Much is imported from Canada and other countries to supply the immense manufacturing-industrial complex that has developed since the Civil War. But the local abundance of minerals was a great help to the development of this complex.

Location as a Resource

As important as the quantity and quality of resources have proved for economic growth in the United States and Canada, the spatial association of agricultural and industrial resource patterns is equally fortuitous. The great Interior Lowlands comprise, with the southeastern Coastal Plain, the second most extensive agricultural region in the world, and they are nearly surrounded by metallic and nonmetallic mineral deposits. Furthermore, the lowlands are underlain by tremendous mineral fuel deposits that are cheaply mined and constitute the primary power source for the numerous urban and industrial clusters in the northeastern United States and southeastern Canada.

These factors—resource quantity, resource quality, general locational associations—however, cannot be well utilized without a developed accessibility network. The two countries of North America are extremely fortunate to possess great natural transportation resources within the very zone containing the spatial association of rich agricultural and industrial resources. Long used by American Indians for interregional trade before the arrival of Europeans, the Great Lakes form an internal waterway broken only at Niagara Falls, between Lakes Ontario and Erie, and at Sault St. Marie, between Lakes Superior and Huron. With these exceptions, the Great Lakes offer over 1600 kilometers (1000 miles) of East–West transportation routes and cover over 1000 kilometers (600 miles) North–South at their greatest extent. They connect the southern margins of the Canadian Shield with the coal-rich and agriculture-rich Interior Lowlands.

Transportation routes are the physical expression of a society's social and economic organization. When these routes are built, as railways or highways, they express either existing needs for interaction or patterns that are desired. A railroad may be constructed to connect a localized resource supply with the industrial consumer of that resource, or it may be built because the region through which it is to pass is believed to possess sufficient economic potential to justify the construction costs. In either case, capital is required and the economic system must have developed to the point where potential can be recognized. The great advantage of internal waterways, whether connected lakes or an extensive system of navigable rivers, lies in their existence exclusive of previous human organization. They are not resources until people use them, to be sure, for "resource" is a human-related concept rather than an absolute. The sequential growth of settlement nd economic interaction in North America, however, definitely follows the pattern of natural accessibility provided by these waterways, especially during the early stages of economic and technological development.

The presence of industrial and agriculture resources adjacent to the Great Lakes lent strong support to the rapid economic and urban development of the northeastern Interior Lowlands. The Great Lakes, of course, are not the only transportation resource. Of direct importance to the long-term growth of the Interior Lowlands was the accessibility provided by the Mississippi River and its many tributaries. The Great Lakes are peripheral to the Interior Plains. The fact that the interior is largely of low relief means that the river systems contained within the region can be extensive; they are not restricted by irregular topography to narrow channels and few tributaries. Similarly, because the region traversed is of relatively even relief, the river system is navigable over great distances by low-draft boats. The agricultural portions of the regional resource potential are therefore well connected by cheap means of movement to the mineral fuel and metallic mineral concentrations which support the industrial base.

Canada

The primary Canadian core immediately north of the United States core—in fact, the Canadian portion of the North American continental core region—has also benefited from the accessibility provided by the internal water transport system. The primary development in Ontario is associated with transport routes on the Great Lakes and the upper St. Lawrence River, while Quebec's development remains closely tied to the lower reaches of the St. Lawrence. So important are the Great Lakes and the St. Lawrence River to Canada, in spite of navigation problems on the St. Lawrence, that it proposed and argued for the development of this waterway into a system capable of handling oceangoing vessels (that is, the St. Lawrence Seaway) decades before the United States acceded and cooperated in its construction. The Canadian core region, therefore, also benefits from the combination of resource quality and quantity and the locational advantages of high accessibility to both major United States markets and the important trading countries of Western Europe.

But Canada's situation is much less favored, overall, than that of the United States. As noted earlier, Canada's population clusters—whether in Ontario or in British Columbia—lie closer to American counterparts just across the U.S. border than they do to each other in Canada itself. This leads to close association expressed in such anomalies as commuting to and from work or shopping across an international boundary, watching foreign television, and so forth. Along the Canada–U.S. border, the pressure toward north–south integration is strong. For this reason, Canada has sought to establish more effective connections with its peripheral regions, cutting across the geographic "grain" of North America in the process.

Canada's core area (its share of the North American core region) occupies just 1 percent of the country's total territory. Yet it contains nearly two-thirds of Canada's population, and has about 80 percent of its manufacturing industries. With so heavy a concentration of population and productive capacity within a small region, there obviously are strong contrasts between the core and peripheral zones of the state. Regional inequalities, major problems for underdeveloped countries, can afflict developed states as well; in Canada's case the comparative remoteness of population clusters outside the core area intensifies local regionalism. In the United States (as Fig. 3-5 shows), linkages between the core area and subsidiary core areas radiate outward in several directions—to Texas, to California, to Oregon–Washington. Canada's connectors are one-directional (westward) and linear.

Agglomeration

The North American core region contains an unusually beneficial combination of economic and geographic resources. The development of industrialization and urbanization in this core during the last 100 years may be illustrated through a reexamination of the concept of *agglomeration* (see page 70) and its effect on cities, on manufacturing, and on the pattern of agricultural production. Agglomeration, we noted, is the clustering or grouping of phenomena from a relatively dispersed spatial pattern to one with little areal extent. Geographers view agglomeration forces as those that lead to this clustering, whether the phenomenon is people (urban places), economic activities (e.g., retail, wholesale, or production concentrations), or political control (nodal structure of political power).

During the last third of the nineteenth century and the beginning of the twentieth, United States and Canadian cities grew, not only as population concentrations, but also, and in general proportion to, the growth of manufacturing. Many factors contribute to the growth or stagnation of an urban area, but during this period one of the more important supports of growth was the increased concentration of industrial activities in metropolitan regions.

It is clear that individuals have long lived together in small areas because of the advantages that accompany this clustering. Economic efforts are concentrated, settlement is more defensible, and unless the cultures of the population are too diverse, the socialization afforded by proximity is generally more pleasant than individual isolation. The advantages related to the clustering of economic activities are of three types: economies of scale, complementarity, and mutual use.

Strictly speaking, economies of scale are a result of large-volume production rather than clustered production. Once the initial investment in basic manufacturing apparatus has been made by a firm, a more complete use of this apparatus permits the initial investment costs to be absorbed more gradually by the many units of output. It is therefore advantageous to use machinery to its productive capacity. In addition, volume production has meant large demands for labor accessible to the production plant and also a large available market for the goods. Large-scale industrial operations, therefore, became established in already developed population clusters, or they drew population to them.

Economies of scale also have meant specialized production. Separate companies manufactured product components in volume and sold them to other firms to use in the assembly of finished items. It proved advantageous to these manufacturers, therefore, to locate close to each other. The complementarity of such activities is often made more efficient by minimizing the cost of interindustry commerce. This clustering of interdependent and interacting industry multiplies the need for labor and markets and reflects the shared growth patterns of urbanization and industrialization.

There are many additional requirements for effective production. Large firms have special needs in water supply, sewage, electrical power, and so forth. Their employees require schools, recreation, food, and clothing, among many other items. Urban places have been best able to supply these functions. Those places already possessing these service functions attract industry, and the arrival of industry requires the expansion of these functions.

Industrial location is much more complicated than presented here, of course, but these "agglomeration economies" have contributed to the clustering of much manufacturing activity in a relatively few metropolitan regions. During the transformation of the North American economies from mercantilist to capitalist-industrial following the United States Civil War and Canadian Confederation, cities in the Northwest, the eastern Interior Plains, and Ontario–St. Lawrence Lowlands grew rapidly as the manufacturing core of the continent developed.

City Size and City Spacing

Cities and towns outside the core region also grew with the national economies. Each urban place possessed a set of site and situational characteristics that changed in importance as conditions altered, leading to the town's rise or its demise. Hence, we should look at the whole range and distribution of North American cities rather than individual towns, for regional patterns reveal significant urban features. First, the size of the population of cities and the number of such urban places are inversely related; that is, there are a great many small towns, fewer large towns, even fewer medium-sized cities, and only a handful of the very largest cities. Second, after taking account of different regional economic intensities, the larger cities tend to be located farther apart than small centers. Both features of North American urbanization reflect aspects of what has come to be known as *central place theory.*

Central place theory had its beginning in the work of the German geographer Walter Christaller (page 71). In 1933, Christaller published *The Central Places of Southern Germany,* in which he outlined a theory dealing with the number, size, and spacing of urban places. Christaller began with a set of assumptions formulated to isolate the most critical factors affecting urban distributions. Environmental differences were "smoothed" by assuming an even distribution of resources. Demographic and income variations were eliminated by assuming an even population density outside the urban places and an equal income distribution among this population. Distance was made important by assuming each person could travel in every direction toward any center with the travel cost proportional only to the distance to the center. Finally, it was assumed that each person desiring something in a town would seek to minimize the expense involved by traveling to the closest center supplying the good or service sought.

With this set of assumptions, each town's activities drew on the surrounding population for its set of customers. Nearby populations would be certain to shop in the closest center. As the radial distance from the urban place increased, a population would eventually be reached that was about the same distance from several towns. To avoid the difficulty of a set of consumers who were indifferent to alternative central places, Christaller argued that competition between equivalent centers for the surrounding population would "fill" the landscape with a series of hexagonal service areas (Fig. 3-8A). If the central places are too far apart to serve the rural population adequately, an intervening center will develop, so the spatial distribution of urban places should be predominantly even.

Any casual inspection of an actual landscape, however, demonstrates the purely theoretical nature of Christaller's proposed hexagonal distribution of central places. Some regions appear to approximate this pattern better than others, but there are several critical explanations for realistic departures from theory. In part, towns are not evenly distributed because resources (environmental and human) are not evenly distributed. Also, social scientists have recently realized that many people do not behave in such a way that their costs are minimized. And finally, as Christaller himself observed, urban places are not all the same size. Most regions therefore contain a varied array of small, medium, and large urban places in complex interrelation with each other.

The basic relationship between urban size and urban spacing can be viewed as an outgrowth of Christaller's initial assumptions and arguments. The simple hexagonal distribution is ideal if the assumptions are met *and* the urban places are about the same size. Some central place functions—activities that take place in urban areas—will not be used as often or by as high a proportion of the total population as will other functions. A grocery store, for example, is likely to be used often and by almost everyone nearby, while a shop that specializes in the sale of high-quality imported glassware will be visited less frequently by the average consumer. Some urban functions can therefore be found in even the smallest central places, while other functions are located only in large towns and cities. Small central places, then, contain relatively few activities—called "lower-order functions"—and large centers contain a greater variety of activities, some very specialized in addition to the lower-order functions. Because higher-order functions must "reach" outward farther to include enough potential customers for the function's economic survival (that is, the "range of the good" is greater) these functions cannot be located too near each other. And because higher-order functions are found in larger central places, fewer such large urban centers can be located in a given area; they will tend to be farther apart than small places (Fig. 3-8B).

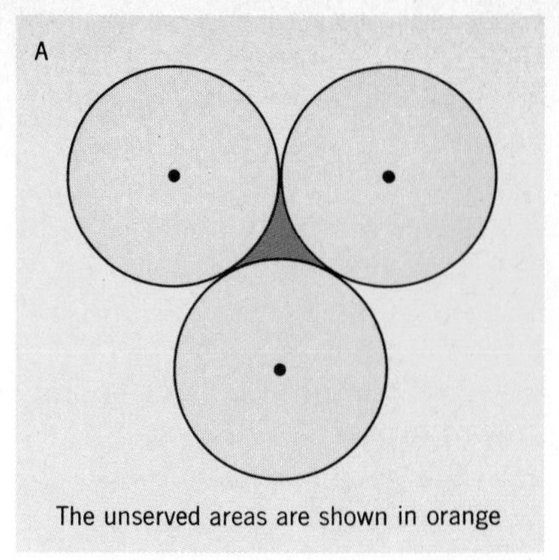

The unserved areas are shown in orange

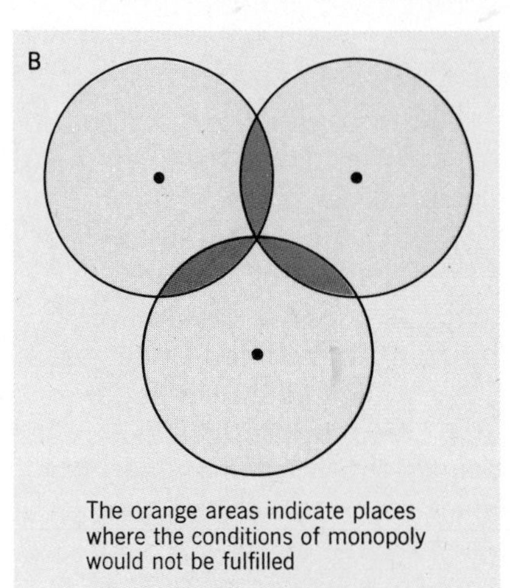

The orange areas indicate places where the conditions of monopoly would not be fulfilled

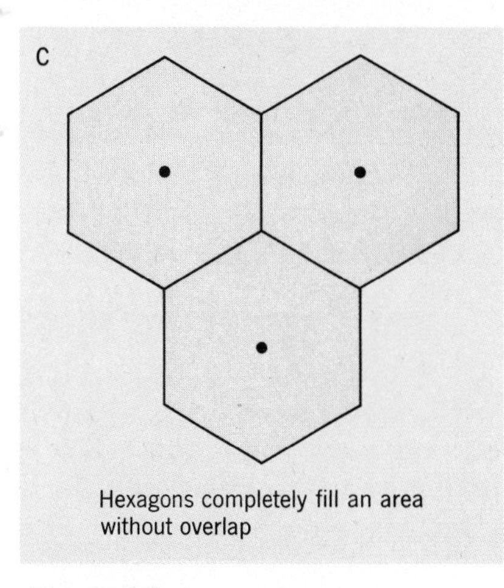

Hexagons completely fill an area without overlap

Figure 3-8

Christaller and other researchers who followed him attempted to extrapolate these conclusions to the identification of an *urban hierarchy.* Because higher-order functions could not survive unless they had access to populations exceeding some "threshold," and because central place theory based the spacing of urban places on the smallest hexagonal lattice of the lowest-order centers, it seemed apparent that the higher-order places would occur in discrete size categories just as they appeared on the basic lattice in discrete spacing patterns.

At first glance, this concept of an urban hierarchy fits with Jefferson's idea of a primate city (page 75) and with the observed pattern of cities in North America. At the highest level, there is only one New York City. Both Chicago and Los Angeles might be placed in the next category; both contain about half

New York's population, fewer functions, and are located at great distances from each other and from New York. The third level in the hierarchy contains even more cities, such as Philadelphia, Montreal, Toronto, Detroit, and others, and so on down the "hierarchy" to the very smallest aggregations containing urban functions. Some researchers, however, have found it very difficult to identify discrete central place categories and the distinct functional thresholds that would create such categories. They argue that the variety of functions occurring in medium and large-size towns creates a system of central places too complex for simple hierarchic categorization.

Regardless of the validity of these alternative positions, many aspects of central place theory are worth considering. Despite the confounding complexities of reality (few places can meet Christaller's simplifying assumptions), the theory does permit identifying and testing the fundamental relationships that govern the growth and distribution of systems of cities. That the general basis of the theory is correct is borne out on the landscape. There *are* fewer and more widely spaced large cities than small ones. Large cities *are* more numerous and closer to each other in the economically intense continental core region than outside that region. Specialized urban activities tend *not* to be found in smaller, closely spaced towns and villages unless the surrounding population is unusually wealthy.

Agriculture's Changing Role

At the same time that many North American cities experienced extremely rapid growth and many others grew substantially from industrial capital and labor inputs, agricultural output declined in relative importance. The key here is the term *relative,* for the total output from the agricultural sector has increased tremendously both in volume and in value. It has not grown as rapidly as the secondary and tertiary industries, however, and agriculture now contributes a much smaller proportion to the Gross National Products (GNP) of the United States and Canada than it did a century ago.

More important than the relative decline of agriculture is the absolute decline in the number of people directly engaged in agricultural production. In 1880, fully 72 percent of the total United States population could be classified as rural, most of these cultivating land for themselves, acting as tenant farmers for others, or practicing animal husbandry. By 1900, this percentage had declined to 53, and in 1920 it was clear that less than half of the population was engaged in rural activities. Except for the period of near economic stagnation during the 1930s, the proportion of population directly involved in agricultural production diminished more each decade so that by 1980, the approximately 220 million inhabitants of the country were fed by the efforts of fewer than 3 million farm families.

An important element of the relative decline of agriculture's contribution to the national economies should not be overlooked. With fewer people actually engaged in agriculture, the absolute increase in agricultural output reflects a tremendous rise in productivity per worker-hour. Today, it takes fewer people working a smaller area of land less time to produce a given quantity of food and plant fiber than even a decade ago, and this increasing efficiency shows few signs of moderation. The efficiency has, in turn, forced marginally productive regions of the country to change their agricultural emphasis. The areas most suited to corn production, for example, are the primary sources of corn.

The great increase in agricultural productivity is a reflection of the many technological innovations that have been accepted and used by North American farmers. Farm machinery, hybrid seeds, and a wide variety of fertilizers, insecticides, and herbicides are some of the more obviously important innovations that have revolutionized agriculture. Also of great impact have been modifications in crop handling and storage, in the care and feeding of farm animals, and in the marketing of farm produce. However, each refinement costs money. With the increased efficiency of production, small farm operations become increasingly difficult to maintain. The farm population has decreased rapidly as average farm size has increased to levels at which the new technology can be used.

Major Crop Regions

By far the most important crops grown in North America, whether measured in value of production or acreage devoted to their growing, are corn and

Mechanization and high productivity are hallmarks of North American agriculture. Wheat harvesting in Washington State. (Grant Heilman)

wheat. Corn production is heavily concentrated in a region extending from western Ohio, through northern Indiana and Illinois, through Iowa, and into southern Minnesota and extreme eastern South Dakota and Nebraska (Fig. 3-9). This region, with most production in Illinois and Iowa, is called the Corn Belt (page 2). It is often assumed that (1) this region is the only source of corn in North America, (2) it produces nothing but corn, (3) most of the corn produced is for human consumption, and (4) corn is grown throughout the region with equal intensity. Each of these assumptions is incorrect. As indicated on Fig. 3-9, corn is also grown throughout the eastern half of the United States, with substantial acreages devoted to its production along the interior Coastal Plain from Maryland and Delaware through the Carolinas into Georgia and southern Alabama. Numerous other crops are grown throughout the Corn Belt, with soybeans, wheat, oats, and alfalfa ranking among the more important. Most of the corn grown in the United States is grown as feed for cattle and hogs, the animals comprising the primary source of meat for the large markets in the northeastern core region. Although Americans do consume corn more readily and in greater volume than many other peoples of the world, corn's primary use is as feed. Corn is used for feed, instead of one of the other feed crops, such as hay, timothy, clover, or alfalfa, because it holds several advantages over most of its crop competitors. The entire plant may be used for silage, and the high-yielding and nutritious grain can be used for feed. The result is greater flexibility and higher value per acre to the farmer.

In contrast to the variety of uses for corn, wheat is primarily consumed by humans. Its areas of production are even more widespread than those of corn, although there are also several regions containing the overwhelming majority of production (Fig. 3-10). A winter wheat region, with wheat sown in the late fall and allowed to remain dormant over a mild winter before early spring growth, occurs in extreme northern Texas and through Oklahoma and Kansas into southern Nebraska. Spring wheat, sown after the harsher winters of the North have moderated into spring, is grown from the Dakotas into the Canadian Plains of western Manitoba, southern Saskatchewan, and into Alberta. Although smaller than these two regions of production, wheat is also grown in the Columbia–Snake River–Palouse region in Washington and Oregon. A substantial amount of wheat is also produced from the mixed farming region in the East, but the winter wheat belt is in fact a virtual one-crop region. Few other crops can approximate wheat's value-per-acre returns in this region's inhospitable environment.

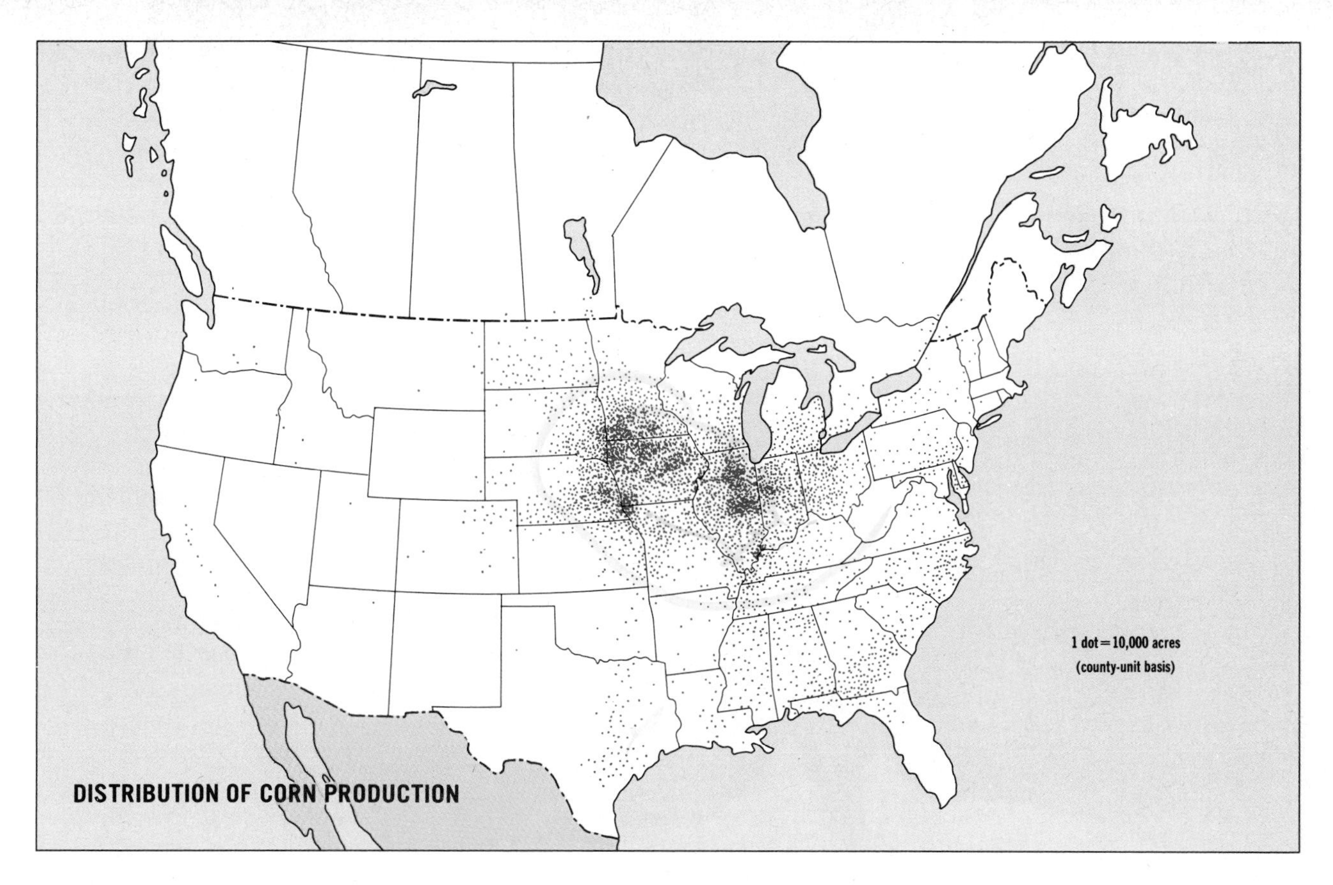

Figure 3-9

Figure 3-10

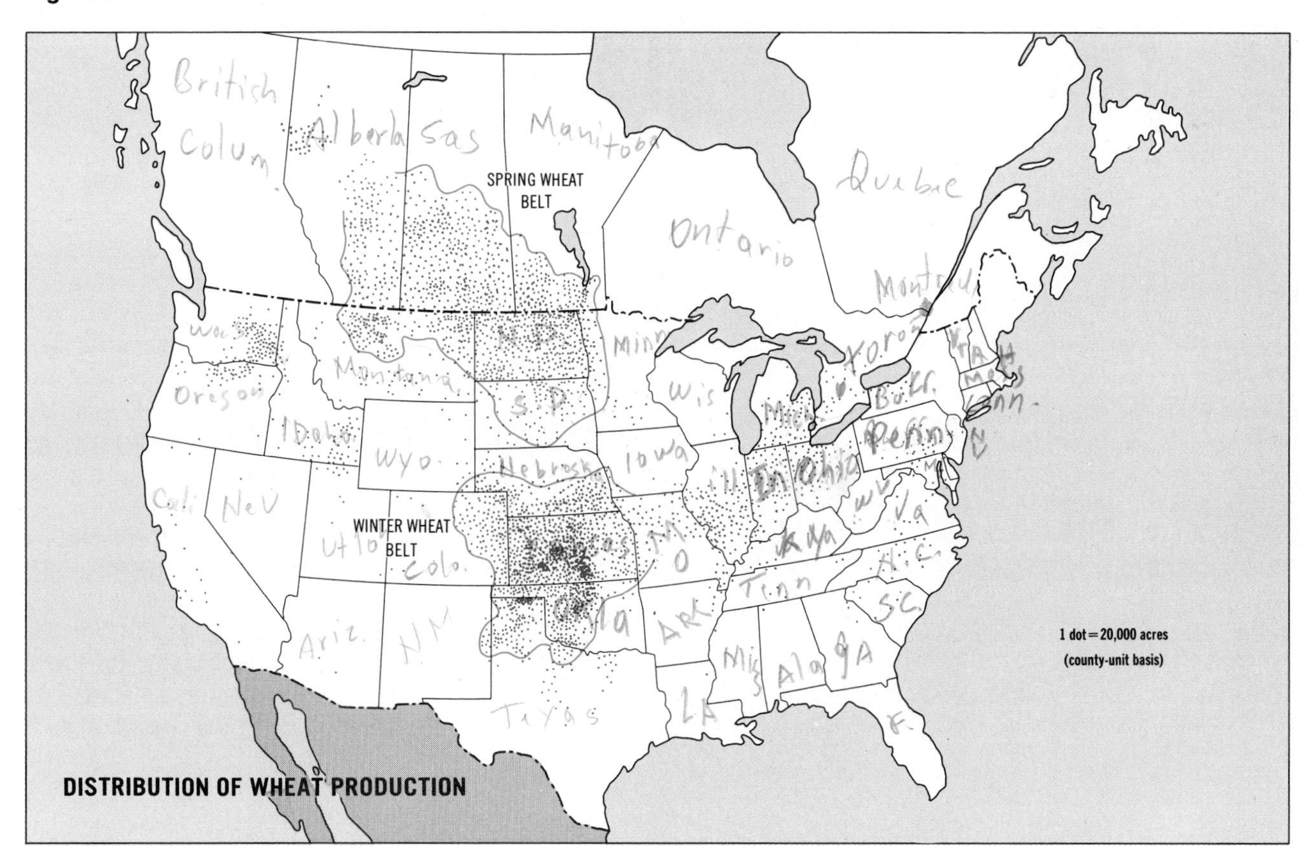

It should be noted, however, that the Canadian section of the spring wheat belt is one of only two major food-producing regions in that country. Food crops are grown on the Ontario peninsula and along the St. Lawrence River to Quebec. Wheat is annually grown on the Canadian Plains in quantities far beyond the needs of the 24 million inhabitants of the country, and so it is an obvious and reasonably dependable source of export funds. Purely in locational terms, the best buyer for this surplus wheat might be the United States, a wealthy country of 220 million people. Of course, the environmental conditions permitting a wheat surplus in Canada are also present south of the border, and so Canada must find buyers for this surplus overseas. Among these buyers are China and the Soviet Union.

Canada is a high-latitude country in which cold, snow, and relatively short growing seasons impose limitations on agriculture. Nevertheless, Canada's wheat lands (such as those of Saskatchewan, upper photo) produce large harvests of grain for export; despite the threat of snow and ice on the pastures (as in Alberta, lower photo) cattle ranching is also a large and thriving industry. (top, ©Craig Aurness/Woodfin Camp; bottom, Gerhard Gscheidle/Peter Arnold)

During most of the first century of the existence of the United States, the populated region east of the Appalachians contributed substantially to the food needs of the country. As agriculture became firmly established in the much more suitable environment of the Interior Plains, the margins of this new agricultural core developed complementary, rather than competitive, products. The land immediately north and east of the mixed farming Corn Belt produces vegetable, fruit, and specialty crops, or *truck* crops—so called because they are sufficiently perishable to require production within trucking distance of the consumption centers. This area also contains a sizable quantity of dairy production. It is no geographic accident that the Wisconsin dairy belt is located adjacent to the Chicago metropolitan region and the agriculturally specialized Corn Belt. Because of storage and shipment requirements, milk production must be, and is, proximate to nearly all major urban centers regardless of the relative environmental suitability.

The patterns of production for several of the speciality crops grown in North America also warrant attention. Cotton and tobacco, especially the former, were long the major agricultural income earners in the

South. However, the primary areas of cotton production have shifted westward, out of much of the region that depended so heavily on its production during the nineteenth century. Examination of the distribution of cotton production (Fig. 3-11) shows the easternmost region of concentrated output in what is called the upper Mississippi Delta; acreage of cotton grown through the eastern states of the former Confederacy is sparse and dispersed across the Piedmont Region. More important than the Piedmont are regions in eastern and northwestern Texas, central California, and southern Arizona along the Gila River. With the exception of eastern Texas, nearly all cotton grown in these western regions is grown under irrigation and is therefore of high quality. Governmental restrictions on production acreages, because of surpluses similar to those of wheat, have led to a sharp reduction in cotton acreage in the last several decades. Production remains high, however, as farmers use the best of their land to grow this lucrative crop. Thus, in spite of a drawing inward of production regions, cotton is still second in value of output only to corn. Tobacco, too, has become more concentrated through stringent government controls in the United States. North Carolina and Kentucky contain most of the tobacco production, as they have since the turn of the century (Fig. 3-11). Unless there is a marked rise in tobacco exports, the production of this crop is very likely to decline in the future. Significant amounts of tobacco are also grown on the Ontario Peninsula in Canada.

Problems on the Farms

Agriculture in North America is a mechanized, highly productive, scientific operation—and yet, during the 1970s, farmers blocked traffic in large cities to protest their low incomes. In 1978 farmers converged on Washington to participate in major demonstrations to underscore their demands for a larger share of the money consumers pay for farm products. What ails agriculture in North America?

Agriculture's problems are structural as well as environmental. The prices of farm products have risen, but the increases have benefited the farmers less than the processors, packagers, and distributors. Farmers in the 1970s reported that the prices their products commanded had fallen

Figure 3-11

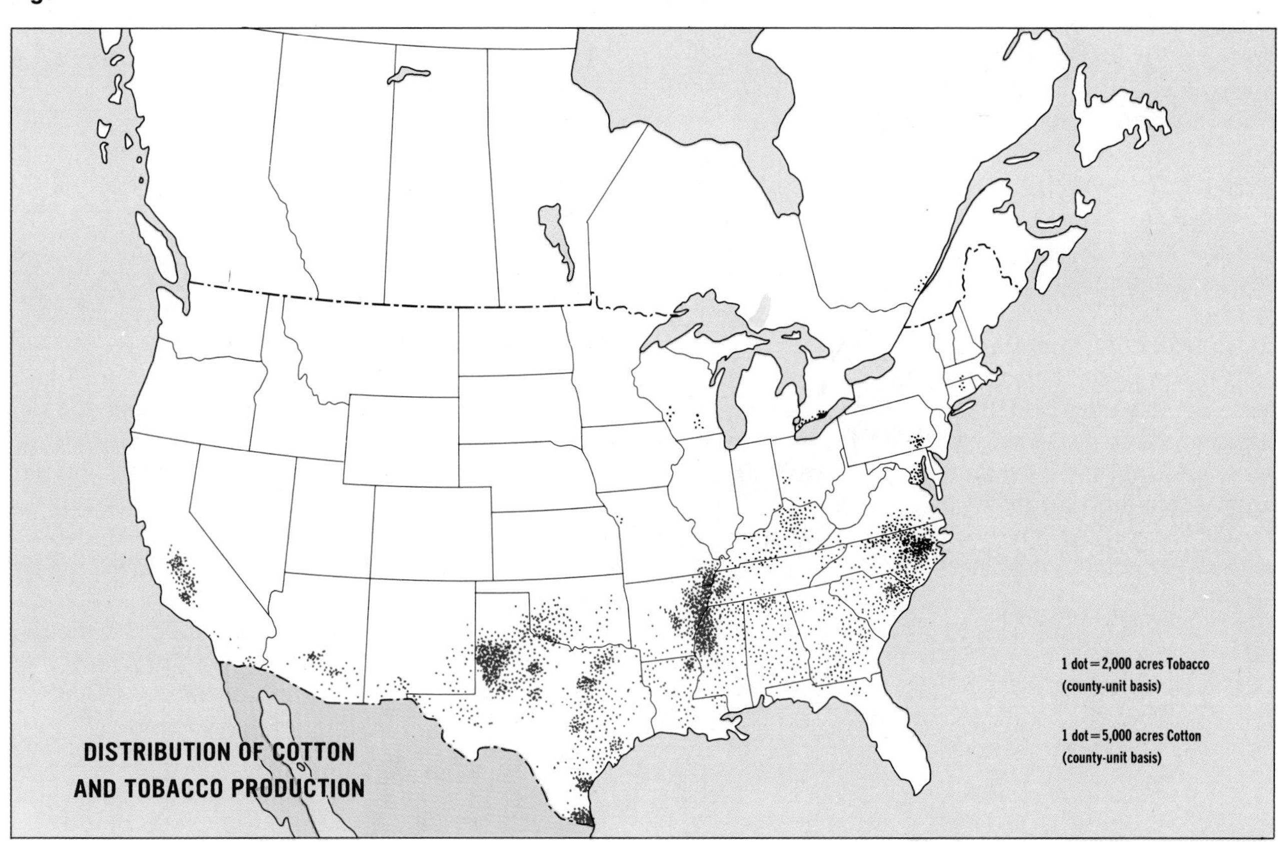

below the actual costs of production. Many, especially the remaining small farmers, faced the loss of their farms.

The apparently huge success of American agriculture, and the surpluses that can be sold to foreign countries, conceal another reality: profitable farming has become big business, requiring land, capital, equipment, and management in quantities beyond the reach of most farmers. In the United States in the late 1970s there were about 2.7 million farm families. Of these, 8 percent (or just over 215,000 farms) accounted for 70 percent of all the farm production annually. The great majority—1.6 million smaller farms, 60 percent of all U.S. farms—generated only 8 percent of the national output. These are the farms in trouble: incomes are small and declining, costs are rising. Many farm families who own small farms must engage in some additional salaried work in order to make ends meet, and every year there are families that lose their farms. Thus, the large operations become larger, and the family farm dwindles. The pattern is not new, but the pressure on the small farmer was never greater than during the 1970s with its inflation and uncertainty.

The family farm in the United States is under pressure. Despite the prosperous appearance of this contour-plowed, well-organized family farm, the farmer's income may not be sufficient to continue the operation and the threat of forced sale to a large farm corporation looms. (Grant Heilman)

Farmers in North America also faced environmental problems during the 1970s. Nothing comparable to the great droughts that brought famine to West Africa and South Asia struck North America, but there were problems nevertheless. In the U.S. West, farming depends heavily upon irrigation; the available water in the spring and early summer in turn depends on the precipitation pattern, especially the winter snows that accumulate for later melting in the high elevations. Repeatedly during the 1970s, precipitation regimes were disrupted and water supply fell below critical levels. Crops withered, a situation that not only threatened the survival of individual farmers but also revealed the vulnerability of dry-area farms to the variability of rainfall regimes.

Patterns of Population Redistribution

The first eighty years of this century have witnessed a substantial redistribution of population in North America, especially, by virtue of its larger numbers, in the United States. Canada has shared in two of these major migration patterns: as in the United States, Canadians have been moving from eastern sections of the country into western areas, and Canada's population has become urbanized as people have moved from the rural areas into the cities. Recently Canada's westward migration stream has been reinforced by the explosive economic growth of Alberta and British Columbia, and another population movement has gained momentum: the exodus of English-speaking people from Quebec. Corporations, businesses, and individual families have made the decision to leave Quebec, partly as a result of the provincial government's internal policies favoring French traditions and culture and partly in anticipation of the effects of Quebec's separation from Canada. Most of these Quebecan emigrants have moved to Ontario, but others have made their way farther west (and some have gone to the United States).

As noted at the beginning of this chapter, North America's population is extremely mobile. This mobility is expressed not only in major and persistent migration streams (such as the east-to-west flow), but also in frequent relocation of people within their region of residence, for example from downtown to the suburbs, or from one suburb to another. The average American family moves its residence every five years; looked at another way, these figures mean that 20 percent of U.S. families relocate every year.

Three major migration streams and two less directional relocation patterns can be identified. The three migration streams are (1) from East to West, (2) from South to North, and (3) from North to South and Southwest. The two relocation patterns are from rural to urban areas and intra-urban movements. To this should be added another migration of recent decades and growing proportions: the entry of Spanish-speaking people from Cuba, Mexico, and other countries into the U.S. Southeast and Southwest. A substantial part of this last stream is illegal and its dimensions are uncertain (especially from Mexico), but millions of immigrants are involved.

Major Migration Streams

Among the three major migration streams identified above, the two oldest ones are those taking people westward and northward. Political and social factors occasionally induced relocation, as did disagreements during the American Revolution and the intense racial hostility experienced by black Americans. But most often the chief motivation for migration has been economic. In the most persistent stream, settlement of the Interior Plains farmland was carried out by families seeking improved agricultural opportunities in the early nineteenth century. People went to the West Coast by ship and by wagon because of the promise of good, cheap land in the Willamette and Sacramento Valleys or quick riches in the California gold fields. The rich Canadian prairies were not settled until they were connected with the eastern Canadian population centers; the transport connection was not needed to carry the migrants, but is was necessary to make the land economically attractive. While the land frontier in the continental United States was filled before the end of the last century and Canada's settlement essentially complete by World War I, western regions continued to grow more rapidly than most others in both countries into the 1970s.

The second major migration stream involved movement out of the Southeastern United States. Hundreds of thousands of families of both races emigrated, but black Americans constituted the most visible and significant element in this population redistribution.

During the twenty-five years following the end of the Civil War, while the North was undergoing a major industrial and urban transformation and the West became settled in at least the outline form that has carried to the present, economic progress lagged in the South. Production increases were below those of the rest of the country mainly as a result of the war. There was some migration from rural to urban areas within the South, a small stream from the South to Northern cities, but most Afro-Americans remained where they had been in 1860. Many became freeholders with small farms, but an even larger number drifted into sharecropping and tenant farming as Southern landowners attempted to meet tax demands levied after the Civil War.

As long as possibilities for personal improvement were formally limited only by the individual's capacity and motivation, the South remained home for the vast majority of Afro-Americans. But as "Jim Crow" laws were introduced in state after state across the old Confederacy during the 1890s, the northward migration of blacks began to increase in volume. By 1914, black Americans were almost completely disenfrachised in the South, thus losing legal political recourse for the correction of racially motivated injustices.

Schools for Afro-Americans were poor or nonexistent, wages low (fifty or sixty cents per *day* in Alabama as opposed to thirty or forty cents per *hour* in the North), and lynchings common (54 occurrences in 1915).

In addition to these undeniably strong push factors, the North vigorously attempted to pull black labor from the South during the years of World War I. National industrialization before World War I depended heavily on millions of European immigrants to meet growing labor requirements. When the war interrupted this flow of labor, an alternative supply was needed. The discrimination-weary Southern black worker met this need nicely. Press campaigns were waged to encourage northward migration and labor recruiters were sent to Southern states to provide the means for this move.

Black Americans continued to leave the South in increasing numbers during the 1920s, primarily for the largest urban centers in the manufacturing core region. There was a decline in black migration during the depression years of the 1930s, but a return to national economic prosperity renewed the population movements during the next three decades. By mid-1973, it was estimated that almost one-half of the Afro-American population lived outside the South, a proportion in stark contrast to only 10 percent in 1900 (Table 3-1).

That black migrants from the South have moved predominantly into cities of the North and West is clear. Over 95 percent of the non-South black population was urban in 1980.

Table 3-1 Black Population of the Coterminous United States by Major Regions, 1890 to 1978, Showing Totals and Percent Distribution

Year	Total	Northeast	North Central	South	West
Population in Thousands					
1978	24,800	4464	4960	12,144	2232
1970	22,580	4335	4560	12,012	1693
1960	18,860	3028	3446	11,312	1074
1950	15,042	2018	2228	10,225	571
1940	12,866	1370	1420	9,950	171
1930	11,891	1147	1262	9,362	120
1920	10,463	679	793	8,912	79
1910	9,828	484	543	8,749	51
1900	8,834	385	496	7,923	30
1890	7,489	270	431	6,761	27
Percent Distribution					
1978	100.0	18.0	20.0	53.0	9.0
1970	100.0	19.2	20.2	53.2	7.5
1960	100.0	16.1	18.3	60.0	5.7
1950	100.0	13.4	14.8	68.0	3.8
1940	100.0	10.6	11.0	77.0	1.3
1930	100.0	9.6	10.6	78.7	1.0
1920	100.0	6.5	7.6	85.2	0.8
1910	100.0	4.9	5.5	89.0	0.5
1900	100.0	4.4	5.6	89.7	0.3
1890	100.0	3.6	5.8	90.0	0.4

Source: C. Horace Hamilton, "The Negro Leaves the South," *Demography,* Vol. 1, 1964, Table 1, *U.S. Census of Population, 1970,* and Bureau of the Census, *Current Population Reports,* series P-23, No. 80, p. 171.

Of equal spatial significance is the pattern of settlement *within* the major cities receiving these migrants. Incoming Afro-Americans have invariably settled in one or two extremely concentrated residential zones, usually in the older and more central portions of the cities where the resident black population already lived. The result since World War I has been both an intensification of habitation and a gradual areal increase in the sections containing the cities' black populations. The higher densities are caused by much greater inflow of migrants than can be accommodated by the slowly expanding residential zones where blacks have been able to buy homes. The de facto residential segregation in Northern and Western cities has been as effective as the de jure separation of racial groups long maintained in the South.

Two important results have followed from the spatial separation of blacks (and other identifiable ethnic groups) and whites in major cities. The high population densities that follow from restricted housing have led to severe structural overcrowding and accompanying environmental deterioration. A Civil Rights Commission report of 1959 states that "if the population density in some of Harlem's worst blocks obtained in the rest of New York City, the entire population of the United States could fit into three of New York's boroughs." When a low-income population lives in old structures under such overcrowding, extremely unfavorable living conditions invariably follow. As the black population has been forced to limit residential expansion almost exclusively to the margins of existing ghettos rather than into an open, widely dispersed residential pattern, the separate cultures are maintained. The black urban culture remains essentially closed to whites just as the white neighborhood remains virtually closed to blacks. This lack of cultural mixing is a concrete example of the maintenance of a plural society in the United States.

The South-to-North migration stream reflects the sharp drop in European immigration and constituted a major population shift within the United States. But it also illuminates other important changes that were taking place in the economy. Urban employment opportunities—in manufacturing, in service, and in government—continued to draw large numbers of workers from rural areas. The increasing use of machinery in agriculture permitted farmers to produce more per worker and per worker-hour. As was noted previously, agriculture has become increasingly efficient, producing more food and industrial crops with less labor and on less land in order to supply the increasingly urban population. Since most rural blacks did not own the land they worked, many moved to urban areas when mechanized farming began to diffuse through the South. Many others remained on the land to subsist without a regular source of income.

The third major migration stream that has shifted U.S. population involves movement from the North and Northeast to the Southeast and Southwest—the "Sunbelt." This movement gained momentum after the end of World War II, and by the decade 1955 to 1965 southward migration was large enough to match the northward flow. Since then, it has actually exceeded the older northward stream and the population of Sunbelt states, notably Florida, has mushroomed.

This southward migration may be of dimensions comparable to the northward population movement, but the people involved are very different. Northward, as noted earlier, went black Americans and poor whites in search of opportunities in northern cities. Southward came older, relatively affluent people (including many recently retired), most of whom moved to Florida and Arizona, and highly skilled younger people who found employment in the space and energy industries, especially in Texas. The people who left the South for Northern destinations mainly left rural areas or small towns and entered the already crowded inner cities of the North. Those who moved southward mainly left good housing in better areas of the cities and found new residences in suburban areas or retirement centers of the South. The southward stream also includes many Canadians who established retirement homes, especially in Florida.

The impact of these movements is far-reaching in both the areas of origin and of destination. We have already noted the problems of inner cities in Chicago, Detroit, Cleveland, and New York (among others), where deteriorating living conditions, unemployment, inadequate services, and other urban ills affect large segments of the population. But even in the Sunbelt the arrival of older Northerners has produced problems. Inflation, rising costs, and fixed incomes have had the effect of forcing hundreds of thousands of retirees into lower-quality housing, and in some instances this has had the effect of changing entire urban landscapes, as in the case of Miami Beach, Florida. Trailer parks in hurricane-vulnerable coastal zones are another reflection of the actual and potential problems of rapid in-migration, even when the immigrants are not poor and are not in search of jobs.

These three regional movements are the major migration streams affecting U.S. and Canadian populations in the 1970s and 1980s, but they are not the only ones that could be identified. People move to the West—but they also leave the West in substantial numbers every year, not only to try their luck in Utah or Colorado, but mainly to return to the Midwest or to the Northeast. A substantial movement carries people from the Southeast to Arizona and California, but a reverse flow nearly three-fourths as large returns to the Southeast.

The two nondirectional movements listed earlier involve rural-to-urban migration and intra-urban relocation. The causes of the rural-to-urban flow already have been underscored; the American farm is changing character and, government support and subsidies notwithstanding, farm families relinquish their land and move to towns and cities. Intra-urban relocation is a major part of America's mobility: the move "upward" to a better home in a more prosperous suburb, the move from one part of the city to another as a result of job relocation, and the decision to leave an apartment downtown for a place within commuting distance. These movements affect the majority of Americans, but they often appear on statistical tables in less dramatic form than the interregional migrations described earlier. Such intra-urban movement *does* affect population numbers in municipal areas, however. Cities have been losing population to outlying suburban areas in spite of continued in-migration to inner city destinations; the population of an entire urban area may be growing, but the "city proper"—the municipality at the heart of the urban area—can be declining. In 1960, Chicago had a population of 3,550,404; ten years later the city had 3,369,359 residents, a decline (according to Census data) of 181,045. But during the same period nearby North Chicago (part of the urban area) more than doubled from 22,938 to 47,275. Detroit, Cleveland, Minneapolis, St. Louis, Pittsburgh, and Philadelphia were among municipalities that recorded a loss of population in this way. It reflects an important shift in the living habits of the majority of urban Americans.

Distribution of Wealth and Poverty

By world standards, the United States and Canada are high-income countries, where annual income per capita exceeds that of all but a few other countries. Visitors to North America often are impressed by the enormous expanses of prosperous suburban developments surrounding American cities, by Americans' capacity to purchase automobiles, television sets, appliances, and other items that, in most of the world, are unattainable luxuries.

Given this considerable wealth, it is only natural that questions are raised about those areas in North America that have failed to share in the realm's overall well-being. Incomes in the East and Northeast (the region centered on megalopolis), in the urban zone of the Great Lakes states, and in the West from California to Alaska are even higher than the national average, higher perhaps than any comparable area in the world. But on this national scale there also are regions where incomes have been lower for many years (Fig. 3-12): the Southeast (except the southern half of Florida), Appalachian areas, and parts of the Southwest. And there are smaller areas where real poverty still prevails, and where America looks more like an underdeveloped country than the highly developed realm it really is. What lies behind this geography of relative well-being?

The regional distribution of low income, in which Southeastern states of the old Confederacy (except Texas and Florida) dominate, reflects the continued regional importance of agriculture during a period of national economic transformation away from such an emphasis. Almost without exception throughout the South, counties with *high* average per capita incomes are those with major urban centers. Concentrating on agriculture is not the only factor that continues low incomes in the South, however; many counties in Ohio, Indiana, and Illinois contain no larger cities and yet have much higher per capita incomes than do Southern counties. The same may be said for much of the High Plains and the Far West. It could be argued that the destruction of the social and economic organization of the South during and after the Civil War lowered the base from which the region had to "rise again"; economically, at least, it has been slow to do so. Perhaps an equally convincing argument is related to the size and racial composition of the South's population.

The South has never had as large a population as the North. The importance of large-scale, cash crop farming in the South—as opposed to the North's emphasis on small, single-family farms—contributed greatly to this pattern. In addition to the large no-income (that is, slave) population in the South, there were many small-scale, essentially subsistence white farmers among the real and aspiring agricultural aristocracy. After 1865, many of the wealthiest aristocrats were taxed into poverty, while freed slaves and low-income whites did not gain much in the way of income. After Reconstruction many whites were able to recover and reaccumulate wealth and power, but many poor white farmers could not. The blacks who remained in the South and who were faced with increasingly repressive "Jim Crow" conditions could achieve even moderate success only under exceptional circumstances. Because black Americans were in the majority in many counties of the South, even in 1970 (Fig. 3-13), the low average per capita incomes there were maintained by the extreme poverty under which the blacks (and many whites) lived. In addition, as long as much of the region's population possessed little income beyond that needed for the barest of necessities, there was an equivalently reduced market for local manufacturers.

It is clear from a comparison of the patterns of income variability and proportion of the population that is black, however, that extreme poverty is not limited to regions with large numbers of black Americans. Wolfe County, Kentucky, for example, in the eastern part of the state, contained no black residents at the time of the 1970 census and yet the median annual family income was only $2694. The "poverty pocket" in eastern Kentucky and adjacent areas of Tennessee and West Virginia, comprising the largest single cluster of low-income counties in Appalachia, has very different conditions behind its poverty than does the South.

Early farmers migrating from the eastern Coastal Plain and Piedmont usually did not settle until they reached the Interior Plains. Some farmers, however, must have been sufficiently attracted to the many small valleys and hollows beteen the low mountains of the Appalachian Highlands to decide to remain there. Four to eight hectares (10–20 acres) of bottomland were enough for subsistence plus sale of whatever hand-manufactured items could support the high transport costs necessary to bring the products to market. Corn mash whiskey, of course, is one such high-value per unit weight product.

Figure 3-12

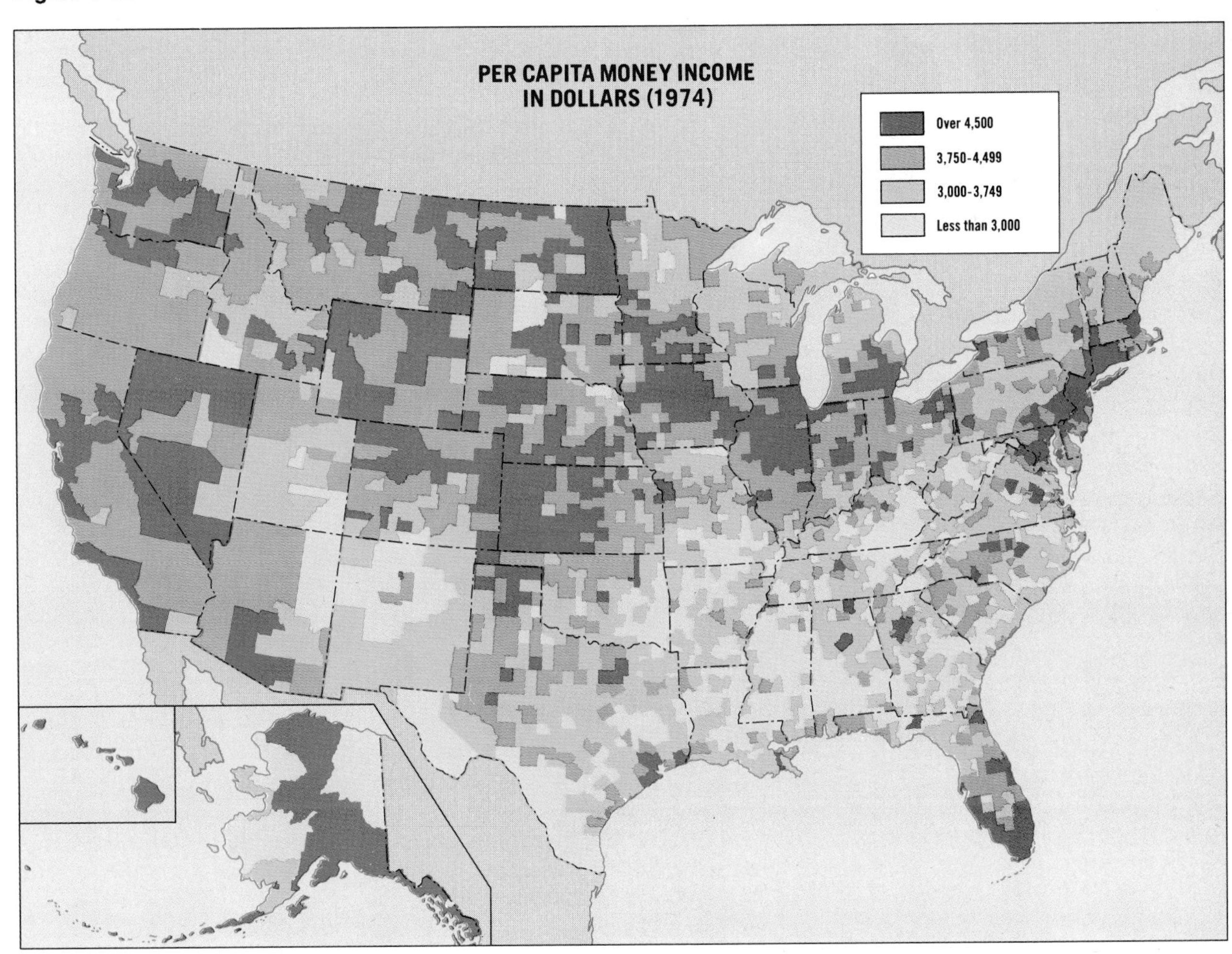

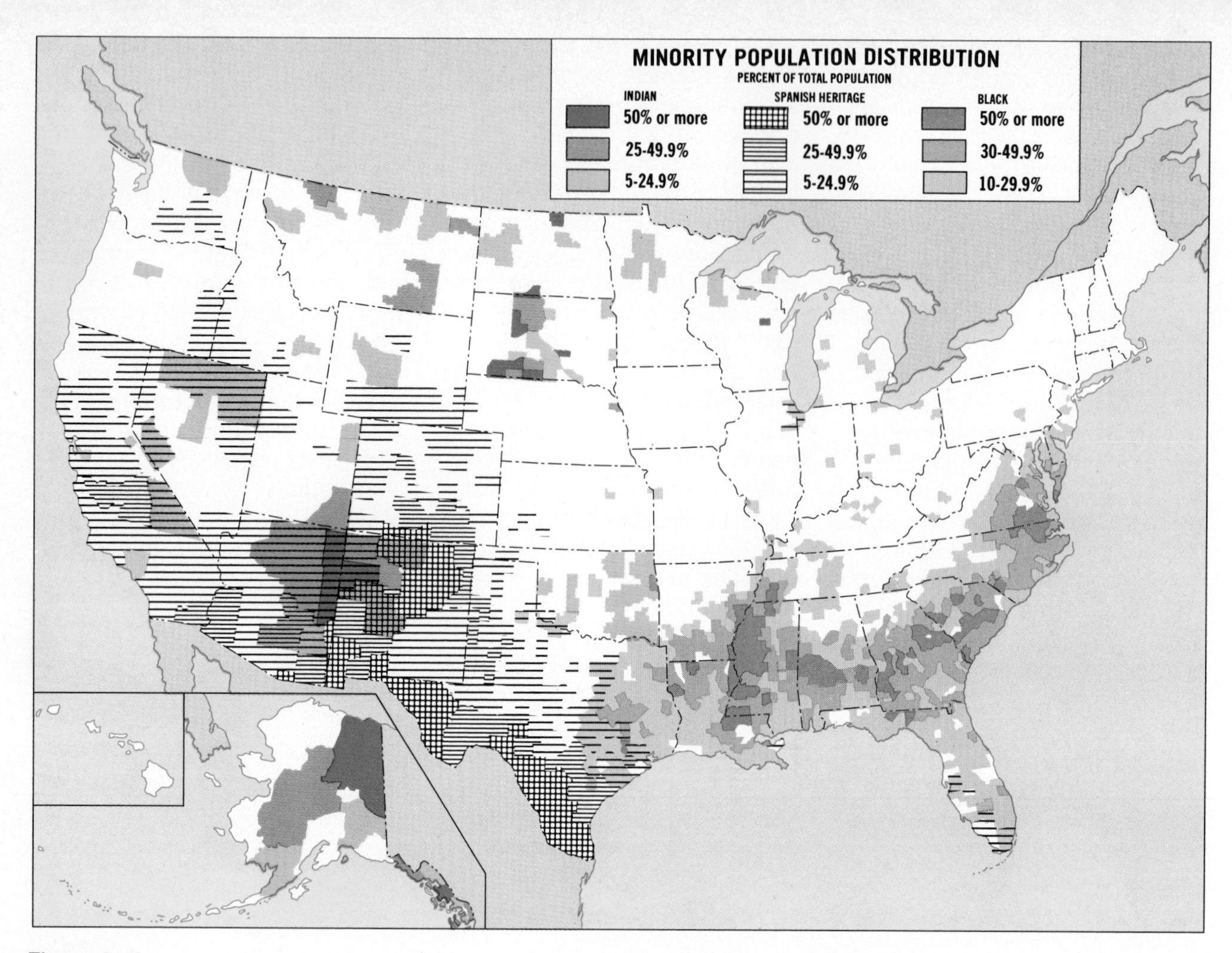

Figure 3-13

As national industrial development proceeded, the coal reserves in this region were mined, introducing much nonfarm population. Communication links with the rest of the country, never good, were improved only in the sense of connecting mine heads and the small mining communities with points or regions of coal consumption. Following the depression in the 1930s, during which the demand for coal was low, the demand skyrocketed through American involvement in World War II, a period of low male labor availability. As in many other industries, this combination of high demand and low labor supply resulted in a sharp rise in production efficiencies by the use of labor-substituting machinery. After the war, highly productive underground mining and strip mining continued to increase regional overpopulation. Farming was also often pushed below the competitive margin as mechanization of agriculture in the Interior Plains, where large farms were possible, proceeded rapidly. To make matters worse, alternative economic pursuits were often not feasible because of the extremely poor transportation network. Roads are rarely built without some economic justification, but development in a modern sense also cannot proceed without these roads. In eastern Kentucky, the road net is sparse, the local terrain rarely permits low-cost road construction, the expected economic support for roads is often indefinite, and the region remains poor. In many respects, except for the role played by coal in Appalachia, the situation is similar in the Ozark and Ouachita Mountains of Arkansas and southern Missouri.

Other small clusters of low-per-capita-income counties exist in the United States, several of them associated with the presence of a distinct ethnic group. In the extreme south of Texas, for example, in the group of counties west and south of San Antonio to the Rio Grande, average annual per capita income is under $1500 in spite of substantial irrigated fruit and vegetable farming in some of these counties. In these counties, at least, the countywide average incomes are lowered by the minimal, often submarginal incomes earned by the numerous Mexican-American laborers living there. Several other regions with a combination of unsuitable agricultural environments and concentrations of Mexican-American or American Indian populations also may be seen on the income map (compare Figs. 3-12 and 3-13).

In Canada, the low-income population is also locationally distinctive. The four Atlantic Provinces—Nova Scotia, New Brunswick, Prince Edward Island, and Newfoundland—contain about 10 percent of the national population but over 45 percent of Canada's poor. The region's development history is similar to that outlined for Appalachia; prior to this century, the population grew slowly and gained income from the relatively strong primary resource base, in this case fishing, timber, and mining. As the economy slowly transformed to one based on secondary and tertiary activities and one more heavily focused in the urban centers of Quebec, Ontario, and recently British Columbia, the Atlantic Provinces became increasingly isolated and their resources relatively less important in the nation's economy. An association between ethnicity and poverty also exists in Canada, most strongly for the Indian and Metis population living mostly in the Prairie Provinces. The Metis are descendants of French fur traders who married into the Plains Indian population and became a distinctive force in the region until these provinces were integrated into the Confederation.

Although quantitatively smaller than the U.S. westward flow of population, migration in Canada also peopled the west, and British Columbia now is the country's third most populous province. The region's leading city is Vancouver, at the heart of a conurbation of nearly 2 million people. (Peter Menzel)

The relationship between income generation and migration in North America is supported by the continent's general population redistribution patterns. Until very recently, the regions with the greatest out-migration were those with the lowest average levels of income—the Southeast, Appalachia, the Ozarks, and sections of the Great Plains. Within Canada, the Atlantic and Prairie Provinces were the major regions of out-migration. In both countries, the strongest flows were from rural areas to the large, economically intensive urban centers. We now examine the geography of North American cities in some detail.

Urban Structure

When entering the city from the surrounding countryside, almost all North American and European travelers would have similar expectations about the urban landscape they would be about to encounter. The first clearly nonfarm residences would still be far from the city center, scattered in small clusters, and occupied by families who commute to the city for work. Closer in, more dense residential development would be found, usually large homes on lots of one-half acre or more, indicating a population of moderately high income. As one traveled through this zone, the homes encountered would be smaller, closer together, and older than those found close to the edge of the metropolitan area. Finally, the two innermost areas expected would be a

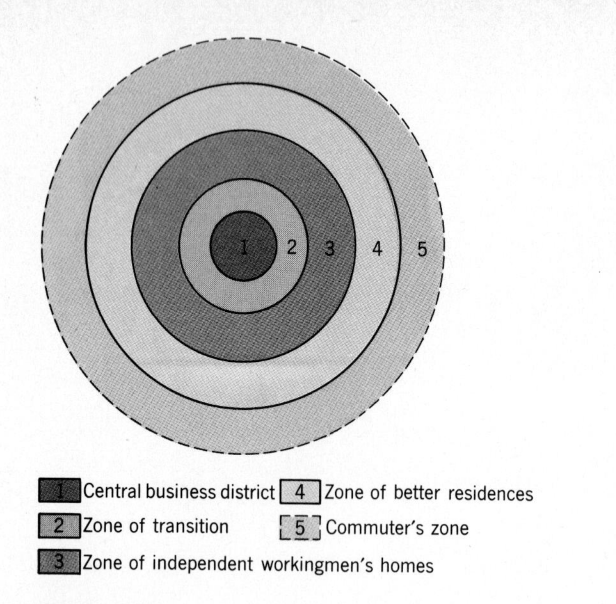

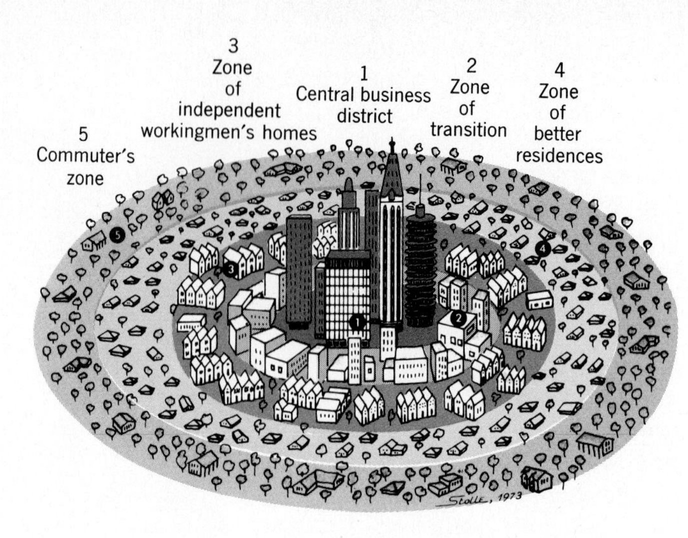

Figure 3-14

functionally complex and intense central business district (CBD) surrounded by a transition zone of light manufacturing and deteriorating residences.

Intuitively pleasing, perhaps because of its simplicity, this sequence of features was first formally proposed by the sociologist E. W. Burgess as a *concentric zone* model of urban spatial structure (Fig. 3-14). Although he described it in specific terms, Burgess recognized that each zone merged with those adjacent to it. Each concentric zone was therefore seen as a multifactor region with the boundaries between them existing as transition zones. The model also included the possibility of growth, for the zone immediately outside the CBD was explicitly subject to encroachment by business, light manufacturing, and wholesaling activities just as every other zone pushed outward into its surrounding environment. Finally, it is no accident that this model is similar to Von Thunen's "Isolated State." While not stated as formally and concerned with social as well as economic factors, some of the assumptions are the same, and the Burgess model describes urban structure as a gradual decline in land use intensity as distance from the city center increases.

Burgess based his model on his observations of urban structure in Chicago (where, because of Lake Michigan, he observed semicircular rather than circular regions as shown in Fig. 3-14). Certainly many cities in the United States and Canada display *elements* of concentricity; frequently the towering skyscrapers of the "downtown" CBD can be seen from the suburbs afar, and a trip into Atlanta, Kansas City, Houston, or Salt Lake City can illustrate just the sequence of zonal changes Burgess postulated. But careful observation will also reveal serious discrepancies. For example, light manufacturing is located in the second zone according to the model, but no provision is made for heavy manufacturing. Heavy industry usually treats bulky raw materials or products and thus tends to locate along major transportation routes. Wholesaling, too, is transport oriented and therefore tends to be located on one or two sides of the CBD instead of in a ring fully surrounding the business district as Burgess suggested.

Major transportation routes are obviously too important an influence on urban land use to be ignored as they are in the concentric zone model. There are economic advantages to location on or near the primary urban arteries for they provide good accessibility to other places in the city. Individuals and activities putting a high value on access are willing to pay more for locations on some of the arterials. Because many cities in the United States and Canada possess major street patterns that focus on the CBD much as the spokes of a wheel focus on the hub, it can be suggested that their land use should approximate a pie-shaped or *sector* model, rather than a concentric pattern.

This, in fact, was the explanation of urban structure offered by the economist Homer Hoyt. Basing this research on residential rent in a variety of cities in the United States, Hoyt

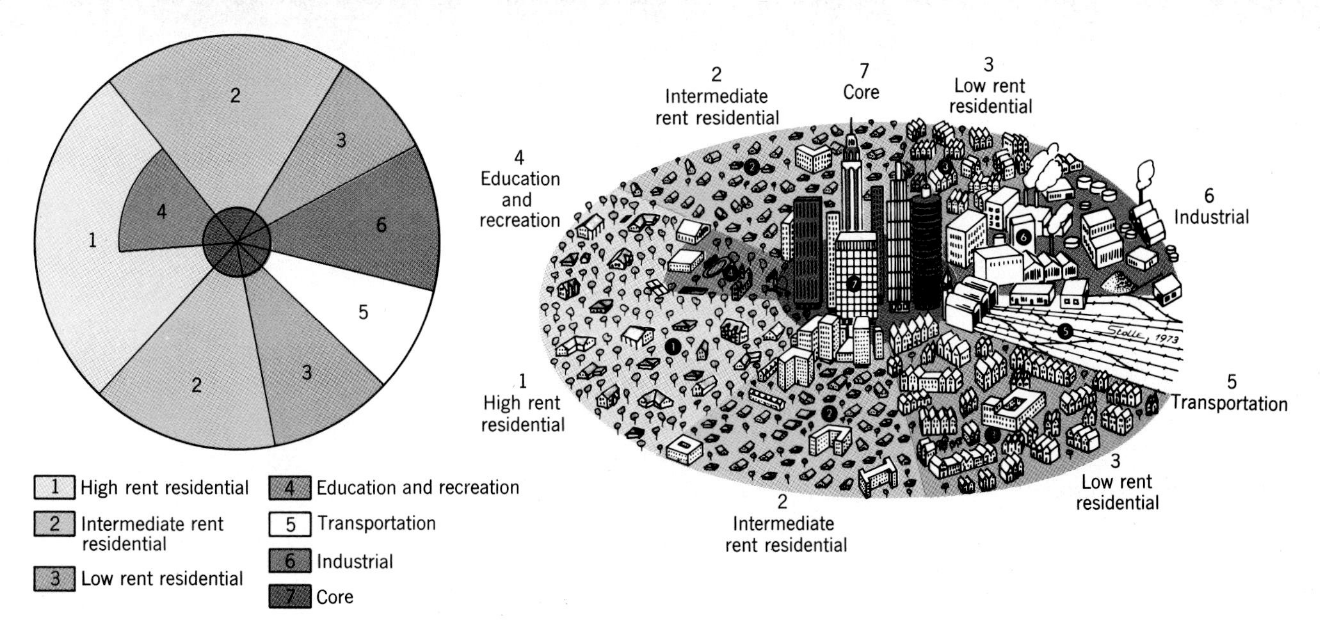

Figure 3-15

argued that rents paid by residents appeared to divide the cities into sectors, not concentric rings, with accessibility to the CBD much more important than simple distance. Thus a low-rent housing district extended from the city center all the way to the suburbs where movement was constrained or inefficient, and high-rent residences could often be found adjacent to the CBD on one side of the city as well as in the outer urban fringes (Fig. 3-15). By adding other urban functions, such as manufacturing and transportation, the general applicability of the sector model is widened beyond a strictly residential description.

Aspects of the sector model can be discerned in the layout of many North American cities. Areas where black workers have settled tend often to adjoin a major transport route leading into the downtown area or to an industrial district. Such axial orientation also affects manufacturing areas, which may follow railroads or waterfront docks. Boston is a major city where sectorial development can be recognized, and Montreal has an axial layout as well.

A third explanation of urban spatial structure was proposed by C. D. Harris and E. L. Ullman. It was observed that the CBD was becoming relatively less important as clusters of business and retail activities moved to sites closer to the urban periphery. Light manufacturing and wholesale activities also were seen to be located in several scattered clusters, or "nucleations," rather than in the regular, CBD-dominated patterns outlined by Burgess and by Hoyt. In fact, Harris and Ullman pointed out, large cities may grow around several nuclei rather than a single one. Therefore, they proposed the *multiple nuclei* model of urban structure (Fig. 3-16). These nuclei were seen as any urban growth stimuli—a government complex, a large research cluster, major shopping centers, a business and financial complex. Urban structure that is a consequence of such growth patterns would have a sprawling, disorganized appearance with no obvious focus. Los Angeles comes immediately to mind as a case in point, but many other North American cities that have grown rapidly since World War II show evidence of a multiple nuclei structure, often blended with aspects of either the concentric or sector patterns. Calgary, for example, grew in a clearly sectoral manner until the early 1960s. Toronto, on the other hand, possesses a dominant business center but also contains more than a dozen clusters of industrial activity.

Megalopolitan Growth

In many respects, the formation of multiple nuclei within and near the periphery of cities has resulted in a great increase in the spatial extent of the urbanized areas, that is, "horizontal" expansion. This aspect of

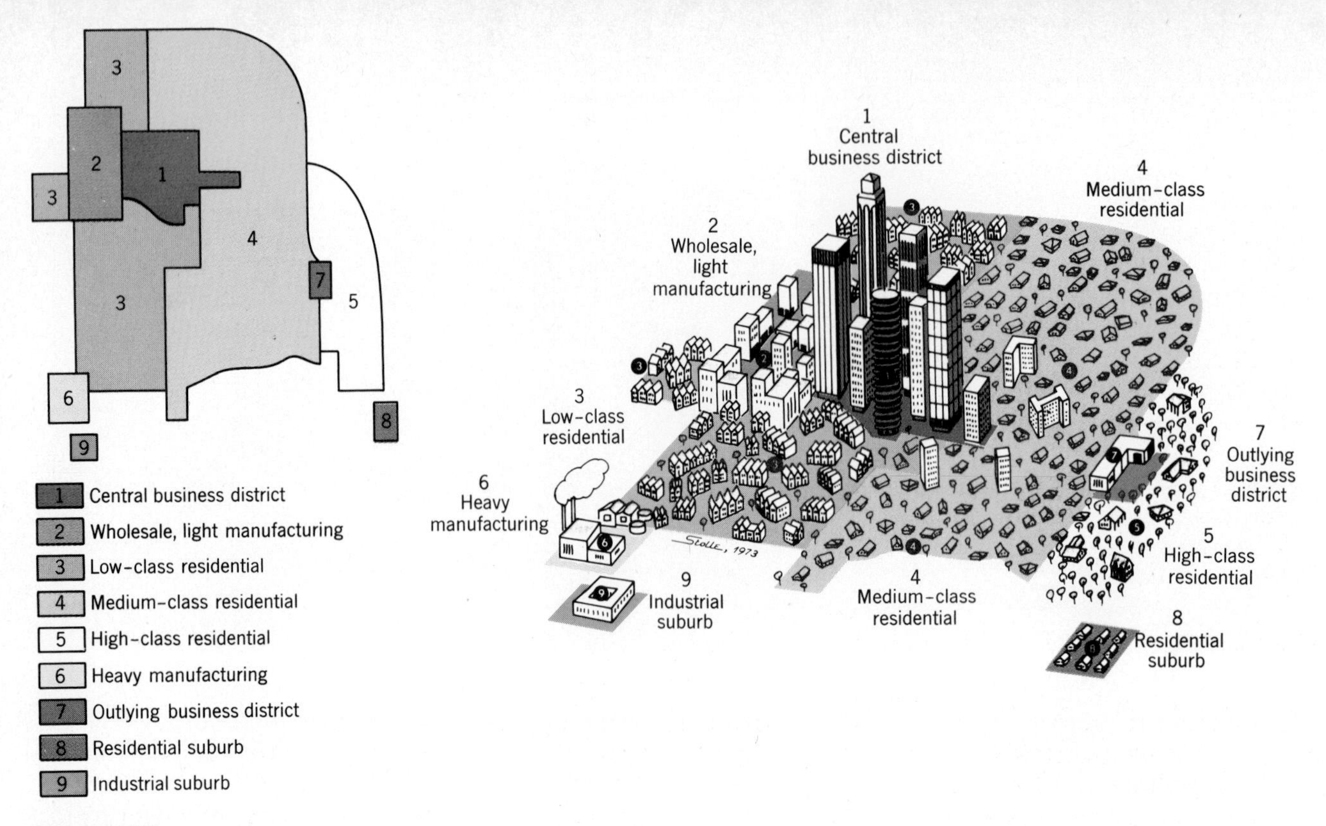

Figure 3-16

urban growth, popularly referred to as "urban sprawl," has proceeded to the point in some sections of the continent that the extensive metropolitan areas of several nearby cities have merged. The result has been the development of a number of "supercities" far larger even than the largest single city on the continent. The largest of these consolidated metropolitan areas, extending from Boston to Washington, D.C., and beyond, has been studied in detail by the French geographer Jean Gottmann. Gottmann chose the name *megalopolis* for this vast urban area, and the term has been applied as a label to other smaller or less well-connected coalesced cities in North America and around the world.

The Boston–Washington megalopolis, being the first and largest such urbanized area formed by the merging of distinct metropolitan units, bears great significance in the geography of the continent. This massive urban region is significant both in terms of its importance relative to the remaining cores of urban development and its indication of future problems that will require solution in other emerging megalopolitan areas. The Northeastern manufacturing core contains a majority of the continent's industrial and financial activity and a sizable proportion of the continent's population. Megalopolis is the eastern "hinge" of this continental core region and therefore of the United States and Canada as a whole. It contains a greater density of high-income population and high-income-producing activities—for example, research and laboratory industries, financial and managerial operations centers—than any other equivalent area in the continental core. Because of this tremendous concentration of population, economic potential, and cultural and intellectual leadership, megalopolis is an area of intense human activity. It is also an urban "laboratory" possessing problems in a variety and seriousness present only on a much smaller scale in other megalopolitan and metropolitan areas. The importance of these problems is magnified by the presence of the United States political and economic foci in megalopolis (Washington and New York City, respectively).

The problems facing megalopolis and other huge urbanized regions are many, but the most crucial may be treated under one of three major headings: political viability, livability, and accessibility. The government operations and organizations of these nodes within megalopolitan regions are restricted by fantastically complex patterns of jurisdiction. Region-wide planning and policy coordination are hampered. As sources of central city revenue diminish and the black population remaining in the city indicates its dissatisfaction with existing occupational and residential discrimination, urban political stability becomes ever more difficult to maintain.

Megalopolitan regions have developed from the coalescence of many cities of varying size. The resulting land-use pattern is extremely complex with innumerable tertiary and secondary activities interspersed among large residential areas. Such is the spatial extent of these various functions that large urban areas possess their own climates, slightly different from the general regional climatic pattern. The efficiency of atmospheric circulation is reduced in such urban areas. Used air is not adequately replaced by fresh air. Air pollution is supplemented by noise pollution to further reduce local environmental quality—that is, the region's "livability."

The atmosphere is not the only polluted portion of the megalopolitan environment. Most internal waterways are fouled by industrial, municipal, and even recreational wastes. Pollution makes more serious the problem of water supply, for great concentrations of people and industry demand large and regular quantities of fresh water for survival.

In addition to environmental pollution and the demand placed on the deteriorating environment by the population and business of megalopolitan regions, livability is also related to the quality and availability of recreational facilities. With the land between urban cores continually being converted to urban uses, less and less space remains available for recreation by the increasing population. Competition for land and water resources between these functions has become sharp, with the less-remunerative use of recreation often losing to either residential, industrial, or retail

At the heart of megalopolis: the center of New York City. Unmatched vertical development has produced an urban landscape without parallel; agglomeration has created locational advantages (see page 70) but environmental problems as well. In the lower photo, air pollution from factories and automobiles chokes the city on an otherwise clear and nearly cloudless day. (top, Owen Franken/Stock, Boston; bottom, Daniel Brody/Stock, Boston)

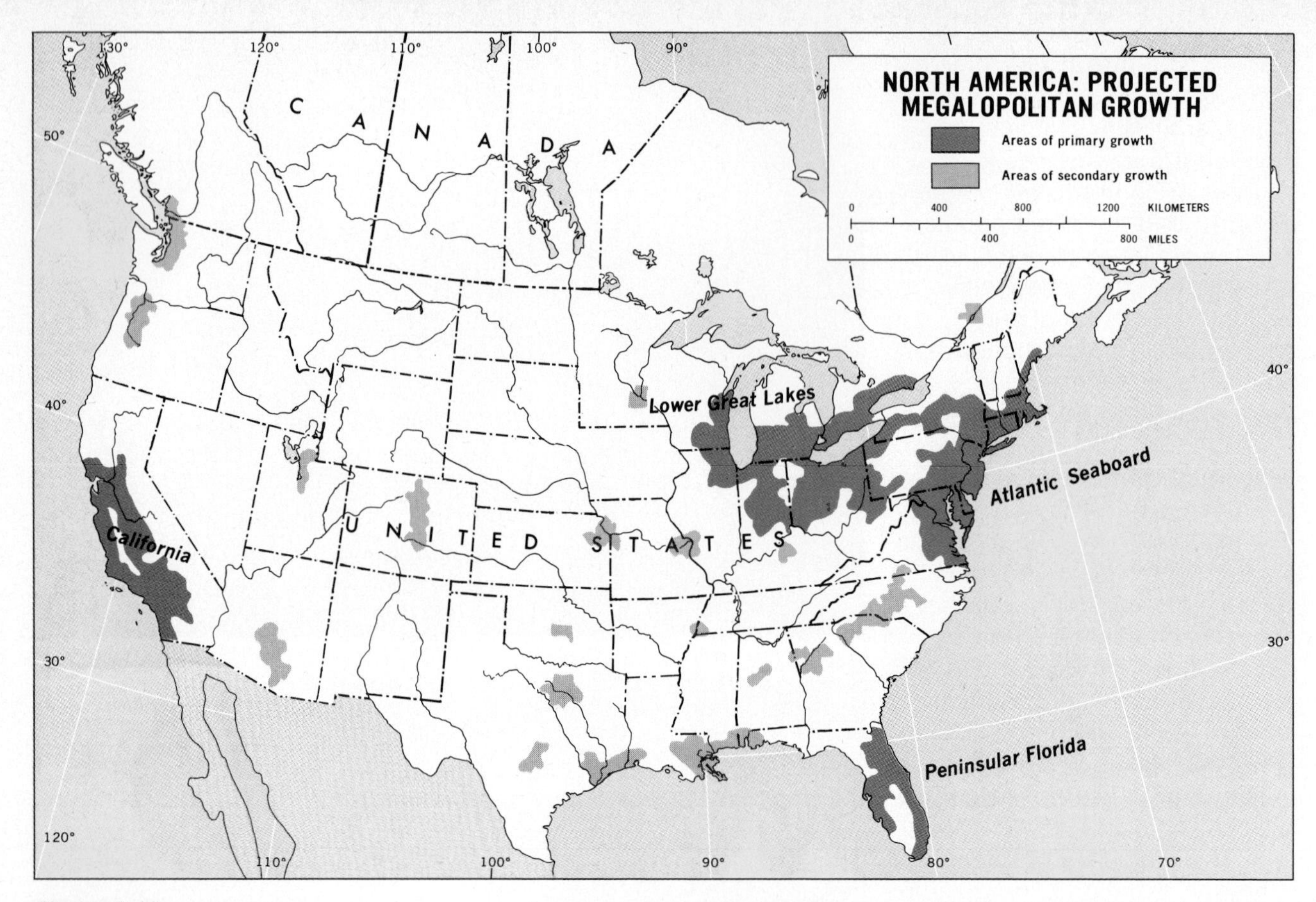

Figure 3-17

activities, or to pollution. In order to obtain the benefits of recreation, great numbers of people from megalopolis and other large urbanized areas must travel increasing distances each year. This, in turn, further encourages the geographic growth of megalopolis.

Many of these factors are related to the problem of constrained accessibility. Cities, or urban agglomerations, exist because advantages accrue to urban functions through a geographic clustering of the functions. Some share inputs, some use the outputs of other activities as input, some compete for the labor or the concentration of personal income found in urban regions, and some desire the advantages of location near a major transportation focus. In each of these cases, accessibility is crucial, for the interdependency described is not possible without the movement of either goods or people from a multitude of origins to a similarly large number of destinations. When the number of people and activity sites is as numerous as in megalopolis, when the distances involved are as great as they have become, and when the primary centers of daily activity continue to increase the volume of interaction within fairly constant areas (e.g., Manhattan, downtown Philadelphia, the governmental complex in Washington, D.C.), then accessibility eventually decreases and movement is constrained. When this happens, the essential purposes for agglomeration are undercut. Tales about one final, permanent, immovable traffic jam in the major urban centers are not total fantasy.

The characteristics and problems in megalopolis are of interest because it is clear from urban expansion elsewhere on the continent (and in the world) that other megalopolitan clusters are in the process of formation (Fig. 3-17). Although it is possible to foresee in the distant future numerous coalesced metropolises in North America approaching megalopolitan dimensions, only three such regions presently warrant consideration. The most extensive of the three lies between Chicago and its nearby cities to Pittsburgh via Cleveland and to Toronto via Cleveland and Buffalo. Some observers also predict a not-too-distant connection of urban settlement between the Chicago–Pittsburgh–Toronto megalopolis and the Boston–Washington megalopolis through the cities of the Mohawk Valley in New York (Rochester, Syracuse, Utica, Schenectady–Troy–Albany). The second major megalopolis that appears to be emerging would result from a future coalescing of the San Diego–Los Angeles urban regions with the rapidly growing urban development extending southward from San Francisco Bay. Some observers argue that these regions are too far apart to merge in the foreseeable future, especially since the demand for municipal, industrial, and agricultural water in Southern California already far exceeds the regional environmental capacity to produce it. This caution is supported by the most recent migration figures that suggest a slowing of westward population movement in the United States. The other possible megalopolis is in peninsular Florida, extending from Jacksonville to Miami along the Atlantic Coast and across the peninsula through Orlando and Lakeland to Tampa Bay. Population growth in Florida, fueled by the search for physical comforts, has slowed only slightly. It has been estimated that by the year 2000, over half of the approximately 270 million people in the United States will live and work in these four megalopolitan regions. This aspect of the continent's changing geography, therefore, wili have great impact in human spatial organization in North America.

Spatial Reorganization and Transport Technology

The apparently contradictory trends of population redistribution in the United States and Canada can ultimately be traced to the same factor. The continent's population has been increasingly concentrated in a relatively few large metropolitan regions, and at the same time, urban populations have been spreading outward from the central cities in a tendency toward dispersal that is at least 30 years old. Both trends are related to the effects of a changing transportation technology, although the more apparent association is with urban sprawl.

The more universal use of the automobile in the United States and Canada after World War II produced a series of changes in human activity and activity patterns that had not been clearly foreseen. Along with increased use of the personal automobile came a reduction in dependence on alternative methods of travel (public transit within the city and railroads between cities). The urban population began to live greater distances from work and shopping areas downtown. Employment and retail functions gradually moved outward toward the more dispersed population, clustering in small shopping centers and industrial "parks." As transport efficiency increased, costs of movement declined and more rapid movement was permitted. People could travel farther to work, to shop, and for recreation with no increase in time or money than they could years earlier. The more than 65,000 kilometers (41,000 miles) of United States interstate highways and equivalent limited-access roadways in Canada have led to a spiral of greater population dispersion→increased demands for transport facilities→improved movement system→greater population dispersion.

It should not be concluded from this that the automobile was the primary generator of urban sprawl. It was the technological means for dispersion, not the motivation. More important, in that they were more basic in urban geographic expansion, were the dispersal of higher incomes among the population, the desire to escape from the unpleasant aspects of city life, and continuing national economic prosperity. Although none of these is specifically geographic, all had direct geographic effects on spatial dispersion of the urban population. "Space-shortening" technology (i.e., the automobile and airplane) allowed the dispersed population to live, in many respects, as though the greater distances did not exist.

Petroleum: Problems of Supply and Demand

The immediate source of North America's "energy crisis" lies in the growth of demand relative to available supplies. The United States and Canada are the world's largest per capita consumers of energy; suppliers of that energy have found it increasingly difficult to meet consumption demands. These difficulties arise from three major, interrelated elements of energy patterns. First, consumption has generally been wasteful. As long as abundant supplies were available, efficiency in energy use was seen as unnecessary. Second, the great growth in energy demand is more and more distant from new supplies. Further, the North American urban manufacturing core region has continued to grow, but, for environmental and technological regions, an increasing share of the total energy consumed in the region (as elsewhere on the continent) has come from the Southwestern and Western petroleum and natural gas reserves. And third, the proven reserves of the fuels in highest demand have not been increased as rapidly as they have been consumed. Even with the addition of the Alaskan North Slope reserves, for example, U. S. domestic production has exceeded new domestic discoveries of oil and natural gas each year since 1970.

There are at least three clear geographic consequences of this changing supply/demand relationship. First, and most directly, the United States as a whole and especially the eastern one-third of the country will be more and more dependent on foreign supplies of these fuels. Only 11 percent of petroleum consumed in the United States in 1957 was imported. This level increased to 19 percent by 1967 and approached 50 percent by 1977. Worried government officials talked of a trade deficit equivalent to 20 percent of the country's gross national product by the mid-1980s just to pay for imported fuel. Second, and less directly, a society dependent on intense spatial interaction would be affected throughout its economy by changes in fuel costs, especially those fuels imported from foreign sources. Third, rising costs of mobility could force a change either in the mode of movement—mass transit instead of individual transportation—or in the location of population and its economic activities. The trend in spatial redistribution of population since World War II has been toward dispersal, but this requires a high consumption of low-cost fuel. With higher movement costs, there may be a reversal of this trend and a return to increased clustering of population and economic activities.

The relationship between travel time, distance, and the process of spatial reorganization was studied by D. G. Janelle in a way that illustrates these changes. He observed that in the past, a demand for accessibility often led to an improvement in transportation, usually through some technological development. Further, the transport innovations, either in the vehicle of movement (railroad, automobile, truck) or in the routeway, improved access between points by reducing the direct cost of travel or the time required to complete the trip. By determining the average travel time between a number ot towns and cities in southern

Energy and its availability have become leading concerns in the United States. Canada to the north and Mexico to the south both export oil, but the United States must import large quantities because its own fields cannot supply domestic demand. Thus the search for new reserves continues. The Alaska pipeline, which carries oil from the distant north to the Pacific coast for shipment south, was opened in 1977. (Courtesy Alyeska Pipeline Service Co.)

Michigan, Janelle found that a trip that required almost 24 hours to complete by land transport in 1840 had been reduced to about 2½ hours by 1900 and about 80 minutes by 1965. If distance is defined as the time it takes to travel a given number of kilometers (or miles), then in effect the trip distance between these places had shrunk greatly; they had become much closer together. Places in the United States and Canada touched by these transport improvements had experienced a *time-space convergence.*

It has been determined, however, that the time-distance between places has not converged at the same rate. The larger the place, the greater the demand for improved access to other locations. Small towns and villages do not generate as much interaction with other places as do major metropolitan concentrations. As an extension of this, the populations of large cities interact most strongly with other large cities. Therefore, improvements in transportation are greatest between these more populated centers. In terms of time-space convergence, large cities are growing closer together more rapidly than are small towns and villages. As discussed earlier, there are economic benefits for activities located close together. Thus, the benefits of time-space shortening have been enjoyed primarily by those places that are already the most dynamic economically—namely, the large metropolitan regions. This has meant, for example, that communications between New York and San Francisco or Toronto and Vancouver are not only faster and cheaper than they were several decades ago, but also the *reduction* in travel time for these pairs of large centers is greater than the improvement for small towns only hundreds of kilometers apart. During the 1970s, at the same time that the rural–urban and South–North migration showed signs of reversing, rapid increases in fuel costs threatened to alter the regularity of this space-shortening process. The implications are not yet clear, but they are unlikely to alter drastically the basic patterns of urbanization, high average annual incomes, and poverty associated with plural societies.

Both the United States and Canada are very large countries, each extending well over 4000 kilometers from east to west.

The dominant physiographic orientation of the continent is north–south. The national integration of each country was achieved in spite of this trend by the early construction of transcontinental railroads. Strengthening circulation patterns have provided the key to the past century of integration. Conversely, to a great extent, the lack of interaction—the exclusion from the dominant circulation patterns—has maintained poverty in east Kentucky, the rural Southeast, the Ozarks, tha Atlantic Provinces, and so forth. The poverty within urban ghettos is similarly accentuated and perpetuated, not by lack of facilities for interaction, but by the absence of residents' participation in the flow of national activity. The reasons for such nonparticipation are many, but they are supported by spatial separation and isolation, whether physical, economic, or social.

Additional Reading

There is a variety of texts on the geography of North America. The most readable are *Regional Landscapes of the United States and Canada* (Wiley, New York, 1978) by S. S. Birdsall and J. W. Florin, and J. H. Patterson's *North America* (Oxford, New York, 1975). You may also want to look at *Regional Geography of Anglo-America* by C. L. White, E. J. Foscue, and T. L. McKnight (Prentice-Hall, Englewood Cliffs, 1974). An excellent treatment of Canadian geography is found in the text edited by John Warkentin, *Canada: A Geographical Interpretation* (Methuen, Toronto, 1968). For alternative treatments of the United States, read W. Zelinsky's *The Cultural Geography of the United States* (Prentice-Hall, Englewood Cliffs, 1973) and the selected regional articles in the June 1972 issue of the *Annals of the Association of American Geographers,* Vol. 62, No. 2.

For an excellent history of the early developments of black–white relationships in America, see Winthrop E. Jordan's *White over Black: American Attitudes Toward the Negro 1550–1812* (University of North Carolina Press, Chapel Hill, 1968). A provocative analysis of slavery is offered in *Time on the Cross* (Little, Brown, Boston, 1974) by R. W. Fogel and S. L. Engerman. A good short article on the motivations for South-to-North black migration is Dewey H. Palmer's "Moving North: Migration of Negroes During World War I," *Phylon,* Vol. 28, No. 1, Spring 1967. Although there have been numerous geographic studies of black America during the last decade, the best overall book remains *The Black Ghetto: A Spatial Behavioral Perspective* (McGraw-Hill, New York, 1971) by Harold M. Rose.

For a discussion of low-income populations in Canada, see *Poverty in Canada* (Prentice-Hall, Scarborough, 1971), edited by John Harp and John R. Lofley. Richard Morrill and Ernest Wohlenberg's *The Geography of Poverty in the United States* (McGraw-Hill, New York, 1971) examines the income distribution in the United States at state and regional levels. A slightly more difficult but worthwhile exposition of noneconomic well-being can be found in David M. Smith's *The Geography of Social Well-Being in the United States* (McGraw-Hill, New York, 1973).

In the area of urban geography, Allan Pred's "Industrialization, Initial Advantage, and American Metropolitan Growth," *Geographical Review,* Vol. 55, No. 2, April 1965, offers a careful introduction to the complexities of the interdependent support of urban and industrial growth. An excellent summary of "Christaller's Central Place Theory" can be found in an article of that title by Arthur Getis and Judith Getis in the *Journal of Geography,* Vol. 65, No. 5, May 1966. Probably the most accessible of E. W. Burgess's statements on the concentric zone model of urban structure is "The Growth of the City," *The City* (University of Chicago Press, Chicago, 1925), edited by R. E. Park, E. W. Burgess, and R. D. McKenzie. A full discussion of the sector model of the city can be found in H. Hoyt, *The Structure and Growth of Residential Neighborhoods in American Cities* (U. S. Federal Housing Administration, Washington, 1939). The article by C. D. Harris and E. L. Ullman, "The Nature of Cities," in the *Annals of the American Academy of Political and Social Sciences,* Vol. 242, 1945, contains a discussion of city size and spacing as well as their multiple nuclei model of urban spatial structure.

Many books have been written in recent years about the problems and characteristics of urban America, but Jean Gottmann's description and analysis of the urbanized northeastern seaboard of the United States, *Megalopolis* (MIT Press, Cambridge, 1961) remains fascinating reading. A shorter and more recent set of studies of megalopolitan centers is Clyde Browning's *Population and Urbanized Area Growth in Megalopolis, 1950–1970* (University of North Carolina, Chapel Hill, 1974). Thirteen urban "vignettes" have also been written by geographers and published in a four-volume series, *Contemporary Metropolitan America,* by the Association of American Geographers (Ballinger, Cambridge, 1976). The concept and implications of time-space convergence are found in "Spatial Reorganization: A Model and Concept," *Annals of the Association of American Geographers,* Vol. 59, No. 2, June 1969, by D. G. Janelle. For those interested in the future, a set of speculative essays has been published in *Human Geography in a Shrinking World* (Duxbury, N. Scituate, 1975) edited by R. Abler, D. Janelle, A. Philbrick, and J. Sommer.

Prodigious Japan: The Aftermath of Empire

IDEAS AND CONCEPTS

Modernization
Relative location
Exchange economy
Industrial location
Areal functional organization
Physiological density measures

In the non-Western, non-European world there is no country quite like Japan. None of the metaphors of underdevelopment applies here: Japan is an industrial giant, an urbanized society, a political power, a vigorous nation. Probably no city in the world is without Japanese cars in its streets; few photography stores are without Japanese cameras; many laboratories use Japanese optical equipment. From tape recorders to television sets, from oceangoing ships to children's toys, Japanese manufactures flood the world's markets. The Japanese seem to combine the precision skills of the Swiss with the massive industrial power of prewar Germany, the forward-looking designs of the Swedes with the innovations of the Americans. How have the Japanese done it? Does Japan have the kind of resources and raw materials that helped boost Britain into its early, revolutionary industrial prominence? Are there locational advantages in Japan's position off the eastern coast of the Eurasian landmass, just as Britain benefited from its situation off the western coast? Could Japan's rise to power and prominence have been predicted, say, a century ago?

Japan consists of four large, mostly mountainous islands (and numerous small islands and islets) whose total area amounts to a mere 370,000 square kilometers (143,000 square miles), a territory smaller than California. Japan's population of 119 million (1980) is crowded on the limited coastal plains, valleys, and lower slopes of these islands, principally on the largest (Honshu) and southernmost (Kyushu and Shikoku) (Fig. J-1). Northernmost Hokkaido Island still has frontier characteristics. And had it not been for Russian imperialism, Japan might also control Sakhalin Island, the fifth large island of Japan's archipelago.

Is Japan a discrete culture realm, or should it be viewed as an insular extension of the Chinese world? Japan was peopled from Asia mostly via the Korean Peninsula, and, over many centuries, it received numerous cultural infusions from the Asian mainland. By the sixteenth century, a highly individualistic culture already had evolved. But for more than a century Japan has taken new directions—technological, cultural, and political—so unlike those prevailing in mainland Asia that ample justification exists for its designation as a culture realm. By the turn of the twentieth century Japan was an industrial force, a colonial power. By a unique blend of tradition and modernization,

Ten Major Geographic Qualities of Japan

1. Japan is an archipelago of mountainous islands off the east coast of Eurasia.
2. Japan buys raw materials and sells finished products the world over; it lies remote from most of the locations with which it has close economic contact.
3. Japan is the prime example in the world of modernization in a non-Western society.
4. Japan was the non-Western world's major modern colonial power; during the colonial era it strongly influenced the development of the Western Pacific.
5. Although recent interaction has been limited, Japan lies poised for a major role in development in East Asia, including Eastern Siberia as well as China.
6. Japan's modern industrialization depends on the acquisition of raw materials (including energy sources) from distant locales, Japan's own mineral resource base is limited.
7. Modern and traditional society exist intertwined in Japan.
8. Japan has the highest physiologic population densities in the world.
9. Japan's agriculture, given conditions of climate, soil, and slope, is the most efficient and productive in Asia.
10. Japan's areal functional organization, based on regional specialization, is fully developed.

borrowing and innovation, discipline and ambition, and with enormous energy and organizational ability, the Japanese transformed their country, overcame defeat and disaster, and created a society unlike any other in Asia.

In the mid-nineteenth century, Japan hardly seemed destined to become Asia's leading power. For 300 years the country had been closed to outside influences; Japanese society was stagnant and tradition-bound. Very early during the colonial period, European merchants and missionaries were tolerated and even welcomed, but as the European presence on the Pacific shores grew greater, the Japanese began to shut their doors. By the end of the sixteenth century Japan's overlords, fearful of Europe's imperialism, decided to expel all foreign traders and missionaries. Christianity (especially Catholicism) had gained a considerable foothold in Japan, but it was now viewed as a prelude to colonial conquest, and in the first part of the seventeenth century it was practically stamped out in a massive, bloody crusade. Determined not to share the fate of the Philippines, which had by then fallen to Spain, the Japanese permitted only the most minimal contact with the Europeans. Thus, a few Dutch traders, confined to a little island near the city of Nagasaki, were for many decades the sole representatives of the European world. And so Japan retreated into a long period of isolation that lasted past the middle of the nineteenth century.

Japan could maintain its aloofness when other areas were falling victim to the colonial tide because of a timely strengthening of its central authority, because of its insular character, and because of its distant position, far to the north along East Asia's difficult coasts. There are other factors, of course—Japan's isolation was far less splendid than that of China, whose silk and tea and skillfully made wares always attracted hopeful traders. But in the eighteenth and nineteenth centuries the modernizing influences brought by Europe to the rest of the world passed by Japan as they did China. And when Japan finally came face to face with the steel "black ships" and the firepower of the United States as well as Britain and France, it was no match for its enemies, old or new.

In the 1850s the United States first showed the Japanese to what extent the balance of power had shifted. American naval units sailed into Japanese harbors and extracted trade agreements through a show of strength; soon the British, French, and Hollanders were also on the scene seeking similar treaties. When there was resistance in local areas to these new associations, the Europeans and Americans quickly demonstrated their superiority by shelling parts of the Japanese coast. By the late 1860s no doubt was left that Japan's lengthy isolation had come to an end.

Figure J-1

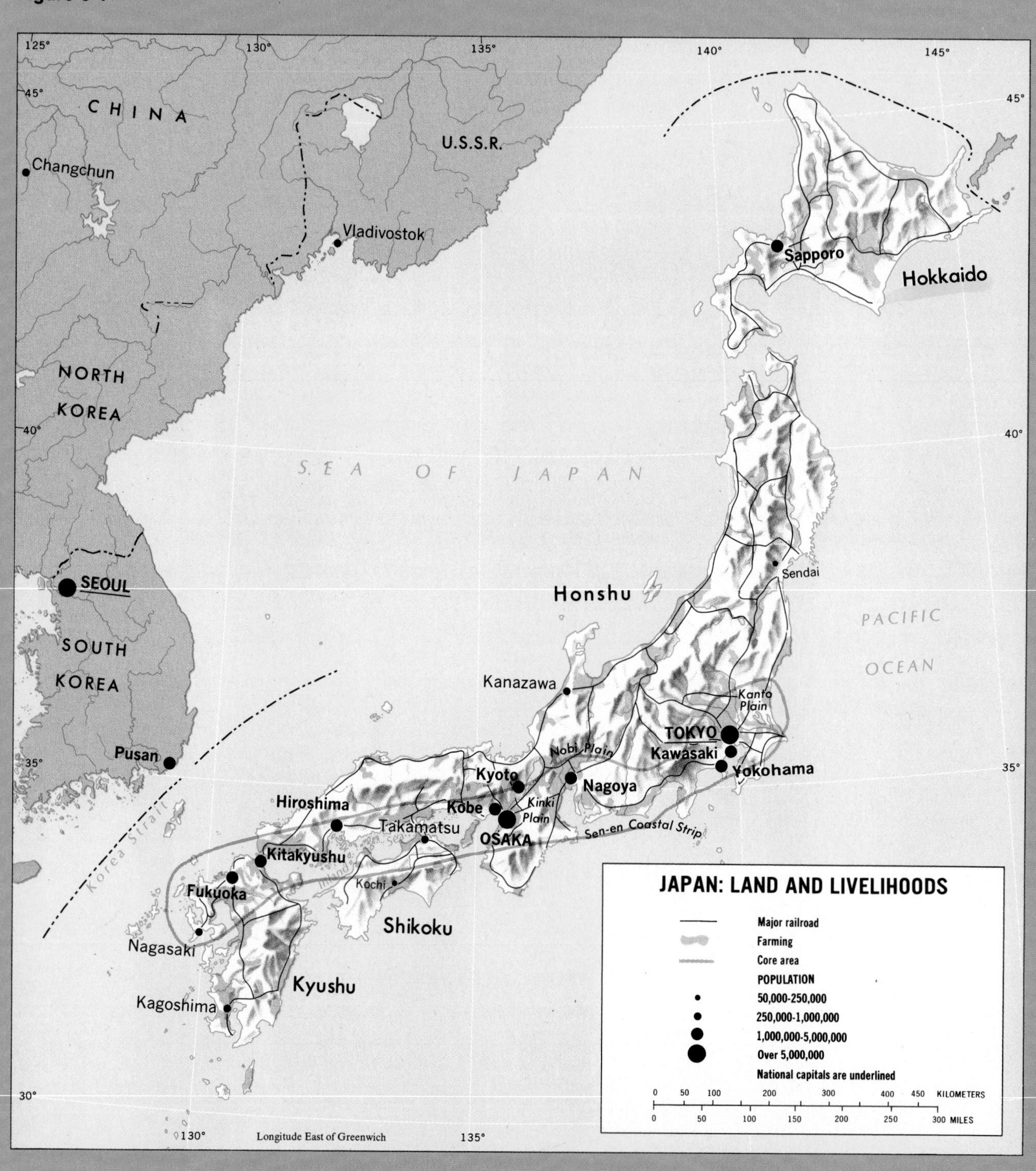

Imperial Japan

In the century that has passed since 1868, Japan emerged from its near-colonial status to become one of the world's major powers, overtaking a number of its old European adversaries in the process. The year 1868 is important in Japanese history, for it marks the overthrow of the old rulers that brought to power a group of reformers whose objective was the introduction of Western ways in Japan. The supreme authority officially rested with the emperor, whose role during the previous, militaristic period had been pushed into the background. Thus, the 1868 rebellion came to be known as the Meiji Restoration—the return of enlightened rule, centered around Emperor Meiji. But, despite the divine character ascribed to the imperial family, Japan in fact was ruled by the revolutionary leadership, a small number of powerful men whose chief objective was to modernize the country as rapidly as possible, to make Japan a competitor, not a colonial prize.

The Japanese success is written on the map; even before the turn of the twentieth century a Japanese empire was in the making, and the country was ready to defeat encroaching Russia. Japan claimed the Ryukyus (1879), took Taiwan from China (1895), occupied Korea (1910), and established a sphere of influence in Manchuria. Various archipelagos in the Pacific Ocean were acquired by annexation, conquest, or mandate (Fig. J-2). In the early 1930s Manchuria was finally conquered, and Tokyo began calling for an East Asia Co-Prosperity Sphere, which would combine—under Japanese leadership, of course—all of China, Southeast Asia, and

Figure J-2

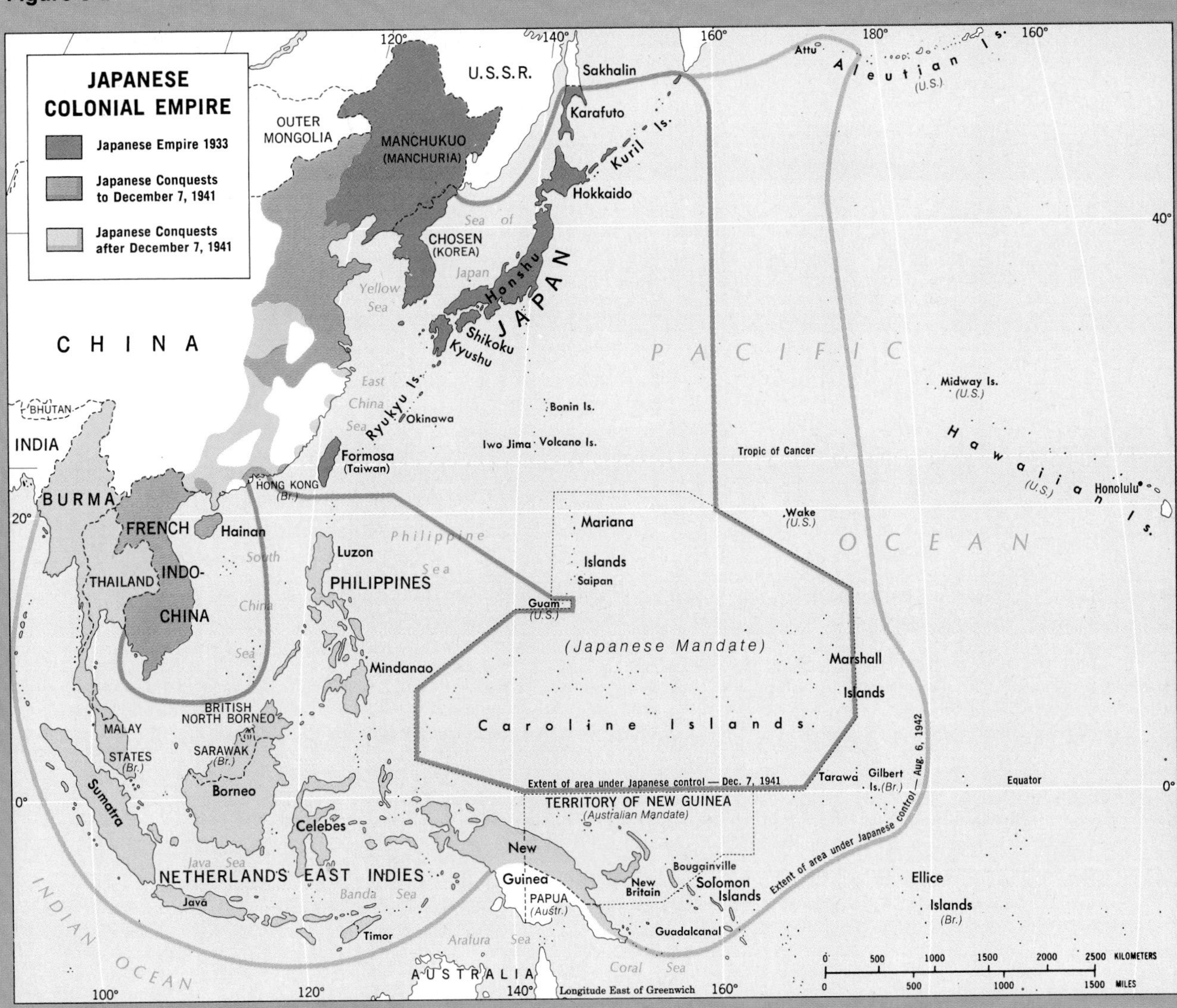

numerous Pacific island territories. By the late 1930s deep penetrations had been made into China, but Japan got its big chance in the early years of World War II. Unable to defend their distant Asian possessions while engaged by Germany at home, the colonial powers offered little resistance to Japan's takeover in Indochina, Burma, Malaya, and Indonesia; neither was the United States able to protect the Philippines. Japan's surprise attack on Pearl Harbor, a severe blow to the United States military installations there, was a major success—and is still admired in Japanese literature and folklore. The Japanese war machine was evidence of just how far Japan's projected modernization had gone: airplanes, tanks, warships, guns, and ammunition all were produced by Japanese industries, and they were a match for their Western counterparts. Japan, we know, was ultimately defeated by superior American power (two cities were devastated by atomic bombs), but this defeat coupled with loss of its prewar as well as wartime empire has still failed to destroy the country's progress. Japan has rebounded with such vigor that its overall economic growth rate currently is among the highest in the world.

With the Meiji Restoration just over a century ago, Japan began its phenomenal growth that brought it an empire, involved it in a disastrous war, and eventually saw the country achieve an industrial capacity unmatched not only in the non-Western world, but in much of the Western world as well. As if to symbolize their rejection of the old, isolationist position of the country, the new leaders moved the capital from Kyoto to Tokyo (or Eastern Capital), the new name for the populous city of Edo in the Kanto Plain. Kyoto had been an interior capital; Tokyo lay on the Pacific Coast, on an estuary that was to become one of the busiest harbors in the world. The new Japan looked to the sea as an avenue to power and empire, and it looked to industrial Europe for the lessons that would help it achieve those ends. While China suffered as conservatives and modernizers argued the merits of an open-door policy to Westernization, Japan's leaders committed themselves—with spectacular results.

Modernization

What has happened in Japan during the past century is a process that is far-reaching, transforming society from "traditional" to "modern"—but the process is complex, difficult to define, and subject to much debate among social scientists. We in the Western world tend to view modernization as identical to Westernization—urbanization, the spread of transport and communications facilities, the establishment of the market (money) economy, the breakdown of local (tribal) communities, the diffusion of formal schooling, the acceptance and adoption of foreign innovations. In the non-Western world, modernization (described perhaps with another word or phrase) is often seen as an outgrowth of colonial imperialism, the perpetuation of a system of wealth-accumulation introduced by alien invaders who were driven by greed. The local elites who have replaced the colonizers in the newly independent states, in that view, only carry on the disruption of traditional societies, not their modernization. The capital cities of Africa and Asia with their crowded avenues and impressive skyscrapers continue to grow—but they do not signal *development*. Traditional societies can be modernized without being Westernized.

Japan's modernization, in this context, is in many ways unique. Having long resisted foreign intrusion, the Japanese did not achieve the transformation of their society by importing a Trojan horse; it was done by Japanese planners, based on the existing Japanese infrastructure, and fulfilled Japanese objectives. Certainly Japan imported foreign technologies and adopted innovations from outside, but the Japan that was built, a unique combination of modern and traditional, was basically an indigenous achievement. Like the European powers they often emulated, the Japanese created a large colonial empire to supply the homeland with raw materials. The ripple effect of the Japanese period still continues in the economies of such former dependencies as Taiwan and Korea—and there the fundamental similarities between imperial Japan and imperialist Europe can still be discerned. But modernization in Japan itself defies the generalizations that apply to the process in the underdeveloped world.

A Japanese cultural landscape: fishing village and terraced farm fields on the shores of the Inland Sea. (Georg Gerster/Rapho-Photo Researchers)

Limited Assets

Japan's success might lead to the assumption that the country's raw material base was sufficiently rich and diverse to support the same kind of industrialization that had characterized England. Britain's industrial rise, after all, was accomplished largely because the country possessed a combination of resources that could sustain an industrial revolution. But Japan is not nearly so well off. Although Japan still has coal deposits in Kyushu and Hokkaido, there is little good coking coal. The most accessible coal seams were soon worked out, and before long the cost of coal mining was rising steadily. There was a time when Japan was a net coal exporter, when coal was shipped to the coastal cities of China, but today Japan has to import this commodity. Nevertheless, Japan did have enough coal, fortuitously located near its coasts, to provide a vital stimulus to industry during the late nineteenth century. Both in northern Kyushu and in Hokkaido, the coal deposits lie so near the coast that transport is not a problem, and by the nature of Japan's landscape the major cities and industrial areas also lie on or near the coast. Thus, coal could be provided quite cheaply and efficiently to every locale of incipient industrial development, and hence the fringes of the Inland Sea were among the first areas to benefit from this.

As for the other main ingredient of major industry—iron ore—Japan has only a tiny domestic supply, and one of small, scattered deposits of variable but generally low quality. In this respect there is simply nothing to compare to what Britain had at its immediate disposal, and neither is Japan rich in the various necessary ferro alloys. The Japanese extract what they can, including ores that are so expensive to mine and refine that it is hardly worth the effort, in view of import alternatives, and they carefully assemble and use all available and imported scrap iron. But iron ore still ranks high on each year's import list, and Japan buys iron all over the world, even as far away as Swaziland in Southern Africa.

Neither does Japan have sizable petroleum reserves. More of this fuel is needed every year, and today Japan already pays about twice as much for its annual petroleum purchases as it does for its iron ores. The few barrels of oil that are produced at home are hardly worth mentioning, and the hydroelectric power that is derived from the country's many favorable sites is not enough to fill its needs. This in itself is a measure of the amount of electricity Japan consumes; with its high mountain backbone, its ample rainfall, deep valleys, and urban markets clustered on the mountains' flanks, all conditions are right for efficient hydroelectric power production and use. But less than one-quarter of Japan's electricity comes from these dams, the demand rises faster than does their production, and the day seems far off when petroleum and coal imports can be reduced.

This, then, answers one of the questions we posed in the first paragraph of this essay. Japan really has nothing to compare to what Britain had during its industrial transformation. In fact, after what has just been said, it would seem that we are dealing with just another of those many countries whose underdevelopment can be attributed largely to a paucity of resources, and whose future is therefore deemed to depend on agriculture. Certainly it would have been difficult in 1868 to predict that the enthusiastic leaders of the Meiji Restoration would really have much success with their plan for the Westernization and strengthening of their country. What, then, gave Japan its opportunity? Obviously foreign trade must have played a great part in it. Was Japan's location the key to its fortune?

Domestic Foundations

It is tempting to turn immediately to Japan's external connections and to call upon factors of location to account for Japan's great industrial growth. The Japanese built an empire, much of it right across the Korea Strait, and this empire contained many of the resources Japan lacked at home. But before we rush into such a deduction, let us consider what Japan's domestic economy was like in the mid-nineteenth century, for, as we shall see, what Japan achieved was in the first place based upon its internal human—and to a large extent natural—resources.

In the 1860s manufacturing—light manufacturing of the handicraft type—was already widespread in Japan. In home industries and community workshops the Japanese produced silk and cotton-textile manufactures, porcelain, wood products, and metal goods. At that time the small metalliferous deposits of the country served their purpose: they were enough to supply these local industries. Power came from human arms and legs and from wheels driven by water; the chief source of fuel was charcoal. Thus there was industry, there was a labor force, and there were known manufacturing skills.

The planners who took over the country's guidance after 1868 realized that this was an inadequate base for industrial modernization, but that it might nevertheless generate some of the capital needed for this process. The community and home workshops were integrated into larger units, thermal and hydroelectric power began to be made available to replace older forms of power and energy, and for the first time Japanese goods—still of the light manufacture variety, of course—began to beat Western products on the world's markets. Meanwhile the Japanese continued to resist any infusion of Western capital, which might have accelerated the industrialization process but would have cost Japan its economic autonomy. Instead, the farmers faced higher taxes, and the money thus earned by the state poured into the industrialization effort.

Now Japan's layout and topography contributed to the modernization process. Coal could be mined and—in the relatively small quantities then needed—transported almost wherever it was needed. Nearly everywhere, a hydroelectric site was nearby, and soon electric power was available anywhere, in the cities as well as the populated countryside. Japan was still managing on what it possessed at home, and the factories and workshops multiplied. Soon the beginnings of a chemical industry made their appearance, and government subsidies and support led to the establishment of the first heavy industry. The period after 1890 was a time of great progress, stimulated by the wars against China (1894–1895) and Russia (1904–1905). Nothing served to emphasize the need for industrial diversification as much as the war, and Japan's victories vindicated the ruthlessness with which industrial objectives were sometimes pursued.

Japan's war effort contributed to its improved economic world position when the imperial period ended. When the Meiji Restoration took place, Britain lay at the center of a world empire and the Europeanization of the world was in full swing; the United States still was a developing country. The wars of the twentieth century saw the Japanese defeating British and French forces in East Asia; the United States and its allies crushed Germany and Italy. Japan came out of World War II a defeated and heavily damaged country, but the world situation had changed dramatically: the United States, Japan's trans-Pacific neighbor, had become the world's most powerful and wealthiest state. Distant Britain was fading, its empire on the verge of disintegration. Japan's *relative location* had changed—its location relative to the economic and political foci of the world. And therein lay much of Japan's postwar opportunity.

Japan's Spatial Organization

Japan is not a very large country, but it was already quite populous when its modern economic revolution began; about 30 million people were crowded into its confined living space. Since then, Japan's population has nearly quadrupled and its economic complex has grown enormously. With a million people, Tokyo (then still named Edo) already may have been the world's largest city in the 1600s, but today the capital is more than 12 times as large and it is part of a still larger urban area that includes Yokohama as well. The whole conurbation remains the largest of its kind, although—if it matters—it can be argued that metropolitan New York remains the largest single urbanized area in the world. That Tokyo should be in New York's league at all is the amazing truth, and a reflection of the phenomenal development of the country.

With such growth, space in Japan has been at a premium, and urban and regional planners have shared with farmers the problem of how to make every square foot count. Before the onrush of urbanization, it was often said that the most densely populated rural areas in Japan were more crowded than any other similar areas in the world. Now the cities, towns, and villages must also vie for the country's limited livable terrain. While Japan's industries were still of the light handicraft variety, the many workshop type of establishments were located in the cities as well as in villages. The small-scale smelting of iron could be carried on in many locales; charcoal was widely available, and textile manufacture could be done in many places. But then the age of the factory arrived, and the rules of industrial location went into effect. We discussed some of these rules earlier in this book: the advantages of treating heavy or bulky raw materials at the spot where they are mined to save transport costs, the transport of lighter and more easily moved raw materials to the areas where heavier, more costly to move resources are found. In Britain during the industrial revolution this led to the rapid growth of some towns into industrial cities, the decline of others, and the founding of still other industrial centers. In Japan, on the other hand, urban industrial growth was straitjacketed; apart from northern Kyushu's coal, no great reserves of raw materials were found, and it soon was clear that these would have to be imported from elsewhere. Internally, the distribution of coal was mainly by water; externally, the iron ore and other requirements, of course, came by sea. Hence the coastal cities, where the labor forces were also concentrated, became the sites of modern factories and industrial plants.

The modern industrial growth of Japan thus tended to confirm the development that was already taking place—Japan did not have the resources to merit any major reorientation. But obviously some cities had advantages over others in terms of their facilities or their position with reference to the source areas of the imported raw materials. Now began the kind of differentiation that has marked regional organization everywhere in the world. A. K. Philbrick in 1957 published an article about this principle, which he called "Areal Functional Organization." Briefly, this idea is based on five concepts. First, human activity has focus; that is, it is concentrated in some locale—a farm or factory or store. Second, such "focal" activity is carried on in certain particular places. Obviously, no two establishments can occupy exactly the same spot on the earth's surface, and so every one of them has an absolute location, which can be measured by latitude and longitude calculations. But what is more relevant, every establishment has a location *relative* to other establishments and activities. Since no human

By some measures Tokyo lies at the heart of the world's largest urban agglomeration. This view of a small part of the urban region overlooks the Sumida River. (Charles E. Rotkin/Photography for Industry)

activity is carried on in complete isolation, the third concept is that interconnections develop among the various establishments. Farmers send crops to markets and buy equipment at service centers. Mining companies buy gasoline from oil firms, lumber from sawmills, and send ores to refining plants. Thus, a system of interconnections emerges and grows more complex as human capacities and demands expand. With their spatial character, these systems form units of area organization. Philbrick's fourth principle is that the evolution of these regions of human organization is the product of human "creative imagination" as people apply their total cultural experience as well as technological know-how when they decide how to organize and rearrange their living space. Finally, it is possible to recognize *levels* of development in area organization, a ranking or hierarchy based on type, extent, and intensity of exchange. We detect here a relationship to Christaller's attempt to discern order in the arrangement of urban centers with different levels of centrality. To quote Philbrick: "The progression of area units from individual establishments to world regions and the world as a whole are formed into a hierarchy of regions of human organization."

In the broadest sense, regions of human organization can be categorized under subsistence, transitional, and exchange types, and Japan's area organization obviously reflects the last of these. But within each type of organization, and especially within the complex exchange type of unit, individual places can also be ordered or ranked on the basis of the number and kinds of activities and interconnections they generate. A map of Japan showing its resources, cities and towns, and surface communications can tell us a great deal about the kind of economy the country has. It looks just like maps of other parts of the world where an exchange type of organization has developed—cities and towns ranging from the largest to the smallest hamlets, a true network of railroads and roads connecting these places, and productive agricultural areas near and between the urban centers. In Japan's case, Figure J-3 shows us something else; it reflects the

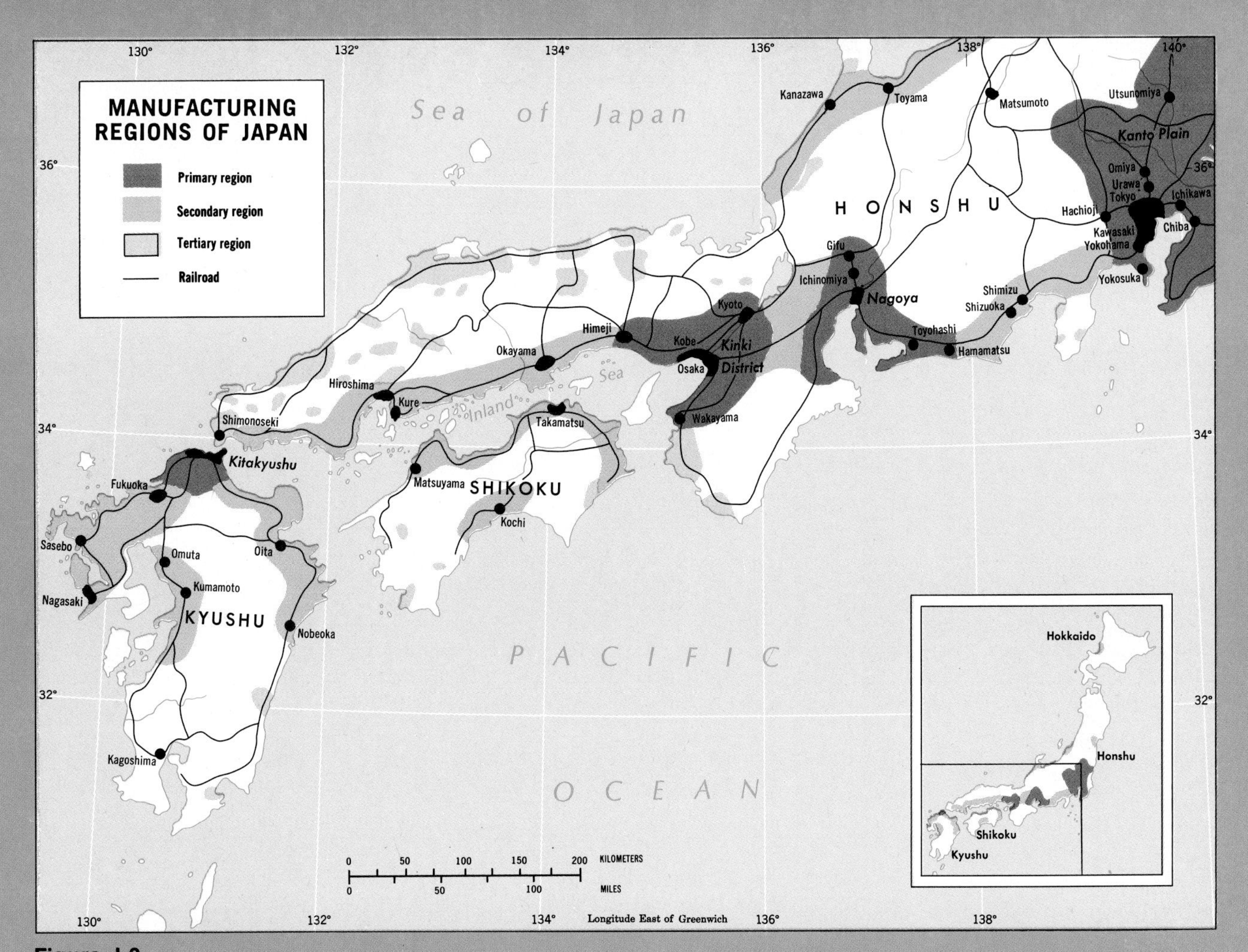

Figure J-3

country's external orientation, its dependence on foreign trade. All primary and secondary regions lie on the coast: of all the cities in the million-size class, only Kyoto lies in the interior, and it lies at the head of the Kinki Plain. If we were to deduce that Kyoto does not match Tokyo, Yokohama, Nagoya, or Kobe-Osaka in terms of industrial development, that would be correct; the old capital remains mainly a center of light manufacture. Actually, Kyoto's ancient character has been preserved deliberately, and large-scale industries have been discouraged. With its old temples and shrines, its magnificent gardens, and its many workshop and cottage industries, Kyoto remains a link with Japan's pre-modern past.

As Figure J-3 shows, Japan's leading primary region of urbanization and industry, along with very productive agriculture, is the Kanto Plain. The Kanto Plain contains not only Tokyo and Yokohama, but also one of Japan's most extensive lowland farming areas. This giant cluster of cities and farms forms the eastern end of the country's elongated, fragmented core area (Fig. J-1), and apart from its flatness it has other advantages: its fine natural harbor at Yokohama, its relatively mild and moist climate, and its central location with reference to the country as a whole. It benefited also from Tokyo's selection as the capital, at a time when Japan embarked on its planned economic growth. Many industries and businesses chose Tokyo as their headquarters in view of the advantages of proximity to the government's decision makers.

The Tokyo–Yokohama conurbation has become Japan's leading manufacturing center, with between one-fifth and one-quarter of the country's output. But the raw materials for all this industry come from far away; Tokyo is among the chief steel producers in Japan, using iron ores from the Philippines,

Malaya, Australia, India, and even Africa. Although electric power comes from the mountain valleys to the northwest and coal is shipped in from Hokkaido, coal also comes from Australia and America, and petroleum from Southwest Asia and Indonesia. As for the supply of food to the huge population, the Kanto Plain cannot produce nearly enough, and imports come from Canada, the United States, and Australia, as well as from some other areas in Japan itself. Thus, Tokyo depends completely on its external trade relations for food, for raw materials, and for markets for its wide variety of products, ranging from children's toys to steel, and from the finest optical equipment to oceangoing ships.

The second-ranking urban-agricultural region in Japan's primary core area is the Kobe–Osaka–Kyoto triangle, at the eastern end of the Seto-Naikai (Seto Inland Sea). The position of the Kobe–Osaka conurbation, with reference to the old Manchurian empire created by Japan, was advantageous. Situated at the head of the Inland Sea, Osaka was the major Japanese base for the exploitation of Manchuria and trade with China, but it suffered when the empire was destroyed, and its trade connections with China were not regained after World War II. Kobe (like Yokohama, Japan's chief shipbuilding center) has remained one of the country's busiest ports, adding the traffic of the Seto Inland Sea to its overseas connections. Among these connections was for many decades the import of cotton for Osaka's textile factories. With its large skilled labor force and its long history of productive settlement, Osaka until very recently was Japan's first-ranked textile producer—before as well as after the Meiji Restoration. Today textiles (mainly synthetics) form only part of Osaka's huge and varied industrial production. Kyoto, as we noted, remains much as it was before Japan's great leap forward; with nearly 2 million people it nevertheless remains a city of small workshop industries.

The Kobe–Osaka–Kyoto region, like the Kanto Plain, is an important farming area (the area is known as the Kinki District) as well as an industrial complex that rivals Tokyo–Yokohama. Rice, of course, is the most intensively grown crop here in the warm, moist lowlands, but the Yamato Plain (one of five fault basins in the Kinki District) is not nearly as extensive as that of Kanto. Here, then, is another cluster of people that requires a large infusion of foodstuffs every year.

The Kanto Plain and the Kinki District, as Figure J-3 suggests, are the two leading primary regions within Japan's core. But between them lies the Nagoya or Nobi Plain, focused on the industrial center of Nagoya (2 million), the city that recently ousted Osaka from first place among Japan's textile producers. The map indicates some of the Nagoya area's advantages and liabilities. The Nobi Plain is larger than that of the Kinki District, and thus its agricultural productiveness is greater, although not as great as that of the Kanto Plain. But Nagoya has neither Tokyo's centrality nor Osaka's position on the Inland Sea; its connections to Tokyo are by the tenous Sen-en Coastal Strip, along which runs the famous Tokaido "bullet train" (Fig. J-1). Its westward connections are somewhat better, and there are signs that the Nagoya area is merging with the Kobe-Osaka region. Still another disadvantage of Nagoya lies in the quality of its port, which is not nearly as good as Tokyo's Yokohama or Osaka's Kobe, and has been plagued by silting problems.

Westward from the three regions just discussed extends the Inland Sea, along whose shores Japan's core area is developing. The most impressive growth has occurred around the western entry to the sea, where Kitakyushu (a conurbation of five north Kyushu cities) constitutes a fourth Japanese manufacturing center. Honshu and Kyushu are connected by road and railway tunnels, but the northern Kyushu area does not have any equivalent on the Honshu side of the Strait of Shimonoseki. The Kitakyushu conurbation includes Yawata, site of northwest Kyushu's large coal mines, and it is on the basis of this coal that the first steel plant in Japan was built there—a complex that for many years was Japan's largest. The advantages of transportation here at the western end of the Inland Sea are obvious. No place in Japan is better situated to do business with Korea and China, and when relations with China are normalized, this area should be quick to reflect the results. Elsewhere on the Inland Sea coast, the Hiroshima—Kure urban area has a manufacturing base that

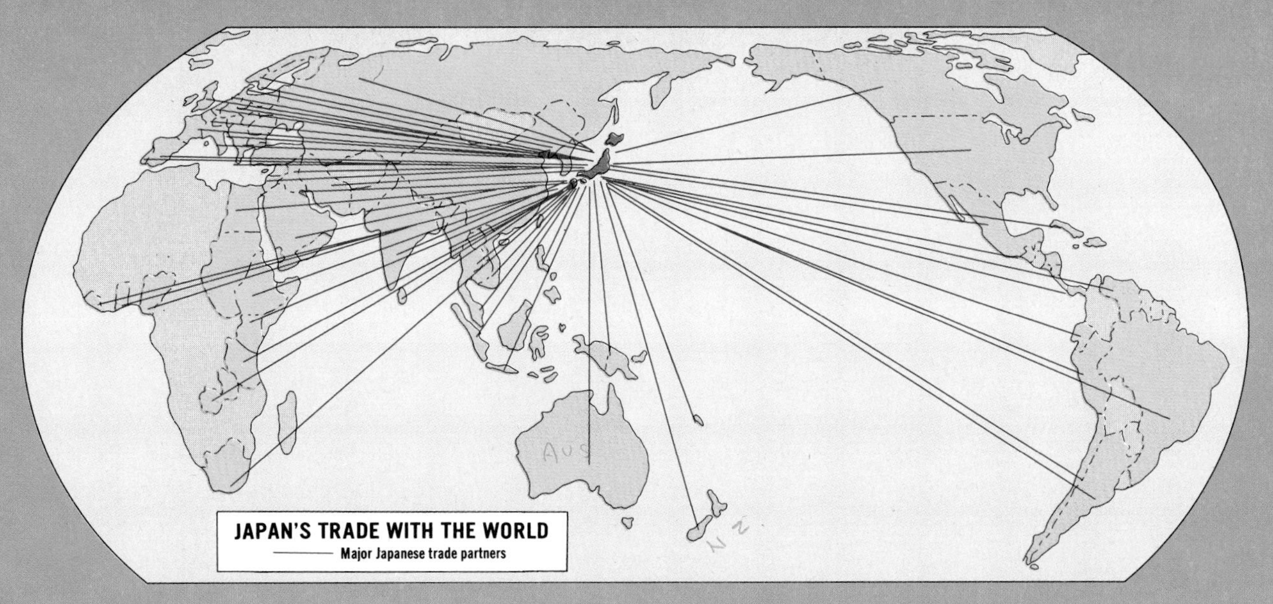

Figure J-4

includes heavy industry, and on the coast of the Korea Strait, Fukuoka and Nagasaki are the principal centers, Fukuoka as an industrial town and Nagasaki as a center of large shipyards.

The map of Japan's area organization shows the country's core area to be constituted largely by four individual primary regions, each of them primary because each duplicates to some degree the contents of the others. Each of the four areas contains iron and steel plants, each is served by one major port, and each lies in or near a large and productive farm area. What the map does not show is that each has its own overseas, external connections for the acquisition of raw materials and the sale of its products. These connections (Fig. J-4) may be stronger than those among the four core sections within Japan itself; only in the case of Kyushu and its coal have domestic raw materials played much of a role in shaping the nature and location of heavy manufacturing. In the shaping of the country's areal functional organization, therefore, more than just the contents of Japan itself is involved. In this respect Japan is not unique: all countries that have exchange-type organizations must to some degree adjust their spatial form and structure to the external interconnections required for progress. But it would be difficult to find a country in which this is more true than in Japan.

Farming and Food

Japan's industrialization so occupies the center stage that it is easy to forget that considerable achievements have been made in the field of agriculture. Japan's planners, no less interested today in closing the food gap than in expanding industries, have created extensive networks of experiment stations to promote mechanization, proper seed-selection, and fertilizer use and information services to diffuse to farmers as rapidly as possible information that is useful in raising crop yields. But although this program has been very successful, Japan faces the unalterable reality of its stubborn topography: there simply is not the necessary land to farm. Less than one-fifth of the country's total area is in cultivation, and although there still may be some land that can be brought into production, it is so mountainous or cold that its contribution to the total harvest would be minimal anyway. Japanese agriculture may resemble Asian agriculture in general, but nowhere else do so many people depend on so little land. Japan's population density overall is about 320 per square kilometer (more than 800 per square mile) to begin with, but when the *cultivable* land is measured it turns out that more than 1800 people depend on the average square kilometer of farmland (4600 per square mile). This measure, persons per unit of cultivable land, is the *physiological* density, and

Japan's is one of the highest in the world. Even Egypt and Bangladesh, notoriously crowded on confined agricultural areas, have a lower physiological density than Japan. The physiological density of India is less than *one-fifth* as high as Japan's!

Japanese agriculture stands apart from farming elsewhere in rice-growing Asia. It is by far the most efficient in all the Asian continent, given the existing conditions of slope, soil, and climate. Rice yields in Japan are among the highest in the world: nearly 60 quintals per hectare (110 bushels per acre) while China has under 35, the United States 51, and South and Southeast Asia under 20.

Japan's population is now approaching 120 million, having tripled over the past century, and although the annual growth rate of 1.3 percent is much lower than those of other Asian countries, Japan no longer enjoys its one-time luxury of self-sufficiency in food. But it is remarkable that the country is somehow able to produce approximately two-thirds of its annual needs—enough to feed nearly 80 million people. It is achieved through turning over more than 90 percent of all farmland to food crops (and 53 percent of this is under rice), through diligent irrigation (more than half of Japan's agriculture is irrigated), and through painstaking terracing, multiple cropping wherever possible (south of the 37th parallel, including the major farm districts), intercropping, intensive fertilization (animal manures, night soil, and chemical fertilizers are bought by the farmers in great volume), and transplanting—that is, the use of seedbeds to raise the rice plants while another crop matures on the paddy.

But the rush of urbanization and postwar industrialization have begun to reverse rural Japan's productivity. Farmers leave the land for the lucrative opportunities in the cities, and farmland lies uncultivated in some areas; the confidence that food can always be purchased with the enormous revenues from manufacturing has begun to erode traditional standards of Japanese agriculture. Also, there have been problems of balance. In recent years Japanese farmers have produced too much rice and too few other crops, and the government has subsidized farmers who would reduce their rice production. Japan's second and third crops, usually wheat and barley, occupy less than one-quarter of the farmland. Wheat is used as a winter crop with rice in the warmer parts of the country, and barley is grown in the north as a summer crop. Sweet and white potatoes, as the major root crops, occupy less than 10 percent of the farmland. Still less (and much of it on hilly slopes) is given over to cash crops, such as tea, grown for local consumption, and mulberry, the plant whose leaves are used to feed the silkworms, the basis of Japan's once-rich silk industry. Of course, there are the vegetable gardens to be expected around the large cities. But Japan's major agricultural effort goes into raising the production of its staple—rice.

As an urbanizing country, Japan is in a class with the United States and Western Europe, in that over 75 percent of the people live in cities and towns (the corresponding figure for the United States is 77 percent, the United Kingdom, 85 percent). Some sources suggest that nearly four-fifths of Japan's population is now concentrated in cities and towns of 20,000 and over, but this seems a high estimate; nevertheless, the percentage of farmers in Japan's population has declined steadily during most of the past hundred years. In the mid-nineteenth century some 80 percent of the Japanese people were farming; today the total number of farmers on the land is not much larger, although the population has nearly quadrupled. But even this is a large number of farmers for so small an agricultural area, and not surprisingly, farms in Japan are quite small—about 8 hectares (20 acres) on the average. This, nevertheless, is twice as large as the average farm was in 1960. Land fragmentation was excessive in postwar Japan, and the American administration in 1946 and 1947 introduced reforms to reverse the process. Then Japan's rapid industrialization and urbanization drew many families away from the land; at present the number of farmers on the land is declining. In addition, studies show that a very large number of farmers are in the upper age groups, so the trend toward sparser rural population and consolidating farms seems likely to continue.

As was the case in precommunist China and in so many other parts of Asia, Japan before its World War II defeat had a very high rate of tenancy in agriculture, with landlords owning great tracts of soil that

Japanese agriculture at its most intensive: tea plantations on steep slopes in interior Honshu. Agriculture in Japan does not invariably present so positive a view, however. The inexorable pull of the urban magnets has drawn farmers from the land, and in parts of Japan farms that could produce needed food crops lie neglected or even abandoned. (Burt Glinn/Magnum)

were farmed by tenants in return for a harvest-rent that went as high as 70 percent of the total yield. Two out of three Japanese farmers were renters, and the system demanded reform. This was also imposed during the American occupation, with the result that today nearly all the land cultivated is farmed by its owners.

Notwithstanding its high-efficiency agriculture, Japan must import foodstuffs every year, including wheat, corn, sugar, and soybeans. Japan pays far more for its imports of petroleum and coal, iron and copper ores, chemicals, and other commodities needed by its industries, but the Japanese would like to reduce their dependence on foreign food supplements.

Fishing

Rice, wheat, barley, and potatoes—staples in Japan—make for a very starchy diet, and the Japanese need protein to balance it. Happily, Japan has the opportunity to secure protein in large quantities, not by purchasing it, but by harvesting it from rich fishing grounds near the Japanese islands. With customary thoroughness, the Japanese have developed a fishing industry that on all accounts is larger than that of the United States or any of the long-time fishing nations of northwestern Europe and now supplies the people with a second staple after rice.

Although mention of the Japanese fishing industry brings to mind fleets of ships and trawlers scouring the oceans and seas distant from Japan, in search of salmon, whales, tuna, and herring, the fact is that most of the catch comes from waters within a few dozen miles of Japan itself. Where the warm Kuroshio and Tsushima Currents meet colder water off Japan's coasts, a rich fish fauna exists, including sardines, herring, tuna, and mackerel in the warmer waters, and cod, halibut, and salmon in the cooler northern seas. Along Japans' coasts there are about 4000 fishing villages, and tens of thousands of small boats ply the waters offshore to bring home catches that are distributed locally and in city markets. Many of the fishermen divide their time between farming and fishing, just as so many Japanese farming families still increase their income by some seasonal or part-time light manufacturing. The Japanese raise freshwater fish in artificial ponds and in flooded paddy fields, seaweeds in aquariums at home, oysters and prawns or shrimps in shallow bays; they are also experimenting with the cultivation of algae for their food potential.

Apart from the wealth of fish in Japan's waters, the industry has other advantages.

Japan's fishing fleets ply all the oceans and harvest an enormous annual catch. In this photograph taken near Tokyo, a tuna boat has just returned to port and the fish are displayed for sale to consumers. (Burt Glinn/ Magnum)

The coastal orientation of Japan's population, and hence the proximity of the markets to the supply points, facilitates both the fishing itself and the quick distribution of the catch. Japan's long coastlines are indented in many places to create good harbors, and offshore, the continental shelf, especially along the coasts of inner Japan, is quite shallow and wide, despite the deep trench that lies off the east coast of the archipelago. With their empire and their shipbuilding industry, it is not surprising that Japanese boats soon ranged far and wide. That advantage—and empire—has vanished. But the Japanese have simply increased the tonnage of their fleets and widened the search to all the world's oceans, and they now bring in more than one-quarter of the combined annual catch of all the fishing countries in the world. Fish, in fact, figures in the Japanese list of exports, as high-priced canned salmon and crab, as well as tuna. The fishing industry is another of Japan's amazing success stories.

Prospects

The energy crisis of 1973, when Arab oil-producing countries began a production cutback, underscored Japan's vulnerability to embargoes and price increases on raw materials the country needs for its continued prosperity. Like other developed countries, Japan in the early 1970s was affected by a severe decline in the economic growth rate and strong inflation. Thus the upward spiral was temporarily halted, and Japanese leaders contemplated the options available.

One asset is Japan's ever wealthier domestic market, now capable of absorbing a substantial part of Japanese production. This growth of Japanese consumerism, however, also increases the need for imports, and the only way Japan can pay for these imports is by selling its products in greater quantities overseas. Thus, there are always risks to face: raw material imports can become more expensive, eventually perhaps prohibitively so, and as time goes on, alternatives may dwindle. Japan's major trade partner is the United States, from where vital imports are drawn, including cotton and scrap iron; other Western, "developed" countries such as Canada, Australia, Britain, and Germany are high on the list. These countries are glad to see Japan buy raw materials, but they are concerned over the competition Japanese products create against their own. And so Japan faces another risk: that the success of its products will eventually mean that countries will erect barriers to their sale, or that states will join in economic unions (such as the European Common Market) to protect themselves against Japanese goods. So far, Japan has not suffered and the country thrives. But its economic base, however fast its current growth, is nevertheless precarious.

Japanese leaders frequently voice the opinion that their country's greatest hope for the future lies in the immense market of China. Major trade with China would greatly reduce the impact of one of Japan's greatest enemies—distance. Ambitious attempts to normalize relations between Japan and China were started in 1972. If they are successful, and China becomes Japan's trading partner once again, the terms will be very different from those that prevailed during the 1930s. In 1975, the value of the two-way trade was a relatively modest $3.7 billion, but this was about one-third more than the 1974 total, and progress is being made.

Another option available to the Japanese relates to the Soviet Far East, but political geography intrudes here: Japan continues to oppose Soviet control in the Kuril Islands. Nevertheless, the Soviets in 1974 sought Japanese assistance in the construction of a second Trans-siberian railroad, and in the same year the Soviet Union and Japan agreed on a joint project for the development of coalfields in southern Yakutsk. Additional future agreements might involve oil and gas reserves in Soviet Sakhalin, gas in Yakutsk, and the Far East's enormous lumber resources (Japanese technology could help develop a paper and pulp industry as well).

The Japanese have found themselves in a difficult position, however, because of the Sino-Soviet quarrel. When Japan and China discussed their "treaty of peace and friendship," one clause approved by Tokyo was

Will success threaten Japan's prosperity? Automobiles are being loaded on a freighter that will carry them to foreign markets. Japanese products sell well overseas because they are competitively priced and their reputation for quality is generally high. Japanese industrialists fear that foreign markets may erect trade barriers against Japanese products; manufacturers of automobiles, television sets and other products are being asked to build factories in countries that also are markets (such as the United States) so that Japanese products can put local labor to work. (J.P. Laffont/Sygma)

the so-called "antihegemony" clause—a reflection of China's continued claims to Soviet Asian territories. As it happens, Japan has some objections to the Soviet presence in some areas as well, and as Japan's involvement with China deepened, Moscow's initiatives cooled. In time Japan may be confronted by a difficult choice involving its relationships not only with China and the Soviet Union, but with Korea and Taiwan as well. There are options for Japan in Asia, but none is without danger.

Still, China and the Soviet Union are proof that Japan's alternatives are not yet exhausted, and it could be that when the time comes that rising costs and closed markets in America and Europe slow the Japanese economy down, Japan will once again have an answer to its problems. If so, the Japanese can be trusted to make the fullest possible use of it, in their unparalleled tradition of adaptation and innovation.

Additional Reading

The most useful bibliography on Japan is *A Selected List of Writings on Japan Pertinent to Geography in Western Languages, with Emphasis on the Work of Japan Specialists,* by David H. Kornhauser, published by the University of Hiroshima in 1979. This 71-page volume constitutes a valuable research resource; Kornhauser's book *Urban Japan: Its Foundations and Growth* (Longman, London, 1976) is an excellent place to start reading about this remarkable region. Arthur E. Tiedemann's *Introduction to Japanese Civilization* (Columbia University Press, New York, 1974) contains a useful chapter by geographer Norton S. Ginsburg on Japan's economic and cultural geography, pp. 423–459. Also see Forrest R. Pitts, *Japan* (Fideler, Grand Rapids, 1974) and Erwin Scheiner, *Modern Japan: an Interpretive Anthology* (Macmillan, New York, 1974). The volume by A. Kolb, *East Asia* (Methuen, London, 1971) in the Advanced Geographies Series contains a detailed and extensive chapter on "Japan: the Island World." Of historical interest is G. T. Trewartha's *Japan: a Geography* (University of Wisconsin Press, Madison, 1949) and E. A. Ackerman, *Japan's Natural Resources and Their Relation to Japan's Economic Future* (University of Chicago Press, Chicago, 1953) which, though dated, proves that a dated book can remain current and provocative. Also see P. Dempster, *Japan Advances: A Geographical Study,* published in New York by Barnes and Noble in 1967, and H. Borton, *Japan's Modern Century,* published in New York by Ronald Press in 1955. The volume by E. O. Reischauer, *The United States and Japan,* a Compass Book of Viking Press, New York, published in 1965, sees Japan through the eyes of one who knows it better than most others. Of interest also is R. B. Hall's *Japan: Industrial Power of Asia,* No. 11 in the Van Nostrand Searchlight Series, published in 1963.

On some specialized topics, see I. B. Taeuber, *Population of Japan,* a 1958 publication of Princeton University Press that traces population patterns from the twelfth century to the mid-twentieth, and T. C. Smith, *The Agrarian Origins of Modern Japan,* published by Stanford University Press, 1959. On the land reform imposed after World War II by the United States, see R. P. Dore, *Land Reform in Japan,* published in New York by Oxford University Press in 1959. The article by A. K. Philbrick, "Principles of Areal Functional Organization in Regional Human Geography," appeared in *Economic Geography,* Vol. 33, No. 4, October 1957. For a further development of the idea, see Philbrick's *This Human World,* a Wiley publication of 1963. A general and very readable book on *The Japanese Way: How They Live and Work* was written by W. Scott Morton and published by Praeger, New York, in 1973.

Part Two

Under-Developed Regions

Contact of cultures, explosion of expression: old and new in the cultural landscape of Mexico City. (Albert Moldvay/Woodfin Camp)

Four

Middle America: Collision of Cultures

IDEAS AND CONCEPTS

- Culture hearth
- Land bridge
- Environmental determinism
- Cultural landscape (2)
- Transculturation
- Mainland-rimland concept
- Altitudinal zonation
- Pluralism (2)

Middle America is a realm of vivid contrasts, matchless variety, turbulent history, and uncertain futures. Physiographically, Middle America extends from the southern borders of the United States to the northern limits of the South American continent, thus including Mexico and the smaller countries from Guatemala to Panama as well as the large and small islands of the Caribbean Sea (Fig. 4-1). Culturally, crowded Middle America spills over into its neighbors—across the Rio Grande into the United States, where Spanish-speaking communities prevail in a large region (see Fig. 3-13), and onto coastal South America, where peoples and ways of life in places resemble Caribbean rather than prevailing South American norms (see Fig. 5-5, page 295).

Middle America's northern boundary quite obviously represents a cultural transition, but some scholars argue that a differentiation between Middle and South America is unnecessary, since both are part of a larger "Latin" American realm. And certainly there are numerous countries in Middle America that share their dominant cultural imprints with South America (notably the Spanish language and Catholic religion). But in Middle America to a far greater degree than in South America, there is variety. Large population groups have African and Asian as well as European ancestries. Nowhere in South America has Indian culture contributed to modern civilization as strongly as it has in Mexico. The Caribbean area is a patchwork of independent states, territories in political transition, and dependencies. The Dominican Republic speaks Spanish, adjacent Haiti uses French. Dutch is spoken in Curaçao and its neighbors, English in Jamaica. Middle America gives definition to concepts of pluralism.

Middle America's cultural diversity is matched by its varied physiography. The numerous Caribbean islands range from nearly flat, coral-fringed platforms to mountainous, volcanic landscapes; some have an area of less than a hectare while others extend over thousands of square kilometers (the largest, Cuba, has 114,000 square kilometers, slightly larger than Tennessee). Most of these islands represent the tops and crests of mountain ranges that rise from the floor of the Caribbean Sea and protrude above the water. Like the mountains of the Middle American mainland, these ranges are susceptible to earthquakes and volcanic eruptions. Because hurricanes also affect Caribbean islands and mainland coasts, the Middle American environment is unusually hazardous.

The Middle American mainland is funnel-shaped, widest in Mexico and then narrowing irregularly to a slim 65 kilometers (40 miles) in Panama. It is a 6000-kilometer connection between North and South America, defined by physical geographers as a *land bridge.* Because sea level varies over time, such land bridges come and go. When the sea level was lower, Australia and New Guinea were connected, as were Korea and Japan, India and Sri Lanka, Malaysia and Indonesia. Geologic processes combine with sea-level changes to create and destroy land bridges, but these links undoubtedly have played crucial roles in the distribution of animals and peoples in various parts of the world. The Middle American land bridge still rises between Atlantic and Pacific waters, and it, too, has been a factor in the dispersal of Indian peoples across the Americas. We return to this theme in Chapter 5, but it is important to note that the Middle American mainland affords no easy transfers. In the north the extensive Mexican Plateau is flanked by major mountain ranges, and Central America is dominated by a mountainous, volcanic, earthquake-prone, lake-studded backbone. Coastlands often are

Figure 4-1

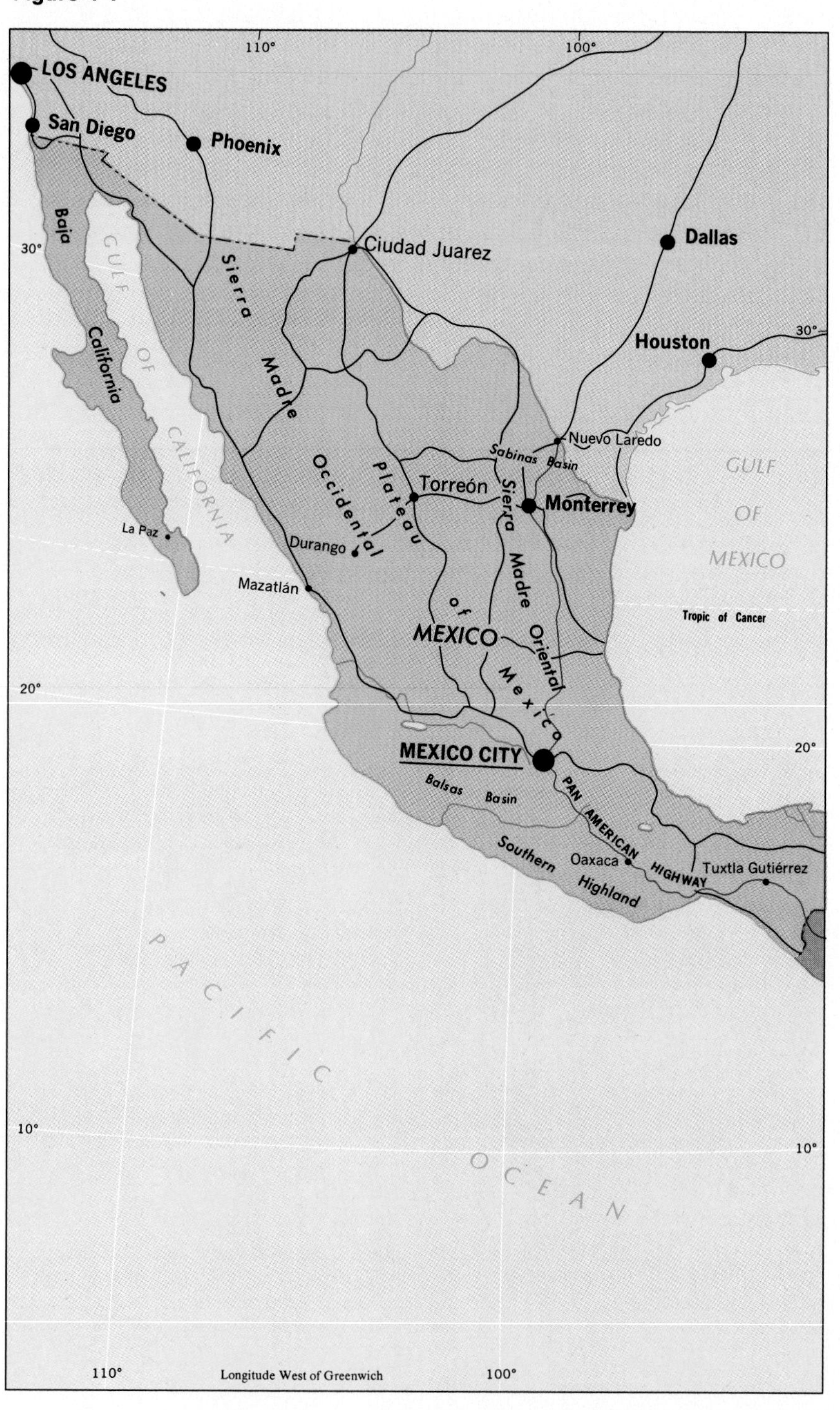

swampy; dense and nearly impenetrable forest covers interior lowlands. Forested Panama long defeated the completion of the intercontinental Pan-American Highway.

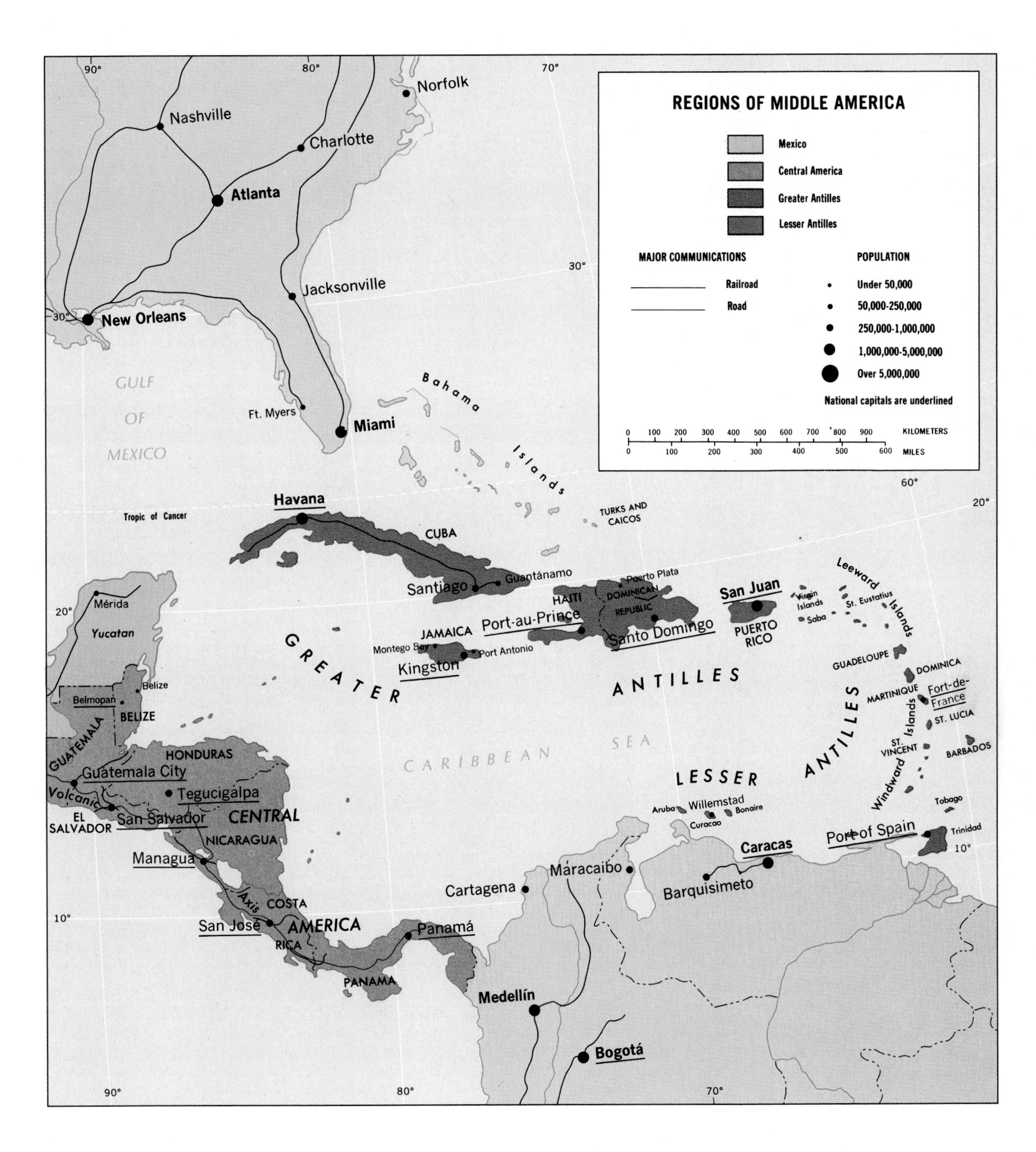

Ten Major Geographic Qualities of Middle America

1. Middle America is a fragmented region consisting of all mainland countries from Mexico to Panama and all the islands of the Caribbean Sea.
2. Middle America's mainland constitutes a crucial barrier between Atlantic and Pacific waters. In physiographic terms, this is an intercontinental *land bridge.*
3. Middle America's tropical location and climatic environments are in important places ameliorated by altitude and its resultant zonation.
4. In terms of area, population, and economic strength Mexico dominates the realm.
5. Middle America's historical geography is replete with involvement by its powerful neighbor, the United States.
6. Middle America's cultural geography is complex. African influences dominate in the Caribbean, while Indian traditions survive on the mainland.
7. Middle America is a realm of intense cultural and political fragmentation. The political geography defies unification efforts and instability recurs.
8. Underdevelopment is endemic in Middle America; the realm contains the Americas' least-developed territories.
9. Out-migration of population for economic and political reasons prevails.
10. The realm (notably Mexico) contains major actual and potential reserves of mineral fuels.

Legacy of Mesoamerica

Mainland Middle America was the scene of the emergence of a major ancient Indian civilization. Here lay one of the world's true culture hearths, an area where population could increase. Agricultural specialization developed, urbanization occurred, transport networks matured, and writing, science, architecture and art, religion, and other spheres of achievement witnessed major advancement. Anthropologists refer to the Middle American culture hearth as *Mesoamerica,* and what is especially remarkable about this development is that it happened in very different geographic environments. In the low-lying, tropical, hot, and humid plains of Honduras, Guatemala, Yucatan, and Belize, the Maya civilization arose. On the higher plateau of present-day Mexico the Aztecs founded their well-organized civilization. In the process the Maya and the Aztecs overcame some serious environmental obstacles. Maya Yucatan may not have been as hot and humid as it is today, but the integration of so large an area was a huge accomplishment. The Aztecs solved problems of distance and remoteness as they forged the largest political unit Middle America has ever seen (Fig. 4-2).

Middle America and Central America

Middle America, as we define it, includes all the mainland and island countries and territories that lie between the United States of America and the continent of South America. Sometimes the term *Central* America is used to identify the same realm, but Central America actually is a region within Middle America. Central America comprises the republics that occupy the strip of mainland between Mexico and Colombia: Guatemala, Honduras, El Salvador, Nicaragua, Costa Rica, and Panama. The self-governing British colony of Belize (formerly British Honduras), adjacent to Guatemala, also is part of Central America.

In the 1920s and 1930s a school of thought developed in geography that favored the view that human cultural, political, and economic progress can occur only under particular environmental circumstances. Pointing to the present concentration of wealth and power in the middle latitudes, especially in the Northern Hemisphere, these geographers theorized that the tropics were simply not conducive to human productivity. Mostly they related their views to one aspect of the environment, the climate. Tropical climates, with their monotonous heat and humidity, were supposed to retard human progress; midlatitude climates, with their variable weather, stimulate human achievement. This school of environmental determinism (or, simply, environmentalism) thus held that the natural environment, and principally climate, to a large extent dictates the course of civilization. It was led by Ellsworth Huntington (1876–1947), who believed that human progress rested on three bases: climate, heredity, and culture. This, of course, is a sensitive area of research, but Huntington did not engage in careless speculation. He wrote no fewer than 28 books and contributed to 30 others (in addition to some 240 articles) in his attempt to measure the influence of climate on civilization. Nevertheless, his conclusions came under severe criticism as ill-founded and supportive of ''master race'' philosophies. But while he may have generalized carelessly and spoken injudiciously, Huntington was posing crucial questions still unanswered today—and now the sociobiologists, going at them from another angle, face similar obstacles. Human societies and natural environments interact, but how? When does a combination of particular environmental circumstances, inherited capacities, and cultural transmissions stimulate a new cultural explosion?

Certainly few environmentalists could easily account for the rise of Mayan civilization in tropical Middle America. Some cultural geographers have suggested that cultural stimuli from ancient Egypt actually reached Middle American shores, and that the pyramid-like stone structures of Mayan cities represent imitations and variations of Egyptian achievements. More likely Maya civilization arose as a spontaneous and independent accomplishment. It experienced successive periods of glory and decline, reaching a zenith in present-day Guatemala from the forth to the seventh century A.D.

The Maya civilization unified an area larger than any of the modern Middle American states except Mexico. Its population probably was somewhere between two and three million; the Mayan language, the local *lingua franca,* and some of its related languages remain in use in the area to this day. This was a theocratic state with a complex religious hierarchy, and the great cities that today lie in ruins, overgrown by the tropical vegetation, served in the first instance as ceremonial centers. Their structures testify to the architectural capabilities of the Maya and include huge pyramids and magnificent palaces. Intricate stone carvings, impressive murals, and detailed ideograms hewn into palace walls tell us that the Maya culture had its artists and writers. We know also that they were excellent mathematicians, had accumulated much knowledge of astronomy and calendrics, and were well ahead of other advanced peoples in the world in these fields. No doubt the Maya civilization had its poets and philosophers, but it had its practical side too: in agriculture and trade the people made major achievements. They grew cotton, had a rudimentary textile industry, and even exported finished cotton cloth by seagoing canoes to other parts of Middle America, in return for, among other things, cacao prized by the Maya. This involvement in foreign trade is another reflection of the degree of organization that was achieved by this empire.

Thus tropical lowland Middle America was the scene of impressive cultural achievements—but no more so than the highlands. At about the same time that the Maya civilization emerged in the forests, the Mexican highland also witnessed the rise of a great Indian culture, similarly focused on ceremonial centers, and likewise marked by major developments in agriculture as well as architecture, religion and ritual, and the arts. Certainly these achievements were comparable to those of the Maya, but they were to be overshadowed by what followed.

One of the important successors to the early highland civilization were the Toltecs, who moved into this area from the north, conquered and absorbed the local Indian peoples, and formed a powerful state centered on one of the first true cities of Middle America, Tula. The Toltecs' period of hegemony in the region was relatively brief, extending for less than three centuries after their rise about A.D. 900. But during this period they conquered parts of the Mayan domain, absorbed many of the Mayas' innovations

Figure 4-2

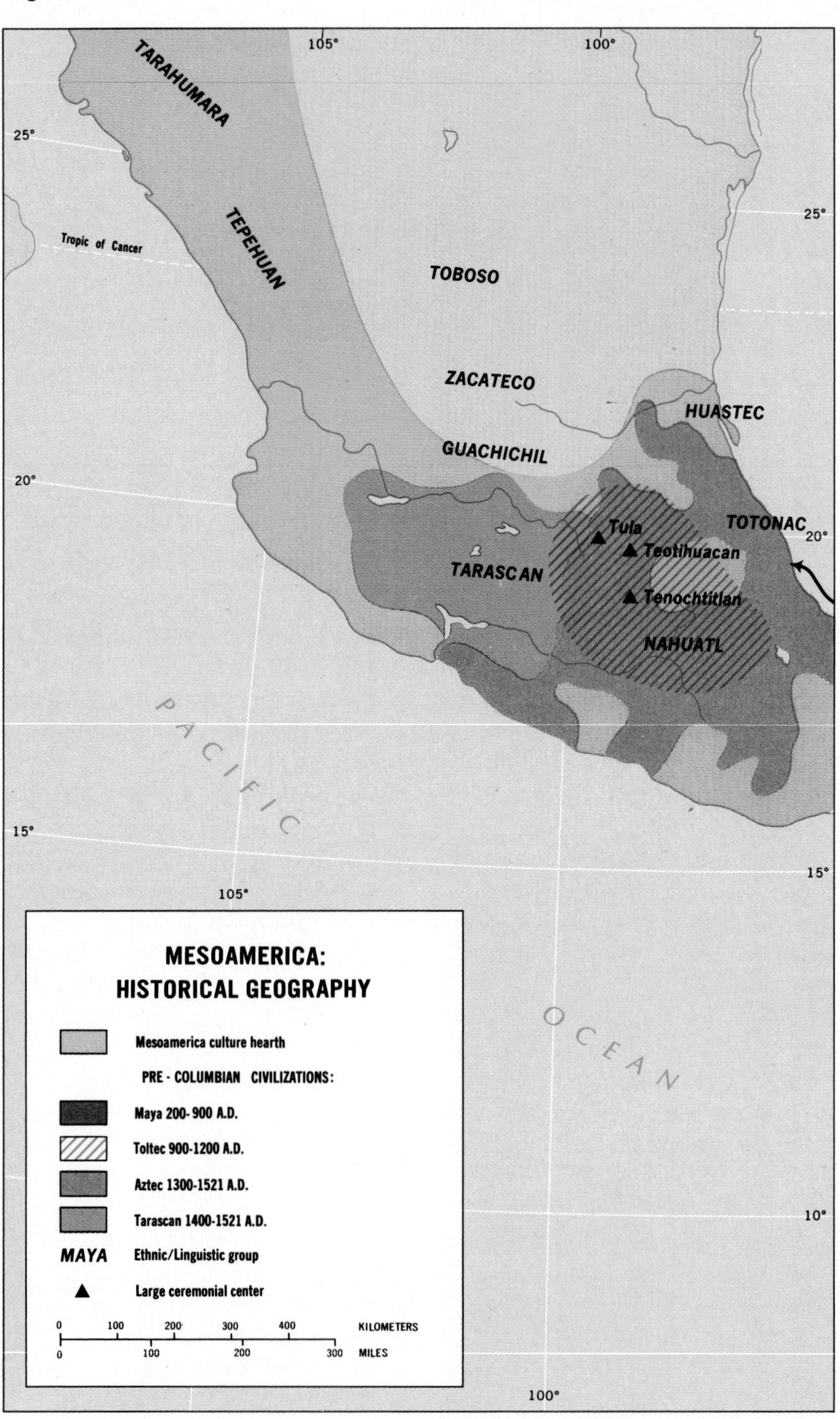

and customs, and introduced them on the highland. When their state was in turn penetrated by new elements from the north, it was already in decay. Still, Toltec technology was not lost; it was readily adopted and developed by their successors, the Aztecs.

The Aztec state, the pinnacle of organization and power in Middle America, is thought to have originated in the early fourteenth century when a community of Nahuatl- (or Mexicano-) speaking Indians founded a settlement on an island in one of the many lakes that lie in the Valley of Mexico. This village and ceremonial center, named Tenochtitlan, was soon to become the greatest city in the Americas and the capital of a large and powerful state. Through a series of alliances with neighboring peoples, the early Aztecs gained control

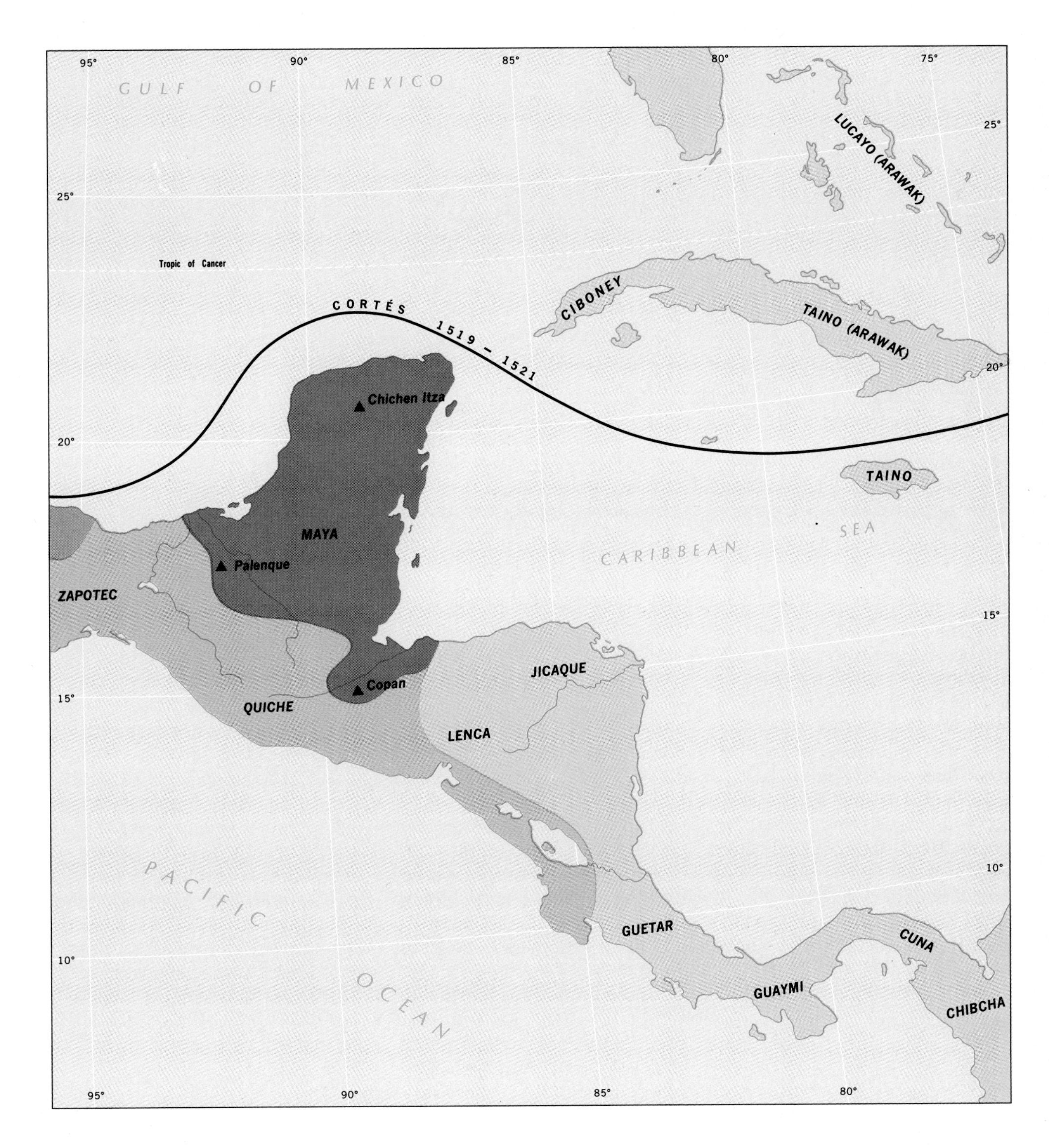

Mayans, Toltecs, Aztects and other peoples of Mesoamerica left their imprints on the landscape. Pyramidal forms are found from the forests of Yucatan to the plateau of Mexico, and the ancient city of Teotihuacan is a remarkable example of the ancient Indians' architectural and organizational capabilities. In the foreground is the Temple of the Sun. (Victor Englebert)

over the whole Valley of Mexico, the pivotal geographic feature of Middle America and today still the heart of the modern state of Mexico. This region is in fact a mountain-encircled basin positioned nearly 2500 meters (about 8000 feet) above sea level. Elevation and interior location both affect its climate; for a tropical area it is quite dry and very cool. A number of lakes lie scattered through the region, and these water bodies formed valuable means of internal communication to the Aztec state. The Indians of Middle America never developed the wheel, and so they relied heavily on porterage and, where possible, the canoe for the transportation of goods and people. The Aztecs connected several of the Mexican lakes by canals and maintained a busy canoe traffic on their waterways. Tens of thousands of canoes carried agricultural produce to the cities and tribute paid by its many subjects to the headquarters of the rulers and nobility.

Throughout the fourteenth century the Aztec state strengthened its position, developing a strong military force and organizing its territory into provinces with their respective governors and district commissioners. By the early fifteenth century the Aztecs were ready to begin with the conquest of neighboring peoples.

The Aztec drive to expand the empire was directed primarily eastward and southward. To the north, the land quickly became drier, inhabited by a sparse nomadic population, and very unproductive. To the west lay a powerful competing state, that of the Tarascans, with whom the Aztecs sought no quarrel in view of the open way to the east and south. Here, then, they penetrated and conquered almost at will. The Aztec objective was not the acquisition of territory, but the subjugation of peoples and towns for the purpose of exacting taxes and tribute. They did not introduce their concepts of religion nor their language; they were not "colonists" in the European sense. On the other hand, they carried off thousands of people for purposes of human sacrifice in the ceremonial centers of the Valley of Mexico, a practice that would have made colonization rather difficult and "pacification" a self-defeating aim. The Aztecs needed a constant state of enmity with weaker people in order to take their human prizes.

As Aztec influence spread throughout Middle America, the volume of goods streaming back to the Valley of Mexico increased. Gold, drawn largely from the Balsas Basin, cacao beans, mostly from coastal areas, cotton and cotton cloth, the feathers of tropical birds, and the skins of wild animals were among the items that were carried back to Tenochtitlan and its surroundings. The state grew ever richer, its population mushroomed, and the cities expanded. Tenochtitlan probably had over 100,000 inhabitants; some put the estimate as high as a quarter of a million. These cities were not just ceremonial centers, but true cities, with a variety of economic and political functions and large populations among which there were labor forces with particular skills and specializations.

Aztec civilization produced an enormous range of impressive accomplishments, although the Aztecs seem to have been better borrowers and refiners than they were innovators. They practiced irrigation by diverting water from streams via canals to farmlands and built elaborate walls to terrace hillslopes where soil erosion threatened. Indeed, when it comes to measuring the legacy of the Mesoamerican Indians to their successors and, indeed, to the world, then the greatest contributions surely come from the field of agriculture. Corn (maize), various kinds of beans, the sweet potato, a variety of manioc, the tomato, squash, the cacao tree, tobacco—these are just a few of the crops that grew in Mesoamerica when the Europeans first made contact.

Collision of Cultures

We in the Western world all too often are under the impression that history began when the Europeans arrived in some area of the world, and that the Europeans brought such superior power to the other continents that whatever existed there previously had little significance. Middle America seems to bear this view out: the great, feared Aztec state fell before a relatively small band of Spanish invaders in an incredibly short time. But let us not lose sight of the facts. At first, the Spanish were considered to be "White Gods," whose arrival was predicted by Aztec prophecy; having entered Aztec territory, the earliest Spanish visitors were able to determine that great wealth had been amassed in the Aztec cities. And Hernán Cortés, for all his 508 soldiers, did not singlehandedly overthrow the Aztec authority. What Cortés brought on was a revolt, a rebellion by peoples who had fallen under Aztec domination and who had seen their relatives carried off for human sacrifice to Aztec gods. Led by Cortés with his horses and artillery, these Indian peoples rose against their Aztec oppressors and followed the Spanish band of men toward Tenochtitlan; thousands of them died in combat against the Aztec warriors. They fed and guarded the Spanish soldiers, maintained connections for them with the coast, carried supplies from the shores of the Gulf of Mexico to the point Cortés had reached, secured and held captured territory while the white men moved on. Cortés started a civil war; he got all the credit for the results. But it is reasonable to say that Tenochtitlan would not have fallen so easily to the Spaniards without the sacrifice of many thousands of Indian lives.

Actually, in America as well as in Africa, the Spanish, Portuguese, Hollanders, and other European visitors considered many of the peoples they confronted to be equals—equals to be invaded, attacked, and, if possible, defeated—but equals nevertheless. The cities and farms of Middle America, the urban centers of West Africa, the great Inca roads of South America all reminded the Europeans that technologically they were, if anything, a mere step ahead of their new contacts. If a gap developed between the European powers and the indigenous peoples of many other parts of the world, it only emerged clearly when the industrial revolution came to Europe—centuries after Columbus, Cortés, and Vasco da Gama.

In Middle America, the confrontation between Spanish and Indian cultures spelled disaster for the Indians in every conceivable way. The quick defeat of the Aztec state (as well as that of the Tarascans) was followed by a catastrophic decline in the population. Whether there were 15 or 25 million Indians in Middle America when the Spaniards arrived (estimates vary), after only one century just 2.5 million survived. The Spanish were ruthless colonizers, but not much more so than other European powers that subjugated other cultures. True, the Spanish first enslaved the Indians and were determined to destroy the strength of Indian society. But biology accomplished what ruthlessness could not have achieved in so short a time. As in the Caribbean islands, the Indians were not immune to the diseases the Spaniards introduced by their presence: smallpox, typhoid fever, measles, influenza, and mumps. Neither did they have any protection against diseases introduced by white people through their African slaves, such as malaria and yellow fever, which took huge tolls in the hotter and more humid lowland areas of Middle America.

Middle America's cultural landscape, its great cities, its terraced fields, and the dispersed villages of the Indians were drastically modified. The Indian cities ceased to function as they had before, and the Spanish brought to Middle America new traditions and innovations in urbanization, agriculture, religion, and other

The Island Indians

Mesoamerican Indian cultures have in some measure survived the European invasion. Indian communities still remain, and Indian languages continue to be spoken by about 3 million people in southern Mexico and Yucatan and another million in Guatemala. But in the islands of the Caribbean the Indian communities were smaller and more vulnerable. At about the time of Columbus' arrival, there probably were about 1.5 million Indian residents in the Caribbean, a majority of them in Cuba, Hispaniola, Jamaica, and Puerto Rico. These larger islands (the *Greater Antilles)* were peopled by the Arawaks, whose farming communities raised root crops, tobacco, and cotton. In the eastern Caribbean, the smaller islands of the *Lesser Antilles* (Guadeloupe, Martinique, St. Lucia, and others) had more recently been peopled by the adventuresome Caribs, who traversed the waters of the Caribbean in huge canoes carrying several dozen persons. When the European sailing ships began to arrive, the Caribs were in the process of challenging the Arawaks for their land, just as the Arawaks centuries earlier had ousted the Ciboney.

The Europeans laid out their sugar plantations and forced the Indians into service. But Arawaks and Caribs alike failed to adapt to the extreme rigors of labor to which they were subjected, and they perished by the thousands; some fled to smaller islands where the European arrival was somewhat delayed. But after just a half century only a few hundred survived in the dense forests of some island interiors. There they were soon joined by runaway African slaves (brought to the Caribbean to replace the dwindling Indian labor force), and with their mixture, the last pure Caribbean Indian strain disappeared.

areas. Having destroyed Tenochtitlan, the Spaniards nevertheless recognized the attributes of its site and situation and chose to rebuild it as their mainland headquarters. But whereas the Indians had used stone almost exclusively as their building material, the Spaniards used great quantities of wood and utilized charcoal for purposes of metal smelting, heating, and cooking. Thus, the onslaught on the forests was such that great rings of deforestation quickly formed around the major Spanish towns. And not only around the Spanish towns; the Indians also adopted Spanish methods of house construction and charcoal use and they, too, contributed to forest depletion. Soon the scars of erosion began to replace the stands of tall trees.

Millions of Indian peasants in Middle America live in isolated villages, remote from the national core areas of the countries under whose jurisdiction they live. Their circumstances differ radically from those of the urban-based elites, and their political consciousness is rising. This particular Indian village lies in the highlands of Mexico, but the scene is repeated in Guatemala, Hunduras, and elsewhere in the region. (Wayne Miller/Magnum)

The Indians had been planters but had no domestic livestock that would make demands of the original vegetative cover. Only the turkey, the dog, and the bee (for honey and wax) had been domesticated in Mesoamerica. The Spaniards, on the other hand, brought with them cattle and sheep—in numbers that multiplied rapidly and made increasing demands not only on the existing grasslands but on the cultivated crops as well. Again, the Indians soon adopted the practice of keeping livestock, putting further pressure on the land. The net effect on food availability was not favorable. Cattle and sheep were avenues to wealth, and the owners of the herds benefited. But the livestock competed with the people for the available food, and they contributed to a gross disruption of the food balance that existed in the region. Hunger became a major problem in Middle America during the sixteenth century and no doubt contributed to the susceptibility of the Indian population to the diseases that threatened them.

The Spaniards also introduced their own crops, notably wheat, and their own farming equipment, of which the plow was the most important. Thus large fields of wheat began to make their appearance alongside the small plots of corn of the Indian cultivators. Inevitably, the wheat fields encroached on the Indians' lands, reducing them further. The wheat was grown by and for the Spaniards, so that what the Indians lost in farmlands was not made up in available food. And neither were their irrigation systems spared. The Spaniards needed water for their fields and power for their mills, and they had the technological know-how to take over and modify the regional drainage and irrigation systems. This they did, leaving the Indian fields either waterless or insufficiently watered, and thus the Indians' chances for an adequate supply of food were diminished further.

The most far-reaching changes in the cultural landscape brought by the Spaniards had to do with their traditions as town dwellers. To facilitate control, the decision was made to bring the Indians from their land into nucleated villages and towns established and laid out by the Spaniards. In these settlements the kind of government and administration to which the Spanish were accustomed could be exercised. The focus of each town was the Catholic church; indeed, until 1565 the resettlement of Indians was mainly the responsibility of missionary orders. The location of each of these towns was chosen to lie near what was thought to be good agricultural land, so that the Indians could go out each day and work in the fields. Unfortunately, the selection was not always a good one, and a number of villages lay amid land that was not suitable for Indian farm practices. Here, food shortages and even famine resulted—but once created, the village must try to survive. Only rarely was a whole village abandoned in favor of a better situation.

Acculturation

In the towns and villages, the Indians came face to face with Spanish culture. Here they learned the white invaders' religion, paid their taxes and tribute to a new master, or found themselves in prison or in a labor gang according to European laws and rules. Packed tightly in a concentrated settlement, they were rendered even more susceptible than they were in their own villages to the diseases that regularly ravaged the people. But despite all this, the nucleated Indian village survived. Its administration was taken over by the civil government, and later by the independent governments of the Middle American states; today it is a feature of Indian areas of the Mexican and Guatemalan landscape. Anyone who wants to see remnants of the dispersed Indian dwellings and hamlets must travel into the most isolated parts of Middle America's remaining Indian areas.

The Spanish towns and cities were administrative centers, located in the interior to function as centers of control over trade, tax collection, labor recruitment, and so on. At times the Spaniards, recognizing an especially salutory location chosen by their Indian predecessors, would build on the same site. Other sites were chosen because they served Spaniards better than preexisting Indian sites would. Within a half century of the Aztec defeat the Spanish conquerors founded more than 50 new towns, some of which have become major modern cities. Along the Caribbean and Pacific coasts, cities emerged that were part of the globe-girdling chain of communications within the rising Spanish empire: Veracruz on the Gulf of Mexico rose to prominence in this way, as did Acapulco on the Pacific coast.

Mining

The Spanish in Middle America had not come simply to ransack the riches in gold, silver, precious stones, works of art, and other valuables that had been accumulated by Indian kings, priests, and nobles. That aftermath of conquest was soon over. Next the Spaniards wanted to organize the sources of this wealth for their own benefit. Mining, commercial agriculture, livestock ranching, and profitable trade were the avenues to affluence, and among these, mining held the greatest promise. The first mining phase, during the first half of the sixteenth century, involved the "washing" of gold from small streams carrying gold dust and nuggets—called *placering*. This is how the Indians had found their gold, but it was (and still remains) a small-scale operation. The Spaniards simply used the Indian placer-workers for their own profit, something that was especially easy during the period of Indian slavery. As the placers produced less and less gold and Indian slavery was abolished, Spanish prospectors looked for other, larger mineral deposits—not just gold, but other valuable minerals as well. They were quickly successful, especially in finding lucrative silver and copper deposits. Between 1530 and 1570 a number of mining towns were founded that have developed into modern urban centers, and a new phase of mining began. The initial focus lay in the southwest and west of the Valley of Mexico, but the most significant finds were made somewhat later and farther to the north, along the eastern foothills of the Sierra Madre Occidental, from Guanajuato and Zacatecas northward to the Chihuahua area.

A small town in Guatemala, founded by the Spanish colonizers, lies at the foot of one of Middle America's numerous volcanoes. The region is vulnerable to volcanic eruptions and, perhaps more seriously, severe earthquakes. (George Holton/Ocelot)

The mining industry set in motion a host of changes in this part of Middle America. The mining towns drew laborers by the thousands. The mines required equipment, timber, mules; the people in the towns needed food. Since many of the mining towns, especially those of the north, were located in dry country, irrigated fields were laid out wherever possible. Mule trains and two-wheeled carts connected farm supply areas to the mining towns, and the mining towns to the coasts. More than a network of administrative towns could have done, the mining towns integrated and organized the Spanish domain in Middle America. More than government and church could have done, the mining towns brought effective and permanent Spanish control to some very far-flung parts of the New Spain. Mining, indeed, was the mainstay of colonial Middle America.

Mainland and Rimland

Outside of Mesoamerica, only Panama, with its twin attractions of transit function and gold supply, became an early focus of Spanish activity. The Spaniards founded the city of Panama in 1519, and apart from their use of the area of the modern Canal Zone as an inter-ocean link, their primary interest lay on the Pacific side of the isthmus. From here, Spanish influence began to extend northwestward into Middle America; Indian slaves were taken in large numbers from the densely peopled Pacific lowlands of Nicaragua, to be shipped to South America via Panama. The highlands, too, fell into the Spanish sphere, and before the middle of the sixteenth century Spanish exploration parties based in Panama met those moving southeastward from Mesoamerica.

But the primary center of Spanish activity was in what is today central and southern Mexico, and the major arena of international competition in Middle America lay not on the Pacific side, but on the islands and coasts of the Caribbean. Here the Spaniards faced the English, French, and Dutch, all interested in the lucrative sugar trade, all searching for quick wealth, and all hoping to expand their empires (Fig. 4-3). Only the English gained a real foothold on the mainland, which otherwise remained an exclusively Spanish colonial domain. Belize (the self-governing former British Honduras) is all that remains of a coastal sphere of British influence that extended from the Yucatan Peninsula to the Nicaragua–Costa Rica boundary. For understandable reasons the Spaniards were more interested in securing

Figure 4-3

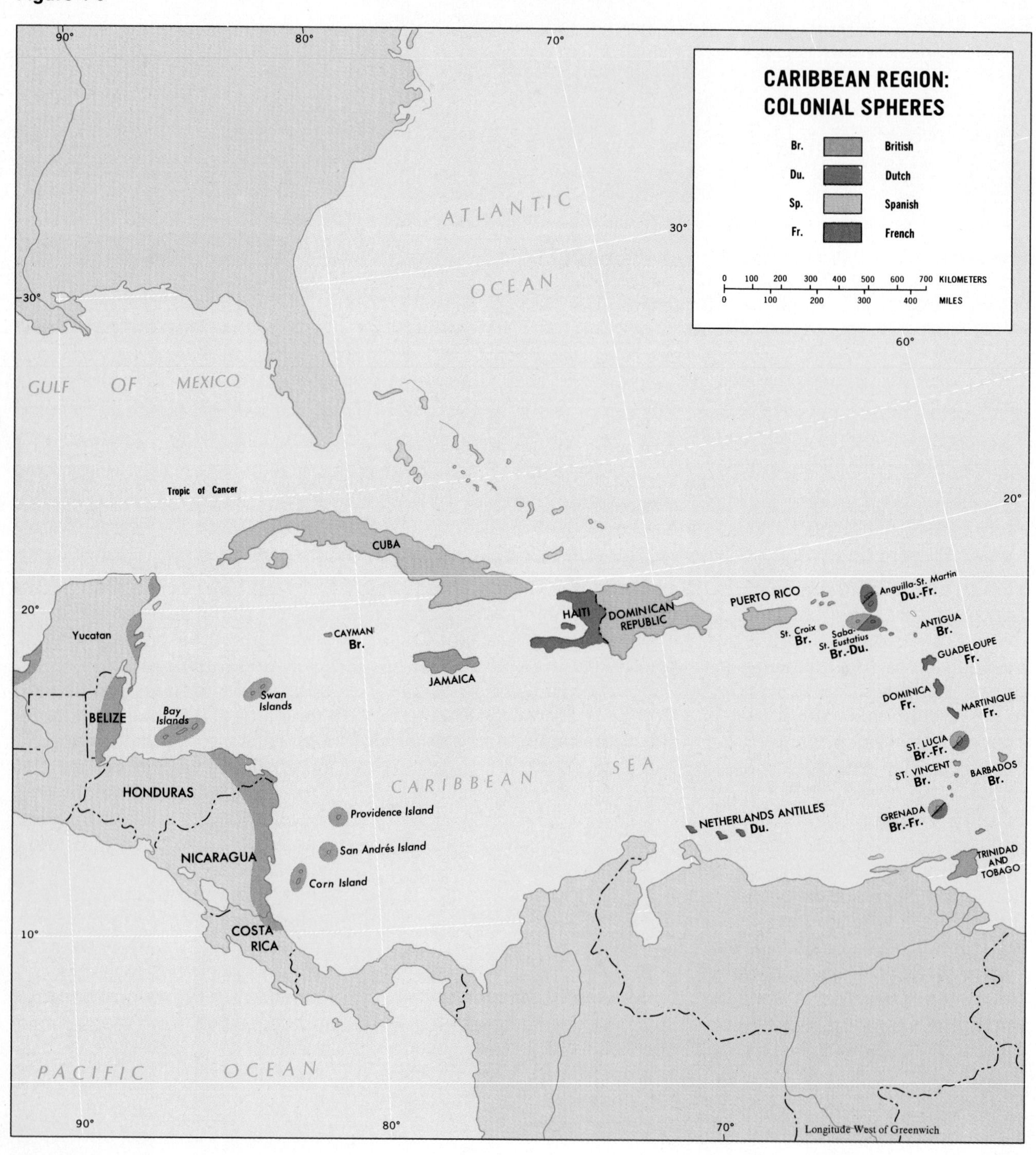

their highland holdings than in the rainy, disease-ridden, forested, swampy coasts, and they formally recognized lowland British interests in 1670. Later, after centuries of European colonial rivalry in the Caribbean, the United States entered the picture and made its influence felt in the coastal areas of the mainland—not through colonial conquest, but by the introduction of large-scale and widespread plantation agriculture. The effects were as far-reaching as was the colonial impact on the Caribbean islands. The economic geography of the Caribbean coastal areas was transformed as hitherto unused alluvial soils in the many river lowlands were planted with thousands of acres of banana trees. Since the diseases the Europeans had brought to the New World had been most rampant in these hot, humid areas, the Indian population that survived was small and provided an insufficient labor force. Tens of thousands of black laborers were brought to the coast from Jamaica and other islands, completely altering the demographic situation. In many physical ways the coastal belt already resembled the islands more than the Middle American plateau, and now the economic and cultural geography of the islands was extended to it.

These contrasts between the Middle American highlands on the one hand, and the coastal areas and Caribbean islands on the other hand, were conceptualized by J. P. Augelli into a Mainland–Rimland framework. Augelli recognized a Euro-Indian *Mainland,* consisting of mainland Middle America from Mexico to Panama but with the exception of the Caribbean coast from Yucatan southeastward, and a Euro-African *Rimland,* comprising this coastal zone and the islands of the Caribbean (Fig. 4-4). The terms *Euro-Indian* and *Euro-African* suggest the cultural heritage of each region: in the Mainland, European (Spanish) and Indian influences are paramount, and in the Rimland, the heritage is European and African. As Fig. 4-4 shows, the Mainland is subdivided into several areas on the basis of the strength of the Indian legacy: in southern Mexico and Guatemala Indian influences are marked; in northern Mexico and parts of Costa Rica, Indian influences are limited. Between these areas

Figure 4-4

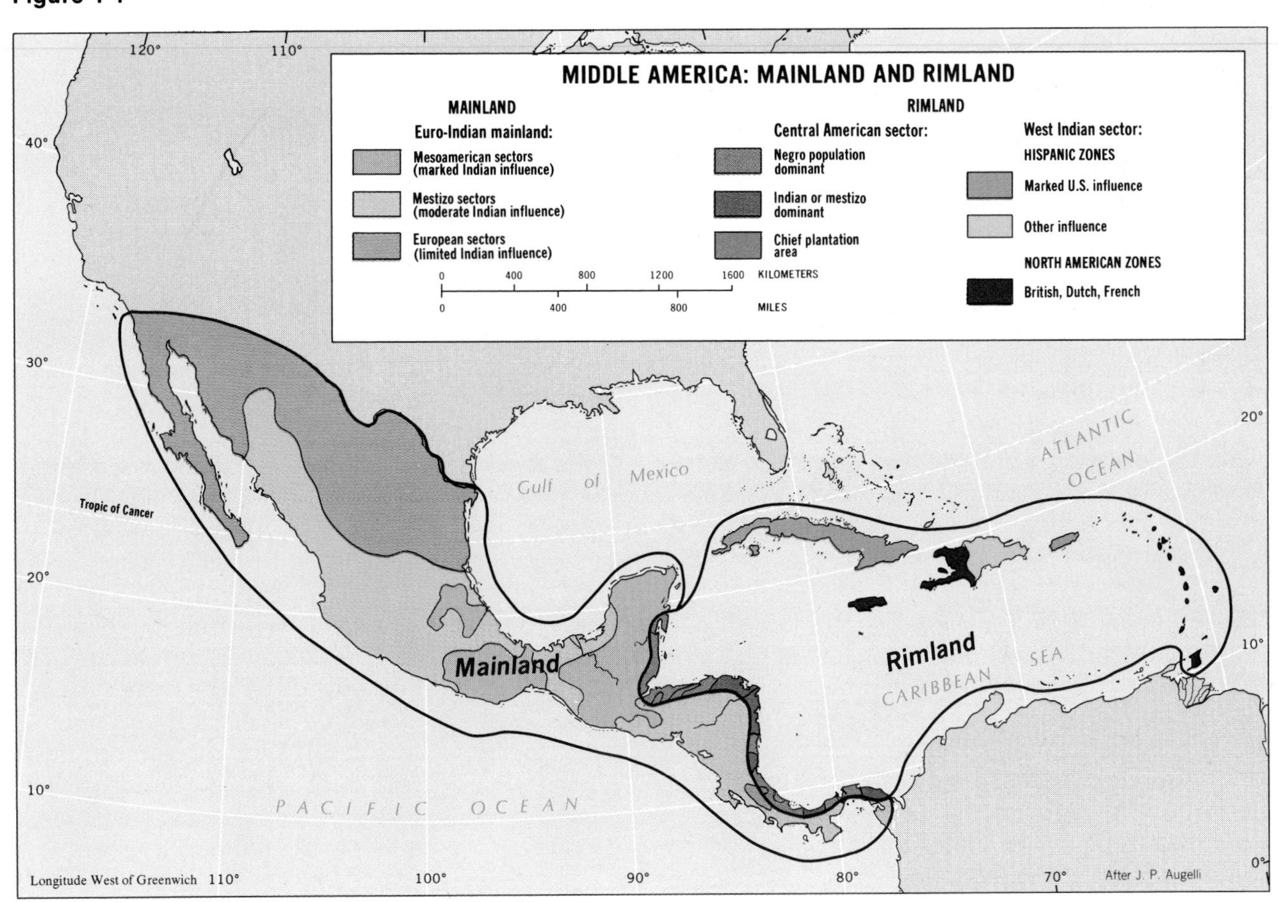

lie sectors with moderate Indian influence. The Rimland, too, is subdivided: the most obvious division is into the mainland-coast plantation zone on the one hand and the islands on the other, but the islands can themselves be grouped according to their cultural heritage. Thus, there is a group of islands with Spanish influence (Cuba, Puerto Rico, and the Dominican Republic on old Hispaniola), and another group with other European influences, including the former British West Indies, the French islands, and the Netherlands Antilles.

The contrasts of human habitat described above are supplemented by regional differences in outlook and orientation. The Rimland was an area of sugar and banana plantations, of high accessibility, of seaward exposure, and of maximum contact and mixture. The Mainland, on the other hand, was farther removed from these contacts; it has been an area of greater isolation, of greater distance from all these forces. The Rimland was the region of the great plantation, and thus its commercial economy was susceptible to fluctuating world markets and tied to overseas capital; the Mainland was the region of the hacienda, more self-sufficient, less dependent on outside markets.

In fact, this contrast between the plantation and the hacienda in itself constitutes strong evidence for a Rimland–Mainland division. The hacienda was a Spanish institution, and the plantation, Augelli argued, was the concept of Europeans of more northerly origin; later it also became an Anglo-American one. In the hacienda the Spanish landowner possessed a domain whose productivity he might never push to its limits: the very possession of such a vast estate brought with it the prestige and the comfort and life-style he sought. The workers would live on the land—it may once have been *their* land—and had plots where they could grow their own subsistence needs. Traditions survived: in the Indian villages incorporated into the early haciendas, in the methods of farming that were used, in the means of transportation of the produce to the markets. All this is written as though it is mostly in the past, but the legacy of the hacienda, with its inefficient use of land and labor, is still there in Middle America.

The plantation, on the other hand, was conceived as something very different. In their volume entitled *Middle America, Its Lands and Peoples,* Robert C. West and J. P. Augelli listed five characteristics of Middle American plantations that always apply. These quickly illustrate the differences between hacienda and plantation: (1) plantations are located in the humid coastal tropical lowlands of the region; (2) they produce for export almost exclusively—usually a single crop; (3) capital and skills are often imported, so that foreign ownership and an outflow of profits occur; (4) labor is seasonal—that its, it is needed in large numbers during harvest time and may be idle at other times—and such labor has been imported because of the scarceness of Indian labor; and (5) with its "factory in the field" characteristic, the plantation is more efficient in its use of land and labor than the hacienda. The objective was not self-sufficiency but profit, and wealth rather than prestige was a leading motive in the plantation's establishment.

During the past century, both hacienda and plantation have changed a great deal. The vast American investment in the Caribbean coastal zones of Guatemala, Honduras, Nicaragua, Costa Rica, and Panama transformed that area and brought a whole new concept of plantation agriculture to the region; in the Mainland the hacienda has been under pressure from governments that view them as political and social (not to mention economic) liabilities. Indeed, some haciendas have been parceled out to smallholders, while others have been pressed into greater specialization and productivity. Still other land has been placed in *ejidos,* and is communally owned by groups of families (see *Mexico).* But both institutions for centuries contributed to the different social and economic directions that have given the Mainland and Rimland their respective individuality.

Political Differentiation

Mainland Middle America today is fragmented into eight different countries, all but one of which (Belize, the former British Honduras) have Hispanic origins. Largest of them all—in fact, the giant of Middle America—is the United Mexican States (Mexico), whose 1,972,000 square kilometers (762,000 square miles) constitute over 70 percent of the whole land area of Middle America (the islands included), and whose more than 70 million people outnumber those of all the other countries and islands of Middle America combined (Fig. 4-1). The cultural variety in Caribbean Middle America is much greater. Here Cuba dominates: its area is larger than that of all the other islands together, and its population of over 10 million is nearly twice that of the next-ranking country (Haiti with about 5 million). But the Caribbean is hardly an area of exclusive Spanish influence: whereas Cuba has an Iberian heritage, its southern neighbor, Jamaica (population 2.3 million, mostly black) has a legacy of British involvement, and Haiti's strongest imprint has been French. The crowded island of Hispaniola is shared between Haiti and the Dominican Republic, where Spanish influence survives; it predominates also in Puerto Rico where, however, it is modified by the impact of the United States.

In the Lesser Antilles, too, the cultural diversity is great. There are American Virgin Islands, French Guadeloupe and Martinique, a group of British-influenced islands including Dominica, Barbados, St. Lucia, and St. Vincent, and Dutch Saba, St. Eustatius, and the ABC islands (Aruba, Bonaire, Curaçao) off the Venezuelan coast. Standing apart from the Antillean arc of islands is Trinidad, another formerly British dependency that, with its smaller neighbor of Tobago, became a sovereign state in 1962.

While the mainland countries (except Belize, which is moving toward independence in the early 1980s) ended their colonial status at a relatively early stage, the Caribbean islands remained colonized for a much longer time. No doubt this was due in large measure to their smaller size and insularity; Spain, which had to yield to demands and struggles for independence on the mainland, could hold Cuba and Puerto Rico, where there was much agitation for reform as well, until the Spanish-American War at the end of the nineteenth century. Spain ceded western Hispaniola to France in 1697, Frenchmen having developed sugar plantations there, and in 1795 gave up the rest of the island to the French. One of the earliest independent republics to emerge in the Caribbean was Haiti, whose black (95 percent) and mulatto (5 percent) population fought a successful slave revolt. In 1804 the Republic of Haiti was established, and the Haitians, having defeated the French—and destroyed a good part of the economic structure of their country—now turned on their Dominican neighbors. From 1822 to 1844 Haiti ruled all of Hispaniola; then, after the Dominicans had fought themselves free, Spain briefly reestablished a colony there (1861–1865). The most recent alien intervention in Hispaniola came during the twentieth century, when the United States occupied both Haiti (1915–1934) and the Dominican Republic (1916–1924).

In the Greater Antilles, then, Spanish influence was strong. By the late nineteenth century, Cuba and Puerto Rico resembled the mainland republics in the composition of their population and in their cultural imprint. But Spain's colonial archrival, Britain, also had a share of the Greater Antilles in Jamaica, and it gained a large number of footholds in the Lesser Antilles. The sugar boom and the strategic character of the Caribbean area brought the European competitors to the West Indies; in the twentieth century the United States of America made its presence felt as well—although the plantation crop was now the banana and the strategic interest even more immediate than that of the European powers. In parts of Caribbean Middle America, the period of colonial control is only just ending. Jamaica attained full independence from Britain in 1962, as did Trinidad and Tobago. An attempt by the British to organize a Caribbean-wide West Indies Federation failed, but other long-time British dependencies, including Barbados, Grenada, St. Lucia, and Dominica, were steered toward a precarious independence nevertheless. France, on the other hand, has made no moves to end the status of Martinique and Guadeloupe as Overseas Departments of the national state, and in 1980 the Dutch ABC islands still were within the Netherlands Empire.

Caribbean Patterns

Caribbean America today is a land crowded with so many people that, as an area, it is the most densely populated part of the Americas. It is a place of poverty and, in all too many locales, much misery and little chance for escape. In many respects Cuba and Puerto Rico constitute exceptions to any such generalizations made about Caribbean America. But in most of the other islands, life for the average person is difficult, often hopeless, and tragically short.

Puerto Rico: Clouded Future?

The largest and most populous United States domain in Middle America is Puerto Rico, easternmost and smallest of the Greater Antilles. The 9000-square-kilometer island, now with about 3½ million inhabitants, fell to the United States during the Spanish-American War, and it was formally ceded to the United States as part of the Treaty of Paris (1898), when the United States also acquired the Philippines and Guam.

Puerto Rico's struggle for independence from Spain was thus terminated (Cuba, on the other hand, achieved sovereignty by the same Treaty of Paris), and United States administration was at first difficult. Not until 1948 were Puerto Ricans allowed to elect their own governor, but in 1952, following a referendum, the island became the autonomous Commonwealth of Puerto Rico. San Juan and Washington share governmental responsibilities in a complicated arrangement in which Puerto Ricans have United States citizenship but pay no federal taxes on local incomes. The political situation fails to satisfy everyone: there still is a small pro-independence segment, a fairly substantial number of voters who favor statehood, and (by most recent count) a majority who want to continue the Commonwealth status, but with certain modifications. But the balance may have shifted in recent years. In December 1976 President Gerald R. Ford declared that he would propose statehood for Puerto Rico for discussion by Congress. Nothing came of his dramatic announcement, however.

Puerto Rico stands in sharp contrast to Hispaniola and Jamaica. Long dependent on a single-crop economy (sugar), Puerto Rico during the postwar period has industrialized rapidly, as a result of tax advantages for corporations, comparatively cheap labor, governmental incentives of various kinds, political stability, and involvement with the United States market. Today textiles—not bananas or sugar—rank as the leading export. Puerto Rico does not have substantial mineral resources, and it lies far from the United States core area: San Juan is about 2600 kilometers (1600 miles) from New York. These disadvantages have been overcome to a great extent by the development program started during World War II. Nevertheless, the Puerto Rican economy responds to and reflects the mainland economic picture, and times of recession and low employment generate an exodus of islanders to the mainland, where they can qualify for welfare support even if unable to find work. In the decade of the 1960s the Puerto Rican population of New York City alone grew by some 450,000, accounting for the comparatively low population growth rate of the island.

Association with the United States has undoubtedly speeded Puerto Rico's development. The question now is whether the island's social order and political system will be able to withstand the pressures that lie ahead.

The availability of food is not always adequate on Caribbean islands, and some countries (most notably Haiti) are chronically food-deficient. Small-market trade in foodstuffs disseminates local staples. This market is in Bridgetown, Barbados. (Jason Lauré/Woodfin Camp)

All this is in almost incredible contrast to the early period of riches based on the sugar trade. But that initial wealth was gained while one whole ethnic group (the Indians) was being wiped off the Caribbean map, while another (the Africans) was being imported in bondage; the sugar revenues always went to the planters, not the workers. Then the economy faced the rising competition of other tropical sugar-producing areas, and it lost its monopoly of the European market. The cultivation of sugar beets in Europe and America also cut into the sales of tropical sugar, and difficult times prevailed. Meanwhile the Europeans helped stimulate the rapid growth of the population, just as they did in other parts of the world. Death rates were lowered, but birth rates remained high, and explosive population increases occurred. With the declining sugar trade, millions of people were pushed into a life of subsistence, malnutrition, and hunger. Many sought work elsewhere; tens of thousands of Jamaican laborers went to the mainland coast when the Panama Canal was constructed and when the United Fruit Company's banana plantations began to be developed there. Large numbers of West Indians went to Britain in search of a better life, while Puerto Ricans have for decades been coming to the United States. But this outflow has failed to stem the tide of population growth; today there are well over 30 million people on the Caribbean islands.

The Caribbean islanders just have not had many alternatives in their search for betterment. Their habitat is fragmented by water and mountains; even on the smaller islands the amount of good, flat, cultivable land may be only a small fraction of the whole area. And although there has been some economic diversification, agriculture remains the area's mainstay, and sugar is still the leading product; it heads the exports of Cuba, the Dominican Republic, and Jamaica. In Haiti, coffee has taken the place of sugar as the chief export (sugar now ranks second), while Trinidad, just off the Venezuelan coast, is fortunate in possessing sizable oil fields (petroleum accounts now for nearly 80 percent of that country's exports). In the Lesser Antilles, sugar has retained a somewhat less prominent position, having been replaced by such crops as bananas, sea-island cotton, limes, and nutmegs. But even here, in Barbados, St. Kitts, St. Lucia, and Antigua, sugar remains the leading revenue producer.

All of the crops grown in the Caribbean—Haiti's coffee, Jamaica's bananas, the Dominican Republic's cacao, the Lesser Antilles' citrus fruits, as well as the sugar industry—face severe competition from other parts of the world and have not become established on a scale that could begin to have a real effect on standards of living. And those minerals that do exist in this area—Jamaica's bauxite, Cuba's iron and chromium, Trinidad's oil—do not support any significant industrialization within Caribbean America itself. As in other parts of the developing world, these resources are exported for use elsewhere.

And so the vast majority of the people in this area continue to eke out a precarious living from a small plot of ground, mired in poverty and threatened by disease. Farm tools are still primitive, and methods have undergone little change over the generations. Inheritance

The Caribbean region is marked by cultural landscapes of varied sources and origins. France, Denmark, Britain, Spain, and the Netherlands have left their imprints. The capital of Curaçao, Willemstad, represents the Dutch colonial impress. (Harm J. de Blij)

customs have divided and redivided peasant families' plots until they have become so small that the owner must sharecrop some other land or seek work on a plantation or estate in order to supplement the harvest. Bad years of drought, cold weather, or hurricanes can spell disaster for the peasant family. Soil erosion threatens; much of the countryside of Haiti is scarred by gulleys and ravines. Where soils are not eroded, their nutrients are depleted, and only the barest of yields are extracted. The good land lies under cash crops, not food crops for local consumption. Those expanses of sugar cane and banana trees symbolize the disadvantageous position of the Caribbean countries, dependent on uncertain markets for their revenues and trapped in an international economic order they cannot change.

With such problems it would be unlikely for Caribbean America to have many large cities; after all, there is little basis for major industry, little capital, little local purchasing power. Indeed the figures reflect this situation, for under one-third of the population is classed as urban. Only Cuba and Puerto Rico have as much as 60 percent of their populations in urban areas. The two largest cities, Cuba's Havana (2 million) and Puerto Rico's San Juan (900,000) owe a great deal of their development to American influences. Jamaica's capital, Kingston, has a population of about a quarter of a million; Santo Domingo (Dominican Republic) and Port-au-Prince (Haiti) each have under a half million. In the late 1970s it was estimated that less than 20 percent of Haiti's population of about 5 million was living in towns larger than 5000.

Living conditions in Haiti are difficult for the great majority of the people. In this trickle of water these residents of Port-au-Prince, Haiti's capital, must do their laundry. Political conditions as well as economic difficulties have driven thousands of Haitians away as "boat people" in search of a better life elsewhere. (Harm J. de Blij)

Tourism: the Irritant Industry. The cities of Caribbean America constitute a potential source of income as tourist attractions and places of call for cruise ships. The tourist industry is growing again after a period of stagnation during the recession of the early 1970s, and places such as Port Antonio (Jamaica) and Puerto Plata (Dominican Republic) have been

The Caribbean's tourist industry has stimulated several local industries to produce for the visitors, as revealed by Nassau's straw market and similar activities elsewhere. One "industry" that has mushroomed with the tourist business is Haiti's visual arts. When the cruise ships are in port, the capital is bedecked with countless paintings and carvings by local artists. (Harm J. de Blij)

added to cruise itineraries that already include San Juan, Port-au-Prince, and Montego Bay. The Caribbean has long been known for its magnificent beaches and spectacular island landscapes, but visitors also are attracted by the night life and gambling of San Juan, the cuisine and shopping of Fort de France (Martinique), and the picturesque architecture of Willemstad (Curaçao).

Certainly tourism is a prospective money-earner for many Caribbean islands; it already ranks at or near the top in such places as the U.S. Virgin Islands, Martinique, Curaçao, and others. But tourism has serious drawbacks. The invasion of overtly poor communities by wealthier visitors at times leads to hostility, even actual antagonism among the hosts. Upon some local residents, tourists have a "demonstration effect" that leads locals to behave in ways that may please or interest the visitors, but is disapproved by the community. Free-spending, sometimes raucous tourists contribute to a rising sense of anger and resentment. Furthermore, tourism has the effect of debasing local culture, which is adapted to suit the visitors' tastes. Anyone who has witnessed hotel-staged "culture" shows has seen this process. And many workers say that employment in the tourist industry is especially dehumanizing. Expatriate managers demand displays of friendliness and servitude that locals find difficult to sustain.

Tourism does generate income in the Caribbean where alternatives are few, but the flood of North American tourists cannot be said to have a beneficial effect on the great majority of Caribbean residents. In the popular tourist areas, the intervention of multinational corporations and governments has removed the opportunities from local entrepreneurs in favor of large operators and major resorts; tourists are channeled on prearranged trips in isolation from the local society. There are some cultural advantages to this, since tourists do not enhance international understanding when they invade Caribbean communities, but on the other hand it deprives local small establishments of potential income.

Tourism, then, is a mixed blessing to underdeveloped Caribbean territories. Given the region's limited options, it provides revenues and jobs where none would otherwise accrue. But there is a negative cumulative effect that intensifies contrasts and disparities: gleaming hotels tower over modest housing, luxury liners glide past poverty-stricken villages, opulent meals are served where, down the street, children suffer from malnutrition. The tourist industry contributes positively to overall economies, but it strains the fabric of local communities.

African Heritage

The Caribbean area is a legacy of Africa, and there are places where cultural landscapes strongly resemble those of West and Equatorial Africa. In the construction of village dwellings, the operation of rural markets, the role of women in rural life, the preparation of certain kinds of food, methods of cultivation, the nature of the family, artistic expression, and a host of other areas the African heritage of Caribbean America can be read.

Nevertheless, in general terms, it is still possible to argue that the European or white person is in the best position in this area politically and economically, the mixed-blood or mulatto ranks next, and the black person ranks lowest. In Haiti, for example, where 95 percent of the population is "pure" black and 5 percent mulatto, it is this mulatto minority that holds the reigns of power. In Jamaica the British have only recently given up control—and the 17 percent "mixed" sector of the population plays a role of prominence in island politics far out of proportion to its numbers. In the Dominican Republic, the pyramid of power puts the 25 percent white sector at the top, the 60 percent mixed group next, and the 15 percent black population at the bottom; here is a country that holds tenaciously to its Spanish-European legacy in the face of a century and a half of hostility from neighboring, Afro-Caribbean Haiti. In Puerto Rico, likewise, Spanish values are adhered to in the face of American cultural involvement and a nonwhite sector counting just under one-fifth of the island's 3.6 million people. In Cuba, too, the 15 percent of the population that is black has found itself less favored than the 15 percent mestizo sector and the 70 percent white Cuban population. In very general terms, it is still possible to see on the map what were the original functions of Caribbean America's black peoples; there is still a correlation between the distribution of sugar-producing areas and black people on the islands.

The composition of the population of the islands is further complicated by the presence of Asians from both China and India. During the nineteenth century, the emancipation of slaves and the ensuing localized labor shortages brought some far-reaching solutions. To Cuba, during the third quarter of the nineteenth century, came over 100,000 Chinese as indentured laborers (their number has dwindled considerably since then), and Jamaica, Trinidad, and Guadeloupe-Martinique saw nearly a quarter of a million East Indians arrive for similar purposes. To the Afro-modified forms of English and French heard in the Caribbean, therefore, can be added several Asian languages; Hindi is especially strong in Trinidad. The ethnic and cultural variety of Caribbean American is endless.

Mexico: Land of Troubled Revolution

Mexico is a land of distinction. This is the giant of Middle America, with a population of more than 70 million (exceeding that of all the other countries and islands of the realm combined) and a territory more than twice as large (Fig. 4-5). In all of Spanish-influenced "Latin" America, no country has even half as large a population as Mexico (Colombia and Argentina are next) and only a few match Mexico's growth rate. So fast was Mexico growing in the 1970s that its population will *double* before the end of the century.

The United Mexican States (the country's official name) consists of 31 states and the Federal District of Mexico City, the capital. About 50 percent of the Mexican population resides in towns and cities over 2500, about 12 million in the conurbation of Mexico City alone. Like all Mexico, the capital is growing at a staggering rate; the country's second city, Guadalajara, has just 2 million inhabitants.

The Indian imprint on Mexican culture is extraordinarily strong. Today, about 56 percent of Mexicans are mestizos, and nearly 30 percent are Indians. Only some 14 percent are Europeans, and there is a Mexican saying that Mexicans who do not have Indian blood in their veins nevertheless have the Indian spirit in their minds. Certainly the Mexican Indian has been Europeanized, but so strong is the Indianization of Mexican society that it would be inappropriate here to speak of acculturation; what happened in Mexico is *transculturation,* the two-way exchange of culture traits between societies in close contact. More than 1 million Mexicans still speak only an Indian language, and as many as 5 million still use Indian languages in everyday conversation although they also speak Mexican Spanish. Mexican Spanish has been strongly influenced by Indian languages, but this is only one aspect of Mexican culture that has an Indian impress. Distinctive Mexican modes of dress, architectural styles, sculpting and painting, and foods reflect the Indian contribution vividly.

Mexico is a large country, but its population is substantially concentrated in a zone that centers on Mexico City and extends across the southern "waist" from Veracruz in the east through Guadalajara in the west. This zone contains over half the Mexican population and is shown on Fig. I-9 (page 26). Half of Mexico's population lives in rural areas, and here lie some of the country's better farmlands. But although some impressive developments have taken place in agriculture, there still are enormous difficulties to be overcome. After achieving independence, Mexico failed for nearly a century to come to grips with the problems of land distribution that were a legacy of the colonial period. In fact, the situation got worse rather than better, and during the 34 years of rule by strongman Porfirio Díaz (1877–1910) matters came to a head. In the first years of the twentieth century the situation was such that 8245 haciendas covered about 40 percent of Mexico's whole area; approximately 96 percent of all rural families owned no land whatsoever and these landless people worked as *peones* on the haciendas. There was deprivation and hunger, and the few remaining Indian lands and the scattered *ranchos* (smaller private holdings, owned by mestizo or white families) could not produce enough to satisfy the country's needs. Meanwhile, thousands of acres of cultivable land lay idle on the haciendas, which occupied just about all the good farmland of the country.

Revolution. The revolution began in 1910 and set in motion a sequence of events that is still going on today. One of its major objectives was the redistribution of Mexico's land, and a program of expropriation and parceling out of the hacienda was made law by the Constitution of 1917. Since then, about half the cultivated land of Mexico has been redistributed, mostly to peasant communities consisting of 20 families or more. Such communaly owned lands are called

Figure 4-5

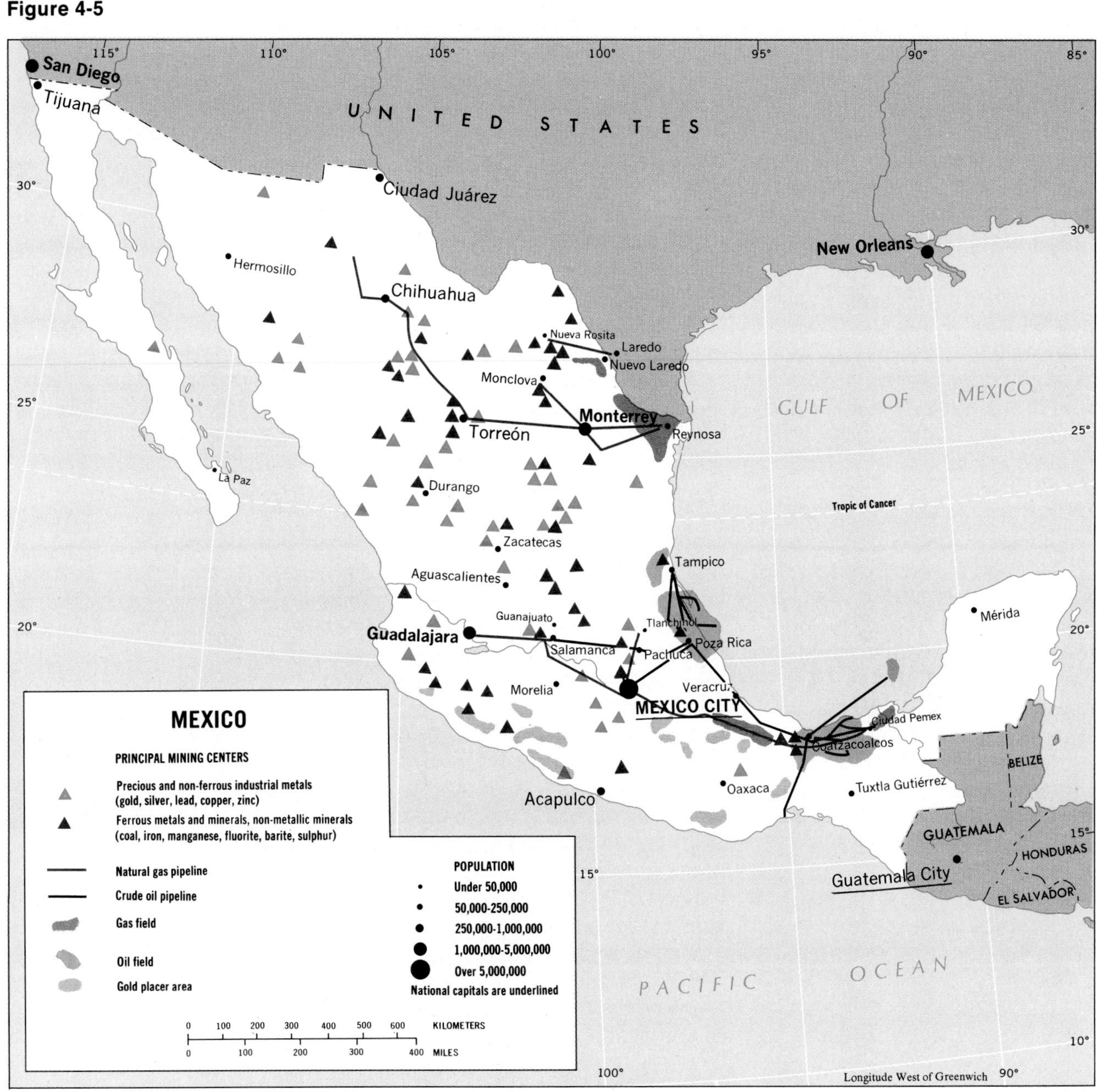

ejidos, and they are an adaptation of the old Indian village farm community; the land is owned by the group, and parcels are assigned to each member of the group for cultivation. Not surprisingly, most of the *ejidos* carved out of haciendas lie in central and southern Mexico, where Indian traditions of land ownership and cultivation survived and where the adjustments were most successfully made. A few *ejidos* are true collectives in that the members of the group do not farm assigned parcels but work the whole area in return for a share of the profits. These are located mainly in the oases of the north, where cotton and wheat are grown, and in some of the former plantations of Yucatan.

With such a far-reaching program it is not surprising that productivity temporarily declined; the miracle is that it has been carried off without a major death toll, and that the power of the wealthy landowning aristocracy could be broken without ruin to the state. Mexico alone among the countries with large Indian populations has made major strides toward solving the land question, although there is still much malnutrition and poverty in the country. But the revolution that began in 1910 did more than that. It resurrected the Indian contribution to Mexican life, and blended Spanish and Indian heritage in the country's social and cultural sphere. It brought to Mexico the distinctiveness that it alone possesses in "Latin" America.

The revolution could change the distribution of land, but it could not change the land itself, nor the methods by which it was cultivated. Corn, beans, and squash continue to form the main subsistence food of most Mexicans, with corn the chief staple and still occupying over half the cultivated land. It used to be more than that; even this proportion suggests that some crop diversification is taking place. But corn is grown all too often where the conditions are not right for it, so that yields are low; wheat might do better, but the people's preference determines the crop, not soil suitability. And if the people's preference has not changed a great deal, neither have farming methods over much of the country.

Economic Geography. Commercial agriculture in Mexico has made major strides in recent decades with respect to both the home market and for export. The greatest productivity is still in the hands of private cultivators, although much of the land has been subdivided into *ejidos.* The central plateau is geared mainly to the domestic market and produces food crops, but in the north, large irrigation projects have been built on the streams coming off the interior highlands. Cotton is cultivated for the home market as well as for export (Mexico leads Middle and South America in this product), and for the home market, wheat and winter vegetables are grown. The boom in large-scale farming in the unlikely area of Mexico's arid north is due in large measure to the adoption of American technology in the fields of irrigation and mechanized agriculture.

The importance of manufacturing in the Mexican economy also is rising. Mexico has a wide range of raw materials, many of which are located in the north and northeast. Quite early, an iron and steel industry was built in Monterrey, using iron ore located near Durango and coking coal from the Sabinas Basin in northern Coahuila. A second plant was later built near the coke, at Monclova. But the majority of Mexico's manufacturing takes place in the cities, and that means that there is a heavy concentration of manufacturing in and around Mexico City. Indeed, by some measures nearly two-thirds of all Mexican industry is located here.

Mexico's metal mining industries are less important today than they once were. The country still exports about one-quarter of the world's silver, but until the steep rise in silver prices other products ranked higher in terms of the revenues they produce, including lead and zinc; some gold is still produced, and some copper as well. The mines lie scattered throughout northern and north central Mexico, but many of the mines that were important in the colonial period—and near which urban centers of some size developed—have been worked out.

On the other hand, a recent success has been the petroleum industry, centered on the Gulf Coast city of Tampico. A series of discoveries of oil and natural gas reserves made Mexico self-sufficient in these energy sources in 1974, based on additional production from wells in the states of Tobasco and Chiapas. In 1975 more oil was found near Veracruz and Campeche and natural gas in Nuevo Laredo and Soto la Marina.

Altitudinal Zonation

Middle America and western South America are areas of high relief and strong local contrasts. People live in clusters in hot, tropical lowlands, in temperate, intermontane valleys, and even just below the snow line in the Andes. In these several zones prevail distinct local climates, soils, crops, domestic animals, and modes of life. As a result, the zones are known by specific names as if they were regions with distinct properties.

The lowest zone, from sea level to about 750 meters (2500 feet), is known as the *tierra caliente,* the "hot land" of the coastal plains and low interior basins where tropical agriculture (including banana plantations) prevails. Above this lowest zone lie the tropical highlands containing Latin America's largest population clusters, the *tierra templada* of temperate land reaching up to about 1700 meters (somewhat over 5500 feet). Temperatures are cooler. Prominent among the commercial crops is coffee; corn (maize) and wheat are staple grains. Still higher, from about 1700 meters to 3000 meters (10,000 feet) or higher (as high as 3500 meters, nearly 12,000 feet) is the *tierra fria,* the cold country of the higher Andes where hardy crops such as potatoes and barley are the people's mainstays. Only small parts of the Middle America highlands reach into the *fria* zone: in Latin America, this environment is much more extensive in the Andes. The highest zone of all is the *tierra helada* or "frozen land," also called the *paramós,* a zone of barren, exposed country above the upper limit of tree growth that reaches to the icy tops of Andean mountains.

Mexico is Middle America's giant, its capital city alone more populous than any country in the region. Smaller cities in Mexico rank with the capitals of other Middle American states. This is a segment of the urban landscape of Taxco. (Albert Moldvay/Woodfin Camp)

Additional discoveries during the late 1970s indicate that Mexico may rank with Saudi Arabia as one of the world's major storehouses of petroleum and natural gas. In 1978 Mexico's President Lopez Portillo announced that reserves of oil and natural gas were estimated to be 200 billion barrels (the 1978 world total was reported as 653.7 billion barrels of oil). Even if some of the "probable" reserves turn out to be smaller than anticipated, Mexico's huge supply is certain to affect world distribution for decades to come. It also will transform the economic geography of Mexico itself; Mexico's government controls the energy industry and its enormous revenues will be invested in industrialization and overall national development.

Mexico is industrializing; it has addressed itself with determination to agrarian reform; it seeks to integrate all sectors of the population into a true Mexican nation. After a century of struggle and oppression the country has taken long strides toward representative government. But its progress is threatened by the explosive growth of its population, for no amount of reform ultimately can keep pace with a growth rate that will, if unchecked, produce a Mexican population of nearly 140 million by the turn of the century. Already, there is local strife over land allocations; tens of thousands of Mexicans move illegally across the United States border in search of a means of survival. Small areas are made unsafe by outlaw bands that prey on highway travelers; they may foreshadow serious regional insurgencies. No political or economic system in the underdeveloped world could long withstand the impact of population change faced in Mexico, and the nation's accomplishments are under threat.

Central American Republics

Crowded on the narrow eastern section of the Middle American isthmus are seven countries, sometimes referred to in combination as the Central American Republics (Fig. 4-6). Territorially they are all quite small—only one, Nicaragua, is larger than the Caribbean island of Cuba. Populations range from Guatemala's 7 million down to Panama's 2 million in the six Hispanic republics, while the sole British territory, Belize (formerly British Honduras), has barely 180,000 inhabitants. As elsewhere in Middle America, the ethnic composition of the population is varied, with Indian and white minorities and a mestizo majority. The exceptions are in Guatemala, where approximately 55 percent of the population remains relatively "pure" Indian (Maya and Quiché) and another 35 percent is mestizo with strong Indian character, and in Belize, where two-thirds of the population is black or mulatto, a situation resembling that prevailing in the Caribbean islands. The least racial complexity exists in Costa Rica, where there is a large white majority of Spanish and relatively recent European immigrants; in a population of over 2 million, fewer than 5,000 Indians are counted, and the black population constitutes less than 2 percent of the total.

The narrow land bridge on which the republics are situated consists of a highland belt flanked by coastal lowlands on both the Caribbean and Pacific sides; from the earliest times the people have been concentrated in the *templada* (temperate) zone. Here, tropical temperatures are moderated by elevation, while rainfall is adequate for the cultivation of a variety of crops. As noted earlier, the Middle American highlands are studded with volcanoes, and areas of volcanic soils are scattered throughout the region (individual areas are too small to be outlined on Fig. I-8). The old Indian agglomerations were located in the fertile parts of the highlands, and the Spanish period perpetuated this human distribution. Today, the capitals of Guatemala (Guatemala City, 800,000), Honduras (Tegucigalpa, 320,000), El Salvador (San Salvador, 350,000), Nicaragua (Managua, devastated by the 1972 earthquake, now recovered to 400,000) and Costa Rica (San Jose, 300,000) all lie in the interior, most of them at 1000 meters or more in elevation; only Panama City (400,000) and Belize (40,000) are coastal capitals in mainland Middle America (Fig. 4-6). The size of those cities, in countries whose population averages about 2.5 to 3 million, is a reflection of their primacy, just as was Mexico City in Mexico. On an average the next ranking town is only one-fifth as large as the capital city.

The distribution of population in the republics, in addition to its concentration in the area's higher sections, also shows greater densities toward the Pacific than toward the Caribbean

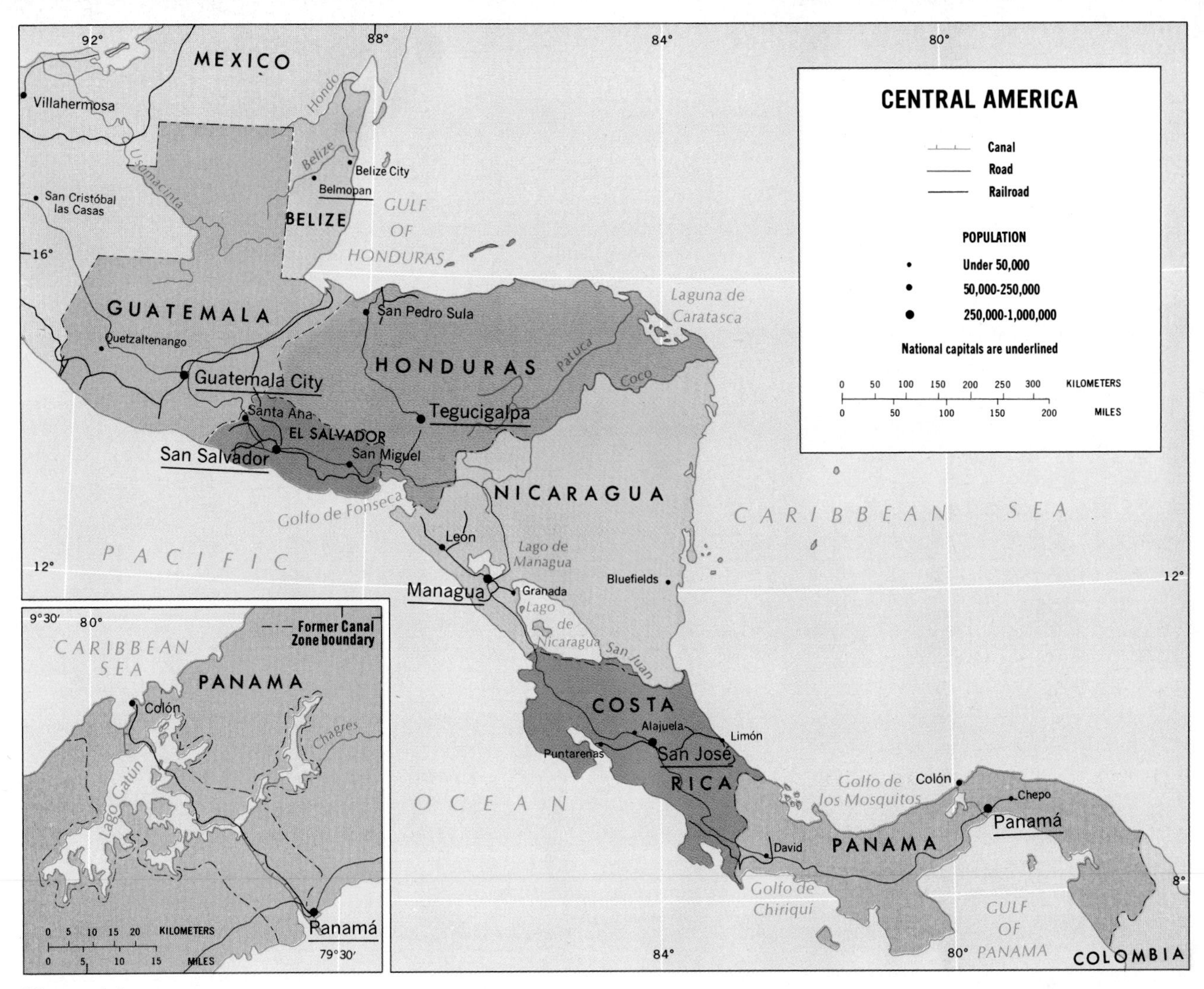

Figure 4-6

coastlands (Fig. I-9, p. 26). El Salvador, with Belize and to a lesser degree Panama, is an exception to the rule that people in mainland Middle America are concentrated in the *templada* zone; most of El Salvador is *tierra caliente,* and the majority of its 4.7 million people are crowded, hundreds to the square mile, in the intermontane plains less than 750 meters (2500 feet) above sea level. In Nicaragua, too, the Pacific areas are the most densely populated; the early Indian centers lay near Lakes Managua and Nicaragua and in the adjacent highlands.

The frequent volcanic activity in this area is accompanied by the emission of volcanic ash, which settles over the countryside and quickly weathers into fertile soils. By contrast, the Caribbean coastal lowlands, hot, wet, and leached, support comparatively few people. In the most populous of the republics, Guatemala, the heartland also has long been in the southern highlands, and although the large majority of Costa Rica's population is concentrated in the central highlands around San Jose, the Pacific lowlands have been the scene of major immigration since banana plantations were established there. Even in Panama there is a strong Pacific concentration: more than half of all Panamanians (and that means over 70 percent of the rural people) live in the southwestern lowlands and adjoining mountain slopes. Another 25 percent live and work in the Canal Zone, and of the remainder, a majority, many of them descendants of black immigrants from the Caribbean, live on the Caribbean side.

Panama and the Panama Canal

The Republic of Panama owes its birth to an idea: the construction of a waterway to connect Atlantic and Pacific waters. In the 1880s, when Panama was still a peninsula of Colombia, a French company directed by Ferdinand de Lesseps tried and failed to build such a canal; thousands of workers died of malaria and other tropical diseases and the company went bankrupt. At the turn of the century U.S. interest in a Panama Canal (which would shorten the distance between the East and West coasts by 8000 nautical miles) rose sharply, and the United States in 1903 proposed a treaty that would permit a renewed effort at construction across Colombia's Panamanian isthmus. When the Colombian Senate refused to approve, Panamanians revolted and the U.S. supported this uprising by preventing Colombian forces from intervening. The Panamanians, at the behest of the United States, declared their independence from Colombia and the new republic immediately granted the U.S. rights in a Canal Zone averaging about 16 kilometers (10 miles) in width and just over 80 kilometers (50 miles) in length.

Soon canal construction commenced, and this time it was a successful project. The Panama Canal was opened in 1914, evidence of the United States' power and influence in the Caribbean and Middle America. The Canal Zone was held by the U.S. under a treaty that granted it "all the rights, powers, and authority" in the area "as if it were the sovereign of the territory. . . ."

This language might suggest that the United States held rights over the Canal Zone in perpetuity, but the treaty nowhere stated specifically that Panama permanently yielded its own sovereignty there. In the 1970s, as the Canal was transferring 20,000 ships per year and generating hundreds of millions of dollars in tolls, Panama sought to terminate U.S. control in the Canal Zone. Delicate negotiations began, and in 1977 agreement was reached on a staged withdrawal by the United States from the territory, first from the Canal Zone and, by the

With the single exception of Costa Rica, Middle America's mainland republics face the same problems as Mexico, only more so; they also share many of the difficulties confronting the Caribbean islands. Although efforts are being made to improve the land situation, the colonial legacy of hacienda and peon hangs heavily over the area. Earlier we referred to the brevity of life for the peoples of the Caribbean: the most recent United Nations statistics indicate that life expectancy in Guatemala, where nearly one-third of the mainland's people outside Mexico live, is 54 years. It is 69 years in comparatively well-off Costa Rica, and 59 in El Salvador. Generations of people have seen little change in their way of life and have had even less opportunity to effect any real improvement: dwellings are still made of mud and straw, sanitary facilities have not reached them, schools are overcrowded, hospital facilities are inadequate. As for their livelihood, each year brings a renewed struggle to extract a subsistence, with no hope that the next year will bring something better. One of the staggering contrasts in Middle America is that between the attractive capital city and its immediate surroundings and the desolation of the distant rural areas; another is that between the splendor, style, and refined culture of the families who own the coffee *fincas* and the destitute, rag-wearing peasant.

A container ship traverses the Panama Canal. (J. P. Laffont/Sygma)

year 2000, from the Canal itself. This agreement took the form of two treaties, and following the signing by the two presidents (Carter and Torrijos) they were ratified in spite of stubborn opposition in the U.S. Congress. Long fragmented (see Chapter 10), Panama will finally become a territorially coherent state.

Coffee—and not bananas—is the most important export crop for all but three of the republics (in Honduras and Panama bananas lead, and in Belize, lumber, naval stores, and chicle). Although introduced as early as the 1780s, major coffee plantings were first made between 1855 and 1860, mostly in Guatemala. This is a *templada* crop that does best on the mountain slopes, and it spread quickly to other countries, especially El Salvador and Costa Rica. During the second half of the nineteenth century, the banana plantation began to make its appearance along the Caribbean coast, and with it came an involvement of American big business in the affairs of Middle America. Black laborers came from Jamaica and elsewhere to take the newly available jobs (as they did when the Panama Canal was dug), and a Caribbean-ward flow of people was stimulated on the mainland.

The companies that introduced banana plantation agriculture in Middle America—among them United Fruit, Standard Fruit and Steamship, and Boston Fruit—were not always benevolent contributors to better days for Middle America. There was a good deal of interference in domestic affairs and considerable resentment in what the Americans sometimes referred to as the "banana republics." But as time went on, positive contributions were indeed made—although the industry has sustained major setbacks. Crop disease caused a large-scale decline in the Caribbean area, and the focus shifted to the Pacific coastal areas, which, in contrast to the always-wet Caribbean shores, are seasonally dry. In recent years, however, new, disease-resistant varieties have been developed and reintroduced in the Caribbean zone (which had not lost its locational advantages with reference to the United States market), and there are signs that the old pattern of Caribbean dominance in the banana industry of Middle America is on the way back.

Additional Reading

The standard geographic work on Middle America is *Middle America, Its Lands and Peoples* by R. C. West and J. P. Augelli, first published in 1966 by Prentice-Hall, Englewood Cliffs, and revised in 1976. On the islands, see H. Blume, *The Caribbean Islands,* a translation from the German by Maczewski and Norton and published in 1974 by Longmans, London. On the Mesoamerican culture hearth, a readable volume is *Middle America: A Culture History of Heartland and Frontiers,* a Prentice-Hall publication of 1975 authored by M. W. Helms. The precolonial period is discussed by E. Wolf in *Sons of the Shaking Earth,* a Phoenix paperback of the University of Chicago Press, published in 1959. Also see a volume edited by H. E. Driver, *The Americas on the Eve of Discovery,* a Prentice-Hall publication of 1964, and R. Wauchope (ed.), *Handbook of Middle American Indians,* University of Texas Press, Austin, 1964. On the problems and potentials of development there are many sources; see E. Staley, *The Future of Underdeveloped Countries,* a Praeger paperback published in 1961, and J. H. Kautsky's *Political Change in Underdeveloped Countries,* published by Wiley in 1962, also in paperback. On plantations, see a publication by the Pan American Union, *Plantation Systems of the New World,* Social Science Monograph No. 7, published in Washington. Another useful work is H. Mitchell's *Caribbean Patterns: A Political and Economic Study of the Contemporary Caribbean,* published in London by Chambers in 1967.

Individual countries and groups of countries have also been discussed by geographers. On Mexico, see D. D. Brand, *Mexico: Land of Sunshine and Shadow,* No. 31 in the Van Nostrand Searchlight Series, and on the Indian peoples, the chapter entitled "The Indians in Mexico" in *Minorities in the New World: Six Case Studies,* by C. Wagley and M. Harris, published by Columbia University Press in 1958. On Central America see Thorsten V. Kalijarvi's *Central America,* No. 6 in the Van Nostrand Searchlight Series, and on the Caribbean islands, consult D. Lowenthal (ed.), *The West Indies Federation: Perspectives on a New Nation,* a Columbia University Press publication of 1961. The perspective has since changed, but the book remains of interest. Also see *The West Indian Scene* by G. Etzel Pearcy, No. 26 in the Van Nostrand Searchlight Series. On Puerto Rico, see the volume by R. Pico entitled *The Geography of Puerto Rico,* published by Aldine, Chicago, in 1974. Puerto Rico is also discussed by E. P. Hanson in another Searchlight Book, No. 7, entitled *Puerto Rico: Ally for Progress.* On El Salvador, there is a book by D. R. Raynolds, *Rapid Development in Small Economies: The Example of El Salvador,* a Praeger publication of 1967. For information of Guatemala, consult the volume by N. L. Whetten, *Guatemala: the Land and the People,* published by Yale University Press in 1961. On other Middle American countries, consult the appropriate editions of *Focus,* published by the American Geographical Society in New York.

J. P. Augelli's article entitled "The Rimland–Mainland Concept of Culture Areas in Middle America" appeared in the *Annals of the Association of American Geographers,* Vol. 52, No. 2, June 1962. On Huntington's viewpoints relating to environmental determinism, the best place to start is with his biography, written by G. J. Martin, *Ellsworth Huntington: His Life and Thought,* published by Archon Books, Hamden, in 1973.

Hope and reality in Bogotá, Colombia. (Victor Englebert)

Five

South America at the Crossroads

IDEAS AND CONCEPTS

- Land alienation
- Isolation
- Culture areas
- Agricultural systems
- Capital cities classification
- Urban structure (2)
- Urban functional classification
- Rural-to-urban migration
- Boundaries

No continent has as familiar a shape as South America, that giant triangle that is connected by Middle America's tenuous land bridge to its sister continent in the north. But South America might be more appropriately called *Southeast* America, for most of it lies not only south, but also east of its northern counterpart. Lima, the capital of Peru, lies farther east than Miami, Florida. Thus South America juts out much farther into the Atlantic Ocean than does North America, and South American coasts lie much closer to Africa and even to southern Europe than coasts of Middle and North America.

Lying so far eastward, South America on its western flank faces a much wider Pacific Ocean than does North America. From its west coast to Australia is nearly twice as far as from San Francisco to Japan, and South America has virtually no interaction with the Pacific world of Australasia—not just because of the vast distances, but also because it lies opposite the insular and less populous south, whereas North America faces Japan and the crowded East Asian mainland.

As if to confirm South America's northward and eastward orientation, the western margins of the continent are rimmed by one of the world's great mountain ranges, the Andes, a giant wall that extends from the very southern tip of the triangle to Colombia in the north (Fig. 5-1). Every map of world environmental distributions reflects the existence of the mountain chain: in the alignment of isohyets (lines of equal precipitation) (Fig. I-5, page 16), in the elongated area of highland climate (Fig. I-6, page 18), and in the vegetation and soil maps. Also, as Fig. I-9 (page 26) indicates, South America's largest population clusters lie along the east and north coasts, overshadowing those of the Andean west.

SOUTH AMERICA: PHYSIOGRAPHY
Punta Gallinas
Pico Cristóbal Colón 5775 m
Orinoco
Llanos
Guiana Highlands
Andes Mountains
Marajo
Equator
Amazon
Gulf of Guayaquil
Selvas
Montaña
Planalto do Mato Grosso
Planalto Central
Brazilian Highlands
PACIFIC OCEAN
Tropic of Capricorn
Campos
Gran Chaco
Paraná
Ojos del Salado 6863 m
Andes Mountains
ATLANTIC OCEAN
Aconcagua 6959 m
Pampas
Rio de la plata
Gulf of San Matias
Chiloé
Gulf of San Jorge
Patagonia
Str. of Magellan
Falkland Is.
Tierra del Fuego
Cape Horn
Longitude West of Greenwich
KILOMETERS
MILES

Figure 5-1

Ten Major Geographic Qualities of South America

1. South America's physiography is dominated by the Andean Mountains in the west and the Amazon Basin in the north; much of the remainder is plateau country.
2. Comparatively low population densities prevail in South America, but growth rates are among the world's highest.
3. South America's population remains concentrated in peripheral zones; much of the interior is sparsely peopled.
4. Half the realm's area and nearly half its population are concentrated in one state, Brazil.
5. Strong cultural pluralism exists in the majority of the realm's countries, and this pluralism frequently is regionally expressed.
6. Interconnections among the states of the realm remain comparatively weak. External ties are frequently stronger.
7. Lingering politico-geographical problems beset the realm; boundary disputes and territorial conflicts continue.
8. Regional contrasts and disparities, both in the realm as a whole and within individual countries, are strong; in general the south is most developed, the northeast least.
9. The Catholic Church dominates life throughout the realm, and constitutes one of its unifying elements.
10. With the exception of three small countries in the north, the realm's modern cultural sources lie in a single subregion of Europe, the Iberian Peninsula. Portuguese (in Brazil) and Spanish are the *linguae franca.*

The Human Sequence

Although modern South America's largest populations are situated in the east and north, there was a time when the Andes Mountains contained the most densely peopled and best organized state on the continent. The origins of the Inca Empire are still shrouded in mystery. It has been accepted that the Incas were descendants of peoples who came to South America via the Middle American land bridge, but even this is not totally beyond doubt. People may have reached Chilean and Peruvian coasts from Pacific islands, and civilization in the Americas may have a northward component as well. In any case, for thousands of years before the first Europeans came on the scene, Indian communities and societies had been developing. About 1000 years ago a number of regional cultures thrived in the valleys between the Andean mountains and along the Pacific Coast. The llama had been domesticated and was a beast of burden, a source of meat, and a producer of wool. Religions flourished and stimulated architecture and the construction of temples and shrines. Sculpture, painting, and other art forms were practiced. It was over these cultures that the Incas extended their authority from their headquarters at Cuzco in the Cuzco Basin (Peru), beginning late in the twelfth century, to forge the greatest empire in the Americas prior to the arrival of the Europeans.

Nothing to compare with the cultural achievements of the Andean area existed anywhere else in South America. In addition to the Andean civilizations, anthropologists recognize three other groups of peoples: those of the Caribbean fringe, those of the tropical forest of the Amazon Basin and other lowlands, and those called "marginal," whose habitat lay in the Brazilian Highlands, the headwaters of the Amazon River, and Patagonia (Fig. 5-2). It has been estimated that the Caribbean, forest, and "marginal" peoples together constituted only about one-quarter of the continent's total population.

The Inca Empire

When the Inca civilization is compared to that of ancient Mesopotamia, Egypt, the old Asian civilizations, and the Aztecs' Mexica Empire, it quickly becomes clear that this was an unusual achievement. Everywhere else, rivers and waterways provided avenues for interaction and the circulation of goods and ideas. Here, the empire was forged out of a series of elongated basins in the high Andes, basins called *altiplanos,* created when mountain valleys between parallel and converging ranges are filled with erosional materials, volcanic debris, etc. The *altiplanos* are often separated from each other by some of the world's most rugged terrain, with high, sometimes snowcapped mountains alternating with precipitous canyons. Individual *altiplanos* accommodated regional cultures;

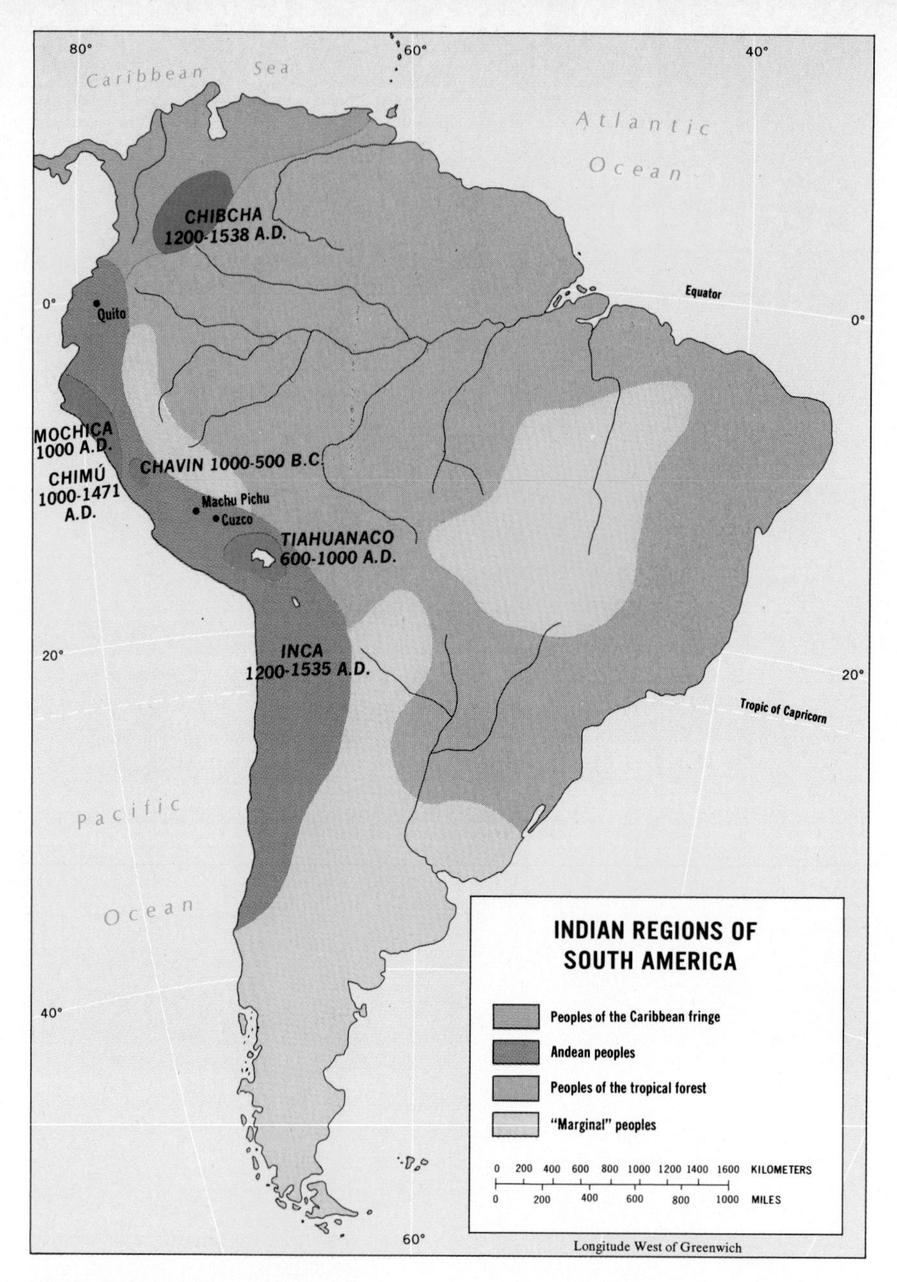

Figure 5-2

the Incas themselves were first established in the intermontane basin of Cuzco. From these headquarters they proceeded, by military conquest, to extend their authority over the peoples of coastal Peru and other *altiplanos.* Their first thrust was southward, and it seems to have occurred toward the close of the fourteenth century. But more impressive than the Incas' military victories was their subsequent capacity to integrate the peoples and regions of the Andean realm into a stable and well-functioning state. All the odds would seem to have been against them; as they progressed farther southward, into northern and central Chile and northwestern Argentina, their domain became ever more elongated, making effective control more and more difficult. The Incas, however, were expert road and bridge builders, colonizers,

In the high Andes the ancient Incas created an empire with impressive and durable achievements in architecture, agriculture, and other fields. Among the most famous ruins remaining from that period are the structures at Macchu Pichu, Peru. (Eric Simmons/Stock, Boston)

and administrators. In an incredibly short time they consolidated their southern territory; shortly before the Spanish arrival (1531) they conquered Ecuador and a part of southern Colombia.

The early sixteenth century was a critical period in the empire, because the conquest of Ecuador and nearby areas for the first time placed stress on the existing administrative framework. Until that time, the undisputed center of the state had been Cuzco, but now it was decided that the empire should be divided into two units—a southern one to be ruled from Cuzco, and a northern sector to be centered on Quito. This decision was related to the continuing problem of control over the dissident and rebellious north and the possibilities of further expansion into central Colombia. Now the empire was beset by a number of internal difficulties: an uncertain frontier in the north, tensions between Cuzco and Quito. And just as the Aztec Empire had been ready for internal revolt when Cortes and his party entered the country, so the Spanish arrival in western South America happened to coincide with a period of stress within the Inca Empire.

When it was at its zenith the Inca Empire may have counted as many as 25 million subjects; estimates range from 10 to 30 million. Of course, the Incas themselves were in a minority in this huge state, and their position eventually became one of a ruling class in a rigidly class-structured society. The Incas, representatives of the emperor in Cuzco, formed a caste of administrative officials who implemented the decisions of their ruler by organizing all aspects of life in the conquered territories. They saw to it that all harvests were divided between the church, the community, and the individual family; they maintained the public granaries; they recognized and reported the need for investments in improved roads, road maintenance, the terracing of hillsides, and the layout of irrigation works. The life of the subjects of the Inca Empire was strictly controlled by this huge bureaucracy of Inca administrators, and there was little personal freedom. Farm yields were predetermined, and there was no real market economy; the produce, like the soil on which it was grown, belonged to the state. Marriages were officially arranged, and families could live only where the Inca supervisors would permit. Indeed, the family (as a productive entity within the community) was considered to be the basic unit of administration, not the individual. Inca rule was amazingly effective, and obedience was the only course for its subjects. So highly centralized was the state and so complete the subservience of its effectively controlled population that a takeover at the top was enough to gain power over it—as the Spaniards proved in the 1530s.

The Inca Empire, which had risen to greatness so rapidly (some argue that it may have taken less than a century and not, as others believe, several centuries) disintegrated abruptly with the impact of the Spanish invaders. Perhaps it was the swiftness of its development that contributed to its weakness. But it left behind many social values that have remained a part of Indian life in the Andes to this day—and that continue to contribute to fundamental divisions between the Iberian and Indian populations in this part of South America. The most basic of these relates to the ownership of land and other property. Even before the advent of the Inca Empire, the Indians of the various regional cultures practiced communal land ownership—if not by the state, then by villages or groups of villages. The Inca period confirmed and even intensified this outlook by rigid state control of all land and resources. Personal wealth simply could not be achieved through the acquisition of land or the control of resources such as minerals or water supplies; not only did the system not permit it, the concept itself was more or less unknown. Even less was prestige associated with land ownership. The Spaniards, we know, held almost precisely opposite views and values. Today, more than four centuries after the fall of the Inca Empire, these conflicting outlooks still divide Iberian and Indian South America.

The Iberian Invaders

In South America as in Middle America, the location of Indian peoples determined, to a considerable extent, the direction of the thrusts of European invasion. The Incas, like the Maya and the Aztec people, had accumulated gold and silver in their headquarters, they possessed productive farmlands, and they constituted a ready labor force. Not long after the defeat of the Mexica Empire of the Aztecs, the Spanish conquerors crossed the Panamanian isthmus and sailed down the west coast. Francisco Pizarro on his first journey heard of the existence of the Inca Empire, and after a landfall in 1527 at Tumbes, located on the northern coast of Peru very near the Ecuador boundary, he returned to Spain to organize a penetration of the Inca realm. He arrived at Tumbes with 183 men and a couple of dozen horses in January 1531, a time when the Incas were occupied with problems of royal succession and strife in the northern provinces. The events that followed are well known; less than three years later the party rode, victorious, into Cuzco. Initially the Spaniards kept the structure of the empire intact by permitting the crowning of an emperor who was, in fact, in their power, but soon the land- and gold-hungry invaders were fighting among themselves, and the breakdown of the old order began.

The new order that eventually emerged in western and southern South America placed the Indian peoples in serfdom to the Spaniards. Land was alienated into great haciendas, taxes were instituted, and a forced labor system was introduced to maximize the profits of exploitation. As in Middle America, the Spanish invaders mostly were people who had little status in Spain's feudal society, but they brought with them the values that prevailed in Iberia—land meant power and prestige, gold and silver meant wealth. Lima, the coastal headquarters of the Spanish conquerors, was founded by Pizarro in 1535, approximately northwest of the Andean settlement of Cuzco. Before long it was one of the richest cities in the world, reflecting the amount of wealth being yielded by the ravaged Inca Empire. Soon it was the capital of the viceroyalty of Peru, as the Spanish authorities in Spain began to integrate the new possession in the colonial empire. Later, when Colombia and Venezuela became Spanish-controlled and Spanish settlement progressed in the coastlands of the Plata Estuary (in present-day Argentina and Uruguay), two viceroyalties were added: New Granada in the north and Rio de la Plata in the south.

Meanwhile, another vanguard of Iberian invasion was penetrating the eastern part of the continent—the coastlands of present-day Brazil. This area had become a Portuguese sphere of influence almost by default. It was visited by Spanish vessels early in 1500, and later that year the Portuguese navigator Cabral saw the coast on his way to the Far East. The Spaniards did not follow up their contact, and the Portuguese in 1501 sent Amerigo Vespucci with a small fleet to investigate further. After this journey the area remained virtually neglected by the Portuguese for nearly 30 years; they were absorbed in the riches of the East Indies. And the Spaniards, we know, had other interests in the Americas. Additionally, the two countries had agreed, in 1494 by the Treaty of Tordesillas, to recognize a line drawn 370 leagues west of the Cape Verde Islands as the boundary between their New World spheres of influence. This north-south border ran approximately along the line of 50 degrees west longitude, thus cutting off a sizable triangle of eastern South America for Portugal's exploitation (Fig. 5-3).

A brief look at the map of South America shows that the Treaty of Tordesillas did not succeed in limiting Portuguese colonial territory to the agreed line of 50 degrees west longitude. True, the boundaries between Brazil and its northern and southern neighbors (French Guiana and Uruguay) both reach the ocean very near the 50 degree line, but then the boundaries of the country bend far inland, to include almost the whole Amazon Basin and a good part of the Parana-Paraguay Basin as well. Portugal's push into the interior was reflected by the Treaty of Madrid (1750) whose terms included a westward shift of the 1494 Papal Line. In fact, Brazil, alone, with over 8.5 million square kilometers (3.3 million square miles), is only slightly smaller than all the other South American countries combined (9.6 million square kilometers). In population, too, Brazil has about half the total of all of South America; history dealt the Portuguese sphere a fairer hand than any treaties did. This enormous westward thrust was the work of many Brazilian elements—missionaries in search of converts, explorers in search of quick wealth—but no group did more to achieve it than the so-called

Figure 5-3

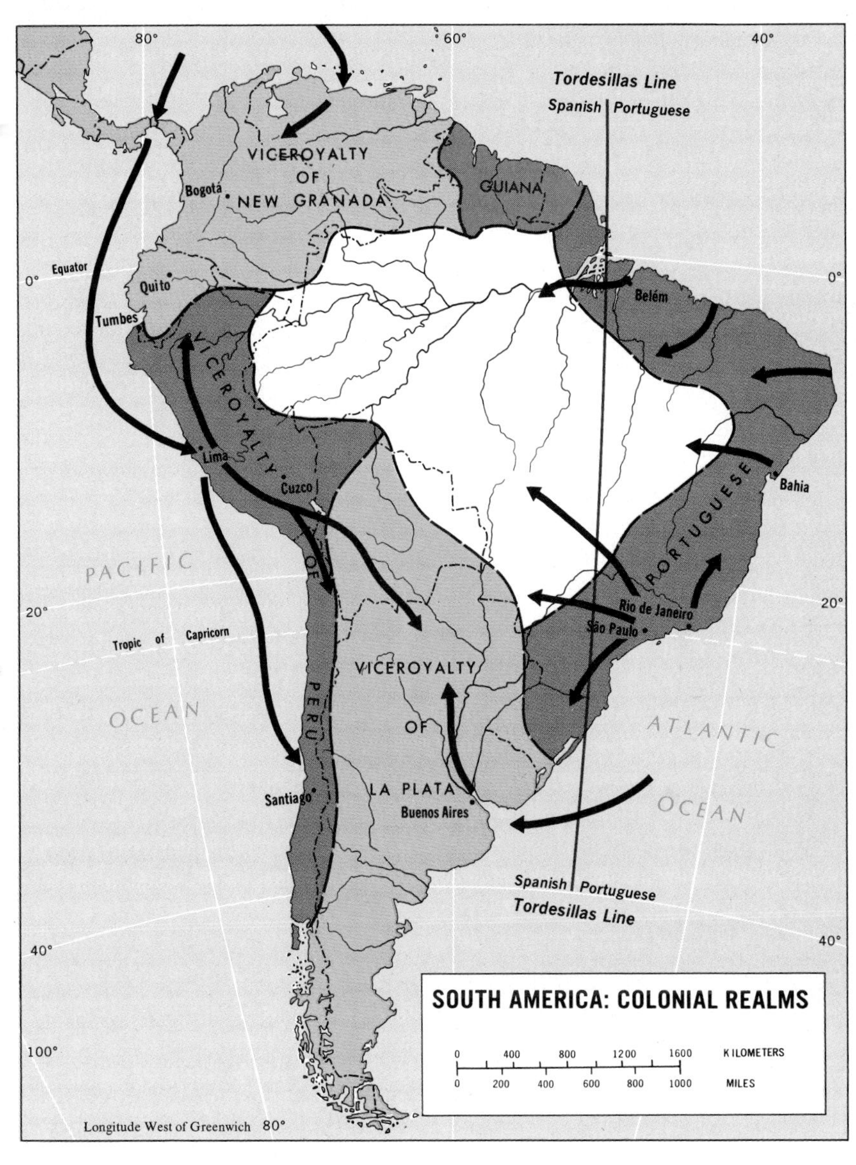

Paulistas, the settlers of São Paulo. From early in its colonial history São Paulo had been a successful settlement, with thriving plantations and an ever-growing need for labor. The *Paulistas* organized huge expeditions into the interior, seeking Indian slaves, gold, and precious stones, and, incidentally, intent on reducing the influence of Jesuit missionaries over the Indian population there.

The Africans

As the map shows, the Spaniards initially got very much the better of the territorial division of the New World—not just quantitatively, but, initially at least, qualitatively as well. No rich Indian states could be conquered and looted in the east, and no productive agricultural land was under cultivation. The Indians were comparatively few in number and constituted no usable labor force. It has been estimated that the whole area of present-day Brazil may have been inhabited by about one million people. When finally the Portuguese began to look with renewed interest to their American sphere of influence they turned to the same lucrative business that was sustaining their Spanish rivals in the Caribbean: the plantation cultivation of sugar for the European market. And they found their labor force in the same region, as millions of Africans were brought in slavery to the Brazilian northeast and north coast. Again, estimates of the total number of African forced immigrants vary, but over the centuries the figure probably exceeded 6 million. Today, with the population of Brazil exceeding 120 million, more than 11 percent of the people are black, and another 30 percent are of mixed African, white, and Indian ancestry. Africans, then, constitute the third major immigration of foreign peoples into South America.

Many hundreds of thousands of African slaves were brought to the plantations of Brazil, and today Brazil has South America's largest black population. This population sector still is strongly concentrated in the poverty-stricken northeast of the country, in Bahia and states to the north. This photograph was taken in the state of Bahia's capital and largest city, Salvador. (Carl Frank/Photo Researchers)

Persistent Isolation

Despite their common cultural heritage (at least so far as their European-Mestizo population is concerned), their adjacent location on the same continent, their common language, and their shared national problems, the states that arose out of South America's Spanish viceroyalties have existed in a considerable degree of isolation from each other. Distance, physiographic barriers, and other factors have played roles in this. To this day, the major population agglomerations of South America lie along the coast, mainly the east and north coast. Of all the continents, only Australia has a population distribution that is more markedly peripheral, but there are only some 15 million people in Australia against over 240 million in South America.

Compared to other world realms, South America may be described as underpopulated, not just in terms of its low total for a continental area of its size, but also in view of the resources available or awaiting development. The continent never drew as large an immigrant European population as did North America. The Iberian Peninsula could not provide the numbers of people that western and northwestern Europe did, and colonial policy, especially Spanish policy, had a restrictive effect on the European inflow. The American viceroyalties existed primarily for the purpose of the extraction of riches and the filling of Spanish coffers; in Iberia there was little interest in developing the American lands for their own sake. Only after those who had made America their permanent home and who had a stake there rebelled against Iberian authority did things begin to change, and then very slowly. South America was saddled with the values, economic outlook, and social attitudes of seventeenth- and eighteenth-century Iberia—not the best equipment with which to begin the task of forging modern states.

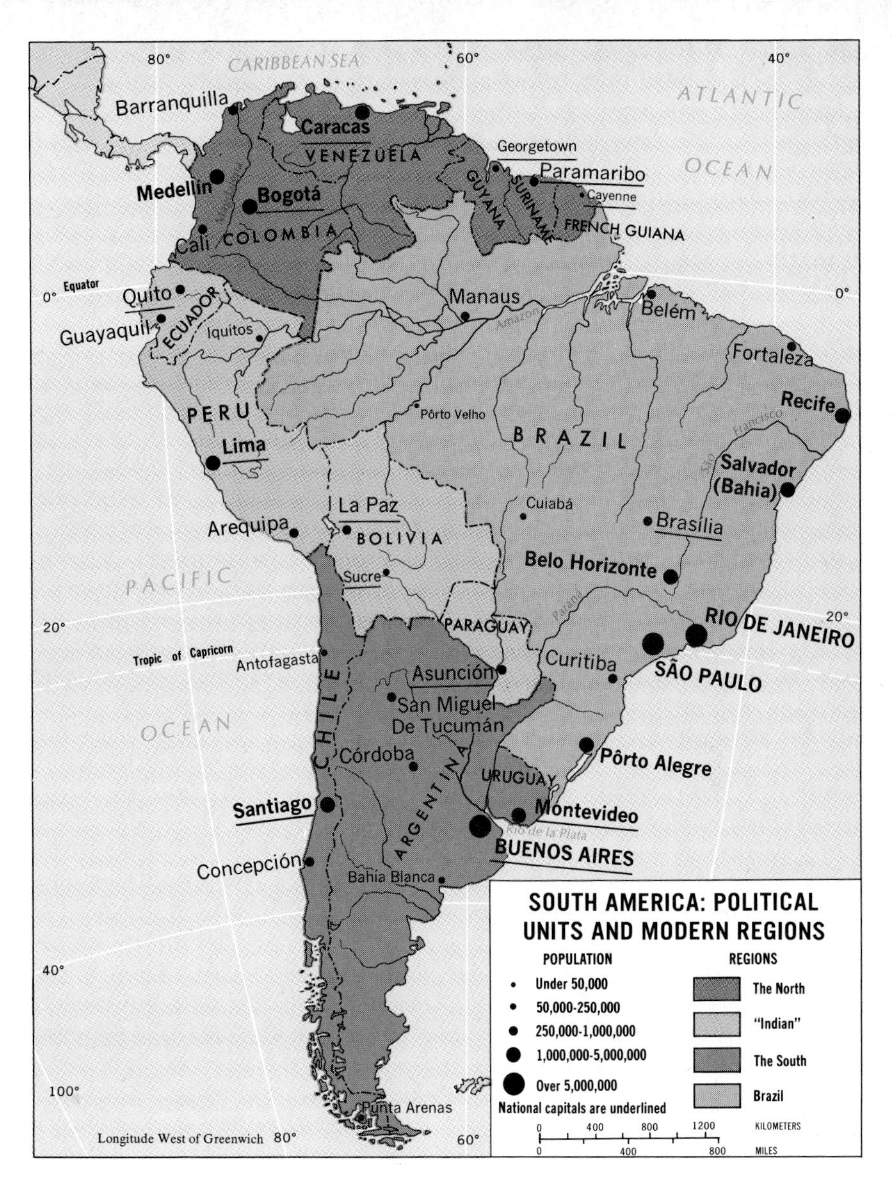

Figure 5-4

Independence

Some isolating factors had their effect even during the wars for independence. Spanish military strength was always concentrated at Lima, and those territories that lay farthest from the center of power—Argentina and Chile—were the first to establish their independence from Spain, in 1816 and 1818 respectively. While the Argentine general José de San Martin led the combined Argentinian and Chilean armed forces to the coast of Peru, Simón Bolivar was leading the north, New Granada, in its fight for independence. Eventually Bolivar organized an assault on the remaining Spanish forces, fortified still in the Andean mountains; in 1824 two decisive battles ended Spanish power in South America. Thus in little more than a decade, the Spanish countries fought themselves free; the significance of their cooperation in this effort can hardly be overstated. But the joint struggle did not produce unity. Nine countries emerged out of the three viceroyalties, including Bolivia, formerly known as Upper Peru and named after Bolivar when it was declared independent in 1826. Bolivar's Colombian Confederacy, which had achieved independence in 1819, broke up in 1831 into Venezuela, Ecuador, and New Granada, which in 1861 was renamed Colombia. Uruguay was temporarily welded to Brazil but it, too, attained separate identity. Paraguay, once a part of Argentina, also appeared on the map as a sovereign state (Fig. 5-4).

It is not difficult to understand why this fragmentation should have taken place; with the Andes intervening between Argentina and Chile, and the Atacama Desert between Chile and Peru, distances seem even greater than they are and the obstacles to contact are very effective. Thus the countries of South America began to grow apart, separated by sometimes uneasy frontiers. Friction and even wars over Middle and South American boundaries have been frequent, and a number of boundary disputes remain unsettled to this day. Bolivia, for example, at one time had a direct outlet to the sea, but lost this access in a series of conflicts involving Chile, Peru, and indirectly, Argentina. Chile and Argentina themselves were long locked in a dispute over their Andean boundary, while Peru and Ecuador both laid claim to the upper Amazon Basin and the town of Iquitos (Peru).

Brazil attained independence from Portugal at about the same time the Spanish settlements in South America were struggling to end overseas domination, although the sequence of events was quite different. In Brazil, too, there had been revolts against Portuguese control—the first as early as 1789; but the early 1800s, instead of witnessing a steady decline in Portuguese authority, actually brought the Portuguese government (Prince Regent Dom João and a huge entourage) from Lisbon to Rio de Janeiro. Thus Brazil in 1808 was suddenly elevated from colonial status to the seat of empire, and it owed its new position to Napoleon's threat to overrun Portugal, which was allied with the British.

At first it seemed that the new era would bring progress and development to Brazil.

While there was some agitation against the regime based in Rio de Janeiro, notably in the Brazilian northeast, where Pernambuco (now named Recife) was the center of a revolt in 1817, the real causes of Brazilian independence lay in Portugal, not Brazil. Dom João did not return to Lisbon immediately after the departure of the French, and by the time he did, it was in response to a revolution there; the Napoleonic period had left behind it a great deal of dissatisfaction with the *status quo.* Worse, the regime in Lisbon wanted to end the status of equality for Brazil and once again make it a colony. Thus Dom João appointed his son, Dom Pedro, as regent and in 1821 set sail for Portugal. It was to no avail: The national assembly of Portugal was determined to undo Dom João's administrative innovations, and Dom Pedro was ordered to return to Lisbon as well. This he refused to do, and in 1822 he proclaimed Brazil's independence and was crowned emperor. He had overwhelming support from the Brazilian people in this decision; the loyalist Portuguese forces still in the country were forced to return to Lisbon.

The postindependence relationships of Brazil to its Spanish-influenced neighbors have been similar to the relationships among the individual Spanish republics themselves. Distance, physical barriers, and culture contrasts serve to inhibit contact and interaction of a positive kind. Brazil's orientation toward Europe, like that of the republics, remained stronger than its involvement with the states on its own continent.

Culture Areas

When we speak of the "orientation" or "interaction" of South American states, it is important to keep in mind just who does the orienting and interacting, for there is a tendency to generalize the complexities of these countries away. The fragmentation of colonial South America into 10 individual states and the nature of their relationships was the work of a small minority of the people in each country. The black people in Brazil at the time of independence had little or no voice in the course of events; the Indians in Peru, numerically vastly in the majority, could only watch as their European conquerors struggled with each other for supremacy. It would not even be true to say that the European minorities *in toto* governed and made policy: it was the wealthy, landholding, upper class element that determined the posture of the state. These were—and in some cases still are—the people who made the quarrels with their neighbors, who turned their backs on wider American unity, and who kept strong the ties with Madrid and Paris (Paris long had been a cultural focus for "Latin" America's well-to-do; many children were sent to French rather than Spanish schools).

So complex and heterogeneous are the societies and cultures of Middle and South America that practically every generalization has to be qualified. Take the one just used: the term *Latin* America. Apart from the obvious exceptions that can be read from the map, such as Jamaica, Guyana, and Surinam, which are clearly not "Latin" countries, it may be improper even to identify some of the Spanish-influenced republics as

"Latin" in their cultural milieu. Certainly the white, wealthy upper classes are of Latin European stock, and they have the most influence at home and are most visible abroad; they are the politicians and the businesspeople, the writers and the artists. Their cultural environment is made up of the Spanish language, the Catholic church, the Mediterranean architecture of Middle and South America's cities and towns. These things provide them with a common bond and with strong ties to Iberian Europe. But in the mountains and villages of Ecuador, Peru, and Bolivia live millions of people to whom the Spanish language is still alien, to whom white people's religion is another element of deculturation, and to whom decorous Spanish styles of architecture are meaningless when a decent roof and a solid floor are still unattainable luxuries.

South America, then, is a continent of plural societies, where Indians of different cultures, Europeans from Spain, Portugal, and elsewhere, Africans from the west coast and other parts of tropical Africa, and Asians from India, Java, and Japan have produced a cultural and economic jigsaw of almost endless variety. Certainly to call this "Latin" America is not very meaningful, but is there a more meaningful way to arrive at a regional differentiation that would represent the continent's cultural and economic spheres even approximately?

One such attempt was made in 1963 by J. P. Augelli, who also authored the *Rimland-Mainland* concept for Middle America. His map (see Fig. 5-5) shows five culture spheres (in effect, culture *areas* or regions of the realm) in South America.

Figure 5-5

The first of these, the *Tropical Plantation* region, in many respects resembles the Middle American Rimland. It consists of several areas, of which the largest lies along coastal northeast Brazil while four others lie along the Atlantic and Caribbean north coasts of South America. Location, soils, and tropical climates favored plantation crops, especially sugar; the small indigenous population led to the introduction of millions of African slave laborers, whose descendants today continue to dominate the racial makeup and strongly influence the cultural expression of these areas. Later the plantation economy failed, soils were exhausted, the slavery system was terminated, and the people were largely reduced to poverty and subsistence—conditions that now mark much of the region mapped as tropical plantation.

The second region on Augelli's map, identified as *European-Commercial*, is perhaps the most truly "Latin" part of

This street scene in Buenos Aires, capital of Argentina, reflects the "European-Commercial" character of the realm's southernmost countries (see map, page 293). (Patti McConville)

South America. Argentina and Uruguay, each with a population that is between 80 and 90 percent "pure" European and with a strong Spanish cultural imprint, constitute the bulk of the European-Commercial region. Two other areas form part of it: the southern section of Brazil's core area and central Chile, the core of that country. Southern Brazil shares the temperate grasslands of the Pampa and Uruguay (see Fig. I-7), and this area has importance as a zone of livestock raising as well as corn growing; Brazil fostered European settlement here at an early stage for strategic reasons. Middle Chile is an old Spanish settlement, and Chile is much more a mestizo country than either Argentina or Uruguay. The one-quarter of the Chilean population that claims pure Spanish ancestry, like 90 percent of all Chilean people, is concentrated in the valleys between the Andes and the coastal ranges, and between the Atacama Desert of the north and the mountainous, forested, sea-indented south. Here, in an area of Mediterranean climate (Fig. I-6), pastoralism (sheep and cattle) and mixed farming are practiced. In general, then, the European-Commercial region is economically more advanced than most of the rest of the continent. A commercial economy prevails rather than subsistence modes of life, living standards are higher, literacy percentages are better, transportation networks are superior, and, as Augelli pointed out in his article, the overall development of this region is well ahead of that of several parts of Europe.

The third region is identified as *Indo-Subsistence,* and it forms an elongated area along the central Andes from southern Colombia to Northern Chile and Argentina, an area that coincides approximately with the old Indian empires. There is a small outlier in southern Paraguay. The feudal socioeconomic structure that was established by the Spanish conquerors still survives. The Indian population forms a large, landless peonage living by subsistence or by working on the haciendas, far removed from the Spanish culture that forms the primary force in the national life of their country. This region includes some of South America's poorest areas, and what commercial activity there is tends to be in the hands of the white or mestizo.

The fourth region, *Mestizo-Transitional,* surrounds the Indo-Subsistence region, covering coastal and interior Peru and Ecuador, much of Colombia and Venezuela, most of Paraguay, and large parts of Brazil, Argentina, and Chile. This is the zone of mixture between European and Indian (or African in Brazil, Venezuela, and Colombia). The map thus reminds us that countries like Bolivia, Peru, and Ecuador are dominantly Indian and mestizo; in Ecuador, for example, these two groups make up nearly 90 percent of the total population, of which a mere 10 percent can be classed as white. The term *transitional* has an economic connotation, because, as Augelli put it, this region "tends to be less commercial than the European sphere but less subsistent in orientation than dominantly Indian areas."

The fifth region on the map is marked as *Undifferentiated,* because it is difficult to identify its characteristics. Some of the Indian peoples in the interior of the Amazon Basin have remained almost completely isolated from the momentous changes in South America since the days of Columbus, and isolation and lack of change are two of the dominant aspects of this region. The Amazon Basin and the Chilean and Argentinian southwest are also sparsely populated and have only very limited economic development; poor transportation and difficult location have contributed to the unchanging nature of this region. The trans-Amazonian highway, now under construction, is a first incursion into the most isolated parts of the area.

The framework of culture spheres described above is obviously a generalization of a very complex situation, but even in its simplicity it underscores the diversity of South American peoples, cultures, and economies.

An Indian market in the highlands of Peru. The heirs to the Inca empire live, often precariously, at high elevations (as much as 11,000 feet or 3500 meters) in the Andes. Poor soils, uncertain water supply, high winds, and bitter cold make farming difficult. (Loren McIntyre/Woodfin Camp)

Agricultural Patterns

Elements of the map of culture spheres (Fig. 5-5) can be discerned in the map of agricultural patterns, on which South America is shown in world context (Fig. 5-6). In the Argentine Pampa and adjacent areas, South America has a grain-producing region comparable to areas of the North American Great Plains (area 4), and in Uruguay and southern Brazil, as well as south-central Chile, areas of mixed livestock and crop farming exist comparable to the Southeastern United States and part of the Midwest (area 3). This is generalized as the European-Commercial sphere on Fig. 5-5. In stark contrast, subsistence modes of life prevail in the forested Amazon Basin (area 11), equivalents of which occur in tropical areas of Africa and Southeast Asia. Tropical Plantation agriculture appears on both maps (area 7) along the eastern and northern fringes of Brazil, Venezuela, Colombia, and the three smaller countries (Guyana, Surinam, French Guiana). Coastal Peru also has intensive agriculture of this kind, but under irrigation in a region that is otherwise desert (Fig. I-6).

Sedentary subsistence cultivation areas (10) in South America on Fig. 5-6 coincide with the Indo-Subsistence culture sphere, extending from western Colombia southward along the Andean altiplanos through Ecuador, Peru, and into Bolivia. Similar farming systems exist in Middle America, in the Ethiopian Highlands in Africa, and in higher interior areas of Southeast Asia, notably Burma. Livestock ranching prevails in much of Augelli's Mestizo-Transitional sphere (area 12). Finally, Chile's central zone constitutes an area of *Mediterranean Agriculture* (area 6). As the world map shows, similar areas occur not only on the margins of the Mediterranean Sea, but also in Southern California, the Cape of South Africa, and Southwestern Australia.

Urbanization

As in other areas of the world, people in South America are leaving the land and moving to the cities. This process speeded up strongly after the Second World War, and today three of the realm's cities (Buenos Aires, capital of Argentina, and Rio de Janeiro and São Paulo in Brazil) are among the world's 25 largest urban centers. A measure of the pace of urbanization is provided by the following indexes: in 1925, about one third of South America's people lived in cities and towns, and as recently as 1950 the percentage was just over 40. But in 1975 the figure had exceeded 60 percent, and in 1980 two out of three persons in South America resided in the ever-growing urban agglomerations. Of course, the percentages hide the actual numbers. Between 1925 and 1950, the continent's towns and cities grew by about 40 million residents as the urbanized percentage rose from 33 to 40. But between 1950 and 1975, more than 125 million people arrived in the urban areas—more than *three times* the total for the previous period.

South America's population of over 240 million has a high growth rate, but nowhere are the numbers increasing more rapidly than in the towns and cities. We usually assume that the populations of rural areas grow more rapidly than urban areas, because farm families traditionally have more children than city dwellers. But overall, the urban population of South America grew by nearly 4.5 percent per year during the past quarter of a century, while the rural areas increased by only about 1.3 percent. These figures reveal the dimensions of the migration stream from the countryside to the towns and cities—still another kind of migration process affecting modernizing societies.

In South America—as in Africa and Asia—people are attracted to the cities and driven from the poverty of the rural areas; both pull and push factors are at work. Land reform has been slow in coming, and many farmers simply give up and leave, seeing little or no change, year after year. The cities attract because they seem to provide opportunity—the chance to earn a regular wage. Visions of education for their children, better medical care, and the excitement of life in a big city draw millions to places such as São Paulo and Bogotá. Road and rail connections continue to improve, so that access is easier and exploratory visits can be made. City-based radio stations beckon the listener to the place where the action is.

But the actual move can be traumatic. South America's cities are surrounded and sometimes invaded by slums *(barrios)* that are among the world's worst, and this is where the uncertain urban immigrant usually finds a first (and often permanent) abode, in a makeshift shack without even the most

Figure 5-6

basic amenities. Many move in with relatives who have already made the transition, but whose dwelling can hardly absorb another family. And unemployment is high, sometimes as high as 25 percent of the available labor force. Jobs for unskilled workers are hard to find and they pay minimally. But the people still come, the crowding in the *barrios* grows worse, and the threat of epidemic-scale disease rises. It has been estimated that in-migration accounts for over 50 percent of urban growth in some countries—and in South America, urban populations have unusually high natural growth rates as well.

Hence South American cities present almost inconceivable contrasts between wealth and poverty, splendor and squalor. The largest city remains Buenos Aires, federal capital of Argentina, with over 9 million residents in 1980; Rio de Janeiro long has been in second place with an estimated 1980

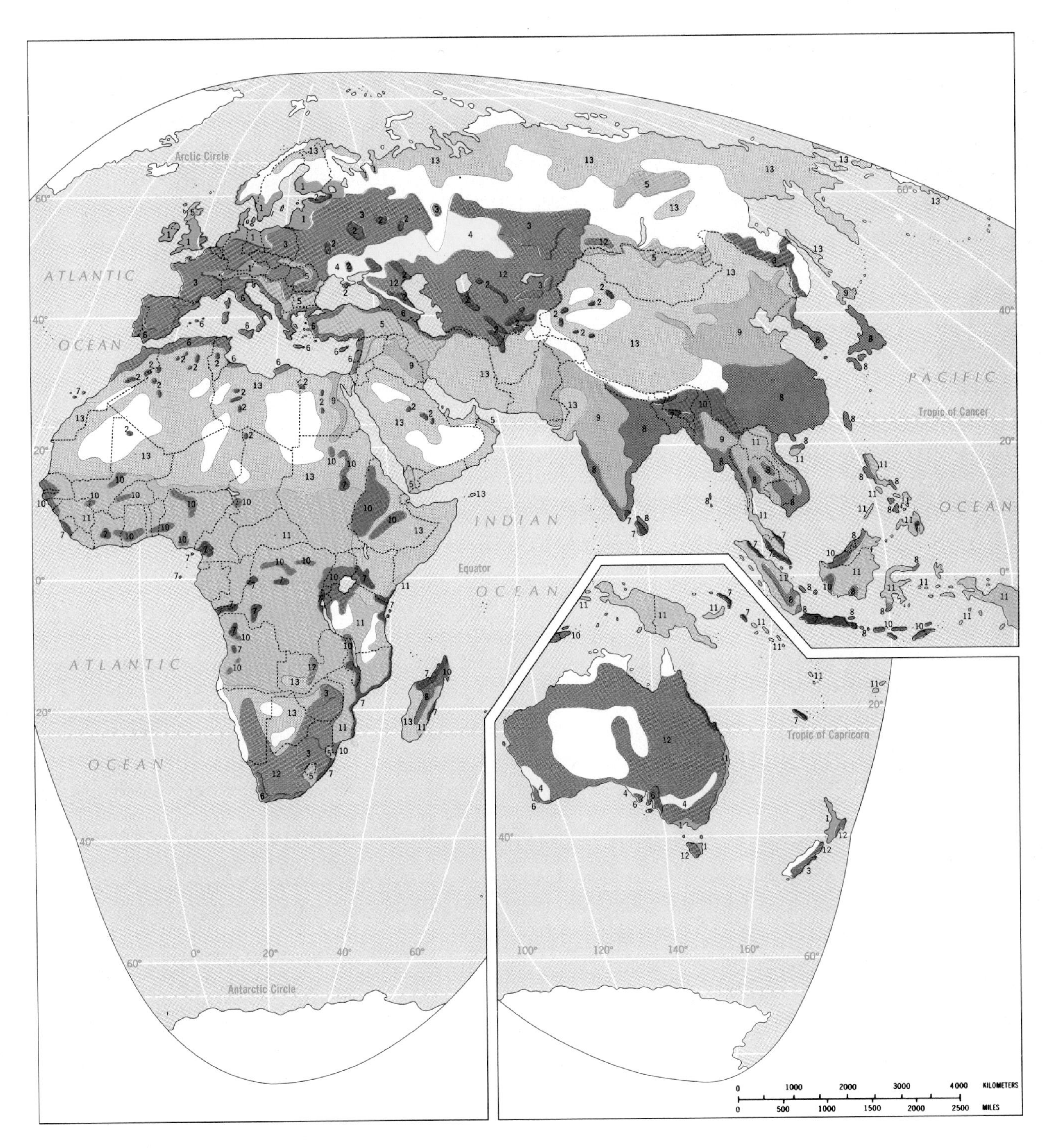

Social contrasts in South America's large urban centers often are harsh. Poorer residents live in substandard housing in the shadow of luxury skyscraper apartments. This photograph is from Caracas, capital of Venezuela. (Owen Franken/Stock, Boston)

population of 8 million. But Rio de Janeiro, long Brazil's capital city, is challenged now by a city that represents more than any other the incredible mass agglomeration, crowding, pluralism, and energy of the realm's urban clusters: São Paulo. Historically Brazil's second city, São Paulo is growing at unprecedented rates; the 1970 census reported just under 6 million inhabitants, but 1980 estimates suggest that the population today may be nearly double this figure. If São Paulo continues to grow at its 1970s pace, it will have 20 million residents by the end of the 1980s. Already it is described as the city with the world's worst traffic jams, greatest air pollution, highest noise levels, and perhaps the lowest quality of life. None of it deters the annual flood of new arrivals. South America's cities are crucibles in which the survival habits of a whole realm's population are being transformed.

Capital Cities

South America's large cities, with few exceptions, would fit Jefferson's concept of the primate city. In all but two of the twelve countries (and in the thirteenth territory, French Guiana, as well), the largest urban center overshadows all other towns, functions as the capital, and is the historic, cultural, political, and economic focus of the state. The exceptions are Brazil, where Rio de Janeiro, long the capital and national center, was replaced as governmental headquarters by newly built Brasilia (see pages 320-331), and Ecuador, where Quito, the capital in the interior Andes, is a smaller city than coastal Guayaquil, the country's port. But in the other countries the capital is the national heart in every way.

Capital cities hold special interest for geographers (particularly political geographers) because they embody national qualities and aspirations, reflect cultural preferences, and represent the essence of the state.

Brasilia, capital of Brazil, lies about 650 kilometers (400 miles) from the coast in Brazil's interior. It was designed not only to succeed Rio de Janeiro as the national headquarters, but also to symbolize the country's modern age, and its westward orientation. Brasilia is indeed a growth pole in Brazil's vast inner frontier, but as a city it hardly represents the vigor and bustle of its older urban places. (Rene Burri/Magnum)

When Japan entered its modern age, its national planners boldly moved the national capital from venerable, revered Kyoto to coastal Tokyo. It was a symbolic as well as practical move, and one of many instances where governments have used their own headquarters to signal a new era, or to underscore a new orientation. Other capitals are seemingly permanent, almost as old as the state itself. London was a village in Roman times, became the capital of England, headquarters of Great Britain, and eventually the center of a world empire.

Capital cities, therefore, can be classified on several grounds. Geographically the most obvious is location relative to the state's territory. Santiago (3 million), capital of Chile, lies in the heart of the country's core area and near the geographical center of the state. In Colombia, Bogotá (3 million) also lies in the interior, nearer the center of the country's population than any coastal capital could be. Ecuador's Quito (600,000) similarly is an interior capital. On the other hand, Montevideo (1.9 million), capital of Uruguay, lies at the very edge of the national territory, as does Buenos Aires, Argentina.

The extent to which a capital's relative location can matter to a state is underscored by Brazil, where coastal Rio de Janeiro long was the primary city as well as the seat of government. Brazil's large population always has been concentrated in a core area that is positioned on its eastern seaboard. Expansion into the vast Brazilian interior has occurred, but slowly—far slower than Brazil's developers would like. In the mid-1950s the Brazilian government announced plans to build a new capital, away from the coast in the interior. Once the decision was made and the site selected, matters moved fast. In 1960, even before the newly planned city was only half finished, Brasilia was inaugurated as Brazil's new headquarters. It was an enormous investment, but it symbolized Brazil's determination to turn "inward" and to open its huge hinterland. By 1980, Brasilia's population (including the surrounding federal district) exceeded 700,000.

In a sense Brasilia represents what political geographers call a "forward capital." There are times when a state will relocate its capital in a sensitive area, perhaps near an area under dispute with an unfriendly neighbor, partly to confirm its determination to sustain its position in the contested zone. A recent example is the decision by Pakistan to move its capital from coastal Karachi to northern Islamabad, near disputed Kashmir. These are called "forward" capitals because of their position in an area that would be first to be engulfed by conflict in case of strife with a neighbor. At one time Berlin was a forward capital, a German headquarters near the margins of Slavic Eastern Europe. Brasilia, of course, does not lie in or near a contested area—but Brazil's interior has been a kind of internal frontier, one to be conquered by a developing nation. In that drive, the new capital has a forward position.

Another way to group capital cities involves their origins and durability. Thus Quito (Ecuador) and Bogotá (Colombia) as well as La Paz (Bolivia) have Indian antecedents, and Quito is the oldest capital city in South America. Bogotá (3 million) began as a Chibcha settlement. As a Spanish center it witnessed the rise and fall of the Viceroyalty of New Granada (Fig. 5-3) and the Confederation of Gran Colombia; since 1835 Bogotá has been the capital of the Republic of Colombia. In another group, Lima (3.5 million), capital of Peru, was founded by the Spanish colonizers in 1535 in preference to Cuzco, the interior Inca city. In Venezuela, Caracas (3.5 million) began as a hamlet built on a colonist's ranch just 11 kilometers (7 miles) from the Caribbean coast, but on the landward side of an Andean mountain spur.

In still another class of capital cities, the administrative functions are divided among more than one headquarters. This peculiar situation has occurred in Libya (to resolve regional quarrels), in Laos (where royal and public capitals emerged), in the Netherlands (where Amsterdam and The Hague share capital roles), and in South Africa (where legislative and administrative functions are handled in Cape Town and Pretoria, respectively). South America, too, has a situation of this kind in Bolivia, where La Paz is the seat of national government but Sucre is the seat of the supreme court and the legal capital. La Paz, the world's highest capital city at 3570 meters (11,700 feet) above sea level, has a population approaching 1 million, 10 times as large as its historic rival.

Urban Structure

In Chapter 3, we viewed North American cities in terms of their spatial structure and in the context of hierarchy. The models of urban structure developed by Burgess and others were based on assumptions (as models normally are), and one of these assumptions is obvious enough: the urban centers showing concentric, sectorial, and nucleated characteristics lie on flat territory, so that terrain will not distort the "ideal" pattern. But when we try to discern familiar patterns in South America's cities, we should note that many of them lie on hilly, even mountainous, sites. Both Rio de Janeiro and São Paulo have been affected in their development by steep slopes: in the former capital, some of these are high and rocky, remote from the main roads, and unserved by city utilities; slums have developed there. In Rio de Janeiro, elevation is less desirable than proximity to the cooling bay and ocean. In São Paulo, on the other hand, the higher elevations soon attracted high-class residential development; this city lies more than 50 straight-line kilometers inland, and here it is height, not seaward location, that affords coolness. Two cities that resemble Chicago somewhat in terms of their site are Buenos Aires—flat, monotonous, and sprawling endlessly—and Montevideo—smaller, and more compact. Both these cities show elements of concentricity in their city plans, and in both, heavy industry is less developed than it is in most large United States cities, so that the pattern is less interrupted. In the case of Buenos Aires, the port stretches for several miles along the Plata Estuary, and the outer suburbs—zone (5)—lie some 30 kilometers from the city center. In Montevideo, the more concentrated port opens into the Old City—the *Ciudad Vieja*—which still remains the core of the urban area; a newer commercial district has developed adjacent to it, and a transition zone to outer residential areas is clearly visible.

A sectorial arrangement appears in Bogotá and Santiago, as well as Rio de Janeiro, induced by the topography of the

Impressive *plazas* still dominate the layout of older sections of South America's cities. A good example is Bogota's Plaza Bolivar, facing the cathedral and named after the great national hero. (Loren A. McIntyre/ Woodfin Camp)

site. Rio de Janeiro is limited by the Atlantic Ocean and the bay on two sides (the south and east); to the west, the twin 1000-meter peaks of Corcovado and Tijuco inhibit both urban growth and land communications. Thus Rio's three major routes to the interior all leave the congested city in a northwesterly direction, and the urban structure has a sectorial rather than a concentric character.

When the Spanish colonizers laid out their cities and towns in South and Middle America, they created a central square or *plaza* dominated by a church or cathedral and flanked by imposing government buildings. Lima's *Plaza de Armas,* Montevideo's *Plaza de la Constitución,* and Buenos Aires' *Plaza de Mayo* are famed examples of this system. Streets led outward from these central squares at right angles, demarcating square or rectangular city blocks. In the process, the Spaniards laid the groundwork for a nucleated urban structure. Early in the city's development the plaza formed the hub and focus of the city, surrounded by shopping streets and arcades. But eventually the city outgrew its old center and new commercial districts formed, leaving the plaza as merely a link with the past. Lima is a prominent case of nucleated development. Thus South America's cities, like those of North America, display elements of all three of the models derived from urban patterns in the United States.

Cities, as we have noted in previous chapters, perform particular functions. Some are capital cities, others are university towns, still others are mining centers. Several geographers have tried to establish a classification of cities based on their primary functions. Perhaps the best known of these efforts is a classification of United States cities by C. D. Harris, proposed in 1943 in an important article in the *Geographical Review.* In this article Harris suggested that a sound way to classify urban centers would lie in the measurement of the activities of the labor force. What are the activities that employ most or a significant portion of the labor force? If in industry, a city would be a manufacturing center. If in selling, it would be a retail center; if in moving goods, a transportation center. Harris also recognized cities where wholesaling prevails; diversified cities, where there is a mix of economic activities; mining towns; university cities; resort and retirement places; and political centers, including national and divisional capitals.

South America's cities, like several major United States cities, in some instances have so diversified an economic base that no single activity dominates. Multifunctional places such as Buenos Aires, Rio de Janeiro, and Caracas serve as industrial, wholesale, retail, and educational centers. On the other hand, Valdivia (Chile), Callao (Peru), La Guaira (Venezuela), and Santos (Brazil) all are transportation centers, serving as ports for larger urban areas located inland. Again, several large mining towns exist, including Cerro de Pasco in Peru, Potosí and Oruro in Bolivia, and Cerro Bolivar in Venezuela. Manufacturing cities are led prominently by São Paulo and Belo Horizonte in Brazil, and include Guayaquil in Ecuador (also an important port) and Medellín in Colombia. Only one primarily political center exists in South America (Brasília), but Sucre in Bolivia and Quito in Ecuador owe their continued prominence in large measure to their political roles in the national state. The necessary data are hard to come by, but an analysis of South American cities over 250,000 suggests that for urban centers in this realm, the Harris classification might well be somewhat modified.

The Republics: Regional Geography

We turn now to the countries of South America and their principal regional-geographic properties. In general terms it is possible to group sets of South America's countries into regional units, because several have qualities in common. Thus the "Caribbean" countries, Venezuela and Colombia, form a regional unit (that might well include adjacent Guyana, Surinam, and French Guiana also). On the basis of their Indian cultural heritage, Andean physiography, and modern population, the republics of Ecuador, Peru, and Bolivia constitute a regional entity, to which Paraguay may be added on grounds to be discussed later. In the south, Argentina, Uruguay, and Chile have a joint regional identity as the realm's midlatitude, most strongly European states. And Brazil, by itself, constitutes a geographic region in South America for the most obvious of reasons: it contains nearly half the land and over half the people of the continent.

The North: "Caribbean" South America

As another look at Fig. 5-5 confirms, the countries of the north coast have something in common other than their coastal location: each has an area of "tropical plantation," signifying early European plantation development, the arrival of black labor, and the absorption of this African element in the population matrix. Not only did black workers arrive: many thousands of Asians (Indians) came to South America's northern shores as contract laborers.

The pattern is familiar: in the absence of large local labor sources, the colonialists turned to slavery and indentured workers to serve their lucrative plantations. But between Venezuela and Colombia on one hand and the "three Guianas" on the other there is this difference: in the former, the population center of gravity soon moved into the interior, and the plantation phase was followed by a totally different economy, while in the Guianas, the plantation economy still dominates.

Venezuela and Colombia have what the Guianas lack (Fig. 5-7). Their territories and populations are much larger, their physiographies more varied, their economic opportunities greater. Each has a share of the Andes Mountains (Colombia's is the larger), and each produces oil from an adjacent joint reserve that ranks among the world's leading fields (here Venezuela is the major beneficiary). Much of what is important in Venezuela is concentrated in the northern and western parts of the country, where the Venezuelan Highlands form a spur of the Andes system. Most of Venezuela's 14 million people are concentrated in the highlands, which include the capital (Caracas), its early rival (Valencia), the commercial and industrial center (Barquisimeto), and San Cristobal, near the Colombian border. The Venezuelan Highlands are flanked by the Maracaibo Lowlands and Lake Maracaibo to the north-

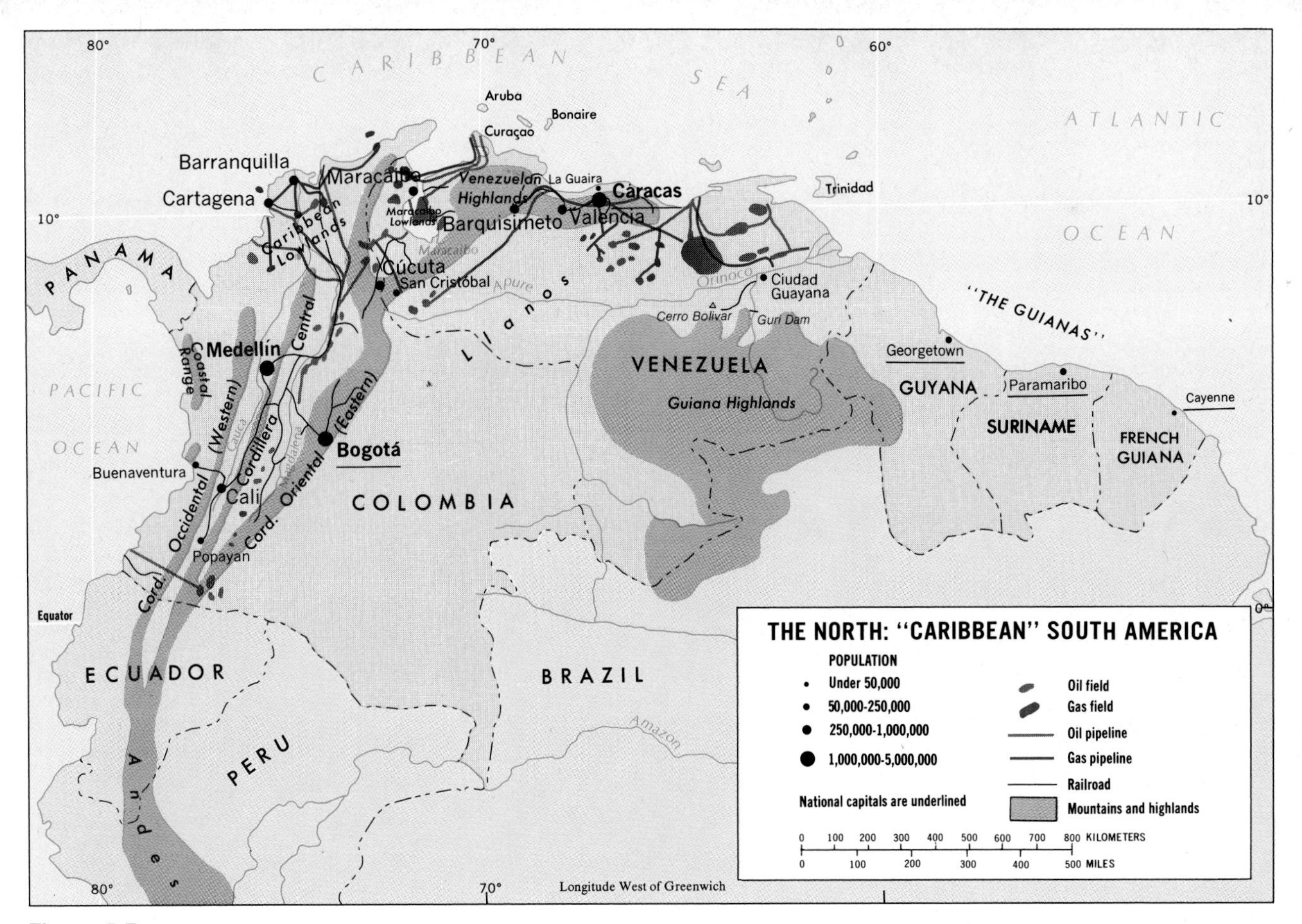

Figure 5-7

west, and by a region of savanna country called the *llanos* to the southeast. The Maracaibo Lowland, once a disease-infested, sparsely peopled coastland, is today one of the world's leading oil-producing areas; much of the oil is drawn from reserves that lie beneath the shallow waters of the lake itself. Actually, "Lake" Maracaibo is a misnomer, for the "lake" is open to the ocean and is in fact a gulf with a very narrow entry. Venezuela's second city, Maracaibo, with over a half million people, is the boom center of the oil industry that has transformed the Venezuelan economy; in the 1970s over 90 percent of the country's annual exports by value were crude oil and petroleum products. Large refineries on the Dutch islands of Curaçao and Aruba for many years have refined the Venezuelan crude oil prior to transportation to United States and European markets, but the capacity of local Venezuelan refineries is steadily increasing.

The llanos, on the south side of the Venezuelan Highlands, and the Guiana Highlands in the country's southeast are two of those areas that contribute to South America's image as "underpopulated" and "awaiting development." Although the llanos do share in Venezuela's oil boom (reserves have been discovered here), the agricultural potential of these savannas—and of the *templada* areas of the Guiana Highlands—has hardly begun to be realized. There are good opportunities for a major pastoral industry in the llanos and for commercial agriculture in the Guiana Highlands. Of course, the llanos and Guiana Highlands are Venezuela's interior regions, and transportation remains a problem. The discovery of rich iron ores on the northern flanks of the Guiana Highlands (chiefly near Cerro Bolivar) has begun to integrate one part of this region with the rest of Venezuela. A railroad was constructed to the Orinoco River, and from there the ores are shipped directly to the steel plants of the coastal eastern United States. Ciudad Guayana, less than two decades old, has over 150,000 people. The Guri Dam on the

The Guianas

Three small countries lie on the north coast of South America, immediately to the east of Venezuela and adjacent to Brazil. In "Latin" South America, these territories are anomalies of a sort: their colonial heritage is British, Dutch, and French. Formerly they were known as British, Dutch, and French Guiana and called the three *Guianas.* But two of them are independent now: Guyana, Venezuela's neighbor, and Surinam, the Dutch-influenced country in the middle. The easternmost territory, French Guiana, still continues under colonial rule.

None of the three countries has a population over 1 million. Guyana is largest with 900,000; Surinam has just over 500,000 inhabitants, and French Guiana, a mere 65,000. Culturally and spatially, patterns here are Caribbean: Asian (Indian) and black people are in the majority, whites a small minority. In Guyana, Asians make up just over half the population, blacks about one-third, and others (mixed and European peoples) small minorities. In Surinam, the population picture is still more complicated, for the colonists brought not only Asian Indians (now nearly 40 percent of the population) and blacks (nearly 33 percent), but also Indonesians (16 percent) to the territory in servitude. Another 10 percent of the Surinam population consists of black communities peopled by descendants of African slaves who escaped from the coastal plantations and fled into the forests of the interior. French Guiana is the most European of the three countries: three-quarters of its small population is French. The majority is of mixed African, Asian, and European ancestry: the *creoles.*

French Guiana constitutes a French *Département* and is represented in Paris by a senator, but this is the least developed of the three Guianas. The capital, Cayenne, has a mere 25,000 inhabitants; there is a small fishing (shrimping) industry, and some lumber is exported. But food must be brought in from overseas. By contrast, Surinam has progressed considerably. The plantation economy has given way to production on small, privately owned farms, where rice, the country's staple crop, is grown in adequate quantities to make Surinam self-sufficient. Citrus fruits are exported, and a wide variety of fruits and vegetables for the home market grow on farms in the coastal zone where once sugar plantations prevailed. Today only two large sugar plantations survive. But Surinam's big income-earner is bauxite, mined in a zone across the middle of the country. The search is on for other minerals, including oil offshore, and the forest resources have barely begun to be exploited. The capital, Paramaribo (150,000), carries the imprint of Dutch architecture; its small industries include a brewery of some reputation.

British-influenced Guyana became independent in 1966 amid much internal conflict that, basically, pitted people of African origins against people of Asian ancestries. Problems of this kind have continued to afflict this country, where the great majority of the people live in small villages in the coastal zone. Plantation products still produce more than half the country's annual revenues, although the contribution of bauxite is rising. Georgetown (200,000) is the major port, largest city, and capital.

Unsettled boundary problems involve all three countries. Venezuela has laid claim to part of western Guyana; Guyana, in turn, claims a corner of southwestern Surinam. Part of the border between Surinam and French Guiana is also under dispute.

Orinoco has been put into service and will soon supply all Venezuela with electricity. Two new paved roads link this eastern frontier with Caracas.

Colombia, too, has a vast area of llanos, covering about 60 percent of the country, and in Colombia as in Venezuela this is comparatively empty land, outside the national sphere, far less productive than it could be. Eastern Colombia consists of the headwaters of the Orinoco and the Amazon Rivers, and it lies partly under savanna and partly under rain-forest conditions. In recent years the Colombian government has begun in earnest to promote settlement east of the Andes. But it will be a long time before any part of eastern Colombia matches the Andean part of the country, or even the Caribbean Lowlands of the north. In these regions live the vast majority of all Colombians, and here lie the major cities and productive areas.

Western Colombia is dominated by mountains, but there is some regularity to this topography. In very broad terms there are four parallel mountain ranges, generally aligned north-south, separated by wide valleys. The westernmost of these four ranges is a coastal belt, less continuous and lower than the other three. These latter are the real Andean mountain chains of Colombia, for in this country the Andes separate into three ranges: the Eastern, Central, and Western Cordillera. The valleys between the Andean Cordillera open into the Caribbean Lowland, where two of Colombia's important ports (Barranquilla and Cartagena) are located.

Colombia's population of over 27 million consists of a set of clusters (there are different ways of identifying them, but in any case there are more than a dozen), some of them in the Caribbean Lowlands, others in the valleys between the great Cordillera, and still others in the intermontane basins within the Cordillera themselves. This was so even before the Spanish colonizers arrived; the Chibcha civilization existed in the intermontane basins of the Eastern Cordillera. The capital city, Bogotá (corrupted from the Chibcha, *Bacatá),* was founded in one of the major basins in this same range, at an elevation of 2,650 meters (8700 feet). For centuries, the Magdalena Valley between the Cordillera Oriental and the Cordillera Central was one link in a cross-continental communication route that began in Argentina and ended at the port of Cartagena, and Bogotá benefited by its position adjacent to this route. Today the Magdalena Valley is still Colombia's major transport route, but Bogotá's connections with much of Colombia still remain quite tenuous, and to a considerable degree the population clusters of this country, like those of Venezuela, exist in isolation from each other.

Colombia's physiographic variety is matched by its demographic diversity. In the south it has a major cluster of Indian inhabitants, and the country begins to resemble its southern neighbors. In the north vestiges remain of the plantation period and the African population it brought. Bogotá is a great "Latin" cultural headquarters whose influence goes beyond the country's borders. In the Cauca Valley (between the Cordillera Central and Occidental), Cali is the urban commercial focus for a hacienda district where sugar, cacao, and some tobacco are grown. Farther to the north, the Cauca River flows through a region comprising the departments (provinces) of Antioquia and Caldas, whose urban focus is the textile manufacturing city of Medellin, but whose greater importance to the Colombian economy lies in a large production of coffee. With its extensive *templada* areas along the Andean slopes, Colombia is one of the world's largest producers of coffee, and in Antioquia-Caldas it is grown on small farms by a remarkably unmixed European population cluster; elsewhere it is produced on the large estates that are so common in Iberian America.

Coffee and oil are Colombia's two leading exports, with coffee accounting for about 60 percent by value and oil about 15 percent. Colombia's oil fields are extensions of Venezuela's reserves, and the oil is piped to its Caribbean ports. The importance of Colombia's window on the Caribbean can be read from the map; two major and several minor ports handle goods brought by land and water from near the southern boundaries of the country, which in total far overshadow the volume of goods transferred at the lone west coast port of Buenaventura, Colombia's leading Pacific seaport.

Venezuela and Colombia both have a marked clustering of population, share a relatively empty interior, and depend on a single product for the bulk of their export revenues. The majority of the people of Colombia and Venezuela subsist agriculturally, and labor under the social and economic inequities common to most of Iberian America.

South America's high-relief coastlines present problems for port development. The majority (though not all) of the realm's major cities lie somewhat in the interior and are served by smaller port-towns. Shown here are the rather confined facilities at La Guaira, gateway to Caracas and Venezuela. (Harm J. de Blij)

"Indian" South America

The second regional grouping of South American states includes Peru, Ecuador, Bolivia, and Paraguay. This is a contiguous group of countries, including South America's only two landlocked entities. Figure 5-8 indicates the common Indo-Subsistence sphere extending along the Andes Mountains and a similar area in southern Paraguay. These are the countries of South America that have large Indian sectors in their populations. Nearly half the people of Peru, numbering over 18 million, are of Indian stock, and in Ecuador and Bolivia, too, the figure is near 50 percent. In Paraguay it may be over 60 percent, but all these percentages are only approximate, since it is often impossible to distinguish between Indian and "mixed" people of strong Indian character. But there are other similarities among the four countries considered here. Their incomes are low, they are comparatively unproductive, and, unhappily, they exemplify the grinding poverty of the landless peonage—a problem that looms large in the future of Ibero-America. These, too, are Iberian South America's least urbanized countries; the capitals of three of the four states have under a million inhabitants, and only Lima, the capital of Peru, ranks with Bogotá and Santiago as a major-scale urban center.

In terms of territory as well as population, Peru is the largest of the four. Its 1.3 million square kilometers (nearly 0.5 million square miles) divide both physiographically and culturally into three regions: (1) the desert coast, the European-Mestizo region, (2) the Andes Mountains or Sierra, the Indian region, and (3) the eastern slopes and the Montaña, the sparsely populated Indian-Mestizo interior (Fig. 5-8). It is symptomatic of the cultural division still prevailing in Peru that the capital, Lima, is located not in a populous basin of the Andes, but in the coastal zone. Here the Spanish avoided the greatest of the Indian empires and chose a site some 13 kilometers (8 miles) inland from a suitable anchorage (now the port of Callao). From an economic point of view the Spanish choice of a coastal headquarters proved to be sound, for the coastal region has become commercially the most productive part of the country. A thriving fishing industry based on the cool, productive waters of the Humboldt (Peru) current offshore contributes a quarter of all exports by value. Irrigated agriculture in some 40 oases distributed all along the coast produces cotton, sugar, rice, vegetables, fruits, and some wheat; the cotton and sugar are important export products, and the other crops are grown mostly for the domestic market.

The Andean region occupies about one-third of the country, and here are concentrated the majority of the country's Indian peoples, most of them Quechua-speaking. But despite the fact that this is one-third of Peru and nearly one-half of its people, the political influence of this region is slight, and its economic contribution (except for the mines) is only minor. In the valleys and intermontane basins the Indian people are concentrated either in isolated villages around which they practice a precarious subsistence agriculture or in the more favorably located and more fertile areas where they are tenants—peons—on white- or mestizo-owned haciendas. Most of the Indian people never get an adequate daily calorie intake or a balanced diet of any sort; the wheat produced around Huancayo, for instance, is sent to Lima's European market and would be too expensive for Indians themselves to buy. Potatoes (which can be grown at altitudes up to 4250 meters or 14,000 feet), barley, and corn are among the subsistence crops, and in the high altiplanos the Indians graze their llamas, alpacas, cattle, and sheep. The major products that are derived from the Sierra are copper, silver, lead, and several other associated minerals. These are taken from mines in a number of districts, of which the one centered on Cerro de Pasco is the chief.

Of the country's three regions, the east—the eastern slopes of the Andes and the Amazon-drained, rain-forest-covered Montaña—are the most

Figure 5-8

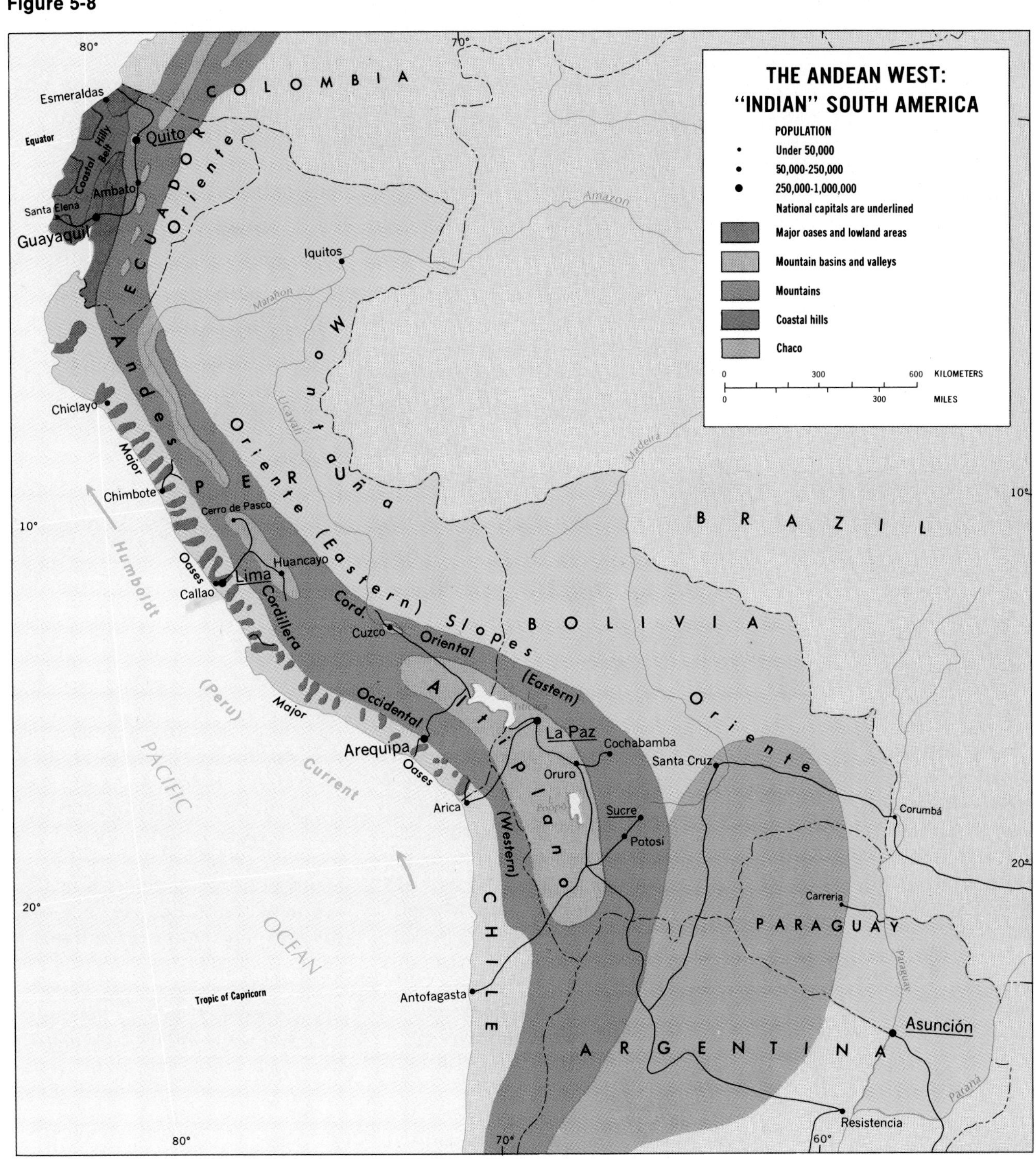

Oil from the Amazon

Peru's limited success in the search for oil in its Western Amazon region is an excellent illustration of the effect of relative location in economic geography. Peru's *Oriente* region became the country's leading source of oil and natural gas as recently as the 1970s; until then, the Talara area in the far northwest had been the chief source (and Peru was an oil exporter until the early 1960s, when the northwestern fields began their decline).

Although yet modest, present known reserves in the Amazonian region amount to about 75 percent of the country's overall total. The major discoveries have been made between Iquitos and the Ecuadorian border, and the river town of Iquitos has become the country's fastest-growing urban center. In 1965 the population was just over 50,000, and in 1980 it was approaching 200,000. In effect Iquitos is Peru's *Atlantic* Ocean port, and it proved simpler to bring equipment and supplies up the Amazon River, 3700 kilometers of waterway, than over the Andes Mountains. Iquitos' small oil refinery provides for local use and sends its petroleum to Brazil on river barges; trucking the oil through the selva and across the mountains would be impractical.

But Peru's core area needs the Amazonian oil, and thus an 850-kilometer pipeline was laid from San Jose de Saramura through the forest, across the Andes (buried below the ground to minimize earthquake danger), and over the narrow coastal zone to the new and expanding refinery at Bayovar in the northwest, not far from the old Talara fields. Since Peru needed Japanese capital to help build this pipeline, it was forced to commit part of the oil to Japan. In any case, a transshipment is needed to move the refined oil from Bayovar to Lima and its hinterland (a feeder pipeline was under construction in the late 1970s). And so Brazil and Japan consume Peru's direly needed domestic energy supplies, in large part because of their situation in Peru's regional-geographic framework.

isolated. A look at the map of permanent (as opposed to seasonal) routes, including railroads, shows how disconnected Peru's regions still are. However marvelous an engineering feat, the railroad that connects Lima and the coast to Cerro de Pasco and Huancayo in the Andes does not even begin to join the country's east and west. The focus of the region, in fact, is a town that looks east rather than west, namely Iquitos, which can be reached by oceangoing vessels sailing up the Amazon River. Iquitos grew rapidly during the Amazon wild rubber boom and then declined, but now it is growing again, reflecting Peruvian plans to begin development of the east. In mid-1977, oil began flowing from wells at San Jose de Saramuro, through an 850-kilometer pipeline across the Andes to the coast at the port of Bayovar (see box). Meanwhile, traders plying the navigable rivers above Iquitos continue to collect such products as chicle, nuts, rubber, herbs, special cabinet woods, and small quantities of coffee and cotton.

Ecuador, smallest of the four republics, on the map looks to be just a corner of Peru. But that would be a misrepresentation. Ecuador has the full range of regional contrasts: it has a coastal belt, an Andean zone that may be narrow (under 250 kilometers or 150 miles) but by no means of lower elevation than elsewhere, and an *Oriente*—an eastern region that is just as empty and just as undeveloped as that of Peru.

As in Peru, the majority of the people of Ecuador are concentrated in the Andean intermontane basins and valleys, and the most productive region is the coastal belt. But here the similarities end. Ecuador's coastal region consists of a belt of hills interrupted by lowland areas, of which the most important one lies in the south between the hills and the Andes, drained by the Guayas River and its tributaries. The largest city and commercial center of the country (but not the capital), Guayaquil, forms the focus for this area. Ecuador's lowland west is not desert country—it is a fertile tropical lowland not bedeviled by excessive rainfall either (see Fig. I-5).

Neither is Ecuador's west really a "European" region as is Peru's, for the white element in the total population of 8.3 million is a mere 10 percent, in part engaged in administration and hacienda ownership in the interior, where most of the 50 percent who are Indians also live. Of the remainder, over 10 percent is black and mulatto, and the rest are mixed Indian-white, many with strong Indian ancestry. The products of this region, too, differ from those of Peru. Ecuador is the world's top banana exporter, and bananas account for 50 percent of its exports by value; small farms owned by black and mulatto Ecuadorians and located in the north, in the hinterland of Esmeraldas, contribute to this total, as do farms on the eastern and northern margins of the Guayas Lowland. Cacao (12 percent) is another lowland crop, and coffee (22 percent) is grown on the coast hillsides as well as in the Andean *templada* areas. Cotton and rice are also cultivated, cattle can be raised, and in recent years the production of petroleum from Santa Elena on the Gulf of Guayaquil has reached substantial proportions. Ecuador now exports petroleum and is a member state of OPEC.

Ecuador is not a poor country, and the coastal region, especially in recent years, has seen vigorous development. But the Andean interior, where the white and mestizo administrators and hacienda owners are outnumbered by Indians by about three or four to one, is a different story—or rather, a story similar to that of other Andean regions. Quito, the capital city, lies in one of the several highland basins in which the Andean population is clustered. Its functions remain primarily administrative; there is not enough productivity in the Andean region to stimulate commercial and industrial development. The Ecuadorian Andes do differ from those of Peru in that they are without known major mineral deposits. Despite the completion of a railroad linking Quito to Guayaquil on the coast, the interior of Ecuador remains isolated and economically comparatively inert.

From Ecuador southward through Peru, the Andes Mountains broaden until, in Bolivia, they reach a width of some 720 kilometers (450 miles). In both the Cordillera Oriental and the Cordillera Occidental, elevations in excess of 6000 meters (20,000 feet) are recorded; between these two great ranges lies the Altiplano proper. On the boundary between Peru and Bolivia, freshwater Lake Titicaca lies at over 3700 meters (12,500 feet). Here, in the west, lies the heart of modern Bolivia; here, too, lay one of the centers of Inca civilization—and, indeed, of pre-Inca cultures. Bolivia's capital, La Paz, is an Altiplano city.

It is Lake Titicaca that helps make the Altiplano livable, for its large body of water ameliorates the coldness on the plateau in its vicinity, and the surrounding lands are cultivable. Grains can be grown in the Titicaca Basin to the amazing elevation of 3850 meters (12,800 feet) and have been for centuries; to this day the Titicaca area, in Peru as well as Bolivia, is a major cluster of subsistence-farming Indians. Modern Bolivia is the product of the European impact, an impact that has passed by some of the Indian population clusters. Of course, the Bolivian Indians no more escaped the loss of their land than did their Peruvian or Ecuadorian counterparts, especially east of the Altiplano. What made the richest Europeans in Bolivia wealthy, however, was not land, but minerals. The town of Potosi in the Eastern Cordillera became a legend for the immense deposits of silver nearby; copper, zinc, and several alloys were also discovered. Most recently Bolivia's tin deposits, among the richest in the world, have yielded some two-thirds of the country's annual export income. Oil and natural gas, however, are contributing a growing share. Bolivia exports gas to Argentina and Brazil; in return Brazil is committed to assist in the development of an economic growth pole (see page 329) in the area of Santa Cruz.

Bolivia has had a turbulent history. Apart from internal struggles for power, the country lost its seacoast in a disastrous conflict with Chile, lost its territory of Acre to Brazil in a dispute involving the rubber boom

A rural scene near Lake Titicaca in Bolivia, showing farm fields and village dwellings. Note that the snow line lies a little higher than the plain in the foreground. (Victor Englebert)

in the Amazon Basin, and then lost 140,000 square kilometers (55,000 square miles) of Gran Chaco territory to Paraguay in the war of 1932–1935. By far the most critical was the loss of its outlet to the sea; although Bolivia has rail connections to the Chilean ports of Arica and Antofagasta, it is permanently disadvantaged by its landlocked situation. Since the Cordillera Occidental and the Altiplano form the country's inhospitable western margins, one might suppose that Bolivia would look eastward and that its Oriente might be somewhat better developed than that of Peru or Ecuador, but such is not the case. The densest settlement clusters occur in the valleys and basins of the Eastern Cordillera, where also the mestizo sector is stronger than elsewhere in the country. Cochabamba, Bolivia's second city, lies in a basin that forms the country's largest concentration of settlement; Sucre, the legal capital, lies in another. Here, of course, lie the chief agricultural districts of the country, between the barren Altiplano to the west and the savannas to the east.

Paraguay is the only non-Andean country in this group, but it is no less Indian. Of 3 million people, perhaps 60 percent are Indian or, by other definitions, mestizo with so strong an Indian element that any white ancestry is almost totally submerged. Although Spanish is the country's official language, Guarani is more commonly spoken. By any measure, Paraguay is the poorest of the four countries of "Indian" South America, although it does have opportunities for pastoral and agricultural industries that have thus far gone unrealized.

One of the reasons for this must be isolation—the country's landlocked position. Paraguay's exports, in their small quantities, must be exported via Buenos Aires, a long river haul from the Paraguayan capital of Asunción. Meat (dried and canned), timber (sold to Uruguay and Argentina), oilseeds, quebracho extract (for tanning leather), cotton, and some tobacco reach foreign markets. Grazing is the

War of the Pacific

Bolivia today is a landlocked country, but its access to the Pacific Ocean may soon be restored—after a century of confinement.

When Bolivia became an independent country in 1825, long stretches of South America's boundaries were only vaguely defined. Bolivia had a sphere of influence along what is today the North Chilean coast. In 1866, the boundary between Chile and Bolivia in this area was defined by treaty (the first step in boundary creation) as lying along the 24th parallel south latitude. After the boundary was delimited (the second step) and put on the map, Chile and Bolivia entered into a complicated agreement whereby Chile would receive part of the revenues derived from the sale of resources found in the Bolivian coastal zone.

What seemed just a stretch of coastal desert turned out to have value far beyond its ports (Antofagasta and Arica served as Bolivian outlets). Guano deposits and nitrate fields attracted Chilean investment, but the Bolivians did not have the capital to exploit their own raw materials. As the Bolivian government saw the Chileans encroach, it sought help from Peru. In 1873 Bolivia and Peru signed a treaty of alliance, and the Bolivians then tried to impose higher taxes on Chilean enterprises operating in their Atacama region. Chile responded in 1879 by attacking both Bolivia and Peru, and by the middle of 1880 the Chileans had captured Bolvia's coastal zone and were on their way into Peru.

The loss of its maritime outlets has had a severely negative impact on Bolivia, and successive Bolivian governments have tried for many years to restore a Bolivian corridor to the sea. In 1976, it appeared that years of quiet negotiation with Chile (interrupted by the political upheavals in Santiago) were about to bear fruit. Chile's government announced its intention to grant Bolivia a zone of land along the Peruvian-Chilean boundary, as yet undefined, to permit Bolivia to reestablish its own, direct connections with the outside world. Political instability in Bolivia (the country had three different presidents in 1978 alone) again slowed the process, and in 1980 the long-awaited Bolivian Corridor still had not materialized. If the Chilean commitment does bear fruit, Bolivia during the 1980s will finally be able to demarcate (the final step in boundary creation) its boundary in an area it lost more than a century ago.

most important commercial activity, but the cattle generally do not compare well to those of Argentina. With respect to its dominantly subsistence economy, Paraguay resembles the Andean Indian clusters of Peru, Ecuador, and Bolivia—with one difference, that the possibilities of attainable alternatives exist here.

The South: Midlatitude South America

South America's three southern countries, Argentina and its neighbors Chile and Uruguay, are grouped in one region. By far the largest in terms of both population and territory is Argentina, whose 2.8 million square kilometers (1.1 million square miles) and 27 million people rank second only to Brazil in South America. Argentina has four major physiographic regions and a great deal of physical variety within its boundaries, but the vast majority of the population—three-quarters of it—is concentrated in one of these, the Pampa. Figure I-9 indicates the degree of clustering of Argentina's inhabitants on the land and in the cities of the Pampa and the relative emptiness of the other three regions, the scrub-forest Chaco in the north, the mountainous Andes in the west (along whose crestlines the boundary with Chile runs), and the arid plateaus of Patagonia south of the Rio Colorado (Fig. 5-9).

Argentina is the product of the last hundred years. During the second half of the nineteenth century, when the great grasslands of the world were being opened up (including those of the United States, Russia, Australia, and South Africa), the economy of the long-dormant Pampa began to emerge. The food needs of industrializing Europe grew by leaps and bounds, and contributions of the industrial revolution—railroads, more efficient ocean transport, refrigerator ships, and agricultural machinery—helped make large-scale commercial meat and grain production in the Pampa not only feasible but very profitable. Large haciendas were laid out and farmed by tenant workers who would clear the virgin soil and plant it with wheat and alfalfa, harvesting the wheat and leaving the alfalfa as pasture for livestock. As the Pampa was soon brought into production, railroads radiated ever farther outward from Buenos Aires. Today Argentina has South America's densest railroad network, and the once-stagnating city is now one of the world's largest. Yet the Pampa has hardly begun to fulfill its productive potential, which might double with efficient and intensive agricultural practices.

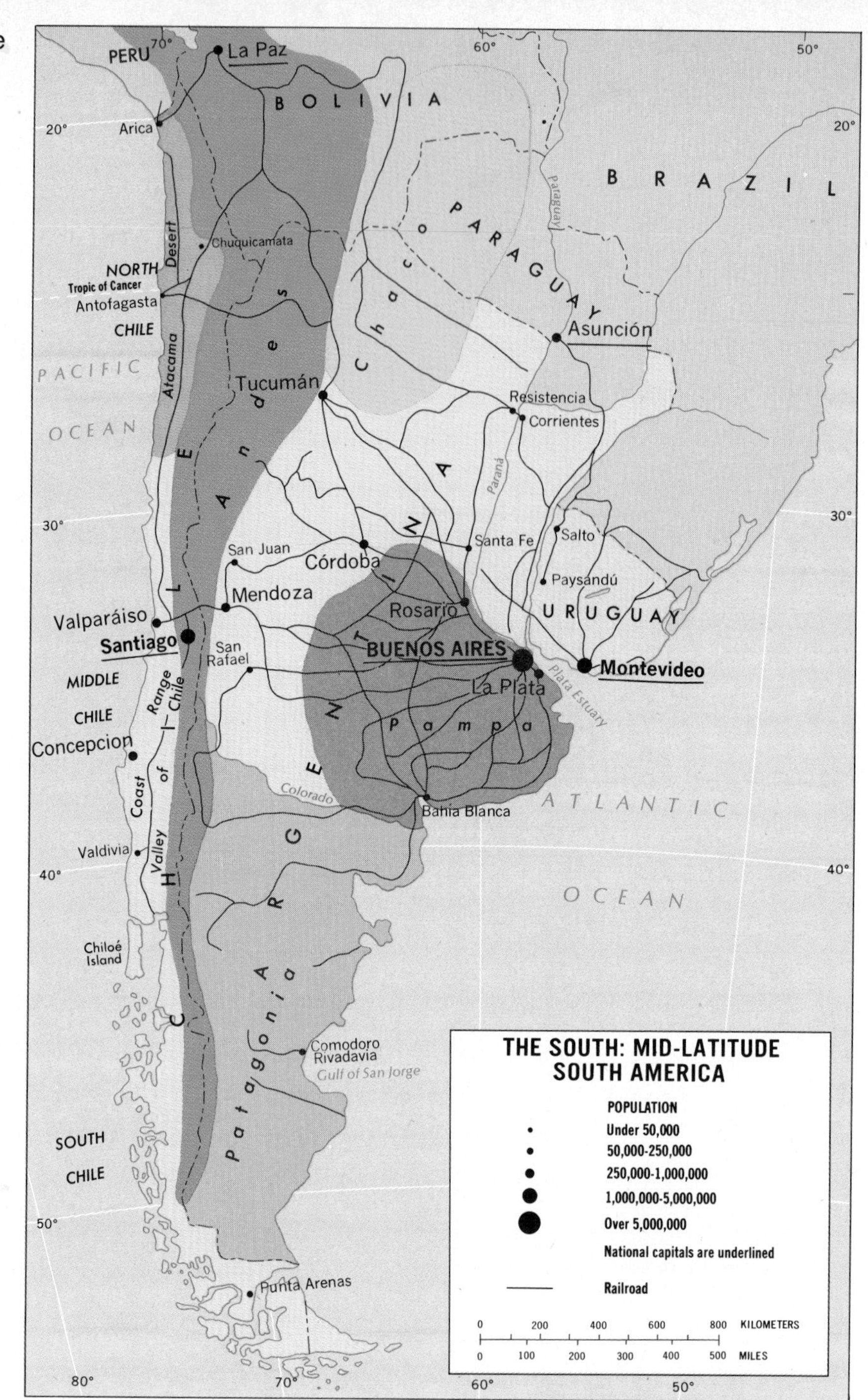

Figure 5-9

Over the decades, the Pampa has developed several areas of specialization. As we would expect, a zone of vegetable and fruit production has become established near the huge conurbation of Buenos Aires. In the southeast is the most exclusively pastoral district, where beef cattle and sheep (for both mutton and wool) are raised. To the west, northwest, and southwest, wheat becomes the important commercial crop, but half the land still remains devoted to grazing. Among the exports, meat usually leads by value, followed by cereals, wool, and some relatively minor exports such as vegetable oils and oilseeds, hides and skins, and quebracho extract.

Buenos Aires is an excellent example of a country's primate city; it dominates life in the country as a whole. (Georg Gerster/Rapho-Photo Researchers)

Argentina's wealth and vigor are reflected in its fast-growing cities; depending on the criteria used, 80 percent of the population may be classified as urbanized, which is an exceptionally high figure for South America. Nearly one-third of all Argentinians live in the conurbation of Greater Buenos Aires alone, and here, too, are most of the industries, many of them managed by Italians, Spaniards, and other recent immigrants. The Cordoba area is also a focus of industrial growth. The majority of the manufacturing in the major cities is the processing of Pampa products and the production of consumer goods for the domestic markets. One of every six wage earners in the country is engaged in manufacturing—another indication of Argentina's advanced economic position.

Argentina's population shows a high degree of clustering and a strong peripheral location. The Pampa region covers only a little over 20 percent of Argentina's area, and with three-quarters of the people here, the rest of the country cannot be densely populated. Outside the Pampa, pastoralism is an almost universal pursuit (except, of course, in the high Andes), but the quality of the cattle is much lower than in the Pampa. In dry Patagonia, sheep are raised. Some of the distant areas are of actual and potential significance to the country: an oilfield is in production near Comodoro Rivadavia, on the San Jorge Gulf, and in the far north, in the Chaco, Argentina may share with Paraguay a larger oil reserve that is yet to be brought into production. Yerba maté, a local tea, is produced in the northeast (Paraguay has a small output of this product as well), and the quebracho extract that figures in both Argentina's and Paraguay's exports comes from the valley of the Paraguay-Paraná River. In addition, the streams that flow eastward off the Andes provide opportunities for irrigation. At Tucuman (325,000), Argentina's major sugar-producing district developed in response to a unique set of physical circumstances and a rapidly growing market in the Pampa cities, which the railroad made less remote; at Mendoza (315,000), and San Juan to the north and San Rafael to the south, vineyards and fruit orchards have been established. But despite these sizable Andean outposts, effective Argentina still remains the area within a radius of 560 kilometers (350 miles) from Buenos Aires.

Uruguay, unlike Argentina or Chile, is compact, small, and quite densely populated. The buffer state of old has become a fairly prosperous agricultural country, in effect a smaller-scale Pampa; Figs. I-2 and I-6 show the similarity of physical conditions on the two sides of the Plata Estuary. Montevideo, the coastal capital with nearly 1.9 million residents, contains well over 60 percent of the country's population, and from here railroads and roads radiate out into the productive agricultural interior.

In the immediate vicinity of Montevideo lies Uruguay's major farming area, and here vegetables and fruits are produced for the city, as well as wheat and fodder crops. Just about all of the rest of the country is used for grazing sheep and cattle; wool constitutes half of the annual exports by value, and meat about a quarter. Of course, Uruguay is a small country, and its 176,000 square kilometers (68,000 square miles), less even then Guyana, do not leave much room for population clustering. But it is nevertheless a special quality of the land area of Uruguay that it is rather evenly peopled right up to the boundaries with Brazil and Argentina. Of all the countries of South America, Uruguay is the most truly "European," without the racial minorities that mark even Argentina and Chile but with a sizable non-Spanish European component in its population.

For 4000 kilometers (2500 miles) between the crestline of the Andes and the coastline of the Pacific lies a narrow strip of land that is the Republic of Chile. On the average just 150 kilometers wide (and only rarely over 300), Chile is the textbook example of what political geographers call an "elongated" state, one whose shape tends to contribute to external political, internal administrative, and general economic problems. In the case of Chile, the Andes Mountains do form a barrier to encroachment from the east, and the sea constitutes an avenue of north-south communication; history has shown the country to be well able to cope with its northern rivals, Bolivia and Peru.

As Figs. I-2 and I-6 as well as Fig. 5-5 indicate, Chile is a three-region country. About 90 percent of Chile's more than 11 million people are concentrated in what is called Middle Chile, where Santiago, the capital and largest city, and Valparaiso, the second city and chief port, are located. North of Middle Chile lies the Atacama Desert, wider and colder than the coastal desert of Peru. South of Middle Chile the coast is broken by a number of fjords and islands, the topography is mountainous, and the climate, wet and cool near the shore, soon turns drier and colder against the Andean interior. South of the latitude of the island of Chiloé no permanent land transport routes of any kind exist, and hardly any settlement.

Figure 5-5 suggests a culture-sphere breakdown for the country that involves three major regions: a mestizo north, a European-Commercial zone in southern Middle Chile, and an undifferentiated south. In addition, a small Indo-Subsistence region in the Northern Andes is shared with Argentina and Bolivia. The Indian element in the two-thirds of the Chilean population that is mestizo largely came from the million or so Indians who lived in Middle Chile.

Despite the absence of a large, landless Indian class, Chile nevertheless had—and continues to have, though in decreasing numbers—tenant farmers on the haciendas and estates. Although they are not peons in the strict sense of the word, in that there is no debt bondage in Chile, these people, the *rotos,* are little if any better off than their counterparts in other South American countries, and there always were few means of escape from the system. The army provided one of these means, and after service in the Atacama Campaign in the late nineteenth century, several thousands of these people were settled on land in the open areas of southern Middle Chile. Another means was to leave the land and head for the cities, to seek work; this migration still goes on. In recent decades the land situation has become a major political issue in Chile, and gradually the large landholdings are being converted into smaller farms.

Some regional differences exist between northern and southern Middle Chile, the country's core area. Northern Middle Chile, the land of the hacienda and of Mediterranean climate with its dry summer season, is an area of irrigated crops that include wheat, corn, vegetables, grapes, and other Mediterranean products; some are grown without irrigation. Livestock raising and fodder crops still take up much of the productive land, and agricultural methods are not particularly efficient; the region could yield a far greater volume of food crops than it does. Southern Middle Chile, into which immigrants from both the north and from Europe (Germany especially) have pushed, is a

A view over the urban landscape of Santiago, capital of Chile, with the Andes in the background. (Rick Merron/Magnum)

moister area where raising cattle predominates but where wheat and other food crops, including potatoes, are also cultivated.

Few of Chile's agricultural products reach external markets, however. Some specialized items such as wine and raisins are exported, but Chile remains a net importer of foods. Thus, despite the fact that nine out of ten Chileans live in what has been defined here as Middle Chile, it is the north, the Atacama region, that provides most of the country's revenues. At first, the mining of nitrates—the Atacama includes the world's largest exploitable deposits—provided the country's economic mainstay, but the industry declined after the discovery of methods of synthetic production (about the time of World War I). Subsequently, copper became the chief export; it is found in several places, but the main concentration lies on the eastern margin of the Atacama near the town of Chuquicamata, not far from the port of Antofagasta. At present, copper in various forms—some of it as pure bars, some as a concentrate refined at Chuquicamanta and some raw ore—constitutes about 70 percent of the value of Chile's exports. Of course, the country is especially vulnerable to fluctuations of prices of this product on the world market.

Copper has been mined for over a century, and American investment in the mining industries of Chile has been heavy. The Chilean government, anxious for a greater share of the profits, expropriated some of the mining properties in the early 1970s. The resulting disagreement over compensation created a political storm that illuminated some of the problems of direct foreign investment in an economy. Large-scale investment by giant multinational firms throughout the Americas and the developing world is a sensitive political issue that could prove troublesome as countries attempt to gain greater control over their economic future.

Additional Reading

Numerous geographies deal with South America as well as Middle America. A good source is a volume edited by H. Blakemore and C. T. Smith, *Latin America: Geographical Perspectives,* published in London by Methuen in 1974. One of the oldest volumes is still full of interest: that of C. F. Jones, *South America,* published by Holt in New York in 1930. Another less durable volume has been F. A. Carlson's *Geography of Latin America,* last printed in 1952 by Prentice-Hall, Englewood Cliffs. The book by R. S. Platt, *Latin America: Countrysides and United Regions,* published by McGraw-Hill, New York, in 1942, consists of field studies and still retains much interest as well. More modern geographies abound. A simple one is G. J. Butland's *Latin America: A Regional Geography,* published in a third edition by Wiley in 1972. J. P. Cole's book, *Latin America: An Economic and Social Geography,* was published by Butterworth in Washington in 1975. P. James's *Latin America* has gone through several Odyssey Press editions. I. Pohl and J. Zepp have edited *Latin America: A Geographical Commentary,* published in London by Murray in 1966. H. Robinson's *Latin America: A Geographical Survey* was published by Praeger, New York, in 1967. Also see C. Wagley, *The Latin American Tradition: Essays on the Unity and Diversity of Latin American Culture,* published by Columbia University Press in 1968. A concise but useful volume is K. E. Webb, *Latin America,* in the Prentice-Hall Foundations of World Geography Series, 1972.

Under the editorship of J. H. Steward, the U.S. Bureau of American Ethnology has published a series of volumes under the general title *Handbook of South American Indians.* There are six volumes and an index, but the best way to study the central issues involved is to consult Steward's summary of these works, entitled *Native Peoples of South America,* edited in cooperation with L. C. Faron and published in 1959 by McGraw-Hill. On individual countries and regions in South America, see T. R. Ford, *Man and Land in Peru,* University of Florida Press,

1962; L. Linke, *Ecuador; Country of Contrasts,* Oxford University Press in New York, 1960; H. Osborne, *Bolivia, a Land Divided,* Oxford University Press in London, 1964; C. Wagley, *An Introduction to Brazil,* Columbia University Press, 1963; G. J. Butland, *The Human Geography of Southern Chile,* published by Philip in London in 1958; and T. F. McGann, *Argentina: The Divided Land,* No. 28 in the Van Nostrand Searchlight Series. Also see a volume edited by A. Taylor, *Focus on South America,* an American Geographical Society collection published in New York by Praeger in 1973.

On urbanization and urban problems in South America, see P. M. Hauser (ed)., *Urbanization in Latin America,* Columbia University Press, 1961, and G. H. Beyer's edited volume, *The Urban Explosion in Latin America: A Continent in Process of Modernization,* Cornell University Press, 1967.

The article by J. P. Augelli, "The Controversial Image of Latin America: A Geographer's View," appeared in the *Journal of Geography,* Vol. LXII, March 1963. C. Harris' article on the "Functional Classification of Cities in the United States" appeared in *The Geographical Review,* Vol. 33, January 1943. A number of South America's continuing boundary problems are discussed by J. R. V. Prescott in *The Geography of Frontiers and Boundaries,* published by Aldine in Chicago in 1965. Also, see P. R. Odell and D. R. Preston, *Economies and Societies in Latin America: a Geographical Interpretation,* published by Wiley in New York in 1973.

Emerging Brazil

IDEAS AND CONCEPTS

Growth pole concept
Development (2)
Multinational influence

Brazil, South America's long-sleepy giant, is stirring. The Brazilian economy is growing at rates that exceed those of underdeveloped countries. The population is increasing at such a pace that, within our lifetimes, Brazil will have more people than the United States. The vast Brazilian interior is finally being penetrated by roads and opened up. The cities are mushrooming. Brazil shows signs of taking off.

With its 8.5 million square kilometers of equatorial and midlatitude South America, Brazil is a giant in world as well as regional terms. Territorially it is exceeded only by the Soviet Union, Canada, China, and the United States. In terms of population Brazil now ranks among the world's half dozen largest countries (see Table I-1). Economically its rank is steadily rising. Brazil seems likely to become a world force of the twenty-first century.

So large is Brazil that it has common boundaries with all other South American countries except Chile and Ecuador (Fig. 5-4). Its environments range from the tropical rain forest of the basin of the Amazon River (nearly all of which lies within Brazil's borders) to the temperate conditions of the Argentine Pampa (Fig. I-6). Its resources are known to include enormous iron ores, extensive bauxite and manganese deposits, large coalfields, and numerous alloys—but Brazil's vast area is yet imcompletely explored, and additional finds will undoubtedly be made. For example, oil and natural gas have long been produced from fields in the state of Bahia (on the central east coast), but never in sufficient quantities to satisfy Brazil's demand. In 1974, a major offshore oil deposit was discovered not far from the coastal city of Campos, less than 200 kilometers northeast of Rio de Janeiro. Self-sufficiency may now be in sight.

Regions

Brazil is a large country, but its landscapes are not as spectacularly diverse as those of several much smaller South American countries. Brazil has no Andes Mountains, and the countryside consists mainly of plateau surfaces and low hills. Even the lower-lying Amazon Basin is not entirely a plain; between the tributaries of the great river lie low but extensive *mesas* sustained by layers of sedimentary rocks. The surface of the great Brazilian plateau rises slowly eastward, but the highest areas fail to reach 3000 meters (under 10,000 feet). Along the coastline, there is a steep scarp leading from plateau surface to sea level, leaving almost no living space along its foot. Thus, although Brazil has some 7500 kilometers (4600 miles) of Atlantic coastline, there is very little coastal plain, and cities such as Rio de Janeiro are crowded between mountain slope and oceanfront. Under the physiographic circumstances, Brazil is fortunate to have several very good natural harbors.

Brazil is a federal republic consisting of 22 states, 4 territories, and the federal district of the capital, Brasilia (Fig. B-1). As in the United States, the smallest states lie in the northeast, the larger ones farther west. The state of Amazonas is the largest, with over 1.5 million square kilometers (more than 600,000 square miles, twice the size of Texas)—but Amazonas' huge area contains barely one million people. Tiny Guanabara, on the east coast near Rio de Janeiro, has a mere 1171 square kilometers (452 square miles) but its population is nearly 6 million. The state with the largest population by far is São Paulo, now with more than 26 million and growing rapidly.

Although Brazil is about as large as the 48 contiguous United States of America, it does not possess the clear physiographic regionalism familiar to us. Apart from the Great Escarpment marking the eastern edge of the Brazilian Plateau, there are no mountain barriers, no well-defined coastal plains, no naturally demarcated desert regions, and no large lakes. Thus the six regions outlined below have no absolute or even generally accepted boundaries; they are subject to debate much like our concept of the United States Midwest as a region. On Fig. B-1, the regional boundaries have been drawn to coincide approximately with the boundaries of states, making identifications easier.

The Northeast

The Northeast was Brazil's source area, its culture hearth. Here began the plantation economy, and to this area came not only the Portuguese planters but also the largest number of African slaves to work in the sugar fields. But the ample, dependable rainfall that occurs along the coast soon gives way, into the interior, to lower and more variable patterns (see Fig. I-6), and today the largest part of the Northeast is poverty-stricken, hunger-afflicted, overpopulated, and subject to devastating droughts. Sugar still remains the chief crop along the moister

Ten Major Geographic Qualities of Brazil

1. Territorially, Brazil ranks among the world's five largest states. The country's area covers nearly half of South America, and Brazil has common boundaries with every South American country except Chile and Ecuador.
2. Brazil is large, but its physiographic diversity does not match its size. The great Andes do not enter Brazil; landscapes consist mainly of plateaus, low hills, and the undulating Amazon Basin.
3. Brazil's population is about as large as that of all other South American countries combined.
4. In recent decades Brazil has had one of the world's highest rates of population growth. Continued growth at this rate will make Brazil the Americas' most populous nation state in the twenty-first century.
5. Brazil's regional development has proceeded most rapidly along its eastern margins and has been slowest in the Amazonian interior.
6. Brazilian governments have strived to focus national attention on the opportunities of the western interior. Brasilia, the new capital, is a manifestation of that effort.
7. Brazil exhibits a strong national culture in which a single language and the domination of one religious faith constitute centripetal forces.
8. Although still ranked as an underdeveloped country, Brazil generates economic indicators that point to "takeoff" conditions. Rapid urbanization and growing industrial strength mark the country.
9. Although it is the giant of South America, Brazil's relationships with its neighbors remain comparatively distant.
10. Brazil is a federation in which eastern states of the modern core area dominate national affairs. Strong central authority and military involvement in government have recently prevailed.

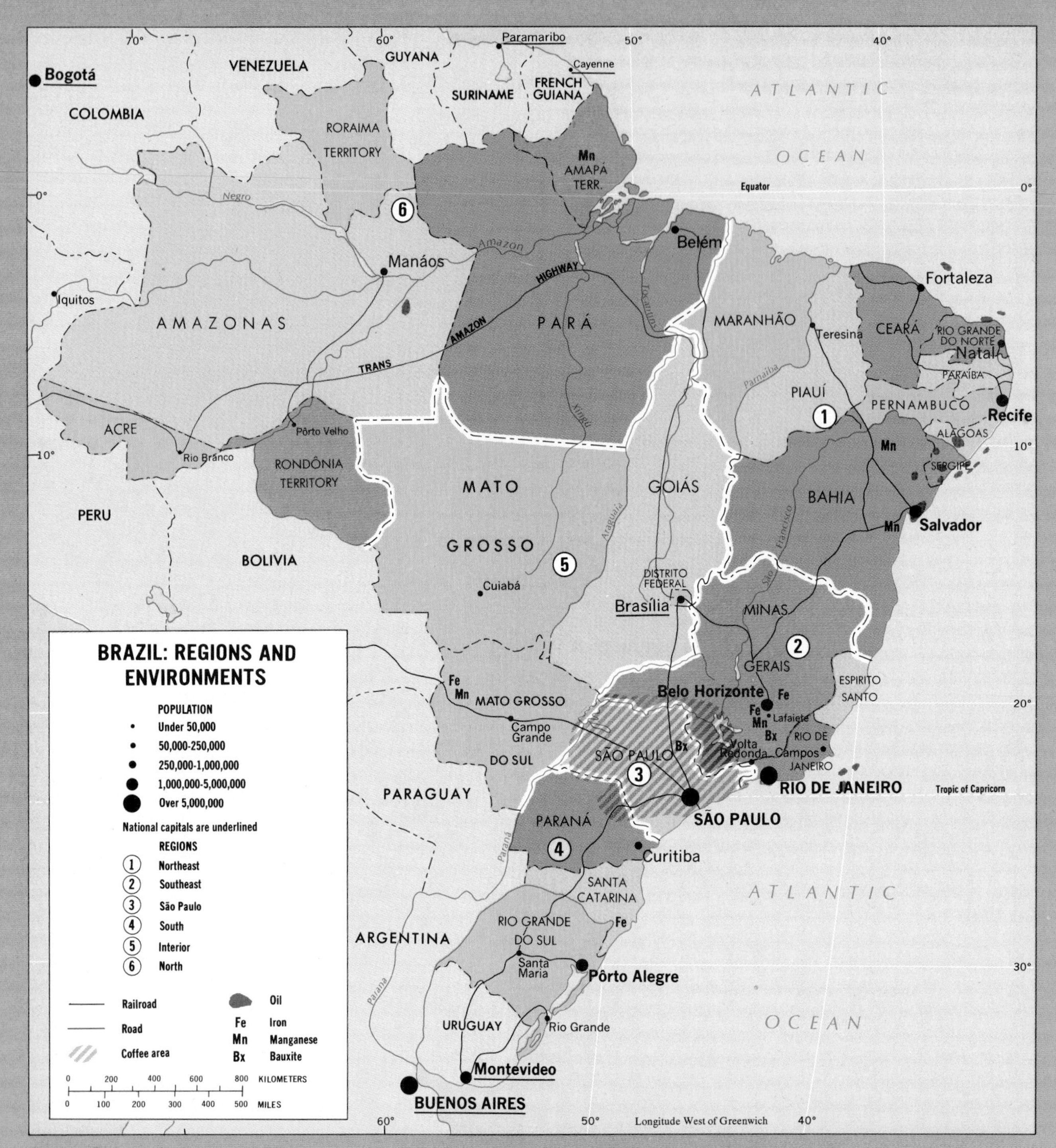

Figure B-1

coast, and livestock herding prevails in the drier *sertão,* with beef cattle in the better grazing zones and goats elsewhere. Human and animal overpopulation have combined to deplete the natural vegetation, and aridity encroaches. The comparatively small areas of successful commercial agriculture (including also cotton in Rio Grande de Norte, sisal in Paraiba, and coffee in Pernambuco) stand in sharp contrast to the patches of shifting subsistence agriculture nearby. The Northeast today is Brazil's great contradiction; in the cities of Recife and Salvador the architecture still bears the imprint of an earlier age of wealth, but thousands of hopeless peasants, driven from the land by deteriorating conditions, arrive constantly in the shantytowns. And few of the generalizations

The Brazilian Northeast is the country's culture hearth. The urban landscape of the capital of the state of Bahia, Salvador, carries the imprint of history. (Jack Fields/Photo Researchers)

about expanding, emerging Brazil yet apply here, in South America's largest most poverty-stricken corner.

The Brazilian government, aware of the Northeast's plight, has directed investments to the region to help diversify its economic geography. The most massive effort so far has been the huge petrochemical complex (the world's largest) built near Salvador. This project, twenty years in the making, became fully operational in 1979, creating thousands of jobs, attracting further (including foreign) investment, and boosting the Northeast's industrial base in general. But even an enormous project such as this has had only limited impact on the overall economic geography of the impoverished Northeast, where subsistence remains the rule rather than the exception.

The Southeast

In the state of Bahia, a transition occurs. The Great Escarpment becomes more prominent, the plateau higher, and the terrain more varied. Annual rainfall increases, and it is seasonally more dependable. The Southeast has been Brazil's modern core area, the scene of successful as well as abortive gold rushes, with its major cities and greatest population clusters. Gold drew many thousands of settlers, and other mineral finds also contributed to the influx (Rio de Janeiro was founded as the endpoint of the "Gold Trail"), but ultimately the region's agricultural possibilities ensured its stabilization and growth. The mining towns needed food, and prices for foodstuffs were high; farming was stimulated and many farmers came (with their slave workers) to Minas Gerais to till the soil. Eventually a pastoral industry came to dominate, with large herds of beef cattle grazing on planted pastures.

Modern times have brought another mineral age to the region, based not on gold or diamonds but on the iron ores at Lafaiete and the manganese and limestone carried to

Brazil's determined industrial push is symbolized by this scene from Volta Redonda, location of large iron and steel plants. The iron ores used here come from la Faiete. (Ellis Herwig/Stock, Boston)

the steel plants at Volta Redonda. Industrial diversification has proceeded apace; Belo Horizonte is the rapidly growing metallurgical center of Brazil, with a 1978 population estimated at 1.9 million, more than three times what it was in the mid-1950s. Rio de Janeiro, of course, continues in second place (after São Paulo) among industrial centers.

São Paulo

The state of São Paulo constitutes part of the southeast region, but it is the focus of Brazilian development today. Largely from the vast coffee *fazendas* concentrated in São Paulo, this state is the largest foreign-exchange earner of the republic, and it leads Brazil in the production of numerous other crops as well. With the help of Japanese technicians (and with the prospect of sale on the large Japanese market), Brazilian farmers have enormously increased their soybean production, surpassing China in the mid-1970s and taking second place among world producers. Matching this prodigious agricultural output is the state's industrial strength. São Paulo (city) is the country's leading manufacturing center today. The state does not have a mineral base to compare to that of Minas Gerais, but it has nevertheless become the leading industrial region not only of Brazil but of all Latin America. The revenues derived from the coffee plantations provided the necessary capital; hydroelectric power from the slopes of the Great Escarpment produced the needed energy; immigration not only from Portugal but also from Italy, Japan, and elsewhere contributed the labor force. São Paulo lay juxtaposed between raw-material-producing Minas Gerais and the states of the South. Communications were improved, notably with the exit port of Santos but also with the interior hinterland, and as the capacity of the domestic market grew the advantages of location and agglomeration secured São Paulo's primacy. São Paulo today is the pulse of Brazil.

The South

Three states make up the southernmost Brazilian region: Parana, Santa Catarina, and Rio Grande do Sul. The contribution of recent-immigrant Europeans to the agricultural development of southern Brazil, as in Uruguay and Argentina, has been considerable. Many came not to the coffee areas of São Paulo, but to the available lands farther south. Here they occupied fairly discrete areas; for example, Portuguese rice farmers clustered in the valleys of the major rivers of Rio Grande do Sul, and the state today produces between one-quarter and one-fifth of Brazil's annual rice crop. The Germans, on the other hand, occupied the somewhat higher areas to the north and in Santa Catarina, where they were able to carry on the type of mixed farming with which they were familiar: corn, rye, potatoes, and hogs, as well as dairying. The Italians selected the highest slopes and established thriving vineyards. The markets for this produce, of course, are the growing cities to the north.

São Paulo is Brazil's largest and fastest-growing city. Its rapid growth is attended by many of the ills of urbanization here as elsewhere: poor internal circulation and resulting congestion, choking air pollution, inadequate housing. (Bruno Barbey/Magnum)

Parana, on the other hand, exports its coffee harvest to overseas markets.

The South never was a boom area, as were the Northeast and Southeast at various times in Brazilian history. It does, however, have a stable, progressive, modern agricultural economy; farming methods here are the most advanced in Brazil. The diversity of its European heritage still is reflected in the regional towns, where German and other European languages are still in use. Some 25 million people live in the three southern states, and the region's importance is increasing. The coal from Santa Catarina and Rio Grande do Sul, shipped north to the steel plants of Minas Gerais, was a crucial element in Brazil's industrial emergence. Local industry is growing as well, especially in Porto Alegre. The contrast with the Northeast could hardly be stronger.

The Interior

Interior Brazil is often referred to as the Central-West—*Centro-Oeste.* This is the region Brazil's developers hope to make a part of the country's productive heartland, and the new capital of Brasilia was positioned on its margins (Fig. B-1). But such integration will take much time. This is a vast upland area, largely a plateau at an elevation over 1000 meters covered with savanna vegetation (see Fig. I-7). Like Minas Gerais, the interior was the scene of gold rushes and some discoveries, but unlike its eastern neighbor, it did not present agricultural alternatives when the mineral age petered out. Today, the two giant states of the interior (Goias and Mato

Grosso) have a combined population of somewhat more than 7 million, so that the average density is under 10 per square kilometer (25 per square mile). Nevertheless, this represents a noteworthy increase, for in 1960 the Centro-Oeste contained just 2.87 million people. Thus the region's population doubled in about 20 years—faster even than Brazil itself.

Still, the Centro-Oeste faces the problems common to savanna regions everywhere: soils are not especially fertile, and the vegetation is susceptible to damage by overgrazing. In the absence of major mineral finds (the region still awaits systematic exploration), pastoralism remains the chief commercial enterprise. There is always the fear that what happened to the Northeast could happen here, and that Brazil's determination to open the interior might create serious problems. But at present that day seems far off. You can fly for hours over the "great forest" and see only an occasional small settlement and a few widely spaced roads (communications are best in the south). What the interior needs is investment—in the clearing and opening of the alluvial soils in the region's river valleys, in experimental farms, in the provision of electric power, in mineral exploration. The first great step was the relocation of the national capital. It was an enormously expensive venture—but it was a mere beginning in view of what the interior really requires.

The North

The largest Brazilian region also is the most remote from the core of Brazilian settlement: the three states (and three federal territories) of the Amazon Basin. This was the scene of the great rubber boom at the turn of the century, when the wild-growing rubber trees in the *selva* (tropical rain forest) produced large profits and the Amazon city of Manaus enjoyed a brief period of wealth and splendor. But the rubber boom ended in 1910, when plantations elsewhere (Southeast Asia) began to produce rubber more cheaply, efficiently, and accessibly.

Since those days the North has been Brazil's stagnant hinterland. Manganese ores are mined in the federal territory of Amapa; a few Japanese settlements near the Amazon mouth are trying to develop a rice culture, and they also grow jute that is bought by a mill in Belem. One pioneering effort has also become an issue of contention: the huge operation of Ludwig that involves the clearing of vast areas of the rain forest and the development of extensive plantations. While bringing jobs and funds to an area that would otherwise lie untouched, such large-scale operations also threaten the forest without regard for (or adequate knowledge of) the possible consequences. A single operation, even at the scale of the Ludwig project, poses no immediate danger. But environmentalists wonder what would happen if this practice expands. However, the North's opportunities have barely been touched. In 1980 the region's total population (in an area half the size of Europe) was about 5 million, and the majority of those lived in the eastern state of Para. The physiographic region of the Amazon Basin still contained under 1 million persons.

The North is not all swampy rain forest, as its popular image sometimes suggests. True, there are low-lying areas at the mouth of the great Amazon River and near the interior streams, but the areas between the rivers are well above the flood-waters, and from the air the aspect is one of undulating, even hilly countryside. The forest cover is dense, because the crowns of the trees interlace, but there is little undergrowth; tree trunks are large and straight. The rubber boom of the late 1800s was not followed by a plantation economy largely because of an absence of labor; the lumbering possibilities have not been exploited because there are too few workers; the agricultural potential remains unrealized because there are few farmers. The Brazilian North is one of the world's few regions that is truly short of people.

Population Patterns

Brazil's population of 122 million (1980) is as diverse as that of the United States, and Bra-

Opening up and exploiting the Amazonian interior is a long-held Brazilian dream, one that generated the great Trans-Amazonian Highway project. The largest project developed in this area is the Ludwig scheme to exploit forest resources and stimulate associated growth. An entire factory was built in Japan and floated on barges to a riverside in Amazonia. (Martin Rogers/Woodfin Camp)

zilian culture has been a melting pot perhaps to an even greater degree than the United States itself. In a pattern that is familiar in the Americas, the original Indian inhabitants of the country were decimated following the European invasion; estimates of the number of Indians who survive today in small communities deep in the Amazonian interior vary, but their total is not likely to exceed 250,000—less than 10 percent of the number thought to have been there when the whites arrived. Africans came in great numbers, and today more than 13 million people in Brazil are black. Significantly, however, there was also much mixing, and about 35 million Brazilians (about 30 percent) have combined European, African, and minor Indian ancestries. The majority, over 70 million, is European.

Until Brazil became independent in 1822, the Portuguese were virtually the only Europeans to settle in this country. But after independence other European settlers were encouraged to come, and many Italians, Germans, and Eastern Europeans arrived to work on the coffee plantations, farm in the south, or try their luck in business. Immigration reached a peak during the decade of the 1890s, when nearly 1.5 million newcomers reached Brazilian shores, and the population was further complicated by the arrival of Lebanese and Syrians, many of whom opened small shops, and Japanese, who formed settlements in Southern São Paulo and Northern Parana in the 1930s.

Brazilian society, to a far greater degree than is true in the United States, has overcome problems of dualism (see Chapter 3). Black Brazilians still, as a group, are the least advantaged among population sectors (except, of course, the Indians), and as in the United States it has proved impossible to overcome the obstacles of history. But the Brazilian ethnic mix is so all-pervasive that hardly any group is unaffected, and official statistics about "blacks" and "Europeans" are rather meaningless in Brazil. What Brazil does have is a true national culture, expressed in an overwhelming adherence to the Catholic faith and the universal use of a modified form of Portuguese as the common language, and in a set of life-styles in which vivid colors, distinctive music, and a growing national pride are fundamental ingredients.

This is not to suggest that Brazilian population patterns do not have regional dimensions. The black component of the population in the Northeast, for example, remains far stronger than it is in the Southeast. After their initial concentration in the Northeast, black workers were taken southward to Bahia and Minas Gerais as the economic core moved in those directions. Today, the black population remains strongest in these areas and in Rio de Janeiro; it is weak in the states of the South, more recently settled and more completely taken over by Europeans.

Nor has Brazil escaped the problems of pluralism also faced in the United States. Black intellectuals, labor leaders, and others in the country have in recent years begun to express publicly their objections against instances of racial discrimination that would, in the past, have gone unnoticed except by the victims and their immediate family and friends. Still, while Brazil may not be the multiracial society it is sometimes portrayed to be, it also does not have the history of overt and legally sanctioned racism that has characterized certain other plural societies.

Development Problems

Until the energy crisis of the early 1970s, Brazil's development had been impressive. The economy was expanding at about 9 percent per year. Since the mid-1960s, the value of Brazilian exports had been doubling every two years. Investments were being made to defeat illiteracy, improve health conditions, assist agriculture, and ameliorate urban living problems. The programs begun after the military regime ended democratic government in 1964 seemed to be succeeding.

The energy crisis severely affected Brazil, and caused a major setback in its economic growth. Plans were made to exploit the country's large oil-shale deposits for oil production, and exploration for oil was intensified—leading to the discovery near Campos. But production from the new sources is still in the future. In the meantime, Brazil has seen its forward drive slowed—a trend shared with many developing countries severely affected by the energy crisis.

Agriculture presents Brazil with great opportunities and serious problems. Apart from the revenues derived from coffee sales overseas, Brazil has also been selling more sugar to the United States, the Soviet Union, and China, and the volume of soybeans harvested is rising as well. The range of crops Brazilian farmers can grow is considerable: corn, rice, and wheat among the staples, beans and vegetables, cotton, tobacco, and grapes. But much land that could be cultivated is not, agricultural methods stilll need improvement, and land reform in the *fazendas* has come slowly. Millions of peasants in Brazil practice shifting agriculture when environmental conditions do not really demand it, so that returns from the soil are only minimal. It has been estimated that the productivity per farm worker in the United States still is almost 50 times as high as it is in Brazil. Mechanization has barely begun, for example, in the Northeast. Notwithstanding the rush to the cities, the labor force engaged in agriculture has been growing, reflecting Brazil's still underdeveloped condition.

In this context we should remember how far Brazil must go from its present takeoff point. Its total area is comparable to the area of Europe, but its total annual production of goods and services in the mid-1970s was only about 15 percent larger than that of the Netherlands and with only 13 million people. As J. P. Cole points out in his book *Latin America* (1975), Norway normally produces 40 percent more hydroelectric power than Brazil (despite Brazil's dependence on hydro power as a major energy source), Venezuela produces more oil in a few weeks than Brazil does in a year, and Brazil's annual coal production "is equal to that of two or three average-sized coal mines in Britain." Brazil's awakening is sometimes compared to the first period of Japan's modernization, but such comparisons may be premature.

Growth Poles

When the military regime took control of Brazil in 1964, it embarked on a development program that is sometimes referred to as the "Brazilian model." Essentially this program involved the government in shared participation with private enterprise, so that public interests, private concerns, and foreign investors would not operate at mutual disadvantage—and at the disadvantage of Brazil as a whole. Agriculture and industry were supported and promoted, but under certain priorities and guidelines.

Among the problems always facing national development planners is the remote region where opportunities lie, but where the investments needed to exploit them are high. If an area is already developing and attracting immigrants, should money be spent to improve already-existing facilities or should a new, distant location be stimulated?

In such decisions, the *growth-pole* concept becomes relevant. The term is almost self-explanatory: a growth pole is a location where a set of industries, given a start, will expand, setting off ripples of development in the surrounding area. Certain conditions must exist, of course: there would be little point in selecting Manaos as a growth pole in the middle of the Amazon Basin where there is no real prospect for development in its hinterland, where there are practically no people and no agricultural or industrial activities to stimulate. Growth-pole theory certainly went into the decision to build Brasilia, for in underpopulated Goias a new market of 700,000 people constitutes a major stimulus for all kinds of activity. We will encounter this concept again as we study the underdeveloped countries of East Africa and South Asia.

Multinational Influences

As we noted earlier, the "Brazilian model" involves government intervention in the economy, including various forms of control over foreign investors. This is a significant dimension of Brazilian development policy, for large corporations have played a major role in Brazil's rapid economic surge. The power of multinational corporations in underdeveloped countries has recently become a matter for concern among the leaders of these countries, because these global corporations, with their enormous financial resources, can influence the economics as well as the politics of entire states.

Brazilian leaders have long welcomed foreign investment, but as Brazil's economic progress accelerated they perceived the risks involved in foreign control over Brazilian firms. Multinational corporations can introduce and spread technological advances, they can provide capital and increase employment—but in the process they gain control over industrial and agricultural export sectors of the economy, and by their efficiency they can throttle local competition and damage business oriented toward local markets. The Brazilian government has, in recent years, prohibited foreign commercial banks from entering the Brazilian economy and has passed regulations forcing multinational corporations active in Brazil to keep more of their huge profits inside the country. It is a measure of Brazil's growing strength that Brasilia was able to make these measures stick—and a source of envy for those underdeveloped countries still unable to control what has been called a new economic colonialism.

Brazil's economic progress has been achieved at a cost to its society. The last elected government (1962–1964) faced food shortages, serious unemployment, high inflation, and achieved little in the way of economic progress. The generals who took power in 1964 provided the stability Brazil needed, and they brought improvement to the economic situation. But dissent was suppressed, sometimes ruthlessly, and Brazil's reputation as an open, good-natured, fermenting society was marred by reports of torture and intimidation, and although the effect of the new policies can be seen in the modernizing cities and near the highways, the lot of the poor in Brazil has hardly begun to improve. The present government anticipates that its policies will ultimately improve the life of every layer of Brazilian society. Whether it can buy the time to make the transition work is the crucial question in South America today.

Additional Reading

Brazil is discussed in volumes that deal with South America as a whole and with Latin America in general. The quotation on page 329 is from J. P. Cole's *Latin America: An Economic and Social Geography,* published by Rowman and Littlefield, Totowa, in 1975, page 214. Also see *Area Handbook for Brazil,* U. S. Government Printing Office, Washington, D. C., 1971. Problems of development are discussed in A. Gilbert's *Latin American Development: A Geographical Perspective,* published by Penguin Books, Baltimore, in 1974. K. E. Webb discusses *The Changing Face of Northeast Brazil,* Columbia University Press, New York, 1974. Also see C. Wagley (ed.), *Man in the Amazon,* published by the University of Florida Press, 1974, and H. O'Reilly Sternberg, *The Amazon River of Brazil,* a Monograph of *Geographische Zeitschrift,* No. 40, published in Wiesbaden, 1975. D. G. Epstein discusses *Brasilia: Plan and Reality,* published by the University of California Press, Berkeley, 1973. On multinational corporations and their influence, see R. J. Barnet and R. E. Muller, *Global Reach: the Power of the Multinational Corporations,* published by Simon and Schuster, New York, in 1974. Several specific references to Brazil are especially instructive.

World of Islam: the mosques of Cairo. (George Holton/Ocelot)

Six

North Africa and Southwest Asia

From Morocco on the shores of the Atlantic Ocean to Afghanistan in Asia, and from Turkey between the Black and Mediterranean Seas to the Somali Republic in the "Horn" of Africa lies a vast realm of enormous historical and cultural complexity. It lies juxtaposed between Europe, Asia, and Africa, and it is part of all three; throughout history its influences have radiated to these continents and to practically every other part of the world as well. This is one of humanity's source areas: on the banks of the Euphrates and the Nile arose civilizations that must have been among the earliest; in its soils plants were domesticated that are now grown from the Americas to Australia; its paths were walked by prophets whose religious teachings are followed by hundreds of millions of people to this day.

It is tempting to characterize this realm in a few words, to stress one or more of its dominant features. It is, for example, often called the "dry world," containing as it does the Sahara and Arabian deserts. But most of the people in the region live where there is water—in the Nile Delta, along the coastal strip (or *tell*) of Tunisia, Algeria, and Morocco, along the eastern and northeastern shores of the Mediterranean Sea, in the Tigris-Euphrates Basin, in the desert oases, and along the mountain slopes south of the Caspian Sea. True, we know this region as one where water is almost always at a premium, where peasants often struggle to make soil and moisture yield a small harvest, where nomadic peoples follow their animals across dust-blown plains, where oases are islands of sedentary farming and trade in a sea of aridity. But it also is the land of the Nile, the lifeline of Egypt, and the great Gezira irrigation scheme, the mainstay of the Sudan, and the crop-covered *tell* of Algeria. This is a dry world—but not everywhere.

IDEAS AND CONCEPTS

- Culture hearth (2)
- Diffusion
- Feudalism
- Urban Dominance
- Nomadism
- Irrigation
- Ecological trilogy
- Cycle theory
- Boundary morphology

North Africa and Southwest Asia also are often referred to as the "Arab world." Again, this implies a uniformity that does not actually exist. In the first place, the name "Arab" is given loosely to the peoples of this area who speak the Arabic language (and some related languages as well), but ethnologists normally restrict it to certain occupants of the Arabian Peninsula, the Arab "source." Anyway, the Turks are not Arabs, and neither are the Iranians, nor the Israelis. Second, while it is true that Arabic is spoken over a wide region that extends from Mauritania in the west across North Africa to the Arabian Peninsula, Syria, Iraq, and southwestern Iran in the east, there are many areas in North Africa and Southwest Asia where it is not used by most of the people. In Turkey, for example, Turkic is the major language, and it has Altaic rather than Semitic or Hamitic roots. In Iran, the Iranian language belongs to the Indo-European family. In Ethiopia, Amharic is spoken by the ruling plateau people, and although it is more closely related to Arabic than is Iranian, it nonetheless remains a distinct language as well. The same is true of Hebrew, spoken in Israel; other "Arab-world" languages that have their own identity are spoken by the Tuareg of the Sahara Desert, the Berbers of Northwest Africa, and the peoples who live in the transition zone between North Africa and Black Africa to the south. In parts of the "Arab world" there are elements that are distinctly non-Arab.

Ten Major Geographic Qualities of North Africa and Southwest Asia

1. This realm contains several of the world's great ancient culture hearths and some of its most durable civilizations.
2. North Africa and Southwest Asia is the source region of several world religions including Judaism, Christianity, and Islam.
3. The realm contains a pivotal area in the "Middle East," where Arabian, North African, and Asian regions intersect.
4. The North African-Southwest Asian realm extends across major parts of two continents, the only geographic world realm to possess this property.
5. Population in North Africa and Southwest Asia is widely dispersed in discontinuous clusters.
6. Natural environments in this realm are dominated by drought and unreliable precipitation; population concentrates where water supply is adequate to marginal.
7. North Africa and Southwest Asia is the "Arab World," but major population components in the realm are not of Arab ancestry.
8. This realm is dominantly (but not exclusively) Islamic, and the faith pervades cultures form Morocco to Afghanistan.
9. North Africa and Southwest Asia is a realm of intense discord and conflict. Geographically these are reflected by territorial disputes and boundary friction.
10. The largest known reserves of petroleum lie in this realm, but this wealth has affected the living standards of a minority of the total population.

Another name given to this region is the "world of Islam." The prophet Mohammed was born in Arabia in A.D. 571, and in the centuries that followed his death in 632, Islam spread into Africa, Asia, and Europe. This was the age of Arab conquest and expansion, during which their armies penetrated southern Europe, caravans crossed the deserts, and ships plied the coasts of Asia and Africa. Along these routes they carried the Moslem faith, converting the ruling class of the states of savanna West Africa, threatening the Christian stronghold in the highlands of Ethiopia, penetrating the deserts of inner Asia, pushing into India and even Indonesia. Islam was the religion of the marketplace, the bazaar, the caravan. Where necessary it was imposed by the sword, and its protagonists aimed directly at the political leadership of the communites they entered. Today, the Islamic religion with its 700 million followers extends well beyond the limits of the region discussed here: it is the major religion in northern Nigeria, in Pakistan, and in Indonesia; it has strength in parts of the Soviet Union and survives even in Eastern Europe, notably in Albania and Yugoslavia (Fig. 6-1). On the other hand, the "world of Islam" is not entirely Islamic either. In Israel Judaism is the prevailing faith. In Lebanon, perhaps as much as half the population adheres to an old form of Christianity. In Ethiopia, the Amharic-speaking ruling class, having managed centuries ago to stave off the Islamic onslaught, also practices an ancient Christian religion. Coptic Christian churches still exist in Egypt. Thus the connotation "world of Islam" for North Africa and Southwest Asia is far from satisfactory: in the first place, the religion prevails far beyond these areas, and, in addition, there are two or three countries within the "Islamic world" where Islam is not the faith of the majority.

Finally, this region is sometimes called the "Middle East." This must sound quite odd to someone, say, in India, who might think of a Middle West rather than a Middle East! The name, of course, reflects its source: the "Western" world, which saw a "Middle" East in Egypt, Arabia, and Iran, and a "Far" East in China and Japan and adjacent areas. Still, the term has taken hold, and it can be heard in general use by members of the United Nations. In view of the complexity of the region, its transitional margins, and its far-flung areal components, at least the name "Middle East" has the merit of being imprecise. It does not make a single-factor region of North Africa and Southwest Asia, as do the terms *dry world, Arab world,* and *world of Islam.*

A Greatness Past

In our discussion of Middle America, first reference was made to the concept of the culture hearth (page 256), a region of cultural growth and development. In those areas of comparative success and progress, clusters of population developed; natural increase was supplemented by the immigration of those attracted from afar. New ways were found to exploit local resources, and power was established over resources located farther away. Farming techniques improved, and so did yields. Settlements could expand, and began to acquire urban characteristics. The circulation of goods and ideas intensified. Traditions emerged in various spheres of life, and these traditions, along with inventions and innovations, radiated outward into the realm beyond. Among the ideas that took hold and developed were political ideas; the theory and practice of the political organization necessary to cope with society's growing complexity.

Such developments occurred on several continents (Fig. 6-2). The Middle East was a prominent source area, where culture hearths lay in the area between the Tigris and Euphrates Rivers (Mesopotamia), in the Nile Valley (Egypt), and in the Indus River basin (in what is now Pakistan). A major culture hearth also was centered on the confluence of the Wei and Huang Rivers, there forming the roots of Chinese culture. In Middle America, as we noted, the Yucatan peninsula and later the Mexican Plateau were significant culture hearths. In South America the Central Andes Mountains witnessed the rise of a major human civilization. And, as Fig. 6-2 shows, in West Africa's savannalands a culture hearth also emerged.

The Middle East, as it now exists as a culture realm, thus included several of the earliest (if not the first) hearths of human culture. Mesopotamia, the land "between the rivers," flanked the Fertile Crescent, one of the places where people first learned to domesticate plants and gather harvests in an organized way. Mesopotamia became a crossroads for a whole network of routes of trade and movement across Southwest Asia, and was a recipient of numerous new ideas and inventions. Many innovations originated in Mesopotamia itself, to be diffused to other regions of development. In addition to the organized and planned cultivation of grain crops such as wheat and barley, the Mesopotamians knew how to make tools and implements from bronze, they had learned to use draft animals to pull vehicles (including plows designed to prepare the fields), and they employed the wheel, a revolutionary invention, and built carts, wagons, and chariots.

The ancient Mesopotamians also built some of the world's earliest cities. This development was made possible by their accomplishments in many spheres, especially agriculture, whose surpluses could be stored and distributed to the city dwellers. Essential to such a system of allocation, of course, was a body of decision-makers and organizers, people who controlled the lives of others—an elite. Such an urban-based elite could afford itself the luxury of leisure, and could devote time to religion and philosophy. Out of such pursuits came the concept of writing and recordkeeping, an essential ingredient in the rise of urbanization. Writing made possible the codification of laws and the confirmation of traditions. It was a crucial element in the development of systematic administration in urbanizing Mesopotamia, and in the evolution of its religious-political ideology. The rulers in the cities were both priests and kings, and the harvest the peasants brought to be stored in the urban granaries was a tribute as well as a tax.

Mesopotamia's cities emerged between 5000 and 6000 years ago, and some may have had as many as 10,000 residents (some archeologists' estimates go even higher). Today these urban places are extinct, and careful excavations tell the story of their significant and sometimes glorious past. Mesopotamia's cities had their temples and shrines, priests, and kings; there were also wealthy merchants, expert craftspeople, and respected teachers and philosophers. But Mesopotamia was no unified political state. Each city had a hinterland

Figure 6-1

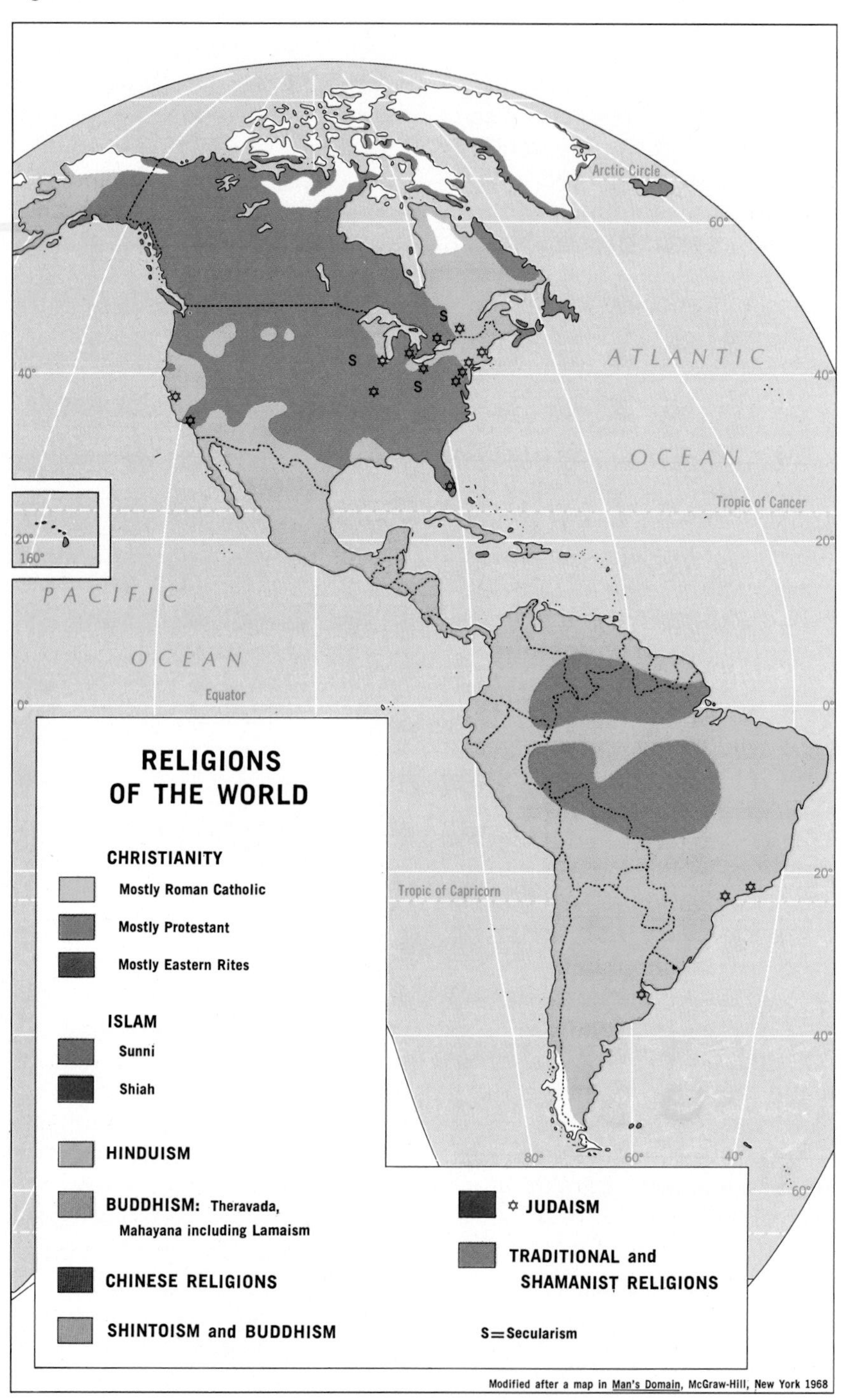

where its power and influence prevailed, and while there were sometimes alliances among cities, there was more often competition and conflict. There was a *feudal* society, a society in which hostility and strife marked political relationships.

Still, Mesopotamia made progress in political spheres as well as other areas, and regional unification eventually was achieved during the fourth millenium B.C. (i.e., between 5000 and 6000 years ago) in at least two early stages named Sumer and Elam.

Another area that witnessed very early cultural and political development was Egypt. Possibly Egypt's evolution started even earlier than Mesopotamia's, but Egypt certainly possessed urban centers 5000 years ago, and there are archeologists who believe that even older remains of urban places may lie buried beneath the silt of the

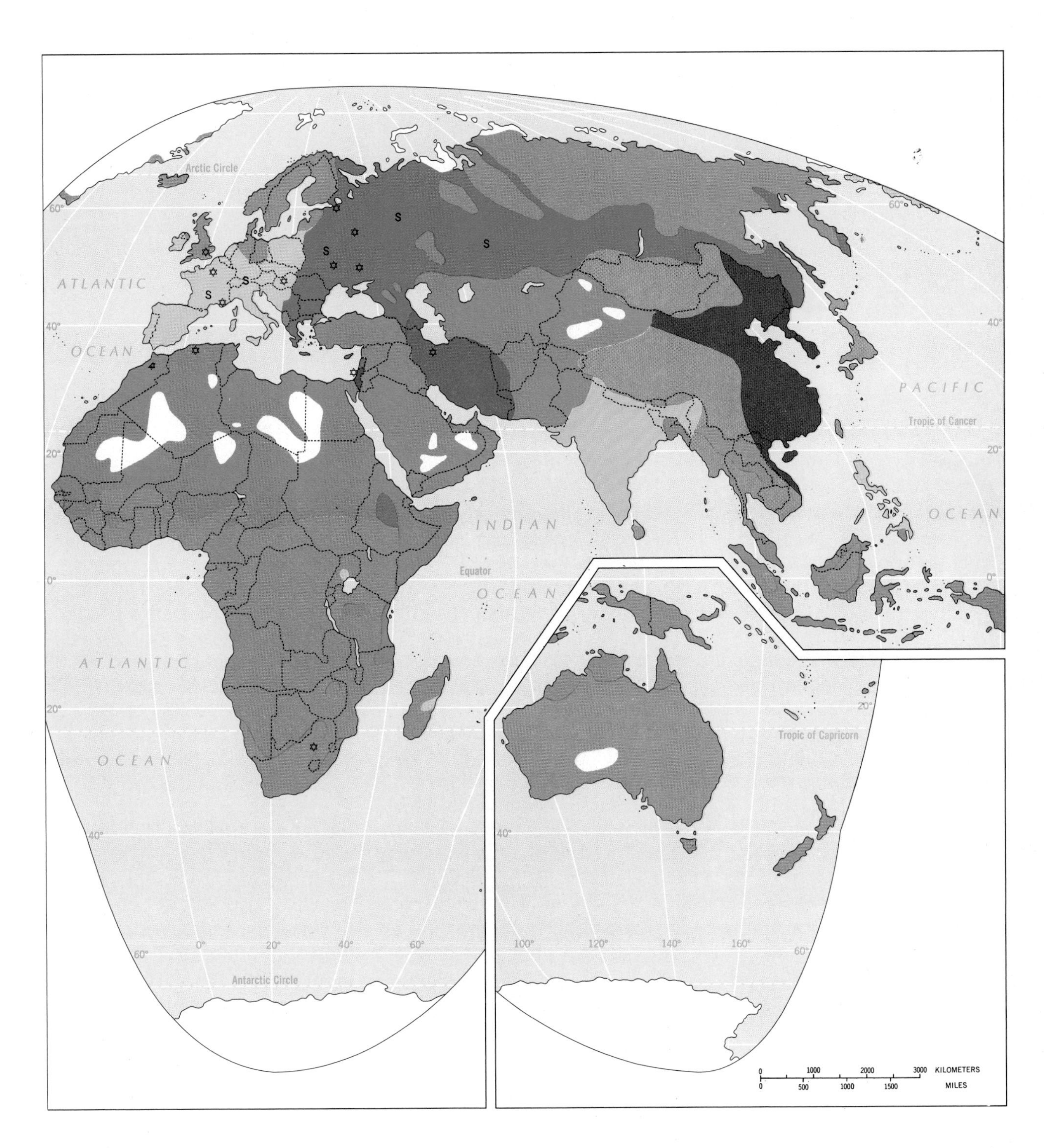

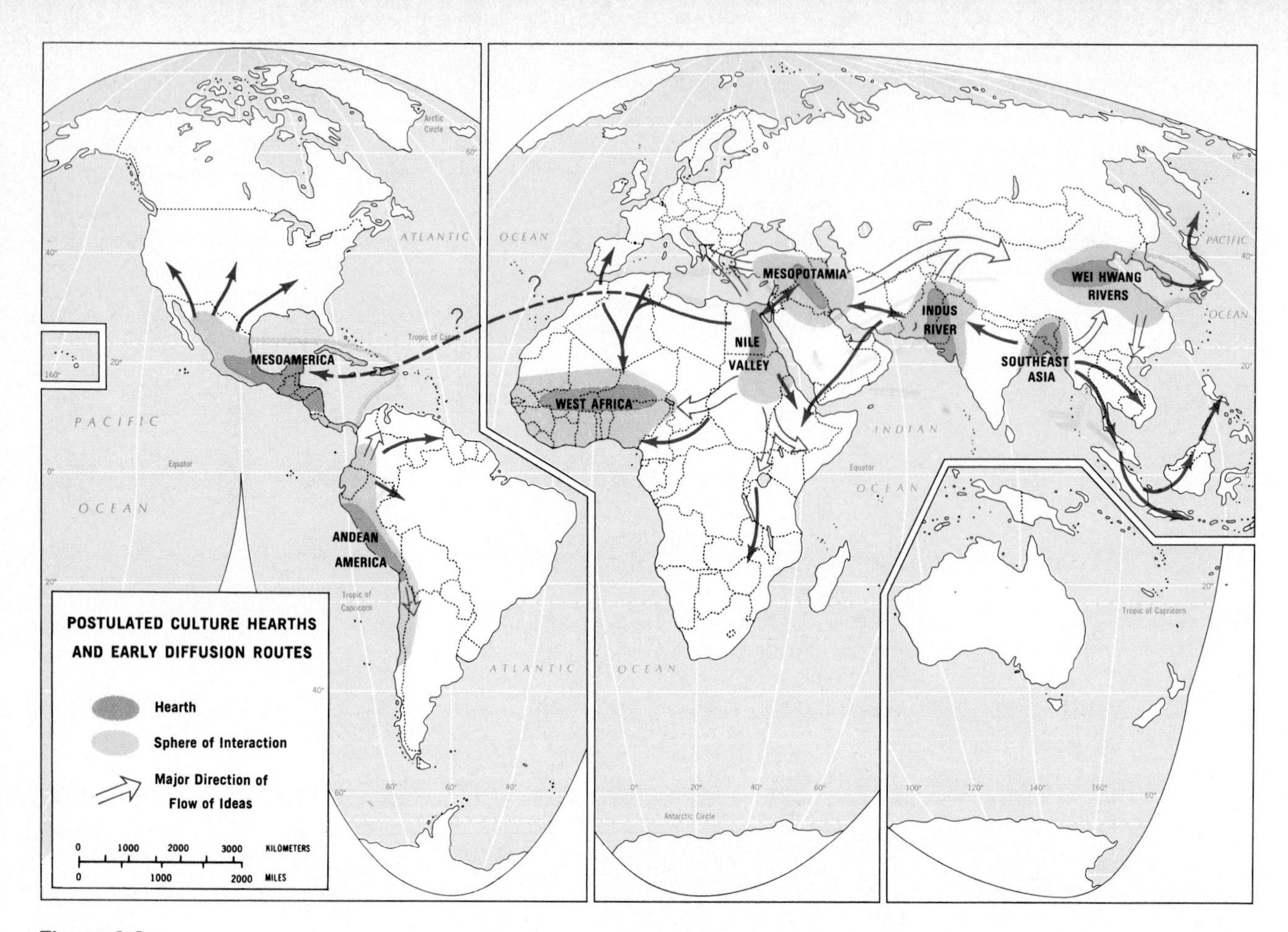

Figure 6-2

Nile Delta. Actually, the focus of the ancient culture hearth of Egypt lay above the delta and below the Nile's first cataract, and this segment of the Nile Valley lies surrounded by rather inhospitable country. The region was open to the Mediterranean, but otherwise it was quite inaccessible by overland contact. In contrast to Mesopotamia, which was something of a marchland, the Nile Valley was a natural fortress of sorts. There, the ancient Egyptians converted the security of their isolation into progress. The Nile waterway was the area's highway of trade and association, its lifeline. It also sustained agriculture by irrigation, and the cyclic regime of the Nile River was a great deal more predictable than that of the Tigris and Euphrates. By the time ancient Egypt finally began to fall victim to outside invaders, from about 1700 B.C. onward, a full-scale, urban civilization had emerged, whose permanence and continuity are reflected by the massive stone monuments its artist-engineers designed and created. The political practices and philosophies of Pharaoic Egypt were diffused far and wide, especially into Africa. Egypt survived as a political entity longer, perhaps, than any other state (China is its only rival); in the process, the state changed from a theocratic to a militaristic one, eventually to fall to colonial status, now to rise again as a modern nation. Ancient Egypt's armies were well disciplined and effective, and the cities were skillfully fortified. Where Egypt's armies ranged, peoples were subjugated and exploited for the state's benefit. Egypt far outlasted its Mesopotamian contemporaries.

Egypt's culture hearth lay on the western fringe of the great Middle Eastern source area. As Fig. 6-2 indicates, Mesopotamia was at the heart of this region, and to the east of the land of Babylon lay a third culture hearth: the civilization of the Indus valley, in what is today Pakistan. By modern criteria, this eastern region lies outside the realm under discussion, but in ancient times it had effective ties with lands to the west. It is believed that Mesopotamia innovations reached the Indus cit-

Scene of one of the worlds earliest experiments with irrigated farming: the banks of the Tigris River, Iraq. (Georg Gerster/Rapho-Photo Researchers)

ies of Harappa and Mohenjo-Daro. When the civilization that was centered on these cities reached its greatest extent, it penetrated the upper Ganges Valley of present-day India and extended as far south as the Kathiawar Peninsula.

In those distant times as today, the key to life in the Middle East was water. The Mesopotamians and Egyptians achieved breakthroughs in the control of river water for irrigation and in the domestication of several staple crops. Today we and most other peoples in the world continue to benefit from the endless list of contributions made by the ancient Mesopotamians and Egyptians: for example, the cereals we eat, such as wheat, rye, and barley; the vegetables we are used to, including peas and beans; the fruits (grapes, figs, olives, apples, peaches, and many more); and the domesticated farm animals—oxen, pigs, sheep, and goats. The horse, too, was first put to human service here. Of course the range of indirect contributions is indeed beyond measure. The ancient Mesopotamians made progress not only in irrigation and agriculture, but also in calendrics, mathematics, astronomy, government and administration, engineering, metallurgy, and a host of other fields. As time went on many of their innovations were adopted and then modified by other peoples and cultures in the Old and eventually even in the New World. Europe was the greatest beneficiary of the legacies of Mesopotamia and Egypt, whose achievements constitute the very foundations of "Western" civilization.

Decline and Rebirth

Today the great cities of the realm's culture hearths are archeological curiosities. In some instances new cities have been built on the sites of the old, but the great ancient cultural traditions of the Middle East went into a deep decline after many centuries of continu-

Where water becomes available, the "dry world" turns green. View over the fertile fields of the Nile Delta. (Robert Azzi/Magnum)

ity. It cannot escape our attention that a large number of the ruins of these ancient urban places are located in what is today desert. Assuming that they were not built in the middle of desert areas, it is tempting to conclude that climatic change (associated with shifting environmental zones in the wake of the last Pleistocene retreat) destroyed the old civilizations. Indeed, some cultural geographers have suggested that the momentous innovations in agricultural planning and in the technology of irrigation may have been made in response to changing environmental conditions, as the river-basin communities tried to survive. The scenario is not difficult to imagine: as outlying areas of the country began to fall dry and farmlands were destroyed, people congregated in the river valleys, already crowded—and every effort was made to increase the productivity of lands that could still be watered. Eventually overpopulation, destruction of the watershed, and perhaps reduced rainfall in the rivers' headwater area combined to deal the final blow. Towns were abandoned to the encroaching desert. Irrigation canals filled with drifting sand. Farmlands dried up. Those who could, migrated to areas reputed to remain productive. Others stayed, the numbers dwindling, reduced to subsistence.

As old societies disintegrated, new power emerged elsewhere. First the Persians, then the Greeks, and later the Romans imposed their imperial designs on the tenuous lands and disconnected peoples of the Middle East. Roman technicians converted North Africa's farmlands into irrigated plantations whose products went by the boatload to Roman shores. Thousands of people were carried in bondage to the cities of the new conquerors. Egypt was a colony; the Arab settlements on the Arabian Peninsula lay more remote, and somewhat more secure in their isolation.

In one of these relatively remote places on the Arabian Peninsula occurred an event that was to change the course of history and affect the destinies of people all over the world. In Mecca, a town 72 kilometers (45 miles) from the Red Sea coast and positioned in the Jabal Mountains, a man named Mohammed (Muhammad) in about A.D. 613 began to receive the truth from Allah (God) in a series of revelations. Mohammed (A.D. 571–632) already was in his early forties when

Islam's impact spread far beyond the Arab World. Millions were converted in Black Africa and in South and Southeast Asia. This is the mosque at Kano, Nigeria, on a Friday afternoon. (Marc and Evelyn Bernheim/Woodfin Camp)

these revelations began, and he had barely 20 years to live. But in those two decades commenced the transformation of the Arab world.

Convinced, after some initial self-doubt, that he was indeed chosen to be a prophet, Mohammed committed his life to the fulfillment of the divine commands he had received. The Arab world was in social and cultural disarray, and in the feudal chaos Mohammed soon attracted enemies who feared his new personal power. He fled Mecca for the safer haven of Medina (but Mecca's place as a holy center was soon assured), and from there he continued his work.

The precepts of Islam constituted, in many ways, a revision and embellishment of Judaic and Christian beliefs and traditions. There is but one god, who occasionally reveals himself to prophets (Islam acknowledges that Jesus was such a prophet). What is earthly and

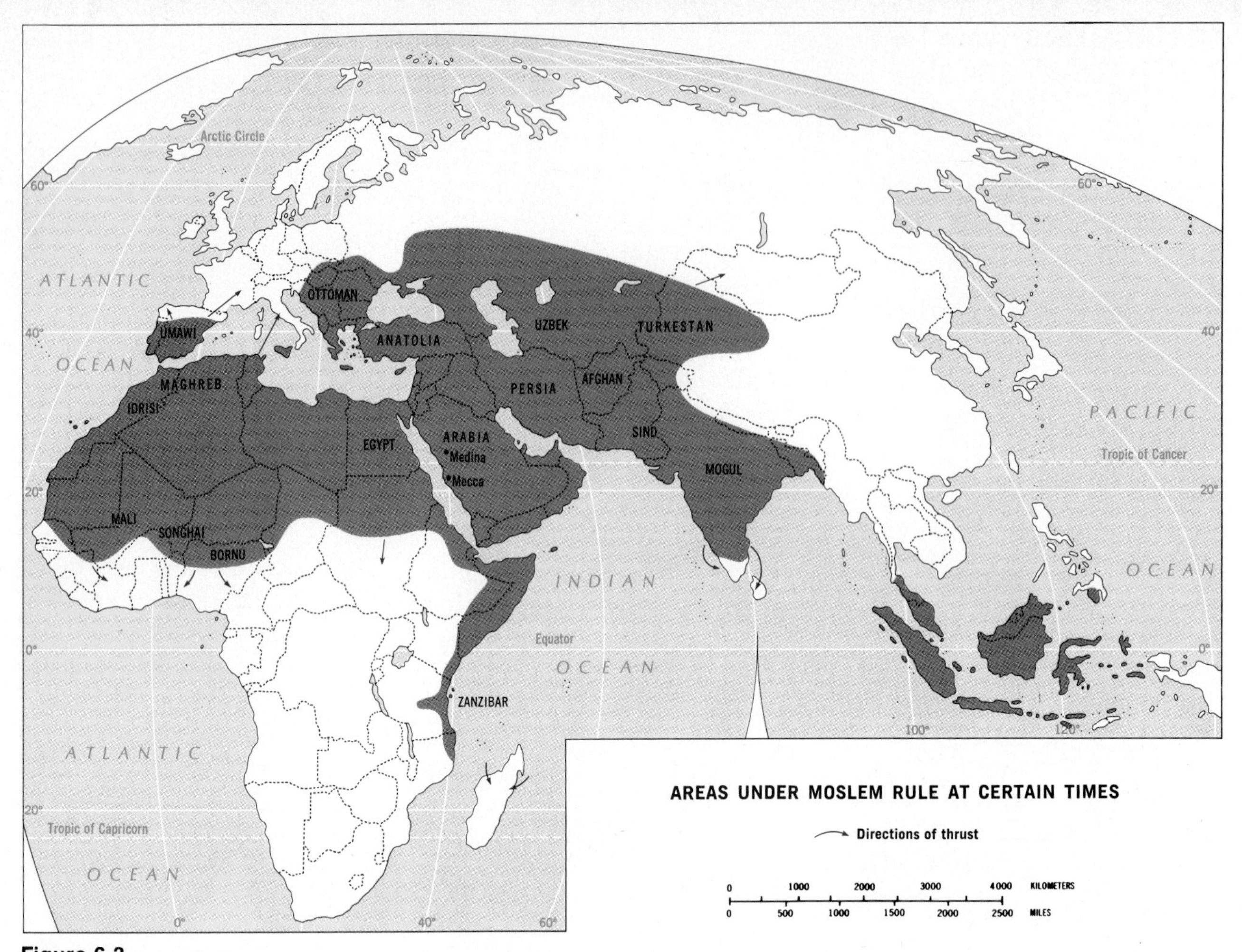

Figure 6-3

worldly is profane; only Allah is pure. Allah's will is absolute; he is omnipotent and omniscient. All humans live in a world created for their use, but only to await a final judgment day.

Islam brought to the Arab world not only a unifying religious faith, but also a whole new set of values, a new way of life, a new individual and collective dignity. Apart from dictating observance of the "five pillars" of Islam (repeated expressions of the basic creed, frequent prayer, a month of daytime fasting, almsgiving, and at least one pilgrimage to Mecca), the faith prescribed and proscribed in other spheres of life as well. Alcohol, smoking, and gambling were forbidden. Polygamy was tolerated, although the virtues of monogamy were acknowledged. Mosques appeared in Arab settlements, not only for the Friday prayer but also as social gathering places to bring communities closer together. Mecca became the spiritual center for a divided, far-flung people for whom a joint focus was something new.

The stimulus given by Mohammed, spiritual as well as political, was such that the Arab world was mobilized almost overnight. The prophet died in 632, but his faith and fame continued to spread like wildfire. Arab armies formed, they invaded and conquered, and Islam was carried throughout North Africa. By the early ninth century, the Moslem world included emirates or kingdoms extending from Egypt to Morocco, a caliphate occupying most of Spain and Portugal, and a unified realm encompassing Arabia, the Middle East, Iran, and most of what is today Pakistan (Fig. 6-3). Moslem influences had penetrated France, attacked Italy, and penetrated what is today Soviet Asia as far as the Aral Sea. Ultimately the Arab empire extended from Morocco to India and from Turkey to Ethiopia. The original capital was at Medina in Arabia, but in response to these strategic successes it was moved, first to Damascus and then to Baghdad. In the fields of architecture, mathematics, and science the Arabs far overshadowed their European contemporaries, and they established institutions of higher learning in many cities, including Baghdad, Cairo, and Toledo (Spain). The faith had spawned a culture; it is still at the heart of that culture today.

Diffusion

The spread of the Islamic faith throughout the realm (and beyond its borders as well) occurred in waves radiating from Medina and the holy city of Mecca. Islam went by camel caravan and by victorious army, it was carried by pilgrim and boatsman, scholar and sultan. Its dissemination, through so vast and discontinuous a realm, is an example of the process of diffusion.

Arab and Moslem horsemen spread the faith over a vast region. These dignitaries and their elaborately decorated horses have come from more than 150 kilometers (100 miles) away to pay their respects to the Emir of Kano. (Thomas D.W. Friedmann/Photo Researchers)

The worldwide diffusion of Islam is still going on. In the United States it is emerging in the religious and nationalist movement commonly called the Black Muslims but officially known as the Nation of Islam. There are Islamic communities in South Africa, in the Philippines, and other far-flung places. This is not the first time that the Middle East generated an idea that affected much of the world. Agricultural methods, metallurgical techniques, architectural styles, and countless other innovations made in this realm have, in the course of history, been taken over by other societies. These inventions reached those distant societies by the same process.

Diffusion processes are of interest to geographers for several reasons. First, like human movement and migration, these are spatial processes in which information flow plays a large role. The diffusion of Islam to West Africa's savanna states, the approximate dates of its arrival, the particular segment of African society that was converted—all these dimensions help us reconstruct trans-Saharan connections between black West Africa and the Arab North. In the case of Islam it is not difficult to identify the source of the innovation, but in other instances an understanding of the nature of the process can help us trace back toward a postulated source. Second, a knowledge of diffusion processes can have a practical side. The spread of a disease (a severe influenza for example) is a diffusion process; if we know how it happens it may be possible to slow it down. On the other hand, if the government wants to disseminate something new (fluoridation of drinking water, for example), it is helpful to be able to predict its adoption, potential areas of resistance as well as places of quick implementation.

Geographers view the diffusion process under two broad rubrics. The spread of Islam is a case, primarily, of *expansion diffusion,* because the faith remained strong at its core and disseminated from there throughout the realm; if you made a series of maps of the converted at intervals of 50 years after 620 the maps would incorporate a larger region every time. Another kind of diffusion process is *relocation diffusion,* whereby whatever is being diffused enters and then leaves the area and populations it successively affects. Thus a disease might affect a particular area, where it affects a large percentage of the population. By the time it has diffused to a new population cluster, it had already run its course in the first area and no longer prevails there. Again, if you were to make a series of maps of this disease, say at bimonthly intervals, its area of prevalence would be different every time, and there would be no overlap. Of course, a combination of the two processes can occur. A Swedish geographer, T. Hägerstrand, studied the diffusion of certain innovations in his home country and developed a basic diffusion model, much in the way Von Thünen, Weber, Christaller, and others created their geographic models. Hägerstrand's model was based on an elaborate set of assumptions (a familiar feature of models). He concluded that four stages mark the diffusion process: a *primary* stage (when the innovation is at its source and being adopted there), a *diffusion* stage (when the innovation radiates rapidly outward, generating new centers of dispersal in remote areas), a *condensing* stage (in which remaining areas are penetrated), and a *saturation* stage (marking the slowdown and end of the diffusion process). Other geographers have confirmed that this wavelike progression characterizes diffusion—Hägerstrand actually called them innovation-waves—and while the Hägerstrand model shows them as curves on graphs, they can also be shown as lines on maps. The lines on our map showing the expansion of Islam represent the great wave of the diffusion stage, when places like Baghdad, Damascus, and Cairo became regional centers of dissemination.

Boundaries and Barriers

If Islam constituted such a strong unifying force, and if Middle Eastern armies could penetrate Europe, Asia, and Africa south of the Sahara—why is the Middle East today a realm of boundaries and barriers, tension and conflict, hostility and disunity?

In part the answer lies in the rise of another religion, one that emerged even earlier than Islam: Christianity. Until the challenge of Islam, the Christian faith (which began as a movement within another Middle Eastern religion, Judaism) had won acceptance in the Roman Empire, could afford the luxury of an east-west ideological split, and diffused to the tribes of Europe north of Rome's boundaries. Then the Arabs rode on Mohammed's teachings into Iberia and threatened Roman Italy itself. Europeans mobilized to meet the challenge, and as soon as Arab power showed signs of weakening (it happened as early as the eleventh century), the first crusade by Christians to the eastern Mediterranean took place. The aftermath of the contest left parts of the area converted to Christianity; still today, nearly half of the population of Lebanon adheres to a Christian faith. Eight centuries later Judaism regained territorial expression as the state of Israel. The city of Jerusalem is a holy place for Jews, Moslems, and Christians alike. Fragmented and embattled, it personifies the region's divisions.

But the realm is vast, and adherents to faiths other than Islam form only small minorities. Most of the boundaries on the map of the Middle East today (Fig. 6-4) were established after the last of the great Islamic empires collapsed. The Moslems' last hurrah in Europe came when the Ottoman Empire, centered on the Turkish city of Constantinople (now Istanbul), extended its power over the Balkans and beyond. Greece, Bulgaria, Albania, Yugoslavia, Hungary, Romania, and the Black Sea area fell to the sultans, whose empire also extended far westward along Mediterranean shores and eastward in Persia. As the twentieth century opened, the Ottoman Empire already was decaying, and when the Turks chose the (losing) German-Austrian side in World War I, its fate was sealed.

From the fragments of the Ottoman Empire, the political divisions of the Middle East emerged. Turkey survived as a discrete entity, but other areas were assigned to the victorious European powers: Syria and Lebanon to France (and the French already had acquired colonies in Algeria, Tunisia, and Morocco); Palestine to Britain (the British also administered Iraq, Transjordan, Sudan, and Egypt). The Italians had established a sphere of influence in Libya and along the Red Sea coast in Eritrea, and they also controlled part of Somalia in Africa's "Horn." A combination of European imperialism and the final breakdown of Moslem power produced a mosaic of political boundaries to suit colonial powers, not Arab interests.

As Figure I-9 (page 27) underscores, this realm's population is clustered, fragmented, strung out in narrow coastal belts and confined, crowded valleys and oases. It is useful to compare Fig. 6-4 and Fig. I-9, for it is clear that thousands of kilometers of political boundaries in North Africa and Southwest Asia lie in virtually uninhabited terrain. Geographers classify boundaries on various grounds, one of which is morphological: does the boundary coincide with a cultural break or transition in the landscape? Does it coincide with a physical feature such as a river or the crest of a mountain range? Is the boundary visibly straight? As our maps show, North Africa has a framework of straightline boundaries, long segments of which lie in the nearly empty Sahara. These are *geometric* boundaries (Egypt in Africa is enclosed by such borders), drawn to coincide with lines of latitude or longitude (or at an angle, from point to point), mostly through territory over which, so it seemed, no one would ever quarrel. That was before some of the region's oil was discovered! Other boundaries, such as the Sudan-Ethiopia border along the foot of the Ethiopian Highlands, and the Israel-Jordan boundary along the Jordan River, coincide with physical features and are *physiographic-political* boundaries. Still another group of boundaries marks changes in the cultural landscape. Essentially the borders between Israel and its Arab neighbors are cultural dividing lines, notwithstanding the large Arab minority within Israel's territory. The boundary between Arab Iraq and Persian Iran may be seen on language maps as well as political maps. These are *anthropogeographic* boundaries.

The North African-Southeast Asian realm may also need some boundaries it does not have. Various proposals to create a state for Palestinians displaced by the creation of Israel have been made; partition has been discussed as a solution for strife-torn Lebanon; Moslem Eritreans and Christian Ethiopians are locked in a conflict that may be solved only by the resurrection of an old colonial boundary.

The United Arab Emirates

Formerly known as British-administered "Trucial Oman" and "The Seven Sheikdoms," this country today is a union of seven emirates on the Persian Gulf coast of the Arabian Peninsula: Abu Dhabi, Dubai, Ajman, Sharjah, Umm al-Qaiwan, Ras al-Khaimah, and Fujairah. By far the largest is Abu Dhabi, with nearly 68,000 square kilometers (26,000 square miles) and a 1980 population of about 120,000. The most populous is Dubai, with 150,000 people but under 4000 square kilometers (1500 square miles). The seven Emirates' total area is only 84,000 square kilometers (32,000 square miles); total population (permanent and transient) is about a half million. Oil output is enormous, in 1980 averaging about $15,000 per person.

Arabian Bonanza

The political fragmentation of this realm into two dozen countries has major ramifications in relation to the distribution of its leading export commodity: oil. Fig. 6-5 displays recent estimates of the world's fuel reserves. A majority of the realm's countries possess substantial oil and natural gas reserves, notably the states and emitrates on the Arabian Peninsula, Saudi Arabia, and Iran (but also Libya and Algeria in North Africa). As new discoveries are made, estimates change, but in combination the states of North Africa and Southwest Asia probably contain over 60 percent of the world's petroleum reserves. Oil has become the realm's most valuable export commodity, providing even small countries such as Kuwait with huge revenues and giving the region a new form of world influence. As the bar graph on Fig. 6-5 indicates, the great consumers of energy are not the countries of the Middle East but the nations of the developed world. Production in the United States and several European countries continues to lag behind consumption, and Western dependence on Middle Eastern oil was underscored by the energy crisis of the early 1970s.

If the oil-exporting states of the North African-Southwest Asian realm were united ideologically, the advantages accruing from their oil wealth would be far greater, and the position of the energy-dependent Western states would be far worse, than they are. But the area's old, traditional disunity has once again worked to its disadvantage. Eight of the thirteen OPEC (Organization of Petroleum Exporting Countries) states lie in this realm, but their efforts to agree on petroleum policy and to develop an effective cartel have been only partly successful. The overthrow of the government of Shah Mohammed Reza Pahlavi in Iran did have the effect of reducing, to some extent, the width of the OPEC policy gap. While in power, the Shah was among the strongest proponents of ever-higher oil prices—but he also

Figure 6-4

permitted the sale of oil to South Africa and Israel. The revolutionary government ended those sales, to which other OPEC members in the realm had always objected.

What has been the impact of the realm's "black gold" on the countries and societies with the good fortune to possess substantial oil reserves? In small states such as Kuwait and the tradition-bound United Arab Emirates of Abu Dhabi and Dubai, a true transformation has taken (and continues to take) place. Kuwait has made its appearance as the country with the world's highest per-capita income. But Kuwait's population is not much over 1 million, and Abu Dhabi and Dubai have just a few hundred thousand inhabitants. In countries with larger populations, the picture is quite

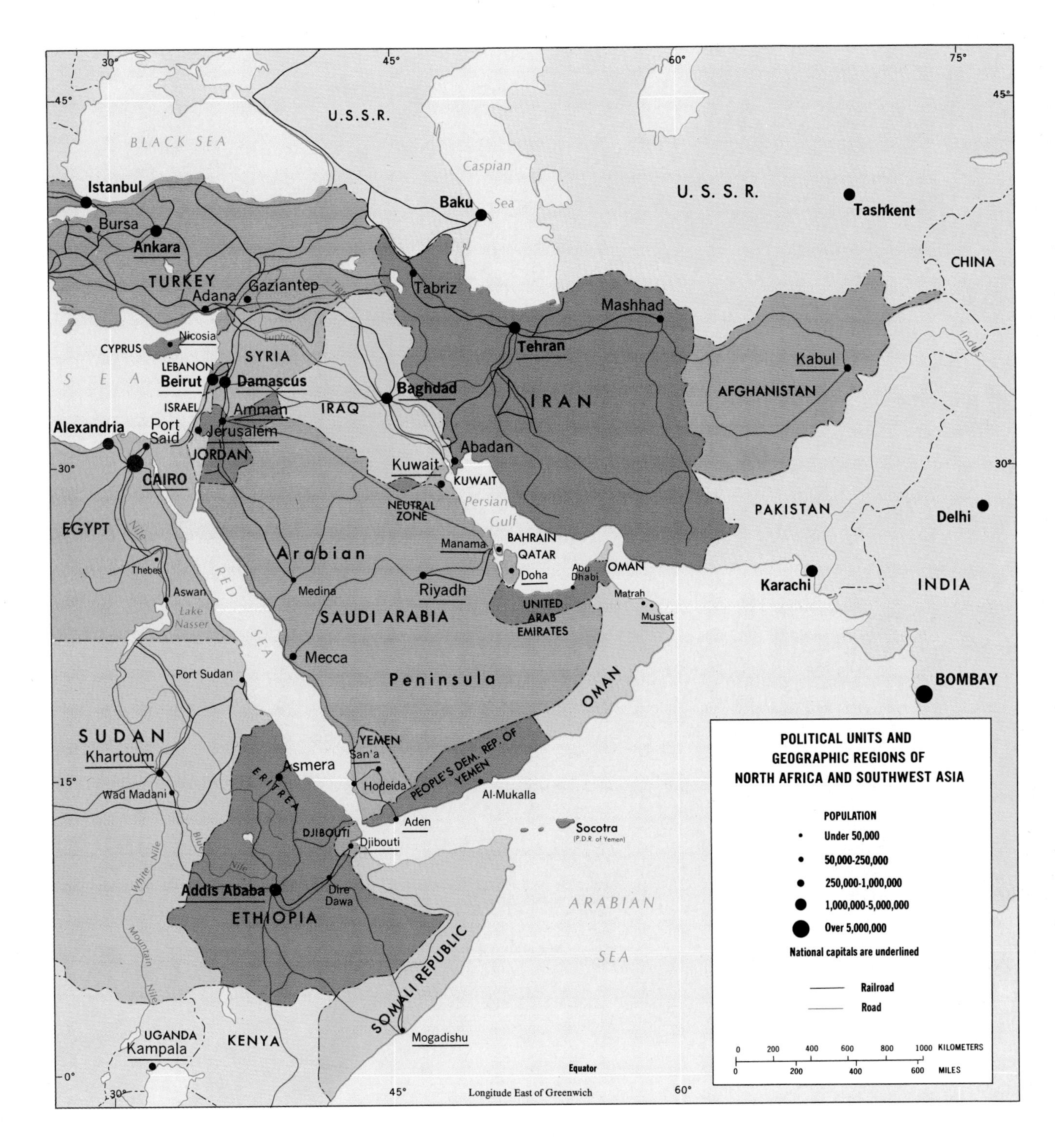

different. In Iran, with a population of 37 million, the Shah's government poured huge amounts of its enormous petroleum-derived income into a major military complex. Huge investments also allegedly disappeared into the personal coffers of the Shah himself and into the accounts of those in favor. The funds that remained were nevertheless sufficient to support a number of major development programs, including an export-oriented group of industries, a series of reforms in agriculture, and an attack on illiteracy. Modernization during this period made its mark on the capital, Tehran, and on other cities (Abadan has long been the oil refinery center and export point). But in the countryside there is much less evidence of change, especially in the east and the northwest. By some counts Iran in the last year of the Shah's government had about 60,000 villages; of these, only some 3000 had running water provided. In the remainder, the old system—wells and pumps—still prevailed. In rural Iran, the presence of the oil fields had changed few lives.

Neither has Iraq (with its 13 million inhabitants) found its petroleum income to be a shortcut to general national welfare. To modernize tradition-bound agriculture is an enormously costly proposition, and Iraq has lacked what Iran's authoritarian government did achieve until 1979: political continuity and stability. Internal political struggles, corruption, and factionalism, together with urban-oriented investment policies that persuaded hundreds of thousands of villagers to leave the countryside and come to Baghdad, deprived Iraq of much of the development its oil revenues should have brought.

Fig. 6-6 shows a system of pipelines and export locations that very much resembles the exploitative railroads in a mineral-rich colony. Such a pattern spells disadvantage for the exporter, whether colony or independent country. In North Africa, Libya again reflects the advantages of scale: its mere 2.8 million people have felt the impact of their oil bonanza far

Figure 6-5

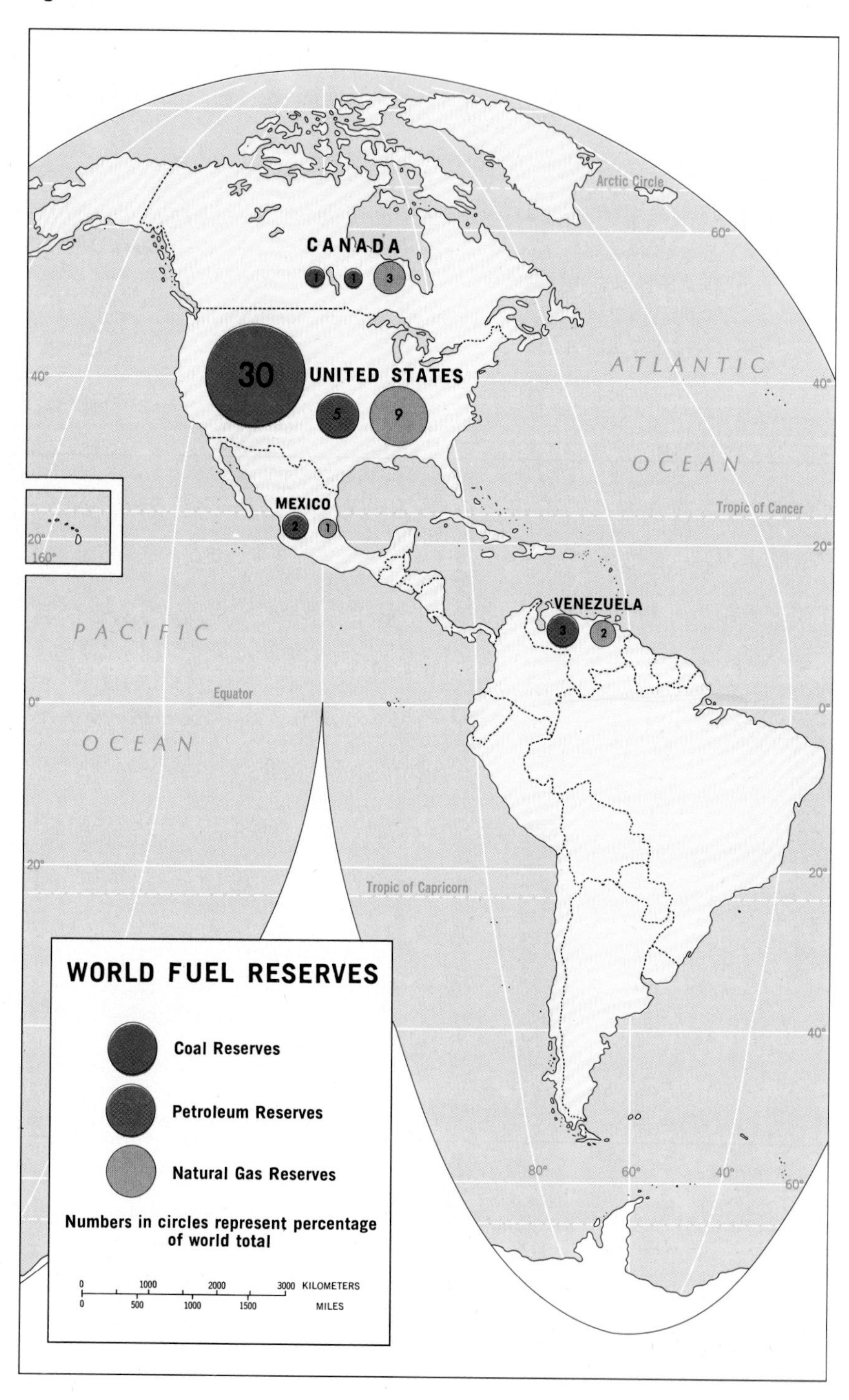

more strongly than Algeria's 20 million, and the Libyan government has used its oil revenues to gain a measure of political influence in the region far beyond its dimensions.

It is evident that the impact of the realm's oil wealth is more evident in the oil-rich small countries than in the more populous states. Furthermore, the oil revenues have had the effect of intensifying urban dominance and regional disparities even where, as in Saudi Arabia, a conservative government implements national development programs funded by oil revenues and designed to limit the abrasive effects of modernization.

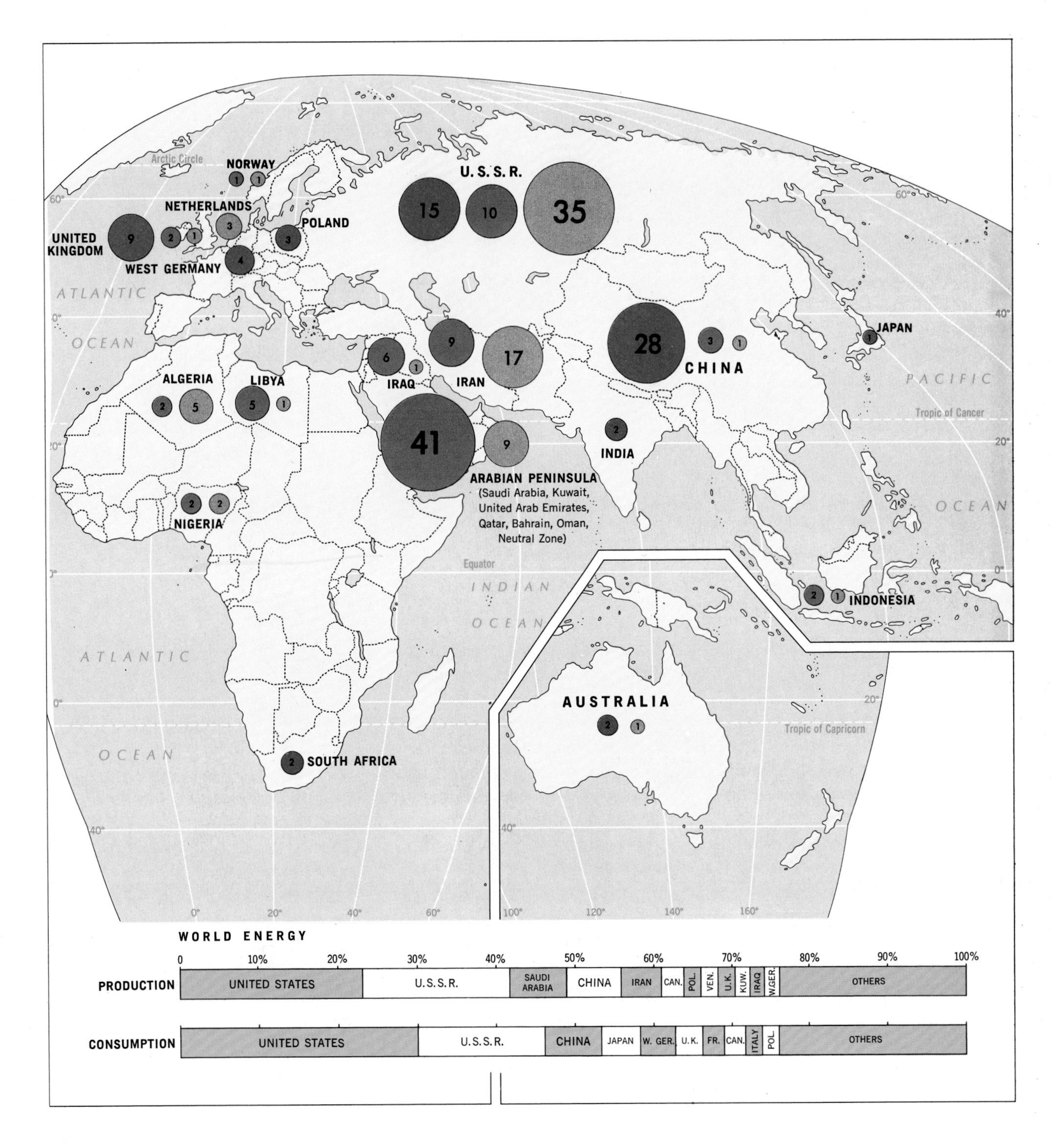

Oil-derived wealth has transformed parts of the Arab World, and it has brought the twentieth century into sometimes rude contact with local traditions. High-rise buildings and modern automobiles vie for space with mud-walled houses and donkey carts. In the desert, refineries shimmer in sun-drenched expanses that until recently were disturbed only by a passing camel train. In this photograph the Petromin Refinery of Saudi Arabia produces petroleum destined for Japan; Japanese technicians are building Saudi Arabia's petrochemical industry. (Alain Nogues/Sygma)

Regions and States

This vast, sprawling realm is not easily divided into geographic regions. Not only are population clusters widely scattered, but cultural transitions—internal as well as external—make it difficult to discern a regional framework. If we constructed a series of maps based on environment (the "dry" factor), religion, ethnic units, and languages, there would be few areas of agreement. Certainly the regionalization that follows is debatable, but it highlights the major geographic components of the realm:

(1) Egypt and the Nile Basin. This region constitutes, in many ways, the heart of the realm as a whole. Egypt is one of the realm's two most populous countries (Turkey exceeds it slightly). It is the historic focus of the area, a major political force. It shares with its southern neighbor, Sudan, the waters of the Nile River.

(2) The Maghreb and the West. The North African Maghreb and the areas that border it also form a region, consisting of Algeria, Tunisia, and Morocco at the center and Libya, Chad, Niger, Mali, and Mauritania along the periphery. The last four of these countries lie astride or adjacent to the African transition zone, where the Arab-Islamic realm merges into Black Africa.

(3) The Middle East. This region includes Israel, Jordan, Lebanon, Syria, and Iraq. In ef-

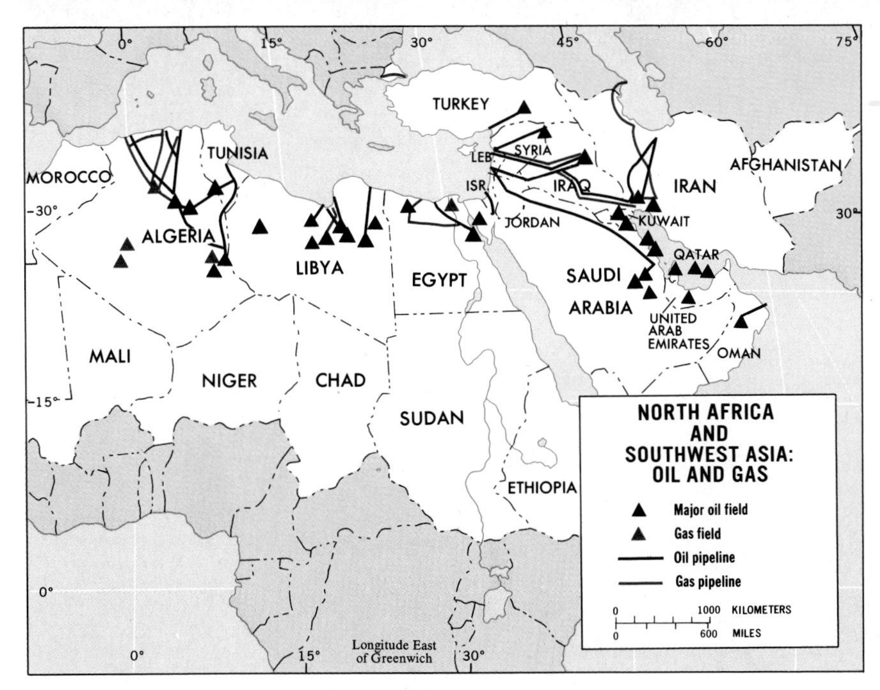

Figure 6-6

fect it is a zone of countries extending from the Mediterranean Sea to the Persian Gulf.

(4) The Arabian Peninsula. Dominated by the giant territory of Saudi Arabia, the Arabian Peninsula also includes the United Arab Emirates, Kuwait, and North and South Yemen. Here lie the source and focus of Islam, the holy city of Mecca; here, too, lie some of the realm's great oil reserves.

(5) The North. Across the northern tier of the realm lie four countries each of which, in one way or another, constitutes an exception to the norms that prevail here. Turkey, Cyprus, Iran, and Afghanistan are not Arab countries; neither the Turks nor the Iranians are Arabs. The great majority of Iranians adhere to a minority Islamic faith, the Shiah, and Shi'ite Moslems do not share certain important values with their Sunni contemporaries.

(6) The African Transition Zone. From Mauritania in West Africa to Ethiopia and Somalia in the east, the realm dominated by Islamic culture interdigitates with the northern region of Black Africa. No sharp dividing lines can be drawn; this is a wide zone of transition south of the Maghreb, Libya, and Egypt. As a result, this is the least sharply defined among the six regions of the North African-Southwest Asian realm.

Egypt and its Neighbors

Egypt occupies, quite literally, Africa's northeastern corner: geometric boundaries separate its 1 million square kilometers (387,000 square miles) from Libya to the west and Sudan to the south. Alone among states on the African continent, Egypt extends into Asia through its foothold on the Sinai Peninsula—a hold that has, in modern times, proved rather tenuous. Egypt lost the entire Sinai region during the 1967 War with Israel, and then regained a toehold across the Suez Canal in the conflict of 1973. The years that followed saw intense diplomatic efforts to persuade Israel to yield more of the occupied Sinai to Egypt, and in 1979 the terms of a peace agreement between the two states included a staged return of Egyptian lands in the Sinai.

Egypt now faces an adversary capable of striking into its very heartland in a matter of minutes, a totally new condition for a country protected for centuries (indeed, millennia) by desert and distance from its enemies. Herodotus described Egypt as the gift of the Nile, but Egypt also was a product of natural protection. The middle and lower Nile lie enclosed by inhospitable country, open to the Mediterranean Sea but otherwise rather inaccessible to overland contact. To the south, the Nile is interrupted by a series of cataracts that begin near the present boundary of Egypt and the Sudan. To the east, the Sinai Peninsula never afforded an easy crossing, and the southwestern arm of the ancient Fertile Crescent lay a considerable distance beyond. To the west, there is the endless Sahara Desert. The ancient Egyptians had a natural fortress, and in their comparative isolation they converted security into progress. Internally, the Nile was then what it remains to this day—the country's highway of trade and association, its lifeline. Externally, the Egypt of the pharaohs had to trade, for there was little wood and metal, but most of

Supranationalism in the Middle East

Notwithstanding their divisions and quarrels, countries of North Africa and Southwest Asia have forged some effective international organizations. One of these is the Arab League, founded in 1945 "to strengthen relationships between members and to promote Arab aspirations." Headquartered in Cairo, the Arab League today counts 20 members:

Morocco	Oman
Algeria	Qatar
Tunisia	Bahrain
Libya	Sudan
Egypt	Somolia
Saudi Arabia	Mauritania
Kuwait	Lebanon
United Arab Emirates	Syria
Yemen (Aden)	Jordan
Yemen (San'a)	Iraq

Member states sometimes have been in conflict; since 1976 Morocco and Algeria have disputed the future of former Spanish Sahara (following partition between Morocco and Mauritania, all of Spanish Sahara became Moroccan area) and Syria and Iraq have argued over the sharing of the waters of the Euphrates River. The League has been effective in the settlement of several of these problems; in 1976 a combined Arab League armed force entered Lebanon to help end the civil war there. In 1979 the League moved its headquarters from Cairo to Tunis, and members voted to suspend Egypt, because of Egypt's peace agreement with Israel.

In another international organization, the OPEC cartel, states from this area have taken a leading role. In 1980, membership in OPEC inclued:

Algeria	Ecuador
Iran	Gabon
Kuwait	Indonesia
Qatar	Nigeria
Saudi Arabia	Venezuela
United Arab Emirates	
Libya	
Iraq	

this trade was left to Phoenicians and Greeks. Egypt's cultural landscape still carries the record of antiquity's accomplishments in the great pyramids and stone sculptures that bear witness to the rise of a culture hearth 5000 years ago.

The Nile. Perhaps 95 percent of Egypt's 42 million people live within a dozen miles of the Nile River or one of its distributaries. The great river is the product of headwaters that rise in two different parts of Africa: Ethiopia, where the Blue Nile originates, and the lakes region of East Africa, source of the White Nile. Before dams were constructed across the river, the Nile's dual origins assured a fairly regular natural flow of water, making the annual floods predictable both in terms of timing and intensity. The Nile normally is at its lowest level in April and May, but rises during July, August, and September to its flood stage at Cairo in October, when it may be more than 7 meters (23 feet) above its low stage. After a rapid fall during November and December, the river declines gradually until its May minimum. In ancient times this regime made possible the invention of irrigation, for the mud left behind by the Nile floods was manifestly cultivable and fertile. Eventually a system of cultivation known as *basin irrigation* developed, whereby fields along the low banks of the Nile were partitioned off by earth ridges into a large number of artificial basins. The mud-rich river waters would pour into these basins during flood time, and then the exits would be closed, so that the water would stand still, depositing its fertile load of alluvium. Then, after six to eight weeks, the exit sluices were opened and the water drained away, leaving the rejuvenated soil ready for sowing. This method, while revolutionary in ancient times, had disadvantages: the susceptible lands must lie near or below the flood level, only one crop could be grown annually, and if the floods were less intense (as they were in some years), some basins remained unirrigated because flood level did not reach them. Still, traditional basin irrigation prevailed all along the Nile River for thousands of years; not until the late nineteenth century did more modern methods develop. Even as late as 1970, as much as one-tenth of all irrigation in Egypt was basin irrigation, most of it practiced, still, in Upper Egypt (the region nearest the Sudan border).

The construction of dams, begun during the nineteenth century, made possible the *perennial irrigation* of Egypt's farmlands. By building a series of barrages (with locks for navigation) across the river, engineers were able to control the floods, raise the Nile's water level, and free the farmers from their dependence upon the seasonality of the river's natural regime. Not only did the country's cultivable area expand substantially but it also became possible to grow more than one crop per year. By the early 1980s all farmland in Egypt will be under perennial irrigation, a transformation that has taken place within one century.

The greatest of all Nile dam projects, the Aswan High Dam, was begun in 1958 and completed in 1971. The High Dam is located some 1000 kilometers (600 miles) upstream from Cairo, in a comparatively narrow, granite-sided section of the valley. The dam wall, 110 meters (364 feet) high, creates Lake Nasser, one of the largest artificial lakes in the world (Fig. 6-7). The reservoir inundates well over 480 kilometers (300 miles) of the Nile's valley not only in Upper Egypt but in the Sudan as well, and the cooperation of Sudan was required since some 50,000 Sudanese had to be resettled. The impact of the High Dam on Egypt's cultivable area was enormous: before construction the Nile's waters could irrigate some 2.53 million hectares (6.25 million acres) of farmland. To this, the Aswan High Dam has added 550,000 hectares. In addition, nearly 400,000 hectares of farmland in Middle and Upper Egypt that were still under basin irrigation could now be converted to perennial irrigation, resulting in increased crop yields. Furthermore, the new dam supplies Egypt with about 50 percent of its energy requirements.

Egypt has often been described as one elongated oasis, and the map—almost any map of the country's human geography—confirms the appropriateness of that description. In Upper and Middle Egypt the strip of abundant, intense green, 5 to 25 kilometers (3 to 5 miles) wide, lies in stark contrast to the barren, harsh, dry desert immediately adjacent, a reminder of what Egypt would be without its lifeline of water. But below Cairo, the great river fans out across its wide delta, 160 kilometers (100 miles) in length and 250 kilometers (155 miles) wide along the Mediterranean coast between Alexandria and Port Said. In ancient times the river's waters reached the sea via several channels, and the delta was flood-prone, inhospitable country; ancient Egypt was Middle and Upper Egypt, the Egypt of the Nile Valley. But today the delta waters are diverted through two controlled channels, the Rosetta and Damietta distributaries. The delta contains twice as much cultivable land as lies in Middle and Upper Egypt, and it has some of Africa's most fertile soils. In 1977, nearly half of Egypt's population (19 out of 42 million) inhabited the delta region.

The increasing use of Nile waters upstream and the diminished flow of the river (and its vital supply of silt) now pose a threat to all the progress that has been made in the delta. There is a danger that brackish water will invade the channels from the north, seep into the soil, and reduce large areas, once again, to their inhospitable and unproductive state. The consequences of such a sequence of events—a real possibility—would, of course, be calamitous.

A Pivotal Location. Egypt has changed substantially in modern times; it is a more populous, more highly urbanized country today than ever before. But Egypt's farmers, the *fellaheen,* still struggle to make their living off the land, as did the peasants in the Egypt of 5000 years ago. In a land that at one time was the source of countless innovations and the testing ground for many others, the tools of the farmer often are as old as any still in use: the handhoe; the wooden, buffalo-drawn plow; the sickle. Water is still drawn from wells by the wheel and bucket; the Archimedean screw can still be seen in service. The peasant still lives in a small mud dwelling that would look very familiar to the farmer of many centuries ago; neither would that distant ancestor be surprised at the poverty, the diseases, the high rate of death among young children, the lack of tangible change in the countryside. Notwithstanding the Nile dams and irrigation projects, Egypt's available farmland per capita has declined steadily during the past two centuries, and rapid population increase keeps nullifying gains in crop harvests.

And yet, of all the countries in the region loosely termed the Middle East, Egypt is not only the largest in terms of population but also, by most measures, the most influential. Its rival, Turkey, has in recent decades

Figure 6-7

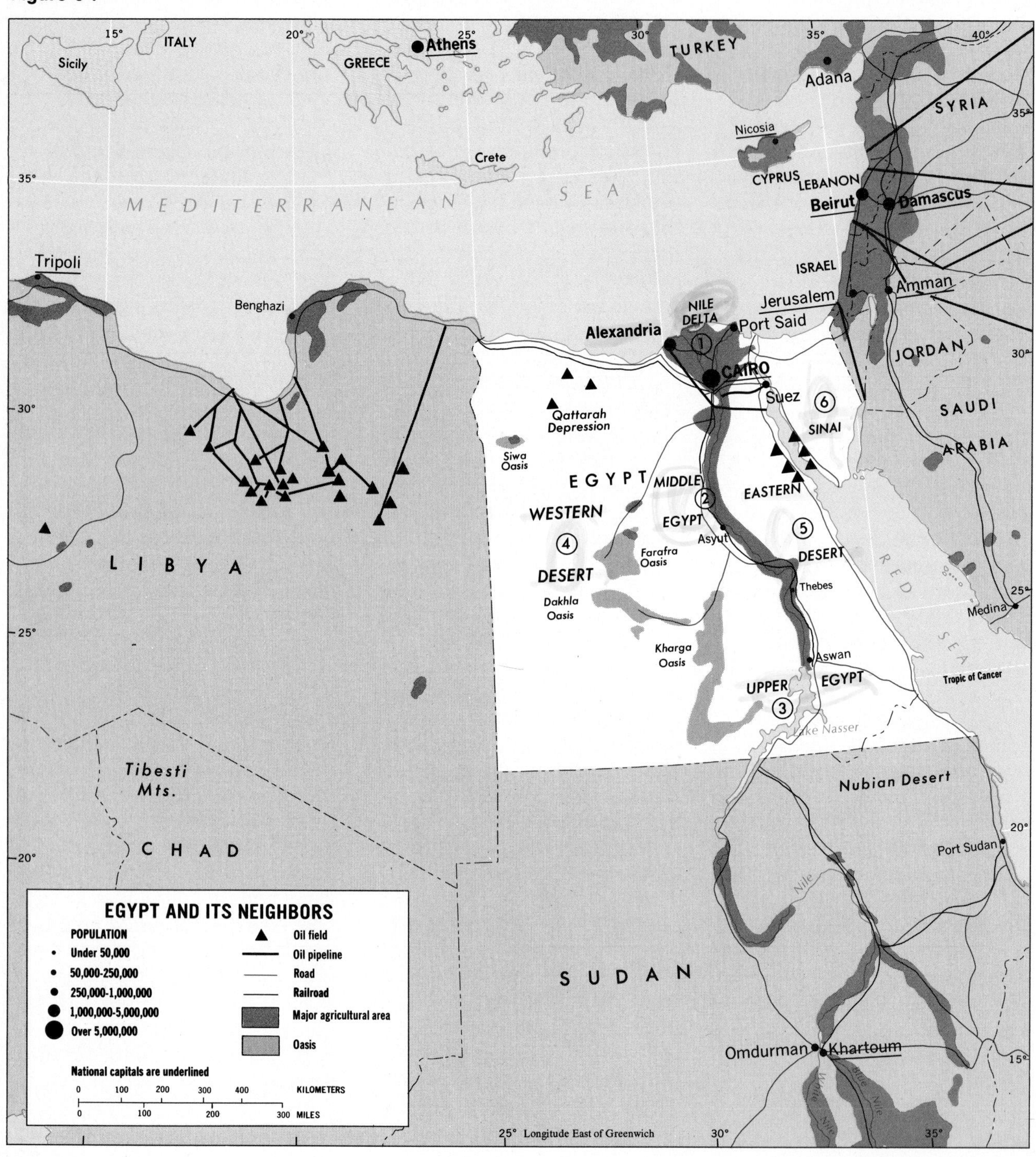

oriented its attentions and energies principally to Europe and the West; in Africa, Egypt has no Islamic competitors to match its status. Indeed, Egypt is continental Africa's second most populous state, after Nigeria. Egypt alone is spatially, culturally, and ideologically at the heart of the Arab world. What factors have combined to place Egypt in this position?

Location is clearly a primary element in Egypt's eminence. The country's position at the southeastern corner of the Mediterranean Sea presented an early impetus as Phoenicians and Cretans linked the Nile Valley with the Levant and other parts of the Mediterranean world. Egypt lay protected but immediately opposite the Arabian Peninsula, and in the centuries of Arab power and empire, it sustained the full thrust of the new wave. The Arab victors founded Cairo in A.D. 969 and made it Egypt's capital (Alexandria, founded in 332 B.C., had functioned as the capital for centuries); they selected a fortuitous site at the junction of valley and delta.

Egypt, too, lies astride the land bridge between Africa and Southwest Asia (and between the Mediterranean and the Red Sea). In modern times this became the major asset to the country as the Suez Canal, completed in 1869, became the vital, bottleneck link in the shortest route between Europe and South and East Asia. Built by foreign interests and with foreign capital, the Suez Canal also brought a stronger foreign presence to Egypt; until 1922, Egypt was a British protectorate. Even when Egyptian demands for sovereignty led to the establishment of a kingdom, British influence remained paramount, and the Suez Canal remained a foreign operation.

The 1952 uprising against King Farouk and the proclamation of an Egyptian Republic in the following year presaged the showdown over the Suez Canal that was now all but inevitable. In July 1956, President Nasser announced the nationalization of the canal and the expropriation of the company that had controlled it; in October, Israel, Britain and France made a miscalculated and strategically disastrous invasion of the Sinai Peninsula. Pressured and pushed into retreat, the invaders abandoned the canal area, and Egyptian stature in the Arab world (and in the Third World generally) was immeasurably enhanced.

The pivotal location of Egypt was expressed in other ways as well. To the west, Algeria was waging a war of liberation against French colonialism, and no Arab country was in a better position to support this campaign than Egypt. To the south, the relationships between Sudan and Egypt had long been strained, but in 1953 the new Egyptian government offered Sudan the option of retaining a political tie with Cairo or full sovereignty. Sudan became fully independent in 1956. And to the east, beyond the Suez Canal and Sinai, lay Israel, its southern flank exposed to the Arab world's strongest partner.

Regions. As Fig. 6-7 reveals, Egypt has six geographic regions: (1) the Nile Delta or Lower Egypt; (2) Middle Egypt, consisting of the Nile Valley from Cairo to Thebes; (3) Upper Egypt, the Nile Valley from Thebes to the Sudan boundary and including Lake Nasser; (4) the Western Desert, including large oases; (5) the Eastern Desert and Red Sea Coast; and (6) the Sinai Peninsula. The great majority of the Egyptian people live in Lower and Middle Egypt.

The Nile Delta covers an area of just under 25,000 square kilometers (10,000 square miles). For thousands of years only the area nearest the Nile Valley was farmed, for the main part of the delta was flood-prone, sandy, lagoon-infested, and excessively salty. Seven Nile distributaries found their way to the Mediterranean Sea. Today, however, the Nile channels are controlled, as modern engineering converted the region into one of perennial irrigation, fertile farmland, and multiple cropping of rice, the staple, and the cultivation of cotton, the chief cash crop. The completion of the Aswan High Dam now makes possible the reclamation and eventual cultivation of an additional 400,000 hectares (1 million acres) of delta land, and by 1990 it is likely that fully half of Egypt's population will reside in this region.

The urban focus of Lower Egypt is Alexandria, Egypt's leading seaport, containing more than 2.5 million residents in 1980. Industrial growth accelerated especially during the period between the two World Wars (today Alexandria is Egypt's leading industrial center), and during the royal period the city served as second capital. Today Alexandria has a canal connection to the Nile, as well as road and rail connections to Egypt's heartland. The city has also become a resort, and, of course, it has shared in modern urban in-migration.

Middle Egypt begins where the Nile Delta ends. Cairo, the capital, lies near the contact between valley and delta. The city's location is quite nodal; its delta connections are good and a railroad extends along the entire length of Middle Egypt and beyond to Aswan. A string of towns lies along this route, and centrally positioned Asyut is Middle Egypt's regional focus.

The metropolitan region of Cairo in 1980 had nearly 10 million residents, and the city's rapid growth of recent decades continued unabated. Cairo is one of the 10 largest urban centers in the world, and it shares with other large cities of the underdeveloped world the problems of crowding, sanitation, health, and education of its huge numbers.

Cairo is a city of stunning contrasts. Along the Nile waterfront, elegant hotel-skyscrapers rise above surroundings that are frequently Parisian and carefully manicured in appearance. But beyond lies the maze of depressing ghettos, narrow alleys, overcrowded slums, a low skyline dominated by the mass of minarets, the towers of mosques pointing skyward. Hundreds of architectural achievements, many of them mosques, shrines, and tombs, are scattered throughout Cairo, but seemingly everywhere one encounters the mud huts and hovels of the very poor. Still, Cairo, a truly cosmopolitan center, is the cultural capital of the Middle East and the Arab world, with a great university, magnificent mosques and museums, renowned zoological and botanical gardens, a symphony orchestra, national theater, and opera. And while Cairo has always been primarily a center of government and administration, it is also an industrial city (textile manufacture, food processing, iron and steel production, assembly plants), and river port.

A street scene in Cairo's less-prosperous area. (Thomas Nebbia/Woodfin Camp)

Countless thousands of small handcraft industries exist in the traditional regions of the city, and the grand *bazaar* throbs daily with the trade in small items.

From the waterfront skyscrapers of Cairo to the monuments of Thebes, ancient capital of the pharaohs, the Nile River gives life to the attenuated oasis that is Middle Egypt. In places the walls of the valley approach the river so closely that the farmlands are interrupted, but along most of the river's course in Middle Egypt the strip of cultivation is between 8 and 16 kilometers (5–10 miles) wide. Occasionally it reaches 24 kilometers (15 miles), and in a few locales the oasis extends only a few hundred yards from the river. But everywhere, the sharply outlined contrast between luxuriant green and barren desert is intense and startling. Clover (*berseem,* a fodder

Modern Cairo has been described as a facade, behind which the rest of the city hides. Certainly these hotels on the Nile River are not representative of Cairo as a whole, but the city has a distinct and characterful cultural landscape and a unique atmosphere. (George Holton/Photo Researchers)

crop), cotton, corn, wheat, rice, millet, sugar cane, and lentils are among the crops that thrive on fields now under perennial irrigation.

Notwithstanding the expansion of its irrigated farmlands, Egypt still must import food to balance local diets better. Although the population growth rate has declined somewhat (from 2.8 percent during the 1960s), development gains are offset by growing numbers. Land reform, increased yields, and expanded cultivation are not enough to overcome Egypt's greatest obstacle to development. Egyptian planners have tried to stimulate industry to help the country escape from its economic treadmill; the most ambitious project is the steel plant at Hulwan (also Helwan) near Cairo, based on the iron deposits found about 50 kilometers (30 miles) west of Aswan, manganese from the Eastern Desert, and local limestone. A search for mineral deposits has yielded some petroleum reserves in the Eastern Desert, on Sinai, and near the Libyan border. But income per person in the late 1970s still was under $400 and President Sadat's policy of accommodation with Israel and alignment with the United States did entail certain political risks. Egypt has, at least, experienced years of political stability; if order were to break down, all hope that a "takeoff" might occur would vanish.

North Africa: the Maghreb

West of Egypt lies Libya, and beyond is the *Maghreb,* the western region of the Arab realm. Together, these four countries (Morocco, Algeria, Tunisia, and Libya) have a population only slightly larger than Egypt's. In 1980, Morocco's population was estimated to be slightly over 20 million; Algeria, the largest Maghreb state territorally, had slightly under 22 million inhabitants; Tunisia, the smallest country in North Africa, had just over 6 million people, and Libya's population was about 2.8 million.

Whereas Egypt is the gift of the Nile, the Atlas Mountains form the physiographic base for the settled Maghreb. The high ranges wrest from the air the moisture that sustains life in the valleys intervening—valleys that contain good soils and sometimes rich farmlands. From the area of Algiers eastward along the coast into Tunisia, annual rainfall averages in excess of 750 millimeters (30 inches), a figure more than three times as high as is recorded at Alexandria and vicinity in Egypt's delta. Even 240 kilometers (150 miles) into the interior the slopes of the Atlas still receive over 250 millimeters (10 inches), of rainfall, and the effect of the topography can be read on the map of precipitation. Where the Atlas terminates, desert conditions immediately begin.

The Atlas Mountains are structurally an extension of the Alpine system that forms the orogenic backbone of Europe and of which Switzerland's Alps and Italy's Apennines are also parts. In Northwest Africa these mountains trend north-northeast and commence in Morocco as the High Atlas with elevations in excess of nearly 4000 meters (13,000 feet). Eastward two major ranges appear, dominating the landscapes of Algeria proper: the Tell Atlas to the north, facing the Mediterranean Sea, and the Saharan Atlas to the south, overlooking the Desert (Fig. 6-8). Between these two mountain chains, each consisting of several parallel ranges and foothills, lies a series of intermontane basins like Andean *altiplanos* but at lower elevations, and markedly drier than the northward-facing slopes of the Tell Atlas. In these valleys the rainshadow effect of the Tell Atlas is reflected not only in the steppelike natural vegetation but also in land-use patterns. Pastoralism replaces cultivation, and strands of esparto grass and bush cover the countryside.

The countries of the Maghreb sometimes are referred to as the Barbary states, in recognition of the region's oldest inhabitants, the Berbers. The Berbers' livelihoods (nomadic pastoralism, hunting, some farming) changed as foreign invaders entered their territory: first the Phoenicians, then the Romans. The Romans built towns and roads, laid out farm fields and irrigation canals, and introduced new methods of cultivation. Then came the Arabs, conquerors of a different sort. They demanded the Berbers' allegiance and their conversion to Islam, radically changed the political system, and organized an Arab-Berber alliance (the *Moors*) that pushed into Iberia and colonized a large part of Southern Europe. After the Moors' power declined, the Ottoman Turks established a sphere of influence along North African coasts. But the most pervasive foreign intervention came during the nineteenth and twentieth centuries, when the European colonial powers (chiefly France, but also Spain) established control and (in the now-familiar sequence of events) delimited the region's political boundaries.

Figure 6-8

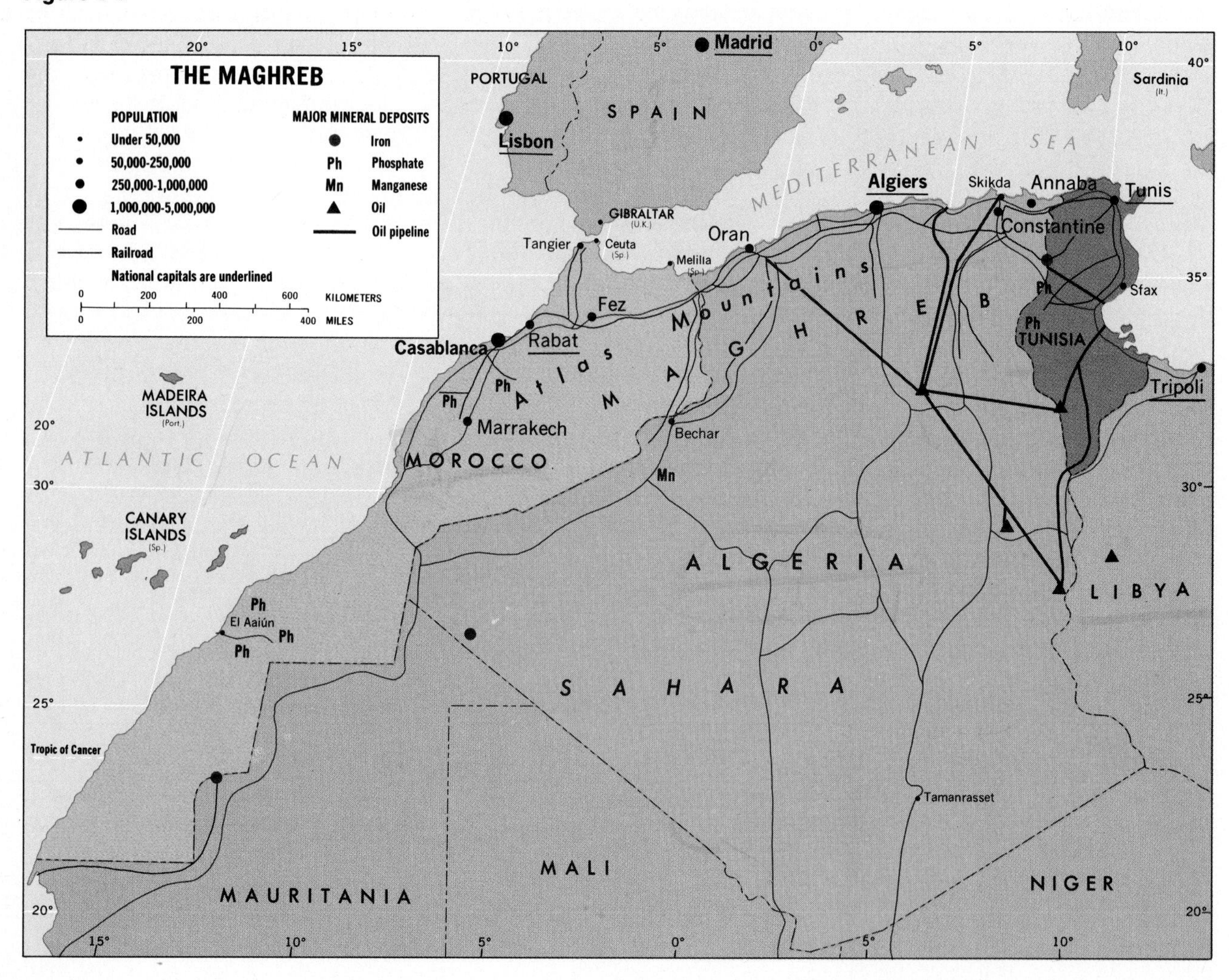

Western Isle

The countries of Northwest Africa are collectively called the *Maghreb,* but the Arab name for them is more elaborate than that: *Djezira-el-Maghreb,* or "Western Isle." The name suggests the physiography of the region, with the great Atlas Mountains rising like a vast island from the waters of the Mediterranean Sea and the sandy flatlands of the Sahara. The Moroccans call their country *El-Maghreb-El Aksa,* "The Farthest West."

During the colonial era, nearly 1.4 million Europeans came to settle in North Africa (most of them French, and a large majority to Algeria), and these immigrants soon dominated commercial life. They stimulated the renewed growth of the region's towns, and Casablanca (nearly 2 million), Algiers (1.6 million), and Tunis (approaching 1 million) rose to become the foci of the colonized territories. But while the Europeans dominated trade and commerce and integrated the North African countries with France and the European Mediterranean world, they did not confine themselves to the cities and towns. They recognized the agricultural possibilities of the favored parts of the *tell,* and established thriving farms. Agriculture here, naturally, is of the Mediterranean variety, and Algeria became known for its vineyards and its wines, its citrus groves, and its dates; Tunisia has long been the world's leading exporter of olive oil; Moroccan oranges went to many European markets. The staples—wheat and barley—long were among the exports.

Despite the proximity of the Maghreb to France and southern Europe, and the tight integration of the region's territories within the French political framework, nationalism emerged as a powerful force in Morocco and Tunisia as well as in Algeria. Morocco and Tunisia secured independence mainly through negotiation, but in Algeria a revolution began in 1954. This costly war did not end until a settlement was reached that took effect in 1962. It was not difficult for the nationalists to recruit followers for their campaign; the justification for it was etched in the very landscape of the country, in the splendid, shining residences of the landlords and the miserable huts of the peasants. But the revolution's success brought new troubles. Hundreds of thousands of French people left Algeria, and the country's agricultural economy fell to pieces; an orderly transition was not possible. Productive farms went to ruin, exports declined, badly needed income was lost.

But there are some bright spots in the Maghreb. Algeria has one resource to compensate for its losses in agriculture—oil, the same oil that led the French in the 1950s to resist Algerian independence. Petroleum has now risen to the top of Algeria's export commodities; it is taken from the Sahara Desert and piped to Algerian as well as Tunisian ports for despatch to Europe. Libya's oil production has risen spectacularly to place the country among the world's leading exporters; again, the fields are located in the Sahara. And, quite unlike most of the remainder of the Middle East, the countries of the Maghreb have a varied set of mineral resources. Chief among these is phosphate of lime, used in the manufacture of fertilizer. Morocco is the world's leading exporter of this commodity. It occurs also in Algeria, and in substantial quantities in Tunisia, where it ranks as the most valuable export as well. Both Morocco and Algeria have Atlas-related iron ores, exported to the United Kingdom from their favorably located sources. Manganese, lead, and zinc are also mined in all three countries, but sufficient coal is not locally available. Despite the new power source in the oil fields (and the associated natural gas), this not inconsiderable range of minerals has stimulated little domestic industrial development.

The oil boom has not yet substantially improved life for the majority—and the majority of the people remain undernourished, illiterate (about 70 percent), poor (barely over $300 income per person annually), unemployed (over 40 percent), and far removed from the foci of change and progress. Tens of thousands of Algerians followed the European emigrants and went to France's cities in search of jobs; the great majority of those who stayed at home remained trapped in the *bidonvilles,* the poverty-striken shantytowns that surround the Maghreb's cities.

The three countries of the Maghreb and Libya reflect the ideological division that so strongly marks the entire realm. Morocco is a kingdom, a relatively conservative force in Arab-world politics. Algeria is nonaligned, anti-imperialist, and socialist in its external and internal political orientations, and a vigorous supporter of Arab causes. Tunisia, poorest of North Africa's states in some ways, became a model of stability under the leadership of its long-time president and national hero, Bourguiba. Tunisia is a country where one-party democracy seems to work without the fanfare that is needed to obscure its failures elsewhere. And Libya, with its oil riches and miniscule population, has become North Africa's radical state, whose leaders can afford to supply military hardware to ideological allies (Soviet MIGs to a former regime of Uganda, for example) while engaging in political subversion in Arab countries not committed to its causes. The political geography of the Arab world's western flank is complex indeed.

Spatial organization of an Israeli kibbutz north of Tel Aviv. (Georg Gerster/Rapho-Photo Researchers)

The Middle East

The Middle East, as a region within the Southwest Asian-North African realm, consists chiefly of the pivotal area positioned between Turkey and Iran to the north and east, Saudi Arabia and Egypt to the south, and the Mediterranean Sea to the west (Fig. 6-4). Five countries lie in this region: Iraq, largest in terms of population (more than 13 million in 1980) as well as territory; Syria, next in both categories; Jordan, Lebanon, and Israel.

Israel thus lies at the very heart of the Arab world. Its neighbors are Lebanon and Syria to the north, Jordan to the east, and Egypt to the southwest—all of them in some measure resentful of the creation of the Jewish state in their midst. Since 1948, when Israel was created as a homeland for Jewish people upon recommendation of a United Nations commission, the Arab-Israel conflict has overshadowed all else in the Middle East.

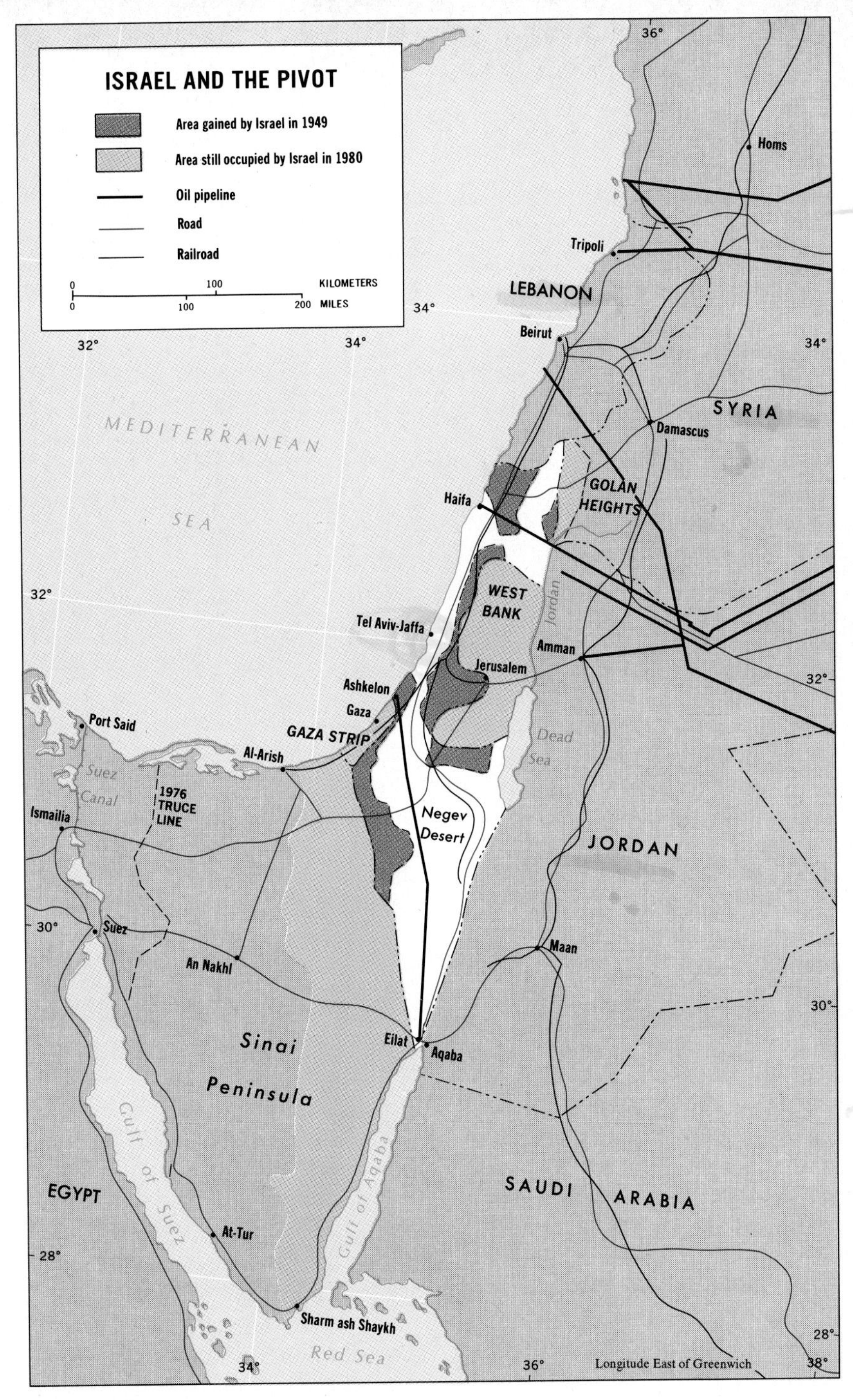

Figure 6-9

Indirectly, Israel was the product of the collapse of the Ottoman Empire. Britain gained control over the mandate of Palestine, and it was British policy to support the aspirations of European Jews for a homeland in the Middle East. These aspirations were embodied in the concept of Zionism. In 1946, the British granted independence to the territory lying east of the Jordan River, and "Transjordan" (now the state of Jordan) came into being. Shortly afterward, the territory west of the Jordan River was partitioned by the United Nations, and the Jewish people got the bulk of it—including, of course, land that had been occupied since time immemorial by Arab and other Semitic peoples. The original United Nations plan proposed to allot 55 percent of all Palestine to the Jewish sector, although only 7 percent of the land was actually owned by Jews (about one-third of the total population of Palestine was Jewish), but this partition plan was never implemented as it was intended. As soon as the Jewish people declared the independent state of Israel, the new state was attacked by its Arab neighbors, who rejected the scheme. In the ensuing battle, Israel not only held its own, but gained some territory in the southern Negev Desert (Fig. 6-9). At the end of the first war (1949), the Jewish population controlled 80 percent of what had been Palestine. Of course, this success was not won by overnight organization: at the time of Israeli independence, there already were three-quarters of a million Jews in Palestine. Indeed, the world Zionist movement had been assisting Jews in their return to Palestine since the late nineteenth century.

For more than 30 years now, a state of latent—and at times actual—war has existed between Israel and the Arab world. In 1967, a week-long conflict resulted in the Israeli occupation of Syrian, Jordanian, and Egyptian territory, including the Sinai Peninsula to the edge of the Suez Canal—a facility the Egyptians had not allowed Israeli vessels to use. In 1973 a brief war led to Israel's withdrawal from the Suez Canal to truce lines in the Sinai, but it

Palestinian Dilemma

Ever since the creation of the state of Israel in what was the British Mandate of Palestine, hundreds of thousands of Arabs who called Palestine their homeland have lived as refugees in neighboring countries. Many have been assimilated in the societies of Israel's neighbors, but a still larger number live—still today—in makeshift refugee camps. The Palestinians call themselves a nation without a state (much as the Jews were before Israel was founded), and they demand that their grievances be heard. Until Palestinian hopes for a national territory in the Middle East are considered by the powers that have influence in the region, a settlement between Israel and the Arab countries may remain unattainable. Estimated Palestinian populations in the Middle East are:

Jordan	900,000
Occupied Jordan (West Bank)	650,000
Egypt and Gaza	400,000
Lebanon	300,000
Syria	200,000
Other Arab states	230,000
Israel	520,000
Total	3,200,000

In recent years militant Palestinian organizations have emerged in Lebanon and Syria. Although the Palestinian Liberation Organization (PLO) has generally represented Palestinian views, several other groups also have done so. Palestinian forces have fought Israeli troops in the Israel-Lebanon border region; Palestinian terrorists have attacked Israeli and non-Israeli targets in Israel and around the world.

Various territorial solutions have been proposed. One suggests that a Palestinian state be created from the presently occupied part of Jordan (west of the Jordan River) plus the Gaza Strip. Another proposes a Palestinian state between Israel and Egypt in Sinai. Still another proposal would partition Lebanon into three segments, one Palestin-

also extended Israel's control over a sector of Syria's Golan Heights, creating a new source of friction. Since then, peace has been sustained largely because of Egypt's willingness to negotiate a staged Israeli withdrawal from the Sinai, but the level of hostility has remained dangerously intense. In addition, the area is a power vacuum, subject to Soviet and United States moves and pressures.

So many issues divide the Jews and the Arabs that a permanent solution (other than the destruction of the state of Israel itself) or even a peace imposed by the great powers does not seem to be attainable. One prominent issue has been the refugee problem: when Israel was created, over 750,000 Palestinian Arabs were forced to leave the area to seek new lives in Jordan, Egypt, Lebanon, or Syria (see box, above). Another problem has been the city of Jerusalem, a holy place for Jews, Moslems, and Christians. In the original United Nations blueprint for Palestine, the city was to have become internationalized; then Jerusalem was divided between the Jewish state and Jordan. This was a point of friction, with neither side satisfied. But the 1967 June War saw Jordanian lands west of the Jordan River fall to Israel, which as a result came to control all of the holy city. A further problem has been the use of the waters of the Jordan River. In 1964, Israel unilaterally diverted over half of the Jordan's flow for its own use. The river's

Baqaa Palestinian refugee camp in Jordan. (F. Sautereau/Rapho-Photo Researchers)

ian, a second Lebanese-Moslem, and a third Lebanese-Christian. But the Palestinians are having difficulty even being heard in international forums dealing with the Middle East; their territorial objectives seem far from realization.

headwaters lie in Lebanon and Syria, and these two countries have joined with Jordan in plans to divert the water completely away from its present course. But the river is crucial to Israel's economy, and the Israelis have threatened to destroy any diversion works or dams built to cut the lower Jordan's water level.

One major irritant in the whole matter has been Israel's rapid rise to strength and prosperity amid the poverty so common to the Middle East. There is nothing particularly productive about most of Israel's land, and neither is it very large (20,000 square kilometers, 8000 square miles, smaller than Massachusetts). But Palestine has been transformed by the energies of the Jewish community and, importantly, by heavy investments and contributions made by Jews and Jewish organizations elsewhere in the world, especially the United States. It is often said that in Israel, desert has been converted into farmland, and it is partly so: the irrigated acreage has been enlarged to many times its 1948 proportions, and water is carried even into the Negev Desert itself.

With 3.9 million people, however, Israel's future obviously does not lie in agriculture—for export or self-sufficiency. Already, the country must import a large volume of wheat. Despite the general intensity of agriculture, and the dairying and vegetable gardens that have grown up around the large cities and towns, the gap between local supply and demand is widening. No effort is spared to maintain as high a degree of self-sufficiency as is possible, but despite the *kibbutzim* (collectivized farm settlements) and other innovations designed to combine communal living, productivity, and defensibility, Israel depends increasingly on external trade and support.

Without an appreciable resource base, industrialization also presents quite a challenge to Israel. Evaporation of the Dead Sea waters has left deposits of potash, magnesium, and salt, and there is rock phosphate in the Negev. But there is very little fuel available within Israel; no coal deposits are known, although some oil has been found in the Negev. To circumvent the Suez Canal, an oil pipeline leads from the port of Eilat on the Gulf of Aqaba to the refinery at Haifa; a second, larger pipeline has been constructed to link Eilat and Ashkelon. The oil formerly came from Iranian fields, but since the Iranian Revolution Israel has been forced to turn elsewhere for its petroleum. One source is Egypt; when Israel yielded a section of the occupied Sinai Peninsula where oil fields had been found and developed, Egypt committed itself to sell oil to Israel in return. The small steel plant at Tel Aviv uses imported coal as well as iron from foreign sources, and was built for strategic reasons; it is not an economic proposition. Thus the only industry for which Israel has any domestic raw materials at all is the chemical industry, and it, in response, has seen considerable growth. But for the rest, Israel must depend mostly on the skills of its labor force. Diamond cutting, for example, is one of the leading industries. Many technicians and skilled craftsmen have been among the hundreds of thousands of Jewish immigrants who have come to Israel since its creation, and the best course, naturally, is to make maximum use of these people.

In effect, then, Israel is a Western type of developed country in the Middle East. It is highly urbanized, with 70 percent of its population living in towns and cities of 10,000 or more residents. Israel's core area includes the two major cities, Tel Aviv-Jaffa (750,000) and Haifa (350,000) and the coastal area between them; in total this core region incorporates over three-quarters of the country's population, although the proportion is less if the territories conquered in the 1967 War are added (including Jerusalem's Jordanian sector of 80,000 population). With the cities surrounded by dairy and poultry farms, vegetable gardens, and dense road networks, the impression given by the core area is indeed Western.

Neighbors and Adversaries. Not only does Israel have the misfortune to be located in the heart of the Arab world, rather than in one of the region's peripheral or transition zones, it also has an inordinate number of neighbors for so small a country with so much coastline—five (counting Saudi Arabia). Of these five, three lie along Israel's northern and eastern boundary: Lebanon, Syria, and Jordan.

Lebanon, Israel's coastal neighbor on the Mediterranean Sea, is one of the exceptions to the rule that the Middle East is the world of Islam: perhaps 40 percent or even more of the population of over 3 million adheres to the Christian rather than Moslem faith. Smaller than Israel (in fact, only a little over half of Israel in terms of territory), Lebanon has a long history of trade and commerce, beginning with the Phoenicians of old—this was their base. Like Israel, Lebanon must import much of its staple food, wheat. The coastal belt below the mountains, although intensively cultivated, normally cannot produce enough grain to feed the entire population.

But normalcy has not prevailed in Lebanon for several years. The country fell apart in 1975, when a civil war broke out between Moslems and Christians. It was a conflict with many causes. Lebanon for several decades had functioned with a political system that divided power between the two dominant communities, but the basis for that system had become outdated. In the early 1930s Moslems and Christians in Lebanon were at approximate parity, but in the 40 years since, Moslems increased in numbers at a much faster rate than the more highly urbanized and generally wealthier Christians. The Moslems' displeasure at an outdated political arrangement (developed during the French occupation) was expressed during several outbreaks of rebellion before the civil war. United States troops entered Lebanon asrecently as 1958 to help suppress anti-Western opposition. In addition, Lebanon had become a base for as many as 300,000 Palestinian refugees (estimates vary). These people, many of them living in squalid camps, were never satisfied with Lebanon's moderate posture toward Israel. When the first fighting between Moslems and Christians broke out in the coastal city of Tripoli, the Palestinians joined the conflict on the

Beirut, once the Middle East's most cosmopolitan and thriving city, was devastated by months of civil war that left areas of the city in chaos. An uneasy peace imposed by outside forces is frequently broken. (Rochot/Sygma)

Moslem side. In the process, Lebanon was wrecked. Beirut, the capital (1 million), often described as the Paris of the Middle East and a city of great architectural beauty, was almost completely destroyed as Christian, Moslem, Palestinian, and eventually Syrian forces fought for control. As the Moslems' strength increased, the Christians concentrated in an area along the coast between Beirut and Tripoli, raising the possibility that the country may be partitioned.

Syria's role as peacemaker in Lebanon has enhanced the country's status since its loss of much of the Golan Heights to Israel. For a long time, Syria was politically unstable, a poor second to Egypt in the so-called United Arab Republic. But the government of President Assad, after taking power in 1971, began to reap the benefits of continuity. In 1976, after a prolonged effort to mediate in the Lebanon civil war, Syrian armed forces intervened and brought a measure of peace to that embattled country.

Syria, like Lebanon and Israel, has a Mediterranean coast where unirrigated agriculture is possible. Behind this densely populated coastal belt, Syria has a much larger interior than its neighbors, but the areas of productive capacity are quite dispersed. Damascus, the capital (1.2 million), was built on an oasis and is one of the Levant's ancient cities; although on the dry, lee side of the coastal mountains, it is surrounded by a district of irrigated agriculture. It lies in the southwestern corner of the country (Fig. 6-9), in close proximity to Israel. In the far north, near the Turkish boundary, lies another old caravan-route center, Aleppo (600,000), now on the little-used railroad from Damascus to Turkey and at the northern end of Syria's important cotton-growing area; here the Orontes River is the chief source of irrigation water. Syria's wheat belt, east along the northern border, also centers on Aleppo. In the eastern part of the country, the Euphrates Valley and the far northeast, known as the Jezira, are being developed for large-scale mechanized wheat and cotton farming with the aid of pump-irrigation systems. Production of these crops is rapidly rising; more than half the Syrian harvest now comes from this region. Recent discoveries of oil deposits add to the potential importance of this long-neglected part of the country.

Southward and southeastward, Syria turns into desert, and the familiar sheep herders and goat herders move endlessly across the countryside. There may be as many as a half million of them, a sizable proportion of Syria's 6 million people. In contrast to Israel, and, to a lesser degree, to Lebanon as well, Syria is very much a country of farmers and peasants; only about 40 percent of the people live in cities and towns of any size. But again, Syria produces adequate harvests of wheat and barley and normally does not need such staple imports: in fact, it exports wheat from its northern wheat zone. It exports barley as well, but its biggest source of external revenue remains cotton. It is, in addition, a country where opportunities for the expansion of agriculture still exist; their realization will improve the cohesion of the state and bring its separate regions into a tighter framework.

None of this can be said for Jordan, the desert kingdom that lies east of Israel and south of Syria. It, too, was a product of the Ottoman collapse, but it suffered heavily when Israel was created—more so, perhaps, than any other Arab state. In the first place, Jordan's trade used to go through Haifa, now an Israeli port, and so Jordan has to depend on Beirut or the tedious route via Aqaba. Second, Jordan's final independence in 1946 was achieved with a total population of perhaps 400,000, including nomads, peasants, villagers, and a few urban dwellers; it was a poor country. Then, with the partition of Palestine and the creation of Israel, Jordan received more than a half million Arab refugees and found itself responsible also for another half million Palestinians who, although living on the western side of the Jordan River, were incorporated in the state. Thus refugees outnumbered residents by more than two to one, and internal political problems were added to external ones—not to mention the economic difficulties. Jordan has survived with United States, British, and other aid, but its problems have not lessened. Many Jordanian residents have little commitment to the country and do not consider themselves Jordanian citizens; they give little support to the embattled king. Extremist groups threaten constantly to drag the country into another war with Israel; the June War of 1967 was disastrous for Jordan. Where hope for progress might lie—as, for example, in the development of the Jordan Valley—political conflicts get in the way. In 1967, Jordan lost its second city, that is, its sector of Jerusalem. Its capital, Amman (600,000), reflects the limitations and poverty of the country. Without oil, without much farmland, without unity or strength, and overwhelmed with refugees, Jordan presents one of the bleakest faces in the Middle East.

By comparison, Iraq is well endowed. Iraq contains the lower Euphrates Valley and also the Tigris River, and its agricultural potential is far greater than what is now used. Iraq is a rarity in the Middle East: it can be described as underpopulated, that is, it could feed a far larger number of people than it now does. This is a leftover from the decline that Mesopotamia went through during the Middle Ages, but the first steps are being taken to improve conditions and to raise standards of living. Iraq is one of the major beneficiaries of the oil reserves of the Middle East, and it needs the income from the petroleum to make the necessary investments in industry and agriculture. Oil accounts for about 90 percent of the country's export revenues; the second export by value, dates, bring in a mere 3 percent, about the same as barley, which usually ranks third.

How can a country that exports food crops and has a huge income from oil have so low a standard of living? There are many answers to this question. For nearly five decades since its independence (in 1932, after a decade as a British mandate), it has suffered from administrative inefficiency, corruption in government, misuse of the national income, inequities in landholding and tenure systems—a set of problems that add up to the concept of underdevelopment. When, in 1958, the Iraqi monarchy was deposed, there was perhaps a promise of better times, but since that event there has been political instability in addition to the other problems.

Very recently the decline of Iran has enhanced Iraq's position, and oil revenues have risen sharply. This has had some effect on living standards, but the vagaries of the past will take many years to erase.

Apart from western Iraq, where nomads herding camels, sheep, and goats traverse the Jordan-Syrian-Iraqi desert, most of the people live in small villages strung along the riverine

People on the Move

Countless thousands of people in Southwest Asia and North Africa are on the move; movement is a permanent part of their lives. They travel with their camels, goats, and other livestock along routes that are almost as old as human history in this area. Most of the time they do so in a regular seasonal pattern, visiting the same pastures year after year, stopping at the same oases, pitching their elaborate tents near the same stream. It is a form of cyclic migration: *nomadism.*

Nomadic movement, then, is not simply an aimless wandering across boundless plains. Nomadic peoples know their domain intimately; they know when the rains have regenerated the pastures they will make their temporary locale, and they know when it is necessary to move on. Some nomadic peoples remain in the same location for several months every year, and their settlements take on characteristics of permanence—until, on a given morning, an incredible burst of activity accompanies the breaking of camp. Amid ear-piercing bleating of camels, the clanking of livestock's bells, and the shouting of orders the whole village is loaded on the backs of animals and the journey resumes. The leaders of the community will know where they are going; they have been making the circuit all their lives.

Among those in the community there will be a division of labor, and some are skilled at crafts and make leather and metal objects for sale or trade when contact is made with the next permanent settlement. A part of the herd of livestock traveling with the caravan may belong to a townsman, who will pay for their care. Nomads are not aimless wanderers—nor are they free from the tentacles of the cities.

Camp is being broken; a camel caravan forms. (Georg Gerster/Rapho-Photo Researchers)

lowland, from the banks of the Shatt-al-Arab (the joint lower course of the Tigris and Euphrates) to the land of the Kurds near the Turkish border. The Kurds, who number over 1 million out of Iraq's 13 million people, have at times been opposed to the Baghdad government, and there still is a serious minority problem of strong regional character. But the general impression of rural Iraq reminds one of rural Egypt, though Egypt is ahead in terms of its irrigation works; the peasants face similar problems of poverty, malnutrition, and disease.

The Arabian Peninsula

South of Jordan and Iraq lies the Arabian Peninsula, physiographically dominated by desert conditions and politically dominated by the Kingdom of Saudi Arabia (Fig. 6-10). With its huge area of 2,150,000 square kilometers (830,000 square miles), Saudi Arabia is the realm's third-largest nation-state; only Sudan and Algeria are somewhat larger. On the peninsula, Saudi Arabia's neighbors include (clockwise from the Persian Gulf coast): Qatar; the United Arab Emirates (see page 345); the Sultanate of Oman; Yemen (Aden), also known as South Yemen or the People's Democratic Republic of Yemen; and Yemen (San'a), also known as the Yemen Arab Republic. Together, these countries on the eastern fringes of the peninsula have fewer than 12 million inhabitants; the largest by far is the Yemen Arab Republic, with nearly 3 million. Boundaries still are inadequately defined in this, one of the world's last frontiers.

Saudi Arabia itself has only about 10 million inhabitants in its vast territory, but the kingdom's importance is reflected by Fig. 6-5: the Arabian Peninsula contains the world's largest concentration of known petroleum reserves, and Saudi Arabia has a major share of it; by some estimates this country alone possesses one-quarter of all the world's remaining oil. The reserves lie in the eastern part of the country, on the Persian Gulf and in the Rub al-Khali (the latter is a southward continuation of the Iraq-Kuwait-Neutral Zone field).

The national state that is Saudi Arabia was consolidated as recently as the 1920s through the energies and organizational abilities of King Ibn Saud. At the time it was a mere shadow of its former greatness as the source of Islam and the heart of the Arab world. Apart from some permanent settlements along the coasts and in scattered oases, there was little to stabilize the country; most of it is desert; annual rainfall almost everywhere is under 10 centimeters (4 inches). The land surface rises generally from east to west, so that the Red Sea is fringed by mountains that reach as high as 3000 meters (nearly 10,000 feet). Here the rainfall is slightly higher, and there are some farms (coffee is a cash crop). The mountains also contain known deposits of gold, silver, and other metals, and at one time these were economic mainstays. Exploration has begun again, and the Saudis hope to diversify their exports by adding minerals from the west to the oil from the east.

As Fig. 6-10 reveals, human activity in Saudi Arabia tends to be concentrated in a wide belt across the "waist" of the peninsula, from the boom town of Dhahran (Az-Zahran) on the Persian Gulf and its neighbors, through the national capital of Riyadh in the interior, to the Mecca-Medina area near the Red Sea. No real and effective internal transport network has yet developed, however: the country's only railroad links the capital to the Persian Gulf town of Ad Damman, and only about 10,000 kilometers (6000 miles) of roads exist in the huge territory. Caravan routes still cross the desert, and in the remote interior Bedouin nomads still traverse their ancient routes. For several decades Saudi Arabia's aristocratic royal families were virtually the sole beneficiaries of their country's incredible wealth, and there was hardly any change in the lives of villagers and nomads. But in recent years the country's rulers have begun to institute reforms in housing, medical care, education, and other spheres. A huge petrochemical industry is being developed for future diversification of the economy. Still, Saudi Arabia remains tradition-bound, and in the villages—even in Medina and Mecca—the modernizing influence of the oil boom seems far away in place as well as time.

Cycle Theory. The new era in the Arab world is one of rejuvenation and reconstruction: rejuvenation in the ouster of colonial regimes and foreign influences and reconstruction with the billions of dollars derived from oil. Historical geographers, noting the rise, decline, and resurrection of nation-states, have attempted to explain the sequence of events in various ways. Environmentalists believe that it all has to do with changing climates and, consequently, changing human energies. Ratzel believed (page 71) that the state functions like a biological organism to grow and die. Another geographer, S. Van Valkenburg, proposed a "cycle in the political development of nations, recognizing four stages, namely youth, adolescence, maturity and old age . . . after completion [a] cycle may renew itself, possibly with a change in political extension, while the cycle can also be interrupted at any time and brought back to a former stage. The time element (the length of a stage) differs greatly from nation to nation. . . ."

In this cycle theory of state evolution, the stage of youth might be called the *organizing* stage, and Tunisia in this realm surely would be the best example. Adolescence may follow as an *expanding* stage since the

Figure 6-10

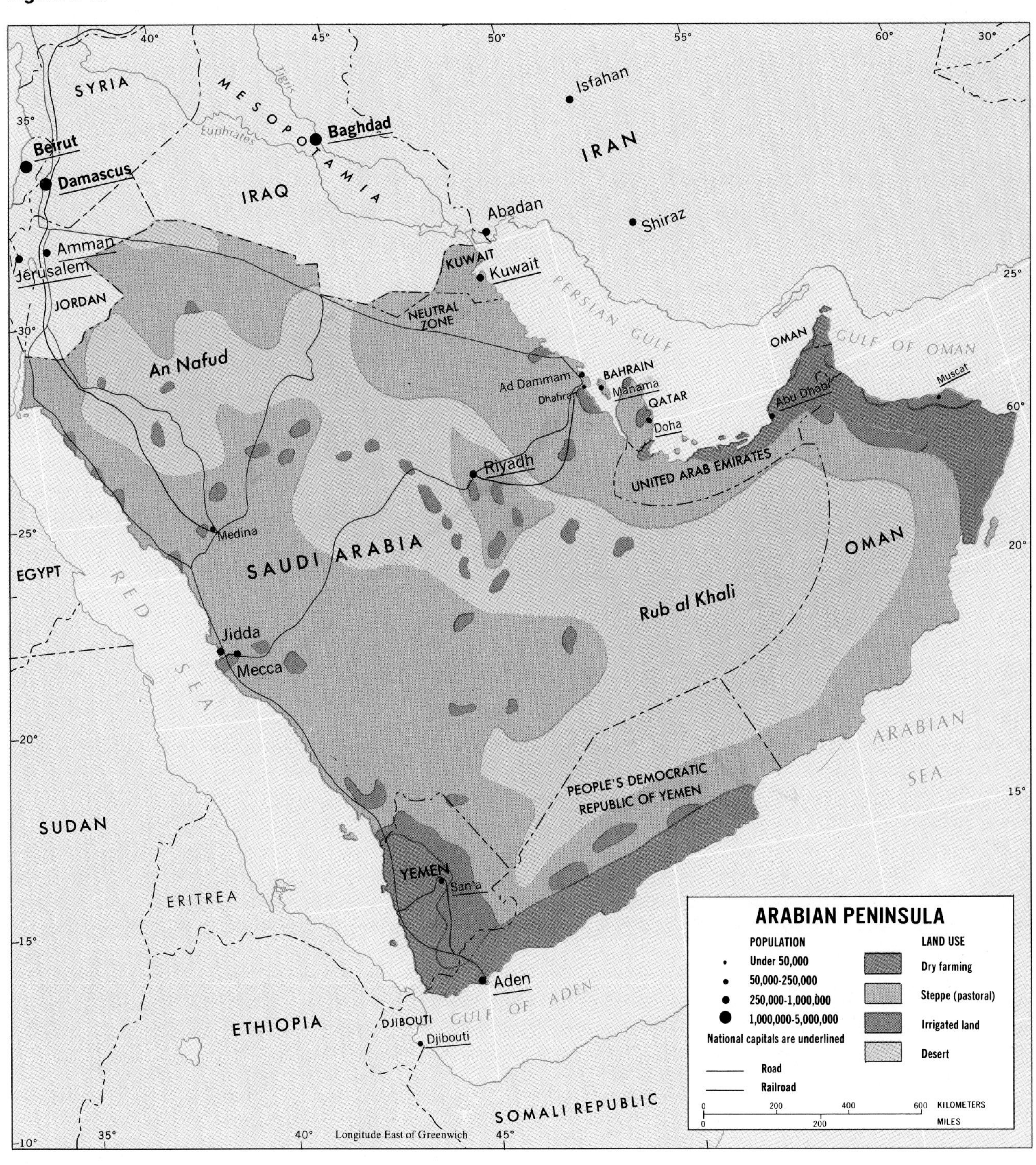

dynamic qualities of the state now emerge, and Iran during the Shah's administration showed such properties: a greater Persian sphere of influence was not beyond Tehran's dreams. Maturity, the *stabilizing* stage, leads the state to concern itself once again with internal problems (as Egypt is doing); external relationships involve moderation and international cooperation. Old age may witness a kind of national senility, but it may also be a *disintegrating* period. Until the cycle was rejuvenated in Saudi Arabia, there was little left of the regional association forged during the rise of Moslem power.

The Non-Arab North

Across the northern tier of the Moslem realm lie three geographically distinct countries: Turkey, Iran, and landlocked Afghanistan. Turkey, although a Moslem country, has been strongly oriented toward Europe in recent decades and has remained aloof from many of the issues that excite Arab countries. Iran also long kept some ideological distance; when the Suez Canal was closed to Israeli shipping by the Arab countries, it was Iranian oil that flowed, on orders of Shah Reza Pahlavi, through Israel's Aqaba pipelines. And Afghanistan, although a Moslem country as well, lies remote and peripheral, vulnerable to foreign invasion, in its mountainous Asian interior.

Turkey, when it entered the twentieth century, was at the center of a decaying, corrupt, and reactionary empire whose sphere of influence had extended across much of the Middle East and Eastern Europe. At that time conditions were such that it would undoubtedly have had to be considered a part of the Middle East, the world of its religious—cultural imprint, Islam. But the Turks are not Arabs (their source area lies in the Sino-Soviet border area in Central Asia), and in the early 1920s a revolution occurred that thrust into national prominence a leader who has become known as the father of Turkey—Mustafa Kemal Ataturk.

The ancient capital of Turkey was Constantinople (now Istanbul), but the struggle for Turkey's survival had been waged from the heart of the country, the Anatolian Plateau, and here Ataturk wanted to place the seat of government. Ankara had many advantages: it would remind the Turks that they were, as Ataturk always said, Anatolians; it lay nearer the center of the country than Istanbul and thus might act as a stronger unifier; and strategically its position was much better than that of Istanbul, located on the western side of the Bosporus (Fig. 6-11).

Although Ataturk moved the capital eastward and inward, his orientation was westward and outward. To implement his plans for Turkey's modernization, he initiated reforms in almost every sphere of life in the country. Islam, the state religion, lost its official status. The state took over most of the religious schools that had controlled education. The Roman alphabet replaced the Arabic. Moslem law was replaced by a modified Western code. Symbols of old—wearing beards, wearing the fez—were prohibited. The emancipation of women was begun (in Moslem society, women have generally been denied access to education, freedom of movement, or social contact, and in a few countries they still must cover their faces when in public), and monogamy was made law. The new government took pains to stress Turkey's separateness from the Arab world, and ever since it has remained quite aloof from the affairs involving the other Islamic states.

What Ataturk wanted for Turkey could not, of course, be achieved in a short time. Turkey is largely an agricultural country, and when three-quarters of the inhabitants of a country are subsistence or near-subsistence farmers, most of whom live in small villages in sometimes quite isolated rural areas, some opposition to change, especially radical change, has to be expected. So it was in Turkey; in fact, the government had to yield to the devoutly Moslem peasants on some issues. Still, Ataturk's directives continue to be pursued more than a half century after he came to power.

Turkey is a mountainous country of generally moderate relief. The highest mountains lie in the east, near the Soviet border, and in Kurdistan. Here elevations reach 2000–3000 meters, but westward the surface declines. The Plateau of Anatolia, in the heart of the country, is dry enough to be identified as a steppe (see Fig. I-6, page 19, the area marked BSk). Here the people live in small villages, grow subsistence cereals, and raise livestock. Sheep and goats bring some revenues, from the wool and hides—and from the mohair of the Angora goat, used in the manufacture of valuable textiles and rugs. But the best farmlands lie in the coastal areas, small as they are. There is little coastal lowland along the Mediterranean, where the

Istanbul, on the threshold of Europe, is Turkey's largest and most varied city. Minarets and mosques rise above a somewhat Eastern European townscape. The Bosporus is in the foreground. (Luis Villota/Image Bank)

areas around Antalya and Adana (northeast of Iskenderun) are the only places where the mountains retreat somewhat from the sea. The Black Sea coast, too, is quite narrow, but it is comparatively moist: it gets winter as well as summer rainfall (though with a winter maximum) totaling 75–260 centimeters (30–105 inches)—uncommonly high for the "dry world." Thus, as Fig. I-9 reflects, these coastal areas, and especially the Aegean—Marmara lowlands, are the most densely peopled zones of Turkey and the most productive ones.

In the farming areas where the export crops are grown, the association is the familiar Mediterranean type, with wheat and barley the staples (they occupy over three-quarters of all the cultivated land in the country), and tobacco, cotton, hazelnuts, grapes, olives, and figs for the external market. Cotton, famous Turkish tobacco, and hazelnuts (the Black Sea coast is the world's chief source of these) form the three top exports by value in a country 90 percent of whose external revenues are derived from the land. As elsewhere in this part of the world, production could be increased, but, unlike Iraq, the problems involved in irrigation expansion are serious: there is little level land and the streams are deeply incised in this mountainous country.

Turkey has the largest population among the countries of the North African-Southwest Asian realm, with 44 million inhabitants. As Fig. 6-11 indicates, this country has the best-developed network of surface communications in the entire realm; it also is the most industrialized, notwithstanding its chiefly agricultural export economy. The production of cotton has stimulated a textile industry, and a small steel industry has been established based on a coalfield near Zonguldak, not far from the Black Sea, and an iron ore deposit several hundred miles away in east central Turkey. In the southeast, Turkey has located some oil, and it may share in the zone of oil fields of which the famed Kirkuk reserve in Iraq is a part. With its complex geology, Turkey has proved to have a variety of mineral deposits if not a large mineral base, and the variety that does exist gives much scope for future development. Spatially, this development seems destined, at present, to confirm the country's westward orientation. Already the Mediterranean coastal zone is the economic focus of the country, and of what remains the western half is again the more developed.

Iran constitutes another exception to the rule that this geographic realm is the Arab world. Iranians are not Arabs; nor are they adherents to the majority (Sunni) Moslem faith. Most Iranians are members of the Shi'ite group, who represent about one-tenth of all Islam, mainly in Iran and neighboring Iraq. Persia (as Iran was formerly called) was a kingdom as long as 2500 years ago; its royal succession faltered repeatedly during the twentieth century. The most recent period of monarchy, the rule of Shah Mohammed Reza Pahlavi, ended in 1979 when an Islamic revolution that had long been intensifying achieved the overthrow of the Shah. Nevertheless, the year 1971 had witnessed the 2500th anniversary of Persia's first monarchy amid celebrations and scenes of royal splendor—and without overt indications of the turbulence that lay ahead.

Iran is a country of mountains and deserts. The heart of the country is a plateau, the Plateau of Iran, that lies surrounded by even higher mountains including the Zagros Mountains in the west, the Elburz Mountains along the southern shores of the Caspian Sea, and the mountains of the Khurasan region to the northeast. The Iranian Plateau, therefore, actually is a high-elevation basin marked by salt flats and expanses of stone and sand. On the hillsides, where the topography wrests some water from the air, lie some fertile soils. Elsewhere only oases break the arid monotony—oases that for countless centuries have been stopping places on the region's

Figure 6-11

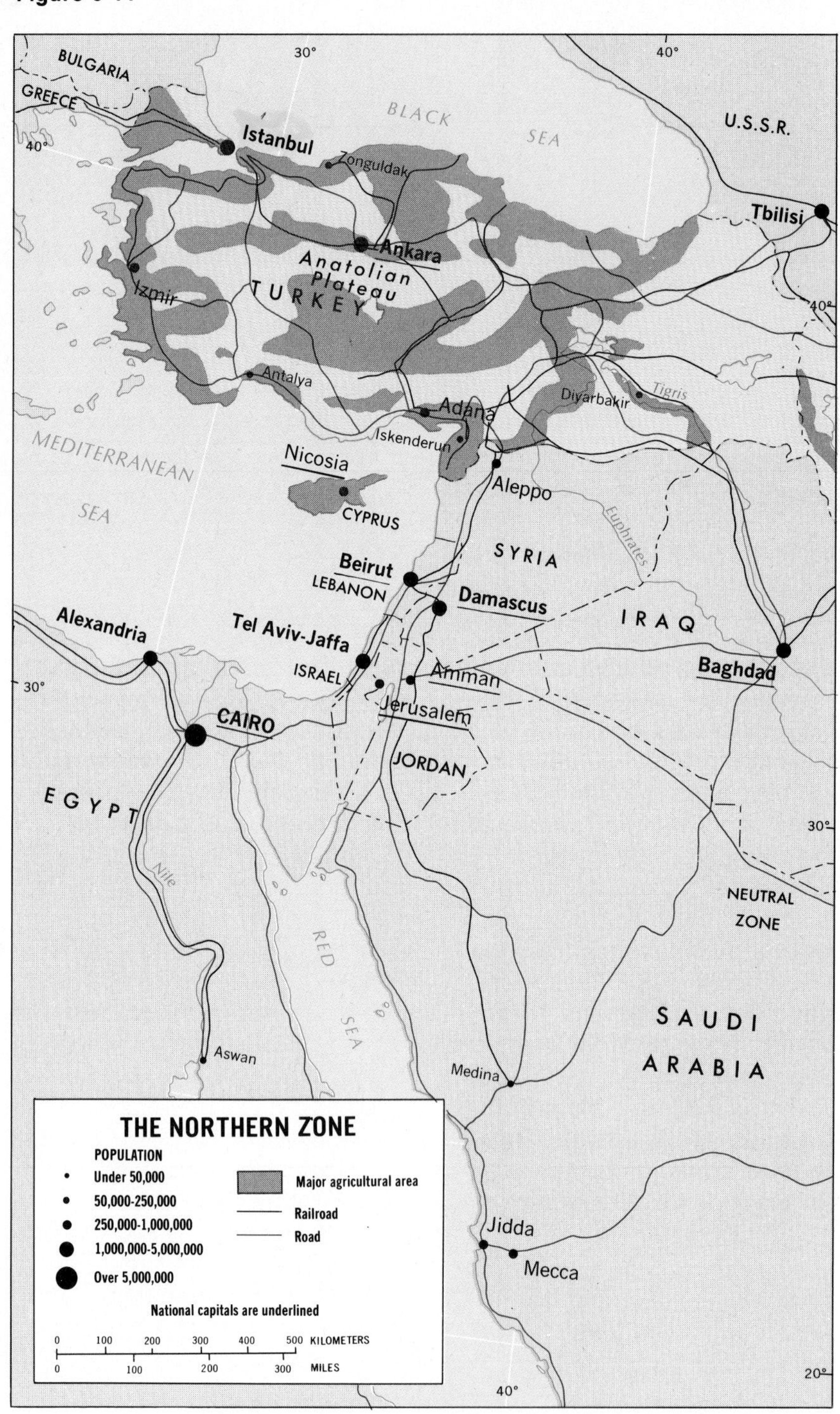

caravan routes. With so little usable land (such as the moist, narrow ribbon along the Caspian Sea coast and the small area along the Shatt-al-Arab in the south) and with most of the people depending directly on agricultural or pastoral subsistence, it is noteworthy that Iran's population is as large as 36 million.

In ancient times Persepolis in southern Iran was the focus for a powerful kingdom. Then, as now, people clustered in and around the oases or depended on *qanats* (underground tunnels from the mountains) for their water supply, as Persepolis did. The focus of modern Iran, Tehran (4.5 million), lies far to the north, on the slopes of the Elburz Mountains. Tehran, a city that also continues to depend to some degree on the *qanat* system for its water, rose from a caravan station to the capital of a modernizing state.

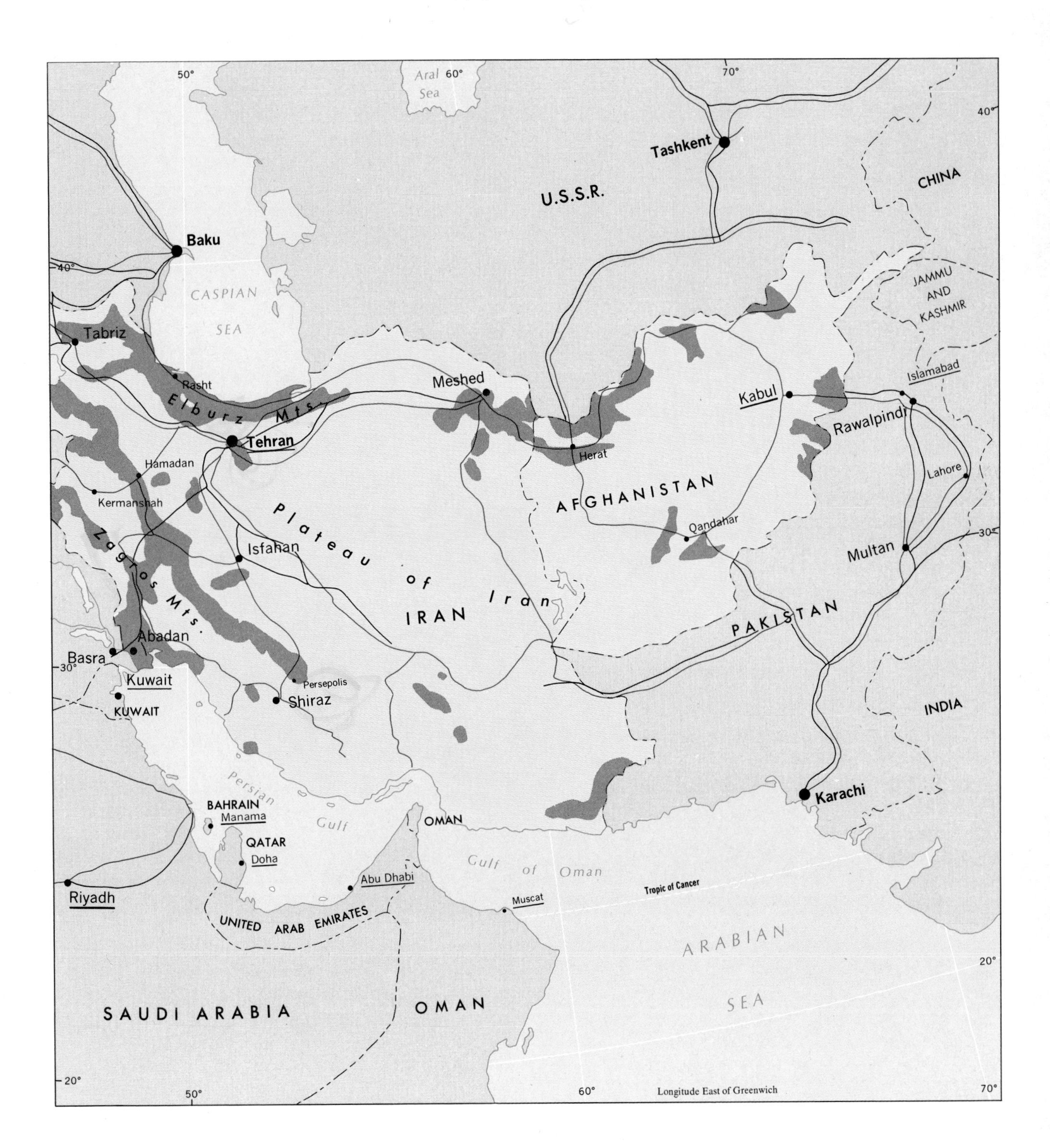

The Problem of Cyprus

In the northeastern corner of the Mediterranean, much farther from Greece than from Turkey or Syria, lies Cyprus—the problem of Cyprus. Since ancient times this island has been dominantly Greek in its population, but in 1571 it was conquered by the Turks, under whose control it remained until 1878. With the decay of the Ottoman state, the British first established an administration on Cyprus under an agreement with the Turks, but when World War I broke out they took full control of it. By the time the British were prepared to offer independence to much of their empire, they had a problem on their hands in Cyprus, for the Greek majority among the 600,000 population preferred *enosis*—union with Greece. It is not difficult to understand that the 20 percent of the Cypriot population that was Turkish wanted no such union; in fact, the answer the Turks gave was partition—the division of the island into a Turkish and a Greek sector. By 1955 the dispute had reached the stage of violence: differences between Greeks and Turks are deep, bitter, and intense. It was really impossible to find a solution to a problem in which the residents of an island country think of themselves as Greeks or Turks first rather than as Cypriots, and yet in 1960 Cyprus was made an independent country with a complicated constitution designed to permit majority rule but also to guarantee minority rights.

The fragile order finally broke down in 1973. As a civil war engulfed the island, Turkish armed forces intervened, and a major redistribution of population occurred. The northern one-third of Cyprus became the stronghold of more than 100,000 Turkish Cypriots; only a few thousand Greeks remained there. The rest of the island was the domain of the Greek majority, now including nearly 200,000 in refugee camps. As in Lebanon, partition became a serious prospect. In 1975 the Turkish part of Cyprus declared itself to be the Turkish Federated State of Cyprus, but the Greek sector refused to recognize it. In 1978 the partition proposal was formalized and submitted to the central government, where it was once again rejected. In 1980 the fundamental differences between Cypriot Greeks and Turks were still unresolved, and a form of *de facto* apartheid existed on the island.

Cyprus is a mostly mountainous country with little good farmland. Nevertheless, more than half of the island's total area is under some form of cultivation; even steep slopes are farmed. Still it is necessary to import grains every year. Few opportunities for irrigation exist, but most Cypriots depend directly on the soil for their subsistence. For its energy sources, too, Cyprus must depend on external supplies. To pay for these requirements, Cyprus exports Mediterranean fruits and wines, potatoes, and untreated copper and iron ores. Unemployment is high, and it has worsened since the civil war severely damaged the tourist industry (tourism has somewhat recovered, but remains far below its former level). Cyprus, a pawn of foreign powers, has been a casualty of history.

Iran indeed did modernize during the Shah's regime, although the villages and nomadic communities are reminders that change does not come quickly or easily to a country as large and tradition-bound as Iran. Thus the process tended to intensify local and regional contrasts, contributing to the success of the revolution that was to come, a revolution that had its roots in the Islamic traditions of the great majority of the people. Iran's national economy is based on the sale of its oil on overseas markets; unlike some other countries in this realm, Iran has been selling oil for 70 years. But never has its annual income been greater than in recent years, and at the direction of the Shah substantial investments were made in industrial diversification projects, power-generating equipment, and the modernization of agriculture—including the official abolition of peasant serfdom and the redistribution of some lands. Yet the real impact of the petroleum age still remained confined to the south between the Zagros Mountains and the Persian Gulf, where the oil reserves lie, as well as the refineries and port facilities—and in Tehran, with its skyscrapers and wide, tree-lined boulevards.

Like Ataturk and his successors in Turkey, Shah Mohammed Reza Pahlavi, who ruled the country from 1941 until his ouster in 1979, sought to bring major reforms to Iran. Apart from the social changes brought by his "White Revolution," the Shah enfranchised women in 1963, introduced profit-sharing in some enterprises, and created a kind of domestic peace corps (aimed principally at medical help and literacy in rural areas). But his reign was sustained by the intervention of foreign interests, and his policies ran counter to the Islamic traditions of the great majority of his people. Moslem leaders opposed him; a network of secret police (SAVAK) protected the almost unchecked privileges of the Shah and his elite. Among exiled religious leaders was Khomeini, and in time he became the symbol of the Islamic revolution that exploded during the late 1970s.

Among the circumstances that brought Khomeini to power in Iran were the searing material inequities that Pahlavi's modernization effort had created. The contrasts between the advantaged elite in Tehran and the poverty of the city's slum-dwellers, the "haves" in places like Isfahan (700,000) and Shiraz and the "have-nots" who remained mired in a web of debt and dependency—all this produced opposition to the Shah's policies. Ruthless political repression, reports of torture, and a lack of avenues for the expression of opposition also contributed to the breakdown of the order the Shah's authoritarian rule had wrought.

The wealth generated by petroleum could not transform Iran in ways that might have staved off the revolution. Modernization remained but a veneer. In the villages away from Tehran's polluted air, the holy men continued to dominate the life of ordinary Iranians. As elsewhere in the Moslem world, urbanites, villagers, and nomads remained enmeshed in a web of production and profiteering, serfdom, and indebtedness that has always characterized traditional society here. This ecological trilogy (as P. W. English called it in a 1967 article in the *Journal of Geography*) is not unique to Iran; it is characteristic of much of this entire realm. It ties the people who live in cities and towns to villagers and nomads; it is the urbanites who dominate because they have the money and own the land and the livestock. Thus, as English writes, "society is divided into three mutually dependent types of communities—the city, the village, and the tribe—each with a distinctive life-mode, each operating in a different setting, each contributing to the support of the other two sectors and thereby to the maintenance of total society . . . the (wealthy, powerful) city dwellers are principally engaged in collecting raw materials from the hinterland—wool for carpets and shawls, vegetables and grain to feed the urban population, and nuts, dried fruit, hides, and spices for export. In return, the urbanites supply peasants and nomads with basic economic necessities such as sugar, tea, cloth, and metal goods as well as cultural imperatives such as religious leadership, entertainment, and a variety of services. This concept of urban dominance is basic to the idea of an interdependent urban trilogy . . . ."

Again, Iran's twentieth-century oil age did little to change these relationships. The peasant tilling the fields near his village still worked on land owned by someone else. The nomadic herdsman moving his flock along centuries-old routes in search of grazing also was not freed from the tentacles of the city; most of the animals belonged to someone living in a far-away town—far away, but in control of his existence nevertheless. In modernizing Iran, new and old were never brought into a harmony that would have averted the turbulence of the 1970s.

Tehran, capital of Iran, with the Elburz Mountains in the background (Ray Ellis/Photo Researchers)

The easternmost country in the realm's northern tier—and also its only landlocked state—is Afghanistan (Fig. 6-11). Afghanistan has a territory that might be described as compact in shape except for a narrow but lengthy land extension, the Wakhan Corridor, that has the effect of adding a significant neighbor, China, to its other contiguous states: the U.S.S.R., Iran, and Pakistan.

Afghanistan was for centuries a battleground for outsiders including Turks and Indians, Tatars and Persians, but the country became a recognized territorial unit because two other competitors—the Russians and the British—agreed to guarantee its integrity. As a buffer state, Afghanistan's high mountains, deep valleys, remote provinces, and inaccessible frontiers were ideal. The nineteenth-century boundary created a country of nearly 650,000 square kilometers (250,000 square miles) that today incorporates about 22 million people, about half of them Pathans (also called Pushtuns and Pakhtuns) and the remainder Tajiks, Uzbeks, Turkmens, Hazaras, and smaller groups. Thus Afghanistan was hardly a nation-state, although virtually the entire population adheres to Islam and are Sunni Moslems.

Although Afghanistan does not possess oil reserves (some natural gas has been found in the north) and remains a country of herders, farmers, and nomads, it does have one asset it cannot sell: its relative location. Southern Afghanistan lies only 400 kilometers (250 miles) from the Arabian Sea, and not far from the mouth of the Persian Gulf. Along Afghanistan's eastern flank lies a province of Pakistan that has displayed separatist tendencies (see Chapter 8). Afghanistan's western border with Iran does not reflect a clear break in the cultural landscape; Dari, a Persian language, has official status inside Afghanistan. And along Afghanistan's northern border lies the Soviet Union which, late in 1979, became the latest of the country's foreign invaders. Afghanistan's relative location may not be saleable, but it is coveted nevertheless.

While foreigners may attempt to erect puppet governments in the capital, Kabul (700,000), the fact remains that Afghanistan is one of the realm's least developed countries. Urbanization is still under 20 percent, communications are minimal [in 1980 there was no railroad in Afghanistan and under 4000 kilometers (2500 miles) of paved roads], agriculture and

In the barren mountains of Afghanistan, firewood is a favorite item of trade. This is the market at Tashkourgan. (Rolland Michaud/Woodfin Camp)

pastoral subsistence remain the dominant livelihoods and there is little national integration. Although the unstable monarchy was overthrown in 1973 and revolutionary governments followed, the cycle of bloody transitions of the royal period continued—now exacerbated by the atheism of the new regimes. The Soviet intervention of 1979–1980 supported a Marxist administration whose modernizing intentions were viewed with suspicion by millions of traditional and conservative Moslem Afghans.

Geographically, Afghanistan not only lies in an historic buffer zone; it also is a transitional region. It differs culturally from Iran as well as Pakistan; its economic geography is unlike either. On various grounds Afghanistan may be viewed as part of each of the three realms here juxtaposed.

African Transition Zone

All across Africa south of the Sahara, from Mauritania and Senegal in the west to Ethiopia and Somalia in the east, Islam's culture yields to the lifestyles of Black Africa. It is a transition zone quite different from the northern tier. Turkey may be Europe-oriented, but it is a Moslem country. So is individualistic, non-Arab Iran. But in Africa, the transition has regional characteristics in several countries. Northern Nigeria is Moslem, the south is not. The northern Sudan is Moslem (including the core area between the White and the Blue Nile, and the capital of Khartoum), but the south is converted to Christianity. Ethiopia's heartland, centered on its capital of Addis Ababa, is the land of the Amharic people, who profess the Ethiopian Orthodox (Coptic Christian) faith—but its coastal province of Eritrea and its eastern region centered on Harar are Moslem. The transition zone is not always peaceful. In 1980, Ethiopia's province of Eritrea still was fighting a war of independence from Addis Ababa. Southern Sudanese throughout the 1960s and 1970s claimed brutal subjugation by their Moslem rulers. A civil war in Chad pitted Islamic northerners against non-Moslem southerners.

This regionalism within African countries came about, of course, because the European colonial invasion came after the Islamic diffusion had penetrated deep into West and East Africa, and the Europeans colonized Moslems and non-Moslems alike. The resulting problems are far from solved. The Somali Republic, where Islam is the state religion, lies adjacent to Ethiopia's Moslem east, and periodically Somalia aspires to a Greater Somali Union. Somali irredentism has even touched Kenya's northeastern territories. Both Ethiopia and Somalia aspire to control the young republic of Djibouti, a former French dependency that became independent in mid-1977 and contains the port of Djibouti—Ethiopia's principal outlet to the sea. Whether Moslem and non-Moslem coexistence can succeed in Nigeria or Sudan still is an unanswered question. That it has failed in Ethiopia is beyond doubt.

Additional Reading

Still probably the best place to start reading about the Middle East is in C. S. Coon's famous book *Caravan: the Story of the Middle East,* published by Holt, New York, in several editions. A book by W. B. Fisher is a standard geography: *The Middle East: a Physical, Social and Regional Geography,* now in its sixth edition from Harper and Row, New York, 1971. An older standard geography is G. B. Cressey's *Crossroads: Land and Life in Southwest Asia,* last published by Lippincott in Philadelphia, 1960. Also see J. Beaumont, et al., *The Middle East: a Geographical Study,* a Wiley publication in London, 1975. Another volume of general interest is J. I. Clarke and W. B. Fisher (eds.), *Populations of the Middle East and North Africa,* published in Cambridge by Heffer in 1972.

On Egypt, see K. M. Barbour, *The Growth, Location, and Structure of Industry in Egypt,* a Praeger, New York, publication of 1972. R. O. Collins and R. L. Tignor produced *Egypt and the Sudan,* published by Prentice-Hall, Englewood Cliffs, in 1967. H. Hopkins' *Egypt the Crucible,* published in London by Secker and Warburg, 1969, is still of interest. On the Maghreb, see I. W. Zartman, editor, *Man, State, and Society in the Contemporary Maghreb,* published by Pall Mall in London, 1973. *Morocco, Algeria, Tunisia* is the title of a volume by R. M. Brace, published in 1964 by Prentice-Hall but still useful. Another general work is edited by M. Brett: *Northern Africa: Islam and Modernization,* published by Cass in London, 1973. J. Wright described *Libya* in a book published by Praeger in 1969.

On Southwest Asia, there are surveys by the International Bank for Reconstruction and Development dealing with Iraq, Syria, Jordan, and Turkey. Also see a volume by R. K. Ramazani, *The Northern Tier: Afghanistan, Iran, and Turkey,* No. 32 in the Van Nostrand Searchlight Series. On Israel, see B. Halpern, *The Idea of the Jewish State,* published by Harvard University Press in 1961, and Y. Karmon, *Israel: a Regional Geography,* published by Wiley, New York, in 1971. P. W. English discussed urban dominance and ecological interconnections in *City and Village in Iran: Settlement and Economy in the Kirman Basin,* published by the University of Wisconsin Press, Madison, in 1966. English's article on "Urbanites, Peasants, and Nomads: the Middle Eastern Ecological Trilogy," appeared in the *Journal of Geography,* Vol. LXVI, No. 2, February 1967. A volume by J. C. Dewdney, entitled *Turkey: an Introductory Geography,* was published by Praeger in New York in 1971. And S. van Valkenburg introduced his cycle theory of politico-geographical evolution in *Elements of Political Geography,* a Prentice-Hall publication of 1939.

A great deal of interest to students of the Middle East occurs in *Arid Lands in Transition* edited by H. E. Dregne, Publication No. 90 of the American Association for the Advancement of Science, Washington, 1970. On the beginnings of agriculture, see E. Isaac, *Geography of Domestication* in Prentice-Hall's Foundations of Cultural Geography Series, 1970.

A market scene in Abidjan, Ivory Coast. (Marc & Evelyn Bernheim/Woodfin Camp)

Seven

African Worlds

Africa is an unusual continent. Even the world map suggests it: no other landmass is positioned astride the equator, reaching as far north as the latitude of Richmond, Virginia, and as far south as Buenos Aires, Argentina. No other continental landmass is without a Pacific Ocean coastline, as Africa is. Only Africa lies at the heart of the Land Hemisphere, with the Americas to the west, Eurasia to the north, Australia to the east, and Antarctica to the south. This means that Africa has a minimum aggregate distance to the world's other continents as well as a central location, antipodal to the Pacific.

Just how unusual Africa's physiography is, however, becomes clear when we study a map of the continent's landscapes (Fig. 7-1). In our discussions of other continental regions, for example, mountain ranges figure prominently. It is hard to visualize a South America without the Andes, North America without the Rockies, Europe without the Alps, Asia without the Himalayas. But Africa, with one-fifth of the land surface of earth, has nothing comparable. The Atlas Mountains of the Maghreb occupy a mere corner in the northwest of the vast African bulk, and in any case, the Atlas Mountains form an extension of Europe's Alps. The Cape Ranges in the far south have local, not regional dimensions. And what about those high mountains of Ethiopia, the great volcanoes of East Africa such as Kilimanjaro, and the sometimes snowcapped Drakensberg of South Africa? These are highlands indeed—but they are not comparable to the Andes or the Himalayas. You can observe the differences even on an ordinary atlas map: no elongated, parallel ranges here, no altiplanos, no mountain "chain." Africa's eastern axis is elevated, true, but its highland scenery is carved by rivers attacking the plateau margins. The highest elevations are those of volcanic peaks that generally stand alone, towering over a surrounding table-landscape.

IDEAS AND CONCEPTS

- Continental drift
- Geography of language
- Complementarity (2)
- Medical geography
- Periodic markets
- Unified field theory
- African socialism
- Irredentism (2)
- Sequent occupance
- Negritude

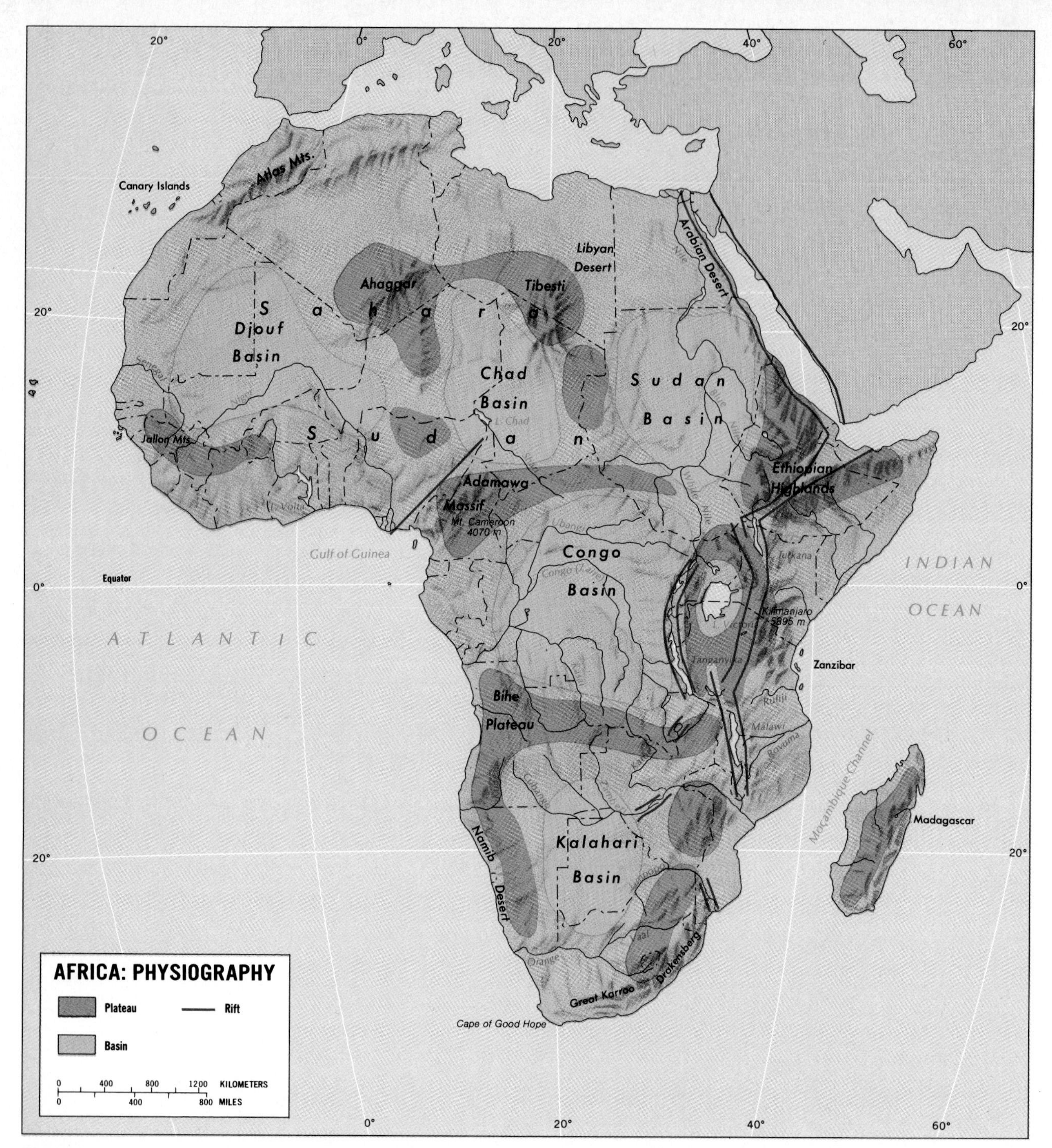

Figure 7-1

This discovery ought to stimulate us to scrutinize other aspects of the African physiography. In East Africa lies a set of great lakes. With the single exception of Lake Victoria, these lakes are markedly elongated, from Lake Malawi (formerly Nyasa) in the south to Lake Turkana in the north. What causes this elongation, and the persistent north-south trend that can be observed in these lakes? The lakes lie in deep trenches cut through the East African plateau, trenches that can be seen to extend *beyond* the lakes themselves. Northeast of Lake Turkana such a trench cuts the Ethiopian massif into two sections, and the Red Sea itself

Ten Major Geographic Qualities of Africa

1. The physiography of Africa is dominated by variable rainfall, soils of low fertility, and persistent environmental problems in farming.
2. The majority of Africa's peoples are dependent upon farming for their livelihood, and live in villages. Urbanization is accelerating, but still remains comparatively limited.
3. The peoples of Africa face a continued high incidence of diseases including malaria, sleeping sickness, and schistosomiasis.
4. Africa's large cities (with some exceptions) were founded by foreigners and continue to represent power and privilege in contrast to less-favored communities.
5. Severe and persistent conditions of underdevelopment affect the great majority of African states.
6. Patterns of raw-material exploitation and export routes, set up during colonial times, still prevail in most of Black Africa. Interregional connections are poor.
7. As a source of vital raw materials, Africa is increasingly drawn into the competition and conflict among the world's major powers.
8. Serious food shortages in areas of Africa notwithstanding, large segments of the best farmlands are still used to produce "cash" crops such as coffee, tea, and cotton for sale overseas.
9. Colonial depredations have been followed, in a number of African countries, by postindependence dislocations involving civil wars, ruthless dictatorships, unprecedented loss of life, and enormous refugee problems.
10. The breakdown of the buffer zone shielding South Africa from Black independent Africa is nearly complete and the political geography of Southern Africa is undergoing fundamental change.

looks very much like a continuation of it. On both sides of Lake Victoria smaller lakes lie in similar huge ditches, of which the western one runs into Lake Tanganyika and the eastern one cuts completely across Kenya and then peters out in Tanzania. The technical term for these trenches is *rift valleys;* as the name implies, they are formed when huge parallel cracks or *faults* appear in the earth's crust, and the in-between strips of land sink or are pushed down to form great valleys. These valleys stretch more than 9600 kilometers (6000 miles) from the north end of the Red Sea to Swaziland in Southern Africa.

In spite of their length, the rift valleys over great distances remain remarkably uniform, considering their dimensions. In general the rifts from Lake Turkana southward are between 30 and 90 kilometers (20 and 60 miles) wide, and the walls, sometimes sheer and sometimes steplike, are well defined. Hence, the rift—almost wherever it may be observed, whether in Swaziland, Kenya, or Ethiopia—is unmistakable in appearance. From the plateau rim, the land falls suddenly to a flat lowland that often possesses climatic and vegetative characteristics quite unlike those above the fault scarp. In the far distance lies the opposite scarp, and the plateau resumes.

Rift valleys fracture the East African landscape from Ethiopia to Malawi and beyond. The photograph shows the well-defined rift within which lies Lake Manyara, Tanzania. The lake has a high salt content; its level is subject to considerable fluctuation, as evinced by the drowned vegetation on the lake shore. (Harm J. de Blij)

Next, Africa's unusual river courses draw our attention. Africa has several great rivers; the Nile and Congo (Zaire) are among the major rivers of the world. But consider their courses. The Niger starts in the far west of Africa, on the slopes of the Futa Jallon Highlands, and then flows *inland* toward the Sahara Desert. Then, after forming an interior delta, it suddenly elbows southward, leaves the desert area, plunges over falls as it cuts through the plateau area of Nigeria, and creates another large delta at its mouth. The Congo begins as the Lualaba River on the Zaire-Zambia boundary, and for some distance it actually flows *northeast* before turning north, then west, then southwest, finally to cut through the Crystal Mountains to reach the ocean. It seems as though the *upper* courses of these two rivers are quite unrelated to the continent's coasts. In the case of the Zambezi River, whose headwaters lie in Angola and northwestern Zambia, the situation is the same; the river first flows south, toward the inland delta known as the Okovango Swamp; then it turns northeast, eventually to reach its delta immediately south of Lake Malawi. We may learn something by looking at the course of the Kafue River, the Zambezi's chief tributary. It flows southwest, also toward the Okovango Swamp, but then abruptly vacates its course to turn due east, as though the Zambezi "captured" and diverted it. Finally there is the famed erratic course of the Nile River, which braids into numerous channels in the Sudd area of the southern Sudan, and in its middle course actually reverses direction and flows southward before resuming its flow toward the Mediterranean delta. With so many course peculiarities, could it be that all of Africa's rivers have been affected by the same event at some time in their history? Perhaps—but let us first look further at our map.

All continents have low-lying areas; witness the coastal plain of North America and the coastal and river lowlands of Eurasia and Australia. But, as the map shows, coastal lowlands are few and of small extent in Africa. In fact, it is reasonable to call Africa a *plateau* continent; except for some low-lying areas in coastal Moçambique and Somalia, and along the north and west coasts, nearly all of the continent lies above 300 meters (1000 feet) in elevation and fully half of it is over 800 meters (2500 feet) high. Even the Congo Basin, Equatorial Africa's tropical lowland, lies well over 1000 feet above sea level, in contrast to the much lower Amazon Basin across the Atlantic.

But although Africa is mostly plateau, this does not mean that the surface is completely flat and unbroken. In the first place, the rivers have been attacking the surface for millions of years and they have made some pretty good dents in it, as the Zambezi's Victoria Falls, 1600 meters (1 mile) wide and 100 meters (over 300 feet) high, can attest. Volcanoes and other types of mountains, some of them erosional "leftovers," stand above the landscape in many areas—even in the Sahara Desert, where the Ahaggar and Tibesti Mountains both reach about 3000 meters (10,000 feet). And in several places the plateau has sagged down under the weight of accumulating sediments—in the Congo (Zaire)

The twin peaks of Kilimanjaro tower above the lava-domed landscape of the Kenya-Tanzania border. Kibo, the taller (and younger) cone, carries permanent snow and ice. Mawenzi, nearer the camera, is a more severely eroded volcanic plug. (Harm J. de Blij)

Basin, for example, rivers for tens of millions of years brought sand and silt downstream and for some reason dropped their erosional loads, apparently in a giant lake the size of an interior sea that seems to have existed here. Today the lake is gone, but the thick sediments that press the African surface into a giant basin are proof that it was there. And it was not the only one. To the south the Kalahari Basin was filling with sediments that today provide the desert's sand, and to the north, in the Sahara, three basins lay in Sudan, in Chad, and farther west in what is today Mali.

The margins of Africa's plateau, too, are of significance. In Africa—especially in Southern Africa—the term *Great Escarpment* is as commonplace as, say, *the Rockies* in the United States. And it is not surprising, for the plateau (the *highveld,* as it is sometimes called) over many hundreds of kilometers drops precipitously down from more than 1.5 kilometers in elevation to a narrow, hilly coastal belt. From Zaire to Swaziland, and intermittently on or near most of the African coastline, a scarp leads to the interior upland. Other parts of the world also have such escarpments: Brazil at the eastern margins of the Brazilian Highlands, India at the western edge of its Deccan Plateau. But Africa, even for its size, has a disproportionately large share of this phenomenon.

Add to this remarkable list of African physiographic peculiarities one other characteristic: the continent's western coastline, with a configuration that matches the east coast of South America. So jigsawlike is this matchup that it was recognized almost as soon as the first maps of Africa and South America were drawn (Francis Bacon drew attention to it in the 1620s!). In combination, Africa's world location relative to the other continents, its distinctive physiography, and its coastal configuration suggest a double-barreled hypothesis: (1) the landmasses at one time were part of a giant, single continent and have since moved apart, and (2) Africa's unique physiography stems from its position at the heart of this ancient supercontinent.

Continental Drift

The hypothesis of continental drift is now generally accepted, and the idea is not new, but it is worth recording the extent of which *geographic, spatial* considerations—direction, distance, shape, relative location—can contribute to the raising of very relevant questions in this connection.

But what can continental drift as a hypothesis do to help explain the general physical characteristics of Africa? First, let us consider the basic idea, which is that Africa, along with South America, Antarctica, Australia, Madagascar, and even southern India at one time formed a supercontinent (called Gondwana or sometimes Gondwanaland), as illustrated by Fig. 7-2. The significant point of this map is that Africa occupied the *central* position in Gondwana and was thus surrounded by the other landmasses. After a long period of unity, it began to break up more than 100 million years ago. The various fragments have moved radially away from Africa and are continuing to do so. Africa itself has moved least of all from its location near the South Pole.

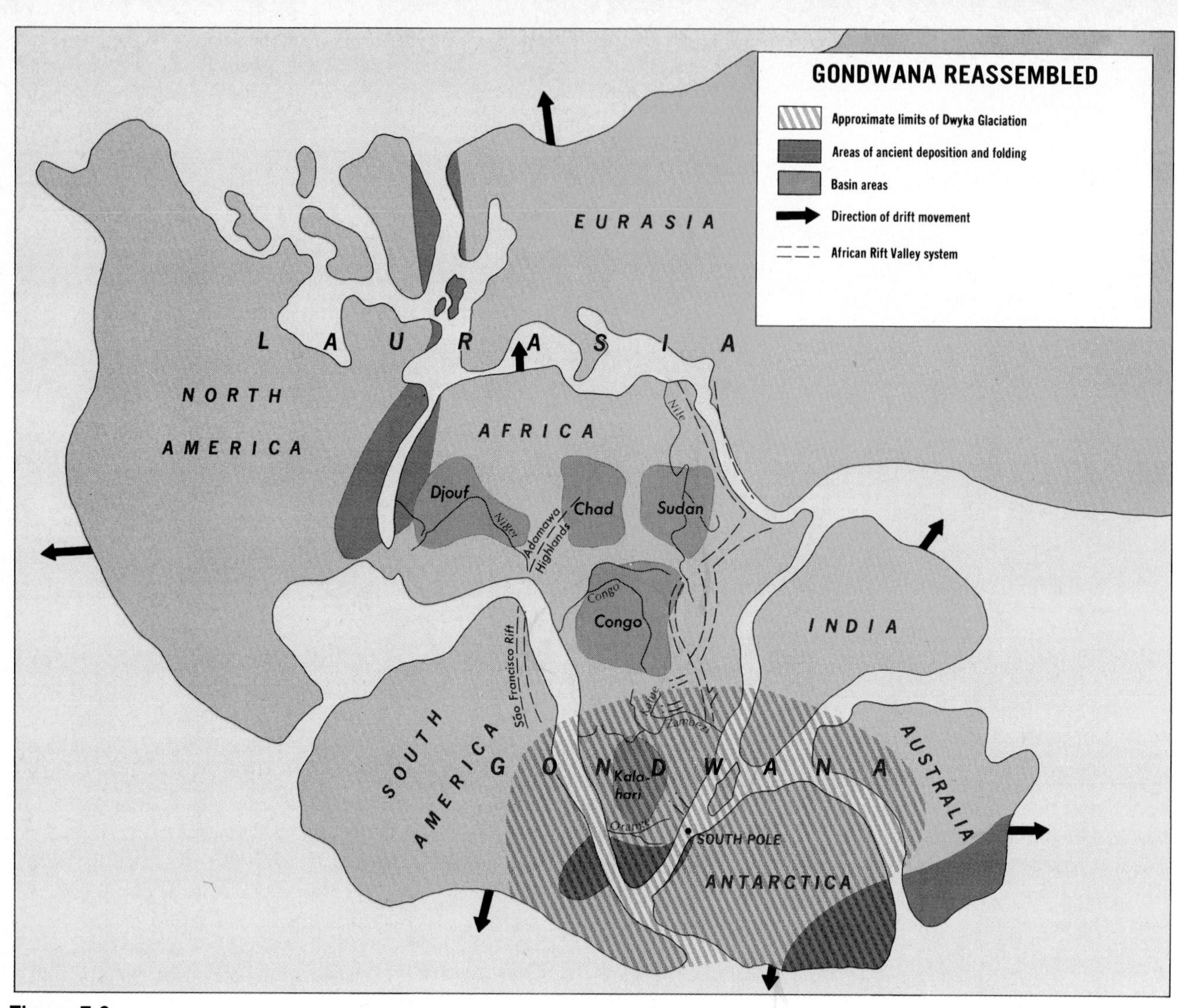

Figure 7-2

In the hypothetical reassembly of Gondwana, then, Africa was the core-shield; it occupied the heart of the supercontinent, and as such it did not have the coasts it has today. The rivers that arose in the interior failed to reach the sea: the upper Niger flowed into Lake Djouf, the Shari River into Lake Chad, the Upper Nile into Lake Sudan, the Lualaba into Lake Congo, and the upper Zambezi into the Okovango Delta on the shores of Lake Kalahari. Other rivers also drained toward the great basins, which filled with sediments. Eventually the great Gondwana landmass, for reasons still in doubt, began to break up. South America began to drift westward, and the West African coast formed. Antarctica moved away to the south, Australia to the east, India to the northeast, and Africa itself drifted slightly northward.

Now the coasts were created all around the continent, where huge escarpments marked the great continental fractures (probably looking a good deal like today's rifts), and rivers began to attack the Great Escarpment. Before long these rivers, by headward erosion, reached the long-isolated lake-basins in Africa's interior. With the huge volumes of lake water, these rivers cut deep gorges and formed fast-retreating waterfalls; it is no accident that Africa, not exactly a moist continent today, still has nearly half of the hydroelectric power potential of the world. According to the drift explanation, then, Africa's major rivers have a double history: the upper courses are predrift and filled the basins, and their lower courses are younger, and resulted from the release of the pent-up lakes.

Victoria Falls on the Zambezi River mark the progress of erosion begun after the breakup of Gondwana. As the photograph (taken from the Zambia side and looking toward Zimbabwe) shows, the water plunges into a deep, lengthy gorge. Weaknesses in the local basaltic rock have been eroded into lengthy gorges that may possibly signal the beginning of rift formation. (George Holton/Photo Researchers)

It is tempting to carry the explanation further, to argue that since landmasses moved away from Africa both east and westward, Africa was under great tensional stress and began to yield along the rift valleys we can observe in the landscape today. In fact it is probably not so simple, although the rifts are certainly related in some way to the drift process, and those who say that these rifts are the new fractures along which Africa will someday split may yet be proved correct. For the absence of mountains, again, a deceptively simple answer appears: all the other landmasses moved many hundreds, even thousands, of miles, and in the process their leading edges were crumpled up into giant folded and crushed mountain arcs, while Africa, which moved only somewhat to the north, shows such evidence only along its northwestern margin. This would explain the location of the American and Australian mountain ranges on the away-side from Africa.

Indeed, consider how the drift hypothesis ties the African—and, indeed, the whole Gondwana—physiography together. It relates the rivers and their peculiarities to the Great Escarpment, the Escarpment to the rift valleys, the rift valleys to the absence of fold-mountains, the absence of fold-mountains to Africa's limited lateral movement, and the factor of lateral movement to the distribution of major world mountain chains. It provides an explanation for the similarities between Africa's west and South America's east, for the stratigraphic (rock-layering) similarities between Africa, southern South America, Madagascar, India, and Australia. It has even made some predictions possible; on the grounds of rock successions in South Africa, geologists predicted that certain beds would be found beneath the ice and the surface in Antarctica—and they were. It suggests that the imbalance inherent in the concentration of the continents in the Land Hemisphere and the virtual emptiness of the Pacific Sea Hemisphere is being corrected by a slow redistribution of the landmasses into the Pacific Basin—and the Ring of Fire marks the forward push of the drift process.

In recent years the problem of a mechanism—what causes the landmasses to "drift"—has come closer to solution with the recognition of seafloor spreading and the identification of tectonic plates in the earth's crust. But before we assume that Africa's physiography has taught us all it could, these new concepts ought to be reviewed in an African light. We know now that Africa's rift valleys are part of a globe-girdling system of *midocean ridges,* magma-producing rifts that are the foci of crustal spreading. The map tells us that these so-called midocean ridges first formed as fractures—rifts—across the great Gondwana landmass; only after the continental pieces began to drift away did ocean water invade as the newly formed, separating crust became "seafloor." Seafloor spreading, then, might better be called *crustal spreading,* and we may see it going on today in East Africa's earthquake-prone, volcanically active rift region. Look again at the African map. Madagascar's separation may represent an old rift stage; the Red Sea rift is just the beginning of what will eventually be a wide portion of ocean. East Africa's rift system probably resembles what the Africa-America contact looked like just before separation began. In West Africa, a line from the

Bodele Depression through Lake Chad, along the Adamawa Highlands to Mount Cameroon may represent a preliminary stage to the whole process. If this is the case, it is probably premature to speak of an "African Plate" as is being done these days. Africa's tectonic plate may well consist of several plates—perhaps as many as six. Whatever the answer, it is difficult to look at the physiographic map of Africa without trying to find another piece to fit into the puzzle. Perhaps no other landmass on earth still carries so faithfully the imprint of its distant past.

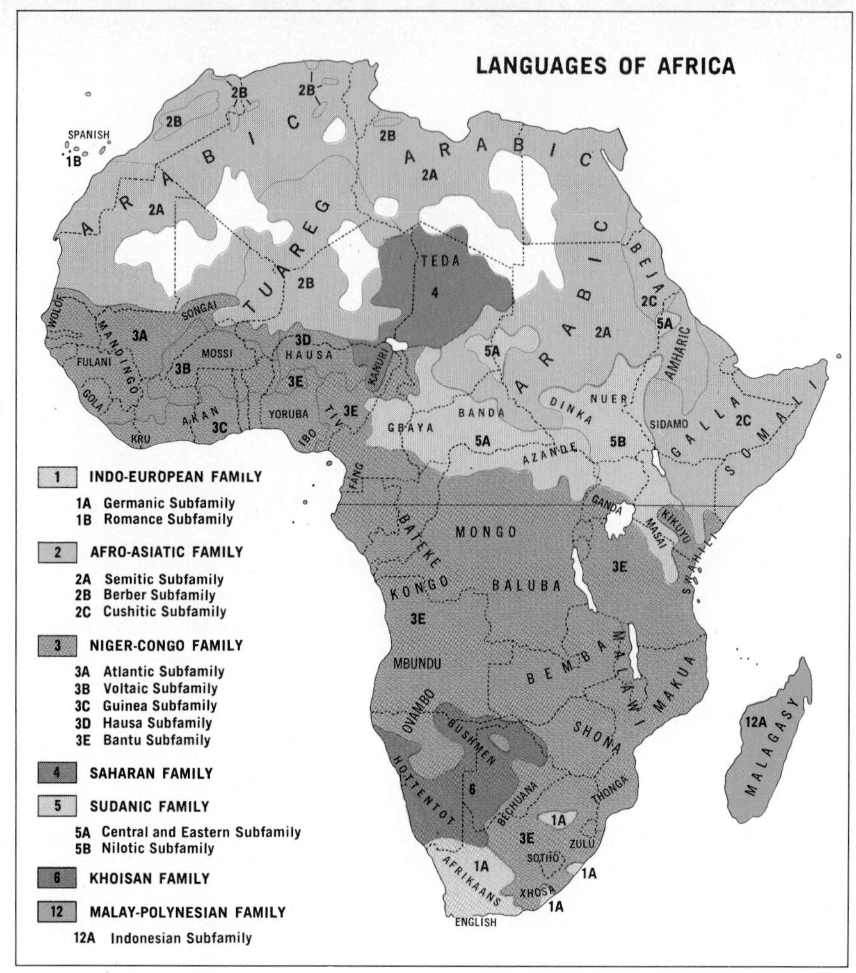

Figure 7-3

The Culture Realm

This chapter focuses on Africa south of the Sahara, the culture realm of Black Africa. Its total population (census data here are still unreliable) was approximately 350 million in 1980, about 8 percent of the world's. But the realm is divided into nearly 50 countries, territories, and island entities, so that it contains almost one-third of the political units in the world today. Black Africa's most populous state is Nigeria, but although Nigeria's population is officially given as over 70 million (Table I-1) much doubt surrounds the accuracy of the 1973 census. Next come Ethiopia (31 million), Zaire (29 million), and South Africa (29 million); at the lower end of the continuum Africa has several countries with populations under 1 million, including Equatorial Guinea, Gambia, and Botswana.

Black Africa's culture realm meets the world of Islam in a wide transition zone that appears on Fig. I-11 (page 41) as a line. We have already discussed the penetration of Islam into Black Africa; the outer limit of the farthest Islamic diffusion wave hardly would constitute a satisfactory boundary for the African culture realm. It would bisect Nigeria! Another criterion lies in the languages of Africa, and in West Africa at least, the linguistic map (Fig. 7-3) comes close to marking the northern limit of the African culture realm. In the "Horn," the close relationships between Amharic and Arabic make the language boundary less meaningful than the religious one.

As Fig. 7-3 shows, most of Africa's languages (as many as 1000 different languages are spoken on the continent) belong in the Niger-Congo family. In the classification of languages, we use terms also employed in biology, so that languages thought to have a shared origin are grouped in a *family;* when their relationship is closer, they belong together in a *subfamily.* The Niger-Congo family of languages, therefore, contains five subfamilies among which the Bantu subfamily is the most extensive. It is interesting to compare Fig. 7-3 and Fig. 1-7 (page 73). Note that the European subfamilies (the Latin-derived Romance languages, for example) are clearly delimited on the map—but the African subfamilies are not. Scholars studying African languages are not yet in agreement over the regional distributions of African language clusters. It *is* agreed that the Khoisan family (6), including the Bushman languages, represents the oldest surviving African languages, spoken over a far larger area of the continent before the Niger-Congo diffusion waves occurred. It also is evident that Madagascar's languages belong to a non-African, Malay-Polynesian family (12), revealing the Southeast Asian origins of a large sector of that island's population. Afrikaans (1A) is an Indo-European language, a derivative of Dutch spoken by a majority of the 4.5 million whites living in South Africa, the realm's southernmost state.

Although the concept of race is under constant debate among anthropologists and other scholars, it is useful to compare Fig. 7-4 and Fig. I-11. It reveals the inclusion of both European and Middle Eastern realms in the area of European (Caucasian) distribution, the wide dispersal of Asian peoples in the Soviet realm, and the clear delimitation of the transition from African to non-African regions. The boundary across West Africa and the "Horn" comes very close to marking the northern limit of the African culture realm as well.

The African Past

Africa may be the cradle of humanity. Research in Kenya, Ethiopia, and Tanzania has steadily pushed back the date of human prototypes' earliest origins by the hundreds of thousands, even millions, of years. It is therefore something of an irony that comparatively little is known about Black Africa from 5000 to 500 years ago—that is, prior to the onset of the colonial era. This is only partly due to the colonial period itself, during which African history was neglected and many misconceptions about African cultures and institutions arose and became entrenched. It is also a result of the absence of a written history over most of Africa south of the Sahara until the sixteenth century, and over a large part of it until much later than that. The best records are those of the savanna belt immediately to the South of the Sahara Desert, where contact with North African peoples was greatest and where Islam made a major impact. But the absence of a written record does not mean, as some historians have suggested, that Africa therefore does not have a history as such prior to the coming of Islam and Christianity. Nor does it mean that there were no rules of social behavior, no codes of law. Modern historians, encouraged by the intense interest shown by the Africans themselves, are trying now to reconstruct the African past, not only where this can be done from the meager written record, but also from folklore, poetry, art objects, buildings, and other such sources. But much has been lost forever. Almost nothing is known of the farming peoples who built well-laid terraces against the hillsides of northeastern Nigeria and East Africa, or of the communities that laid irrigation canals and constructed stone-walled wells in Kenya; and very little is known with any certainty about the people who, perhaps a thousand years ago, built the great walls of Zimbabwe. Porcelain and coins from China, beads from India, and other goods from distant sources have been found at Zimbabwe and other points in East and Southern Africa; but the trade routes within Africa itself, and even the products that moved on them and the people who handled them, still remain the subject of guesswork.

Livelihoods

A great many Africans today live as their predecessors did, by subsistence farming, herding, or both. The oldest means of survival, hunting and gathering, still sustain the Bushmen of the Kalahari and the Pygmies of Zaire, but for many centuries the majority of Africans have been farmers. Not that the African environment is particularly easy for the farmer who tries to grow food crops or raise cattle; tropical soils are notoriously unproductive, and much of Africa suffers from excessive drought. It seems strange, on a continent located astride the equator, most of it in the tropics, flanked by two oceans, that water supply can possibly be a problem. Consider Fig. I-5 (page 17); only small parts of West Africa and interior Zaire have really high annual precipitation totals, and the rainfall drops off very quickly both northward and southward. In West Africa only a comparatively narrow coastal zone is well watered; East Africa is actually quite dry, with steppe conditions (compare with Fig. I-6) penetrating far into Kenya; and in Southern Africa the rainfall is a modest 50 to 100 centimeters (20 to 40 inches) only in the east and south, while westward the Kalahari steppe and desert take over. And what the map does not tell us is that the hot tropical sun, by evaporating a good part of the rainwater that does reach the ground, reduces the moisture available for plant growth even further. Moreover, much of the rainfall of Africa outside the wettest areas comes concentrated in one or perhaps two seasons, and the intervening periods may be bone-dry; before long the herds of livestock have used up the reserve from the last wet season and their owners must drive them in search of water somewhere else. This is not just a problem faced by the Masai pastoralists of Kenya and Tanzania; it also confronts the modern ranchers in South Africa. The Masai drive cattle in a never-ending search for pasture and water; the South African ranchers may rail cattle out of the drought area before it is too late.

The vast majority of African families still depend on subsistence farming for their living. Only very few still survive on hunting and gathering alone; even the Pygmies trade for vegetables with their neighbors. Along the coast, especially the west coast, and the major rivers, some communities depend primarily on fishing. But otherwise the principal mode of life is farming—of grain crops in the drier areas and root crops in moister zones. In the methods of farming, the sharing of the work between men and women, the value and prestige attached to herd animals, and other cultural aspects, subsistence farming gives a great deal of insight into the Africa of the past. Moreover, the subsistence form of livelihood was little changed by the colonial impact; tens of thousands of villages all across Africa never were fully brought into the economic orbit of the European invaders, and life in these settlements went on more or less unchanged.

Africa's herders more often than not mix farming with their pastoral pursuits, and very few of them are actually "pure" herders. In Africa south of the Sahara there are two belts of herding, one extending along the West African savanna-steppe and connecting with the East African area (where the famous

Figure 7-4

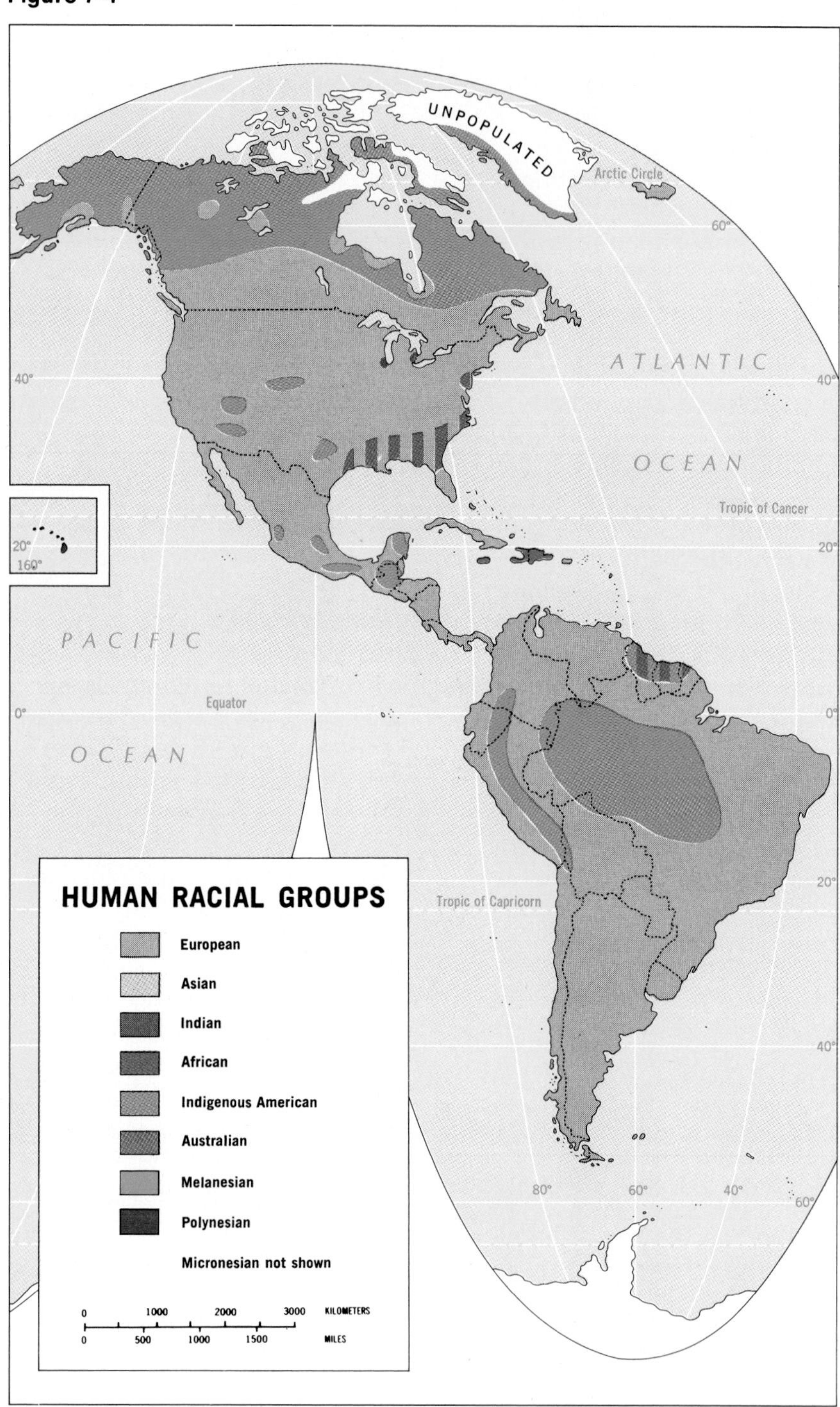

Masai drive their herds), and the other centering on the plateau of South Africa. Especially in East and South Africa, cattle are less important as a source of food than they are as a measure of the wealth and prestige of their owners in the community; hence African cattleowners in these areas have always been more interested in the size of their herds than in the quality of the animals. As far as the staple food in these areas is concerned, the grain crops are those with which the herding areas overlap. Probably a majority of Africa's cattle owners are sedentary farmers, although some peoples—such as the Masai—engage in a more or less systematic cycle of movement, following the rains and seeking pastures for their livestock. In West Africa the pastoralists who

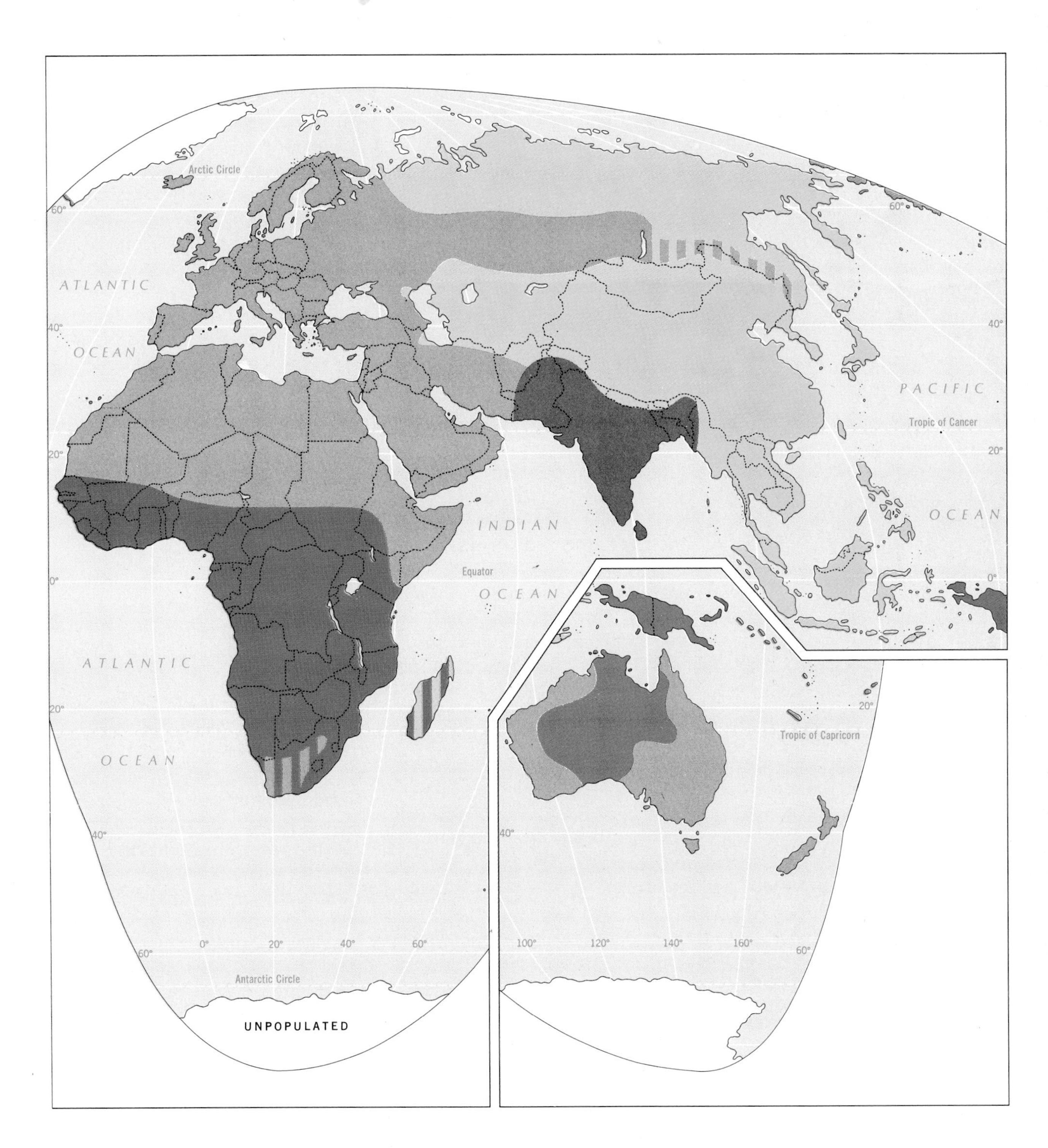

Cattle are the essence of existence for East Africa's Masai. To a considerable extent the Masai have managed to resist modernization and integration in the new society; their traditions have proven durable. (Marvin Newman/Woodfin Camp)

will sell their cattle for the meat face the problem of the considerable distances they are located from the major coastal markets. If the animals are railed or taken by truck the several hundred miles from the savanna-steppe margins to the coast, the cost is high but they do not lose much weight between farm and market; on the other hand, if the animals are driven to the markets on the hoof, the cost of moving them is low but their weight loss is considerable. And the cattle herder, too, faces environmental problems. Not least of these is the dreaded tsetse fly, which carries sleeping sickness to humans and which ravages Africa's animals—wild as well as domestic. Large parts of Africa that might otherwise be usable cattle country are infested by the tsetse fly and are rendered practically useless for pastoralism.

Cattle, of course, are not the only livestock in Africa. There are millions of goats everywhere—in the forest, in the savanna, in the steppe, even in the desert; they always seem to survive, and no African village would be complete without a few of these animals. Where conditions are favorable they multiply very fast, and then they denude the countryside and promote soil erosion; in Swaziland, for example, they constitute both an asset to their individual owners and a serious liability to the state. Nevertheless, goats are an important and valuable property.

States and Peoples

Africa on the eve of the colonial period was in many ways a continent in transition. For several centuries the habitat in and near one of the continent's most culturally and economically productive areas—West Africa—had been changing. For 2000 years and probably more, Africa had been creating as well as adopting ideas. In West Africa, cities were developing on an impressive scale; in Central and Southern Africa, peoples were moving, readjusting, sometimes struggling with each other for territorial supremacy. The Romans had penetrated to the Southern Sudan, North African peoples were trading with West Africans, Arab *dhows* were plying the eastern coasts, bringing Asian goods in exchange for gold, copper, and a comparatively small number of slaves.

Consider the environmental situation in West Africa as it relates to the past. As Figs. I-5 to I-8 indicate, the environmental regions in this part of the continent have a strong east-west orientation. The isohyets run parallel to the coast (Fig. I-5); the climatic regions, now positioned somewhat differently from where they were two millenia ago, still trend strongly east-west (Fig. I-6). Soil regions are similiarly aligned (Fig. I-8), and the vegetation map, although very generalized, also reflects this situation (Fig. I-7), with a coastal forest belt yielding to savanna (tall grass in the south, short grass in the north), which in turn gives way to steppe and desert.

Essentially, then, the situation in West Africa was such that over a north-south span of not too many hundreds of kilometers there was an enormous contrast in environments, economic opportunities, modes of life, and products. Obviously the people of the tropical forest produced and needed goods that were quite different from the products and requirements of the peoples of the dry, distant north. To give an example, salt is a prized commodity in the forest, where the humidity precludes its formation, but salt is in plentiful supply in the desert and steppe. Hence the desert peoples could sell salt to the forest peoples. What could the forest peoples offer in exchange? Ivory and spices could be sent north; there were elephants in the forest, and certain plants that yield valuable condiments. Thus, there was a degree of *complementarity* between the peoples of the forests and the peoples of the dry lands. And the peoples of the savanna—the in-betweens—were beneficiaries of this situation, for they found themselves in a position to channel and handle the trade, and that activity is always economically profitable.

The markets on which these goods were exchanged prospered and grew, and so, in the savanna belt of West Africa, there arose a number of true cities. One of these old cities, now an epitome of isolation, was once a thriving center of commerce and learning and one of the leading urban places in the world: Timbuktu. Others, predecessors as well as successors of Timbuktu, have declined, some of them into oblivion. Still other savanna cities continue to have considerable importance, like Kano in the northern part of Nigeria.

States of impressive strength and truly amazing durability arose in the West African culture hearth. The oldest state about which anything at all concrete is known is Ghana. Ancient Ghana was located to the northwest of the coastal country that has taken its name in modern times, covering parts of present-day Mali and Mauritania, along with some adjacent territory. It lay astride the Upper Niger River and included gold-rich streams coming off the Futa Jallon, where the Niger has its origins. For a thousand years, perhaps longer, old Ghana managed to weld various groups of people into a stable state. The country had a large capital city, complete with markets, suburbs for foreign merchants, religious shrines, and, some miles from the city center, a fortified royal retreat. There were systems of tax collection for the citizens and extraction of tribute from subjugated peoples on the periphery of the territory, and tolls were levied on goods entering the Ghanaian domain. An army kept control, and even after the Moslems from the northern dry lands invaded Ghana in about 1062, the state continued to show its strength: the capital was protected for no less than 14 years. A decade after its fall, a successful rebellion brought the old royal dynasty back. However, the invaders had ruined the farmlands of the country, and the trade links with the north were destroyed. Ghana could not survive, and it finally broke up into a number of smaller units.

In the centuries that followed, the center of politico-territorial organization in the West African culture hearth shifted almost continuously eastward—first to the successor state of Mali, which was centered on Timbuktu and the Middle Niger River, and which became consolidated in the thirteenth century, and then to Songhai, whose focus was Gao on the Niger, and which is still on the map today. Eventually certain states in northern Nigeria rose to prominence. One possible explanation for this eastward movement may lie in the increasing influence of Islam; Ghana had been a pagan state, but Mali and its successors were Moslem and sent huge, rich pilgrimages to Mecca along the savanna corridor south of the desert. Indeed, hundreds of thousands of citizens of the modern state of Sudan trace their ancestry to Nigeria, their forefathers having settled down there while journeying to or from Mecca. Whether Islam is indeed a major cause for West Africa's eastward shift is not certain; quite possibly the answer may lie in some other area, for example, that of intervening opportunity—the development to the east of shorter and better trade routes and larger trade volumes to the northern markets.

In any event, the West African savanna was the scene of momentous cultural, political, and economic developments for many centuries—but in its progress it was not alone in Africa. In what is today southwestern Nigeria, a number of urban farming communities became established, the farmers being concentrated in these walled and fortified places for reasons of protection and defense; surrounding each "city of farmers" were intensively cultivated lands that could sustain thousands of people clustered in towns. In the arts, too, southern Nigeria produced some great achievements, and the bronzes of Ife and Benin are true masterworks. In the region of the Congo (Zaire) mouth, a large state named Kongo existed for centuries. In East Africa, trade on a large scale with China, India, Indonesia, and the Arab world brought crops, customs,

An outpouring of artistic expression always has been part of life and culture in West Africa. This Benin bronze head from Nigeria, perhaps five centuries old, represents visual arts that also included sculpture and carving; today West African playwrights and authors produce a prodigious volume of plays, novels, short stories, poems, and other literature. (Marc & Evelyn Bernheim/Woodfin Camp)

and merchandise from these distant parts of the world to the coast, to be incorporated in African cultures. In Ethiopia and Uganda, populous kingdoms emerged. And much of what Africa was in those earlier centuries has yet to be reconstructed. But with all this external contact, it was clearly not isolated, as the lack of historical records might suggest.

Colonial Transformation

The period of European involvement in Black Africa began in the fifteenth century, a period that was to alter irreversibly the entire cultural, economic, political, and social makeup of the continent. It started quietly enough, with Portuguese ships groping their way along the west coast, rounding the Cape of Good Hope not long before the turn of the sixteenth century, and finding a route across the Indian Ocean to the spices and riches of the Orient. Soon other European countries sent their vessels to African waters, and a string of coastal stations and forts sprang up. In West Africa, the nearest part of the continent to European spheres in Middle and South America, the initial impact was strongest. At their coastal points the Europeans traded with African middlemen for the slaves that were wanted on American plantations, and for the gold that had been flowing northward across the desert, and for ivory and spices. All of a sudden the centers of economic activity lay not in the cities of the savanna, but in the foreign stations on the coast. As the interior declined, the coastal peoples thrived. Some of the small forest states rose to power and gained unprecedented wealth, transferring and selling slaves captured in the interior to the white men on the coast. Dahomey (now called Benin) and Benin (now part of Nigeria) were slave-trade-built states, and when the practice of slavery eventually came under attack in Europe, abolition was vigorously opposed by those who had inherited the power and riches it had brought.

Although it is true that slavery was not new to West Africa, the kind of slave trading introduced by the Europeans certainly was. In the savanna states, African families who had slaves usually treated them comparatively well, permitting marriage, affording adequate quarters, and absorbing them into the family. The number of slaves held in this way was quite small, and probably the largest number of persons in slavery in precolonial Africa were in the service of kings and chiefs. In East Africa, however, the Arabs had introduced—long before the Europeans—the sort of slave trading that the white man brought first to the west: African middlemen from the coast raided the interior for slaves, marched them in chains to the Arab *dhows* that plied the Indian Ocean, and there, packed by the hundreds in specially built vessels, they were carried off to Arabia, Persia, and India (Fig. 7-5). It is sad but true that European, Arab, and African combined to ravage the black continent, forcing perhaps as many as 30 million persons away from their homeland in bondage, destroying families, whole villages, and cultures, and bringing to those affected an amount of human misery for which there is no measure.

The European presence on the West African coast brought about a complete reorientation of trade routes, for it initiated the decline of the interior savanna states and strengthened coastal forest states, and it ravaged the population of the interior through its insatiable demand for slaves. But it did not lead to any major European thrust toward the interior, nor did it produce colonies overnight. The African middlemen were well organized and strong, and they managed to maintain a standoff with their European competitors, not just for a few decades, but for centuries. Although the European interests made their appearance in the fifteenth century, West Africa was not carved out among them until nearly four centuries later; the British could not establish control over the Ashanti group of states (in present-day central Ghana) until the first years of the twentieth century. Even the Portuguese, earliest of Africa's European colonizers, just a century ago had effective control over only about one-quarter of their African provinces.

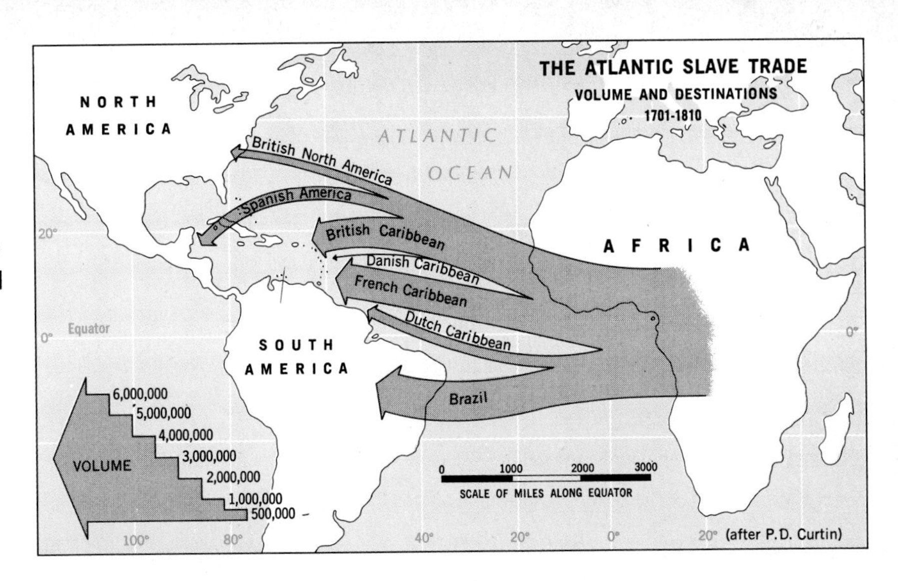

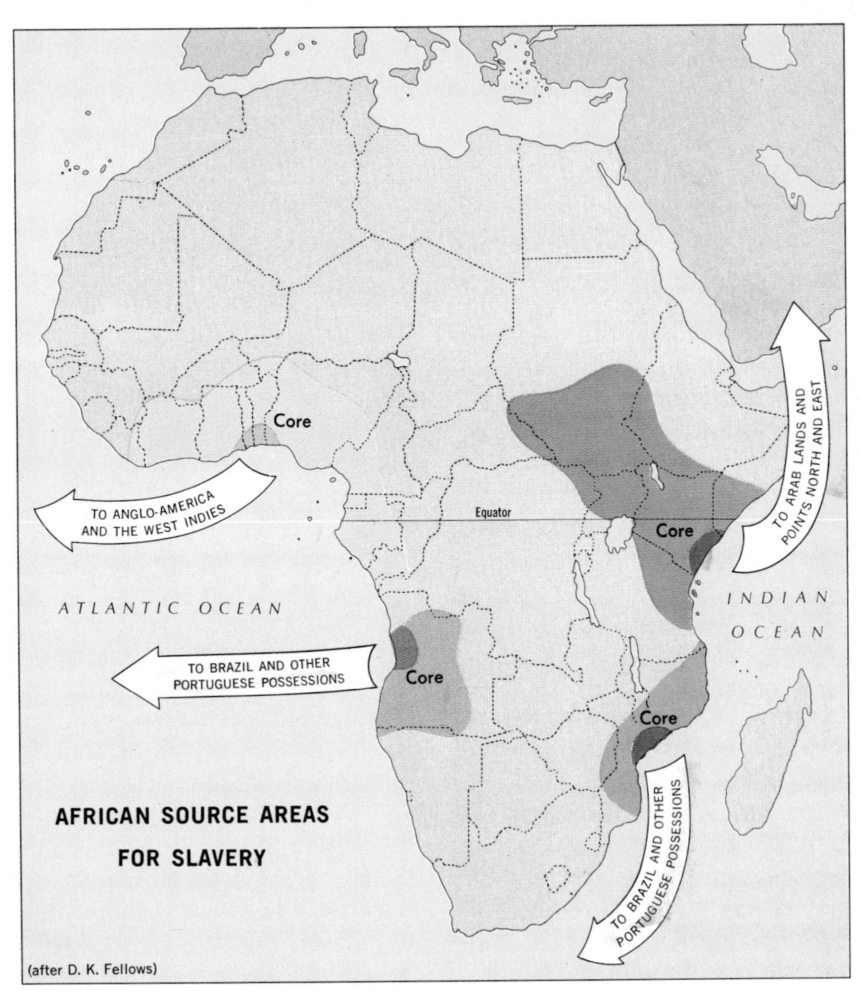

Figure 7-5

As fate would have it, European interest was to grow strongest—and ultimately most successful—where African organization was weakest. In the middle of the seventeenth century the Hollanders chose the shores of Table Bay, where Cape Town lies today, as the site for a permanent settlement, although not for purposes of colonization. The objective, rather, was to establish a revictualing station on the months-long voyage to and from Southeast Asia and the Indies; the tip of Africa, where the Atlantic and Indian Oceans meet, was the obvious halfway house. African considerations hardly entered into the choice, for there was no intent to colonize there. Southern Africa was not known as a productive area, and the more worthwhile East lay in the spheres of the Portuguese and the Arabs. Probably the Hollanders would have elected to build their station at the foot of Table Mountain whatever the indigenous population of the interior, but as it happened they picked a location about as far away from the major centers of Bantu settlement as they could have found. Only the Bushmen and their rivals, the Hottentots, occupied Cape Town's hinterland. When conflicts developed between Amsterdam and Cape Town and some of the settlement's residents decided to move into this hinterland, they initially faced only the harassment of groups of these people, not the massive resistance that might have been offered by the Bantu states. To be sure, a confrontation eventually did develop between the advancing Europeans and the similarly mobile Bantu Africans, but it began several decades after Cape Town was founded and hundreds of miles from it. Unlike some of the West African way stations, Cape Town was never threatened by African power, and it became a European gateway into Southern Africa.

Elsewhere in Black Africa, the European presence remained confined almost entirely to the coastal trading stations, whose economic influence was very strong, of course. No real frontiers of penetration developed; individual travelers, missionaries, explorers, and traders went into the interior, but nowhere else in Africa south of the Sahara was there an invasion of white settlers comparable to Southern Africa's.

The Scramble

After more than four centuries of contact, Europe finally laid claim to all of Africa during the second half of the nineteenth century. Adventurous men such as Livingstone, Park, Speke, Burton, and Grant had "explored" parts of the continent. Now individual representatives of European governments sought to expand or create African spheres of influence for their homelands: Rhodes for Britain, Peters for Germany, De Brazza for France, and Stanley for the king of Belgium were among those who helped shape the destiny of the continent. The British talked of an all-British axis from Cairo to the Cape; the French desired to create a vast colonial empire across West and North Africa from Dakar to Cape Guardafui in the east, or at least to Djibouti. The Portuguese suddenly sought to connect their Angolan and Moçambique possessions across South-Central Africa, and the Germans, coming late to the colonial scene, showed up in part to acquire colonies, as they did in West, South, and East Africa, but also to obstruct the colonial designs of their European rivals. In some areas such as along the lower Congo (Zaire) River and in the vicinity of Lake Victoria, the competition between the European powers was very intense. Spheres of influence began to crowd each other; sometimes they even overlapped. And so, late in 1884, a conference was convened in Berlin to sort things out. At this conference the groundwork was laid for the now-familiar boundary lines of Africa.

As the twentieth century opened, Europe's colonial powers were busy organizing and exploiting their African dependencies. The British, having defeated the Boers, came very close to achieving their Cape-to-Cairo axis: only German East Africa interrupted a vast empire that stretched from Egypt and the Sudan through Uganda, Kenya, Nyasaland, and the Rhodesias to South Africa (Fig. 7-6). The French took charge of a vast realm that reached from Algiers in the north and Dakar in the west to the Congo River in Equatorial Africa. King Leopold II of the Belgians held personal control over the Congo. Germany had colonies in West Africa (Togo), Equatorial Africa (Kamerun), South Africa (South West Africa) and East Africa (German East Africa, later Tanganyika). The Portuguese controlled two huge territories in Angola and Moçambique, both much larger than the original spheres of influence along the Atlantic and Indian coasts; in West Africa, Portugal also held a small entity known as Portuguese Guinea. Italy's posses-

Maize (corn) is Africa's leading staple. Here the cobs are shelled in the village of Chikumbi, near Lusaka, in Zambia. The corn will be pounded into a flour and cooked to be consumed as a porridge. (Marc & Evelyn Bernheim/Woodfin Camp)

sions in tropical Africa were confined to the "Horn," and even Spain was in the act with a small dependency consisting of the island of Fernando Poo and a mainland area called Rio Muni. The only places where the Europeans showed some respect for African independence were Ethiopia, which fought some heroic battles against Italian forces, and Liberia, where Afro-Americans retained control.

The two world wars had some effect on this politicogeographical map of Africa. In World War I, defeated Germany lost its African possessions altogether, and they were placed under the administration of other colonial powers by the League of Nations' Mandate System. In World War II, fascist Italy launched a briefly successful campaign against Ethiopia, but the ancient empire was restored to independence when the Allied forces won the war. Otherwise the situation in colonial Africa in the late 1940s—after a half century of colonial control and on the eve of the "Wind of Change"—was still quite similar to the one that arose out of the Berlin Conference.

Colonial Policies

Geographers—especially political geographers—are interested in the way in which the philosophies and policies of the colonial powers were reflected in the spatial organization of the African dependencies. These colonial policies can be expressed in just a few words. For example, Britain's administration in many parts of its vast empire was referred to as "indirect rule," since indigenous power structures were sometimes left intact and local rulers were made representatives of the crown. Belgian colonial policy was called "paternalism" in that it tended to treat Africans as children, to be tutored in Western ways, although slowly. While the Belgians made no real efforts to make Belgians of their African subjects, the French very much wanted to create an "Overseas France" in their African dependencies. French colonialism has been identified as a process of "assimilation"—the acculturation of Africans to French ways of life. France

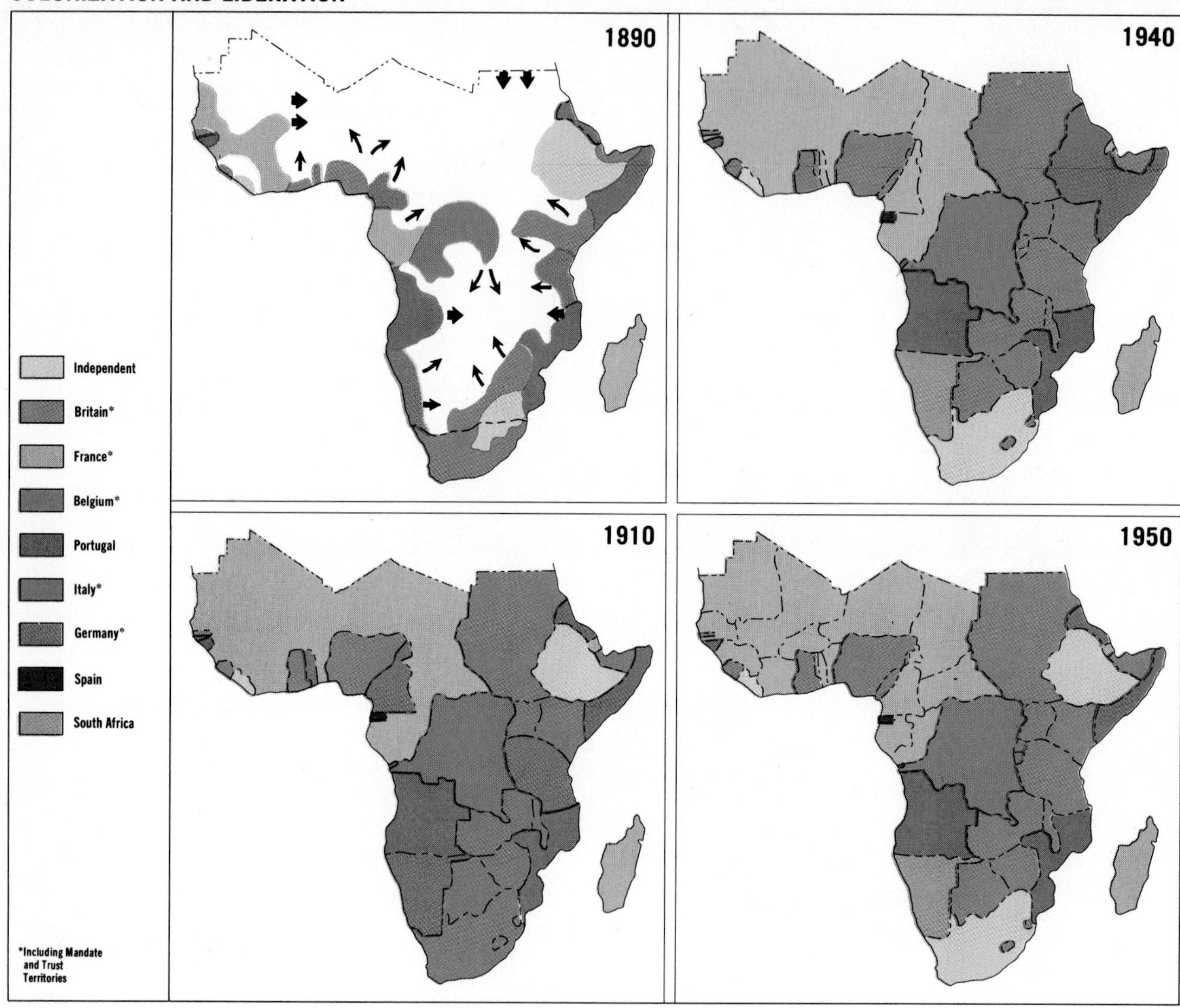

Figure 7-6

made a strong cultural imprint in the various parts of its huge colonial empire. Portuguese colonial policy had similar objectives, and the African dependencies of Portugal were officially "Overseas Provinces" of the state. If you sought a one-word definition of Portuguese colonial policy, however, the term *exploitation* would probably emerge most strongly. Few colonies have made a greater contribution (in proportion to their known productive capacity) to the economies of their colonizing masters than have Moçambique and Angola.

Colonial policies have spatial expression as well, and the spatial organization fostered by the colonial powers has become the infrastructure of independent Africa. As the map (Fig. 7-6) shows, Britain possessed the most widely distributed colonial empire in Africa. British colonial policy tended to adjust to individual situations: in *colonies,* white settler minorities had substantial autonomy (Rhodesia, Kenya); in *protectorates,* the rights of African peoples were guarded more effectively; in *mandate* (later *trust*) territories, the British undertook to uphold League of Nations (later United Nations) administrative rules; and in one case the British shared administration with another government, as in Sudan's *condominium.* Britain's colonial map of Africa was a patchwork of these different systems, and the independent countries that emerged reflect the differences. Nigeria, which had been a colony in the south and an indirectly ruled protectorate in the north, became a federal state.

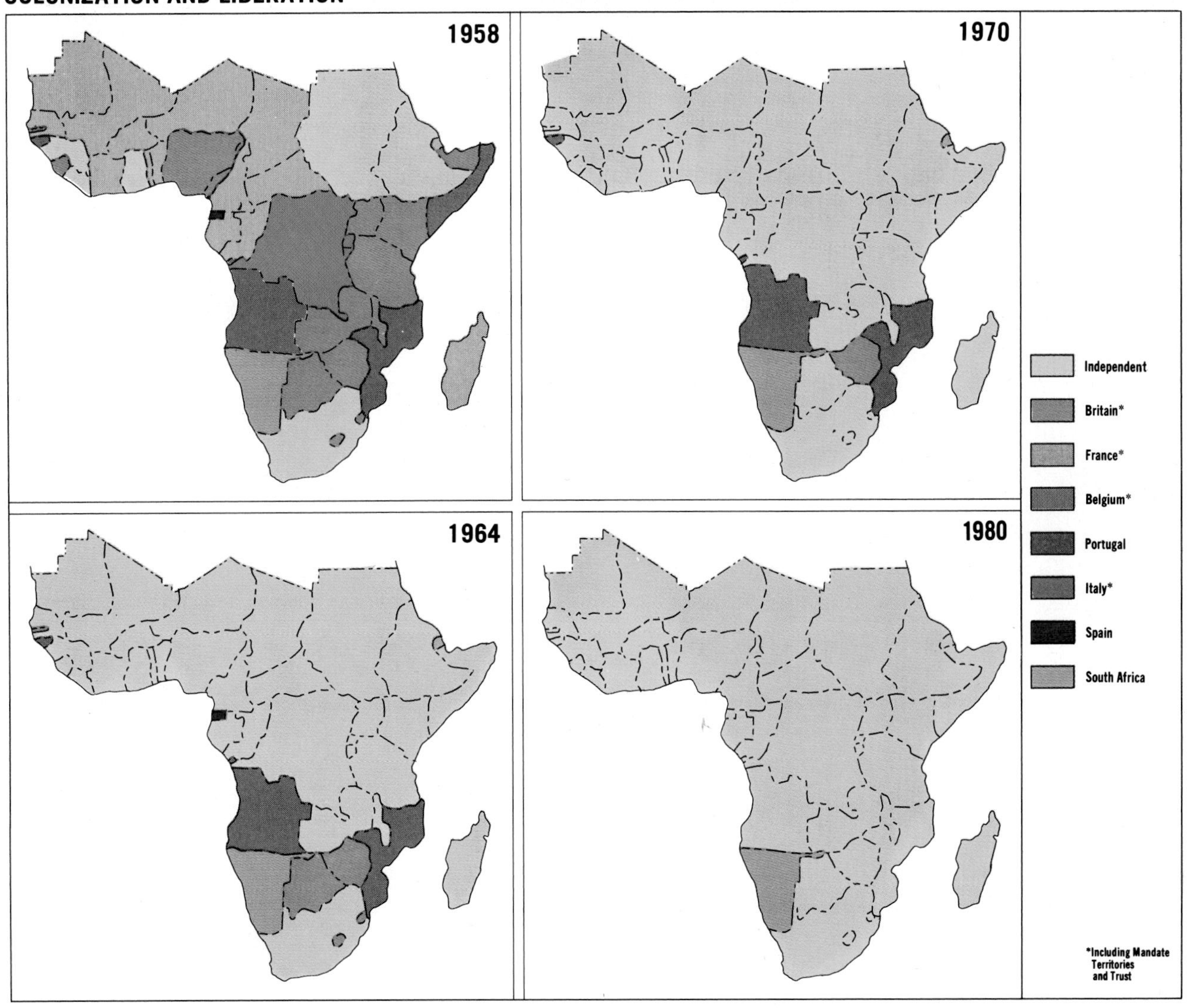

Kenya, the former colony, became a highly centralized unitary state after the Africans wrested control of the productive core area from the whites.

In contrast to the British, the French put a cloak of uniformity over their colonial realm in Black Africa. Contiguous and vast, although not very populous, France's colonial empire extended from Senegal eastward to Chad and southward to the former French Congo. This huge area was divided into two units, French West Africa (centered on Dakar) and French Equatorial Africa (whose headquarters was Brazzaville). After World War I, France was granted a mandate over the former German colony of Kamerun in Equatorial Africa and over a part of Togo in West Africa; its only other dependencies were French Somaliland, the gateway to Ethiopia, and the island of Madagascar. France itself, we know, is the textbook example of the centralized, unitary state, whose capital is the cultural, political, and economical focus of the nation, overshadowing all else. The French brought their concept of centralization to Africa as well. In France all roads led to Paris: in Africa all roads were to lead to France, to French culture, to French institutions. For the purposes of assimilation and acculturation, French West Africa, half the size of the entire United States of America (although with a population of only some 30 million in 1960), was divided into administrative units, each centered on

The Berlin Conference

In November 1884 the imperial chancellor and architect of the German empire, Bismarck, convened a conference of 14 powerful states (including the United States) to settle the partition of Africa. Bismarck wanted not only to expand German spheres of influence in Africa; he sought to play Germany's colonial rivals off against each other, to the Germans' advantage.

The major colonial contestants in Africa were the British, who held beachheads along the West, South, and East African coasts; the French, whose main sphere of activity was in the area of the Senegal River and north of the Congo; the Portuguese, who now desired to extend their coastal stations in Angola and Moçambique deep into the interior; King Leopold II of Belgium, who was amassing a personal domain in the Congo (Zaire); and Germany itself, active in areas where the designs of other colonial powers might be obstructed as in Togo (between British holdings), Cameroon (a wedge into French spheres), South West Africa (taken from under British noses in a swift strategic move), and East Africa (where the effect was to break the British design for a Cape-to-Cairo axis).

When the conference convened in Berlin, over 80 percent of Africa still was under traditional African rule, but the colonial powers' representatives nevertheless drew their boundary lines—through known as well as unknown regions, haggling over pieces of territory, erasing and redrawing boundaries, and exchanging pieces of African real estate at the urging of their governments. In the process, African peoples were divided, unified regions were ripped apart, hostile societies were thrown together, hinterlands were disrupted, and migration routes were closed off. All this was not felt in the beginning of course—but such was the effect when the colonial powers began to consolidate their holdings, and those boundaries on paper became barriers on the African land. As Fig. 7-6 shows, that process was under way by 1890, but was not really completed until the early 1900s.

The Berlin Conference was Africa's undoing in more than one way. Not only did the colonial powers superimpose their domains on the African continent; when independence returned to Africa, the realm also had a legacy of political fragmentation that could neither be eliminated nor made to work satisfactorily. The African politico-geographical map is a permanent liability resulting from three months of ignorant, greedy acquisitiveness during Europe's insatiable search for minerals and markets.

the largest town, all oriented toward the governor's headquarters at Dakar. As the map shows, great lengths of these boundaries were delimited by straight lines across the West African landscape; history tells us to what extent they were drawn not on the basis of African realities but on grounds of France's administrative convenience. The present-day state of Upper Volta, for example, existed as an entity prior to 1932, when, because of administrative problems, it was divided up among Ivory Coast, Soudan (now Mali), and Dahomey. Then, in 1947, the territory was suddenly re-created. Little did the boundary makers expect that what they were doing would one day affect the national life of an independent country.

During their period of tenure the French created a French-speaking, acculturated

African elite, trained at French universities and often experienced in French politics through direct representation in Paris. From this elite corps, headquartered in the colonies' capital cities, the governments of the now-independent states were forged. These have retained their ties with France, and virtually every modern institution, including the political machinery, is based on French models. Heavy investments are made, where possible, to make the capital a true primate city—Ivory Coast's Abidjan is a good example—and to improve the connections between the various regions of the country and the capital and core area. France continues, through aid programs, loans, educational assistance, and various other programs to maintain its presence in what is now called Francophone Africa.

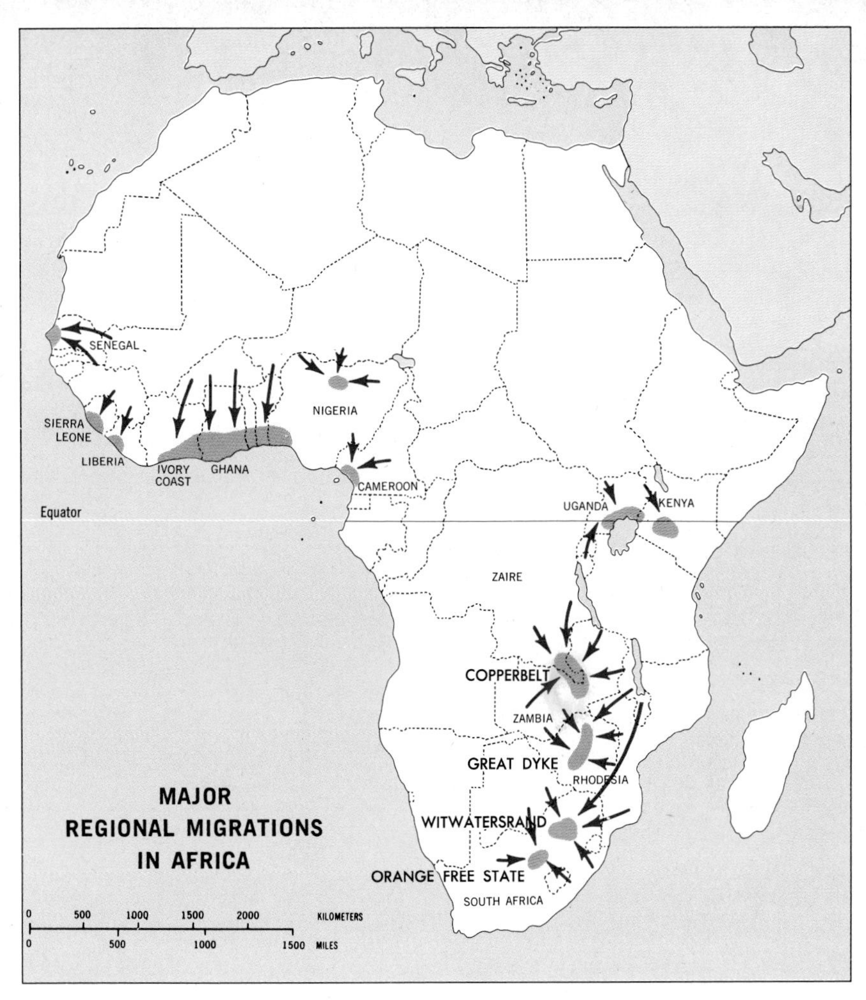

Figure 7-7

Belgian administration in the Congo (now Zaire) provides another insight into the results, in terms of spatial organization, of a particular set of philosophies of colonial government. Unlike the French, the Belgians made no effort to acculturate their African subjects. The policies identified under the catchword "paternalism" actually consisted of rule in the Congo by three sometimes competing interest groups: the Belgian government, the managements of huge mining corporations, and the Catholic church. As it happened, each of these groups had major regional spheres of activity in this vast country. As the map shows, Zaire has a corridor to the ocean along the Zaire River, between Angola to the south and the former French Congo to the north. The capital, long known as Leopoldville (now Kinshasa), lies at the eastern end of this corridor; not far from the ocean lies the country's major port, Matadi. Here has developed the administrative and transport core area of Zaire: this was the place from which the decisions made in Brussels were promulgated by a governor-general; here now rules the government. But the economic core of the Congo (and now of Zaire) lay in the country's southeast, in the province then known as Katanga and centered on the copper-mining headquarters of Lubumbashi (formerly Elizabethville). This is the northernmost extension of Southern Africa's great mining complexes, and it continues into Zambia's Copperbelt. From hundreds of miles around, African workers in search of jobs streamed to the mines (Fig. 7-7). These Southern African mining regions set up patterns of regional migration that persisted for decades and still exist today.

Between the former Belgian Congo's economic and administrative core areas lay the Congo Basin, once the profitable source of wild rubber and ivory. The colony's six administrative subdivisions were laid out in such a way that each incorporated part of the area's highland rim and part of the forested basin. When independence approached, it briefly seemed that each of these Congolese provinces, centered on its administrative capital, might break away and become an independent African country (as each French-Africán dependency in West and Equatorial Africa did). But in the end the country held together, and Leopoldville became Kinshasa, capital of Black Africa's largest state in terms of territory.

Portugal's rule in Angola and Moçambique was designed to exploit four assets: (1) labor supply to the interior mines (especially in Moçambique), (2) transit functions and port facilities (Moçambique from South Africa and Rhodesia; Angola from the Copperbelt and Katanga), (3) agricultural production (especially cotton from northern Moçambique), and (4) minerals (mainly from diamond- and oil-rich Angola). In this effort the Portuguese created a system of control that involved forced labor and the compulsory farming of certain crops. In Moçambique, especially, the country was split into a large number of small districts that were tightly controlled. Movement and communication, even within a single African ethnic area, were kept to a minimum. Portuguese colonial rule often was described as the most harsh of all European systems, and for a long time it seemed unlikely that an independence movement could be mounted. But when independence came to Angola's and Moçambique's neighbors (Zaire and Tanzania), Portugal's days in Africa were numbered. As elsewhere in recently colonial Africa, however, the colonial imprint in former Portuguese Africa remains strong, a pervasive element in the regional geography of the continent.

Hazards and Diseases

Africa is not a densely populated realm. The total population of approximately 350 million (1980) includes several large concentrations (Fig. I-9) in Nigeria, in the region of Lake Victoria, and in Southern Africa, but much of Africa south of the Sahara remains sparsely peopled.

African environments are difficult. In the 1970s parts of Ethiopia, West Africa (the Sahel), Tanzania, and Moçambique suffered from regional or local famines. Diets are not well

Figure 7-8

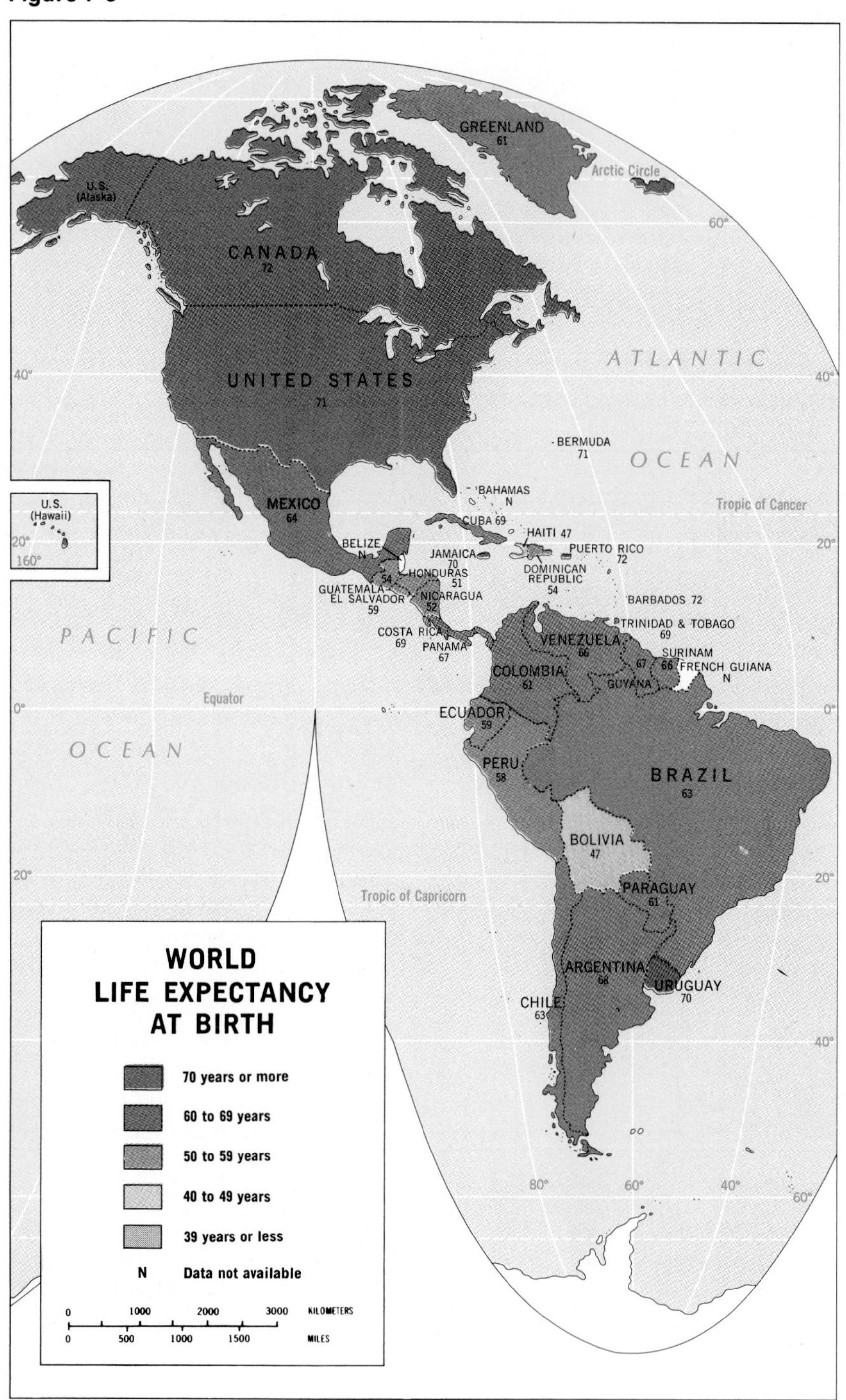

balanced, and life expectancies are nowhere lower (Fig. 7-8). Millions of Africans are unwell their entire lives, always fighting illnesses that afflict them almost from the day they are born. Children do not eat sufficient proteins, and kwashiorkor and marasmus are common, especially in areas where people depend on starchy root crops for their staples.

Diseases strike populations in different ways. When a sudden outbreak occurs, leading to a high percentage of afflictions in a population (and perhaps a large number of deaths), the outbreak is an *epidemic.* An epidemic is a regional phenomenon, or even a local one; the "Legionnaire's Disease" that struck a large number of conventioneers in a Philadelphia hotel in 1976 was an epidemic among that concentrated population. When an outbreak that

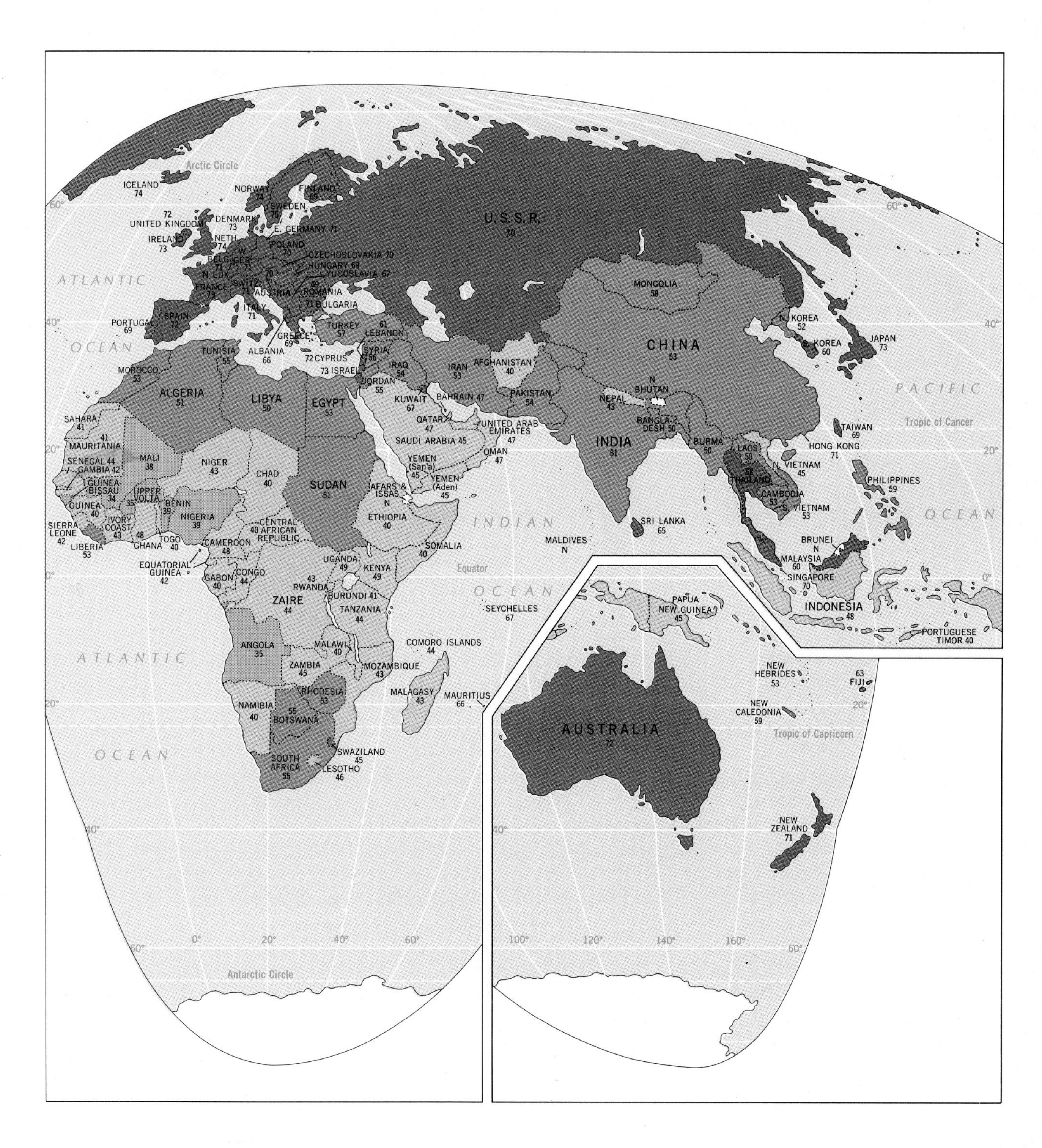

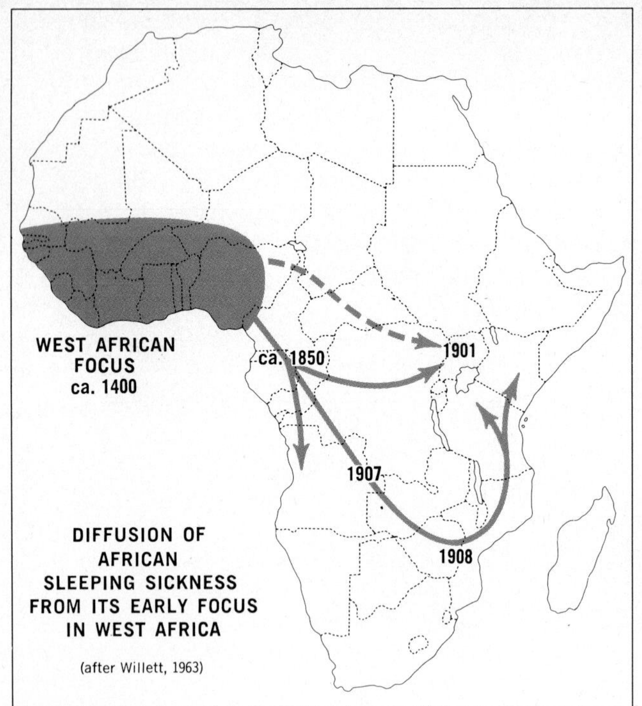

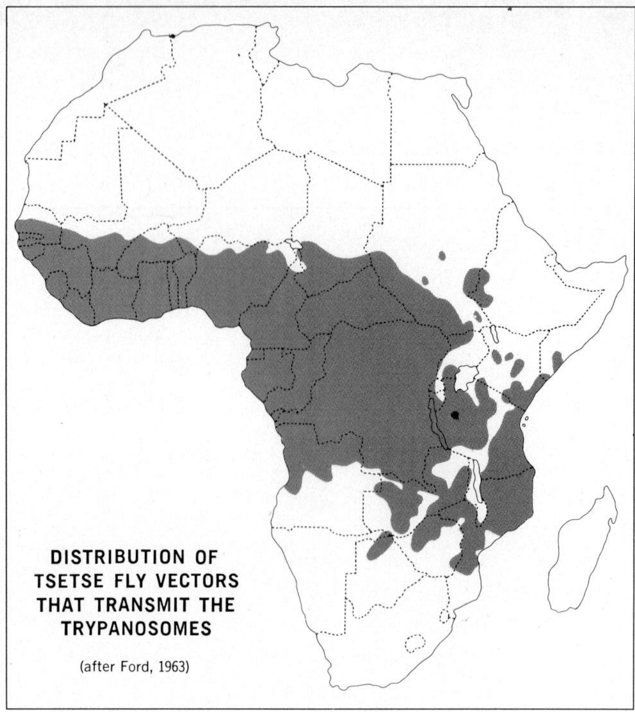

Figure 7-9

begins locally spreads regionally and then worldwide as various forms of influenza have done in modern times, it is described as having *pandemic* proportions. But a disease can inhabit a human population in still another way. Some illnesses do not strike in a violent attack, but exist in a sort of equilibrium with the population. Many people will have it, and the disease saps energies and shortens lifetimes. These, nevertheless, are the survivors of the first attack during childhood or later, now able to withstand the parasites. Such *endemic* diseases (syphilis and mononucleosis are examples in the United States) affect many millions of Africans.

Undoubtedly malaria, killer of small children and endemic to most of Africa, is the worst. In recent years malaria has again been on the increase. The malaria mosquito is the vector (carrier) of the parasite, and the mosquito prevails in almost all of inhabited Africa. Africans who survive childhood are likely to suffer from malaria in some degree, with a debilitating effect. African sleeping sickness is transmitted by the tsetse fly (Fig. 7-9) and now affects most of tropical Africa; the fly infects not only people, but also their livestock. Its impact on Africa's population has been incalculable. The disease appears to have originated in a West African source area about A.D. 1400, and since then it has inhibited the development of livestock herds where meat and milk would have provided crucial balance to seriously protein-deficient diets. It channeled the migrations of cattle-herders through fly-free corridors in East Africa (Fig. 7-9), destroying herds that moved into infested zones. But most of all it ravaged the human population, depriving it not only of potential livelihoods but also of its health. Still another serious and widespread disease is yellow fever, also a mosquito-transmitted malady. Yellow fever is endemic in moister tropical zones of Africa, but it sometimes appears in other areas in epidemic form, as in 1965 in Senegal when 20,000 cases were reported (there undoubtedly were thousands more). This is another disease that strikes children and, if they survive, they have a certain level of immunity. To this depressing list must be added schistosomiasis, also called bilharzia, transmitted by snails. The parasites enter via body openings when people swim or wash in slow-moving or standing water infested by the snails. Internal bleeding, loss of energy, and pain result, but schistosomiasis is not a killer disease. It is endemic today to over 200 million people, most of them living in Africa.

These are major and more or less continent-wide diseases, and there are numerous others of regional and local distribution. The dreaded river blindness, caused by a parasitic worm transmitted by a small fly, is endemic in the savanna belt south of the Sahara from Senegal to Kenya, and in northern Ghana it blinds a large percentage of the adult villagers. And animals and plants as well as people are attacked by Africa's ravages. Besides sleeping sickness, rinderpest also afflicts livestock herds. Crops and pastures are stripped by swarms of locusts that number an average of 60 million insects each and travel thousands of kilometers, devouring the vegetation of whole countrysides. Add these problems to those of widespread poor soils and inadequate precipitation, and Black Africa's population total of 350 million appears a great deal more impressive.

Various regions of Africa suffered from severe droughts during the 1970s. The impact of aridity can be seen in the dying vegetation of the East African savanna (above) and in the desperate search for individual kernels of grain by inhabitants of West Africa's Sahel (below). In East Africa and the Horn, hundreds of thousands of people and animals died of starvation. A similar fate befell countries of Sahel—a name that became synonymous with drought. (top, Harm J. de Blij; bottom, Alain Nogues/Sygma)

African Regions

On the face of it, Africa would seem to be so massive, so compact, and so unbroken that any attempt to justify a regional breakdown appears doomed to failure. No deeply penetrating bays or seas create peninsular fragments, as in Europe; no major islands (other than Madagascar) provide the sort of fundamental regional contrasts we see in Middle America; nor does Africa taper southward to the peninsular proportions of South America. Africa is not cut by an Andes or a Himalaya range. Does any clear regional division nevertheless exist?

Indeed it does. As we noted earlier, the Sahara Desert, extending as it does from west to east across the entire northern part of the continent, constitutes a broad transition zone between Arab Africa and Black Africa. While busy trade routes have crossed it for centuries, and millions of people who live south of the Sahara share the religious outlook of the Arab world, Black Africa retains a cultural identity quite distinct from that of the Arab realm.

Within Africa south of the Sahara, regional identities exist as well, although the regional boundaries involved are not easily defined and have at times been the subjects of debate (Fig. 7-10). Three such regions are in common use: *West* Africa, which includes the countries of the west coast from Senegal to Nigeria, *East* Africa, by which is normally understood the three states of Kenya, Uganda, and Tanzania, and *South* Africa (or *Southern,* to get away from the political connotation), of which South Africa and Zimbabwe are parts. Less clear has been the use of such regions as *Equatorial* Africa, by which is generally meant Zaire and the countries that lie between it and Nigeria. Although the appellation *Equatorial* presumably has to do with the location of these countries on or near the equator, the similarly positioned countries of East Africa are practically never referred to in this way. In fact, *Equatorial* has come to mean hot, tropical, and low-lying rather than equatorially located; thus Gabon is a part of Equatorial Africa, while Uganda is not. The regional divisions shown on

Figure 7-10

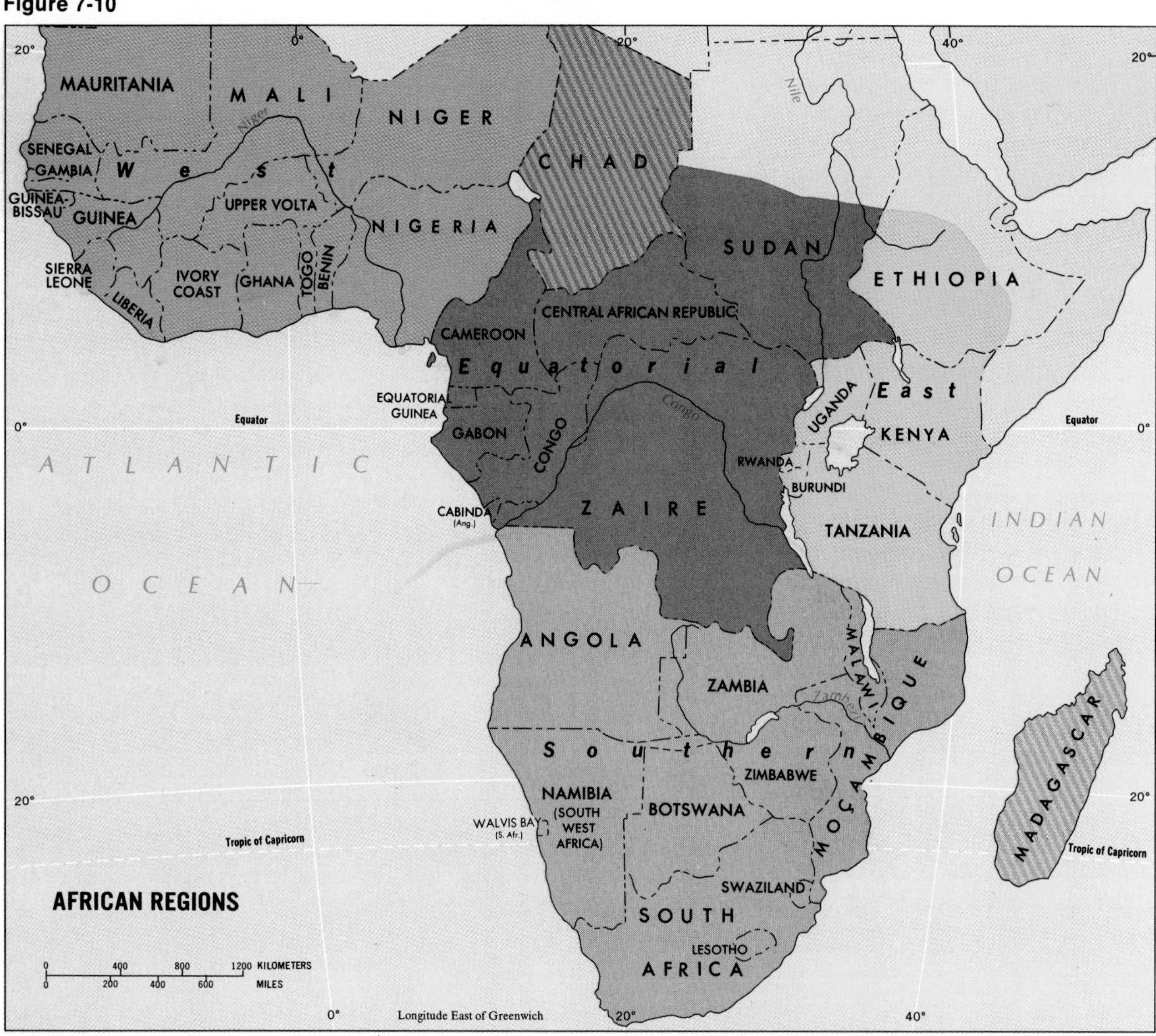

Fig. 7-10 should therefore be viewed in general terms and as a matter of orientation; their boundaries, as will be clear from the following discussion, are not beyond dispute in certain areas.

West Africa

West Africa extends from the margins of the Sahara Desert south to the coast, and from Lake Chad to Senegal. Politically, the broadest definition of the region includes all those states that lie to the south of Morocco, Algeria, and Libya, and west of Chad (itself sometimes included) and Cameroon. Apart from Portuguese Guinea and long-independent Liberia, West Africa comprises only former British and French dependencies: four British and nine French. The British-influenced countries (Nigeria, Ghana, Sierra Leone, and Gambia) lie separated from each other; Francophone West Africa is contiguous. As Fig. 7-11 shows, political boundaries extend from the coast into the interior, so that from Mauritania to Nigeria the West African habitat is parceled out among parallel-oriented states. Across these boundaries, especially across those between former British and former French territories, moves very little trade. For example, in terms of value, Nigeria's trade with Britain is about one hundred times as great as its trade with nearby Ghana. The countries of West

Figure 7-11

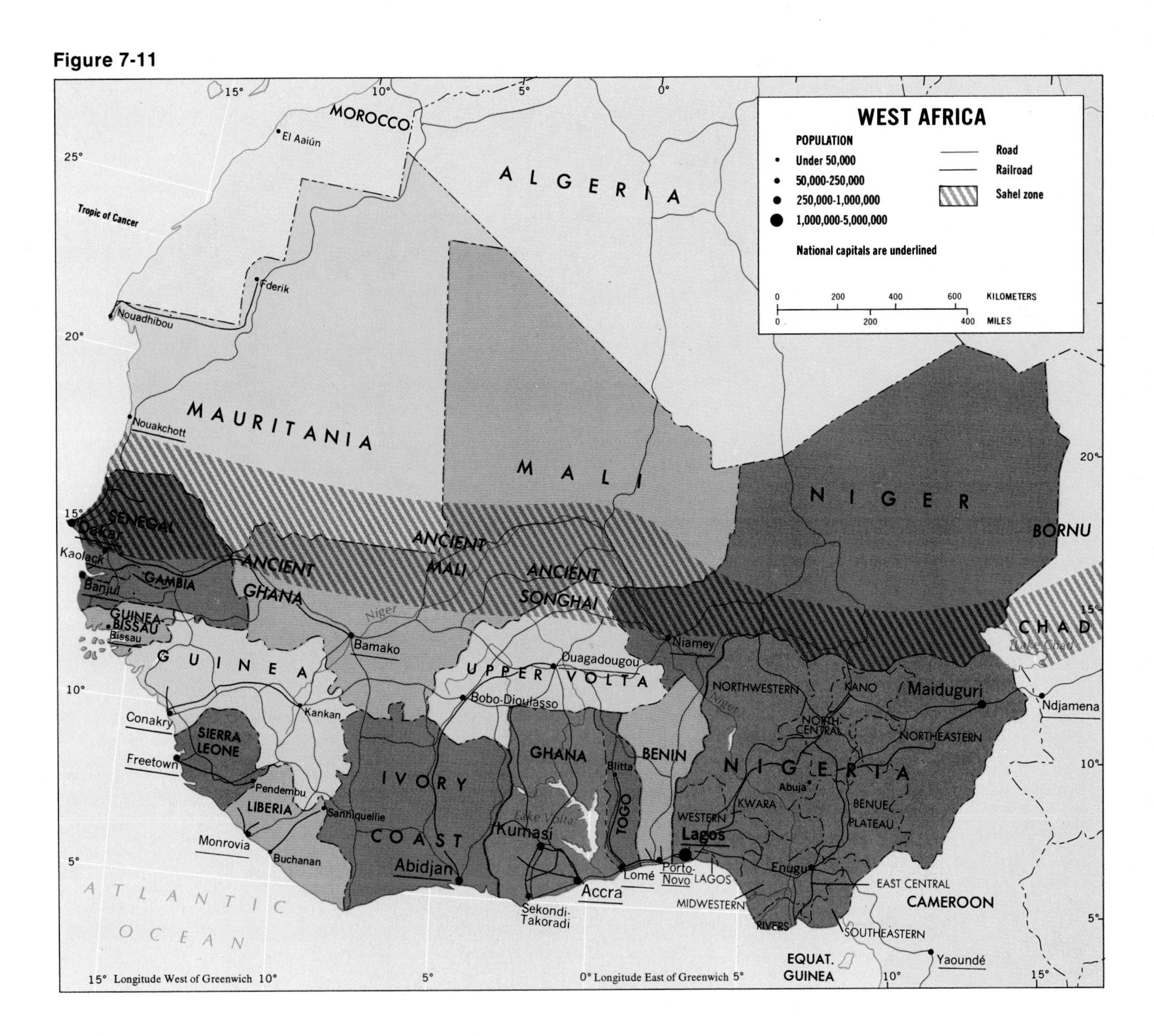

Lagos, long Nigeria's capital, has been called the "Calcutta of Africa." The coastal city exemplifies Nigeria's varied population, its enormous energy, bustle, and rapid change. Streets are filled with thousands of people, buses, and automobiles; internal circulation is poor as crowding causes huge traffic jams. Congestion, pollution, inadequate housing and other facilities do not deter Nigerians from moving to Lagos, where the action is; but the Nigerian government has decided to remove the capital functions from here to a city more centrally positioned. (Bruno Barbey/Magnum)

Africa are not interdependent economically, and their income is to a large extent derived from the sale of their products on world markets. But the African countries do not control the prices their goods can demand on the world markets, and when these prices fall, they face serious problems.

If the economic, and in some cases the political, contacts between the former colonial powers and the West African countries remain stronger than the ties between the African states themselves, what are the justifications, if any, for the concept of a West African region? First, there is the remarkable cultural and historical momentum of this part of Africa. The colonial interlude failed to extinguish West African vitality, expressed not only by the old states and empires of the savanna and the cities of the forest, but also by the vigor and entrepreneurship, the achievements in sculpture, music, and dance of peoples from Senegal to Iboland. Second, West Africa from Dakar to Lake Chad carries a set of parallel ecological belts, so clearly reflected by Figs. I-5 to I-8, whose role in the development of the region is pervasive. As a map of transport routes in West Africa indicates, connections within each of these belts, from country to country, are quite poor; no coastal or interior railroad was ever built to connect the coastal tier of countries. Communication and contact across these belts, on the other hand, is better, and some north-south economic exchange does take place, importantly in the coastal consumption of meat from cattle raised in the northern savannas. Third, West Africa received an early and crucial impact from European colonialism, which with its maritime commerce and the slave trade transformed the region from one end to the other. This impact was felt all the way to the heart of the Sahara, and it set the stage for the reorientation of the whole area—out of which emerged the present patchwork of states.

The effects of the slave trade notwithstanding, West Africa today is Black Africa's most populous region (Fig. I-9). In these terms, Nigeria, with perhaps 90 (but certainly over 80) million people, is Africa's largest state, and Ghana, with 11 million, also has a high ranking. As Fig. I-9 shows, West Africa also has regional identity in that it constitutes one of Africa's five major population clusters (the others are northern Morocco and Algeria, The Nile Valley and Delta, the Lake Victoria environs, and eastern South Africa). The southern half of the region understandably carries the majority of the people. Mauritania, Mali, and Niger include too much of the unproductive Sahel's steppe and desert of the Sahara to sustain populations comparable to Nigeria, Ghana, or Ivory Coast.

Négritude and Nationalism

Négritude is a concept and a philosophy, a literary movement, and an appeal to African values. In the decades beginning with the 1930s, black writers in Caribbean America and African scholars working in France began to pour out their feelings about dehumanization of black peoples caught up in the web of French colonialism. In poems, short stories, and novels these writers—Aime Césaire in Martinique, Leon Damas in French Guiana, and Léopold Senghor in Senegal prominent among them—emphasized the contradictions between French colonial policies of "assimilation" and the realities of European superiority and discrimination.

In Paris the strands of négritude coalesced, and African and French Caribbean intellectuals urged their countrypeople in the Francophone black world to reassess the power of their own culture with its intense personal relationships, its closeness to nature, its special contact with ancestors, its disdain for material wealth. They appealed to Africans to reject the arrogance and acquisitiveness of European culture—to borrow the good but to reject the bad. As a philosophy, négritude also supported African political aspirations (and eventually Léopold Senghor became the first president of Senegal), and it became the cultural focus of French-speaking West Africa from Senegal to Chad.

Négritude encompasses all that is good in being black, including African history, arts, and music; it is a state of mind, a self-confidence that comes with self-knowledge. It recognizes the kinship, the spoken and unspoken ties among black people everywhere.

This is not to say that only the coastal areas of West Africa are densely populated. True, from Senegal to Nigeria there are large concentrations of people in the coastal belt, but the interior savannalands also contain sizable clusters. The peoples along the coast reflect the new era brought by the colonial powers: they prospered in their new-found roles as middlemen in the coastward trade. Later they were in position to undergo the changes the colonial period brought: in education, religion, urbanization, agriculture, politics, health, and many other fields they adopted new ways. The peoples of the interior, on the other hand, retained their ties with a very different era in African history; distant and often aloof from the main scene of European colonial activity, they experienced very much less change. But the map reminds us that Africa's boundaries were not drawn to accommodate such differences. Both Nigeria and Ghana have population clusters representing the coastal as well as the interior peoples, and in both countries, the wide gap between north and south has produced political problems.

When, in 1960, Nigeria achieved full independence, it was endowed with a federal political organization consisting of three regions based on the three major population clusters within its borders—two in the south and one in the north. Around the Yoruba core in the southwest lay the Western Region. The Yoruba are a people with a long history of urbanization, but they are also good farmers; in the old days they protected themselves in walled cities around which they practiced intensive agriculture. The colonial period brought coastal trade, increased urbanization, cash crops (the mainstay, cocoa, was introduced from Fernando Poo in the 1870s), and, eventually a measure of security against encroachment from the north. Coastal Lagos, the country's first federal capital (a new capital, Abuja, is under construction in a more central location), grew on the region's south coast and now is a conurbation of an estimated 3.2 million people. Ibadan, one of Black Africa's largest cities, evolved from a Yoruba settlement founded in the late eighteenth century and now has about 2 million inhabitants. At independence the Western Region counted some 10 million inhabitants, and more than any other part of Nigeria it had been transformed by the colonial experience. East of the Niger River and south of the Benue, the Ibo population formed the core of the Eastern Region. Iboland, although coastal, lay less directly in the path of colonial change, and its history and traditions, too, differed sharply from those of the West. Little urbanization had taken place here, and even today, while over one-third of the former Western Region's people live in cities and towns over 20,000, a mere 10 percent of the Eastern population is urbanized. With some 12 million people, the rural areas of eastern Nigeria are densely peopled. Many Ibo over the years have left their crowded habitat to seek work elsewhere, in the west, in the far north, in Cameroon, and even in Fernando Poo. The third federal region at independence was at once the largest and the most populous: the Northern Region, with some 30 million people. It extended across the full width of the country from east to west, and from the northern border southward beyond the Niger and Benue Rivers. This is Nigeria's Moslem North, centered on the Hausa-Fulani population cluster, where the legacy of a feudal social system, conservative traditionalism, and resistance to change hangs heavily over the country. Nigeria's three original regions, then, lay separated—not only by sheer distance, but also by tradition and history, and by the nature of colonial rule. In Nigeria, even physical geography and biogeography conspire to divide south from north: across the heart of the country (and across much of West Africa in about the same relative location) stretches the so-called Middle Belt—the poor, unproductive, disease- and tsetse-ridden country that forms in effect an empty barrier between the regions.

Nigeria's three-region federal system failed. Following attacks on Ibo families living in Northern Nigeria and a mass movement of Ibos back to the Eastern Region, the East declared its secession from the Nigerian Federation, and civil war broke out in 1967. The war continued until 1971, accompanied by enormous destruction and loss of life. Even as it continued, Nigerian politicians were trying to devise a system that would prevent such disasters in the future. The latest of these, a federal plan that divides the country into 19 states, was adopted in 1976.

Nigeria today is a cornerstone of the new Africa, a country on the move whose economic windfall has been the discovery of substantial oil reserves in the area of the Niger Delta. In the late 1970s over 90 percent of Nigeria's export revenues were derived from the sale of petroleum and petroleum products (cocoa beans and peanuts were next), and cities from Port Harcourt to Lagos reflected the oil boom of the south. Overcrowded Lagos has been described as the Calcutta of Africa, the scene of intramilitary rivalries for control over the country and costly corruption in government. But Nigeria is changing. Urbanization, barely 20 percent 15 years ago, is accelerating, surface communications are being improved, and national integration is proceeding. Nigeria's old regional disparities still exist, of course, and the former Eastern Region's war of independence may not have been the last civil conflict to disrupt the country. But Nigeria nevertheless overshadows all other countries in West Africa.

In the swamplands of the Niger Delta lies the source of Nigeria's modern transformation: its oil. Nigeria is Black Africa's leading oil-exporting country, and oil earns over 90 percent of the country's external revenues. (Georg Gerster/Photo Researchers)

Periodic Markets

The great majority of the people of West Africa are not involved in the production of exports for world markets, but subsist on what they can grow and raise—and trade. Their local transactions take place at small markets in villages. These village markets are not open every day, but operate at intervals. In this way, several villages in a region get their turn to attract the day's trade and exchange, and each benefits from its participation in the network of communications. People come to these *periodic markets* on foot, by bicycle, on the backs of their animals, and by whatever other means are available.

A periodic market in Mali, not far from the ancient city of Timbuktu. (George Holton/Photo Researchers)

Periodic markets are not exclusively a West African phenomenon. They occur also in interior Southeast Asia, in China, and in Middle and South America, as well as in other parts of Africa. The intervals between market days vary. In much of West Africa, village markets tend to be held on every fourth day, although some areas have two-day and eight-day cycles; in Ghana's northern region, markets are held in a particular village every third day. In China, various periodic market cycles developed, including three, five, six, ten, and even twelve days.

Periodic markets, then, form a sort of interlocking network of exchange places that serves even areas where there are no roads; as each market in the network gets its turn, it will be near enough to one section of the area so that the people who live in the vicinity can walk to it, carrying what they want to sell or trade. In this way, small amounts of produce filter through the market chain to a larger regional market, where shipments are collected for interregional or perhaps even international trade. What is traded, of course, depends on where the market is located. A visit to a market in West Africa's forest zone will produce impressions that are very different from a similar visit in the savanna zone. In the savanna, sorghum, millet, and shea-butter (an edible oil drawn from the shea-nut) predominate, and you will see some Islamic influences here; in the south, such products as yams, cassava, corn, and palm oil change hands. In the southern forest zone, too, one is more likely to see some imported manufactured goods passing through the market chain, especially near the relatively prosperous cocoa, coffee, rubber, and palm oil areas. But in general the quantities of trade are very small—a bowl of sorghum, a bundle of firewood—and their value is low; these markets serve people who mostly live near the subsistence level.

This does not mean that West Africa's markets fail to generate excitement, color, and jostling crowds. In fact, the markets are the main arenas not only for trade, but also for social and cultural exchange. In preliterate society they take the place of newspapers, radio, and television: here the latest news can be heard. Elders meet together on market day to discuss matters of local traditional government. Young men and women customarily use market days for courtship. Men attend market, especially in the savanna zone, to sit around and drink millet beer; the women do the bulk of the trading. All day, on market day, people barter, discuss, gossip, argue, eat, and drink—and the market place is deserted until, three or four days later, the market has another turn.

East Africa

Despite the fact that they are neighbors, the three formerly British territories of East Africa—Kenya, Uganda, and Tanzania—and the former Belgian wards of Rwanda and Burundi display numerous differences. These differences have arisen less out of any contrasts in colonial rule than out of the nature of the East African habitat, the course of African as well as European settlement, and the location of the areas of productive capacity. This, of course, is highland, plateau Africa, mainly savanna country that turns into steppe toward the northeast. Great volcanic mountains rise above a plateau that is cut by the giant rift valleys; the pivotal physical feature is Lake Victoria, where the three major countries boundaries come together (Fig. 7-12), and on whose shores lie the primary core area of Uganda and secondary cores of both Kenya and Tanzania. With limited mineral resources (the chief ones are diamonds in Tanzania, not far south of Lake Victoria, and copper in Uganda, to the west of the lake), most people depend on the land—and on the water that allows crops to grow and livestock to live. In much of East Africa, rainfall is marginal or insufficient. The heart of Tanzania is dry, tsetse- and malaria-ridden, and occasionally still the scene of local food shortages. Eastern and northern Kenya consist largely of dry steppe country. As Fig. I-5 shows, the wettest areas lie spread around Lake Victoria, and Uganda receives more rainfall than its neighbors.

Tanzania is the largest of the East African countries. Its area exceeds that of the four other countries combined, although its population of more than 17 million is only slightly larger than Kenya's (16 million). Uganda, with nearly 14 million, ranks next; and even the small countries of Burundi and Rwanda together exceed 10 million, so that East Africa contains some densely peopled areas. Although all these states lie in the same region, they display some strong differences. Tanzania, for example, has been described as a country without a primary core area, because its areas of productive capacity (and its population, Fig. I-9) lie spread about, mostly near the country's margins on the east coast, near the shores of Lake Victoria in the north, near Lake Tanganyika in the west, and near Lake Nyasa (Malawi) in the south. Kenya, on the other hand, has a strongly concentrated core area centered on its capital, Nairobi (900,000). Again, Tanzania is a country of many ethnic groups, none with a numerical or locational advantage to permit domination of the state, but Kenya is dominated by the Kikuyu, the country's largest national group whose traditional domain includes much of the productive farmland of the core area. In terms of political and politico-geographical philosophy, Tanzania has gone the route of African socialism while Kenya has become a capitalist state. Although latent problems always remain, both Kenya and Tanzania have come to terms with their non-African minorities, but in different ways. Kenya had some 70,000 white settlers, over 200,000 Asians, and perhaps 30,000 Arabs; Tanzania never had foreign minorities as large (the white population never reached 20,000). In African-socialist Tanzania, Asians and other minorities have been integrated in a new economic and social order, but in Kenya, many of the old conditions still prevail: Asians, for example, still

Figure 7-12

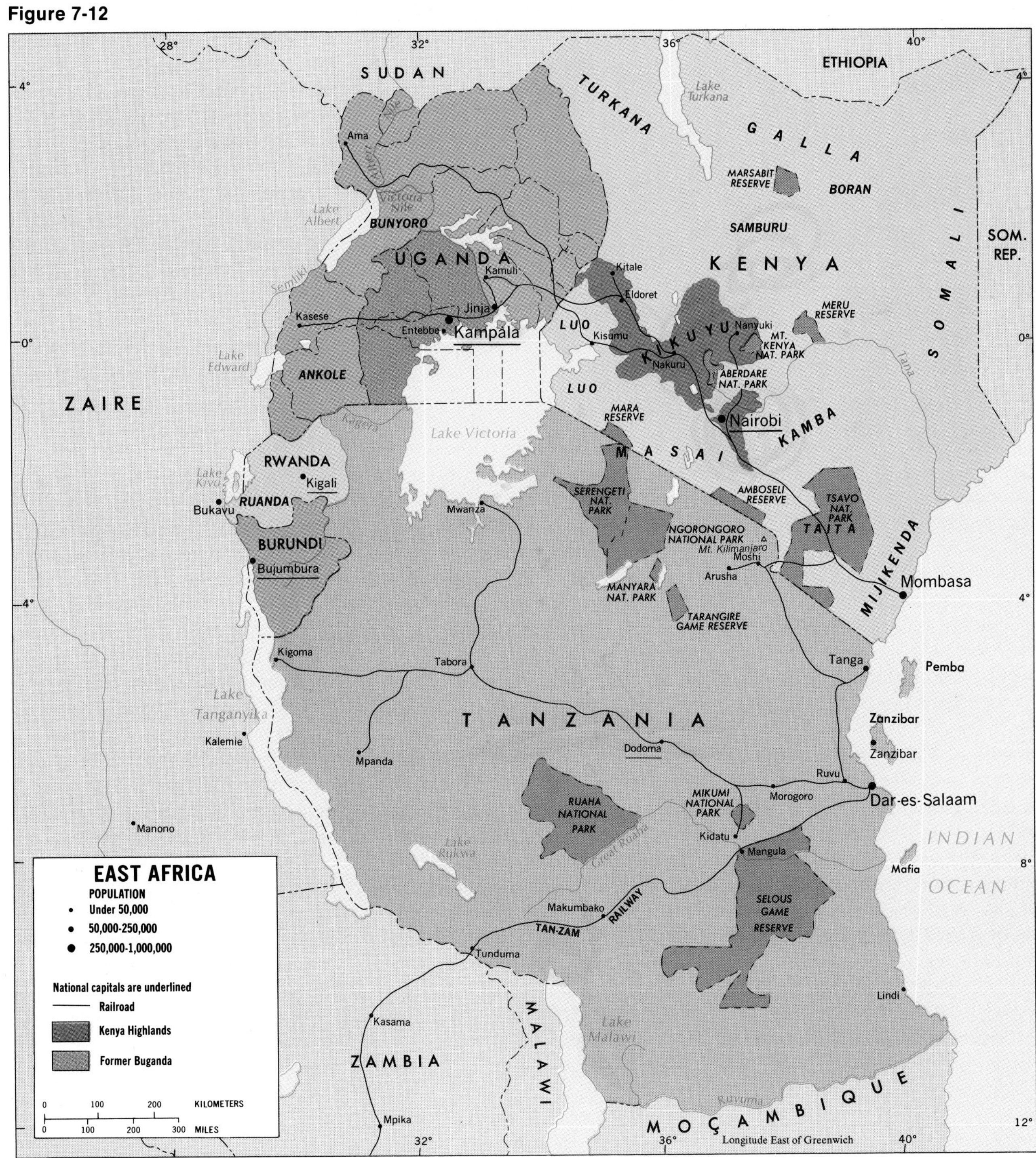

On the highway from Kenya to Ethiopia. Road construction, on a continent so vast and with resources so limited, proceeds slowly. No tarmac or other all-weather road yet links any of Africa's major regions. (Courtesy American University Field Staff)

have a prominent role in commerce. Neither Tanzania nor Kenya took the drastic step ordered by the Amin regime in Uganda, where all 75,000 Asians were ordered to leave the country within three months in 1972. It was a modern case of forced migration (page 128); the emigrants left behind $400 million in assets and a commercial system that soon fell apart.

Tanzania has become an important example to the rest of Africa. Notwithstanding its limited resources and fragmented population, the country achieved political stability in the face of pressures generated by its transition to self-help African socialism. Tanzania has some commercial agriculture (sisal plantations along the north coast, coffee farms on the slopes of Mount Kilimanjaro near the Kenya border, tea in the southwest, near Lake Malawi), but the great majority of Tanzanians are subsistence farmers. The government of president Julius Nyerere undertook a major reorganization of agriculture, creating new villages, forming cooperatives, and supporting improved farming methods. There was opposition, of course, but the new order prevailed. All this occurred while Tanzania was a haven for FRELIMO (in English, Front for the Liberation of Moçambique) insurgents fighting the Portuguese in Moçambique, a difficult merger with Zanzibar was accomplished, and the capital was moved from coastal Dar es Salaam (400,000) to Dodoma in the interior. At the same time the Chinese were in Tanzania, working on the TanZam (or TAZARA) Railway. Even when Tanzania faced the impact of the same drought that caused such ravages in the West African Sahel, things did not fall apart.

Kenya's comparative prosperity is reflected by the tall skyscrapers of Nairobi and the productive farms in the nearby highlands. But Kenya's development is concentrated, not spatially dispersed like Tanzania's, so that the evidence of postindependence development is strong in the core area, but very limited in the sparsely peopled interior. As the map shows, both Tanzania and Kenya have a single-line railroad that traverses the whole country from the major port in the east (Mombasa in the case of Kenya) to the far west, and feeder lines come from north and south to meet this central transport route. In Kenya, that central railroad and its branches in the highlands really represent the essence of the country, but in Tanzania the railroad lies in the "empty heart" of the territory, and the branches lead to the productive and populated peripheral zones. The new TAZARA Railway gives Tanzania a link to a neighboring landlocked country as well.

Kenya does not possess major known mineral deposits, but in addition to its coffee and tea exports it receives substantial revenues from a tourist industry that grew rapidly during the 1960s and 1970s, until political difficulties in East Africa in the late 1970s began to reduce the flow of visitors. In the mid-1970s the total of foreign visitors to Kenya was approaching a half million, and the industry became the largest single foreign exchange earner. Kenya's famous wildlife reserves lie mostly in the drier and remote parts of the country, but population pressure is a growing problem even in those areas—posing still another threat to tourism's future.

If Tanzania is an example to Africa, Kenya is as well, but of

Sequent Occupance

The African realm affords numerous opportunities to illustrate a geographic concept introduced by D. Whittlesey in 1929: *sequent occupance*. This concept involves the study of an area that has been inhabited—and transformed—by a succession of residents, each of whom leaves a lasting cultural imprint. A place and its resources are perceived differently by peoples of different technological and other cultural traditions. These contrasting perceptions are reflected in their respective cultural landscapes. The cultural landscape today, therefore, is a collage of these contributions, and the challenge is to reconstruct the contributions made by each community.

The idea of sequent occupance is applicable in rural as well as urban areas. The ancient Bushmen used the hillsides and valleys of Swaziland to hunt and to gather roots, berries, and other edibles. Then the cattle-herding Bantu found the slopes to be good for grazing, and in the valleys they planted corn and other food crops. Next came the Europeans to lay out sugar plantations in the lowlands, but after using the higher slopes for grazing they planted extensive forests, and lumbering became the major industry.

The Tanzanian city of Dar es Salaam provides an interesting urban example. Its site was first chosen for settlement by Arabs from Zanzibar to serve as a mainland retreat. Next it was selected by the German colonizers as a capital for their East African domain, and it was given a German layout and architectural imprint. When the Germans were ousted following their defeat in World War I, a British administration took over in Dar es Salaam, and the city began still another period of transformation; a large Asian population left it with a zone of three- and four-story apartment houses that seem to be transplanted from Bombay. Then in the early 1960s Dar es Salaam became the capital of newly independent Tanzania, under African control for the first time. Thus Dar es Salaam in less than one century experienced four quite distinct stages of cultural dominance, and each stage of the sequence remains imprinted in the cultural landscape. Indeed, a fifth stage has begun, since the capital functions were recently moved from Dar es Salaam to the interior town of Dodoma.

Because of Africa's cultural complexity and historical staging, numerous possibilities of this kind exist. The Dutch, British, Union, and Republican periods can be observed in the architecture and spatial structure of Cape Town and other urban places in South Africa's Cape Province. Mombasa in Kenya carries imprints from early Persian, Portuguese, Arab, British, and several African communities. African traditional farmlands taken over by Europeans and fenced and farmed are now being reorganized by new African proprietors, and contrasting attitudes to soil and slope are etched in the landscape.

another kind. Unlike Tanzania, Kenya chose a capitalist route in its quest for accelerated development. In the process, Kenya in many respects forged ahead of its larger neighbor, exceeding it in gross national product, per capita income, and in a host of other indicators. Certainly this was accompanied by a less equable division of wealth (in Nairobi a class of people owning expensive automobiles are referred to as *Wa-Benzi*), but entrepeneurship and

An air view of the modern center of Nairobi, East Africa's largest urban agglomeration. (George Gerster/Rapho-Photo Researchers)

initiative had fuller reign in Kenya. In 1978 Kenya passed a major test when its president, Jomo Kenyatta, died and, in an orderly succession, the vice president, Daniel arap Moi, assumed the state's highest office.

Landlocked Uganda depends on coastal Kenya for its exit to the ocean, but relations between Uganda and Kenya have not been good in recent years. For several years after independence Uganda seemed set on a course similar to that of many other African countries, but a military coup (an event occurring with growing frequency in postcolonial Africa) brought to power the unpredictable Idi Amin. The Amin regime soon claimed a large part of western Kenya as Ugandan territory, and Amin also spoke of a Ugandan corridor to the sea through northern Tanzania. In a sense it was the final blow to concepts of a greater East African international community, already endangered by deteriorating relationships between Kenya and Tanzania.

Uganda contained the most important African political entity in the region when the British entered the scene during the second half of the nineteenth century. This, the Kingdom of Buganda, faced the north shore of Lake Victoria, had an impressive capital in Kampala, and was stable—as well as ideally suited for indirect rule over a large hinterland. The British established their headquarters at Entebbe on the lake (thus adding to the status of the kingdom) and proceeded to solidify their hold over their Uganda Protectorate, using Buganda representatives among other peoples.

What happened next in Uganda forms an excellent illustration of a concept in political geography, the *unified field theory* proposed by S. Jones. According to this idea, whenever we see a politically organized area, we are looking at the end result of a number of "hubs" of activity that have followed each other over a period of time. The first of these is an idea—a political idea, or an idea with political implications. In the end there will be a politically organized area. His five-step model looks as follows:

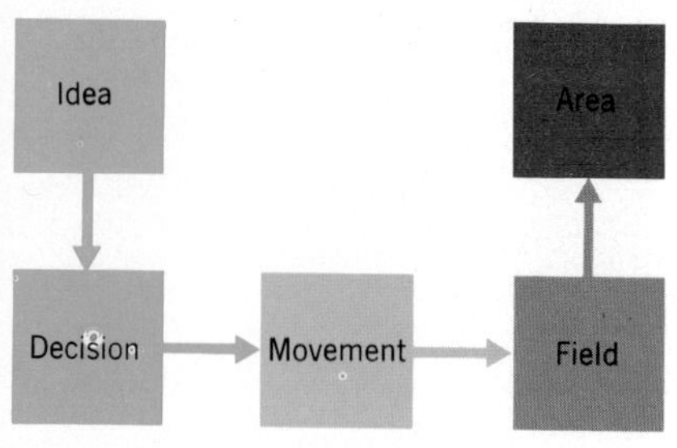

This became known as the "idea-area chain," but you should not assume that it is simply a sequence in which one step gives rise to the next. Rather, the several elements in the model interact with each other all the time, backward as well as forward. In Uganda, the fundamental political idea involved the welding of the Buganda Kingdom to various smaller kingdoms nearby (and more distant peoples as well) under the general philosophy of indirect rule. The specific decision was to use Buganda people to administer the resulting protectorate on behalf of the British crown. The movement phase witnessed Buganda dispersal into non-Buganda areas of Uganda, but it also meant the movement of ideas, new goods, and practices. Roads, police posts, water traffic on rivers and lakes, post offices, and other kinds of movement-related innovations were brought to Uganda's interior. The "field" phase of the model produced the actual boundaries of

Uganda, the area within which these changes had effect. Finally, there emerged the politically organized area: Uganda as it was until the British departed.

In modern Uganda, the Buganda always were in a minority, and when the British departed they endowed the country with a complicated federal constitution that was designed to perpetuate Buganda primacy. But the system failed under pressure of new ideas and new decisions, and in the new Uganda, movement has been restricted, modernization and interaction reversed.

These circumstances, extending over most of the 1970s, brought ruin to once-prosperous Uganda. Submerged animosities were revived and waves of retribution and revenge flooded the countryside. After the ouster of the Asians and the collapse of the economy, Uganda largely reverted to a state of fragmentation. Markets closed, subsistence living reappeared. Wildlife in the national parks was hunted out. Thousands escaped to neighboring Kenya but many more were killed; some estimates put the death-by-violence toll as high as 300,000. Ultimately Amin began a conflict with Tanzania over an area along Uganda's border near Lake Victoria, and the Tanzanian army invaded Uganda and drove the despotic ruler into exile. For Uganda, the 1970s constituted a decade of almost incalculable setbacks, from which the country may never fully recover.

Equatorial Africa

As the term is used *Equatorial* Africa lies to the west of highland East Africa, and consists of the Zaire Republic, the People's Republic of Congo (capital Brazzaville), Gabon, the Central African Republic, Cameroon, and Equatorial Guinea (Fig. 7-10). Chad, to the north of the Central African Republic, was an administrative part of French Equatorial Africa and often is still included in this region, but it lies between Niger and Sudan, in the West-Africa-related savanna belt, and only by the most tenuous of arguments can it be considered a part of Equatorial Africa.

The giant of Equatorial Africa, and by far its most developed (rather, least underdeveloped) state is Zaire, the former Belgian Congo. With nearly 2.5 million square kilometers (well over 900,000 square miles) and almost 30 million inhabitants, Zaire is one of Black Africa's largest politico-geographical units. It has the bulk of Equatorial Africa's human and natural resources, its largest cities, its best-developed communications, and undoubtedly its greatest opportunities and potential. The regional geography of Zaire was outlined on page 401, and the fragments of former French Equatorial Africa add comparatively little to the region's contents. France never managed to create an interconnected whole in its equatorial posessions, and the *total* population of the five former French dependencies today is barely over 10 million. Independence has not made the economic picture much brighter, although some new mineral resources have been brought into production (including Gabon's oil reserves). But Equatorial Africa remains a stagnant region of subsistence and raw-material exports; only in the growth areas of Zaire can the beginnings of a new age be detected.

Southern Africa in Crisis

IDEAS AND CONCEPTS

Buffer zone (2)
Separate development
Domestic colonialism

In Southern Africa, a struggle for power between white minorities and African peoples is under way. What began as a revolution in French Algeria in the 1950s and continued in West, East, and Equatorial Africa as European powers yielded to African demands for self-determination—the "Wind of Change," as British Prime Minister Harold Macmillan called it—has now touched the largest white stronghold of all, South Africa itself. In 1979, the combatants in a long and costly war that afflicted Rhodesia for years were brought to a London conference table, and a transition to black rule was achieved; Rhodesia became Zimbabwe. In South West Africa (Nambia), armed insurrection and diplomatic maneuvers proceeded. In Moçambique and Angola, five centuries of Portuguese colonial control came to a convulsive end. The apparently inevitable crisis in South Africa now appears at hand.

As a geographic region, Southern Africa consists not only of the Republic of South Africa and its immediate neighbors, but of

The ruins of the ancient city of Zimbabwe reflect an earlier age of African political power in this region. Following a prolonged political and armed struggle, the former British colony of Southern Rhodesia (and briefly the breakaway white-dominated state of Rhodesia) in 1980 became the black-ruled independent state of Zimbabwe, named after this historic place. (Harm J. de Blij)

Zambia and Malawi as well (Fig. S-1). Not long ago these two countries (then named Northern Rhodesia and Nyasaland) were locked into the ill-fated Central African Federation, dominated by white (Southern) Rhodesian interests. For many decades Zambia, whose economic core area is the Copperbelt, has looked southward for its electric power supply and its fuel, and for its outlets. The TanZam (or TAZARA) Railway was built, in large measure, to redirect this trade, but Zambia's involvement with Southern African countries set a pattern that will probably emerge again after the political struggles are over. Malawi, too, also has had to look south for its outlets. Like Zambia, Malawi is a landlocked country, and its core area lies (as it always has) in the southern part of its elongated territory.

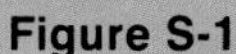

Figure S-1

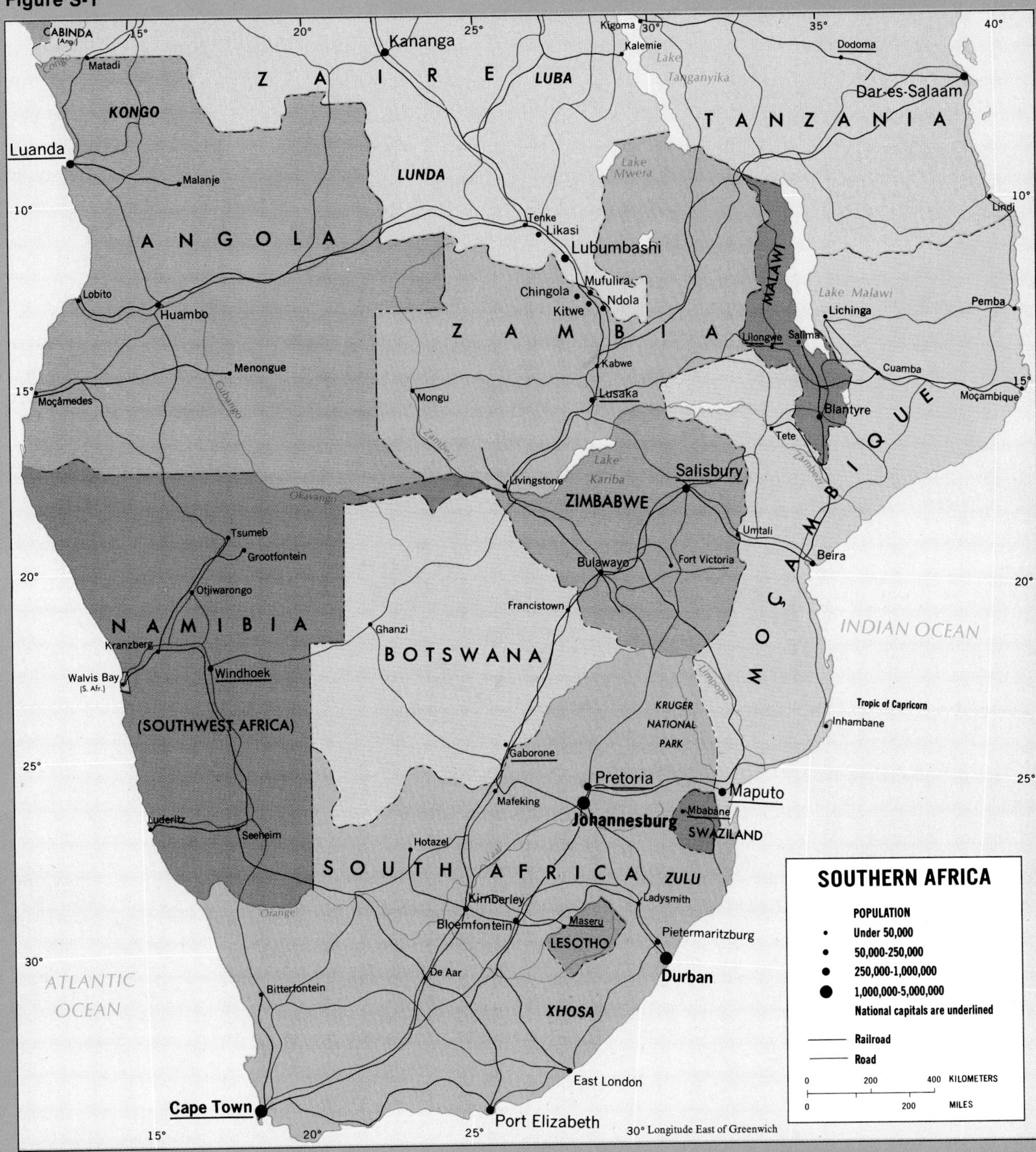

Ten Major Geographic Qualities of Southern Africa

1. South Africa's relative location at the southern tip of the African continent assigns the Republic considerable strategic importance.
2. South Africa is Africa's only true temperate-zone country; latitude and elevation combine to produce a range of natural environments unmatched on the continent.
3. South Africa's physiography is dominated by a plateau, the *highveld.*
4. South Africa's plural society consists of four major cultural components, each itself divided.
5. The historical geography of South Africa involves the in-migration of population sectors from other parts of Africa, from Europe, and from Asia.
6. South Africa's cultural pluralism is strengthened by its regional expression. A mosaic of "traditional" regions marks the Republic.
7. South Africa's white population is larger than the white settler populations of all other black African countries combined, even during the height of the colonial period.
8. South Africa's political geography is evolving as a racial mosaic under the white government's policy of *separate development.*
9. The buffer zone that has long had the effect of separating South Africa from black independent "front line states" is breaking down.
10. Although South Africa is often identified as Africa's only developed country, the republic consists of juxtaposed, sharply contrasted developed and underdeveloped regions.

Southern Africa is the richest region, in material terms, in Africa. A great zone of mineral deposits extends from Zambia's Copperbelt, through Zimbabwe's Great Dyke and South Africa's Bushveld Basin and Witwatersrand, to the gold fields of the Orange Free State in the heart of the Republic of South Africa. The range and volume of minerals mined in this belt are enormous, from the copper of Zambia and the chrome and asbestos of Zimbabwe to the gold, chromium, diamonds, platinum, coal, and iron ore of South Africa. Not all of Southern Africa's mineral deposits lie in this central backbone; there is coal in western Zimbabwe at Wankie, and in central Moçambique near Tete. In Angola, petroleum from fields along the north coast heads the export list, but diamonds are mined in the northeast and manganese and iron on the central plateau. Namibia (South West Africa) produces copper, lead, and zinc from a major mining complex centered on the town of Tsumeb in the north, while diamonds occur along the beaches facing the Atlantic Ocean in the south.

This is a mere summary of Southern Africa's mineral wealth, and it is matched by the variety of crops cultivated in the region. Vineyards drape the valleys of the Cape Ranges in South Africa; apple orchards, citrus groves, banana plantations, and fields of sugar cane, pineapples, and cotton reflect South Africa's diverse natural environments. In Zimbabwe, tobacco has long been the leading commercial crop, but cotton also grows on the plateau, and tea thrives along the eastern escarpment slopes. Angola is a major coffee producer, and Moçambique has huge cashew coconut plantations. The staple crop for the majority of Southern Africa's peoples is corn (maize), but wheat and other cereals are also grown. Even the pastoral industry is marked by variety. Not only are dairying and beef cattle herds large, but South Africa also is one of the world's leading wool exporters. And when it comes to livestock, Southern Africa's almost infinite variety appears again; in Nambia, karakul pelts are a high-value product, and in the Cape Province of South Africa, an ostrich farm is both a commercial enterprise and a tourist attraction.

But this mass of diverse resources has not been enough to improve the lives of all of Southern Africa's peoples—nor do the region's 10 countries share equally in these

This sweeping view over the city of Cape Town, South Africa, shows Table Mountain (background), the city center (left) and Lion's Head (right). Here landed the first Dutch settlers in 1652. (Richard & Mary Magruder/Image Bank)

assets. By far the richest country in Southern Africa is South Africa; indeed, South Africa is the wealthiest country in all of Black Africa. On the other hand, Lesotho, the mountainous country that is completely surrounded by South Africa, is quite poor by any standards. Botswana—landlocked and bounded by South Africa, Namibia, and Zimbabwe—is mainly desert and steppe country, and in 1980 this Texas-sized state still had under 1 million inhabitants. Moçambique remains an example of a seriously underdeveloped entity, its 10 million people only 8 percent urbanized in the late 1970s, their life expectancy 43 years, literacy perhaps 19 percent, annual income per person under $300. Malawi, Zambia, and Angola also suffer from these symptoms, albeit to a lesser degree. The overwhelming majority of the people in these countries live a life of often difficult subsistence, tilling patches of land near their villages that may or may not produce an adequate crop to sustain them during the approaching dry season, eating meals that are ill-balanced at best.

Even in South Africa and in Zimbabwe, skyscrapered cities, air-polluting industrial complexes, and modern mechanized farms still adjoin areas of almost unchanged rural Africa, where all those manifestations of modernization seem far away and of little relevance to village life. On world maps showing developed and underdeveloped countries, South Africa often appears as one of the developed countries—but these are two South Africas. The cities and their factories, the mines, and the commercial farms generate the incomes and other measures of development that place South Africa in the category of developed countries. But it is white South Africa that is developed South Africa, with a minority of African peoples who have been drawn into the sphere of development. Much of black South Africa remains underdeveloped, rural, remote, poverty-stricken. South Africa's wealth is substantial, but it has been neither sufficient nor adequately distributed among its multiracial population to produce a truly developed plural society.

Buffer Zone's Breakdown

In Chapter 2 we discussed the role of the frontier as a divider between expanding, ultimately competitive power cores. Southern Africa affords a modern example of the same phenomenon, but under different politico-geographical circumstances. A frontier developed in this region, but it did not consist of unclaimed country; rather, it extended across a belt of bounded political units. As the "Wind of Change" pushed African independence southward and toward South Africa, its rush in the late 1950s and 1960s was halted at the northern borders of Angola and Moçambique, and in what was then Rhodesia. Zambia and Malawi became independent states as early as 1964, shortly after the dissolution of the Central African Federation and just seven short years after Ghana's independence began the dramatic sweep of decolonization. But there the drive was slowed, and predictions that Rhodesia would soon follow Zambia proved in error. In northern Moçambique and the north of Angola, insurgent movements began to challenge Portuguese authority, and in Rhodesia, the white minority (now about 200,000 among nearly 7½ million Africans) declared the colony an independent, settler-ruled state. As the Portuguese struggled against the nationalists and the Rhodesian whites fortified their political position, South Africa lay protected by a buffer zone that kept the contest away from its own borders. Ultimately, just as happened in the frontiers of old, South African power made an incursion into the area as Pretoria's armed forces entered Angola. Briefly there was a prospect of partition in Moçambique, yielding the area to the north of the Zambezi River to the African nationalists and claiming the south as white-controlled country. Events overtook these efforts, however, and in 1975 Angola and Moçambique became independent African states.

A look at the map (Fig. S-1) reveals the significance of the developments of the 1970s and, especially, Zimbabwe's independence of 1980. True, Namibia still remains an area of contention, and the great majority of its African peoples (numbering under 1 million) inhabit the far north of the territory, a desert removed from the South African heartland. Adjacent Botswana, we noted, is virtually an empty country; many Botswana families make ends meet because a son or brother works in South Africa and supports those at home. This kind of independence obviously limits Botswana's political options. But in 1980 Rhodesia became Zimbabwe, and an African government led by one of the leaders of the independence movement, Mugabe, took control in Salisbury. This development is far more consequential than Moçambique's earlier sovereignty. The Swaziland ministate and the vacuum of the elongated Kruger National Park create separations along the Moçambique border, and in any case, South Africa continued to cooperate with Maputo in the operation of its major port facilities. Additionally, Moçambique's attentions for more than a decade were diverted to the Rhodesian armed struggle, since Moçambique served as a sanctuary for Zimbabwe insurgents.

Independence in Zimbabwe has transformed the politico-geographical map of Southern Africa. It marks the breakdown of the very heart of the buffer zone and exposes South Africa, finally, to a potentially strong black African state. As the old and the new Africa come face to face, uncertainty prevails.

Emergent Zimbabwe

The protracted struggle for control in Britain's former dependency of (Southern) Rhodesia illustrates the regional-geographic problems of decolonization in Africa. The country's core area is well defined, lying astride the mineral-rich Great Dyke and extending, approximately, from the environs of the capital, Salisbury (600,000) to the vicinity of the second city, Bulawayo (400,000). This is the stronghold of the former colony's white population, although white-owned farms and plantations lie scattered far beyond its limits. Land division in colonial Rhodesia always has been an issue of contention between black and white. In 1930 the white settlers imposed a Land Apportionment Act that allotted half of the colony's nearly 400,000 square kilometers (150,000 square miles) to the small European population, one-third to the African majority, and the remainder to the government for its disposition. That Act was revised repeatedly, but as recently as 1970 the whites (then a quarter of a million) and Africans (over 5 million) were awarded equal amounts of land—an unacceptable arrangement to the African majority. In addition, the white-owned lands contained the great majority of the good soils and well-watered areas of Rhodesia. Furthermore, the government stipulated that Africans could not grow certain crops on their land, thus leaving white farmers to reap added commercial advantage from their properties. Political pressures led the white regime to abandon these stipulations in 1977, but by then the damage had been done.

The drive for independence by Rhodesia's Africans was not simply a white-black confrontation, however. If there are strong animosities between white and black, there are also serious divisions among the African peoples of Rhodesia, and these divisions have regional expression. To begin with, Rhodesia's African population is fragmented into about 40 different peoples. These may be grouped, on cultural, ethnic, and locational bases, into two major units: the Shona-speaking Mashona peoples of the east and northeast and the Nguni-speaking Ndebele (or Matabele) of the southwest and west. The seven principal Shona-speaking peoples have created a loose federal alliance, and their territory extends across the Zimbabwean border into neighboring Moçambique. The Mashona are Rhodesia's oldest modern inhabitants, and the famous relic city of Zimbabwe is an ancient cultural center of the nation. Shona society is quite egalitarian and comparatively passive. The Mashona outnumber all other peoples in Zimbabwe combined, so that they have the capacity to control the country's government.

In contrast, the Ndebele peoples are recent arrivals in Rhodesia. When the British arrived to colonize the area during the second half of the nineteenth century, the Ndebele had just invaded the west and southwest of Zimbabwe, and they were in the process of colonizing Shona communities. The European intrusion stopped this process, but Ndebele society retains its stratified character, with a "pure" Nguni-speaking elite as the upper class, a somewhat absorbed and acculturated middle sector, and an overpowered and colonized lower class consisting of Shona (and other) groups in the process of transformation under Ndebele cultural dominance.

Like other African countries dominated by one or two national groups, Zimbabwe has a number of smaller minority peoples, but the Mashona and Ndebele constitute the bulk of the population, now exceeding 7 million. Even as Ndebele and Shona leaders were caught up in the struggle for independence, they clearly differed on ideological grounds, and saw Zimbabwe's post-independence future in different ways.

The intervention of the British government, which brought all parties to the conference table, ended the escalating war of independence, and transferred sovereignty to the elected representatives of the African majority, was a major achievement. It brought both prominent African leaders (Mugabe and the Ndebele representative, Nkomo) into a government of reconciliation that also included a substantial minority of whites. But independent Zimbabwe will continue to face its old divisions, and it will also confront new pressures as the latest "frontline" state—a state that adjoins the continent's last white-ruled country.

South African Prospects

South Africa has the greatest material wealth in the entire African realm. It is rich in natural resources as well as in the diversity of its cultures. The Republic's 29 million people (1980) included about 21 million Africans representing several nations; 4.5 million whites of whom the majority are Afrikaners while most of the remainder have British ancestries; over 2.5 million "Coloured" people of mixed African-white ancestry mainly concentrated in the southernmost Cape Province; and nearly 1 million Asians, mostly of Indian descent and largely concentrated in the port city of Durban and in rural Natal.

South Africa's modern history has set it apart from the rest of Black Africa. The old

Boer republics were founded before the colonial scramble elsewhere really gathered momentum; the diamonds of Kimberley attracted thousands of white men long before individual European travelers first saw much of tropical Africa. While African laborers from Lesotho made their way in the 1860s and 1870s to the mines of Kimberley to seek work and wages, Asians from India were arriving by the thousands on the shores of Natal, under contract to work on the sugar plantations there. Together, black, white, Asian, and Eurafrican contributed to the emergence of Africa's most developed country. Whites located the mineral resources of South Africa and provided the capital for their exploitation, but blacks did the work in the mines at wages low enough so that their extraction was an economic proposition. Whites laid out the farms and plantations that produce so wide a range of crops, but Africans and Asians constituted the essential labor force to make them successful. Whites built the factories of Johannesburg, but blacks form the majority of the wage earners that work in them. Whites built or bought the fishing boats that annually bring in a catch large enough to rank South Africa among the world's top 10 fishing nations, but the nets are manned mostly by Coloureds.

With such interracial cooperation in common economic pursuits, one might expect that South Africa's plural society over the years would have become more and more integrated. But in fact this has not been the case. South Africa stands today as a world symbol of racial discrimination, an outpost of white minority rule on a black continent. It acquired this reputation largely after World War II. Although much of the rest of the world reflected on the racial injustices involved in that war and the decolonization period of the 1950s transformed the international political scene, South Africa became known for its official policy of racial subjugation and separation as summarized by one word: *apartheid*. With this policy, South Africa moved in a direction that was more or less opposite to those ideals that were championed—if not necessarily practiced—in other parts of the world.

The platform promulgated after World War II by the new government, dominated by the conservative Afrikaners (successors to the Boers), included an enunciation of the *apartheid* doctrine. But within South Africa and in the world at large, the concept has often been incorrectly defined. *Apartheid* involved not only the separation of the black peoples of South Africa from the white, but also the Coloured from the white, and the Asian from the African. Behind it lay a vision of South Africa as a community of racial states, in which each racial sector would have its own "homeland", although residents of one such "homeland" could travel to another to work. What gave South Africa its international reputation was less this scheme than the *ad hoc,* "petty" *apartheid* that was designed to set the wheels of ultimate segregation in motion. The number of "Europeans Only" signs on public places, park benches, bus stops, and so forth, multiplied. Discriminatory pass laws appeared. An "Immorality Act" was passed, designed to stop interracial sexual contact. Violators were severely punished and suspects harrassed by decoys. Persons who were not white began to disappear from the white universities. Police powers increased; persons of liberal political views lost their freedom. African political organizations were banned, their leaders exiled. Eventually, a protest at Sharpeville led to a mass killing of Africans by police—an event that evoked world condemnation.

Nevertheless, South Africa, with the view that the end justifies the means, is proceeding in its program of *separate development*—now the official name for the *apartheid* scheme. The first of the African "homeland states," the Transkei, has been functioning for some years; it was formally declared independent in 1976. Bophutatswana followed in 1978, and Venda in 1979; and a total of 10 homelands were scheduled for "independence" by the late 1980s (Fig. S-2). Initially there were plans for a massive relocation of industry, from the core areas of the country to the "borderlands," the boundaries between the Bantustans and the white areas. If industries could be located there, it was argued, then skills and capital could come from white South Africa and labor from the black states, without any need for racial mixing. This part of the program has proved unworkable on any large scale, but other aspects of it are being vigorously pursued. Taking into consideration the spatial characteristics of the country, it is not surprising that a program that might seem totally unworkable in most other countries is viewed in South Africa as the ultimate and

The Transkei was the first African "Homeland" to achieve identity as a state under the Separate Development program. Border formalities are similar to those at any international boundary. Transkei, however, is not a member of the United Nations nor have countries other than South Africa recognized it as an independent state. (Harm J. de Blij)

inevitable solution to racial problems in plural societies. South Africa has a history of regional racial separation. Quite apart from the obvious associations between Bantu peoples and certain areas (such as the Zulu in Natal, the Xhosa in the Eastern Cape, and so on), there are other, pervasive regional contrasts. The Coloured people, for example, are largely concentrated in Cape Town and its environs; they are, indeed, called the Cape Coloured in South Africa. South Africa's Asians, on the other hand, came mainly to Natal and to the city of Durban. Many of them went to other parts of the country, of course, but Natal remains the South African province with the largest Asian population. Not unexpectedly, South African planners have considered a Coloured "homeland" in the Western Cape and an Asian one in Natal.

But the majority of the people in South Africa are black, and the separate development program's first objective has been the allocation of "homelands" for these people. When the Boers penetrated the Highveld, they left behind them the Xhosa and other peoples of Southern Natal and the Eastern Cape, who were hemmed in by the Great Escarpment on one side and the ocean on the other. On the plateau itself they found relatively little effective opposition, and before long virtually the whole of the Highveld was white country, with the Africans confined in reserves around the drier Kalahari margins, the rugged Northern and Eastern Transvaal, and below the escarpment in Natal and the Eastern Cape. Thus the horseshoe arrangement of reserves existed before the separate development program came into being; the invasion of white people had jelled the Bantu migrant peoples in their domains.

What made the whole separate development program viable was the agricultural, mining, and manufacturing development of the Highveld. From Swaziland, Zululand, Lesotho, the Transkei, and every other African territory, black laborers have come to the farms, the factories, and the mines of the Highveld. They came by the hundreds of thousands; the population of Johannesburg proper today is about 1.7 million, nearly two-thirds of it African. All too often African families would find themselves in the miserable shantytowns, without adequate shelter, food, medical care, schools, or other necessary facilities. Building programs could not keep pace with the influx; an ugly juvenile delinquency problem arose. Yet the government could not prohibit the immigration of workers, for they were always needed. This situation was high on the list of problems the separate development program was designed to solve. If an African man from an autonomous "homeland"—say, the Transkei—wanted to work in a white-owned mine or factory, he would apply for a permit to enter white South Africa on a temporary visa, would live in a dormitory type of residence in the city, near the mine or factory, and would send his wife and children part of his income for their subsistence. Thus the municipal government would no longer be faced with the task of making available the facilities necessary for the man's family, while the economy would not lose its labor supply.

Naturally, such a program could not be implemented overnight. The homelands must be established and given the trappings of semiautonomous states, governments installed, arrangements for the accommodation of repatriated families must be made—and, of course, people who may have been

Figure S-2

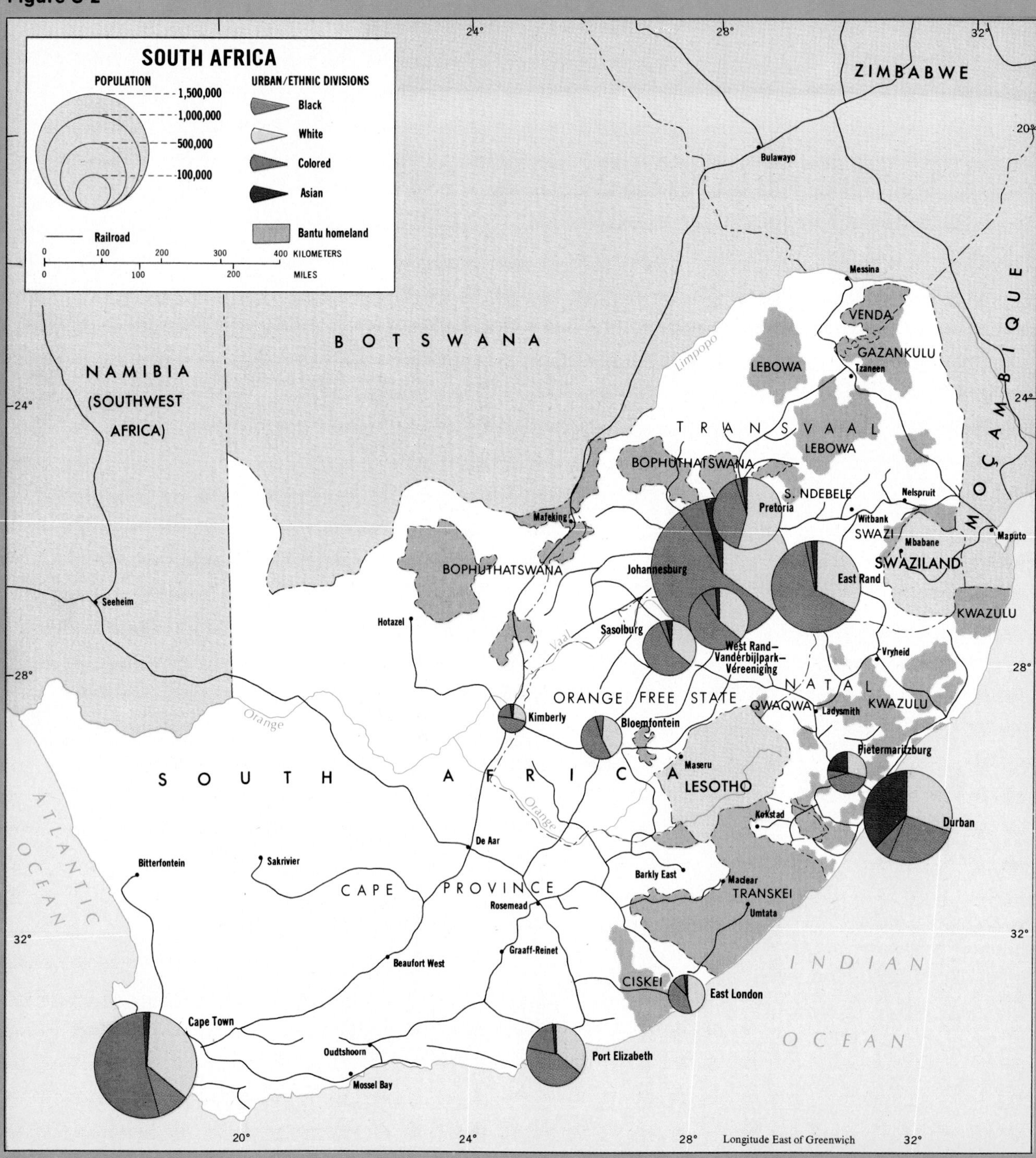

living in the big cities for generations must be classified according to their "race." In this process, once again, the South African government paid little heed to the sensitivities of many of its subjects. People were classified and—regardless of their race—were made to carry identity cards to prove their racial heritage. In many cases, of course, there was little doubt; a man's language quickly tells whether he is a Zulu or a Xhosa. But race in South Africa has long been correlated with status, and if a man can prove he is white rather than Coloured, his chances for a decent life are immeasurably improved; if a man can prove that he is Coloured, not black, he may be substantially better off than otherwise. Some of the saddest tales ever to come from Africa tell of people desperately trying to disavow their own ancestries, hoping to climb the social ladder one rung, to give themselves and their children a better chance in life.

Separate development is designed ultimately to solve all this. Instead of struggling for advancement in the white society, black or Coloured people would do so in their own society, without racial barriers, job restrictions, or other racially imposed obstacles. Thus separate development is *parallel* development, and eventually all the states of South Africa are destined to have every facility, including universities, hospitals, transport services, and so forth. But the critical flaw in the design appears to lie in the allocation of the country's areas of productive capacity to the homelands. of South Africa's 1¼ million square kilometers (nearly 500,000 square miles) of territory, only about one-sixth is presently being considered for the African homelands; yet the blacks constitute well over two-thirds of the total population. Four-fifths of the territory of the country will become "white" South Africa—to be controlled by the one-sixth of the population that is of European ancestry. Obviously, these figures by themselves mean very little; in many of the countries of the world, most of the productive lands lie on less than one-sixth of the total area. But a look at maps of the distribution of resources, cities, transport lines, and the other assets of South Africa quickly shows that the African homelands, far from having a real share of South Africa's prosperity, are among the poorest and most isolated parts of the country. No major mineralized areas lie in the homelands: from 1965 to 1975 only about 3 percent of South Africa's income from its mineral resources came from the Bantu areas. As far as agricultural areas are concerned, there is good farmland in the eastern homelands—in Natal and the Eastern Cape—but these areas, especially the Transkei and Zululand, also face a serious overpopulation problem. Those reserves that lie in the Northern Cape and on the Western and Northern Transvaal fringe suffer from drought and terrain problems. In terms of industrialization, the homelands have hardly begun to change; secondary industries are practically absent there. They have few resources, have never had much capital, and lie far from the markets and sources of power supply. South Africa's railway network and the country's major roads skirt the homelands or cross them only to connect cities of white South Africa; internal communications remain poor. And none of South Africa's major cities lies in a present or potential Bantu homeland. Indeed, when the separate development program was instituted, plans were actually laid to create—from the ground up—several dozen towns within the African homelands, to relieve the pressure on the land and to serve as dormitory settlements for the border industries scheme.

But towns as well as states need *raisons d'être,* and in the foreseeable future the African homelands of South Africa appear to be little other than a political gamble. Despite the optimistic implications of the term, it is hard to conceive of anything parallel between the development of white South Africa and that of the impoverished Transkei or KwaZulu—and these may be the best endowed of the projected black states. With such limited opportunities, the labor forces of the homelands will have to seek work in the mines and factories of white South Africa; this, of course, strengthens the white sector's hand in its dealings with the "autonomous" black states. And while it is possible that the African homelands will be nearly exclusively Bantu in their population, it is inconceivable that white South Africa will ever find itself without a resident (as opposed to a temporary migrant) black population sector. Today, 8 out of 10 farm laborers in "white" South Africa are black Africans, as are 9 out of 10 miners and 9 out of 20 factory workers.

South Western Townships (Soweto for short) present a rather bleak appearance in this air photo. This black suburb of Johannesburg has been a focus of resistance against South African authority. Note the mine dumps in the background. (Georg Gerster/Rapho-Photo Researchers)

Signs of the coming crisis in South Africa emerged shortly after Angola and Mocambique achieved independence and as Rhodesia's problems deepened. In mid-1976, urban riots broke out in several of South Africa's largest cities. The disorders began in the African towns that contain the black labor force working in the factories, mines, and businesses, but they soon spilled into downtown shopping streets. Patterns of separate living broke down as young blacks refused to attend school, prevented older workers from boarding trains to their jobs, and roamed streets in defiance of a police force that had orders to restore control at any cost. Hundreds died, and in thousands of families the seeds of future defiance were sown. Ancient animosities between African peoples crowded into the black towns led to further violence.

There is some evidence that leaders of the white South African minority recognize the untenable nature of the separate development doctrine. In 1978, following months of deteriorating negotiations over the issue of territorial expansion into adjacent areas, the Transkei broke diplomatic relations with South Africa (still the only country in the world to recognize the Transkei's status as a sovereign republic). In 1979, following the creation of Venda, there were predictions that the "independence" of Venda might well be the last to be formalized. Zulu opposition to the creation of KwaZulu as a national homeland has long been intense and has been effectively expressed by the Zulu leader, Chief Buthelezi. But while government leaders responded by opening dialogues with black leaders and by showing (as in the case of Crossroads, the reprieved black community in Cape Town scheduled for demolition) a capacity for compromise, the white electorate confirmed its conservative preferences. When the government proposed to create more skilled jobs for blacks in the mining industry, white mineworkers promptly struck in protest. In local elections to fill vacated parliamentary seats, the more conservative candidates were elected. If the white South African government is indeed prepared to moderate its racial policies, there is as yet little evidence that the voters are prepared to follow such a lead. And yet the future seems to demand a new course for South Africa. Geography and history may have conspired to divide its peoples, but economic realities have thrown them irretrievably together.

Additional Reading

A very good place to start reading about Africa is a book edited by C. G. Knight and J. L. Newman, *Contemporary Africa: Geography and Change,* published in 1976 by Prentice-Hall, Englewood Cliffs. Another recent volume was written by A. C. G. Best and H. J. de Blij, *African Survey,* published in 1977 by Wiley in New York. Also see W. A. Hance, *The Geography of Modern Africa,* undoubtedly the most comprehensive study on the geography of Africa available today, published by Columbia University Press, New York, in 1975. Another volume by Hance,

Population, Migration, and Urbanization in Africa was also published by Columbia in 1970. Among standard geographies published and/or revised in the 1960s and 1970s, see those by R. J. H. Church, A. T. Grove, R. M. Prothero (ed.), and Stamp and Morgan. The book by W. Fitzgerald, *Africa,* was published by Methuen, London, in numerous editions and it is now a classic. General introductions to the African culture realm include the readable *Africa* and *Africans* by P. Bohannan and P. Curtin, published by Natural History Press (Doubleday) in New York in 1971, the now-dated but still interesting two-volume *Tropical Africa* by G. H. T. Kimble, published in New York by Twentieth Century Fund in 1960, and G. P. Murdock's *Africa: Its Peoples and Their Culture History,* published by McGraw-Hill, New York, in 1959. J. N. Paden and E. Soja edited a three-volume *African Experience,* published in Evanston by Northwestern University Press in 1970.

Regional discussions abound. R. J. H. Church wrote *West Africa,* published by Wiley, New York, in several editions. B. Floyd's *Eastern Nigeria: a Geographical Review* appeared in 1969 from Macmillan in London. Also see A. L. Mabogunje, *Urbanization in Nigeria,* a London University Press publication of 1968. Still a valuable work is V. Thompson and R. Adloff, *The Emerging States of French Equatorial Africa,* published by Stanford University Press in 1960; another older work by A. Merriam, *Congo, Background to Conflict,* published by Northwestern University Press, Evanston, in 1961, also retains much interest. On East African countries, see J. Hatch, *Tanzania: a Profile,* published in New York by Praeger in 1972, S. H. Ominde, *Land and Population Movement in Kenya,* a Northwestern University Press publication of 1968, and P. M. Gukiina, *Uganda: a Case Study in African Political Development,* published by Notre Dame University Press in 1972. Also see a volume edited by W. T. W. Morgan, *East Africa: Its Peoples and Resources,* recently revised by Oxford University Press, Nairobi. In the early 1960s the International Bank for Reconstruction and Development published volumes on all three of the major East African countries, and they are dated but still of interest. On Southern Africa, see G. Kay's *Rhodesia: a Human Geography,* published by Africana Publishing Corporation in New York in 1970 and P. O'Meara, *Rhodesia: Racial Conflict or Coexistence?* published by Cornell University Press in Ithaca, 1975. On Portuguese colonial policy, a benchmark was a small paper by M. Harris, *Portugal's African Wards,* published by the American Committee on Africa in New York in 1958. Another commentary on Portugal in Africa is E. de S. Ferreira, *Portuguese Colonialism in Africa: the End of an Era,* printed by UNESCO in Paris in 1974. No reading on Portuguese Africa would be complete without a look at J. Duffy, *Portugal in Africa,* a Penguin Book, 1962, and A. Moreira's reply on behalf of Portugal: *Portugal's Stand in Africa,* published by University Publishers in 1962. On South Africa, a large compendium by M. Cole is *South Africa,* published by Methuen in London in 1966. L. Marquard's *Peoples and Policies of South Africa,* Oxford University Press, is a readable volume that has gone through several editions. A recent volume is L. Thompson and J. Butler (editors), *Change in Contemporary South Africa,* published by the University of California Press in Berkeley, 1975. Still of interest is D. Niddrie's *South Africa: Nation or Nations?* in the Van Nostrand Searchlight Series, 1968. Undoubtedly Namibia will attract world attention in the future, and a geography in depth is J. H. Wellington, *South West Africa and its Human Issues,* published by Oxford University Press in London in 1967.

On Africa's physical geography, see another volume by J. H. Wellington, *Southern Africa,* published by Cambridge University Press in 1955 but still productive, and L. C. King, *South African Scenery,* published by Oliver and Boyd, Edinburgh, in 1963. No reading would be complete without A. L. du Toit's classic *Our Wandering Continents,* another Oliver and Boyd publication of 1937. For an update, see J. T. Wilson, *Continents Adrift,* a Freeman, San Francisco, publication of 1970. S. B. Jones' article introducing the unified field theory appeared in the *Annals of the Association of American Geographers,* Vol. XLIV, 1954. On the problems of disease in Africa and elsewhere, see J. M. Hunter (editor), *The Geography of Health and Disease,* published in 1974 by the University of North Carolina and available from the Department of Geography. And D. Whittlesey's article entitled "Sequent Occupance" appeared in the *Annals of the Association of American Geographers,* Volume 19, in 1929.

Festival of faith: source of sustenance for millions. (Jehangir Gazdar/
Woodfin Camp)

Eight

India and the Indian Perimeter

The south of Asia consists of three protruding landmasses. In the southwest lies the huge rectangle of Arabia. In the southeast lie the slim peninsulas and elongated islands of Malaysia and Indonesia. And between these there is the familiar triangle of India, a subcontinent in itself. Bounded by the immense Himalaya ranges to the north, by the mountains of eastern Assam to the east, and by the rugged, arid topography of Iran and Afghanistan to the west, the Indian subcontinent is a clearly discernible physical region (Fig. 8-1).

This chapter focuses on India's huge human population in a world context. India alone, with 660 million people in 1980, has a larger population than South America, Black Africa, and Australia combined. Neighboring Bangladesh (90 million) and Pakistan (82 million) also rank among the world's 10 most populous countries. With over 865 million inhabitants, this realm constitutes one of the great human concentrations on earth. The scene of one of the world's oldest known civilizations, it later became the cornerstone of the British colonial empire.

The giant of South Asia, India, lies surrounded by the smaller countries. Bangladesh and Pakistan flank India on the east and west. To the north lie the disputed territory of Kashmir and the landlocked states of Nepal and Bhutan. Sikkim, formerly an Indian protectorate, has been incorporated into India as that federal country's 22nd state. And off the southern tip of the Indian triangle lies the island state of Sri Lanka (formerly Ceylon). The landlocked countries of the north lie in the historic buffer zone between the former British sphere of influence in South Asia and its Eurasian competitors; India, Pakistan, Bangladesh, and Sri Lanka all formed part of what was once known as British India.

IDEAS AND CONCEPTS

Boundary genetics (2)

Boundary superimposition

Green revolution

Demographic cycle and transition

Population explosion

Exponential growth

Physiologic and arithmetic density

Social stratification

Centripetal and centrifugal force

Irredentism (2)

Ten Major Geographic Qualities of South Asia

1. South Asia is a well-defined physiographic region extending from the southern slopes of the Himalayas to the island of Sri Lanka.
2. The South Asian realm's politico-geographical framework results from the European colonial period, but important modifications took place after the European withdrawal.
3. India lies at the heart of the world's second largest population cluster.
4. Two river systems (the Ganges-Brahmaputra and the Indus) form crucial lifelines for hundreds of millions of people in this realm. The annual monsoon is a critical environmental element.
5. Strong cultural regionalism marks South Asia. Hindu precepts dominate life in India; Pakistan is an Islamic state; Buddhism thrives in Sri Lanka.
6. India constitutes the world's largest and most complex federal state.
7. All the states of South Asia suffer from underdevelopment. Food shortages exist, nutritional imbalance prevails, and famines occur.
8. The great majority of South Asia's peoples live in villages and subsist directly from the land.
9. Agriculture in South Asia, in general, is comparatively inefficient and not as productive as it is in other parts of Asia.
10. No part of the world faces demographic problems with dimensions and urgency comparable to South Asia.

British colonial rule threw a false cloak of unity over a region that has enormous cultural variety, and even before the British departed, in 1947, plans had been developed for the separation of an Islamic state, Pakistan, from Hindu-dominated India. Islamic Pakistan formed a state that consisted of two territories, West Pakistan (now Pakistan) and East Pakistan (now Bangladesh). Sri Lanka became an independent state in 1948, and in 1972, following a bitter war, Bangladesh freed itself from its ties with Pakistan. Nevertheless, it is something of a miracle that a region as large and populous as South Asia has generated only four postcolonial states. Within India itself, regional differences remain strong, but the country has held together for more than 30 years. In Pakistan, persistent separatist movements affect Pakhtunistan in the northwest and Baluchistan in the southwest, but hitherto the pressure has been accommodated and the integrity of the state is sustained. In Sri Lanka, the Buddhist Sinhalese majority has from time to time been at odds with the minority Hindu Tamils, but Sri Lanka is no Cyprus.

Physiographic Regions

The Indian subcontinent is a land of immense physiographic variety, but it is possible to recognize several rather clearly defined physical regions. In the most general terms, there are three such regions: the

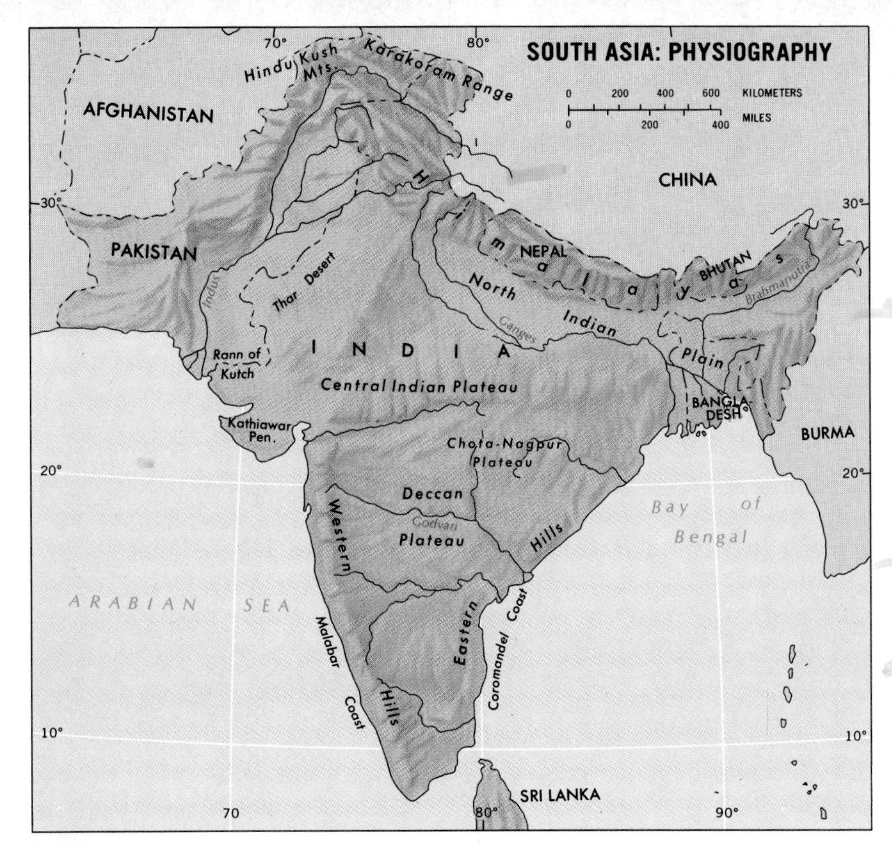

Figure 8-1

northern mountains, the southern peninsular plateau, and between these two, the belt of river lowlands. In this division, the northern mountains consist of the Baluchistan, Kashmir, Himalayan, and Assam uplands. The southern plateau is constituted mainly by the huge Deccan Plateau. And the tier of river lowlands extends from Sind (the lower Indus Valley) through the Punjab and middle Ganges to the great delta and on into Assam's Brahmaputra Valley (Fig. 8-1).

But this set of regions only introduces the varied Indian landscape. Depending on the degree of detail employed, it would be easy to subdivide each of these three into more specific physiographic subregions. Thus, in the mountain wall that all but surrounds India and Pakistan, we may recognize the desert ranges of Baluchistan and the Afghanistan border, the snowcapped ranges of the Himalayas, and, in the east, the jungle-clad mountains of the Assam-Burma margin. And, of course, these northern mountains do not simply rise out of the river valleys below: there is a persistent belt of foothills between the lofty peaks and the moist river basins.

The belt of river lowlands that extends from the Indus to the Brahmaputra also is anything but uniform. This is sometimes called the North Indian Plain, and some geographers consider it to include the Kathiawar Peninsula and part of the Thar Desert as well as the zone of river lowlands. But even without these this is a region full of contrasts, well reflected, incidentally, by Figs. I-5 to I-8. The Indus, which has its source in Tibet and which outflanks the Himalayas to the west in its course to the Arabian Sea, receives its major tributaries from the Punjab ("land of five rivers").

The physiographic region of the Punjab, then, extends into Pakistan as well as India. The region of the lower Indus is characterized by its low precipitation, its desert soils (Fig. I-8), and its irrigation-based cluster of settlement. This is the heart of Pakistan, and the farmers grow wheat for food and cotton for sale. Some 70 percent of all cultivated land in Pakistan is irrigated, most of it by an elaborate system of canals; wheat occupies more than one-third of the cropped land.

Hindustan, the region extending from the vicinity of the historic city of Delhi to the delta at the head of the Bay of Bengal, is wet country; the precipitation exceeds 250 centimeters (100 inches) in sizable areas and 100 centimeters (40 inches) almost everywhere (Fig. I-5). Deep alluvial soils cover much of the region (Fig. I-8), and the combination of rainfall, river flow, good soils, and a long growing season make this India's most productive area. As Fig. I-9 proves, this is also India's largest and most concentrated population core; not only do rural densities in places approach 400 people per square kilometer, but also, from Delhi to Calcutta, lies a series of the country's leading urban centers, connected by the densest portion of its rail and road transport network. In the moister east, rice is the chief food crop; jute is the main commercial product, notably in the delta, in Bangladesh. To the west, toward the Punjab and around the drier margins of the Hindustan region, wheat and such drought-resistant cereals as millet and sorghum are cultivated.

The easternmost extension of the North Indian Plain comprises the Brahmaputra Valley in Assam. This valley is much narrower than the Ganges Basin, it is very moist, and suffers from frequent flooding, and the river has only limited use for navigation. Up on the higher slopes, tea plantations have been developed, and in the lower sections rice is grown. Assam is just barely connected with the main body of India through a corridor of only a few miles between the southeast corner of Nepal and the northern boundary of Bangladesh, and it remains one of Asia's frontier areas.

Turning now to the peninsular part of India, referred to previously as plateau India, we find once again that there is more variety than first appearances suggest. There are physiographic bases for dividing the plateau into a northern area (the Central Indian Plateau would be an appropriate term) and a southern sector, which is already well known as the Deccan Plateau. The dividing line can be drawn on the basis of the roughness of terrain (along the Vindhyas) or on the basis of rock type (much of the Deccan is lava-covered), and the Tapti and Godavaria Rivers—in the west and east, respectively—form a clear lowland corridor between the two regions. The Deccan (meaning "south") has been tilted to the east, so that the major drainage is eastward; the plateau is marked along much of its margin by a mountainous, steplike descent formerly called the *Ghats* but now, simply, *Hills.* Parallel to the coast in both east and west are the Eastern and Western Hills; they meet near the southern tip of the subcontinent on the Cardamon Upland, recognized by some as a distinct physiographic region.

Peninsular India possesses a coastal lowland zone of varying width. Along the southwest coast lies the famous Malabar Coast, and, north of it, the Konkan; along the eastern shore is the Coromandel Coast. These physiographic regions lie wedged between the interior plateau and the Indian Ocean; the Malabar Coast is the more clearly defined since the Western Hills are more prominent and higher in elevation than the Eastern Hills. Thickly forested and steep, the Malabar escarpment dominates the west coast and limits its width to an average of less than 80 kilometers (50 miles); the Coromandel coastal plain is wider and its interior margins are less pronounced. While not very large in terms of total area, these two regions have had and continue to have great importance in the Indian state. Figure I-5 indicates how well watered the Malabar Coast is; this rainfall supply and the warm, tropical temperatures have combined with the coast's fertile soils to create one of India's most productive regions. On the lowland plain, rice is grown; on the adjacent slopes stand spices and tea. Of course, this combination of favorable circumstances for intensive agriculture led to the emergence of southern India's major population concentration (Fig. I-9). Along these coasts the Europeans made their first contact with India, beginning with the Greeks and Romans. Later these regions became spheres of British influence, and it was during this period that two of India's greatest cities, Bombay and Madras, began their growth.

"India"

Our use of the name *India* for the heart of the South Asian realm derives from the Sanskrit word *sindhu,* used to identify the ancient civilization in the Indus Valley. This word became *sinthos* in Greek descriptions of the area, and then *sindus* in Latin. Corrupted to *indus,* which means "river," it was first applied to the region that now forms the heart of Pakistan, and subsequently it was again modified to *india* to generally refer to the land of river basins and clustered peopled from the Indus in the west to the Brahmaputra in the east.

Human Sequence

The Indian subcontinent is a land of great river basins. Between the mountains of the north and the uplands of the peninsula in the south lie the broad valleys of the Ganges, the Brahmaputra, and the Indus. In one of these—the Indus—lies evidence of the region's oldest urban civilization, contemporary to and interacting with ancient Mesopotamia (see box). Unfortunately much of the earliest record of this civilization lies buried beneath the present water table in the Indus Valley, but those archeological sites that have yielded evidence indicate that here was a quite sophisticated urban culture, with large and well-organized cities in which houses were built of fired brick and consisted of two and sometimes even more levels. There were drainage systems, public baths, and brick-lined wells. As in Mesopotamia and the Nile Valley, considerable advances were made in the technology of irrigation, and the civilization was based on the productivity of the Indus lowland's irrigated soils. In the pottery and other artistic expressions of this literate society lies evidence of contact—and thus trade—with the lowland civilizations of Southwest Asia (Fig. 8-2).

But the Indus Valley did not escape the invasion of the Aryans any more than did Europe or the Mediterranean area. After about 2000 B.C., peoples began to move into the Indus region from Western Asia, through what is today Iran and through passes in the mountains to the north. Culturally these peoples were not as advanced as the Indus Valley inhabitants, and they brought destruction to the cities of the Indus. But they also adopted many of the innovations of the Indus civilization, and pushed their frontier of settlement out beyond the Indus Valley into the Ganges area and southward into the peninsula. They absorbed the tribes they found there, through conquest and enslavement, and the language they had brought to India, Sanskrit, began to differentiate into the linguistic complex that is India's today. Their village culture spread over a much larger part of the subcontinent than the urban civilization of the Indus had done.

In the centuries during and following the Aryan invasion, Indian culture went through a period of growth and development. From a formless collection of isolated tribes and their villages, regional organization began to emerge. Towns developed; local rulers became something more as surrounding areas fell under their control. Arts and crafts blossomed once again, and trade with Southwest Asia was renewed. Hinduism emerged from the religious beliefs and practices brought to India by the Aryans; tribal priests took on the roles of religious philosophers shaping a whole new way of life. Social stratification evolved, with the ruling Brahmans, administered by powerful priests, at the head

Figure 8-2

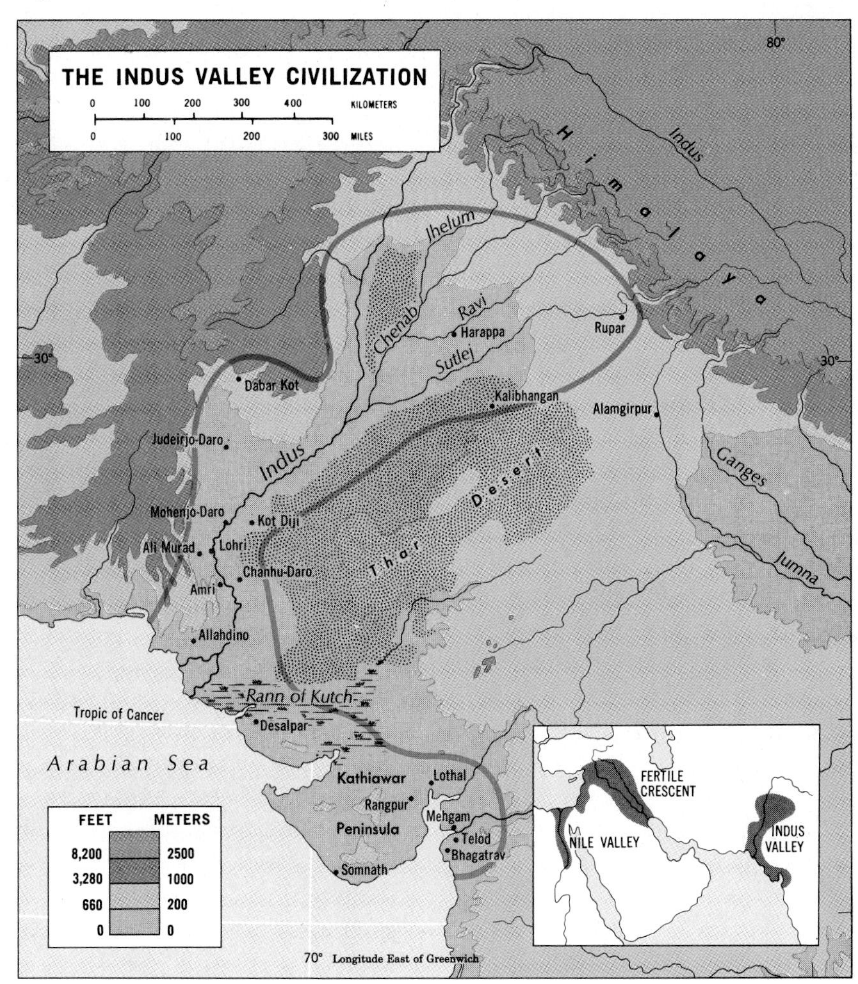

Bathing in the holy river, the Ganges, is an element of Hindu tradition that draws huge crowds to Benares every year. (Jehangir Gazdar/Woodfin Camp)

of a complex bureaucracy—a caste system—in which soldiers, artists, merchants, and others all had their place. Aggressive and expansion-minded kingdoms arose, always competing with each other for greater power and control. It was in one of these kingdoms, in northeastern India, that Prince Siddharta, or Buddha, was born in the sixth century B.C. Buddha voluntarily gave up his princely position to seek salvation and enlightenment through religious meditation. His teachings demanded a rejection of earthly desires and a reverence for all life. But although Buddha had a substantial following during his lifetime, the real impact of Buddhism was to come during the third century B.C., after an interval marked by the end of Persian domination in the northwest and a brief but intense intervention by Alexander the Great (326 B.C.). Alexander's Greeks pushed all the way to the Indian heartland in the Ganges Valley.

Thus Northern India was the theater of cultural infusion and innovation. The south lay removed, and protected by distance, from much of this change. Here a very different culture came into being. The darker skins of the people today still reflect their direct ties with ancient forebears such as the Negritos and black Australians, who lived here even before the Indus civilization arose far to the northwest. Their languages, too, are distinctive and not related to those of the Indo-Aryan region. Both the peoples and the languages of southern India are known collectively as *Dravidian.* The four major Dravidian languages—Telugu, Tamil, Kanarese (Kannada), and Malayalam—all have long literary histories, and Telugu and Tamil are the languages of 18 percent or nearly one-fifth of India's 640 million inhabitants (Fig. 8-3).

Mauryan Era and Islam

The first Indian Empire to incorporate most of the subcontinent emerged with the decline of Hellenic influence, shortly before 300 B.C. Its heartland lay in the middle Ganges Valley, and quite rapidly it extended its power over India as far west as the Punjab and the Indus Valley, as far east as Bengal, and as far south as modern Mysore. The state was led by a series of capable rulers, among whom the greatest no doubt was Asoka, who ruled for nearly 40 years during the middle period of the third century B.C. and who was a convert to Buddhism. In accordance with Buddha's teachings, Asoka diverted the state's activities from conquest and expan-

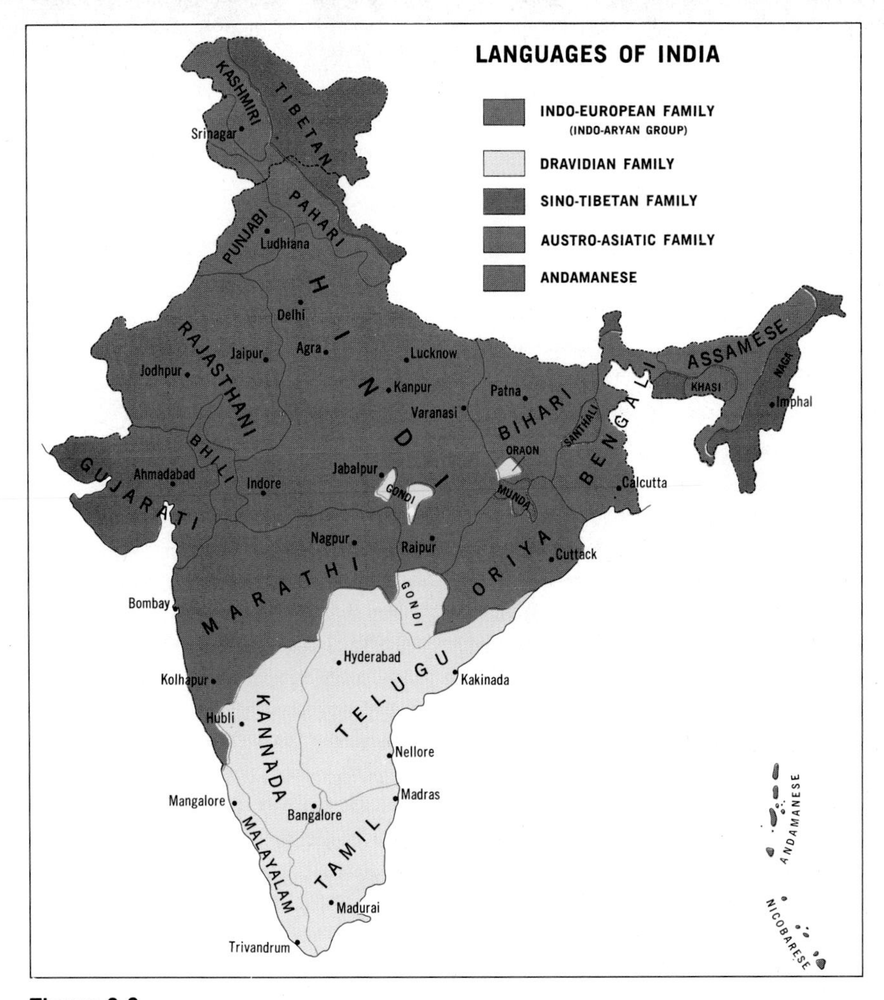

Figure 8-3

sion to the attainment of internal stability and peace; had he not done so it is likely that all of the subcontinent would have fallen to the Maurya's rule. A vigorous proponent of Buddhism, Asoka sent missionaries to the outside world to carry Buddha's teachings to distant peoples; in so doing he contributed to the further spread of Indian culture. Thus Buddhism became permanently established as the dominant religion in Ceylon (now Sri Lanka), and it achieved temporary footholds even in the eastern Mediterranean lands—although it eventually declined to minor status in India itself.

The Mauryan state, which represented the culmination of Indian cultural achievements up to that time, was not soon to be repeated. When the state collapsed late in the second century A.D., the old forces—regional-cultural disunity, recurrent and disruptive invasions, and failing central authority—again came to the fore. Of course, the cultural disunity of India was a reality even during the Mauryan era; the Mauryans could not submerge it. And while India did not see anything like the Aryan invasions again, there were almost constant infusions of larger and smaller population groups from the west and northwest. Persians, Afghans, and Turks entered the subcontinent, mostly along the obvious avenue: across the Indus, through the Punjab, and into the Ganges (Ganga) Valley. Thus beginning late in the tenth century, the wave of Islam came spreading like a giant tide across the subcontinent from Persia and Afghanistan to the northwest. In the Indus region there was a major influx of Moslems and a nearly total conversion to Islam of the local population. In the Punjab perhaps two thirds of the population was converted; Islam crossed the bottleneck in which Delhi is positioned and spread into the Indian heartland of the Ganges (Ganga) Valley—Hindustan. While the Moslem impact in Hindustan was much less, about one eighth of the population there became adherents to the new faith. In the delta region of the Ganges, where up to three quarters of the people became Moslems, Islam seems to have spread at the expense especially of Buddhism. Southward into the peninsula, the force of Islam was spent quite quickly, and the extreme south was never under Moslem control.

Islam was an alien faith to India, as it was to southwestern Europe, and it brought great changes to existing ways of life. It was superimposed by political control; early in the fourteenth century a sultanate centered at Delhi controlled all but extreme southern, eastern, and northern margins of the subcontinents. There were constant struggles for control, as challengers to existing authority came from the Afghan empire to the west and from Turkestan and inner Asia to the northwest. Out of one of these challenges arose the largest unified Indian state since Asoka's time, the so-called Mogul or Mughal Empire, which in about 1690 under the rule of

Aurangzeb comprised almost the whole subcontinent from Baluchistan to the Ganges Delta and from the northern foothills to Madras. But Islam in India was neither the monopoly of the invaders from outside nor the religion only of the rulers and the new aristocracy. Islam provided a welcome alternative to Hindus who had the misfortune of being of low caste; it was an alternative also for Buddhists and others who faced absorption into the prevailing Hindu system. In the majority of cases the Moslems of the Indian subcontinent are indistinguishable racially from their non-Moslem neighbors; there is no correlation between race and religion. Most of the subcontinent's Moslems today are descendants of converts rather than descendants of any invading Moslem ruling elite.

European Intrusion

Into this turbulent situation of religious, political, and linguistic disunity still another element intruded: European powers in search of raw materials, markets, and political influence. The Europeans profited from the Hindu-Moslem contest; they were able to exploit local rivalries, jealousies, and animosities. British merchants gained control over the trade to Europe in spices, cotton, and silk goods, and in time they ousted their competitors, the French, Dutch, and Portuguese. The British East India Company's ships also took over the intra-Asian sea trade between India and Southeast Asia, long in the hands of Arab, Indonesian, Chinese, and Indian merchants. In effect the East India Company became India's colonial administration.

As time went on, the East India Company faced problems it could not solve; its commercial activities remained profitable, but it became entangled in an ever-growing effort to maintain political control over an expanding Indian domain. It proved an ineffective governing agent at a time when the increasing Westernization of India brought to the fore new and intense kinds of friction. Christian missionaries were challenging Hindu beliefs, and many Hindus thought that the British were out to destroy the caste system. Changes came also in public education, and the role and status of women began to improve. Aristocracies saw their positions threatened; landowners had their estates expropriated. Finally, in 1857, a rebellion occured that changed the entire situation. It took a major military effort to put down, and from that time the East India Company ceased to function as the government of India. The administration was turned over to the British government, the company was abolished, and India became formally a British colony. It retained this status until 1947.

Colonial Transformation

Four centuries of European intervention in India greatly changed the subcontinent's cultural, economic, and political directions. Certainly the British made positive contributions to Indian life, but colonialism also had serious negative consequences. In this respect there are important differences between the Indian case and that of Black Africa, for when the Europeans came to India they found a considerable amount of industry, especially in metal goods and textiles, and an active trade to both Southwest and Southeast Asia in which Indian merchants took a leading part. The British intercepted this trade, and in the process the whole pattern of Indian commerce changed. India now ceased to be Southern Asia's manufacturing area: soon India was exporting raw materials and importing manufactured goods—from Europe, of course. India's handcraft industries declined, and after the first stimulus the export trade in agricultural raw materials also suffered as other parts of the world were colonized and tied in trade to Europe. Thus the majority of India's people, who were farmers then as now, suffered a setback as a result of the colonial economy. Therefore, although in total *volume* of trade the colonial period brought considerable increases, the *kind* of trade India now supported was by no means a way to a better life for the people.

Neither did the British manage to accomplish what the Mauryans and the Moguls had also tried to do: unify the subcontinent and minimize its cultural and political divisions. When the crown took over from the East India Company, nearly 2 million square kilometers (about 750,000 square miles) of Indian territory still were outside the British sphere of influence, and slowly the British extended their control over this huge, unconsolidated area, including several pockets of territory already surrounded but never integrated into the company's administration. Also, the British government found itself with a long list of treaties that had

The colonial imprint on the urban landscape of Calcutta. (Cary Wolinsky/ Stock, Boston)

been made by the company's administrators with numerous Indian kings, princes, regional governors, and feudal rulers. These treaties guaranteed various degrees of autonomy for literally hundreds of political entities in India ranging in size from a few acres to Hyderabad's more than 200,000 square kilometers (80,000 square miles). The British crown saw no alternative but to honor these guarantees, and so India was carved up into an administrative framework in which, in the late nineteenth century, there were over 600 "sovereign" territories in the subcontinent. These "Native States" had British advisors; the large British provinces such as Punjab, Bengal, and Assam had British governors or commissioners who reported to the Viceroy of India who, in turn, reported to the Parliament and British crown. It was a near-chaotic wedding of modern colonial control and traditional feudalism, reflecting and in some ways deepening the regional and local disunity of the Indian subcontinent. While certain parts of India quickly adopted and promoted the positive contributions of the colonial era, other units rejected and repelled them, thus adding still another element of division to the already complex picture.

Indeed, colonialism did produce assets for India. As a glance at a world map of surface communications shows, India was bequeathed one of the better road and railroad transport networks of the colonial realm. British engineers laid out irrigation canals through which millions of acres of land were brought into cultivation. Settlements that had been founded by the British developed into major cities and bustling ports, as did Calcutta (9.5 million), Bombay (7 million), and Madras (3.3 million), three of India's largest urban centers. Modern industrialization, too, was brought to India by the British on a limited scale. In education an effort was made to combine English and Indian traditions; the Westernization of India's elite was supported through the education of numerous Indians in Britain. Modern practices of medicine also were introduced. In addition, the British administration tried to eliminate features of Indian culture that were deemed undesirable by any standards, such as the burning alive of widows on the funeral pyres of their husbands, female infanticide, child marriage, and the caste system. Obviously the task was far too great to be successfully achieved in less than four generations of rule, but India itself has continued the efforts initiated by the British where necessary.

Partition

Even before the British government decided to yield to Indian demands for independence it was clear that British India would not survive the coming of sovereignty in one piece. As early as the 1930s the idea of a separate Pakistan was being promoted by Moslem activists, who circulated pamphlets arguing that India's Moslems were a distinct nation from the Hindus and that a separate state consisting of the Punjab, Kashmir, Sind, Baluchistan, and a section of Afghanistan should be created. The first formal demand for partition was made in 1940, and the idea had the almost universal support of the region's Moslems as subsequent elections proved. As the colony moved toward independence, a

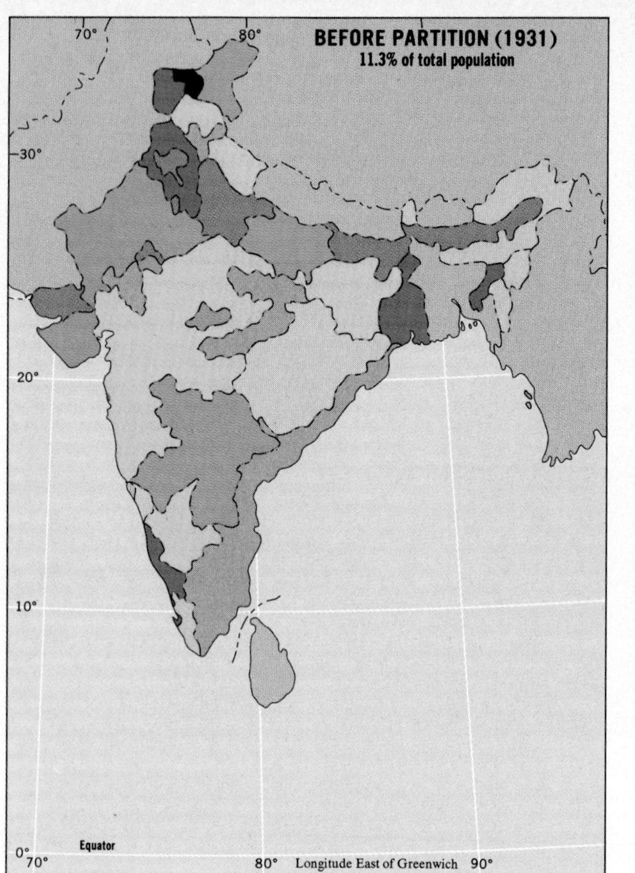

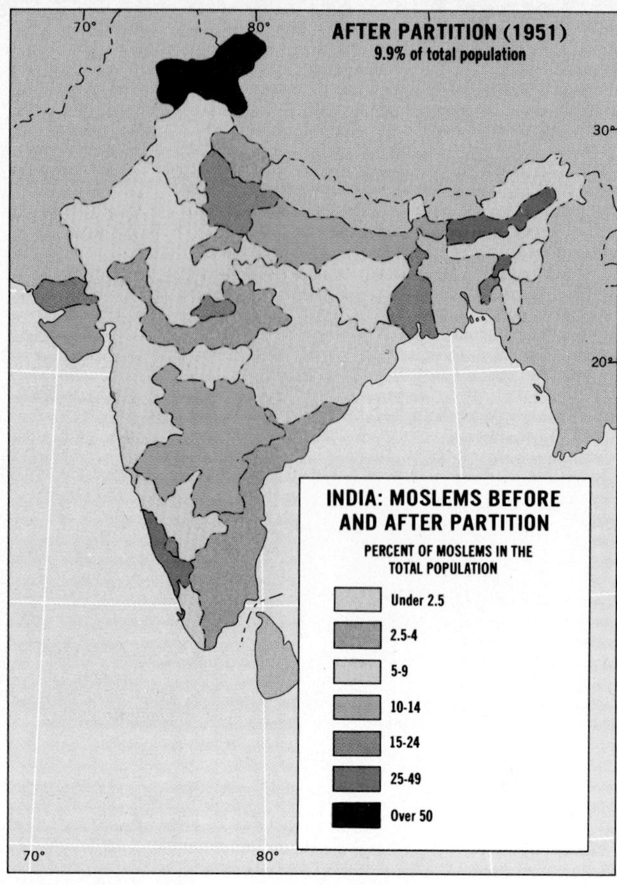

Figure 8-4

crisis developed: the majority Congress Party would not consider partition, and the minority Moslems refused to participate in any unitary government. But partition was not simply a matter of cutting the Moslem areas off from the main body of the country; true, Moslems were in the majority in what is today Pakistan and Bangladesh, but other Moslem clusters were scattered all over the subcontinent (Fig. 8-4). The boundaries between India and Pakistan would have to be drawn right through transitional areas, and people by the millions would be dislocated.

In the Punjab, for example, a large number of Sikhs—whose leaders were intensely anti-Moslem—faced incorporation into Pakistan. Even before independence day, August 15, 1947, the Sikh leaders had been talking of revolt, and there were some riots, but no one could have anticipated the horrible killings and mass migrations that followed independence and official partition.

Just how many people participated in these migrations will never be known, but the most common estimate is 15 million, representing a mass of human suffering that is indeed incomprehensible. And even that huge number of refugees hardly began to purify either India of Moslems or (former East and West) Pakistan of Hindus. After the initial mass exchange, there were still tens of millions of Moslems in India, and East Pakistan (now Bangladesh) remained about 20 percent Hindu. Facing the difficult alternatives, many people decided to stay where they were and make the best of the situation. And for most it turned out to be a wise choice.

The actual process of partition was done quite quickly, and of necessity rather arbitrarily, by a joint commission whose chairman was a neutral, a British representative. Using data from the 1941 census of India, Pakistan's boundaries were defined in such a manner that the Moslem state would incorporate all contiguous civil divisions and territories in which Moslems formed a majority. The commission had to make decisions it knew in advance would be

This remarkable photograph, taken in 1947, shows thousands of Moslem refugees jamming the railroad cars of a train leaving the New Delhi area for Pakistan, the Islamic republic. (Wide World)

highly unpopular, though no one foresaw the proportions of the Sikh-initiated violence.

What happened in India was the *superimposition* of a political boundary on a cultural landscape in which such a boundary had hitherto not existed or functioned. Political geographers consider boundaries in this light—that is, with reference to the stage of development of the cultural landscape at the time boundaries were established. On this basis a genetic classification of boundaries has been developed (as opposed to the morphological classification discussed in the context of the Middle East), whose terminology was proposed by R. Hartshorne in 1936.

Boundaries that were defined and delimited before the main elements of the present-day cultural landscape began to develop are identified as *antecedent.* For example, a boundary may be drawn, even geometrically, through a desert area occupied, if at all, only by some nomadic groups of people. Then, however, oil may be discovered in the area through which the boundary lies, and a whole new pattern of settlement emerges. As the cultural landscape evolves, it must adjust to the reality of the boundary.

Subsequent boundaries display a certain degree of conformity to the main elements of the cultural landscape through which they lie. They come about as part of the process of spatial organization in the regions they now fragment. Not infrequently this sort of boundary conforms to linguistic, religious, or ethnic breaks or transition zones in the landscape. In the partition of the Indian subcontinent, the newly created boundaries between India and the two sections of Pakistan over parts of their course are subsequent in nature, in that they confirmed a strong cultural break already existing in the cultural landscape. But British India has been a political and economic whole, and Hindus and Moslems had coexisted for several centuries, so that a strong case can be made for the argument that the Pakistan-India boundary is largely *superimposed* on the landscape—that is, it was established after a period of development during which Hindus and Moslems had lived in close contact; its effect was to reverse a process of cultural accommodation. In the east, for example, in the area where the boundary between former East Pakistan and India was delimited, cotton fields and jute plantations cover the countryside. For many decades cotton and jute farmers had been taking their produce to cotton mills and jute factories nearby—but when the boundary was created, the majority of the mills and factories were in India, and most of the farmlands in East Pakistan. Here the boundary was superimposed on a functional region.

Federal India

India today is the world's most populous federal state. Including newly incorporated Sikkim, the Indian federation consists of 22 states and 8 union territories (Fig. 8-5). A comparison between Fig. 8-5 and the map of languages in India (Fig. 8-3) indicates a considerable degree of coincidence between linguistic and state boundaries. The present framework is not the one with which India was born as a sovereign state in 1947, and it is likely to be modified again in the future. Soon after the British withdrew, the several hundred "princely states," whose rights had been protected during the colonial period, were absorbed into the states of the federation although their rulers' privileged

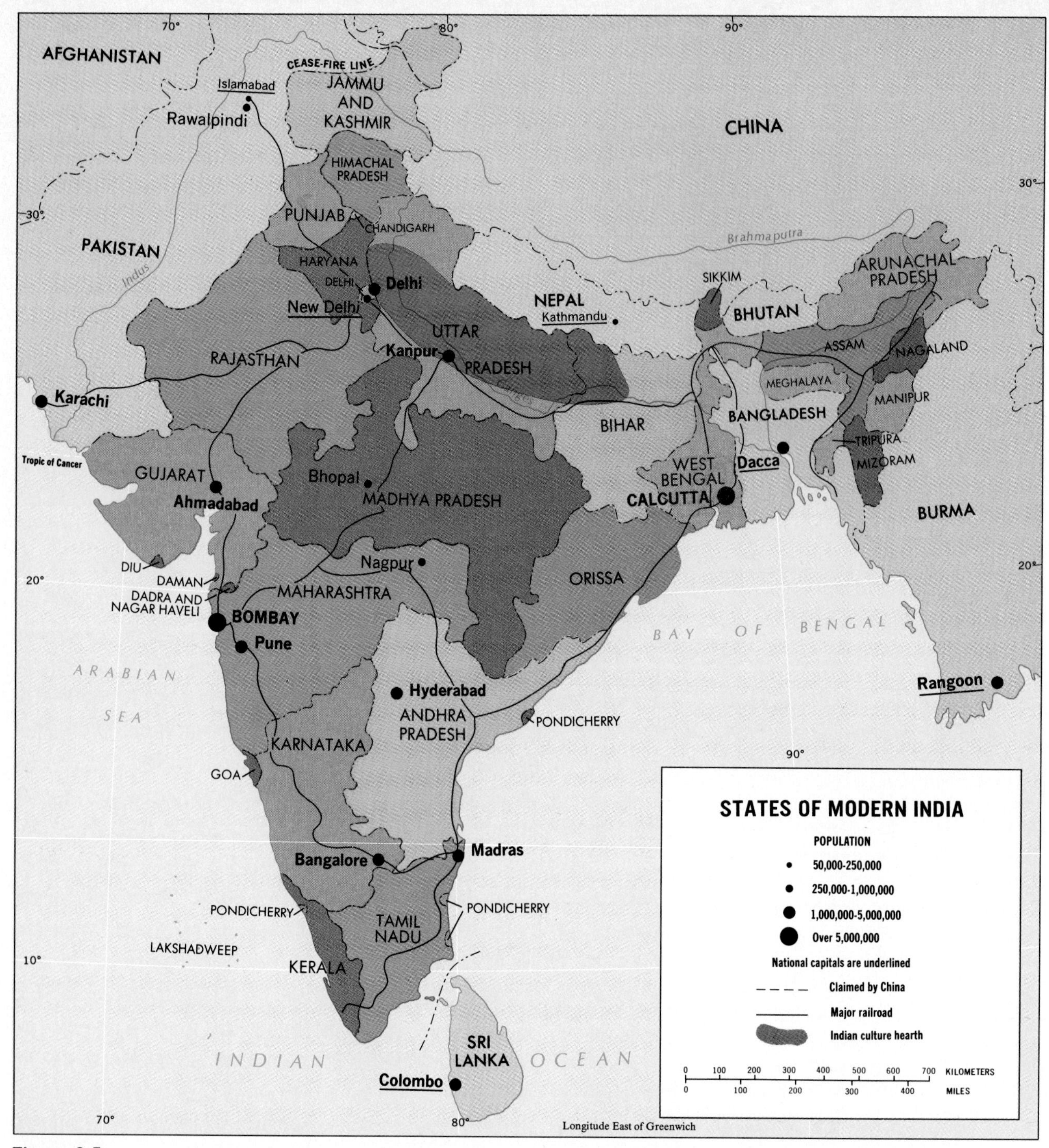

Figure 8-5

"princely orders" were not terminated until 1972. Next the Indian government reorganized the country on the basis of its regional languages, and Hindi, spoken by nearly one-third of all Indians, was designated the official language. Hindi was one of fourteen languages given national status by the Indian constitution—ten of them spoken in the northern and central part of the country and four in the Dravidian south. English, it was anticipated, would remain a lingua franca when Hindi could not serve as a medium of communication at government and administrative levels.

The politico-geographical framework based on the major regional languages proved to be unsatisfactory to many communities in India. In the first place, many more languages are in use than the fourteen that were officially recognized—even after

Sindhi (spoken in the Rann of Kutch and the Kathiawar Peninsula adjacent to Pakistan) was added to the list. As early as 1953, the government yielded to demands for the creation of a Telugu-speaking state from Tamil-dominated Madras, and thus the state of Andhra Pradesh was formed. In 1960 the state of Bombay was fragmented into two linguistic states, Gujarat and Maharashtra (Fig. 8-5). In the distant eastern borderlands, the Naga peoples, numbering less than a half million, put up a struggle against federal authority and against Assamese administration; after Indian armed forces were sent into the area, Nagaland was established as an Indian state. In the west the Sikhs demanded the breakup of the original state of Punjab into a Sikh-dominated west (now Punjab) and a Hindu east (now Haryana). Pressure for greater regional autonomy continues in several other areas, and it is remarkable that the Indian state has been able to survive such centrifugal forces.

India's unity is all the more remarkable in view of its enormous population and the problems inherent in its demographic condition. The country's most populous states (Uttar Pradesh, approaching 100 million; Bihar, nearly 70 million; Maharashtra, over 60 million) have populations larger than the great majority of the world's countries. No country contains greater cultural diversity than India, and variety in India comes on a scale unique in the world. Even today, after the religious partition between Hindu and Moslem, India remains 11 percent Moslem. That might seem to be a rather small minority—except that in India, an 11 percent minority represents over 70 million people.

The language issue has constituted a centrifugal force in India for decades. Here, demonstrators demand the creation of their own state in the Punjab. (Marc Riboud/Magnum)

A still-pervasive aspect of Indian society is its stratification into *castes,* a dimension of Hinduism. Castes are layers in society whose ranks are based on ancestries, family ties, and occupations. The caste system may have developed from society's early differentiation into priests and warriors, merchants and farmers; it may also have a racial foundation (the Sanskrit term for caste is color). Over centuries its complexity grew until India possessed several thousand castes, some with a few hundred members, others with millions. In village as well as city, communities were thus divided into castes ranging from the highest (priests, princes) to the lowest (the untouchables). A person was born into a caste based on his or her actions in a previous existence; hence it would not be appropriate to counter such ordained caste assignment by permitting movement (or even contact) from a lower caste to a higher one. Persons in a particular caste could do only certain jobs, wear only certain clothes, worship only in certain ways at particular places. They or their children could not eat with, play with, or even walk with people more fortunate. The untouchables were the most debased, wretched members of a rigidly structured social system.

Although the British ended the worst excesses associated with the caste system and modern Indians, including Gandhi and Nehru, have worked to modify it, centuries of class consciousness are not wiped out in a few decades. In traditional India, caste provided stability and continuity. In modernizing India, it constitutes an often painful and difficult legacy.

What centripetal forces have helped keep India unified? Without question the dominant binding force in India is the cultural strength of Hinduism, its holy writings (read in all languages), its holy rivers, and its many other influences over Indian life. Hinduism is a way of life as much as it is a faith, and its diffusion over virtually the

Centrifugal and Centripetal Forces

Political geographers use the terms *centrifugal* and *centripetal* forces to identify forces that, within the state, tend respectively to pull the system apart and to bind it together. *Centrifugal* forces are divisive. They cause deteriorating internal relationships. Religious conflict, racial strife, linguistic division, and contrasting outlooks are among major centrifugal forces. During the 1960s the Indochina War became a major centrifugal force in the United States, one whose aftermath will be felt for years to come. Newly independent countries find tribalism a leading centrifugal force, sometimes strong enough to threaten the very survival of the whole state system, as the Biafra conflict did in Nigeria.

Centripetal forces tend to bind the state together, to unify and strengthen it. A real or perceived external threat can be a powerful centripetal force, but more important and lasting is a sense of commitment to the system, a recognition that it constitutes the best option. This commitment is sometimes focused on the strong charismatic qualities of one individual—a leader who personifies the state, who captures the population's imagination. (The origin of the word *charisma* lies in Greek, where it means "divine gift.") At times such charismatic qualities can submerge nearly all else; in India, Gandhi and Nehru had personal qualities that went far beyond political leadership, and their binding effect on the nation extended far beyond their lifetimes.

The degree of strength and cohesion of the state depends on the excess of centripetal forces over the divisive centrifugal forces. It is difficult to measure such intangible items, but some attempts have been made in this direction; for example, by determining attitudes among minorities, and by evaluating the strength of regionalism as expressed in political campaigns and voter preferences. When the centrifugal forces gain the upper hand and cannot be checked (even by external imposition) the state will break up—as Pakistan did in the early 1970s.

entire country (Moslem, Christian, and Sikh minorities notwithstanding) brings with it a national coherence that constitutes a powerful antidote to regional divisiveness. In addition, communications in much of densely populated India are better than they are in many other recently independent countries, and the circulation of people, ideas, and goods helps bind the nation together. Before independence, opposition to British rule was a shared philosophy throughout the country, but India remained divided and separated internally. Independence brought with it a first taste of national planning, national political activity, and national debates over priorities. Furthermore, India generated some strong leaders whose very personalities constituted a binding force. Gandhi and Nehru did much to forge the India of today. And finally, federal India has proved capable of accommodating far-reaching change, thereby avoiding the alternative of secession. Although political leaders in some states have on occasion openly endorsed the prospect of secession, India's flexibility on the language issue, its ability to tolerate individuality in its states, and its capacity to modify and remodify the federal map all give evidence that change *is* possible and that political, economic, educational, and other objectives can in fact be achieved within the Indian framework. This, too, has served to sustain the most complex federation in the world.

Indian Development

If India has faced problems in its great effort to achieve political stability and national cohesion, these are more than matched by the difficulties that lie in the way of economic progress and development. The large-scale factories and power-driven machinery of the colonial powers wiped out a good part of India's indigenous industrial base. Indian trade routes were taken over. European innovations in health and medicine sent the population growth rate soaring, without introducing solutions for the many problems this entailed. Surface communications were improved and food distribution systems became more efficient, but local and regional famines occurred nevertheless. In the 1970s India was struck by droughts and crop failures, as were Africa's Sahel, Ethiopia, and other parts of the world, and food shortages were widespread.

Agriculture

India's underdeveloped condition is nowhere more evident than in its agriculture. Traditional farming methods continue to prevail; yields per hectare and per worker remain low for virtually every crop grown in the region. As the total population grows, the amount of cultivated land per person declines; in the late 1970s the physiological density was estimated at 325 per square kilometer (850 per square mile). This is nowhere near as high as it is in neighboring Bangladesh (where it is *four times* greater) or in Egypt, but Indian farming is so inefficient that this is a deceptive comparison. More than two-thirds of India's huge working population depends directly on the land for its livelihood, but the great majority of Indian farmers are poor, unable to improve their soils, equipment, or yields. Those areas where India has made substantial progress in the modernization of its agriculture (as in the Punjab's wheat fields) still remain islands in a sea of agrarian stagnation.

This stagnation has persisted in large measure because India failed, after independence, to implement a much-needed nationwide land reform program. In the late 1970s, nearly one third of India's entire cultivated area still was owned by only 4 percent of the country's farming families; nearly half the rural families either owned as little as a half hectare (1¼ acres) or less—or no land at all. Independent India inherited inequities from the British colonial period, but the individual states of the union would have had to cooperate in any national land reform program. The large landowners had much political influence, and so the program never got off the ground, although several states in recent years have placed upper limits on the holdings families may possess, and some redistribution of land has taken place. In addition, much of India's farmland is badly fragmented as a result of local rules of inheritance, inhibiting cooperative farming, mechanization, shared irrigation, and other opportunities for progress. Land consolidation efforts have had only limited success except in the states of Punjab, Haryana, and Uttar Pradesh, where modernization has gone farthest. Certainly official agricultural development policy, at federal as well as state levels, has contributed to India's agricultural malaise also. Unclear priorities, poor coordination, inadequate dissemination, and other failures have been reflected in the country's output.

It is useful to compare Fig. 8-6, showing the distribution of crop regions and water supply systems in India, with Fig. I-5 (page 17), indicating mean annual precipitation in India and the world. In the comparatively dry northwest, notably in the Punjab and neighboring areas of the Upper Ganges, wheat is the leading cereal, and in this region India has made major gains in annual production through the introduction of high-yielding varieties developed experimentally in Mexico. This innovation was part of the so-called Green Revolution of the 1960s, when strains of wheat and rice (the latter in the Philippines) were developed that were so much more productive than other varieties that they were called "miracle" crops. Introducing these new seeds also led to the expansion of cultivated areas, the development of new irrigation systems, and the more intensive use of fertilizer (a mixed blessing, for fertilizers are expensive and the "miracle" crops are heavily dependent on fertilizers).

Toward the moister east and in the monsoon-drenched south, rice takes over as the dominant staple. Over one quarter of all of India's farmland lies under rice, most of it in the states of Assam, West Bengal, Bihar, Orissa, and eastern Uttar Pradesh. This area has over 100 centimeters (40 inches) of rainfall, and irrigation supplements precipitation where necessary;

Concepts of Density

The last two columns of Table I-1 (pages 28–32) list the population *density* of individual countries of the world per unit area (square kilometers in this instance). For the countries of South Asia, the table records that Bangladesh, with 555 persons per square kilometer in 1978, has double the population density of Sri Lanka (227) and more than 2½ times that of India (210).

These figures represent the *arithmetic* population density, derived by simply dividing the area into the population total. But the total area of a country inevitably includes uninhabitable and unproductive terrain as well as farmland and pastures. More significant, therefore, is a measure of the number of people per unit area of agriculturally productive land. This is the *physiological* population density (other terms also are applied to it), and now the South Asian comparisons look very different. India's physiological density is 325, about 1½ times its arithmetic density, but in Bangladesh it is nearly 1350 per square kilometer, which amounts to some 3500 persons per square mile of productive land (or less than one-fifth of an acre per person)—fully 2½ times the arithmetic density.

Statistics on physiological density tend to vary, because "productive land" can be variously defined and specialists disagree on criteria. Nevertheless, these figures can explain much. Bangladesh's physiological population density is four times greater than India's—one crucial reason why hope may remain for India in the struggle for food supply while Bangladesh appears headed for disaster.

Agriculture in India suffers from numerous problems. Ancient methods continue to prevail, as in this area of Madras State, where water is brought to the surface not by a pump but by oxen pulling a bucket from a well. The animals walk endlessly back and forth along the short path (center), but are capable of bringing only a comparative trickle of water to the surface (left). (Jacques Jangoux/Peter Arnold)

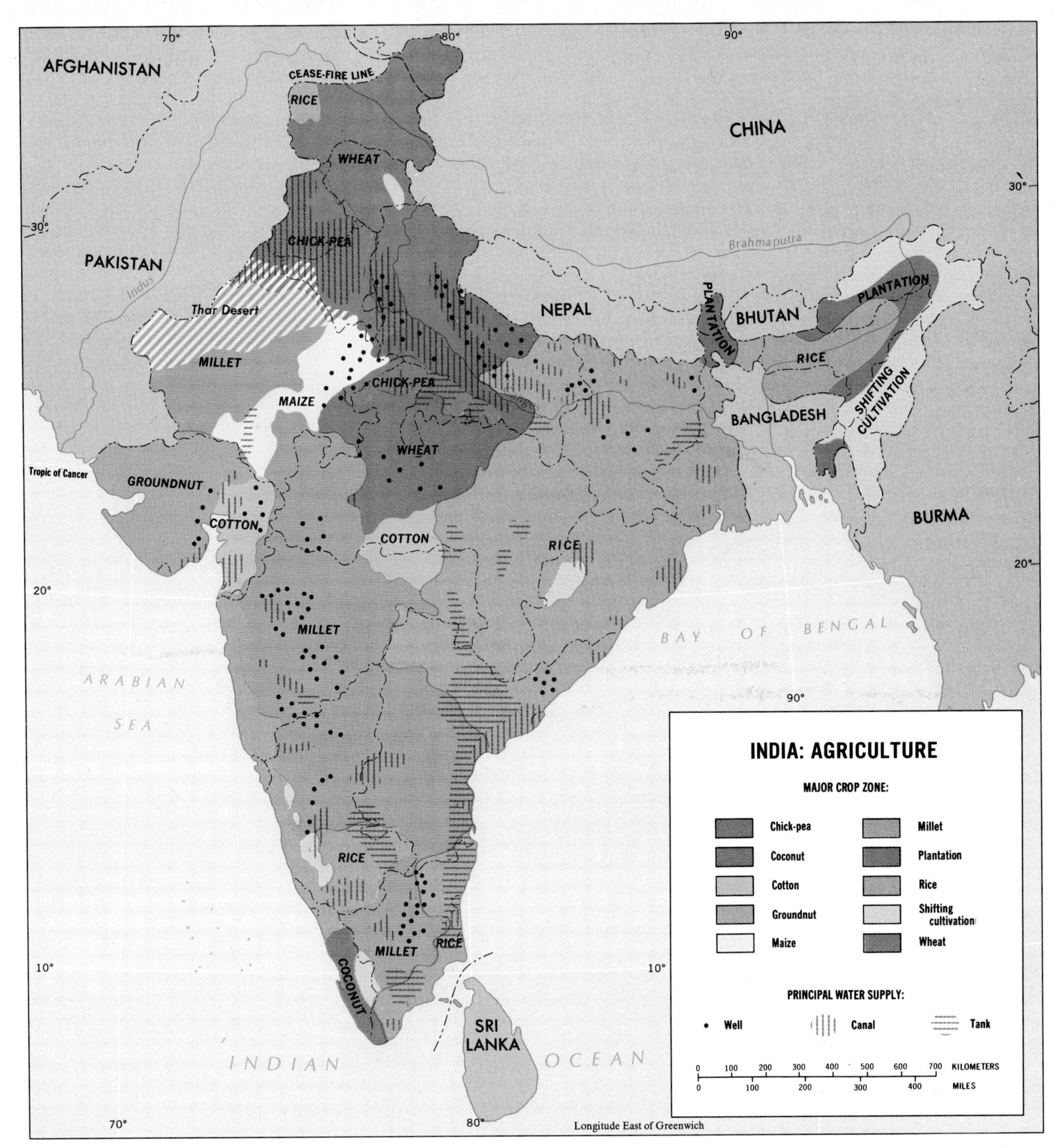

Figure 8-6

India has the largest acreage of rice among all the world's countries. But notwithstanding the introduction of "miracle rice" in parts of India, the average yield per hectare still is below 1000 kilograms (900 pounds per acre). The population map (Fig. I-9) reveals a high degree of coincidence between India's rice areas and its most densely peopled zones, but rice yields per unit area in India are among the lowest in the world.

As the map of agricultural zones indicates, India's varied environments support the cultivation of millet, barley, corn (maize), and other cereals. But subsistence remains a way of life for countless millions of Indian villagers, who cannot afford to buy fertilizers, cannot cultivate the new strains of wheat or rice, and cannot escape the cycle of abject poverty.

Perhaps as many as 115 million Indians do not even have a plot of land and must live as tenants, always uncertain of their fate. This is the background against which optimistic predictions of better food conditions in the India of the future must be seen. True, wheat production increased over 50 percent in five years during the Green Revolution, and rice production nearly doubled between 1950 and 1972. But the gap beteen population and food supply has not closed. When the worldwide droughts of the mid-1970s struck India, there was famine and malnutrition. In 1974 people in the state of Gujarat rioted when food supplies ran out, and the government was overthrown; the federal administration was forced to take control. People by the hundreds of thousands left the land and headed for the cities and towns, and the specter of hunger faced India once again.

Industrialization

Notwithstanding the problems faced by the farmers, agriculture must be the basis for development in India. Agriculture employs the vast majority of the workers (about 70 percent), generates most of the government's tax revenues, contributes its chief exports by value (cotton textiles, jute manufactures, tea, and tobacco all rank high), and produces most of the money the country can spend in other sectors of the economy. Add this to the compelling need to grow ever more food crops, and India's heavy investment in agriculture is understandable.

In 1947 India inherited the mere rudiments of an industrial framework. After a century of British control over the economy, only 2 percent of Indian workers were involved in industry, and manufacturing and mining combined produced perhaps 6 percent of the national income. Textile and food-processing industries dominated; although India's first iron mill had been opened in 1911 and the country's first steel mill began production in 1921, the first major stimulus for heavy industry came after the outbreak of the Second World War. Manufacturing was concentrated in the major cities; Calcutta led, Bombay was next, and Madras ranked third.

The pattern of industrialization today still reflects these beginnings, and industrialization and urbanization in India have proceeded slowly even after independence (Fig. 8-7). Calcutta now forms the center of India's eastern industrial region, the Bihar-Bengal area, where the jute industry dominates but engineering, chemical, and cotton industries also exist. In the interior Chota Nagpur, coal mining and iron and steel manufacture have developed, and Jamshedpur is the nucleus of heavy industry. On the opposite side of the peninsula, two industrial areas combine to form the western industrial region: one centered on Bombay and the other focused on Ahmadabad. This, the Maharashtra-Gujarat area, specializes in cotton and chemicals, with some engineering and food-processing. The cotton industry long has been an industrial mainstay in India, and it was one of the few industries to derive some benefit from the new economic order brought by the British. With the local cotton harvest, the availability of cheap yarn, abundant and cheap labor, and the power supply from the Western Hills' hydroelectric stations, the industry thrived and today outranks Britain itself in the volume of its exports. Finally, the southern industrial region consists chiefly of a set of linear, city-linking corridors centered on Madras, smallest of the three large cities and also a textile and light engineering area.

When India achieved independence, its government immediately set out to develop its own industries, to lessen its dependence on imported manufactures, and to cease being an exporter of raw materials to the developed world. In the process, emboldened by the early results of the Green Revolution, Indian planners actually overspent on their industrial programs—but the problem of rapid population growth bedeviled industry as it did agriculture. Unemployment was to be reduced as industrialization progressed; instead, unemployment rose. Per-capita incomes would rise as high-value products came from new assembly lines; but incomes rose very little and today they still are barely over $100 per year. India has limited coking coal deposits in the Chota Nagpur and large lower-grade coalfields, and the country ranks among the world's 10 largest coal producers, but in the absence of known major petroleum reserves (some oil comes from Assam, Gujarat, off-shore from Bombay, and Punjab), it must spend heavily on energy every year. Major investments have been made in hydroelectric plants, especially multipurpose dams that provide electricity, make irrigation possible, and facilitate flood control. India's iron ores in Bihar and Mysore may be the largest in the world, but despite the emergence of

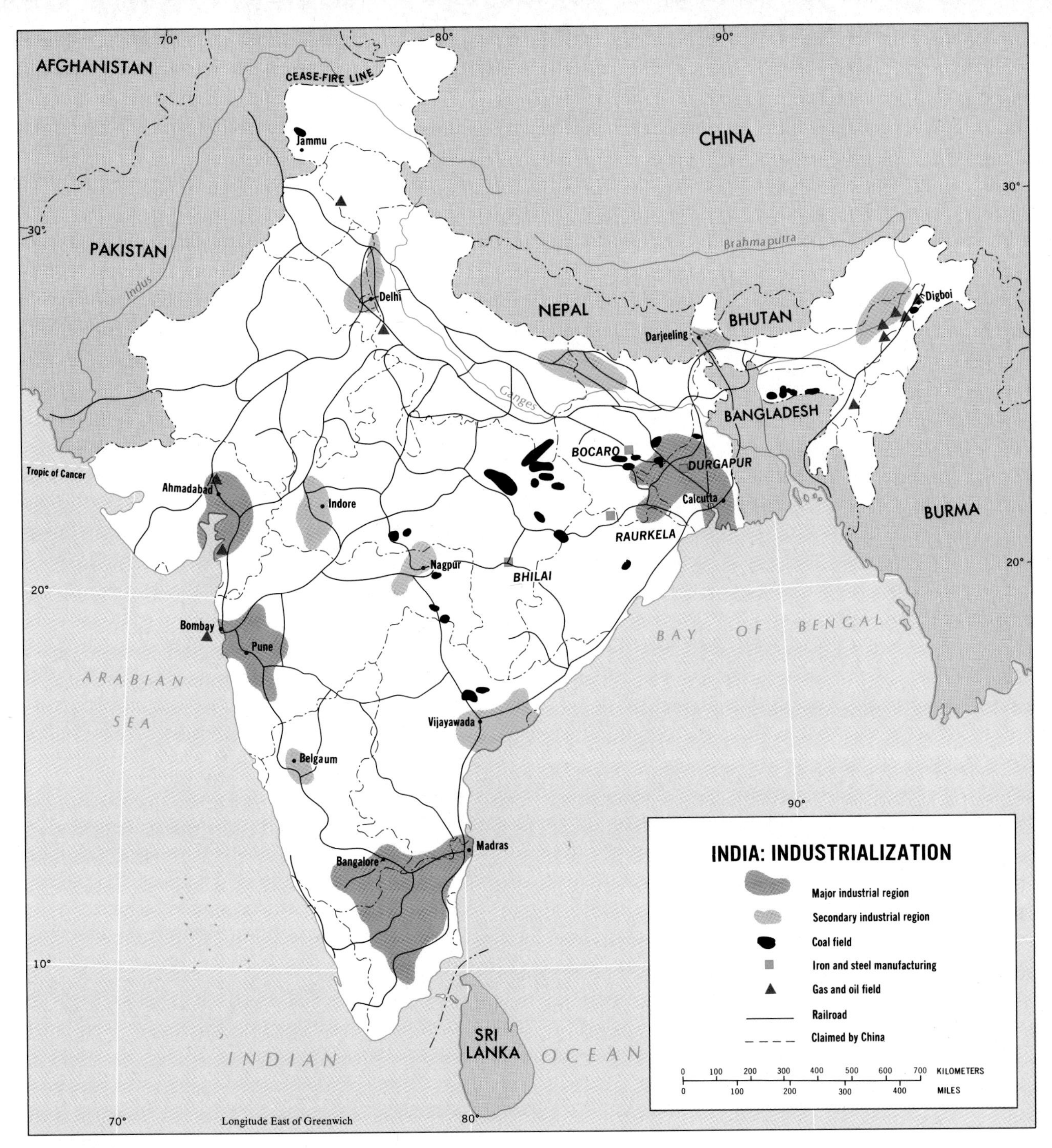

Figure 8-7

Jamshedpur and environs as India's Pittsburgh, India still exports iron ore as raw material to developed countries. In underdeveloped countries, entrenched patterns are difficult to break.

Urbanization. In the late 1970s only about one-fourth of India's population resided in urban areas. India may be known in part for its teeming, crowded cities and its homeless, sidewalk-sleeping urbanites, but urbanization has proceeded very slowly. The largest city in a country of 640 million, Calcutta, in 1978 still had under 10 million inhabitants; Bombay, the second city, had under 8 million, and Delhi's conurbation fewer than 5 million. Once again we must take note of India's massive proportions, and realize that 25 percent of its population involves 160 million people. In-

People sleep on the streets in the shadow of modern Bombay. (J. P. Laffont/Sygma)

dia's urban crisis has enormous dimensions.

India's modern urbanization also has its roots in the colonial period, when the British selected Calcutta, Bombay, and Madras as regional trading centers and as foci for their colony's export and import traffic. All were British military outposts by the late seventeenth century: Madras, where a fort was built in 1640, lay in an area where the British faced little challenge. Bombay (1664) had the advantage of being located closest of all Indian ports to Britain, and Calcutta (1690) was positioned on the margin of India's largest population cluster and had the most productive hinterland. Calcutta lies some 130 kilometers (80 miles) from the coast on the Hooghly River, and a myriad of delta channels connect it to its hinterland, a natural transport network that made the city an ideal colonial headquarters. But the British had to contend with Indian rebelliousness in the region, and in 1912 they moved the colonial capital from Calcutta to the safer interior city of New Delhi, built adjacent to the old Mogul headquarters of Delhi (today the urban areas have coalesced).

Indian urbanization reveals several regional patterns. In the northern heartland, the west (the wheat-growing area, approximately) is more urbanized than the east (where rice forms the main staple crop). This undoubtedly relates to the differences between wheat farming and the labor-intensive, small-plot cultivation and multiple-cropping of rice. In the west, urbanization may be as high as 37 percent (high by Indian standards); in the east less than 10 percent of the population resides in urban centers. Further, India's larger cities (over 100,000) are concentrated in three regions: (1) the northern plains from Punjab to the Ganges Delta, (2) the Bombay-Ahmadabad area, and (3) the southern peninsula including Madras and Bangalore. The only interior cities with populations over 1 million not located within one of these zones are centrally positioned Nagpur (2.6 million) and the capital of Andhra Pradesh, Hyderabad (2.1 million).

In contrast to many other former colonial and now underdeveloped countries, India has a relatively well-developed network of railroads with over 100,000 kilometers (63,000 miles) of track. Once again the British colonizers built the first railroads (the first train ran 34 kilometers from Bombay to Thana in 1853), and as usual they bequeathed India with a li-

Indian cities are places of great social contrasts. This is the waterfront of Bombay. (Ira Kirschenbaum/Stock, Boston)

ability as well as an asset. The railroads were laid, of course, to facilitate exploitation of interior hinterlands and to improve India's governability; this effective transport system permitted the British to move their capital from coastal Calcutta to the deep interior at Delhi. But the railroad system never developed as a unified whole. Different British colonial companies constructed their own systems, and no fewer than four separate railway gauges came into use. The two widest gauges now constitute about 90 percent of the whole system, but many transshipments are still necessary, and the country must buy different sets of equipment. India's planners have begun to standardize the system when parts of the track must be replaced, but the expense is enormous and progress remains slow. Still, India's railway network connects most parts of the country and in terms of overall length it ranks as the world's fourth largest by country.

Crisis of Numbers

The population of the South Asian realm now exceeds 825 million and constitutes one-fifth of all humanity. This reality is alarming enough, especially because so many millions of these people are unable to secure adequate food and there still is death by starvation. But even more frightening is the rate at which the population continues to grow. In the late 1970s, 19 million people were being added to the South Asian population every year—over 14 million in India alone and over 2 million each in Pakistan and Bangladesh. Here lies the region's greatest obstacle to progress—whatever the gains in crop production and industrial output, the demands of an ever-larger population consume them.

The situation in India reflects an equally disturbing world pattern. Not only is the world's population growing; the *rate* at which it is growing increases all the time. This is dramatically illustrated when we take a backward look. At the time of the birth of Christ, the world population probably was about 250 million. It took until about A.D. 1650 to reach 500 million. Then, however, the population grew to 1 billion by 1820, and in 1930 it was 2 billion. In 1975 it passed the 4 billion mark, and if current growth rates continue the world's population will be 8 billion by 2010.

We can look at these figures in another way. It took nearly 17 centuries for 250 million to become 500 million, but then that 500 million grew to 1 billion in just 170 years. The billion doubled to 2 billion in 110 years, and 2 billion became 4 billion in only 45 years. The population's *doubling time* at present growth rates is only 35 years. The shorter the doubling time, the higher the rate of population growth. The doubling time has been decreasing steadily, and the growth rate has been rising without letup. This is what mathematicians call *exponential* growth—growth along an upward curve rather than along a straight line.

Importantly, the world's population is not increasing at the same rate in all regions. The overall growth rate, producing a doubling time of 35 years, is 2 percent. But, as Fig. 8-8 shows, the population of many Middle and South American countries is growing more rapidly than this—in excess of 3 percent (which produces a doubling time of just 23 years). In Africa, too, a number of countries have very high birth and growth rates.

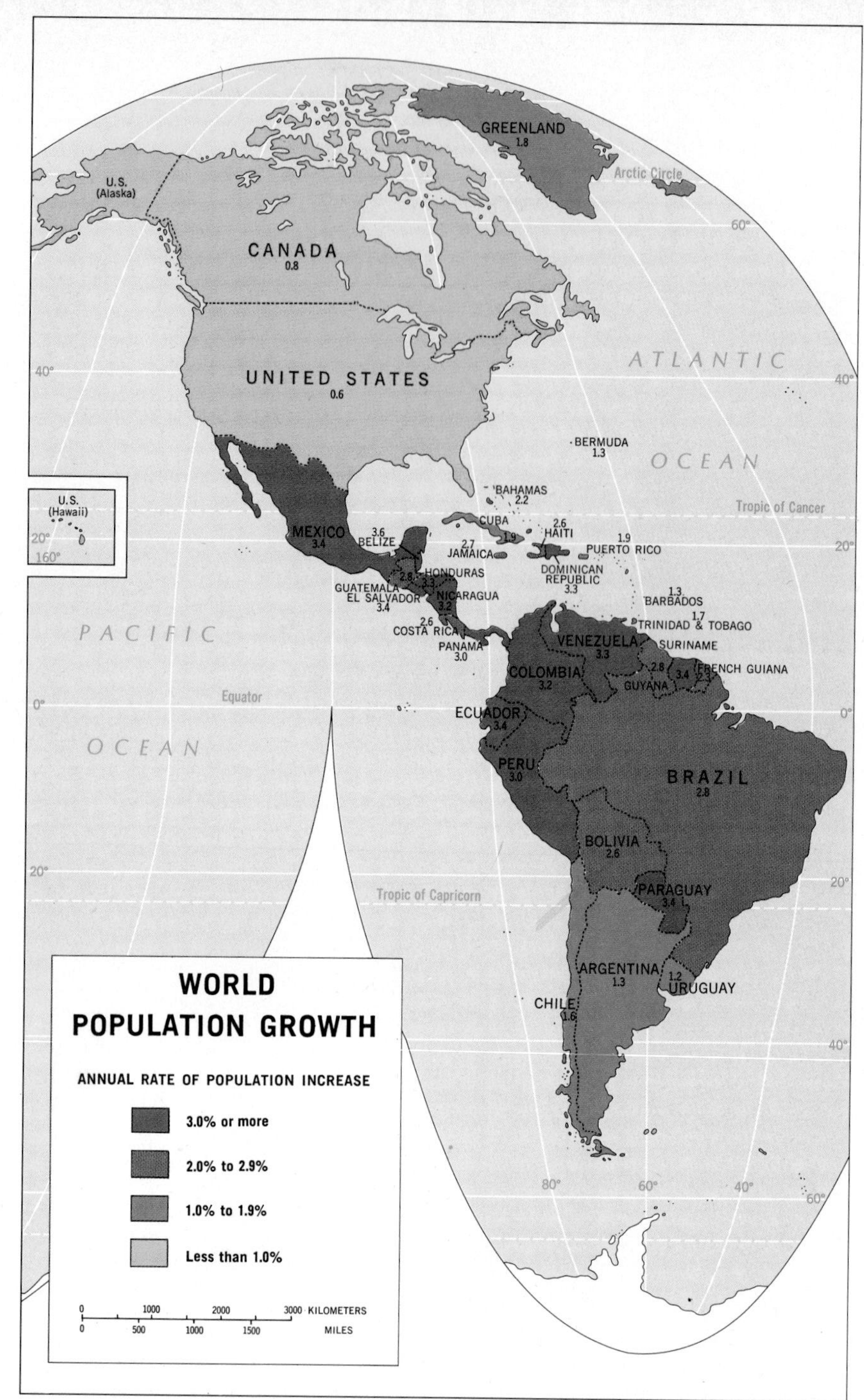

Figure 8-8

Annual population growth is calculated as the number of excess births over deaths recorded during a given year, usually expressed as a percentage, as on Fig. 8-8. In Bangladesh, for example, the annual growth rate is given as 2.7. The birth rate in Bangladesh is currently 44 per thousand, and the death rate 17 per thousand. Thus, each year for every thousand persons in Bangladesh there are 27 excess births over deaths, or 2.7 percent. In the United States, the population growth rate in the late 1970s is 0.6 percent, with 15 births per thousand population and 9 deaths. To compare what these figures mean in terms of doubling time, the doubling time of Brazil's population is just 25; for the United States, it is 116. Some populations are doubling even faster than Brazil's: Mexico, with a population approaching 70 million, has an annual growth rate of 3.4 percent and a doubling time of only 20 years. In just two decades, at present growth rates, there will be twice as many people in Mexico as there are now.

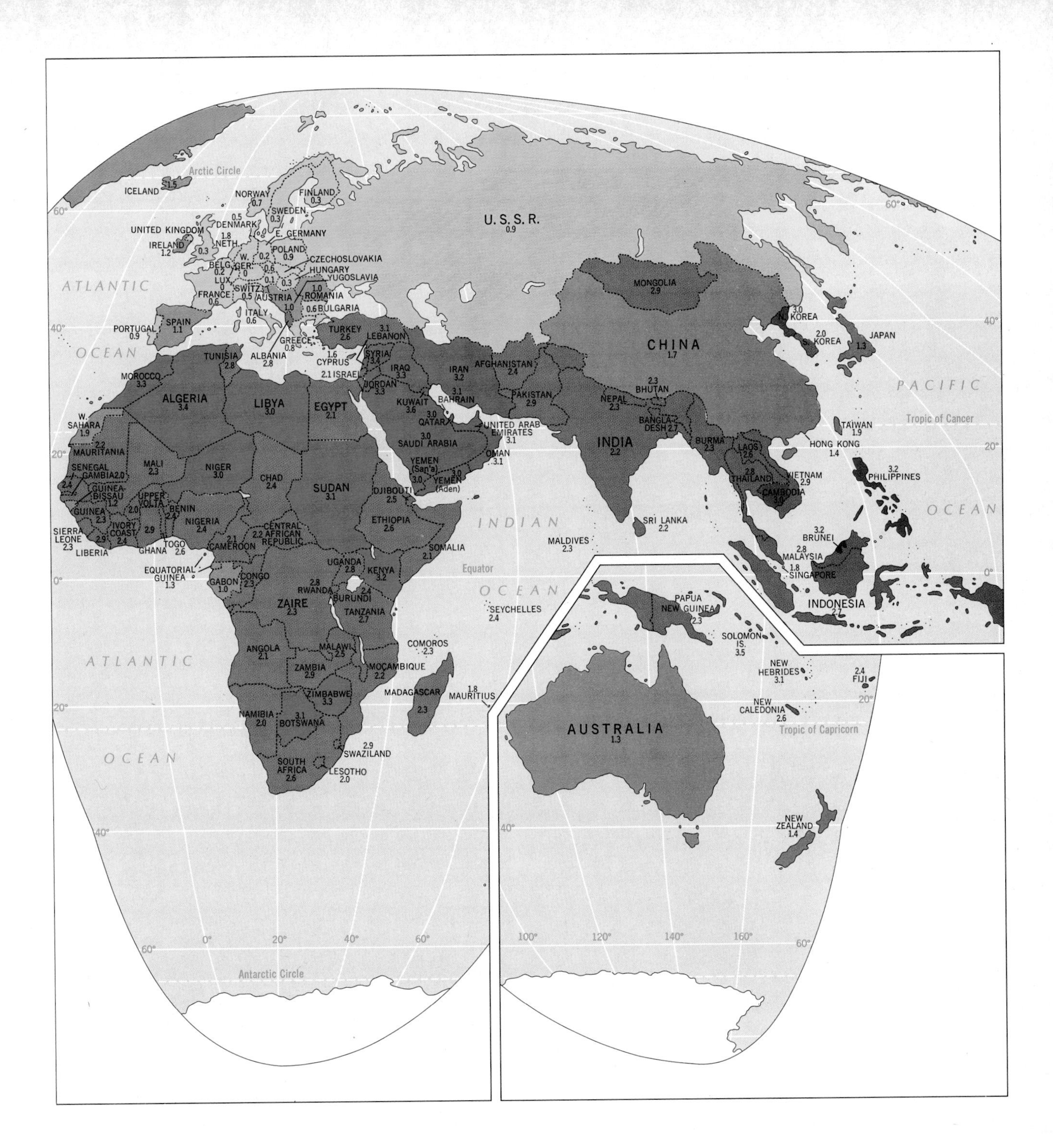

The Dilemma in India

At the beginning of the twentieth century, India had under 250 million inhabitants. This population fluctuated with the vagaries of climate, disease, and war, but students of population problems believe that the average had remained almost stationary for centuries before the Europeans intervened—and did not change significantly until the impact of Europe's industrial revolution was diffused to the colonies. Then the gap between birth rates and death rates began to widen. Birth rates remained high, but death rates declined as medical services became available, food distribution systems improved, food production expanded, and

Predicting the Nightmare

In 1650 the world's human population was approximately 500 million; not until 1820 did the world accommodate one billion inhabitants. Even then, a century and a half ago, some scientists realized that the world was headed for trouble. The most prominent among these was Thomas Malthus, an English economist who sounded the alarm as early as 1798 in an article entitled *An Essay on the Principle of Population as it Affects the Future Improvement of Society.* In this essay Malthus reported that population in England was increasing faster than the means of subsistence; he described population growth as geometric and the growth of the means of subsistence as arithmetic. Inevitably, he argued, population growth would be checked by hunger. Malthus' essay caused a storm of criticism and debate, and between 1803 and 1826 he revised it several times. He never wavered from his basic position, however, and continued to predict that the gap between the population's requirements and the soil's productive capacity would ultimately lead to hunger, famines, and the cessation of population growth.

We know now that Malthus was wrong about several things. The era of colonization and migration vastly altered the whole pattern of world food production and consumption, and worldwide distribution systems made possible the transportation of food from one region of the world to another. Malthus could not have foreseen this; nor was he correct about the arithmetic growth of food production. Expanded acreages, improved seed strains, and better farming techniques have produced geometric increases in world food production, but Malthus was correct in his prediction that the gap between need and production would widen. It has—and there are areas in the world in the 1970s where population growth rates are being checked in the way he predicted, as famines claim the lives of hundreds of thousands in Africa and Asia.

Even today there are scholars and specialists who adhere essentially to Malthus' position, modified, of course, by modern knowledge and experience. These people are sometimes referred to as prophets of doom by those who take a more optimistic view of the future. They are also called neo-Malthusians, tying their concerns to those of that farsighted Englishman who first warned that there are limits to this earth's capacity to sustain our human numbers.

costly wars were suppressed. The last time death rates (from famine and outbreaks of disease) approached the birth rate was in 1910–1911, when births were near 50 per thousand and deaths in the high 40s. Since then, the birth rate has declined to 37 (it first went below 40 as recently as 1970), but the death rate in India now is only 15. The population doubled to about 500 million by the mid-1960s and now, scarcely a decade later, another 140 million have been added. These are staggering figures.

In terms of population growth—as in culture, economy, and urbanization—there is not just one India but several distinct and different Indias. The growth of population is most rapid in Assam and in West Bengal, both adjacent to Bangladesh (Fig. 8-9). In Nagaland and Manipur it actually exceeds 3.5 percent, and it is over 2.5 percent (as it is in Bangladesh) throughout much of India's great eastern population cluster in the lower Ganges-Brahmaputra region. In Uttar Pradesh, however, it has been below 2.0 percent. In the southern peninsula, the growth rate tends to lie

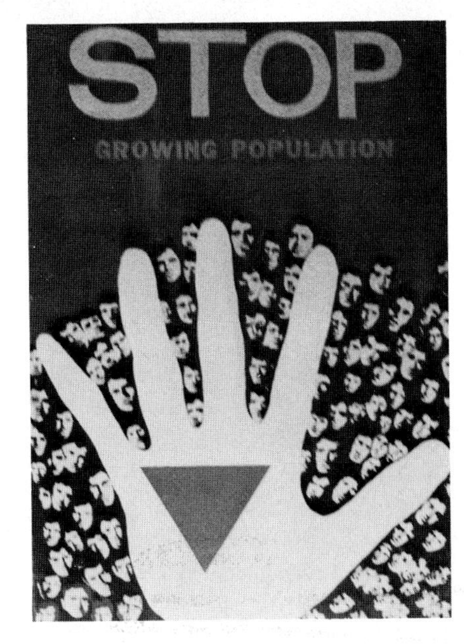

The Indian government, and governments of individual Indian states, have attempted to stem the tide of population growth through legislation and through propaganda campaigns. This poster was seen throughout Bombay and surrounding towns. (Sygma)

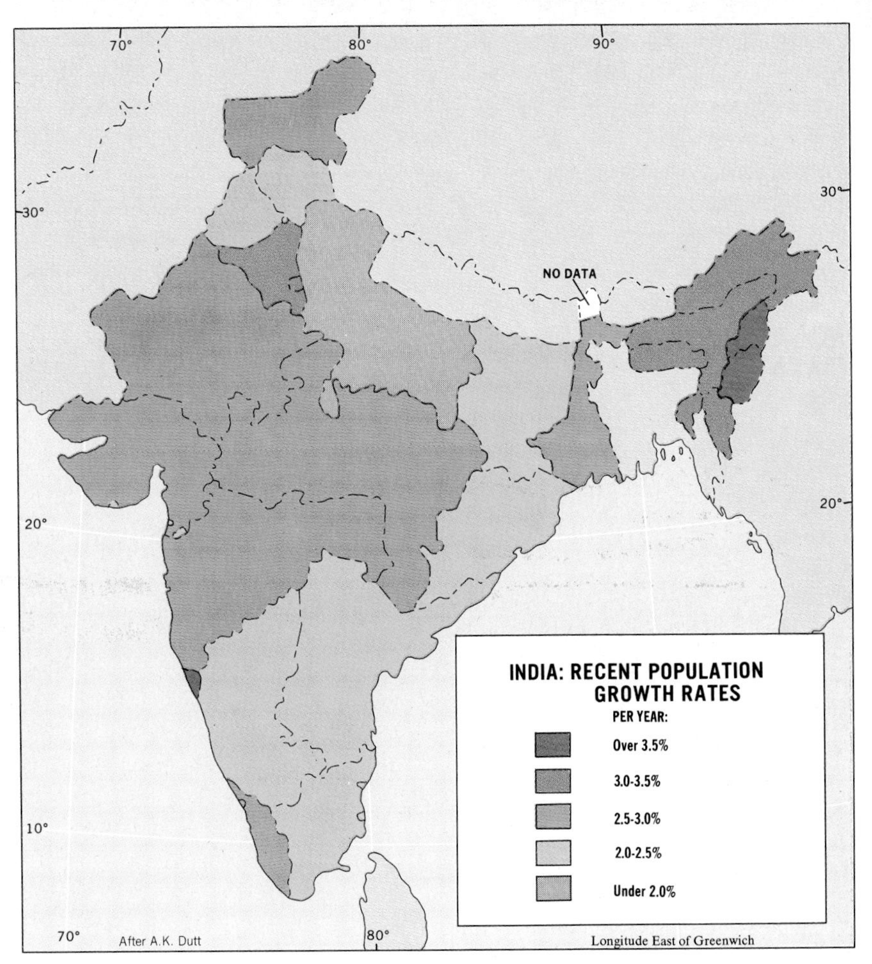

Figure 8-9

near the national average, but in the west it is somewhat higher. All this mirrors India's regional food shortages; hunger and starvation can afflict the crowded east when food supply is adequate in Punjab.

India obviously must reduce its birth rate, but in a country as vast and heavily rural as this, such a national task confronts virtually insurmountable obstacles. Indian governments have endorsed family planning as part of official development programs, but it has proved difficult to disseminate the practice throughout tradition-bound Indian society. Not all the states of the union cooperated equally in the effort, and when, in the mid-1970s, a program of compulsory sterilization of persons with three children or more was instituted in Maharashtra state, there were riots. After more than 3.7 million people were sterilized (bringing the 1977 total in India to about 22.5 million), the government was forced to state that sterilization would no longer be compusory. Successive Five-Year Plans project the reduction of the birth rate to levels that have yet to be reached (it was to have been 32 per thousand by the end of the 1969-1974 Plan). Meanwhile, India remains on the demographic treadmill.

Demographic Cycles

Some people predict that India will overcome its population crisis because other countries, too, went through periods of explosive growth only to reach a demographic near-equilibrium. Fig. 8-8 indicates that the low-growth countries are the developed countries, many of them with an annual population increase of less than 1 percent. In the nineteenth century, many of those countries also experienced explosive population expansion—but when they became industrialized, urbanized, and modernized, their rapid growth declined. Thus it is possible to discern three major stages in the "ideal" demographic cycle. First, birth as well as death rates are high, and the population fluctuates within a

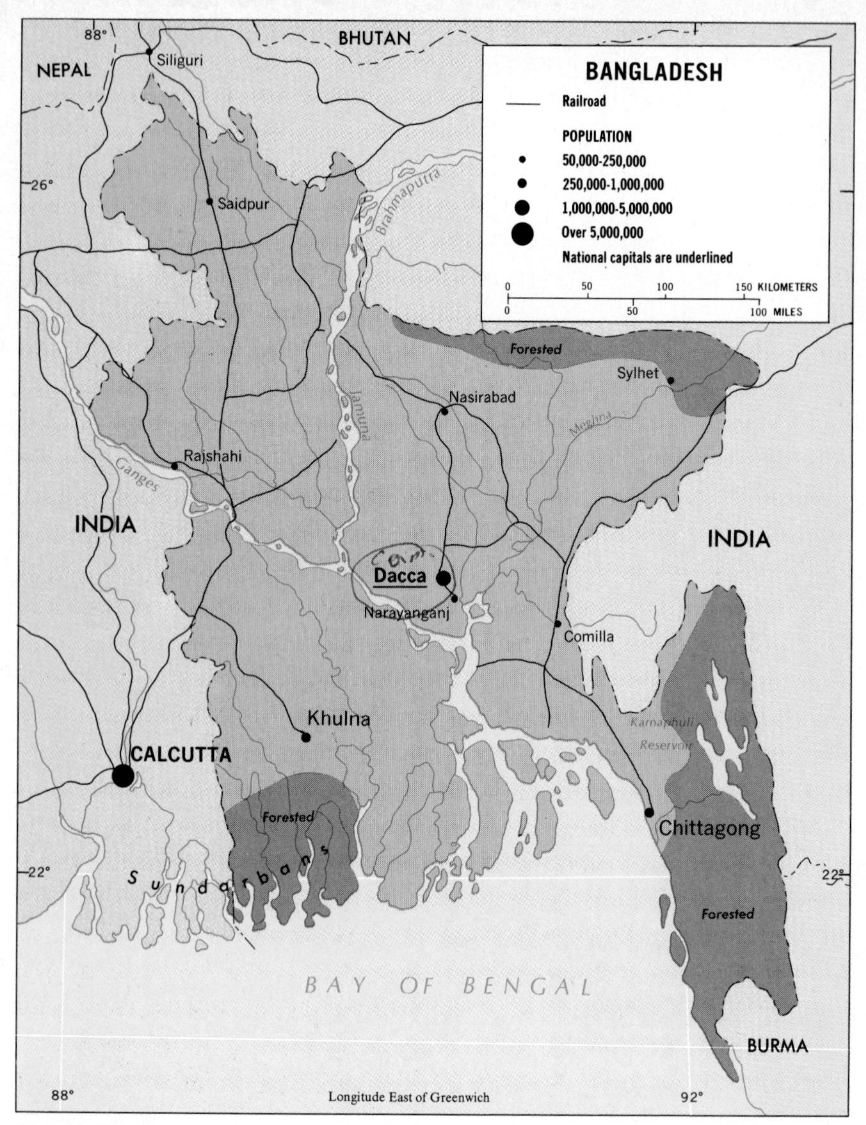

Figure 8-10

certain range. Next, as industrial, medical, and technological innovations are introduced, the death rate declines but the birth rate stays high and begins to decline much later (this is the period of "explosion"). Finally, birth rates decline more sharply, and the overall growth rate is reduced below 1 percent. This is essentially what occurred in Western Europe and, some scholars insist, underdeveloped countries also will experience such a sequence.

But the European model may not fit countries of the modern underdeveloped world. First, the population of a country such as the United Kingdom was under 10 million when the rapid-growth stage began. Two doubling periods still would produce only 40 million on this base, and while England went through its explosive period there was mass emigration to alleviate the population pressure further. Present population totals in countries such as Pakistan and Bangladesh (let alone India) are of a very different order. In two doubling periods (our remaining lifetime, with any luck) Pakistan would go from 77 to 308 million, Bangladesh from 80 to 320 million, and India from 640 million to over 2.5 billion. Furthermore, many underdeveloped countries do not at present appear to have the raw materials to sustain the kind of development that Europe witnessed. As we noted earlier, all the oil in the Middle East has not generated genuine development, even though it has produced money with which the products of development can be bought. The Indian dilemma is far from solution.

Bangladesh

The state of Bangladesh was born in 1972 following a war of independence against Pakistan— of which it had for nearly 25 years been a part. With its economy shattered by many months of conflict, Bangladesh commenced its sovereign existence impoverished, ill-fed, and overcrowded. Bangladesh is a comparatively small country (143,000 square kilometers, 55,000 square miles), somewhat smaller than the state of Florida. Its population, now over 90 million, constituted 55 percent of formerly united Pakistan. The territory, nearly surrounded by India (a short stretch of boundary adjoins Burma to the east), is essentially the floodplain of the Ganges-Brahmaputra system, which drains into the Bay of Bengal through numerous channels (Fig. 8-10). Only in the extreme east and southeast, in the hinterland of Chittagong, does the country's topography rise into hills and mountains.

Periodic floods affect the low-lying farmlands of Bangladesh. (Albert Moldvay/Photo Researchers)

The land of Bangladesh is extremely fertile. Practically every cultivable foot of soil is under crops: rice for subsistence, jute and tea for cash. The jute industry made a major contribution to the economy of Pakistan prior to Bangladesh's secession, producing (with other products) well over half of the country's annual export revenues. It was always a bone of contention that Bangladesh's share of the Pakistani budget was only about 40 percent, so that Bangladesh served as an exploited colony to West Pakistan in the eyes of many of its people.

The staple food for Bangladesh's enormous population is rice, grown on fields whose fertility is renewed by the silt swept down by the river systems' annual floods. In most of the country three harvests of rice per year are possible, and the country normally was able to produce some 80 percent of its people's food requirements. Then the dislocation brought on by the war of secession threatened widespread starvation as harvests rotted and crops were abandoned. Large-scale emergency imports saved much of the situation, but Bangladesh never recovered. The country remains the epitome of Malthusian forecasts, in a constant state of hunger and local starvation, dislocation, and misery.

Bangladesh's land is fertile because it is low-lying and subject to river flooding—but this has negative aspects, too. To the south, the country lies open to the Bay of Bengal, where destructive tropical storms are born and sometimes make landfalls. With much of southern Bangladesh less than four meters above sea level, the penetration of cyclones (as these hurricane- and typhoon-type storms are called there) can do incalculable damage. In November 1970, a cyclone hit Bangladesh, and the rising waters and high winds exacted a toll of 600,000 persons. It was perhaps the greatest natural disaster of the twentieth century, but it is unlikely that it was the last such assault upon the land and people of Bangladesh. Crowding on low-lying farmlands, inadequate escape routes and mechanisms, and insufficient warning time continue to exist.

The people of Bangladesh in recent years have faced disaster of several kinds. The indifference of the Pakistani government during the aftermath of the November 1970 cyclone was one of the leading factors in the outbreak of open revolt against established authority. The war of secession brought unspeakable horror to villages and towns; millions left their homes and streamed across the border into neighboring India. Then followed hunger and rampant disease. It is impossible to be certain of the consequences, but estimates of the total loss of life (and not counting those permanently disabled and ruined) run as high as 3 million.

Again in 1974, Bangladesh was struck by destructive flooding, and the country was forced to seek aid in the form of grain shipments from the developed countries. The United States, which had sent 850,000 tons of grain to Bangladesh following the civil war, shipped another 1 million tons of wheat and rice to Bangladesh in 1975 and 1976. Already, one in five inhabitants of Bangladesh depends on food imports, paid for by foreign aid receipts. If the overseas grain harvest fail or the foreign aid stops, Bangladesh will be the scene of starvation on a scale not witnessed in modern times.

Rice yields per unit area are low in Bangladesh (about one-third of the world average and 10–15 percent of the yield Japanese rice farmers wrest from their fields). Grain production could be tripled if high-yielding varieties were used everywhere, fertilizers were made available, and traditional farming methods were changed. But 40 percent of the country's peasants are landless villagers, and those who do own land possess, on the average, less than 1 hectare (2.5 acres). With little or no capital, these farmers would have to borrow money at annual interest rates of 200–300 percent from the local moneylender in order to buy fertilizer and better rice plants. If the crop failed, the farmer would lose the family's land. Few take the risk.

Among the world's 10 most populous countries, none suffers from underdevelopment as severely as Bangladesh. It was the people's shared adherence to the Islamic faith that led to the merger with Pakistan, and still today Bangladesh's population is 80 percent Moslem; 18 percent follow the Hindu faith and did so even while Bangladesh was part of the "Islamic Republic" of Pakistan. There is something symbolic in this, for religious animosities never appeared as strong here as they were in the west. But after centuries of British colonialism Bangladesh reaped a harvest of Pakistani neglect, and now the country is overpopulated, malnourished even where people do have food (protein deficiency is a serious problem), and economically in general disorder. The jute industry has suffered severely, the already-inadequate transport system is in disarray, Governmental instability, corruption in administration, and terrorism have further bleakened Bangladesh's first years of independence.

Dacca, the centrally positioned capital (1.4 million), and the port of Chittagong (500,000) are urban islands in a country where only 9 percent of the people live in towns. A measure of Bangladesh's economic problems can be gained from a look at the transport system. There is still no road bridge over the Ganges anywhere in the country, and only one railroad bridge; there is neither a road nor a railroad bridge across the Brahmaputra anywhere in Bangladesh. Road travel from Dacca to the eastern town of Comilla involves two ferry transfers over the same river (distributaries of the Meghna); there is hardly a means of surface travel from Dacca to the western half of the country, except by boat. Thousands of boats ply Bangladesh's numerous waters, which still form more effective interconnections in the country than do the roads. In Bangladesh, survival is the leading industry; all else is luxury.

Pakistan

The Islamic Republic of Pakistan (Fig. 8-11) also is an underdeveloped country—but it differs in almost every respect from its former federal partner. Indeed, so strong are the contrasts between former East Pakistan and the state that now carries the name that the generation-long survival of divided Pakistan was something of a miracle. Islam was the unifier, but there were differences even in the paths whereby the faith reached west and east. In dry-world West Pakistan it arrived over land, carried eastward by invaders from the west and northwest. In monsoon-wet East Pakistan it came by sea, brought to the delta by Arab traders. In fact, it is difficult to find any additional area of correspondence or similarity between former East and West Pakistan. The two wings of the country lay in different culture realms, with all that this implied. Pakistan lies on the fringes of Southwest Asia; Bangladesh adjoins Southeast Asia. In Pakistan the problem has always been drought, and the need is for irrigation. In Bangladesh the problem is floods, and the need is for dikes. Pakistanis build their houses of adobe and matting, eat wheat and mutton, and grow cotton for sale; the people of Bangladesh build with grass and thatch, eat rice and fish, and grow jute.

Livelihoods

Contrary to the predictions made at the time Pakistan became a separate state, it has managed to survive economically and has even made a good deal of progress. This is so despite the country's poverty in known mineral resources; apart from natural gas in Baluchistan (which may be related to oil reserves) and chromite, Pakistan has only minor iron deposits that are being used in a small plant at Multan. So the country must make up in agricultural output what it lacks in minerals, and since independence it has made considerable strides. Favored by high prices generated by the Korean War, the economy got an early boost from jute and

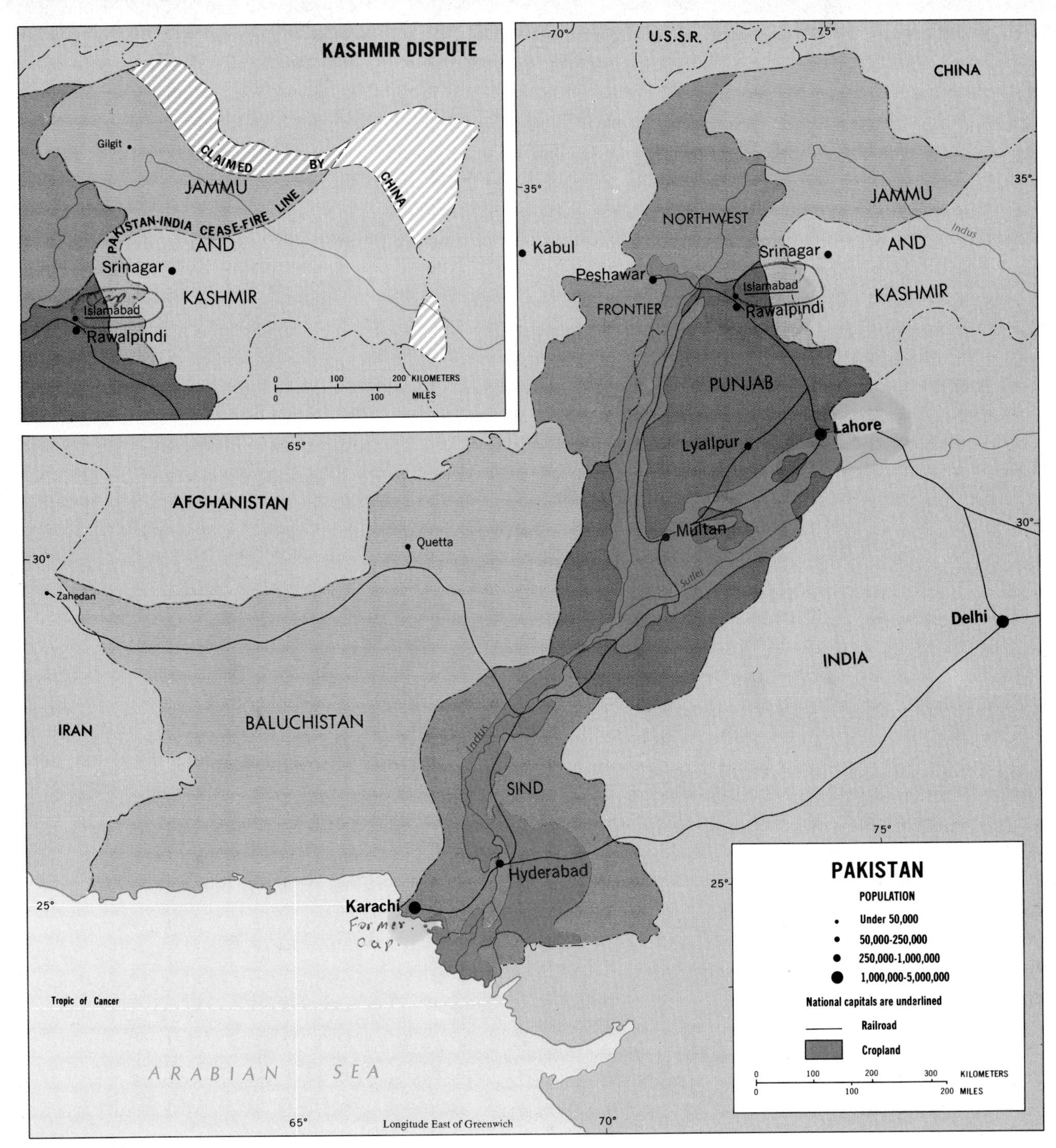

Figure 8-11

cotton sales. Then the mid-1950s brought a period of stagnation, made worse by the political conflicts with India. But during the 1960s and 1970s a considerable expansion of irrigated acreages, land reform, the improvement of marketing techniques, and the success of the textile industry all contributed to considerable progress.

Certainly there was room for such progress; Pakistan shares with India the familiar low yields per hectare for most crops, per capita annual incomes are very low (around $250), and manufacturing is only now emerging as a significant contributor to the economy. The low yields for vital crops such as wheat and rice have long been due to the inability of the peasants to buy fertilizers, to the poor quality of seeds, and to inadequate methods of irrigation. In many areas there has not been enough irrigation water. In still other places there is excessive salinity in the soil. In Sind, where large estates farmed by tenant farmers existed before the British began their irrigation programs in the Punjab, yields are kept down by outdated irrigation systems and by the low incentives to landless peasants.

Pakistan's virtually new textile industry, based on the country's substantial cotton production, has developed rapidly. It now satisfies all of the home market, and textiles have quickly risen to become the top contributor to exports. Other industries also have been stimulated; increasingly Pakistan is able to produce at home what formerly it had to import from foreign sources. A chemical industry is emerging, and an automobile assembly plant has opened at Karachi, where a small steel-producing plant now operates as well. Thus Pakistan is trying to reduce its dependence on foreign aid, investing heavily in the agricultural sector and encouraging local industries as much as possible.But there is a long way to go, and a difficult one. There are few natural resources, and the export products are subject to sudden price fluctuations on world markets and to the increasing pressure of competition. And the political crisis of the late 1960s and early 1970s also took an inevitable toll.

Pakistan still faces a host of unsolved problems. Literacy is low. Family planning is government-approved, but its dissemination and acceptance are just beginning. Many millions of people still live in poverty, subsist on an ill-balanced diet, have a short life expectancy, and experience little tangible progress during their lifetimes. Of course there are successful farmers in the Punjab, in Sind, and elsewhere. But they are still far outnumbered by those for whom life remains mainly a struggle for survival.

The Cities

It is understandable that the young state of Pakistan chose Karachi as its first capital. Clearly it was desirable to place the capital city in the Moslem stronghold of the country, in West Pakistan, but the outstanding center of Islamic culture, Lahore, lay too exposed to the nearby, sensitive Indian boundary.

Lahore, with 2.5 million residents, grew rapidly as a result of the Punjab's partition. In 1950 its population was only about one-third as large. Founded in the first or second century A.D., Lahore became established as a great Moslem center during the Mogul period. As a place of royal residence the city was adorned with numerous magnificent buildings, including a great fort, several palaces, and mosques that remain to this day monuments of history, with their marvelous stone work and excellent tile and marble embellishments. The site of magnificent gardens and an old university, Lahore also was the focus of a large area in prepartition times, when its connections extended south to Karachi and the sea, north to Peshawar, and east to Delhi and Hindustan. Its Indian hinterland was cut off by the partition, but Lahore has retained its importance as the center of one of Pakistan's major population clusters, as a place of diverse industries (textiles, leather goods, gold and silver lacework), as the unchallenged historic headquarters of Islam in this area, and as an educational center.

Karachi (4 million) grew even faster than Lahore. After independence it was favored not only as the first capital of Pakistan, but also as West Pakistan's only large seaport. Its overseas trade rose markedly; new industries were established; new regional and international interconnections developed between Karachi and former East Pakistan and between Karachi and other parts of the world. A flood of Moslem refugees came to the city, presenting serious problems—there simply was not enough housing or food available to cope with the half million immigrants who had arrived by 1950. Nevertheless, Karachi survived and in its growth has reflected the new Pakistan. As a result of the expansion of cultivated acreages in the Pakistani Punjab and in Sind, Karachi's trade volume has increased greatly, especially in wheat and cotton. Imports of oil and other critical commodities required construction of additional port facilities.

But Karachi never was the cultural or emotional focus of the nation. It lies symbolically isolated, along a desert coast, almost like an island; the core of Pakistan still lies far inland. In 1959, after just over a decade as the federal capital of Pakistan, Karachi surrendered its political functions to an interior city, Rawalpindi. This was a temporary measure: an entirely new government headquarters is being completed a small distance from Rawalpindi, at Islamabad. This new town, as the

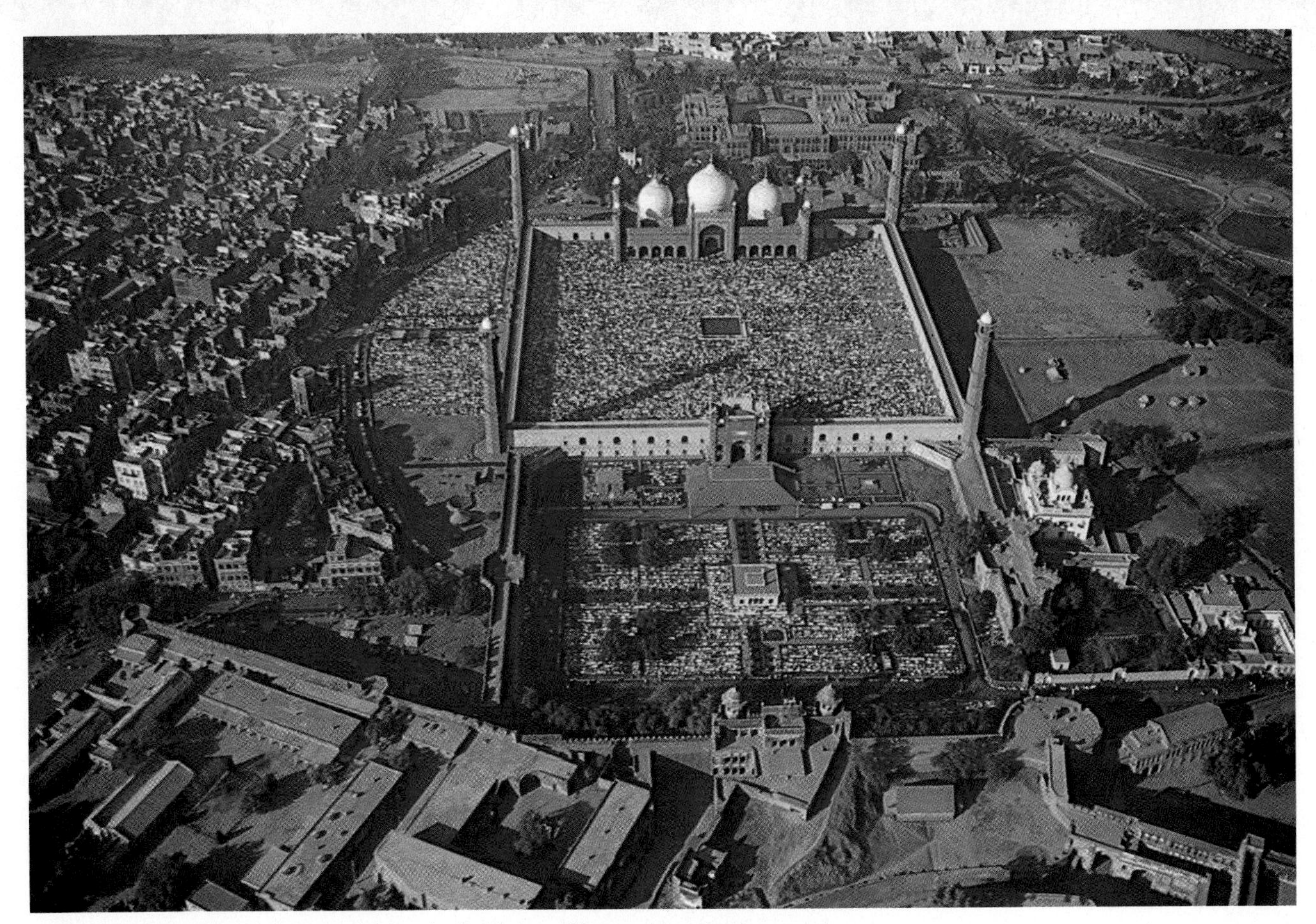

Lahore is the cultural focus of Islamic Pakistan. Shown here is the great Badshai Mosque, with a half million of the faithful in attendance. (Franke Keating/Photo Researchers)

map shows, lies near the boundary of Kashmir. It confirms not only the internal position of Pakistan's cultural and economic heartland, but also the state's determination to emphasize its presence in the contested north. It is evidence of a sense of security Pakistan did not yet possess when it chose Karachi over exposed Lahore. In this context, Islamabad is a prime example of the principle of the *forward* capital.

Irredentism

For many years Pakistan has been beset by problems in its marginal areas (Fig. 8-11). The boundary with India in Kashmir is a truce line that appears to be stabilizing, but in the northwest (along the boundary with Afghanistan) and in the west (in Baluchistan) Pakistan has faced more recent difficulties. In Baluchistan, still a remote area where the Iran-Pakistan border has little practical meaning, government efforts to exercise its authority were met by opposition from local chiefs. Baluchistan is a region of rugged mountains, severe desiccation, scattered oases, and ancient nomadic routes resembling neighboring Iran and quite unlike the settled farmlands of the Indus Valley. Baluchistan's peoples were not ready to be controlled by Pakistan's government, and in their rebelliousness they were supported by the government of Afghanistan in Kabul. Afghanistan's majority population, the Pakhtuns (also called Pashtuns, Pathans, and Pushtuns), constitute about 51 percent of that country's 22 million people and are concentrated in the region centered on the capital, Kabul—positioned opposite the Khyber Pass less than 200 kilometers (120 miles) from the Pakistan border. Kabul's government has also encouraged Pakhtuns living in Pakistan's northwest to demand their own Pakhtunistan (Pathanistan) state. Pakistan's moderate response has been to hasten the integration of its northwestern areas into the national state through improved

The Problem of Kashmir

Kashmir is a territory of high mountains bounded by Pakistan, India, China, and, along a few kilometers in the far north, by Afghanistan (Fig. 8-11). Although known simply as Kashmir, the area actually consists of several political divisions, including the state properly referred to as Jammu and Kashmir (one of the 562 Indian states at the time of independence) and the administrative areas of Gilgit in the northwest and Ladakh (including Baltistan) in the east. Ladakh gained world attention as a result of Chinese incursions there in the early 1960s, but the main conflict between India and Pakistan over the final disposition of the territory has focused on the southwest, where Jammu and Kashmir are located (Fig. 8-11).

When partition took place in 1947, the existing states of British India were asked to decide whether they would go with India or with Pakistan. In most of the states, this issue was settled by the local authority, but in Kashmir there was an unusual situation. There were about 5 million inhabitants in the territory at that time, nearly half of them concentrated in the so-called Vale of Kashmir (where the capital, Srinagar, is located). Another 45 percent of the people were concentrated in Jammu, which leads down the foothill slopes to the edge of the Punjab. The small remainder of the population is scattered through the mountains, including Pathans in Gilgit and other parts of the northwest. Of these population groups, the people of the mountain-encircled Vale of Kashmir are almost all Moslems, while the majority of Jammu's population is Hindu. But the important feature of the state of Jammu and Kashmir was that its rulers were Hindu, not Moslem, although the overall population was more than 75 percent Moslem. Thus the rulers were faced with a difficult decision in 1947—to go with Pakistan and thereby exclude themselves from Hindu India, or to go with India and thus incur the wrath of the majority of the people. Hence, the Maharajah of Kashmir sought to remain outside both Pakistan and India, and to retain the status of autonomous, separate unit. This decision was followed, after partition of India and Pakistan, by a Moslem uprising against Hindu rule in Kashmir. The maharajah asked for the help of India, and Pakistan's forces came to the aid of the

communications, education, and other facilities, but Afghan irredentism continues. Thus Islamabad is situated between pressure areas in Kashmir *and* Afghanistan. The Soviet invasion of Afghanistan, and its resultant refugee problem, have further aggravated its regional difficulties. Pakistan's territorial problems did not end with the secession of Bangladesh.

Sri Lanka

Sri Lanka (formerly Ceylon), the compact, pear-shaped island located off the southern tip of the peninsula of India, is the fourth independent state to have emerged from the British sphere of influence in South Asia. Sovereign since 1948, Sri Lanka has had to cope with political as well as economic problems, some of them quite similar to those facing India and Pakistan, and others quite different (Fig. 8-12).

Good reasons exist for Sri Lanka's separate independence. This is neither a Hindu nor a Moslem country; the majority—some 70 percent—of its 15 million people are Buddhists. And, unlike India or Pakistan, Sri Lanka is plantation country (a legacy of the European period), still the mainstay of the external economy.

Moslems. After more than a year's fighting and through the intervention of the United Nations, a cease-fire line was established that left Srinagar, the Vale, and most of Jammu and Kashmir in Indian hands, including nearly four-fifths of the territory's population. In due course this line began to appear on maps as the final boundary settlement, and Indian governments have proposed that it be so recognized.

Why should two countries whose interests would be served by peaceful cooperation allow a distant mountainland to trouble their relationship to the point of war? There is no single answer to this question, but there are several areas of concern for both sides. In the first place, Pakistan is wary of any situation whereby India would control vital irrigation waters needed in Pakistan, and as the map shows, Kashmir is traversed by the Indus River, the country's lifeline. Moreover, other tributary streams of the Indus originate in Kashmir, and in the Punjab Pakistan learned the lessons of dealing with the Indians for water supplies. Second, the situation in Kashmir is analogous to the one that led to the partition of the whole subcontinent—Moslems are under Hindu domination. The majority of Kashmir's people are Moslems, so Pakistan argues that free choice would deliver Kashmir to the Islamic Republic, and a free plebiscite is what the Pakistanis have sought and the Indians have thwarted. Furthermore, Kashmir's connections with Pakistan prior to partition were much stronger than those between Kashmir and India (although India has invested heavily in improving its links to Jammu and Kashmir since the military stalemate). In addition, Pakistan argues that it needs Kashmir for strategic reasons, in part to cope with the Pakhtun secession movement that extends into this area.

As the 1970s drew to a close it appeared likely that the cease-fire line would indeed become a stable boundary line between India and Pakistan. In February 1975 the incorporation of the state of Jammu and Kashmir into the Indian federal union was confirmed when India was able to reach agreement with Sheikh Muhammad Abdullah, the state's chief minister and leader of the majority.

The majority of Sri Lanka's people are not Dravidian, but rather Aryans who link their history to ancient northern India. After the fifth century B.C. they began to migrate to Sri Lanka, a migration that took several centuries and that brought the advanced culture of the northwestern part of the subcontinent to this southern island. Part of that culture was the Buddhist religion; another part of it was a knowledge of irrigation techniques. Today the descendants of these early invaders, the Sinhalese, speak a language related to the Indo-Aryan language family of northern India, especially to Bengali.

The Dravidians from southern India never came in sufficient numbers to challenge the Sinhalese. They introduced the Hindu way of life, brought the Tamil language to Sri Lanka, and eventually came to constitute a substantial minority (18 percent) of the country's population. Their number was strengthened substantially during the second half of the nineteenth century, when the British brought many thousands of Tamils from the mainland to Sri Lanka to work on the plantations that were laid out. Sri Lanka has sought the repatriation of this element in its popu-

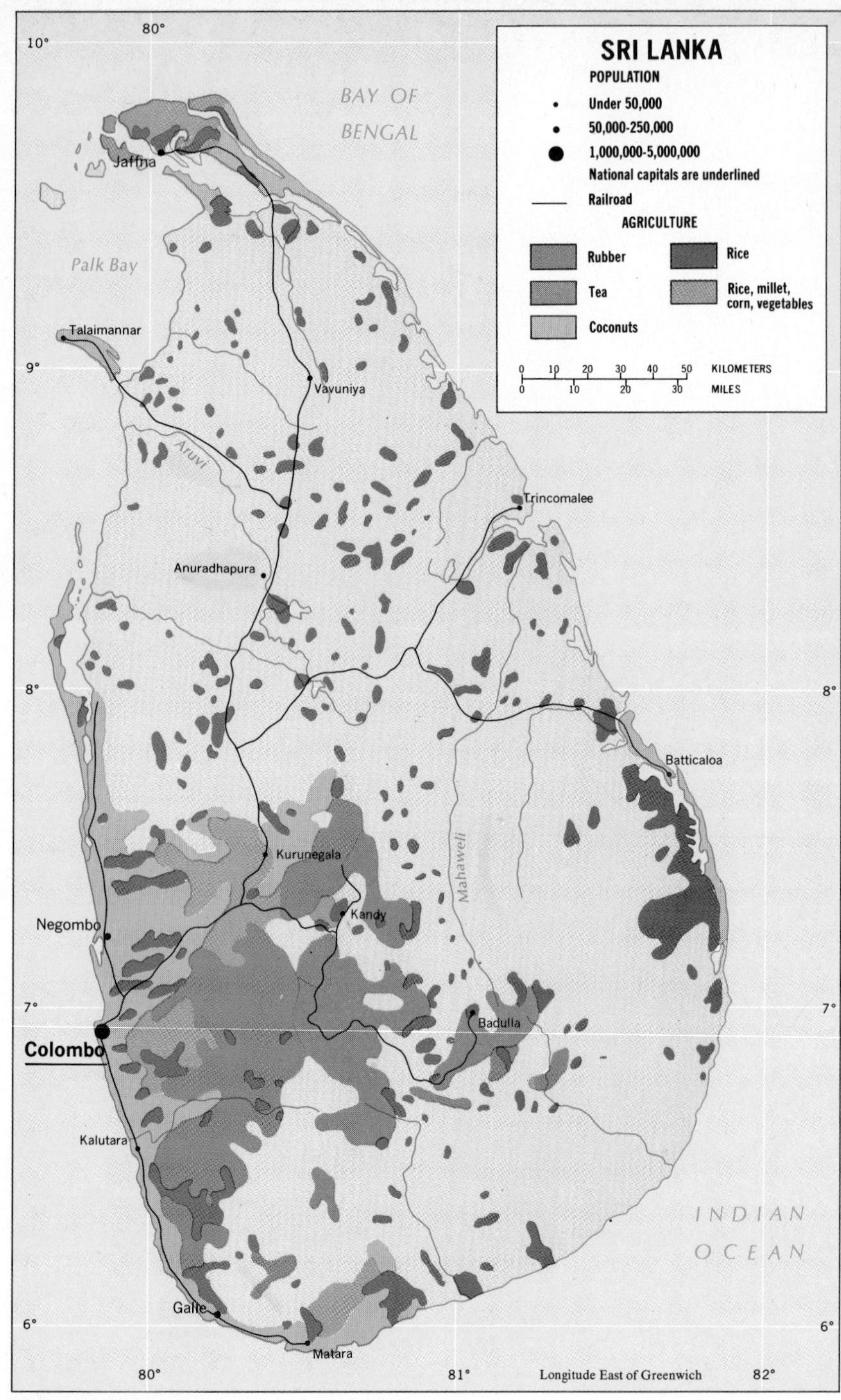

Figure 8-12

lation, and an agreement to that effect was even signed with India. In 1978, however, Tamil was granted the status of a national language of Sri Lanka.

Sri Lanka is not a large island (66,000 square kilometers, 25,000 square miles), but it has considerable topographic diversity. The upland core lies in the south, where elevations reach 2500 meters (over 8000 feet) and sizable areas exceed 1500 meters (5000 feet). Steep, densely overgrown slopes lead down to a surrounding lowland, most of which lies below 300 meters (1000 feet); northern Sri Lanka is entirely low-lying. The rivers, the sources of the irrigation waters for the rice paddies, flow radially from the highland across this lowland rim.

Since the decline of the Sinhalese empire, centered on Anuradhapura, the moist southwestern uplands have been the major areas of productive capacity (the plantations are concentrated here) and the population core as well. Three plantation crops have been successful: coconuts in the hot lowlands, rubber up to about 600 meters (2000 feet), and in the highlands above, tea, the product for which Sri Lanka is famous. Tea continues in most years to account for two-thirds of Sri Lanka's exports by value.

Sri Lanka's plantation agriculture is very productive and quite efficient, but the same cannot be said for the island's rice lands. In the 1950s and early 1960s it was necessary to import half the rice consumed in Sri Lanka, a situation that was detrimental to the general economic situation. Thus the government made it a priority to reconstruct the lowland irrigation systems, to repopulate the lowlands (until its recent eradication, malaria was an obstacle to lowland settlement), and to intensify rice cultivation. The result has been to increase rice production substantially, but Sri Lanka is still far from self-sufficient. Population growth remains 2.2 percent annually, and Sri Lanka spends one-fifth of its import expenditures on grains.

A freighter under sail leaves the harbor of Colombo, Sri Lanka. (Harm J. de Blij)

In a country so heavily agricultural, it is not surprising to find very little industrial development except for factories processing farm and plantation products. Sri Lanka appears to have very little in the way of mineral resources; graphite is the most valuable mineral export, although gemstones (sapphires, rubies) once figured importantly in overseas commerce. The industries that have developed, other than those treating agricultural products, depend on Sri Lanka's relatively small local market. Predictably, these include cement, shoes, textile, paper, china and glassware and the like. The majority of these establishments are located in the country's major port, largest city, and capital: Colombo (750,000); there are indications that plans to build a small steel mill will go ahead. But Sri Lanka is an agricultural country; it is still in a position to close the gap between the demand for staple foods the local supply capacity. Its chief concern must be to increase productivity of its soil. In this respect it shares the objective of all South Asia.

Plantation type + zonation agriculture.

[illegible] a warm temperature.

Industry tied in with agriculture.

G.N.P – 220 p.c. reflective of the agriculture.

Additional Reading

Probably the most comprehensive and best-documented geographical publication on South Asia is Joseph E. Schwartzberg (ed.), *A Historical Atlas of South Asia* (Chicago: University of Chicago Press, 1978). Still a standard geography on South Asia is the two-volume *India and Pakistan: A General and Regional Geography* by O. H. K. Spate and A. T. A. Learmouth, a revision of earlier editions, published in 1971 by Methuen in London. Also see the appropriate chapters in J. E. Spencer and W. L. Thomas, *Asia, East by South: A Cultural Geography,* published in a second edition by Wiley in New York in 1971. For a very comprehensive bibliography on the realm, no source is more useful than B. L. Sukhwal's *South Asia: A Systematic Geographic Bibliography,* published in Metuchen by Scarecrow Press. Although they are somewhat dated now, G. Myrdal's three volumes on the *Asian Drama: An Inquiry into the Poverty of Nations* still contain much food for thought; they were published in New York by Pantheon in 1968. A collection of *Focus on South Asia* issues was prepared by A. Taylor of the American Geographical Society and published by Praeger in New York in 1974.

India is discussed by A. K. Dutt and several colleagues in a volume entitled *India: Resources, Potentialities, and Planning,* published by Kendall-Hunt, Dubuque, in 1972. Also see the volume by J. H. Hutton, *Caste in India: Its Nature, Function and Origins,* published in a fourth edition by Oxford University Press, New York, in 1963; it contains a wealth of detail and insights. In the Van Nostrand Searchlight Series, No. 24 is *India: the Search for Unity, Democracy, and Progress,* by W. C. Neale. Two collections of essays are R. R. Platt, et al., *India: A Compendium,* published by the American Geographical Society in New York in 1962, and P. Mason, *India and Ceylon: Unity and Diversity,* published in New York by Oxford University Press in 1967. The Indian population question is discussed in R. O. Whyte, *Land, Livestock and Human Nutrition in India,* a Praeger volume published in 1968. For a sympathetic and easily read volume see P. Griffiths, *Modern India,* published in a third edition by Praeger in 1962. Indian federalism is discussed in detail by R. D. Dikshit in *The Political Geography of Federalism: An Inquiry into Origins and Stability,* published by Wiley in 1975. Two other volumes on India are worth your attention: one by J. R. McLane entitled *India: A Culture Area in Perspective,* published in Boston by Allyn and Bacon in 1970 and S. Wolpert's *India,* a Prentice-Hall publication of 1965.

Writings on Bangladesh before 1972 are either incorporated in general geographies of Pakistan or listed under East Pakistan. Among relatively recent discussions is N. Ahmed, *An Economic Geography of East Pakistan,* published by Oxford University Press in London in 1968, but the newest volume is by B. L. C. Johnson, *Bangladesh,* published by Barnes and Noble in New York in 1975.

On Pakistan, see H. Feldman, *Pakistan: An Introduction,* published by Oxford University Press in 1968, and R. P. Long, *The Land and People of Pakistan,* published by Lippincott in Philadelphia in 1968. The volume by D. P. Singhal, *Pakistan,* was published by Prentice-Hall in Englewood Cliffs in 1972.

The volume by R. D. Campbell, *Pakistan: Emerging Democracy* is No. 14 in the Van Nostrand Searchlight Series. And, of course, the appropriate chapter in the Spate-Learmouth volume is an excellent source on this country.

Among the countries of the perimeter, Sri Lanka is discussed in detail by R. F. Myrop in the *Area Handbook for Ceylon,* published by the United States Government Printing Office, Washington, in 1971. L. Boltin's *Ceylon,* a Doubleday, New York, publication of 1966, still contains much valuable detail. H. N. S. Karunatilake's *Economic Development in Ceylon* was published in New York by Praeger in 1971.

P. P. Karan has published several volumes on the countries of the northern interior, including *Nepal: A Cultural and Physical Geography* from the University of Kentucky Press, Lexington, in 1960, and a similar volume on Bhutan that appeared in 1967. In collaboration with W. M. Jenkins, Jr., Karan wrote *The Himalayan Kingdoms: Bhutan, Sikkim, and Nepal,* No. 13 in the Van Nostrand Searchlight Series.

R. Hartshorne makes his recommendations for the naming of boundaries in "Suggestions of the Terminology of Political Boundaries," in an abstract of a paper in the *Annals of the Association of American Geographers,* Vol. 26, March 1936.

On the population problem, you will want to consult the publication of the Population Reference Bureau, Washington, D.C., whose *Bulletins* include statistics, maps, and commentaries on current questions. Also see S. Johnson (ed.), *The Population Problem* (Wiley, New York, 1973), and John I. Clarke's excellent introduction to the field of *Population Geography* (Pergamon, New York, 1972). Some interesting readings were assembled by Q. H. Stanford in *The World's Population: Problems of Growth* (Oxford, New York, 1972), and you will want to spend an evening with the Ehrlichs' *Population, Resources, Environment* (Freeman, San Francisco, 1975), a book whose first edition aroused considerable debate. If you are writing a paper and need recent statistics, remember the United Nations Yearbooks in your library.

A general introduction to world population distribution is G. T. Trewartha's *A Geography of Population: World Patterns* (Wiley, New York, 1969), and you will find some disturbing details in a book by D. H. Meadows; D. L. Meadows, J. Randers, and W. W. Behrens entitled *The Limits to Growth* (Universe Books, New York, 1972).

Supreme Harmony Hall in Peking's Forbidden City. (Terry Madison/Image Bank)

Nine

China of the Four Modernizations

More than one-fifth of all humanity exists within the borders of a single country: China. The Chinese national territory is almost as large as that of the United States (Table I-1), but China's population is between four and five times larger. The Chinese themselves have never been very open with statistical information about their country, and estimates of the Chinese population range considerably. It probably exceeds 900 million and may be as high as 950 million; in any case, China will reach 1 billion people during the 1980s.

China today represents the modern expression of what may be the world's oldest continuous national culture and civilization. The present capital, Beijing (Peking) lies near the same region where ancient China first emerged and where cities existed four thousand years ago. Over the course of its lengthy history, China's political geography changed many times as the mushrooming state gained territory and then lost it again. But in the heartland there was always China, hearth of culture, source of innovations, focus of power. Over countless centuries the Chinese created for themselves a world apart, a society with strong traditions, values, and philosophies that could repel and reject influences from the outside. In the past, Westerners sometimes used the offensive term *dark continent* to describe Africa's unknown, uncertain qualities. But more even than Africa, China remained a land little known and less understood—distant, remote, and aloof.

IDEAS AND CONCEPTS

- Confucianism
- Extraterritoriality
- Central place theory (2)
- Action space
- Regional complementarity (2)
- Boundary superimposition
- Geomancy

This is not to suggest that the Chinese themselves did not deliberately contribute to the isolation of their country from foreign contacts and influences. Indeed, a certain degree of isolation is itself an integral part of Chinese tradition, and spatially the Chinese certainly were in a position to sustain this posture. To China's north lie the great mountain ranges of eastern Siberia and the barren country of Mongolia. To the northwest, beyond Xinjiang (Sinkiang), the mountains open—but into the vast Kirghiz Steppe. To the west and southwest lie the Tien Shan, Pamirs, the forbidding Plateau of Xizang (Tibet), and the Himalayan wall. And to the south there are the mountain slopes and tropical forests of Southeast Asia—the same that mark India's eastern flank with Burma (Fig. 9-1). But more telling even than these physical barriers to contact and interaction is the factor of distance. China has always been far from the modern source areas of industrial innovation and technological change. True, China itself was such an area, but China's contributions to the outside world were very limited, and European capacities and ideas did not change China as they transformed other societies. China interacted to some extent with Japan and with areas of Southeast Asia. Compare this to the Arabs, who ranged far and wide and who brought their religion and political influence to areas from southern Europe to Indonesia,

Figure 9-1

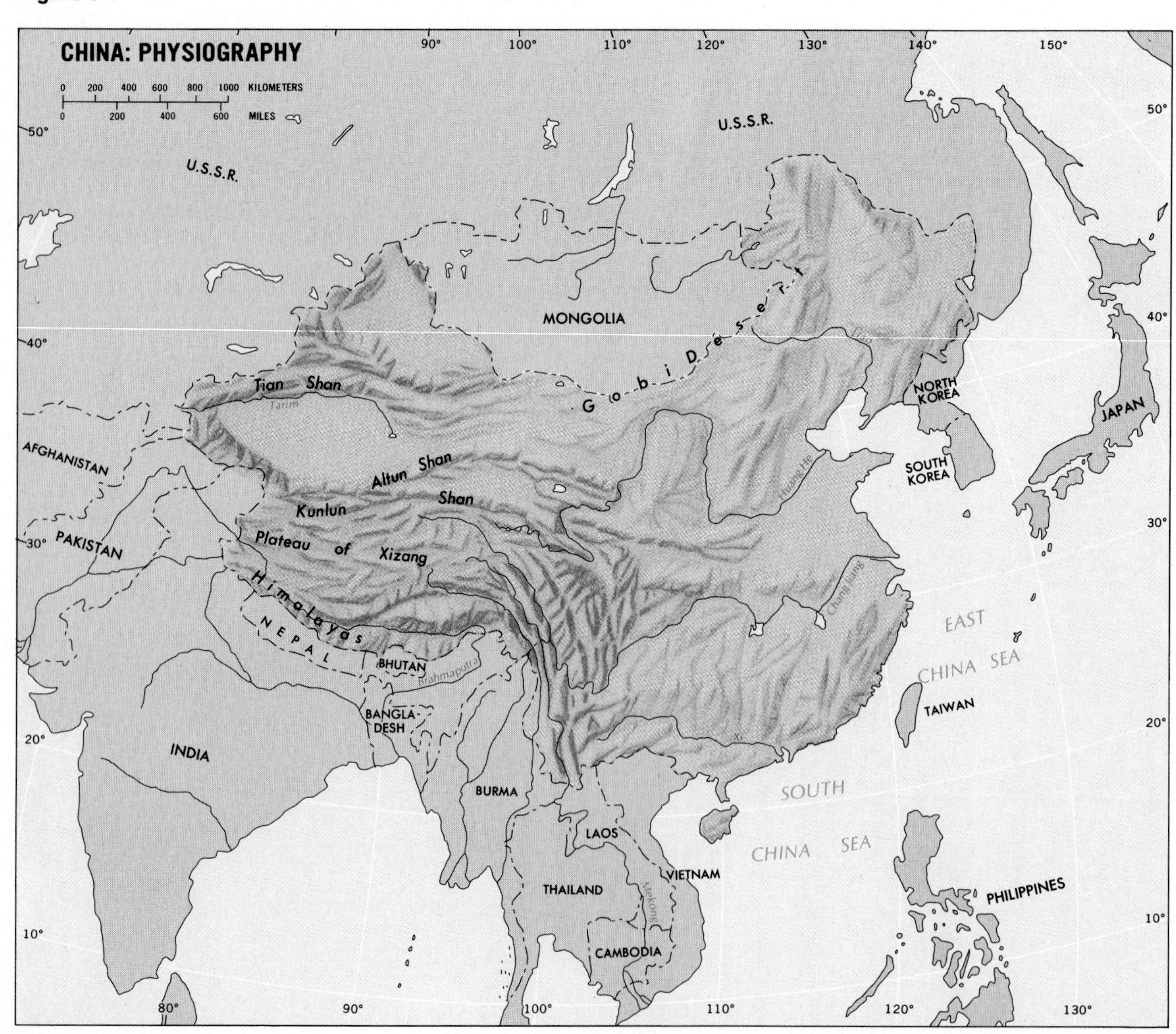

India to East Africa. When Europe became the center of change, China lay farther away by sea than almost any other part of the world, farther even than the coveted Indies.

Today, modern communications notwithstanding, China is still distant from almost anywhere else on earth. Going by rail from the heartland of China's Eurasian neighbor, the Soviet Union, is a long and tedious journey of several days. Direct overland communications with India are virtually nonexistent. Communications with Southeast Asia are still tenuous. And China's long Pacific coastline adjoins the world's largest ocean; Japan is China's only nearby, industrialized trading partner.

Ten Major Geographic Qualities of China

1. China's population represents over one-fifth of all humanity; territorially China ranks among the world's three largest states.
2. China occupies the eastern side of Eurasia; its sphere of influence was reduced by Russian expansionism in East Asia.
3. China is one of the world's two oldest continuous civilizations.
4. China's recent modernizing drive notwithstanding, the country remains a dominantly rural society with limited urbanization and industrialization.
5. China's enormous population is strongly concentrated in the country's eastern regions; western zones remain comparatively empty and open (they also are more arid and less productive).
6. The Chinese state and national culture evolved from a core area that emerged in the north, near the present capital of Beijing (Peking), and there has lain China's cultural hearth ever since.
7. China's civilization developed over a long period in considerable isolation, protected by physiographic barriers and by sheer distance from other source areas.
8. Foreign interventions had disastrous impact on Chinese society, from European colonialism to Japanese imperialism; intensified regionalism and territorial losses are only two of many resulting afflictions.
9. China's communist-designed transformation after World War II involved unprecedented regimentation and the imposition of effective central authority, with results that are perhaps permanently imprinted on the cultural landscape.
10. China's role in the political geography of East Asia remains uncertain. Boundary disputes, external minorities, relations with Taiwan, and ideological competition in Southeast Asia are among the factors involved.

The Pin-yin System

If a Western traveler in China were to stop and ask for directions to "Peking," the response would quite probably be a shake of the head. The Chinese call their capital "Beijing," not "Peking." For centuries the romanization of Chinese has led to countless errors of this kind, in part because Chinese dialects are also inconsistent and the Chinese in northern China may pronounce a place name somewhat differently from those living in the south.

In 1958 the Chinese government adopted the Pin-yin system of standard Chinese—not to teach foreigners how to spell and pronounce Chinese names and words, but to establish a standard form of the Chinese language throughout China. The Pin-yin system is based on the pronunciation of Chinese characters in Northern Mandarin, the Chinese spoken in the region of the capital and the north in general.

During the 1970s China's contact with the outside world expanded, but the Pin-yin system was not in use elsewhere. Of course the Chinese preferred the new standard, which represented not only a linguistic modernization but also something of a psychological break with the past. From 1979 all press reports from China contained only the Pin-yin spelling of Chinese names. In the United States and in other parts of the world, newspapers and magazines began to use the system. Atlases, wallmaps, and other geographic materials (including textbooks) were revised. In some instances the change was resisted, and the old spellings persist. But Pin-yin undoubtedly will become the world standard for romanized Chinese.

Strange as the familiar Chinese place names seem when transcribed into Pin-yin, the new system is not difficult to learn or understand. Most of the sounds are clear to someone who speaks English, except the q and the x. The q sounds like the ch in *cheese;* the x is pronounced sh as in *sheer.*

Some familiar place names, spelled the old and the Pin-yin way, are listed below:

Peking	Beijing
Shanghai	Shanghai
Canton	Guangzhou
Tibet	Xizang
Sinkiang	Xinjiang
Yangtze Kiang	Chang Jiang (River)
Nanking	Nanjing
Sechwan	Sichuan
Hwang Ho	Huang He (River)
Tientsin	Tianjin

And some personal names:

Mao Tse-tung	Mao Zedong
Chou En-lai	Zhou Enlai
Teng Hsiao-ping	Deng Xiaoping

In this chapter both the old and new Pin-yin spellings are given, so that it will be easy to become familiar with the new names. The maps have been revised to conform to the Pin-yin system. In the text, the Pin-yin version is given first, and the old spelling is in parentheses.

China in the World Today

For all its isolation and remoteness, China has in recent years moved to the center stage of world attention. It was Napoleon who long ago remarked that China was asleep, and whoever would awaken the Chinese giant would be sorry. Today China is awake. It was stung by Japanese aggression in the 1930s and 1940s, and after the end of World War II the growing communist tide took power following a bitter civil war in which the United States supported the losing side. European colonialism, Japanese imperialism, and a communist ideology combined to stir China into action. The foreigners were ousted, China's old order was rejected and destroyed, and since 1949 the communist regime has been engaged in a massive effort to remake China in a new image of unity and power.

This effort has taken China from a position of backwardness and weakness to one of considerable strength. China is not yet a third power equal to the United States or the Soviet Union. But China is on the move, and by almost any combination of measures it now ranks in third place.

What is most remarkable about all this, of course, is that after three decades of communist control, China has been able to place itself in a position of serious potential contention for world power, something that hardly seemed possible at the end of World War II. Chinese assistance is going to countries in Asia and America; Chinese technicians are building railroads in Africa. And at a time when the two greatest powers in the world are attempting to achieve a world in which they can coexist, China looms as a third-world threat to this joint monopoly over ultimate power.

China's potential for world influence was underscored by its explosion of a nuclear device in the early 1960s and its subsequent development of a nuclear arsenal as well as an expanding capacity to deliver these weapons over a long range. After the communist takeover the Chinese were assisted by the Soviets in their industrial and military development, but an ideological quarrel ended Sino-Soviet cooperation and the Soviet technicians were sent home. The Chinese accused the Soviets of "revisionism"—the dilution of communist ideology with doses of capitalism. So it was that China, awakened to communism by the revolutionary example of Russia, took on the mantle of the "purest" of communist systems and rejected the modern Soviet version. When anti-Soviet feeling in China reached its peak in the late 1960s and Soviet and Chinese forces exchanged shots across the Amur and Ussuri Rivers, it was not difficult to hear the echo of Napoleon's famous words.

Momentous changes came to China in the 1970s. Following China's entry into the United Nations in 1971 (replacing the delegation from the nationalist holdout island of Taiwan), a United States president visited Beijing (Peking) in 1972. Diplomatic relations with Japan were established and several trade agreements followed; China was an important oil supplier to Japan during the energy crisis. In 1976 both Premier Zhou Enlai (Chou En-lai) and the architect of modern communist China, Mao Zedong (Mao Tse-tung), died and a power struggle ensued after Mao's death. This struggle involved "moderate" and "radical" factions of the Communist Party and had been building even before Chou's death. The basic issues centered on the course of economic development and the future of education in China. The pragmatic moderates wanted to speed economic growth by offering workers some material rewards; they also proposed that the schools spend less time on ideological teaching and more on practical technical training. The radicals feared that such changes would weaken China's "pure" communism and produce new elite people, setting back three decades of class struggle. Eventually the pragmatists prevailed, and before the end of the decade of the 1970s a new China had begun to emerge. Thousands of American visitors were admitted; diplomatic relations returned to normal. In the 1979 municipal elections, noncommunist as well as communist candidates were allowed to compete for office. And China's planners laid out a new course for the new China: the "Four Modernizations." In industry, defense, science, and agriculture China opened an era of modernization as rapid as the country could achieve.

China is vulnerable to severe earthquakes. In 1976, perhaps as many as 750,000 people lost their lives when an earthquake devastated Tangshan; damage extended to the port of Tientsin and the capital, Beijing. The photograph shows some of the damage in Beijing in July, 1976. (UPI)

On July 28, 1976, two devastating earthquakes struck the northeast of China, killing perhaps as many as 750,000 people, wrecking the important industrial city of Tangshan in Hebei (Hopeh) Province and damaging the port of Tianjin (Tientsin, China's third largest city) as well as the capital, Beijing (Peking). It was by far the worst disaster of its kind in the past four centuries, and it dealt a staggering blow to China's economy, already impeded by the ideological struggle and associated labor problems. The 1970s in China were turbulent years indeed.

China's expanding role in the world is a major portent for the future. Beijing's (Peking's) government is opposed to what it views as the socialist imperialism of the Soviet Union as well as the capitalist imperialism of the United States, but it has come to consider the Soviet threat the greater. A major reorientation took place when the Chinese joined the United Nations, invited United States leaders for discussions, normalized relations with Japan—all this notwithstanding the continuing Indochina War (then still in progress) and the United States presence in Taiwan and South Korea. But the map of Eastern Asia helps explain some of China's recent actions. As the Indochina War came to a close, the Soviet presence in Southeast Asia strengthened even while the Americans withdrew. Although there are large Chinese minorities in Southeast Asian countries (see page 524) and the Chinese had given assistance to the communist insurgents during the war in Vietnam, much anti-Chinese hostility exists in the region. The Soviets capitalized on the situation to begin developing a sphere of influence, a prospect that concerned China greatly. And, of course, China lies an ocean away from California's beaches, but only a stone's throw from Soviet soil. It was Russian imperialism that cost China dearly, not American imperialism. It is Soviet territory that China claims on historic grounds as its own, not American soil. And as China's capacity for war grows, it is Russia that will lie in its reach first; even in these modern times of intercontinental missiles, an ocean provides a time cushion and therefore constitues an advantage to the technologically superior foe. That was proved by the Cuban missile crisis during the Kennedy administration: rockets poised to strike the United States from Russia was one thing, but rockets aimed at America from Cuba was something else. China is testing rockets and bombs, but it still does not have the capacity to strike effectively—it does not as yet possess a sufficient arsenal. When the time comes, will the Soviet Union accept missiles poised to strike it from Xinjang (Sinkiang) any more than the United States did in the case of Cuba? Thus China's approaches to the United States are a matter of self-interest, consistent with the policies of the pragmatic moderates who are trying now to speed the country's economic development along revisionist lines.

Evolution of the State

China may have developed in comparative isolation for more than 4000 years, but there is evidence that China's earliest core area, which was positioned about the confluence of the

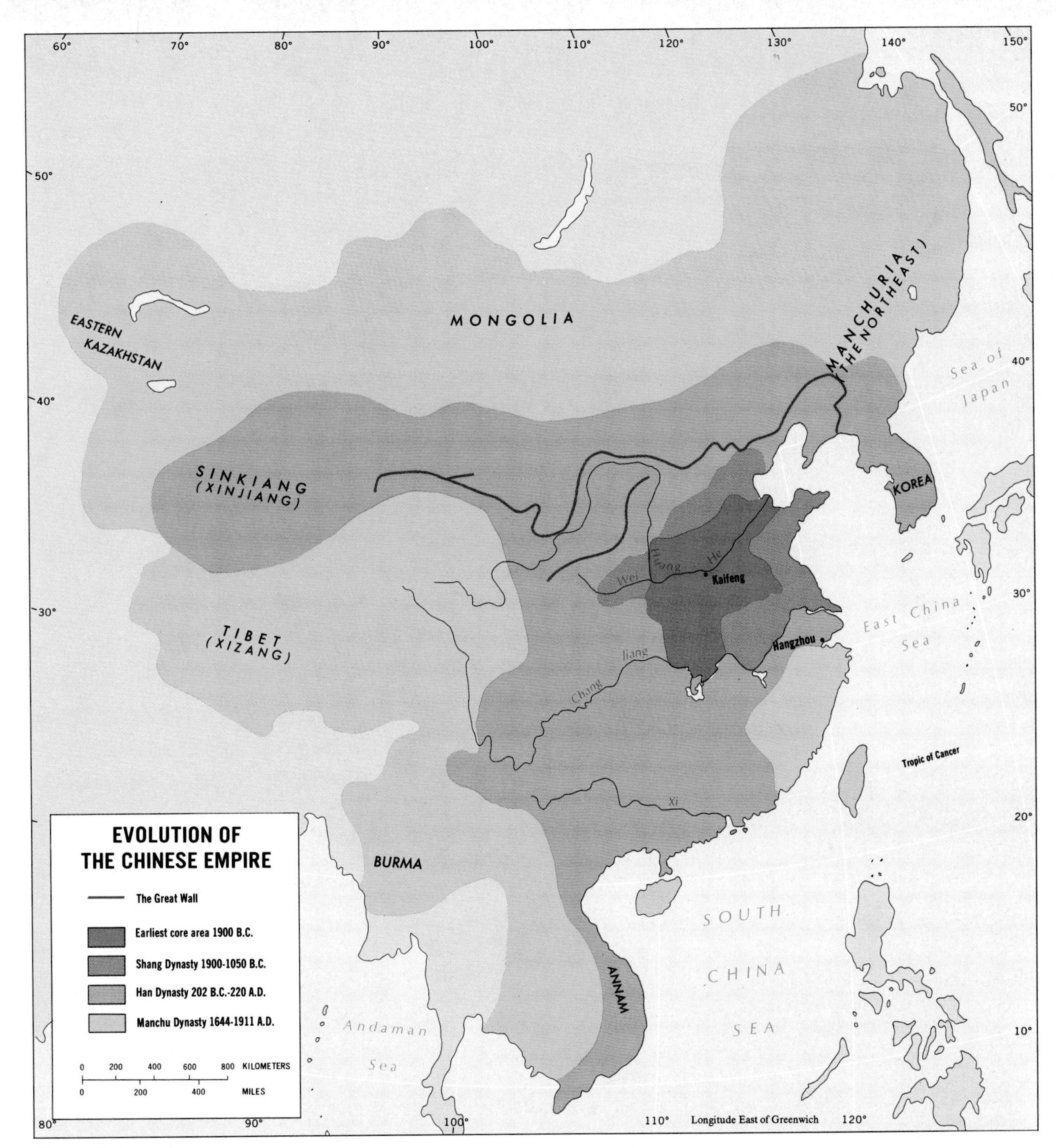

Figure 9-2

Huang (Yellow) and the Wei Rivers (Fig. 9-2), received stimuli from other, distant, and possibly slightly earlier civilizations: the river cultures of Southwest and South Asia. From Mesopotamia and the Indus, techniques of irrigation and metalworking, innovations in agriculture, and possibly even the practice of writing reached the Yellow River basin, probably overland along the almost endless route across desert and steppe. The way these early Chinese grew their rice crops gives evidence that they learned from the Mesopotamians; the water buffalo probably came from the Indian subcontinent. But quite soon the distinctive Chinese element began to appear, and by the time the record becomes reliable and continuous, shortly after 2000 B.C., Chinese cultural individuality was already strongly established.

The oldest dynasty of which much is known is the Shang Dynasty (sometimes called Yin), centered in the Huang-Wei confluence from perhaps 1900 B.C. to about 1050 B.C. Walled Cities were built during the Shang period, and the Bronze Age commenced during Shang rule. For more than a thousand years after the beginning of the Shang dynasty, North China was the center of development in this part of Asia. The Chou Dynasty (about 1050 to 250 B.C.) sustained and consolidated what had begun during the Shang period.

Eventually agricultural techniques and population numbers combined to press settlement in the obvious direction—southward, where the best opportunities for further expansion lay. During the brief Chin Dynasty the lands of the Chang (Yangtze) Jiang (*jiang* means "river") were opened up, and settlement spread even as far southward as the lower Xi Jiang (Hsi Jiang or West River). A pivotal period in the historical geography of China lay ahead: the Han Dynasty (202 B.C. to A.D. 220). The Han rulers brought unity and stability to China, and they enlarged the Chinese sphere of influence to include Korea, the Northeast, Mongolia, Xinjiang (Sinkiang), and, in Southeast Asia, Annam (located in what is today Vietnam) as well. This was done to establish control over the bases of China's constant harassers—the nomads of the surrounding steppes, deserts, and mountains—and to protect (in Xinjiang) the main overland avenue of westward contact between China and the rest of Eurasia. The Han period was a formative one in the evolution of China. Not only was Chinese military power stronger than ever before, but there were also changes in the systems of land ownership as the old feudal order broke down and private, individual property was recognized, and the silk trade grew into China's first external commerce. To this day, most

The Great Wall of China in winter. (George Holton/Ocelot)

Chinese, recognizing that much of what is China first came about during this period, still call themselves the "people of Han."

Han China was the Roman Empire of East Asia. Great achievements were made in architecture, art, the sciences, and other spheres. Like the Romans, the Han Chinese had to contend with unintegrated, hostile peoples on their empire's margins. In China these peoples occupied the mountainous country south of the national territory, and the Han rulers built military outposts to keep these frontiers quiet. Again, like the Roman Empire, the China of Han went into decline and disarray, and more than a dozen states arose after A.D. 220 to compete for primacy. Not until the Sui Dynasty (510–618) did consolidation begin again, to be continued during the Tang Dynasty (618-906), another period of national stability and development.

It is often said that the year 1949, when Communist Party Chairman Mao Zedong proclaimed the People's Republic of China at Beijing's (Peking's) historic Gate of Heavenly Peace, marked the beginning of a new dynasty that is not so different from the old—communist doctrine notwithstanding. Certainly some of China's old traditions continue in the new communist era, but in many more ways the new China is a totally overhauled society. Benevolent or otherwise, the rulers of old China headed a country where—for all its splendor, strength, and culture—the fate of the landless and the serf was often indescribably miserable, where famine and disease could decimate the population of entire regions, where local lords could repress the people with impunity, where children were sold and brides bought. Contact with Europe, however peripheral, brought ruin to Chinese urban society where slums, starvation, and depravation were commonplace. Mao's long-term and apparent omnipotence may remind us of the dynastic rulers' frequent longevity and absolutism, but the new China he left behind is radically different from the old.

Following the Tang period, when China once again was a great national state, the Sung Dynasty (960–1279) carried the culture to unprecedented heights, notwithstanding continuing problems with marauding nomadic peoples. This time the threat came from the north, and in 1127 the Sung rulers had to abandon their capital at Kaifeng on the Huang He in favor of Hangzhou (Hangchou) in the southeast. Nevertheless, China during the Sung Dynasty was the world's most advanced state; it had several cities with more than 1 million people, paper money was in use, commerce intensified, literature flourished, schools multiplied in number, the arts thrived as never before, and the philosophies of Confucius, for many centuries China's guide, were modernized, printed, and mass-distrubuted for the first time.

Eventually Sung China fell to conquerors from the outside, the Mongols led by Kublai Khan. The Mongol authority, known as the Yuan Dynasty (1279–1368) made China a part of a vast empire that extended all the way across Asia to Eastern Europe. But it lasted less than a century and had little effect on Chinese culture. Instead, the Mongols adopted Chinese civilization. In 1368 a Chinese local ruler led a rebellion that ousted the Mongols, and this signaled the beginning of another great indigenous dynasty, the Ming Dynasty (1368–1644). Under the Ming rulers China's greatness was restored, its territory once again consolidated from the Great Wall in the north to Annam in the south. Finally, in 1644 another northern, foreign, nomadic people forced their way to control over China. They came from the far Northeast, but unlike the Mongols they sustained and nurtured Chinese traditions of administration, authority, and national culture. The Manchu (or Ching) Dynasty (1644–1911) extended the Chinese sphere of influence to include Xinjiang and Mongolia, a large part of southeastern Siberia, Xizang, (Tibet), Burma, Vietnam, eastern Kazakhstan, and Korea (Fig. 9-2). But the Manchus also had to contend with the rising power of Europe and the West and the encroachment of Russia. Large parts of the Chinese spheres in Siberia, interior Asia, and Southeast Asia were lost to China's voracious competitors.

Manchuria: the *Northeast*

Three provinces, Liaoning, Jilin (Kirin), and Heilongjiang (Heilungkiang) constitute China's Northeast, a region bounded by the Soviet Union in the north, by Mongolia in the west, and by North Korea in the east. This was the home of the conquering Manchus and, later, the scene of foreign domination. Russians and Japanese struggled for control over the area, and ultimately Japan created a dependency here called *Manchukuo.*

Later the region came to be called *Manchuria,* but Manchuria is not an accepted regional appellation in China. Chinese geographers refer to the region encompassed by the three northeastern provinces, simply, as the *Northeast.* This practice is followed in the present discussion.

A Century of Convulsion

China long withstood the European advent in East Asia with a self-assured superiority based on the strength of its culture and the continuity of the state. There was no market for the British East India Company's rough textiles in a country long used to finely made silks and cottons. There was little interest in the toys and trinkets the Europeans produced in the hope of bartering for Chinese tea and pottery. And even when Europe's sailing ships made way for steam-driven vessels, and newer and better factory-made textiles were offered in trade for China's tea and silk, China continued to reject the European imports that still were, initially at least, too expensive and of too poor quality to compete with China's handmade materials. Long after India had fallen into the grip of mercantilism and economic imperialism, China was able to maintain its established order. This was no surprise to the Chinese; after all, they had held a position of undisputed superiority among the countries of Eastern Asia as long as could be remembered, and they had dealt with invaders from land and from the sea before.

The nineteenth century shattered the self-assured isolationism of China as it proved the superiority of the new Europe. On two fronts, the economic and the political, the European powers destroyed China's invincibility. In the economic sphere, they succeeded in lowering the cost and improving the quality of their manufactured goods, especially textiles, and the handcraft industry of China began to collapse in the face of unbeatable competition. In the political sphere, the demands of the British merchants and the growing British presence in China led to conflicts. In the early part of the nineteenth century, the central issue was the importation into China of opium, a dangerous and addictive intoxicant. As the Manchu government moved to stamp out the opium trade in 1839, armed hostilities broke out, and soon the Chinese sustained their first defeats. Between 1839 and 1842 the Chinese fared very badly, and the first "opium war" signaled the end of Chinese sovereignty. British forces penetrated up the Chang Jiang (Yangtze) and controlled several areas south of that river; Beijing hurriedly sought a peace treaty. As a result, leases and concessions were granted to foreign merchants. Hong Kong Island was ceded to the British, and five ports (Guangzhou, formerly Canton, and Shanghai among them) were opened to foreign commerce. No longer did the British have to accept a status that was inferior to the Chinese in order to do business; henceforth, negotiations would be pursued on equal terms. Opium now flooded into China, and its impact on Chinese society was

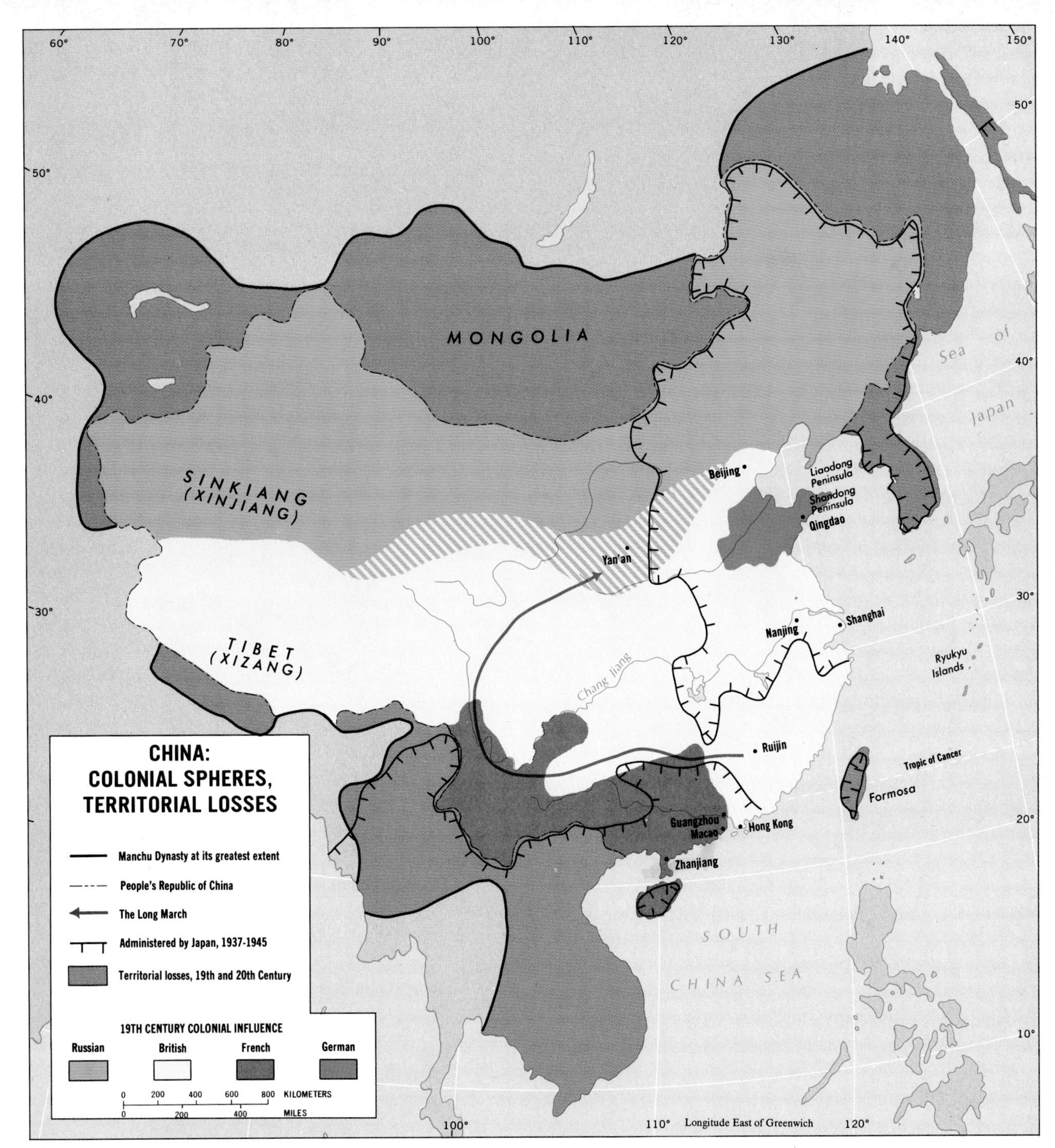

Figure 9-3

devastating. Fifteen years after the first "opium war," the Chinese tried again to stem the disastrous tide—and again they were bested by the foreign parasites that had attached themselves to their country. Now the cultivation of the opium poppy in China itself was legalized. Chinese society was disintegrating, and the scourge of opium was not defeated until the revival of Chinese power in the early twentieth century.

But before China could reassert itself, much of what remained of China's sovereignty was steadily eroded away. The Germans obtained a lease of Qingdao (Tsingtao) in 1898, and in the same year the French acquired a sphere of influence at Kwanchowan, now Zhanjiang (Fig. 9-3). The Portuguese took Macao, the Russians obtained a lease on Liaodong (Liaotung) in Manchuria and railway concessions as well, and even Japan got in the act by annexing the Ryukyu Islands and more importantly, Formosa (Taiwan) in 1895. After millenia of cultural integrity, economic security, and political continuity, the Chinese world lay open to the aggressions of foreigners whose innovative capacities China had denied to the end. But now, ships flying European flags lay in the ports of China's coasts and rivers; the smokestacks of foreign factories rose above the urban scene of its great cities. The Japanese were in Korea, which had nominally been a Chinese vassal; the Russians were in China's Northeast. The Foreign invaders even took to fighting among themselves, as did Japan and Russia in Manchuria (as the foreigners called the Northeast) in 1904.

Rise of a New China

In the meantime, organized opposition to the foreign presence in China was emerging. The twentieth century opened with a large-scale revolt against all outside elements. Bands of revolutionaries roamed the cities as well as the countryside, attacking not only the hated foreigners but also Chinese citizens who had adopted some form of Western culture. Known as the Boxer Rebellion (after a loose translation of the Chinese name for these revolutionary groups), the 1900 uprising was put down with much bloodshed. But simultaneously another revolutionary movement was gaining strength, aimed against the Manchu leadership itself. In 1911 the emperor's garrisons were attacked all over China, and in a few months the two and a half-century old dynasty was overthrown. It was, indirectly, still another casualty of the foreign intrusion, and it left China divided and disorganized.

The end of the Manchu era and the proclamation of a republican government in China did little to improve the country's overall position. The Japanese captured Germany's holdings on the Shandong (Shantung) Peninsula, including the city of Quingdao (Tsingtao) during World War I, and when the victorious European powers met at Versailles to divide the territorial spoils they confirmed Japan's rights in the area. This led to another significant demonstration of Chinese reassertion as nationwide protests and boycotts of Japanese goods were organized in what became known as the May 4 Movement.

Extraterritoriality

A sign of China's weakening during the second half of the nineteenth century was the application in its cities of a European doctrine of international law, *extraterritoriality.* This principle originated with the French jurist Ayraut (1536–1601); it denotes a situation in which foreign states or international organizations and their representatives are immune from the jurisdiction of the country in which they are present. This, of course, constitutes an erosion of the sovereignty of the state hosting these foreign elements, especially when the practice goes beyond the customary immunity of embassies and persons in diplomatic service. In China, extraterritoriality reached perhaps unprecedented extremes. The best residential suburbs in the large cities, for example, were declared to be "extraterritorial" parts of foreign countries and made inaccessible to Chinese citizens. Soon the Chinese found themselves unable to enter their own public parks and many buildings without permission from foreigners. Christian missionaries fanned out into China, their bases fortified with extraterritorial security. To the Chinese, this involved a loss of face that contributed to the bitter opposition against the presence of all foreigners, a resentment that exploded in the Boxer Rebellion of 1900.

One participant in these demonstrations was a young man named Mao Zedong. Nevertheless, China after the First World War remained a badly divided country. By the early 1920s there were two governments—one in Beijing (Peking) and another in the south, in Guangzhou (Canton), where the famous Chinese revolutionary Dr. Sun Yat-sen was the central figure. Neither government could pretend to control much of China. The Northeast was in complete chaos, petty states were emerging all over the central part of the country, and the Guangzhou (Canton) "parliament" controlled only a part of Guangdong (Kwangtung). Yet it was just at this time that the power groups that were ultimately to struggle for supremacy in China were formed. While Sun Yat-sen was trying to form a viable Nationalist government in Guangzhou (Canton), the Chinese Communist Party was formed by a group of intellectuals in Shanghai. Several of these intellectuals had been leaders in the May 4 Movement, and in the early 1920s they received help from the Communist Party of the Soviet Union. Mao Zedong was a prominent figure in these events.

Initially there was cooperation between the new Communist Party and the nationalists led by Sun Yat-sen. The nationalists were stronger and better organized, and they hoped to use the communists in their anti-foreign (especially anti-British) campaigns. By 1927 the foreigners were on the run; the nationalist forces entered cities and looted and robbed at will while aliens were evacuated or, failing that, sometimes killed. As the nationalists continued their drive northward and success was clearly in the offing, the luxury of internal dissension could be afforded. Soon the nationalists were as busy purging the communists as they were pursuing foreigners, and the central figure to emerge in this period was Chiang Kai-shek. Sun Yat-sen died in 1925, and when the nationalists established their capital at Nanjing (Nanking) in 1928, Chiang was the country's leader.

Three-way Struggle

The post-Manchu period of strife and division in China was quite similar to other times when, following a lengthy period of comparative stability under dynastic rule, the country fragmented into rival factions. In the first years of the Nanjing government's hegemony, the campaign against the communists intensified and many thousands were killed. Chiang's armies drove them ever deeper into the interior (Mao himself escaped the purges only because he was in a remote rural area at the time), and for a time it seemed that Nanjing's armies would break the back of the communist movement in China.

The Long March. A core area of communist peasant forces survived in the zone where the provinces of Jiangxi (Kiangsi) and Hunan adjoin (Fig. 9-4) and defied Chiang's attempts to destroy them. Their situation grew steadily worse, however, and in 1933 the nationalist armies were on the verge of encircling this last eastern communist stronghold. The communists decided to avoid inevitable strangulation by leaving. Nearly 100,000 people—armed soldiers, peasants, local leaders—gathered near Juichin and started to walk westward in 1934. It was a momentous event in modern China; among the leaders of the column were Mao Zedong (Mao Tse-tung) and Zhou Enlai. The nationalists rained attack after attack on the marchers, but they never succeeded in wiping them out completely; as the communists marched, they were joined by sympathizers. The Long March, as this drama has come to be called, first took them to Yunnan Province, where they turned north to enter western Sichuan (Szechwan); they traversed Gansu (Kansu) Province and eventually reached their goal, the mountainous interior near Yan'an (Yenan) in Shaanxi (Shensi) Province. The Long March covered nearly 10,000 kilometers (6000 miles) of China's most difficult terrain, and the nationalists' continuous attacks killed an estimated 75,000 of the original participants. Only about 20,000 survived the epic migration, but among them were two leaders who were convinced that a new China would emerge from the peasantry of the rural interior to overcome the urban easterners whose armies could not eliminate them. The two were Mao and Zhou.

The Japanese. While the Nanjing government was pursuing the communists, foreign interests made use of the situation to further their objectives in China. The Soviet Union held a sphere of influence in Mongolia and was on the verge of annexing a piece of Xinjiang (Sinkiang). Japan was dominant in Manchuria, where it had control over ports and railroads. The

Figure 9-4

Nanjing government tried to resist the expansion of Japan's sphere of influence, but failed, and the Japanese even set up a puppet state in the region. They appointed a ruler and called their new dependency Manchukuo.

The inevitable war between Japanese and Chinese broke out in 1937. There were calls for a suspension of the nationalist-communist struggle in the face of the common enemy, but after a brief armistice the factional conflict arose again while both sides fought the Japanese. China became divided into three regions: the areas taken by the Japanese during their quick offensive of 1937-1938 (when they took much of Eastern China including the principal ports), the nationalist zone centered on China's wartime capital of

Chongqing (Chungking), and the communist-held areas of the interior west. The Japanese, by engaging Chiang's forces, gave the communists an opportunity to build their strength and prestige in China's western regions.

Victory. When Japan was defeated by the Western powers, principally the United States, the full-scale nationalist-communist struggle was quickly resumed. The United States, hopeful for a stable and friendly government in China, sought to mediate in the conflict; but did so while recognizing the nationalist faction as the legitimate government. The chances of mediation were impaired by this position, and also by the military aid given to the nationalists' forces. By 1948 it was clear that Mao Zedong's armies would defeat the forces of Chiang Kai-shek and that the final victory was only a matter of time. Chiang kept moving his capital—back to Guangzhou (Canton, where Sun Yat-sen had assembled the first nationalist government), and then to Chongqing (Chungking), where the nationalists had held out during World War II. But late in 1949, following disastrous defeats in which hundreds of thousands of nationalist forces were killed, the remnants of Chiang's faction fled to the island of Taiwan. In Beijing, Mao Zedong proclaimed the birth of the People's Republic of China.

Regions of China

The human drama of modern China's evolution was played out on a physical stage of immense variety of regional as well as local dimensions. As we noted earlier, China's territorial proportions are almost identical to those of the United States including Alaska, and latitudinally China lies in the same general range as the contiguous United States (although it extends somewhat farther both to the north and south, as Fig. 9-4 shows). There are some other similarities. In China as well as the United States, the core area lies in the eastern part of the national territory. Both China and the United States have lengthy east coasts. But on the other hand there is nothing in China to compare to California—and neither is there a west coast; no subsidiary core areas or large conurbations are emerging in China's far west. Yet China contains well over four times as many people as the United States, a context in which the distribution shown on Fig. I-9 (page 27) should be seen.

In the following paragraphs we consider China's general physiographic layout and then focus on the country's human regions. In very general terms we can observe still another similarity to the United States: politically (Fig. 9-4) China has small provinces in the east and Texas-sized units toward the west. And physiographically, China divides into about the same number of regions as the United States, ranging from the snow-capped mountains of Xizang (Tibet) to the warm river lowlands of Guangdong and from the deserts of Xinjiang to the wheat fields of the North China Plain.

Physiographic Regions

We have already come to know China as a land of river basins, fertile alluvium, and loess (Figure I-8), temperate to continental climates (Figure I-6), severe dryness in the north and west (Figure I-5), and great mountain chains (Figure I-2). To get a picture of the spatial arrangement of things in China, a general frame of reference is useful. As in the case of India, we can identify several major regions and then break them down into more detailed subregions. Thus China's four major regions are (inset, Fig. 9-5): (1) the river basins and highlands of east and Northeast China, (2) the plateau steppe of Mongolia, (3) the high plateaus and mountains of Xizang (Tibet), and (4) the desert basins of Xinjiang (Sinkiang). Strictly speaking, of course, these are not physiographic *provinces;* they are broader than that. Similar "regions" in the United States would be, say, the Midwest—which includes several different kinds of landscapes—or the Rocky Mountains, diverse enough to be divided into several distinct regions as well.

Region 1, River Basins and Highlands of East China and the Northeast, contains the greater length of the course of China's three most important rivers: the Huang (Yellow), Chang, and Xi (West), marked as (A), (B), and (C), respectively, on the larger map (Fig. 9-5). All

East China is a land of river basins and valleys separated by highlands, the lowlands covered by farm fields. This view is over the Li River and its fertile banks. (Terry Madison/Image Bank)

three of these rivers rise on eastern slopes of the Xizang (Tibet)-Yunnan Plateau and flow eastward, the Yellow River through the most circuitous and longest course, the West River through the most direct and shortest. The upper courses of the Yellow and Chang Rivers lie in close proximity, but in two distinct physiographic provinces; their lower reaches however lie in an area that can with justification be identified as a single region, the Eastern Lowland. In human terms this Eastern Lowland is China's most important region, including as it does the North China Plain and the cities of Beijing and Tianjin (Tientsin), and the productive lower Chang area with Nanjing and China's largest city, Shanghai. In every respect this is China's core area with its greatest population concentration, largest percentage of urbanization (35 percent, as opposed to approximately 20 percent for the country as a whole), enormous agricultural production, growing industrial complexes, and intensive communications networks. China's heartland, as Fig. 9-5 shows, extends into the Northeast, where the lowland of the Liao River and the city of Shenyang form part of it.

The Huang He, in its upper basin, makes an immense bend and in the process almost encircles one of China's driest areas, the Ordos Desert. Below the Ordos, the Yellow River enters the loess plateau, and here conditions are quite different. Loess is a wind-borne deposit whose origin in this area is related to the nearby deserts (possibly the Ordos) and the Pleistocene glacial epoch, during which the deposits were laid down in a mantle up to 75 meters (250 feet) thick, covering the preexisting landscape. This loess is quite fertile, and, unlike ordinary soil, its fertility does not decrease with depth.

The Huang (Yellow) River has always been marked by violent floods and frequent changes in course. Alternately it has drained to the north into the Yellow Sea or to the south of the Shandong (Shantung) Peninsula, with numerous distri-

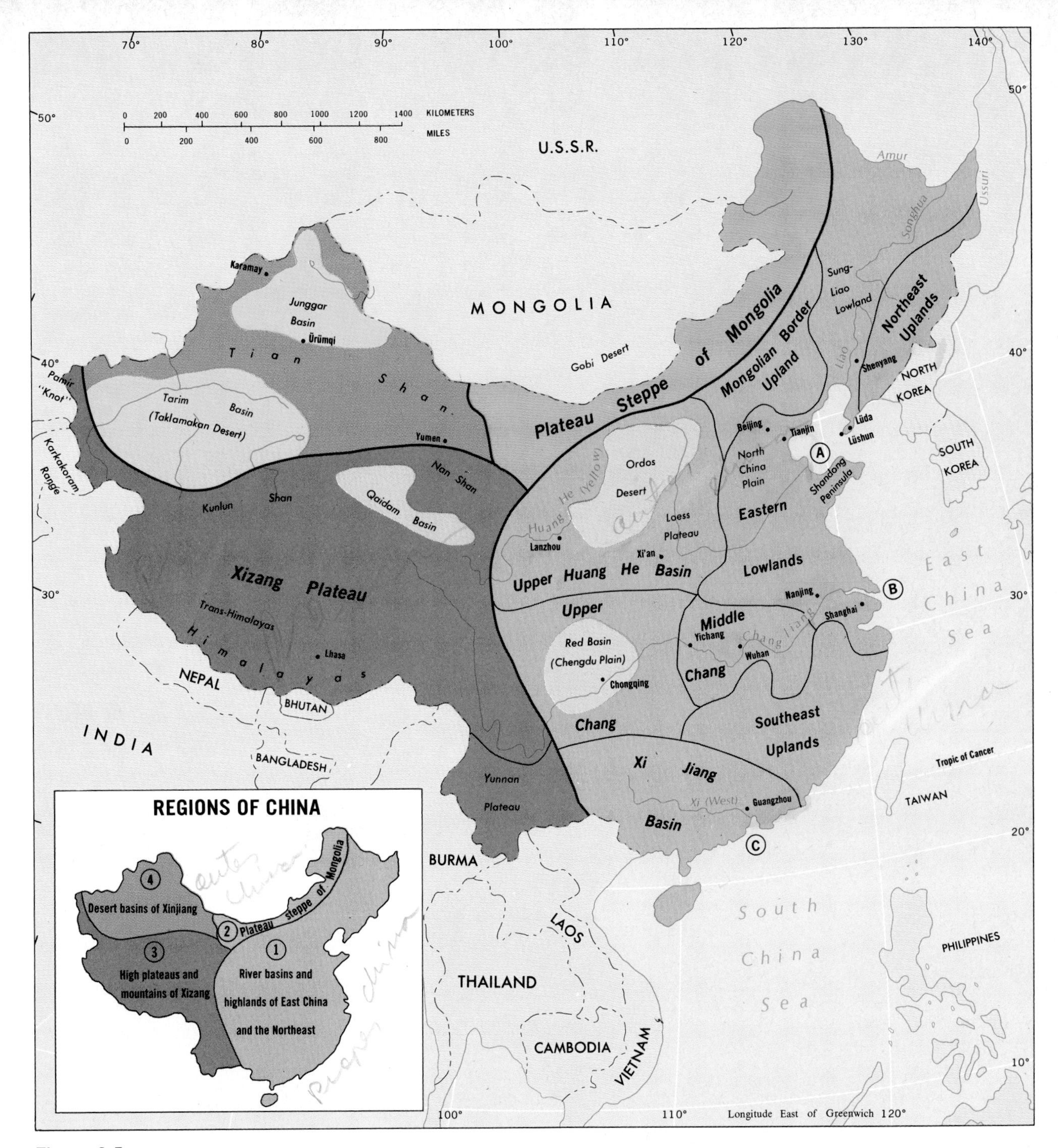

Figure 9-5

butaries forming and shifting position over time. Flowing as it does through the loess plateau, the river brings enormous quantities of silt to the North China Plain, one of the world's most fertile and productive farming regions. There it deposits all this sediment and then proceeds to flow over the new accumulation; for uncounted centuries the local inhabitants have tried through dikes and artificial levees to keep the river's various channels to some extent stable. All that was needed for a disastrous flood was a season with a particularly high volume of water, enough to overflow and breach the dikes. It has happened dozens of times, and the lives lost directly in these floods or subsequently in the inevitable famine must number in the tens of millions.

Confucius

Confucius (K'ung-fu tsu in Chinese) was China's most influential philosopher and teacher, whose ideas dominated Chinese life and thought for over 20 centuries. Confucian ideals were considered incompatible with communist doctrine by the leaders who took control of China in 1949, and the elimination of Confucian principles was one of Mao Zedong's primary objectives. Confucius left his followers a wealth of "sayings," many of which were frequently quoted as a part of daily life in China. Mao's "Red Book" of quotations was part of the campaign to erase the Confucian lifestyle.

Confucius was born in 551 B.C. and died in 479 B.C. He was one of many philosophers who lived and wrote during China's classical age, and during his lifetime he was prominent, but no more so than a number of other philosophers. The Confucian school of thought was one of several to arise during this period; the philosophies of Taoism also emerged at this time. Confucius was appalled at the suffering of the ordinary people in China, the political conflicts, and the harsh rule by feudal lords. In his teaching he urged that the poor assert themselves and demand the reasons for their treatment at the hands of their rulers (thereby undermining the absolutism of government in China), and he also tutored the indigent as well as the privileged, giving the poor an education that had hitherto been denied them and thus ending the aristocracy's exclusive access to the knowledge that constituted power.

Confucius was therefore a revolutionary in his time—but he was no prophet. Indeed, he had an aversion to supernatural mysticism and argued that human virtues and abilities should determine a person's position and responsibilities in society. In those days it was believed that China's aristocratic rulers had divine ancestors and governed in accordance with the wishes of these godly connections. Confucius proposed that these dynastic rulers give the reins of state to ministers chosen for their competence and merit. This was another Confucian heresy, but in time the idea came to be accepted and practiced.

Notwithstanding his earthly philosophies, Confucius took on the mantle of a spiritual leader after his death. His ideas spread to Korea, Japan, and Southeast Asia; temples were built in his honor all over China. As so often happens, Confucius was a leader whose teachings were far ahead of his time. His thoughts emerged from the mass of philosophical writing of his day to become a guiding principle during the formative Han Dynasty. Confucius had written that the state should not exist for the pleasure and power of the aristocratic elite; it should be a cooperative system, and its principal goal should be the well-being and happiness of the people. As time went on, a mass of Confucian writings evolved, most of which Confucius never wrote but which were attributed to him nevertheless. At the heart of this body of literature lie in the Confucian Classics, 13 texts that became the focus of education in China for 2000 years. In fields of government, law, literature, religion, morality, and in every conceivable way the Confucian Classics were the Chinese civilization's guide. The whole national system of education (including the state examinations, where everyone, poor or privileged, could achieve entry into the arena of political power) was based on the Confucian Classics. Confucius was a champion of the family as the foundation of Chinese culture, and the Classics prescribe a respect for parents and the aged that was a hallmark of Chinese society. It has been said that to be Chinese, whether Buddhist, Christian, or communist, one would have to be a Confucian; hardly any conversation of substance could be held without reference to some Confucian principle.

When the Western powers penetrated China, Confucian philosophy came face to face with practical Western education, and for the first time a segment of China's people (initially small) began to call for reform and modernization, especially in teaching. Confucian principles could guide an isolated China, but they were found wanting in the new age of competition. The Manchus resisted change, and during their brief tenure the nationalists under Sun Yat-sen tried to combine Confucianism and Western knowledge into a neo-Confucian philosophy. But it was left to the communists, beginning in 1949, to attempt to substitute an entirely new set of principles to guide Chinese society. Confucianism was attacked on all fronts, the Classics were abandoned, ideological indoctrination pervaded the new education, even the family was assaulted during the early days of communization. But it is difficult to eradicate two millenia of cultural conditioning in three decades. The dying spirit of Confucius will haunt physical and mental landscapes in China for years to come.

The course of China's middle river, the Chang (Yangtze) Jiang, is usually divided into three basins, of which the westernmost, the Red Basin, contains one of China's largest population clusters. The 1980 population of Sichuan (Szechwan) Province was estimated to be 106 million; about 76 million of these are clustered in the Red Basin, where lies one of the most intensively cultivated areas in the world. Apart from the Chengdu (Chengtu) Plain, where there is relatively level land, the slopes of the basin's hilly country have been transformed by innumerable terraces, where rice grows in summer and wheat in winter; apart from other cereals, including corn, such crops as soybeans, sweet potatoes, sugar cane, and a wide range of fruits are also grown; on the warmer slopes, tea flourishes. The Chang's middle course begins in the vicinity of Yichang, where the river emerges from a series of gorges that limit navigation to Chongqing (Chungking) to small motorized vessels. It enters an area of more moderate relief and flows through middle China's lake country—agriculturally one of the most productive parts of the nation. This is the southern part of the Eastern Lowland, and along or near the Chang lie the large cities of Wuhan (a three-city conurbation of 5 million), Nanjing (2.5 million), Hangzhou (Hangchow) (1.5 million), and China's leading urban center, Shanghai (13 million).

China's southernmost major river is actually called the West (Xi) River, and neither in length nor in terms of the productivity of the region through which it flows is this river comparable to the Chang or the Huang. In the Xi Jiang Basin there is much less level land; in the hills and mountains of South China remain many millions of local indigenous peoples not yet acculturated to the civilization of the people of Han. Hence the western part of the Xi Jiang Basin is constituted largely by the Guangxi Zhuangzu Zizhiqu (Kwangsi-Chuang Autonomous Region). To the east lie Guangdong (Kwangtung) province, the populous delta, and Guangzhou (Canton), the south's largest and most important city (3.7 million).

Between the deltas of the Xi and Chang Rivers lie the Southeast Uplands, a region of rugged relief which for a long time in Chinese history remained a refuge for southern tribes against the encroachment of the people of Han from the north. With its steep slopes and narrow valleys, this region has very limited agricultural potential, although a massive land development program was begun here in the 1960s. For many years this has been one of China's most outward-looking regions, with a considerable emigration (via Guangzhou) to the Philippines and Southeast Asia, substantial overseas trade in the leading commercial product, tea, and a seafaring tradition for which the people of Fujian (Fukien) Province have become famous.

In the Northeast, the Sung-Liao Lowland, while not comparable to the three great rivers in some respects (for example, this is largely an erosional rather than a depositional basin), is attaining population densities and productivities that are in a class with those of the heartland provinces farther to the south. The Northeast Uplands, comparable in terms of relief to those in the southeast, are not coastal; the Northeast's nearest coastline is silt-plagued Liaodong (Liaotung) Gulf, and its most effective direct outlet, Luda (Dalian) lies on the tip of the Liaodong Peninsula.

Region 2, the Plateau Steppe of Inner Mongolia, actually constitutes the southern rim of the basin of the Gobi Desert. Near the physiographic boundary line, elevations in a series of ranges reach an average of 1700 meters (5600 feet), but toward the heart of the Gobi the land lies much lower, as low as 750 meters (2500 feet). As is suggested by Fig. I-5 and I-7, the moistest parts of this region lie on the slopes of those southern, higher hills, although even there the vegetation is only poor steppe grass with some scrub; some of the mountains are rocky and barren and the whole aspect of the area is one of drought. Not surprisingly, this is a difficult environment in many ways; summer temperatures are often searingly hot, while winters are bitterly cold. Vicious winds blow up sand and dust. Altogether this is unlike the image we have of populous, productive, rice-paddy China. Despite the government's efforts to spread sedentary agriculture here through river dams and irrigation projects, as a whole, the Plateau Steppe remains one of the country's more sparsely populated areas, with an average of just 12 people per square kilometer (30 per square mile) in a country where 300 people per rural square kilometer is no rarity. Only Xinjiang (Sinkiang) and Xizang (Tibet) are less populous.

Region 3, the High Plateaus and Mountains of Xizang (Tibet), consists of one of the world's greatest assemblages of high, snow-covered mountain ranges and high-elevation plateaus. The southern margins of this region are the great Himalaya Mountains themselves, and in the north lie the Kunlun Mountains and the Nan Shan. From the Pamir "Knot" in the west, these great mountain chains spread out eastward in a series of vast arcs that eventually converge again and turn southward into the Yunnan Plateau. The *average* elevation in the region as defined on Fig. 9-5 is probably around 4500 meters (15,000 feet), with the higher mountain ranges standing 3000 meters (10,000 feet) above this level and the valleys—where most of the people live—going down to about 1500 meters (5000 feet) above sea level. The central plateau of Xizang is desolate and barren, cold, windswept, and treeless; here and there some patches of grass sustain a few animals. In terms of human development, there are two areas of interest. The first of these is the area between the Himalayas in the south and the Trans-Himalayas that lie not far to the north. In this area there are some valleys below 2000 meters (under 7000 feet) where the climate is milder and cultivation is possible. Here is Xizang's main population cluster; the capital of Lhasa is situated at the intersection of roads leading east-west along the valley and northward into Qinghai and China Proper. The Chinese government, since its confirmation of control in Tibet, has made investments to speed up the snail's-pace of development that prevailed under the previous administration. The southern valleys contain excellent sites for hydroelectric power projects, and some of these have been put to use in a

The mountainous country of Xizang (Tiber) towers over one of the area's grassy but treeless valleys. (Brian Brake/Rapho-Photo Researchers)

few light industries. More importantly, the south of Xizang may be an area of considerable mineral wealth, and despite the enormous difficulties of distance and terrain, any such minerals may become vitally important to the developing China of the future.

One other part of the Xizang region is of political economic importance—the Qaidam (Tsaidam) Basin, in Qinghai on the edge of Xinjiang (Sinkiang). This basin lies hundreds of meters below the surrounding Kunlun and Nan Mountains, and as such it has always contained a cluster of nomadic Xizang (Tibetan) pastoralists—in fact, there were times when these nomads plundered the trade route to the west that skirted the Nan Shan to the north. Recently, however, exploration has revealed the presence of oil fields and coal reserves below the surface of the Qaidam Basin, and the development of these resources has already begun.

Region 4, the Desert Basins of Xinjiang (Sinkiang), comprises two huge, mountain-enclosed basins and several smaller basins. The two largest of Xinjiang's basins are separated by the Tien Shan, a mountain range that stretches across the region from the Kirghiz border in the west to Mongolia in the east. Climatically these are dry areas; the Tarim Basin is in fact a desert (the Turkestan or Taklamakan Desert), while the basin of Junggar (Dzungaria) is steppe country. Both the Junggar and the Tarim Basins are areas of internal drainage; that is, the rivers that rise in the adjacent mountain slopes and flow to the basin floor do not continue to the sea. The rivers flowing off the adjacent mountains are the chief supply of water for the region's rough gravels that have been washed from the mountains; they disappear below the surface as they reach these coarse deposits. But then they reappear where the gravels thin out, and there, oases have long existed. Along the southern margin of the Taklamakan Desert lie a string of these oases that at one time formed stations on the long trade route to the west.

Since 1949 the Chinese have made a major effort to develop the agricultural potential of the Tarim area. Canals and *qanats* were built, oases enlarged, and the acreage of productive farmland has probably quadrupled by now from what it was 30 years ago. Especially the northern rim of the Tarim Basin, long neglected, has been brought into the sphere of development, and fields of cotton and wheat now attest to the success of the program.

Junggar (Dzungaria), although it contains perhaps just one-quarter of Xinjiang's more than 10 million inhabitants, has

A rural scene in Xinjiang. An Uighur village near Urumqi lies within sight of the great Tan Shan, which here reach as high as 5500 meters (17000 feet). (George Holton/Ocelot)

a number of assets of importance to China. First, it has long been the site of strategic east-west routes. Second, the main westward rail link toward the Kazakh S.S.R. and Soviet Russia goes from Xian (Sian) in China Proper via Yumen in Gansu and Urumqi (Urumchi) in Junggar, which is its present terminus. Third, Junggar has proved to contain sizable oil fields, notably around Karamay, not far from the Soviet border. Pipelines have been laid all the way to Yumen and Lanzhou (Lanchou), where refineries have been built. Thus it is not altogether surprising that the region's capital of Urumqi (350,000) lies in the less populous but strategically important northern part of Xinjiang.

Human Regional Geography

Interior China (the China of Xinjiang and Xizang and Inner Mongolia) still is frontier China, where China faces neighbors across unsettled boundaries and where human occupation is still spotty and tenuous. Xinjiang's population of over 10 million now is nearly one-half Chinese as a result of Peking's determination to integrate this distant province into the national framework; just 25 years ago the population was only about 5 percent Chinese. The majority of the people in Xinjiang are Moslem Uighurs, Kazakhs, and Kirghiz, with cultural affinities across the borders in the Soviet Asian Republics. The Uighurs, most numerous (about 4.5 million) among these peoples, have for centuries been concentrated in the oases of the Tarim Basin, and the mobile Kazakhs and Kirghiz moved along nomadic routes across the Soviet borders. In recent decades Chinese from Eastern China have come to build and develop towns and roads, to secure the Soviet boundaries, and to exploit the region's resources. In the process they have accomplished the Sinicization of an area that always was beyond China's national sphere, and before the end of the century the majority of Xinjiang's population will be Chinese.

Tibet (now Xizang) was pressed into the Chinese fold in the 1950s, first through frontier settlement and economic interference, and in 1958 by force of arms after the Tibetan villagers tried to resist the Chinese presence. Tibetan society had been organized around the fortress-like monasteries of monks who paid allegiance to the Dalai Lama. The Chinese wanted to modernize this feudal system, but the Tibetans clung to their traditions. In 1958 they proved no match for the Chinese forces, the Dalai Lama was ousted, and the monasteries were emptied. Since then, Xizang has been administered as a Chinese Autonomous Region, but although it is large (1,222,000 square kilometers, 472,000 square miles) its population remains only about 3 million. The Chinese presence is far weaker than in Xinjiang, but Xizang may yet come to play a crucial role in China's development because of its enormous mineral potential.

Inner Mongolia lies closer to China's eastern heartland—and it also has a history of closer association with Han China than either Xinjiang or Xizang. Here (and in adjacent Mongolia, now an independent country in the Soviet sphere) Chinese farmers long competed with nomadic horse- and camel-riding Mongols for control over an area that consists of vast expanses of steppe and desert and riverine ribbons and oases of farmland. Inner Mongolia's population of perhaps 10 million now is overwhelmingly sedentary and Chinese (fewer than 1 million Mongols concentrate in the northern border area) and depends largely on irrigation systems built in the basin of the great bend of the Yellow River.

The three regions of interior China cover a huge territory—some 3.3 million square kilometers (1.3 million square miles), an area larger than all of India—but they contain fewer than 25 million of China's over 950 million people. Small wonder that China's east, where the great majority of Chinese are concentrated, is known as the "real" China, China Proper. More even than in the United States, "the east" is China's historic and present heartland, where the civilization of the People of Han had its roots and where the modern core of the state evolved. China's east consists of China Proper and the Northeast, and of the two regions, China Proper still retains its primacy in every way. This is humid China, agricultural China, the China of wheat fields and rice paddies, villages and cities, workshops and factories. Above all it is the China of the masses, the throngs whose regimentation on the farms and in the towns has brought the country an era of comparative stability and order.

China Proper

China Proper is the China south of the Great Wall, the China of the three rivers (Fig. 9-6). Of the three, the middle river—the Chang—and its basin are in almost every respect the most important. From the Red Basin of Sichuan in its upper catchment area to its delta in the Eastern Lowlands (Fig. 9-6) the Chang traverses China's most populous and most productive areas. As a transportation route it is China's most navigable waterway; oceangoing ships can sail over 1000 kilometers (600 miles) up the river to the Wuhan conurbation (Wuhan is short for Wuchang, Hanyang, and Hankow), and boats of up to 1000 tons can even reach Chongqing (Chungking). Several of the Chang's tributaries are also navigable, and—depending, of course, on the size of the ships capable of using these various stretches—over 30,000 kilometers of water transport routes exist in the Chang Basin. The Chang Jiang thus is one of China's major transport arteries, and with its tributaries it attracts the trade of a vast area including nearly all of middle China and large parts of the north and south. Funneled down the Chang, most of this enormous volume is transferred at Shanghai, whose size reflects the productivity and great population of this hinterland.

Early in China's history, when the Chang's basin was being opened up and rice and wheat cultivation began, a canal was built to link this granary to the northern core of old China. Over 1600 kilometers (1000 miles) in length, this was the longest artificial waterway in the world, but during the nineteenth century it fell into disrepair. Known as the Grand Canal, it was dredged and rebuilt during the period when the nationalists held control over Eastern China, and after 1949 the communist regime continued this restoration effort. During the 1960s the southern section of the canal was opened once again to barge traffic, supplementing the huge fleet of vessels that transports domestic interregional trade along the east coast.

The bulk of China's internal trade in agricultural as well as industrial products is either derived from or distributed to the Chang region. International trade, also, goes principally through Shanghai, whose port normally handles half the country's overseas tonnage. The rest is split among China's other ports, including Tianjin (Tientsin) on the northern coast and Guangzhou (Canton) in the south.

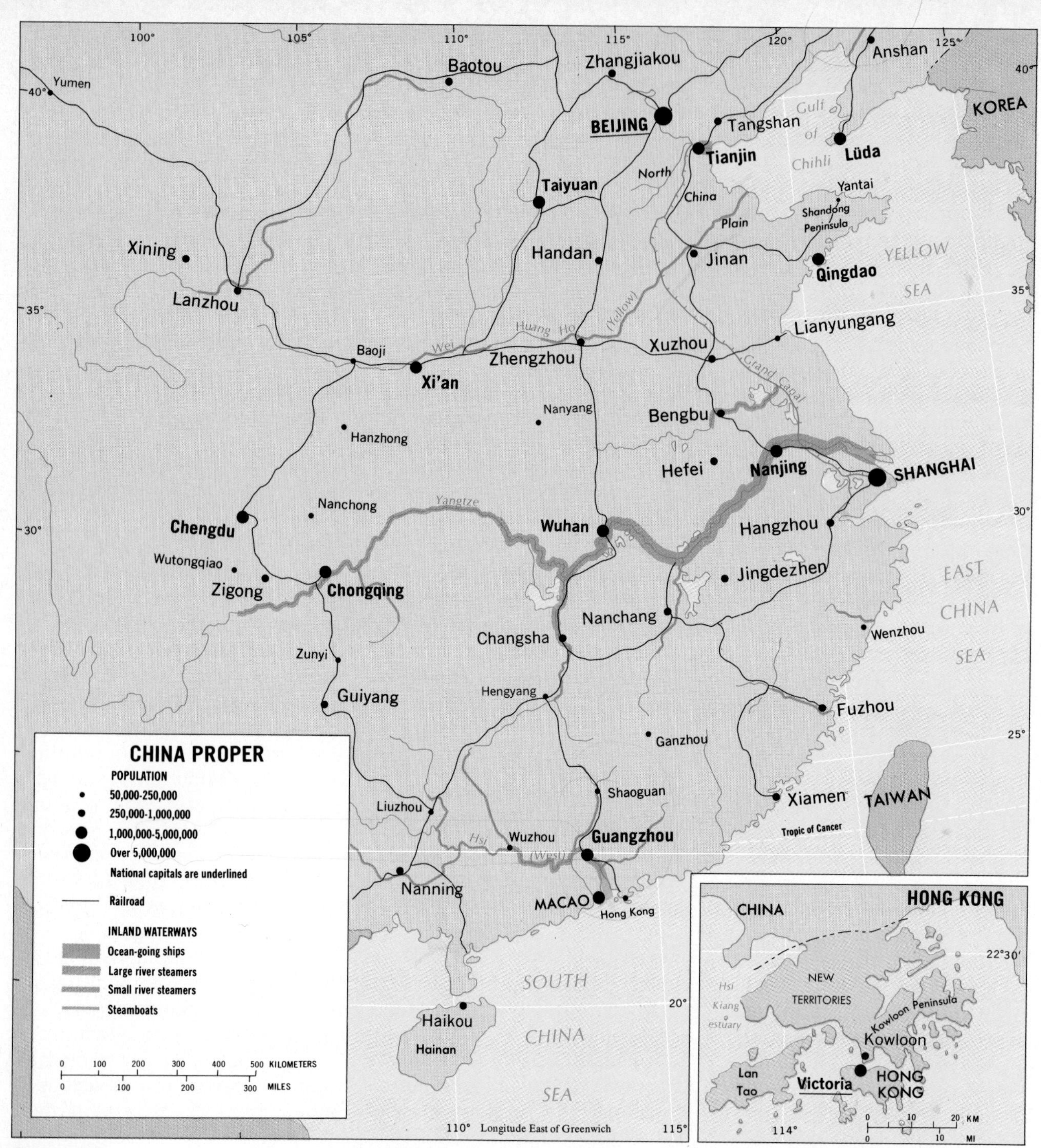

Figure 9-6

Shanghai, just a regional town until the mid-nineteenth century, rose to prominence as a result of its selection as a treaty port by the British, and ever since then its unparalleled locational advantages have sustained its position as China's leading city in almost every respect. The city lies at one corner of the Chang Delta, an area of about 50,000 square kilometers (20,000 square miles) in which live perhaps as many as 50 million people. Some two-thirds of these are farmers who produce food as well as silk and cotton for the city's industries. Thus Shanghai has as its immediate hinterland one of the most densely populated areas in the world, and beyond the delta lies what must be the most populous region in the world to be served by one major outlet. During the nineteenth century and up to the war with Japan, the

The Grand Canal, the longest artificial waterway in the world, connects four Chinese provinces. (George Holton/Ocelot)

Faces in a Chinese crowd along Najing Road, Shanghai. (Richard Choy/ Peter Arnold)

principal exports to pass through Shanghai were tea and silk; large quantities of cotton textiles and opium were imported. At that time Shanghai's position was undisputed; it handled two-thirds of all of China's external trade. But its fortunes suffered during and after World War II. First the nationalists blockaded the port (1949) and made bombing raids on it; then the Beijing government decided to disperse its industries upcountry, thereby reducing their vulnerability to attack. Meanwhile safer Tianjin (Tientsin) had taken over as the leading port. But Shanghai's situational advantages promised a comeback, and it came. In the 1960s the port regained its dominant position, and the industrial complex (textiles, food processing, metals, shipyards, rubber, chemicals) continued to expand.

The vast majority of the Chiang Basin's more than 300 million people are farmers on cooperatives and collectives that produce the country's staples and cash crops. We have already seen the amazing variety of crops grown in Sichuan there, as in the Chang's lower basin, rice predominates (Fig. 9-7). However, a look at Fig. I-5 and I-6 will suggest why the Sichuan-Chang line is more or less the northern margin for rice cultivation. To the north, temperatures go down and rainfall diminishes as well; even in the irrigated areas wheat rather than rice is the grain crop. In the Chang Basin, rice rotates with winter wheat, and in effect this is the transition zone; a line drawn from southern Shaanxi to mid-coastal Jiangsu approximates the northern limit of rice cultivation.

As Fig. 9-6 shows, the Eastern Lowland merges northward into the lower basin of the Yellow River and the North China Plain. Here spring wheat is grown in the northern areas, while winter wheat and barley are planted in the south; in the spring other crops follow the winter wheat. Millet, sorghum (kaoliang), soybeans, corn, a variety of vegetables, tobacco, and cotton are cultivated in this northern part of China Proper; on the Shandong Peninsula's higher ground, fruit orchards do well. This part of China was marked by very small parcels of land, and the communist regime has effected a major reorganization of landholding here; at the same time an enormous effort has been made to control the flood problem that has bedeviled the Huang He Basin for uncounted centuries, and to expand the land area under irrigation.

The North China Plain is one of the world's most overpopulated agricultural areas, with nearly 500 people per square kilometer of cultivated land. Here the ultimate hope of the Beijing government lay less in land redistribution than in raising yields through improved fertilization, expanded irrigation facilities, and the more intensive use of labor. A series of dams on the Huang He now reduces the flood danger, but outside the irrigated areas the ever-present problem of rainfall variability and drought recurs. The North China Plain has not produced any substantial food surplus even under normal circumstances, and when the weather is unfavorable the situation soon becomes precarious. The specter of famine has receded, but the food situation is still uncertain in this northern part of China Proper.

The layout of villages, farmlands, and market areas in Northern China affords an opportunity ot reexamine a concept introduced in Chapter 1 and also discussed in Chapter 3: *central place theory.* You will recall that Christaller developed his ideas relating to a hierarchy of market centers (central places) in southern Germany, and that one of his assumptions involved a flat, uninterrupted plain that could be traversed in all directions with equal ease: The North China Plain comes close to just such a circumstance. Over centuries of human settlement, China's rural population has become distributed among nearly 1 million farm villages, many of them in the North China Plain. These villages each have populations of several hundred, and they are surrounded by their own farm fields. Groups of these villages (averaging about 18) lie in the hinterlands of larger market towns. Each market town lies within walking distance (6 to 7 kilometers, about 4 miles) of these market towns. For centuries the market towns (Christaller's *K Places)* were the crucial places where peasant and merchant met and goods and ideas were exchanged. Here the peasant would sell the farm's produce, buy goods and services, and talk to people from other villages. The action space of the Chinese peasant always was very limited, but the market town was part of the network of still larger market centers, including, at the top of the hierarchy, China's largest cities. In the market towns the peasants were exposed to the Chinese world beyond their own villages, so that these places served as more than markets for the exchange of goods—they also functioned to integrate the Chinese nation.

The Chinese network and hierarchy of central places in particular areas strongly resembles elements of the Christaller model (page 208) as G. William Skinner describes in an article entitled "Marketing and Social Structure in Rural China" published in 1965. But even on the North China Plain there are distortions; for example, villages tend to cluster on higher ground and away from streams in order to reduce the flood danger. Still, the Chinese pattern confirms that Christaller's concepts, based as they were on observations in a totally different cultural realm, have universal application.

The two major cities of the Yellow River Basin are the historic capital, Beijing, and the port city of Tianjin, both positioned near the northern edge of the plain. In common with many of China's harbor sites, that of the river port of Tianjin

Figure 9-7

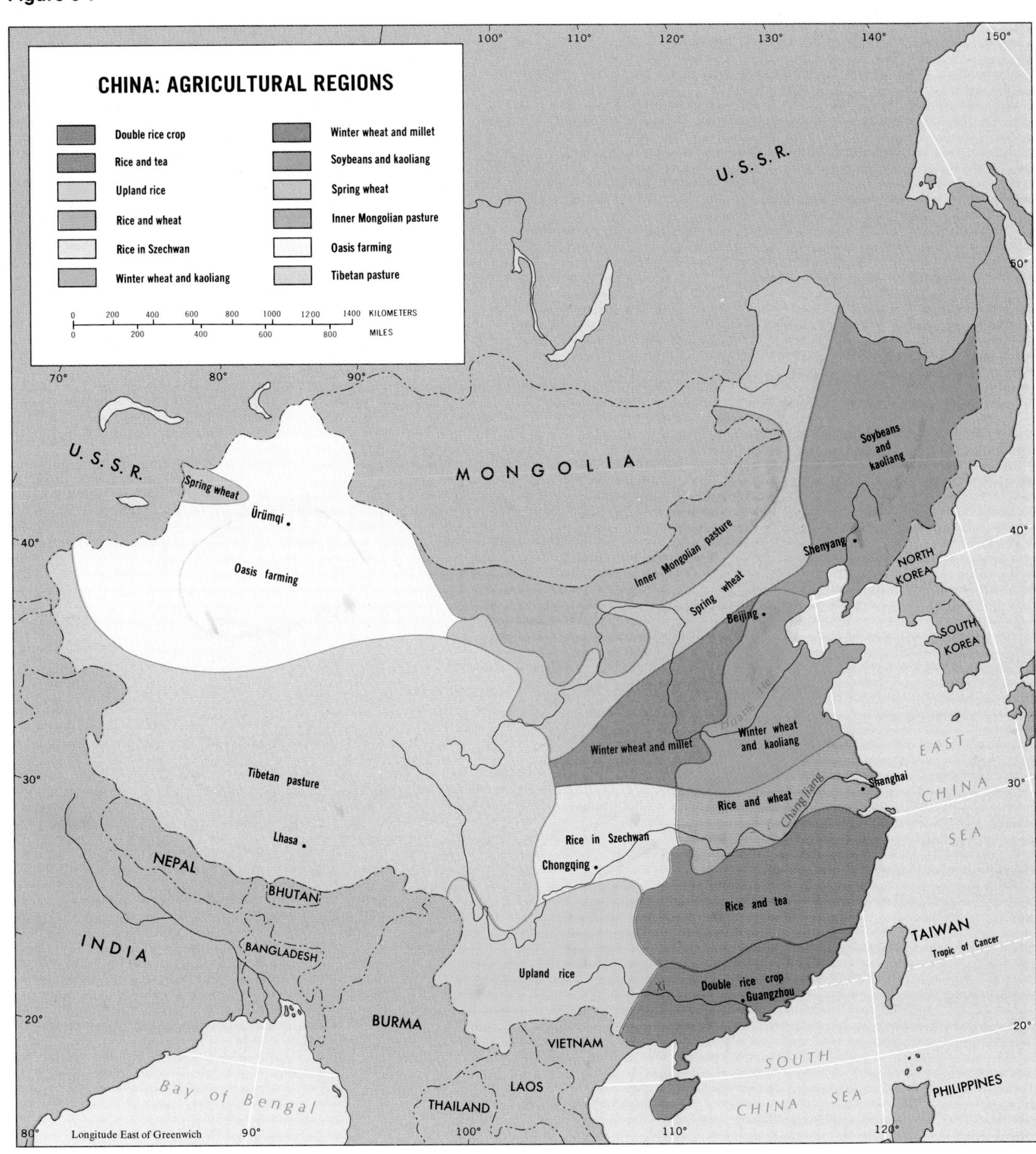

is not particularly good, but the city is well situated to the northern sector of the Eastern Lowlands, the capital (now with a population of about 9 million), the Loess Plateau to the west, and Mongolia beyond. Like Shanghai, Tianjin had its modern start as a treaty port, but the city's major growth awaited communist rule. For many decades it had remained a center for light industry and a flood-prone port, but after 1949 a new artificial port was constructed and flood canals were dug. More importantly, Tianjin was chosen as a site for major industrial development, and large investments were made in the chemical industry (in which Tianjin now leads China), in heavy industries, especially iron and steel production, in heavy machinery manufacture, and in the textile industry. Today, with a population of over 5 million, Tianjin is the center of the country's third-largest industrial complex, after the Northeast's Liaoning Province and its old competitor, Shanghai. Beijing, on the other hand, has remained mainly the political, educational, and cultural center of China; although industrial development occurred here also after the communist takeover, this has not been on a scale comparable to Tianjin. The communist administration did, however, greatly expand the municipal area of Beijing, which is not controlled by the province of Hebei (Hopeh) but is directly under the central government's authority. In one direction Beijing was enlarged all the way to the Great Wall—50 kilometers (30 miles) to the north—so that the "urban" area includes hundreds of thousands of farmers.

The upper basin of the Yellow River—and that of the Wei River—includes the Loess Plateau, an area of winter wheat and millet cultivation whose environmental problems were discussed earlier. Beneath the loess lie major coal deposits (Fig. 9-8), and already Taiyuan in Shanxi (Shansi) is the site of a large iron and steel complex, machine manufacturing plants, and chemical industries. As the map indicates, the upper Yellow River area is not especially well positioned for the transportation of raw materials, although rail connections do exist to the Northeast's source area.

Southern China is dominated by the basin of the Xi Jiang—the West River. While Northern China is the land of the ox and even the camel, Southern China uses the water buffalo. Northern China grows wheat for food and cotton for sale; Southern China grows rice and tea. Northern China is rather dry and continental in its climate; Southern China is subtropical and humid. And the people of northern China are much more clearly Mongoloid in their appearance; Northern China has long looked inward, to interior Asia, while Southern China has oriented itself outward, to the sea and even to lands beyond. With its very mixed population and multilingual character, Southern China carries strong Southeast Asian imprints (Fig. 9-9).

The map suggests that the Xi Jiang is no Chang, and this impression is soon verified. Not only is the West River much shorter, but for a great part of its course it lies in mountainous or hilly terrain. Above Wuzhou (Wuchou) the valley is less than a half mile wide and, confined as it is, the river is subject to great fluctuations in level. Except for the delta, whose 8500 square kilometers (3100 square miles) support a population of some 15 million, there is little here to compare to the lower Chang Basin. Cut by a large number of distributaries and by levee-protected flood canals, the delta of the Xi Jiang is southern China's largest area of flat land and the site of its largest population cluster; here, too, lies Guangzhou (Canton), the south's leading city.

Subtropical and moist, southern China provides a year-round growing season. In the lower areas rice is double-cropped; one planting takes place in mid- to late winter with a harvest in late June or shortly thereafter, and a second crop is then planted in the same paddy and harvested in mid-autumn. It is even possible to raise some vegetables or root crops between the rice plantings! As in Sichuan, whole areas of hillsides have been transformed into a multitude of artificial terraces, and here, too, rice is grown, although the higher areas permit the harvesting of only one crop. But, again, there is time for a vegetable crop or some other planting before the next rice is put down. Fruits, sugar cane, tea, corn—wherever farming is possible southern China is tremendously productive, and the range of produce is almost endless. If only there were more level land—and fewer people; this is one of China's food-deficient regions, always requiring imports of grain.

With its 4 million people, Guangzhou (Canton) is the urban focus for the whole region, and despite its rather narrow valley the West River is navigable for several hundred miles (Fig. 9-6), so that hinterland connections are effective. Hong

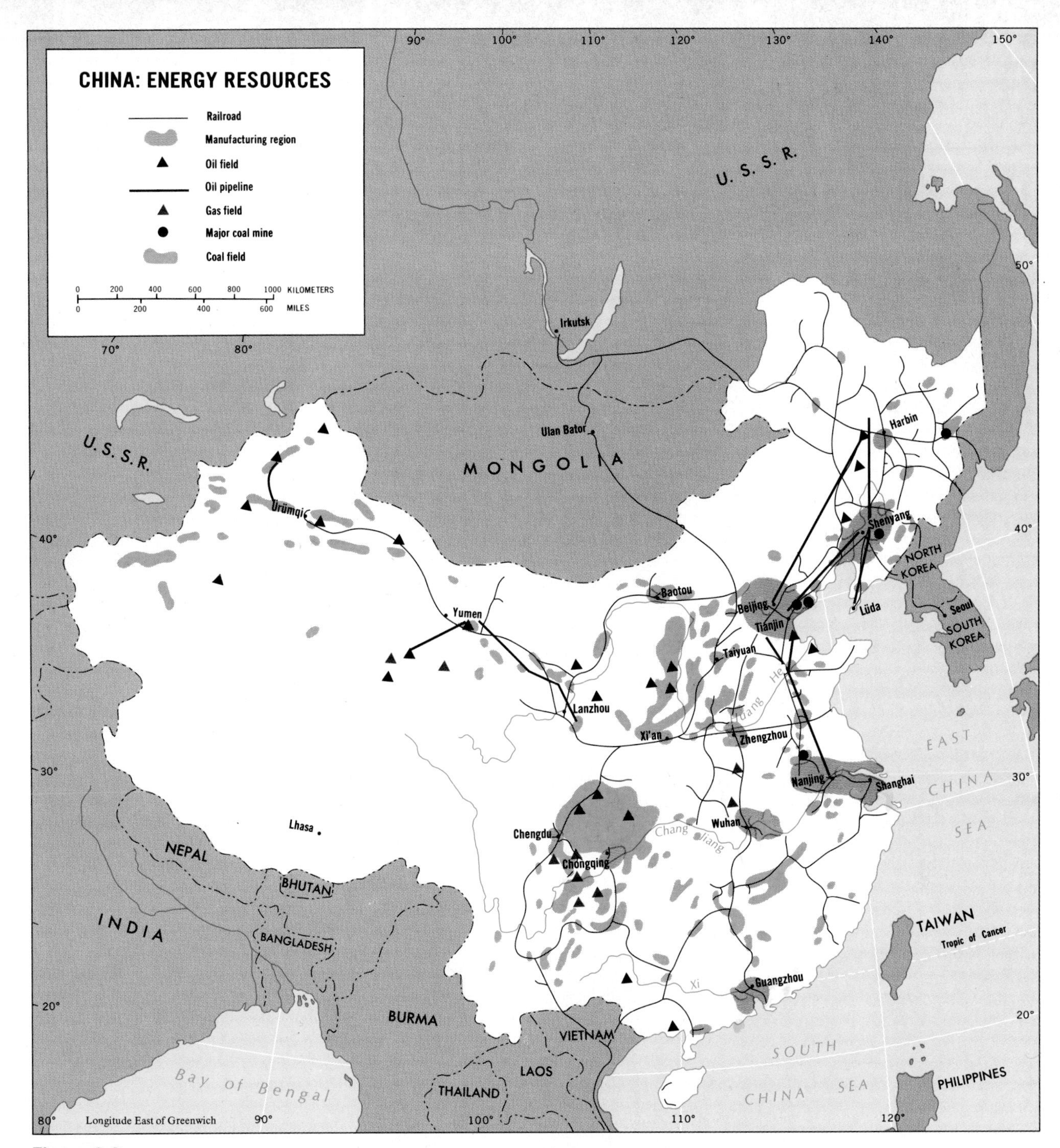

Figure 9-8

Kong, just over 160 kilometers (100 miles) away, overshadows Guangzhou in size as well as trade volume. Ghangzhou has faded as an industrial center; there was a time when its factories and production made it a competitor for Shanghai, but in modern times the paucity of natural resources in the West River basin has caused its industrial development to lag. In current development plans Ghangzhou's harbor is undergoing improvement, and the port facilities are being expanded (with an eye to the competition of British-owned Hong Kong), but industrial growth is envisaged as relating to the agricultural production of the hinterland—sugar mills, textile factories, fertilizer plants. There is a fuel problem in Southern China that has long been countered by coal imports from

Northern China, but now the hydroelectric potential of the West River basin, which is good, is being put to use.

The Northeast

China's northeastern region, home of the Manchus and embattled frontier, has become a vital part of the new Chinese world. The Manchus were the last to impose a dynastic rule on China, but since its collapse (in 1911) and after the elimination of foreign interests in the Northeast, massive Chinese immigration into the region and vigorous development of its considerable resources have combined to make this an integral and crucial part of the new China.

Northeast China has lengthy (and in places sensitive) foreign boundaries. Korea lies to the east, the Soviet Union to the east and north. Chinese Inner

Figure 9-9

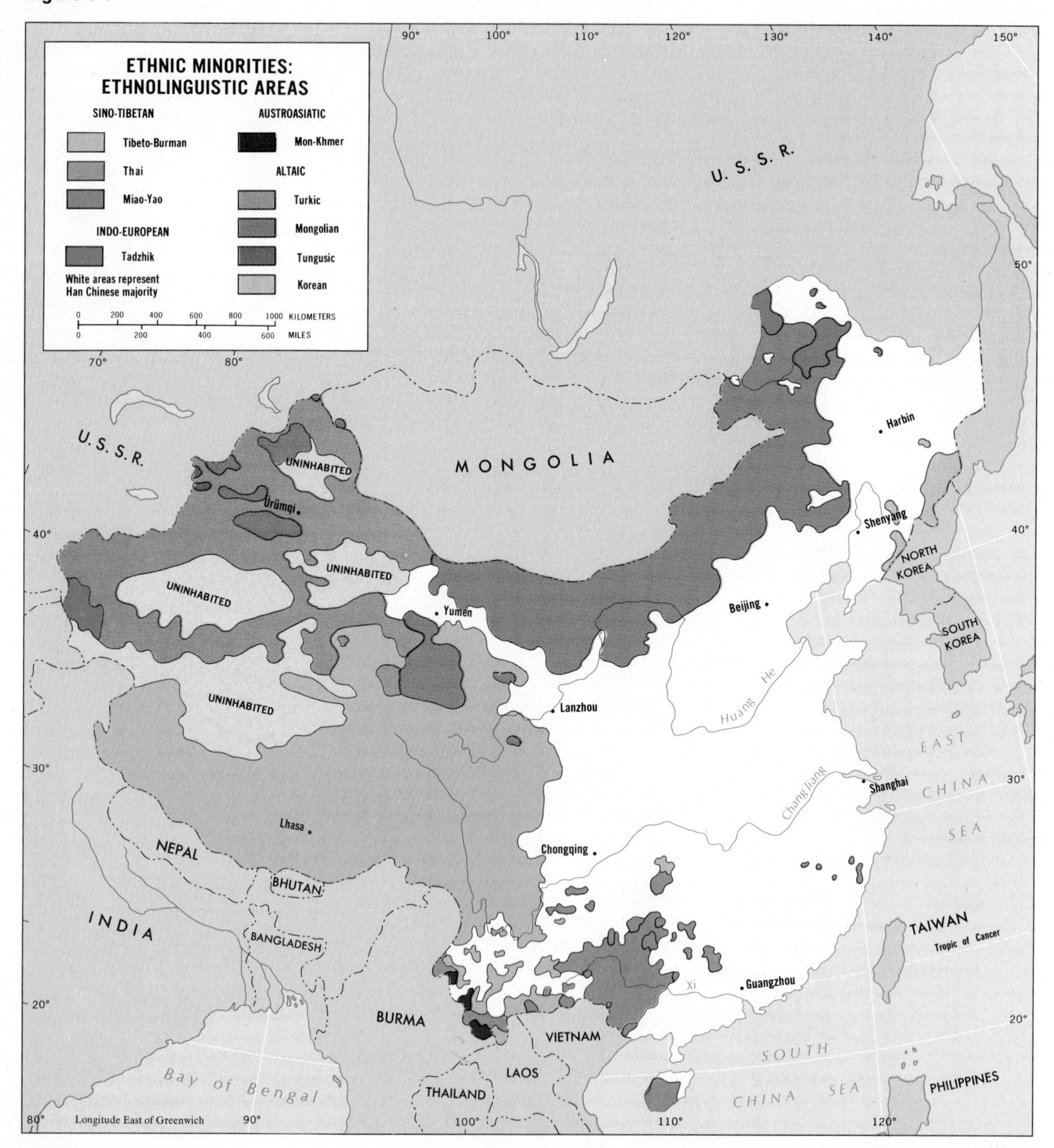

Mongolia and Soviet-sphere Mongolia lie to the west (Fig. 9-10). As the map shows, the shortest route from Vladivostok to the Soviet heartland is right across this region via Harbin, China's northernmost large city. This is the railroad in which the Soviets long had an interest, finally relinquished by treaty as late as 1950. With the increasing tension between the two countries, the Soviets have come to rely on the route skirting the Chinese border, via Khabarovsk—a trip that is 480 kilometers (300 miles) longer.

Japan also had imperial designs on the Northeast, but the aftermath of Japan's defeat in World War II was the confirmation of Chinese hegemony in the region. The Japanese contributed to the legacy by leaving behind railroads, factories, and a general framework for the region's development. The Soviets removed machinery, railroad rolling stock, and a large quantity of hoarded gold. China controlled an area of over 1.2 million people, with enormous economic possibilities. Today the three provinces of Heilongjiang, Jilin, and Liaoning contain over 90 million people. The old Manchu culture has been completely submerged.

The physical and economic layout of the Northeast is such that the areas of greatest productive capacity (and consequently largest population) lie in the south, where they form an extension of the core area of China Proper. The lowland axis formed by the basin of the Liao River and the Songhua (Sungari) Valley extends from the Liaodong Gulf north and then northeastward. Although the growing season here is a great deal shorter than in most of lowland China Proper, it is still long enough in the Liao Valley to permit the cultivation of grains such as spring wheat, kaoliang (sorghum), barley, and corn; and large areas are planted to soybeans as well. Although the Northeast has drought problems and severe winters, its southern zone has some excellent fertile soils. Toward the northern margins, extensive stands of oak and other hardwood forests on higher slopes have great value in timber-poor China.

Han Chinese have been moving into the Northeastern provinces for centuries, driven by hunger and devastating Huang He floods, dislocated by the ravages of war, and attracted by the region's oppor-

This petrochemical complex in Liao-Yang in Northeast China symbolizes the emergence of the region as China's industrial heartland. European and American companies are assisting the new China in its modernization effort; the French helped build this facility. (Andanson/Sygma)

Hong Kong

The mouth of China's southernmost major river, the Xi Jiang, opens into a wide estuary below the city of Guangzhou (Canton). At the head of this estuary (on the northeast side) lies the British colony of Hong Kong; the Portuguese dependency of Macao long occupied the southwestern side.

Hong Kong consists of three parts: the island of Hong Kong, 82 square kilometers (32 square miles), the Kowloon Peninsula on the mainland opposite this island, and, adjacent to the west, the so-called New Territories (inset, Fig. 9-6). The total area of the British dependency is only just over 1000 square kilometers (400 square miles), but the population is very large; in 1980 it had exceeded 5 million, 99 percent of them Chinese. Hong Kong Island and the Kowloon Peninsula were ceded permanently by China to Britain in 1841 and 1860 respectively, but the New Territories were leased on a 99-year basis in 1898.

With its excellent deep-water harbor, Hong Kong is the major entrepot of the western Pacific between Shanghai and Singapore. Undoubtedly the Chinese could have recaptured the territory during the successful campaign of the late 1940s, but both London and Beijing saw advantages in the status quo. During its period of isolation and

Crowded Hong Kong. (Peter Arnold)

communist reorganization, Hong Kong provided a convenient place of contact with the Western world without any need for long-distance entanglement. Even when hundreds of thousands of Chinese from Guangdong and Fujian Provinces fled to Hong Kong and the city was a place of rest and recuperation for American forces during the Indochina War, the Chinese government continued to tolerate the British colonial presence. Indeed, the colony depends on China for vital supplies, including fresh water and food. After Japan, China is Hong Kong's chief source of imports.

The colony is incredibly crowded. Hong Kong Island, where the capital of Victoria is located, is beyond the saturation point, and urban sprawl now extends onto the Kowloon Peninsula where high-rise apartments continue to go up. Until the early 1950s Hong Kong was a trading colony, but then the Korean War and the United Nations' embargo on trade with China cut the dependency's connections with its hinterland and an economic reorientation was necessary. This came about in an amazingly short time; in just a few years a huge textile industry and many other light manufacturing industries developed. Today textiles and fabrics make up over 40 percent of the exports by value (and their manufacture employs nearly half the labor force), but the volume of electrical equipment and appliances is growing. These consumer goods go to the United States, the United Kingdom, West Germany, and Japan; China is no major importer of Hong Kong's products.

Hong Kong is a hub of capitalist enterprise on China's very doorstep, a dependency on the very coast of the bastion of anti-colonialism, an outlet for Chinese emigrants, a window on a world to which China long turned its back. It is an anomaly full of potential dangers, but Hong Kong survives.

tunities of land and space. Since 1949 the settlement and development of the region has been promoted as a matter of official policy, and millions more have come to the Northeast. Not all of them came to farm the soil, because this region has become a Chinese industrial heartland. The largest city, Shenyang (Mukden on Western maps of China), has a population of 5 million and has emerged as a Chinese Pittsburgh. All this is based on large iron ore and coal deposits, which—like the best farmland—lie in the area of the Liao Basin. Shenyang lies in the middle of it: there is coal 160 kilometers (100 miles) to the west and less than 50 kilometers (30 miles) to the east of the city (at Fushun); about 100 kilometers (60 miles) to the southwest, near Anshan, there is iron ore. Since the iron ore is not very pure, the coal is hauled from Fushun (and also from northern China Proper) to the site of the iron reserves; this is the cheapest way to convert these low-grade ores to finished iron and steel. Anshan has thus become China's leading iron- and steel-producing center, but Shenyang remains the Northeast's largest and most diversified industrial city. Machine building and other engineering works in Shenyang supply the entire country with various types of equipment including drills, lathes, cable, and so forth. As the Northeast's most productive farming area, Shenyang has other functions as well; it is the leading agricultural-processing center of the Northeast.

The southern area of the Northeast, however, is not the only area where coalfields and iron ores exist. From near the

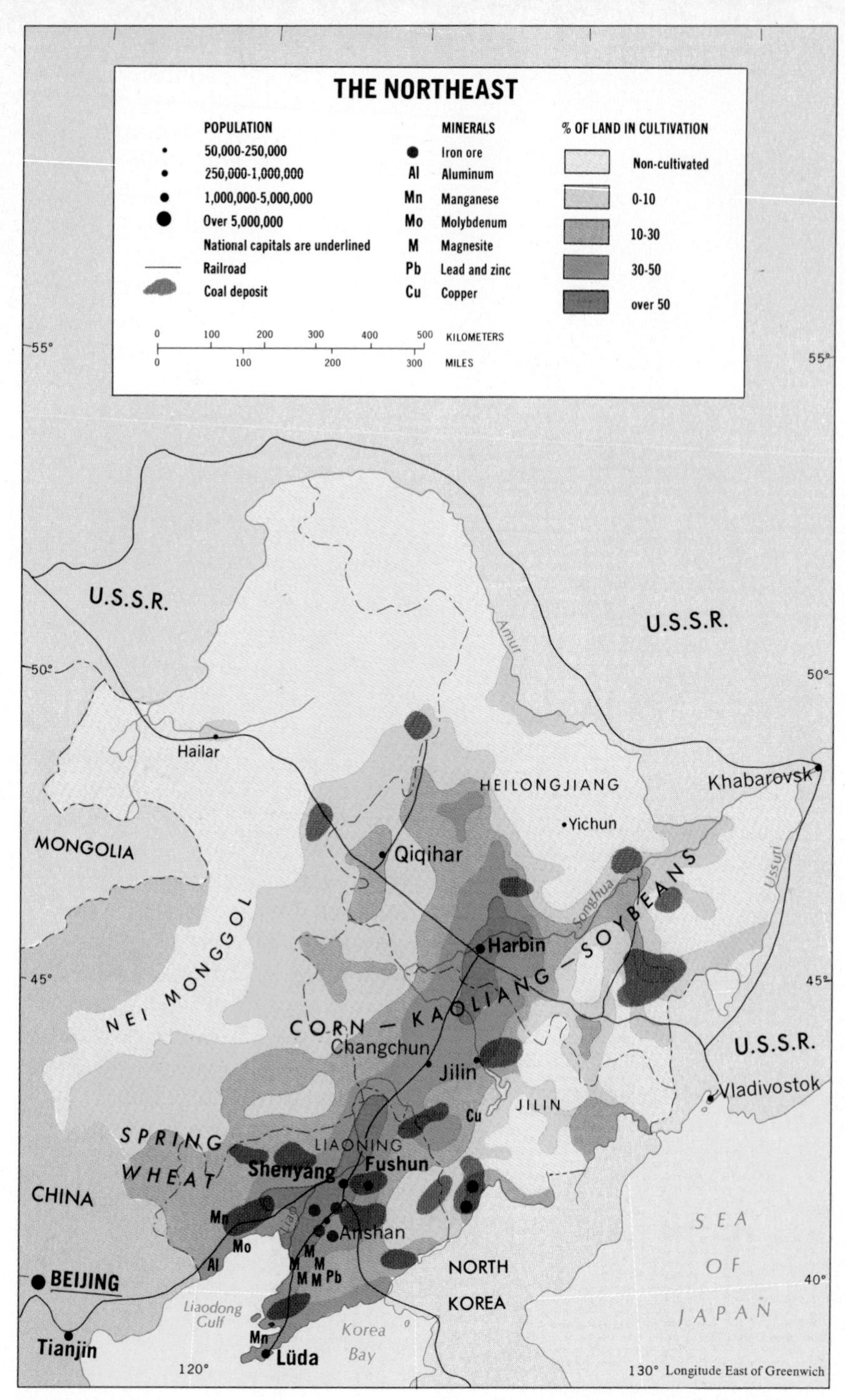

Figure 9-10

Korean border (where the iron ore is considerably more pure) to the northeast, where the coal is of best quality, the region has natural resources in abundance, and there is also aluminum ore, molybdenum, lead, zinc, and limestone, an important ingredient in the manufacture of steel. Reflecting the northward march of development in the Northeast, the railroad crossroads of Harbin has become a city of 3 million, with large machine manufacturing plants (especially agricultural equipment such as tractors), a wide range of agricultural processing factories, and such industries as leather products, nylons, and plastics. Not only does Harbin (Ha-Erh-Pin) lie at the convergence of five railroads; it also lies at the head of navigation on the Songhua (Sungari) River, connecting the city to the towns of the far northeast.

Communist doctrine in this Northeast frontier always has been somewhat relaxed; wages have always been higher than in China Proper in order to attract the skills the industries needed. In terms of planning, the rebuilt industries of Shenyang, Harbin, and Anshan in some instances are models of everything the communist regime would like them to be; schools, apartments, recreation facilities, hospitals, and even old-age homes are all part of the huge industrial plants, and having a job means access for the worker to all these. But most important, in the northeastern provinces the large-scale and efficient agricultural development stands in contrast to the parceled chaos of old China. Here the farms have been laid out more recently; they have always been larger, more effectively collectivized, and more quickly mechanized. In many ways the Northeast is the image in which Chinese planners would like to remake all China.

The New China

China at midcentury was a country torn by war almost as long as the people could remember. Foreign interference and exploitation and civil strife had been added to the problems China already faced, stemming from its huge population, its rapid population growth, the limited fertile land in the country, and the low levels of output. In the absence of any effective central government, the worst aspects of human nature could freely run their course; feudal warlords held parts of the country, village landlords mercilessly exploited their victims, corruption was rife, the cities had the world's most terrible slums where rats thrived by the millions, children were dying of starvation by the thousands, beggars were everywhere. Of course, in so large a country, there were variations. There were areas where things were not quite so bad, and elsewhere they were worse. But when the new government took control, it faced an enormous task of reconstruction, one at which the nationalist regime of the 1920s and 1930s had only begun to nibble.

In 1949, the communist government initiated a program of reform in China. The central theme of this program, initially, was the reorganization of agriculture, following the doctrines applied earlier in the Soviet Union; but conditions in China were different. In the Soviet Union, there was a distinction—in terms of productivity—between the poor peasants who formed the majority of the inhabitants of thousands of rural villages, and the subsistence farmers; when the Soviet communist government expropriated the landowning minority it thereby acquired most of the productive land. In China, on the other hand, no such clear distinction existed. There were poor peasants, rich peasants, and "middle" peasants; almost every peasant family sold some produce on the market. As in Russia there was a system of tenancy, which the new government was determined to wipe out. But in China the reorganization of agriculture was not simply a matter of expropriating the rich. Proportionally fewer peasants lived a pure subsistence existence; the risks to production of collectivization were therefore greater.

Nevertheless, the reform program was initiated almost immediately in every village of Han China (the minorities were at first excluded), involving the redistribution of the land of the landlords among all the landless families of the village. This distribution was based on the number of people per family and was done strictly according to these totals; the landlords themselves were included and received their proportionate share of what had been their own land. Those landlords who had been guilty of the worst excesses were executed; most were absorbed into the new system. Meanwhile the villagers who were not landlords but who did own the land they worked were allowed to keep their properties; hence the reform program did not produce a truly egalitarian situation.

Collectivization

During the early 1950s, the whole pattern of land ownership in China was being radically changed. In response, there was an increase in agricultural output, but not as great as the authorities had hoped. True, the taxes the new landowners had to pay on their acquired land were only on the order of one-third of what they previously had to pay their landlords, and so they had some money available to buy fertilizer and improve yields. But in the absence of any prospect of large-scale mechanization of agriculture, the Chinese leaders now began to seek a method whereby still greater productivity could be achieved. The year 1952 produced exceptionally good harvests and encouraged the government to press ahead with a mutual aid program. This program envisaged the creation of countrywide mutual-aid teams, in which all the peasants in a neighborhood were encouraged to render assistance to each other during planting and harvest seasons. In 1955 the first stage of collectivization started whereby all the peasants in a village pooled their land and their labor. Compensation would be in proportion to the size of each share and to the labor one had contributed; the farmers retained the right to withdraw their land from the cooperative if desired. In the beginning participation was slow; by early 1955 only about 15 percent of China's farmers were in these new Agricultural Producers' Cooperatives (APC). But then the government pressed for greater participation, and by coercion and often quite brutal methods compliance was raised to nearly 100 percent by 1956. In 1956 the cooperatives were being turned into collective farms, in which it was not the share of the land, but the amount of the farmer's labor that determined returns; as in Soviet collectives the farmer was able to hold on to a small plot of land for private cultivation. Thus by the end of 1956 virtually the whole countryside was organized into socialist collectives; the Chinese peasants, who had briefly held some land under the Agrarian Reform Law of 1950, lost it again to the collectives into which they had been pressed.

Geomancy

To us in the Western world, especially in the United States, intense concern for the preservation of the natural environment is something comparatively new. Environmentalists' objections to strip mining, highway construction, dam building, and other manifestations of "progress" became commonplace only in the 1960s.

In China, such concerns are almost as old as the civilization itself. The Chinese believed that powerful spirits reside in the landscape, in hills and mountains, soil and plants, water and air. These spirits (or *feng shui*) must not be angered by human endeavors, and any assault on the landscape must be done with the greatest of care. There always was much disapproval of mining, an obvious offense against *feng shui;* even the burial of the dead must be planned with circumspection. The towns and villages of China all had *geomancers,* philosopher-teachers who studied the spirits and whose help was needed when mining, building, burial, or some other significant activity was to take place. One's place in a burial mound, for example, was determined after much consideration of one's rank and status in the community, and in consultation with the spirits through the geomancers.

When Western interests penetrated China and their ruthless exploitation of its resources began, there was great resentment among Chinese citizens because of the violations of the *feng shui.* Much later the Chinese communist planners, determined to pull China out of its dated past, leveled burial mounds to enlarge farm fields and declared geomancy a dead issue. In so doing they prodded many villagers into opposition against communist modernization—a costly misjudgment of the strength of an ancient attachment to a spiritual landscape.

Communization

In 1958 the program of collectivization was carried still one step further. In an incredibly short time—less than a year—over 120 million peasant households, most of them already organized in collectives, were reorganized into about 26,500 People's Communes numbering about 20,000 people each. This was not just a paper reorganization: it was a massive modification of China's whole socioeconomic structure. This was to be China's Great Leap Forward from socialism to communism. Teams of party organizers traveled throughout the country; opposition was harshly put down while the new system was being imposed. In effect, the new communes were to be the economic, political, as well as social units of the Chinese communist state. But most drastic was the way these communes affected the daily life of the people. The adults, male and female, were organized into a hierarchy of "production teams" with military designations (sections or teams, companies, battalions, brigades). Communal quarters were built, families were disrupted through distant work assignments, children were put in boarding schools, and households were viewed as things of the past. (Later, the Chinese Communist Party rescinded this order and allowed families to stay together.) The private lands of the collective system were abolished, and even the private and personal properties of the peasants were ruled communal, although this rule was soon relaxed. The wage system of the collectives was changed in favor of an arrangement whereby the farmer or worker received free food and clothing plus a small salary—another step toward the communist ideal.

The impact of the commune system on the face of China was enormous and immediate. Workers by the thousands tackled projects such as irrigation dams and roads; fences and hedges between the lands of former collectives were torn down and the fields consolidated. Villages were leveled, others enlarged. New roads were laid out to serve the new system better. Schools to accommodate the children of parents in the workers' brigades were newly built. Each commune was given the responsibility to maintain its own budget, to make capital investments, and to pay the state a share of its income to replace the taxation formerly levied on the collectives and the individual peasants.

As might be expected, there was opposition to the introduction of the communes. Peasants in some areas destroyed their crops; elsewhere they were left to be overgrown with weeds. But there was also peasant support for the concept, for most Chinese farmers know that some form of communal organization is necessary in their heavily overcrowded country. Nevertheless, the commune system faltered—not so much because of peasant rebellion but because of two other factors, one human and one environmental. The human factor must be obvious from the preceding description of the communes' introduction: it was all done with too much haste, too little planning (although there had been pilot communes in Henan (Honan) Province and elsewhere), and too little preparation for what was truly an immense switch from private, family living to a communal existence. But more devastating was the environmental factor; China had several successive bad years from 1959 to 1961, with severe droughts and destructive floods in various sections of the country. In combination with the negative effects of such resistance as the peasants did offer, it was enough to cause a moderation of the communization program.

Cultural Revolution

It was natural that China's revolutionary leaders would attempt to transform agriculture before turning their attention to other programs of change. China's earth must produce far greater yields if China's people were to be fed; from China's land would have to come the revenues to sustain industrial and technological development; on China's farms there prevailed much of the worst corruption and exploitation in China's precommunist society. But Mao Zedong and his colleagues also wanted to rid Chinese society of its Confucian traditions and prescriptions and to substitute the principles of communism as a philosophical basis for the new Chinese order. Further, the new leaders intended to give China industrial strength based on domestic resources and generated by heavy investment in this sector.

The industrial growth of China has been spectacular, helped at first by Soviet technicians and loans and sustained by new mineral discoveries resulting from intensified exploration. The energy picture brightened as oil reserves were located in the far Northeast, Sichuan, and Xinjiang; China's coal reserves extend from the Northeast to the Chang Basin and Sichuan. China's coal production in recent years has ranked third in the world (460 million tons in 1977 as against 722 million for the Soviet Union and 611 million for the United States), and in terms of iron ore and other vital raw materials China's domestic reserves are as good as those of the United States or the Soviet Union. China's planners have sought to diversify the country's industrial base and to disperse industry into the interior (the east always has been China's industrial heartland), and to a considerable extent they have succeeded. Compared to the slow progress of agriculture, China's industrial march has been powerful.

China's first five-year plan (1952–1957) gave an indication of the aspirations of China's new leaders: more than half the state's investments were poured into industry, against only 8 percent into agriculture. Of the enormous allotment to the industrial sector, about 80 percent went to heavy industry. True, when the second five-year plan (1957–1962) faltered as a result of the problems associated with the communization program, these priorities were reconsidered and agriculture got a much larger share. But the first plan gives a good indication of Chinese priorities. A strong military establishment needs a heavy industrial base, and the Beijing government was determined to create it. Recent problems notwithstanding, those priorities still exist.

Wall posters carrying news, information, and sometimes personal accusations have constituted a crucial means of communication in China. The authorities have at times sought to terminate the practice, but the posters keep reappearing, occasionally with political consequences. (George Holton/Ocelot)

China's major difficulties in the sociopolitical arena have involved the so-called cultural revolution and the reorientation of China's population from Confucian to communist precepts. Mao Zedong, apparently fearing that the China of the 1960s and 1970s might abandon its revolutionary fervor and become "revisionist" like the modern Soviet Union, initiated a "cultural revolution" to rekindle the old enthusiasm and to recoup and conceal losses sustained during the ill-fated Great Leap Forward. The cultural revolution centered initially on education, and programs of adult education were begun in the communes. University and college teachers joined workers on assembly lines and peasants in the fields; the idea was to eliminate the old Confucian distinction between mental and manual labor and to prevent the emergence of an educated elite in the urban areas, living in comfort off the labors of the peasants. But serious problems arose. In June 1966 China's schools were closed on the grounds that the entry system was unfair to the mass of the students and the teachers perpetuated bourgeois principles. In this way, millions of young people found themselves at loose ends—ready to be recruited into the Red Guards, the organization that was to become the heart of the revolution. But just as the Great Leap Forward simply did not mobilize enough of the needed energy in China, so the cultural revolution failed to get the necessary commitment from the people. The Red Guards met with opposition, and there was even fighting between them and cadres organized to support and protect local party leaders. Once again the country was badly dislocated by a wholesale, revolutionary program, and it did not stop here. People were encouraged to criticize their superiors if they thought they might be corrupt or incompetent. Further factions were created; even the army was threatened. Violent battles occurred between workers and Red Guards and even between rival pro-Mao groups.

Since 1967 the party, as well as the country, has at times been badly divided; suspicions and accusations had been brought to the surface and could not easily be buried again. After Mao's death these divisions led to a power struggle that nearly plunged China into civil war. But worst of all, China had once again been set back by the instability the whole affair had caused—the disruption of education, the diversion of the activities of workers who should have been on their jobs, the loss of managerial personnel accused and forced out by the Maoists, the interruption of the country's train services as millions of people moved about for no other than political reasons. That China could be able to absorb the collectivization drive, the Great Leap Forward, and the Cultural Revolution—and still could show substantial progress in agriculture, industry, and several other spheres (such as, for example, the production of nuclear weapons)—is proof of the amazing capacity of this huge country and its people to overcome enormous odds.

The "Four Modernizations"

Historical geographers of the future will undoubtedly point to the late 1970s as a crucial time in the development of the modern Chinese state. With the ideological struggle between the radicals and the moderates resolved in favor of the pragmatic moderates, the new Chinese leadership could turn its attentions to the country's urgent needs in several spheres. Premier Hua Guofeng managed to enlist Deng Xiaoping, previously dismissed as too "moderate," as the country's deputy premier. Deng's pragmatic approach to China's future soon made its mark on government policies.

In a country where exhortations, slogans, and other calls for action (including huge billboards and numerous wallposters) are commonplace, Deng introduced his plea for "Four Modernizations." Briefly, these involved (1) the rapid modernization and mechanization of agriculture; (2) the immediate upgrading of the defense forces; (3) the modernization and expansion of industry; and (4) the development of science, technology, and medicine. Importantly, Deng proposed that Mao Zedong's long-prevailing policy of self-reliance be abandoned through the expansion of foreign trade, the purchase of foreign technology and machinery, and the use of foreign scientists to help China in its modernization drive. Further, capitalist-type incentives to spur production would be used, and education would return to practices where success in examinations rather than political attitudes and associations counted most.

It was not difficult to see, in this new determination to modernize, the reflection of Japan just slightly more than a century ago. Japan, too, broke its long-term isolation (see page 238) and embarked on an intensive modernizing drive—one that in a remarkably short time produced an imperial power whose colonial acquisitions encompassed much of East Asia.

China's initial target year is 2000. By that year, its planners hope, agriculture will be substantially mechanized and self-sufficiency, even in years of drought, assured; the gap between China and the Soviet Union in military capacity will no longer exist; industrialization will rank China as one of the world's three leading producers; and the temporary dependence on foreign technology and science will have ended.

These are optimistic predictions, but in 1980 the groundwork was being laid. Japanese technicians were in China planning some of 120 large industrial complexes to be operational by 1985 (according to the new regime's first 10-year plan, announced in 1978 by Premier Hua Guofeng). In return, Japan was to receive Chinese oil and coal. United States researchers were in China to help in mineral exploration; an American corporation was contracted to develop additional Chinese iron mines. China also signed a major trade agreement with the European Community. The need for foreign exchange was among the factors that led China to open its doors to hundreds of thousands of tourists. As the Maoist doctrines of the previous 30 years faded, the modernization effort accelerated.

All this must be seen against the reality of a China that, in the late 1970s, still was far from a developed country. Grain imports still were needed to supplement inadequate production at home. Annual income per person was well under $400. Urbanization hovered around 20 percent. More than 70 percent of the huge labor force worked in agriculture. Such statistics emphasize the magnitude of the effort on which the new China is embarked, and they suggest that more than two decades will be required to achieve its goals. On the other hand, there is no doubt that China has the human and material resources to bring its "four modernizations" within ultimate reach.

Taiwan

Less than 200 kilometers (120 miles) off the coast of China lies Taiwan, the island where the defeated Chiang Kai-shek and the remainder of his nationalist faction took refuge in 1949. In all, about 2 million refugees reached Taiwan following the end of World War II and during China's civil war. Assisted by the United States, Taiwan has survived as "Nationalist China" on the very doorstep of the People's Republic.

Mountainous Taiwan is one island in a huge belt of islands and archipelagos that stretches along Asia's eastern coasts from the Kurils north of Japan to Indonesia south of the Philippines. Unlike many of the other islands, Formosa (as the Portuguese called the island when they tried to colonize it) is quite compact and has rather smooth coastlines; few good natural harbors exist. In common with several of its northern and southern neighbors, the island's topography has a linear, north-south orientation and a high, mountainous backbone. In Taiwan this backbone lies in the eastern half of the island (Fig. 9-11); elevations in places exceed 3000 meters (10,000 feet). Eastward from this forested mountain backbone, the land

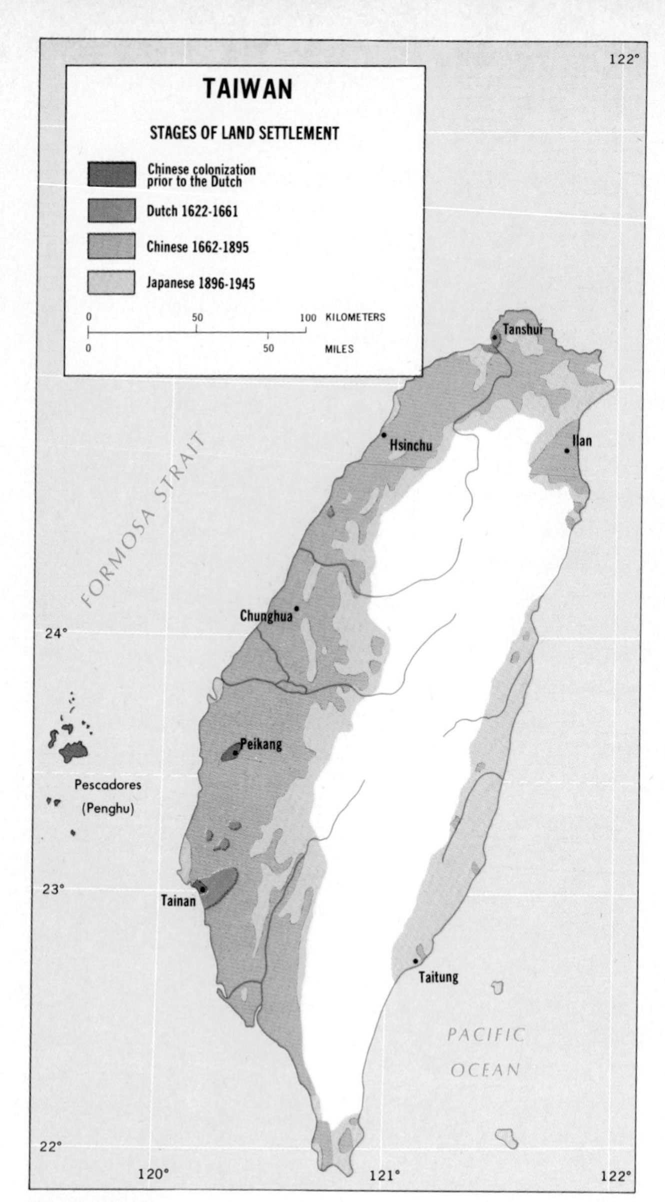

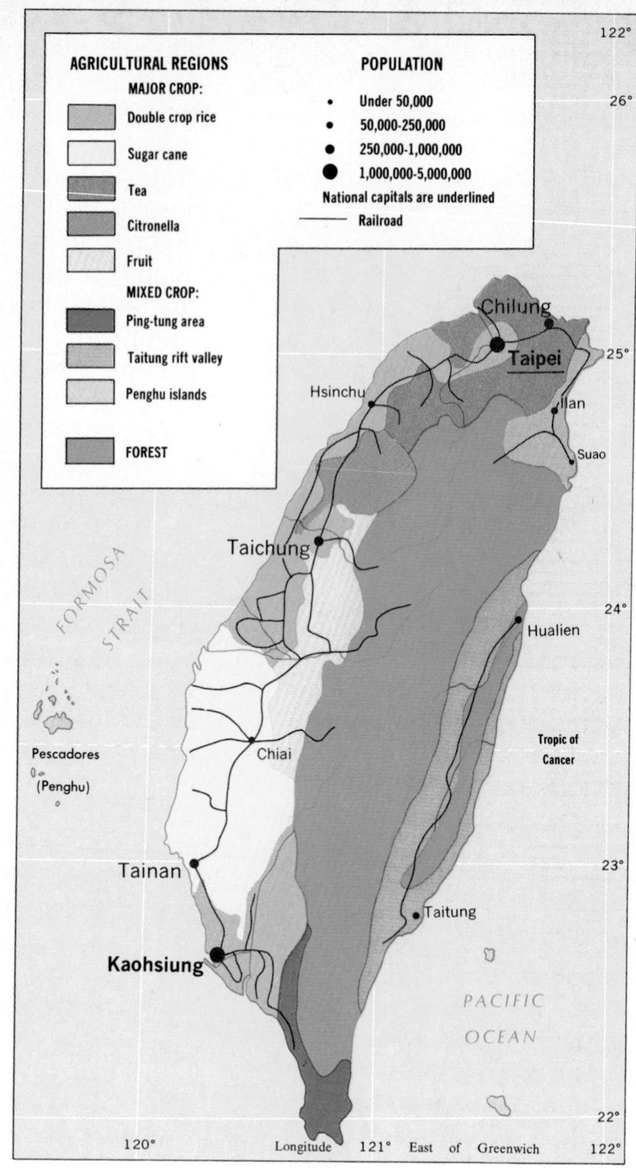

Figure 9-11

drops very rapidly to the coast and there is very little space for settlement and agriculture, but westward there is an adjacent belt of lower hills and, facing the Taiwan Strait, a substantial coastal plain. The overwhelming majority of Taiwan's 18 million inhabitants live in this western zone, near the northern end of which the capital, Taipei (2.2 million), is positioned.

There have been times when Taiwan was under the control of rulers based on the mainland, but history shows that its present, separate status is really nothing new. Known to the Chinese since at least the Sui Dynasty, it was not settled by mainlanders until the 1400s. For a time there was intermittent Chinese interest in Taiwan, and then the Portuguese arrived. During the 1600s there was conflict among the Europeans, including the Hollanders and the Spaniards, but in 1661 a Chinese general landed with his mainland army and ousted the foreign invaders. This was during the decline of the Ming Dynasty and the rise of the Manchus, and, much like the nationalists of the 1940s, hundreds of thousands of Chinese refugees fled to Taiwan rather than face Manchu domination. But before 1700 the island fell to the Manchu victors and became administratively a part of Fujian (Fukien) Province. By then the indigenous population, of Malayan stock, was already far outnumbered and in the process of retreating into the hills and mountains.

Taipei, Taiwan's capital, contrasts strongly against mainland Chinese cities of similar size. Busy traffic, ample consumer goods, and other aspects of Taiwan's capitalistic economics produce a strikingly different atmosphere. (James R. Holland/Stock, Boston)

Taiwan did not escape the fate of China itself during the second half of the nineteenth century. In 1895 it fell to Japan as a prize in the war of 1894–1895, and for the next half century it was under foreign rule. The Japanese saw more of the same possibilities in Formosa as they saw in Manchuria: the island could be a source of food and raw materials and, if developed effectively, a market for Japanese products. Thus Japan engaged in a prodigious development program in Taiwan, involving road and railroad construction, irrigation projects, hydroelectric schemes, mines (mainly for coal), and factories. The whole island was transformed; farm yields rose rapidly as the area of cultivated land was expanded and better farming methods were introduced; in the sphere of education the Japanese attacked the illiteracy problem and oriented the entire system toward their homeland.

Japanese rule ended with Japan's collapse (1945) in World War II, and briefly the island was to return to Chinese control, but before long it became the last stronghold of the nationalists and once again its mainland connections were severed. A large influx of refugee nationalists arrived, and they were fortunate; here was one of the few parts of China where they could have found a well-functioning economy, productive farmlands, the beginnings of industry, good communications—and the capacity to absorb an immigrant population of some 2 million or more. True, United States assistance made a major difference. But Taiwan had much to offer.

Nationalist "China"

From 1945 until 1949, Taiwan was officially part of China, torn as it was by the civil war. This period was one of quite ruthless and generally unpopular government on the island, which had to do with the degree to which the Japanese had transformed its life and culture, and the large residue of sympathy they had left behind. But 1949 brought a new phase with the installation of Chiang Kai-shek's regime in Taipei and the arrival of some 2 million immigrants, who constituted about 25 percent of the total population in 1950. In some ways the problems Chiang's regime faced were similar to those of the new central government. The Japanese had achieved much, but the war had brought destruction; as on the mainland there was a need for land reform and increased agricultural yields. While United States assistance helped reconstruct Taiwan's transport network and industrial plants, the Taipei government initiated a program whereby the farm tenancy system was attacked. In 1949 most Taiwanese farmers still were tenant farmers who paid rent amounting to as much as 70 percent of the annual crop. By law, the maximum allowed rent was reduced to 37.5 percent; the land of large landowners was bought by the government in exchange for stock holdings in large government-owned corporations, and this land was sold to farmers at low interest rates. Through incentives, seed improvement, additional irrigation layouts, more fertilizers, and new double-crop rotations, it has been possible to double the yields per acre of Taiwan's major crops.

These efforts to raise the volume of the grain harvest are familiar to anyone who has studied the Beijing regime's attempts to do the same on a much larger scale. Taiwan lies astride the Tropic of Cancer, and thus its lowland climate is comparable to that in the Xi Jiang Basin; rice is the leading staple, and about two-thirds of the harvest is from double-cropped land. Wheat and sweet potatoes also are important staples; sugar cane in the lower areas and tea higher up are grown for cash. But in at least one respect Taiwan faces problems even more serious than China itself. In recent decades the population's growth rate has been very high; in the 1950s and early 1960s it was 3.4 percent, one of the highest in the world. In the early 1970s it averaged 2.4 percent, still a high rate of increase, so that annual grain harvests must rise continuously just to keep pace. Recent figures show a decline to about 2 percent. The Green Revolution has helped Taiwan overcome its shortage of good farmland, but there are limits to what can be accomplished; in parts of the coastal plain, the rural population density exceeds 1500 per square kilometer (4000 per square mile).

In contrast to overwhelmingly rural China, about two-thirds of Taiwan's inhabitants live in urban areas. The Japanese began an industrialization program that temporarily faltered during the nationalist takeover, but the 1960s and 1970s have witnessed the island's industrial resurgence. In the late 1970s agriculture accounted for about one-third of Taiwan's gross national product, and income per person was over $1300 per year, statistics no other East, Southeast, or South Asian country could match (except, of course, Taiwan's model, Japan). All this has been accomplished despite the limitations of Taiwan's raw material base. Mineral resources are very limited (although power sources are ample); there is no large iron ore reserve, for example. Along the western flank of the mountain backbone there are coal deposits, and numerous streams on this well-watered isle provide good opportunities for hydroelectric power development. There even is some petroleum and natural gas. But the textile industry, Taiwan's major foreign-revenue earner, depends on imported raw cotton; the aluminum industry gets its bauxite from In-

Taipei is the capital city of Taiwan and the island's chief industrial center. Taiwan's economic strength derives in part from its industrial successes; shown here is one of numerous large-scale textile factories. (J.P. Laffont/Sygma)

donesia; and ores and minerals rank high on the list of imports. On the other hand, the Taiwanese in 1976 concluded an agreement with Saudi Arabia whereby Chinese technicians would help develop the Arabian economy in return for large loans (and oil shipments) to support Taiwan's vigorous industrialization program. In recent years Taiwan's planners have moved toward the establishment of a domestic iron and steel industry, nuclear power plants, shipyards, a larger chemical industry, and an electrified and more comprehensive railroad network.

Figure 9-12

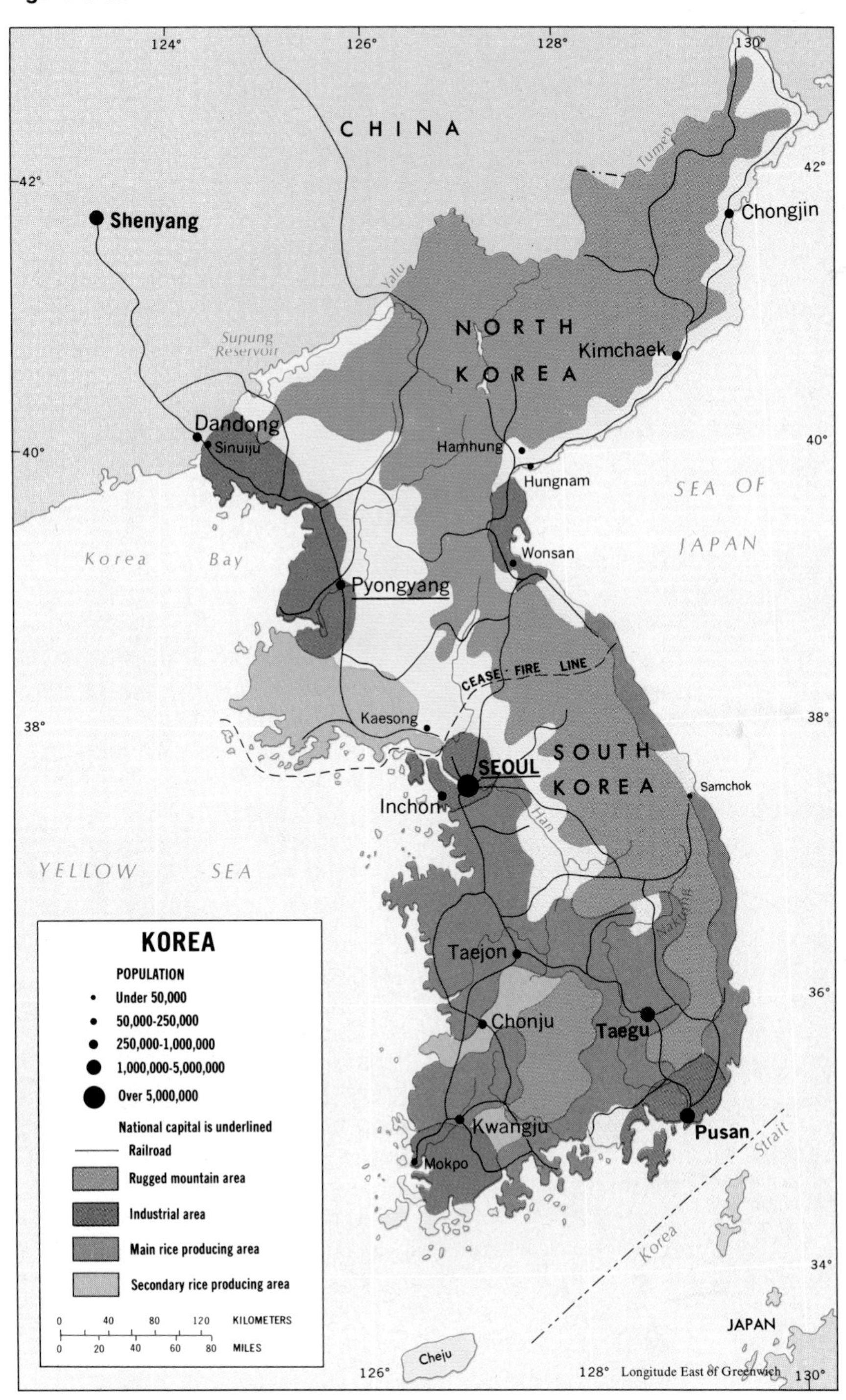

In the international arena, Taiwan's position has become more difficult as a result of China's emergence from its isolation. Taiwan was expelled from the United Nations to make way for China; the United States, Taiwan's vital ally, closed its embassy and since 1979 has been under pressure to lower its commitments to Taipei in favor of more productive relationships with Beijing. In 1976 the Canadian government refused to permit Taiwan to participate in the Montreal Olympic Games. But Taiwan remains in effect a United States protectorate, a bastion of support for the American presence in East Asia. A military alliance between the United States and Taiwan was signed in 1954, during the Eisenhower administration. In any case, mainland China has opposed a policy whereby the United States would recognize both Beijing and Taipei, so that a two-China policy is no solution to this persistent problem.

Korea

The peninsula of Korea reaches from the East Asian mainland toward Japan. It has had a turbulent history. For uncounted centuries it has been a marchland, a pawn in the struggles among the three powerful countries that surround it (Fig. 9-12). Korea has been a dependency of China and a colony of Japan; when it was freed of Japan's oppressive rule in 1945, it was divided for administrative purposes by the victorious powers. This division gave North Korea beyond the 38th parallel to the Soviet Union forces, and South Korea to the United States forces. In effect, Korea traded one master for two others. The country never was reunited, as North Korea fell under the communist ideological sphere and South Korea, with massive American aid, became part of Asia's noncommunist perimeter of which Japan and Taiwan are also parts. Once again it was the will of external powers that prevailed over the desires of the Korean people themselves. In 1950 North Korea sought to reunite the country by force and invaded the South across the 38th parallel, an attack that drew a United Nations response led by the United States. This was the beginning of a devastating conflict in which North Korea's forces pushed far to the south, only to be driven back into their own half of Korea nearly to the Chinese border, whereupon Chinese armies entered the war to drive the United Nations forces southward again. Eventually a cease-fire was arranged in mid-1953, but not before the people and land of Korea had been ravaged in a way that was unprecedented even in its violent past.

The Two Koreas

The fragmentation of Korea into two political units happens, to a considerable extent, to coincide with the regional division any physical geographer might suggest as an initial breakdown of the country. As Fig. I-5 shows, South Korea is moister than North; Fig. I-6 emphasizes the temperate maritime climatic conditions that prevail over most of the South as opposed to the more continental, "snow" character of the North. Fig. I-8 shows much of Korea as a whole to lie under mountain soils, but the most extensive belt of more productive soils is in a zone that lies largely in South Korea. Fig. I-7 proves South Korea to possess a sizable area of broadleaf evergreen trees like that of Southern Japan; the remainder of the country has a deciduous forest as its natural vegetation. And Fig. I-9 shows that the majority of Korea's 57 million people live in a zone in the western part of the country, a zone that widens southward so that the great majority of the population is in South Korea, in a triangle roughly to the west of a line drawn from Seoul to Pusan.

Thus a great number of contrasts exist between North and South Korea; the North is continental, the South peninsular; the North is more mountainous than the South; the North can grow only one crop annually and depends on wheat and millet while in much of the South multiple cropping is possible and the staple is rice; the North has significantly fewer people than the South (18 million against 39 million), but the North has a large food deficit while the South comes close to feeding itself (it has had surpluses in the past). But perhaps the most striking contrast lies in the distribution of Korea's raw materials for industry. North Korea has always produced vastly more coal and iron ore than the South, and the overwhelming majority of all other Korean production also comes from the North. Similarly, North Korea has maintained its great lead in hydroelectric power development, an advantage that was initiated by the Japanese and has been sustained. In recent years several discoveries of coal and iron have been made in South Korea, but the overall balance relating to the bases for heavy industry remains strongly in favor of North Korea.

Although it is practically impossible to obtain data concerning North Korea's external trade, one thing is certain: the superimposed political boundary (can there be a better example?) between the two Koreas has been virtually watertight, and, in effect, no trade has passed across it. North Korea's trade connections have been with China and the Soviet Union, those of South Korea with the United States, Japan, and West Germany. Yet it must be clear from what has just been said that North Korea, the focus of heavy industry in Korea as a whole, could do a great deal of business with South Korea, the more agriculturally productive region of the country. The two Koreas are interdependent in so many ways; even within the industrial sector itself there are complementarities. While North Korea specializes in heavy manufacturing, light industries (cotton textiles, food processing), still prevail in South Korea, although heavy industries are now developing, especially around Pusan, since the discovery of iron ores and of Samchok's coal (Fig. 9-12). North Korea's chemical industries produce fertilizer; South Korea needs it. South Korea long exported food to what is today North Korea. North Korea has much electric power; the transmission lines that used to carry it to the South were cut soon after the postwar division

The demarcated boundary between North and South: symbol of divided Korea. (Chauvel/Sygma)

Seoul, capital of South Korea and largest population cluster in the Koreas. (Dennis Stock/Magnum)

of the country. South Korea has the largest part of the domestic market as well as the largest cities, the old capital of Seoul (7 million) and the southern metropolis of Pusan (2.5 million). Pyongyang, the North Korean headquarters, is just approaching 2 million. Seoul is the country's primate city, whose location in the waist of the peninsula, midway between the industrialized northern and agriculturally productive southern regions of Korea, has been an advantage. Pusan is closest to Japan and grew in phases during the twentieth century, first under Japanese stimulus and later as the chief United States entry point. Pyongyang was also developed by the Japanese, and it lies at the center of Korea's leading industrial region, which is merging with the mining-industrial area of the northwest.

The Koreans, like the Vietnamese, are one people, with common ways of life, religious beliefs, historic and emotional ties, and with a common language. In the entire twentieth century the Koreans have barely known what self-determination in their own country would be like, but they have not forgotten their aspirations. After seemingly endless suffering through conflicts and wars, most of them precipitated by the "national" interests of other states, Korea today, ideologically and politically divided as it is, still retains the ingredients of a cohesive national entity. In 1972 a first effort was begun to thaw the hostilities between North and South Korea through the reopening of trade connections, but this attempt failed. In the late 1970s, however, there were signs that a new initiative was in the making.

Perhaps this period, with great power "spheres of influence" in a state of flux, will witness a unified, progressive Korea functioning as a cultural and economic whole.

Additional Reading

China, like India, is discussed in detail in several geographies that deal with Asia as a whole or with all of non-Soviet Asia. The volumes by Spencer and Thomas, Ginsburg, and Myrdal have been quoted on page 466. A. Kolb's *East Asia* was published by Methuen in London in 1971. An especially interesting treatment of China is Yi-Fu Tuan's *China* in the Aldine World's Landscapes Series, published in Chicago in 1969. R. R. C. de Crespigny published *China: The Land and its People* through St. Martin's Press, New York, in 1971. An argumentative but nevertheless readable volume is B. Appel's *Why the Chinese Are the Way They Are,* published by Little Brown in Boston in 1973 in a revised edition. J. Robottom wrote *Twentieth-Century China* for Putnam, published in New York in 1971. An older but still worthwhile book is L. Fessler's *China,* a Time-Life publication of 1963. K. Buchanan's book on the *Transformation of the Chinese Earth* was published by Bell in London in 1970. Also see T. Shabad, *China's Changing Map: National and Regional Development, 1949–1971,* a sourcebook published in New York by Praeger in 1972. A somewhat dated but basic geography is T. R. Tregear's *A Geography of China,* published by Aldine in Chicago in 1965. V. P. Petrov's volume on *China: Emerging World Power* is part of the Van Nostrand Searchlight Series. Another volume in this series is by Chiao-min Hsieh, *China: Ageless Land and Countless People,* published in 1967.

Despite its alleged shortcomings, Edgar Snow's *Other Side of the River—Red China Today,* published in 1962 by Random House in New York, remains one of the most challenging and informative works currently available. Also see the collection of essays edited by Ruth Adams, *Contemporary China,* a Vintage paperback published by Random House in 1966.

On the interaction between China and its Asia neighbors, see A. Lamb, *The China-India Border: The Origins of the Disputed Boundaries,* Oxford University Press, 1964, published in New York; O. Lattimore, *Pivot of Asia: Sinkiang and the Inner Asian Frontiers of China and Russia,* published in Boston by Little Brown in 1950; and H. J. Wiens, *China's March Toward the Tropics,* published by Shoe String Press, Hamden, in 1954; G. S. Murphy, *Soviet Mongolia: A Study of the Oldest Political Satellite,* published by the University of California Press, Berkeley, in 1966; and C. P. Fitzgerald's *Chinese View of Their Place in the World,* published by Oxford University Press in New York in 1964.

Historical perspectives are obtained by consulting Albert Herrmann's *Historical Atlas of China,* edited by N. Ginsburg and published by Aldine, Chicago, in 1966, and through the book by K. S. Latourette, *The Chinese: Their History and Culture,* the fourth edition of which was published in 1964 by Macmillan, in New York. Also see O. and E. Lattimore, *China, a Short History,* published in New York by Norton in 1947. Another useful atlas is Chiao-min Hsieh, *Atlas of China,* published in New York in 1973 by McGraw-Hill. *China in Maps* was produced by Denoyer-Geppert, Chicago, in 1968, edited by H. Fullard. It contains an excellent, clear series of historical maps.

On Korea, a standard work remains S. McCune, *Korea's Heritage: a Regional and Social Geography,* published by Tuttle in Rutland in 1956. Also see McCune's *Korea: Land of Broken Calm,* published by Van Nostrand, Princeton, in 1966. A more recent volume that deals with South Korea only was written by P. M. Bartz, *South Korea,* a Clarendon Press publication, Oxford, 1972.

G. W. Skinner's two-part article on "Marketing and Social Structure in Rural China" appeared in the *Journal of Asian Studies,* Volume 24, Numbers 1 (1964) and 2 (1965).

Crowded waterfronts, teeming cities: Djakarta personifies Southeast Asia's cultural pluralism. (Georg Gerster/Rapho-Photo Researchers)

Ten

Southeast Asia: Between the Giants

IDEAS AND CONCEPTS

- Spatial morphology
- Territoriality
- Insurgent states
- Domino theory
- Maritime boundaries
- Economic zones
- World lake concept

Southeast Asia, Asia's southeastern corner of peninsulas and islands, is bounded by India to the northwest and by China to the northeast; its western coasts are washed by the Indian Ocean and to the east stretches the vast Pacific. From all these directions, Southeast Asia has been penetrated by outside forces. From India came traders and from China came settlers. Across the Indian Ocean came the Arabs to engage in commerce and the Europeans in pursuit of empires. And from across the Pacific came the Americans. Southeast Asia has been the scene of countless contests for power and primacy; the competitors have come from near and far. Like Eastern Europe, Southeast Asia is a region of great cultural diversity; again like Eastern Europe, it is also a shatter belt.

Southeast Asia is nowhere nearly as well-defined a culture realm as India or China. It does not have a single, dominant core area of indigenous development. Culturally it is diverse in every respect; there are numerous ethnic and linguistic groups, various religions, and different economies. Even spatially the region's discontinuity is obvious; it consists of a peninsular mainland section and hundreds of islands that form the archipelagos of Indonesia and the Philippines. The mainland area, smaller than India, is divided into seven political entities (Fig. 10-1). This underscores not only the cultural complexity of this realm, but also the historical geography of European intervention. Except Thailand, none of the states on the map of Southeast Asia today was independent when the Second World War came to an end.

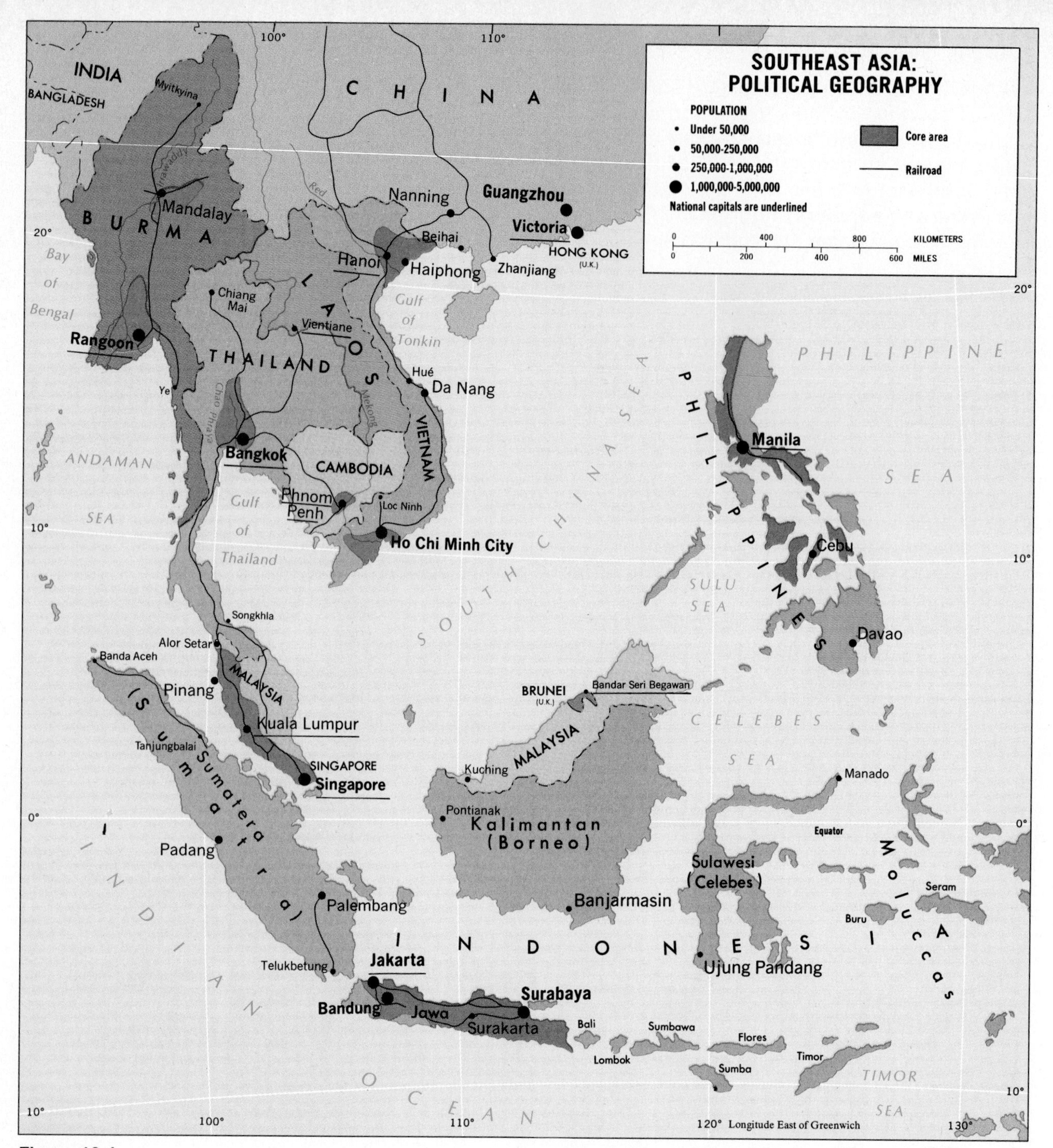

SOUTHEAST ASIA: POLITICAL GEOGRAPHY
POPULATION
Under 50,000
50,000-250,000
250,000-1,000,000
1,000,000-5,000,000
National capitals are underlined
Core area
Railroad
KILOMETERS
MILES
INDIA
BANGLADESH
CHINA
BURMA
LAOS
THAILAND
CAMBODIA
VIETNAM
MALAYSIA
BRUNEI (U.K.)
SINGAPORE
PHILIPPINES
INDONESIA
HONG KONG (U.K.)
Myitkyina
Mandalay
Rangoon
Ye
Chiang Mai
Vientiane
Hanoi
Haiphong
Nanning
Guangzhou
Victoria
Beihai
Zhanjiang
Hué
Da Nang
Bangkok
Phnom Penh
Loc Ninh
Ho Chi Minh City
Songkhla
Alor Setar
Banda Aceh
Pinang
Kuala Lumpur
Tanjungbalai
Singapore
Kuching
Bandar Seri Begawan
Pontianak
Kalimantan (Borneo)
Banjarmasin
Sumatera
Padang
Palembang
Telukbetung
Jakarta
Bandung
Jawa
Surakarta
Surabaya
Bali
Lombok
Sumbawa
Sumba
Flores
Timor
Sulawesi (Celebes)
Ujung Pandang
Manado
Moluccas
Buru
Seram
Manila
Cebu
Davao
Bay of Bengal
ANDAMAN SEA
Gulf of Thailand
Gulf of Tonkin
SOUTH CHINA SEA
SULU SEA
CELEBES SEA
PHILIPPINE SEA
TIMOR SEA
INDIAN OCEAN
Red
Irrawaddy
Chao Phraya
Mekong
Equator
120° Longitude East of Greenwich

Figure 10-1

Ten Major Geographic Qualities of Southeast Asia

1. The Southeast Asian realm is fragmented into numerous peninsulas and islands.
2. The regional boundaries of Southeast Asia are problematic. Transitions occur into Indian, Chinese, and Pacific realms.
3. Southeast Asia, like Eastern Europe, exhibits shatter-belt characteristics. Pressures upon this realm from external sources have always been strong.
4. Compared to neighboring regions, mainland Southeast Asia's physiologic population densities remain relatively low.
5. Population in Southeast Asia tends to be strongly clustered, even in the rural areas.
6. Southeast Asia exhibits intense cultural fragmentation, reflected by complex linguistic and religious geographies.
7. Southeast Asia's politico-geographical traditions involve frequent instability, fragmentation, and conflict.
8. Intraregional communications in Southeast Asia remain poor. External connections often are more effective than internal ones.
9. The legacies of powerful foreign influences (Asian as well as non-Asian) continue to mark cultural landscapes in Southeast Asia.
10. Explosive population growth has prevailed in the island regions of Southeast Asia, mainly in Indonesia, during the middle part of the twentieth century.

Population Patterns

Compared to the population numbers and densities in habitable regions of India and China, figures for the countries of Southeast Asia seem almost to suggest a sparseness of people (see Table I). Laos, territorially quite a large country, has a population under 4 million and an average population density that is less than desert-dominated Iran. Much of interior mainland Southeast Asia has population densities similar to those of savanna Africa, the margins of South America's Amazon Basin, and Soviet Central Asia (Fig. I-9). Even the more densely inhabited coastal areas of Southeast Asia have fewer people and smaller agglomerations than elsewhere in South and East Asia. There is nothing in this realm to compare to the immense human clusters in the Ganges (Ganga) Basin or the Huang He and Chang Jiang (Yangtze) Lowlands. The whole pattern is different: Southeast Asia's few dense population clusters are relatively small and lie separated from each other by areas of much sparser occupancy.

Why the difference? When there is such population pressure and such land shortage in adjacent regions, why has Southeast Asia not been flooded by waves of immigrants? Several factors have combined to inhibit large-scale invasions. In the first place, the overland routes into Southeast Asia are not open and unobstructed. In discussing the Indian subcontinent, we noted the

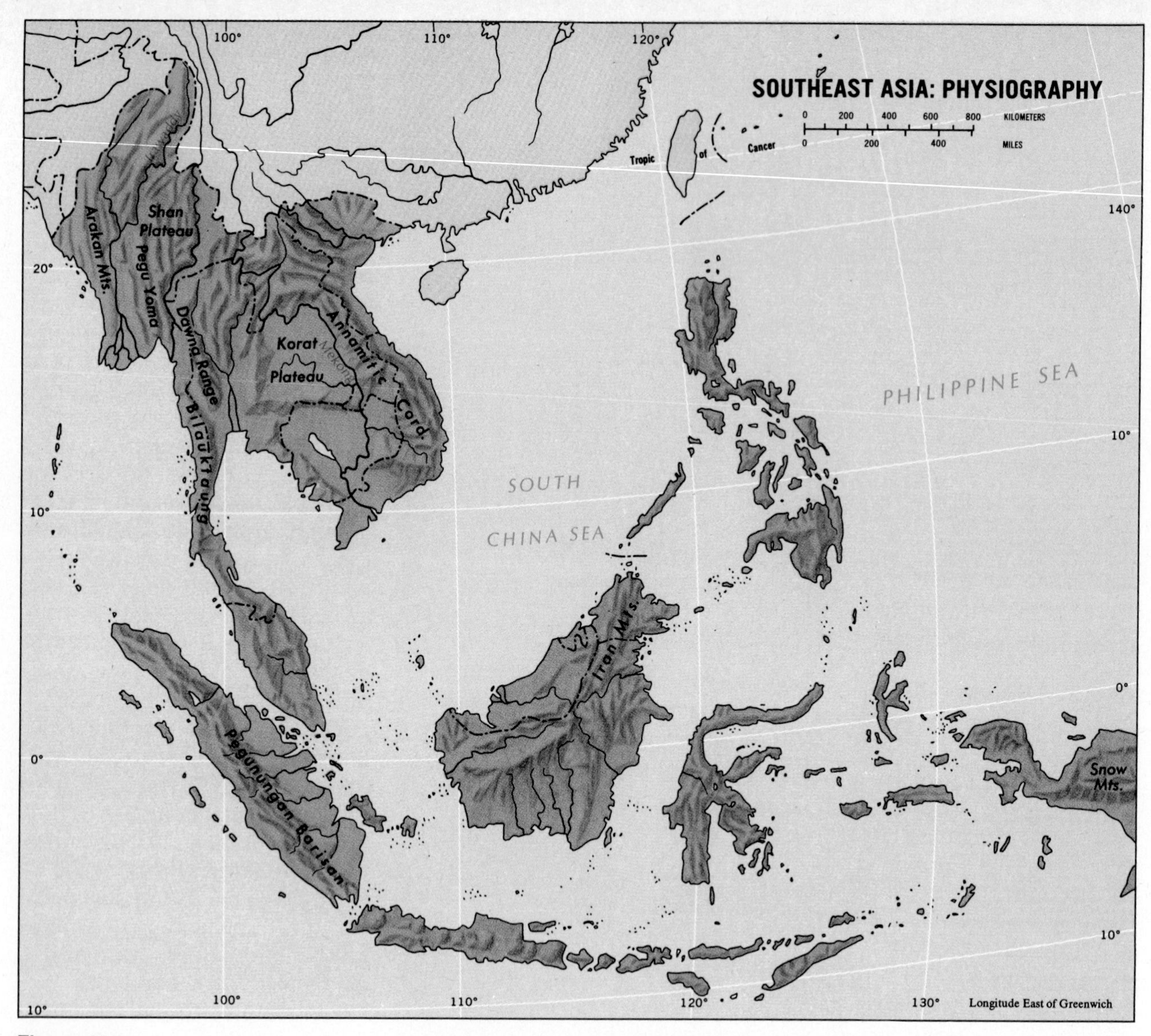

Figure 10-2

barrier effect of the densely forested hills and mountains that coincide with the border between Indian Assam and western Burma. North of Burma lies forbidding Tibet, and northeast of Burma and north of Laos is the high Yunnan Plateau. Transit is easier into northern Vietnam from southeastern China, and along this avenue considerable contact and migration has indeed occurred. Neither is contact within Southeast Asia itself helped by the rugged, somewhat parallel ridges that hinder easy east-west communications between the fertile valleys within the region (Fig. 10-2). Second, and the population map reflects this, Southeast Asia is not an area of limitless agricultural possibilities and opportunities. Much of the region is covered by dense, tropical monsoon forest and rain forest, parts of an ecological complex whose effect on human settlement we have previously observed in other areas of the world. Except in certain locales, the soils of mainland Southeast Asia are excessively leached by the generally heavy rains. In the areas of monsoonal regimes and savanna climate (the latter prevails over much of the interior) there is a dry season, but the limitations imposed by savanna conditions on agriculture are all too familiar. High evapotranspiration rates, long droughts, hard-baked soils, slight fertility, high runoff, and erosional problems add up to anything but a peasant's paradise in these parts of Southeast Asia.

The Clusters

Population in Southeast Asia has become concentrated in three kinds of natural environments. First, there are the valleys and deltas of Southeast Asia's major rivers, where alluvial soils have been formed. Four major rivers stand out. In Burma, the Irawaddy rises near the border with Tibet and creates a delta in the Bay of Bengal. In Thailand, the Chao Phraya traverses the length of the country and flows into the Gulf of Siam. In southern Vietnam lies the extensive delta of the great Mekong River, which rises in the high mountains on the Qinghai-Xizang (Tibet) border in interior China and crosses the whole peninsula. And in northern Vietnam lies the Red River, whose lowland probably is the most densely settled area in Southeast Asia (the Tonkin Plain). Each of these four river basins contains one of mainland Southeast Asia's major population clusters—in effect the core areas of those four countries.

Second, Asia is known for its volcanic mountains—at least in the archipelagos. In certain parts of the islands the conditions are all right for the formation of deep, dark, rich volcanic soils, especially in much of Jawa (Java).[1] The population map indicates what this fertility means; the island of Jawa is one of the world's most densely populated and most intensively cultivated areas of the world.

[1]As in Africa, names and spellings have changed with independence. In this chapter, the modern spellings will be used, except when reference is made to the colonial period. Thus Indonesia's four major islands are Jawa, Sumatera, Kalimantan (the Indonesian part of Borneo), and Sulawesi. The Dutch called them Java, Sumatra, Borneo, and Celebes.

On Jawa's productive land live about 90 million people—nearly two-thirds of the inhabitants of all the islands of Indonesia and more than a quarter of the population of the whole Southeast Asian realm.

Another look at Fig. I-9 indicates that one area remains to be accounted for: the belt of comparatively dense population that extends along the western coast of the Malay Peninsula, apparently unrelated to either alluvial or volcanic soils. This represents a third basis for population agglomeration in Southeast Asia—the plantation economy. Actually, plantations were introduced by the European colonizers throughout most of insular Southeast Asia, but nowhere did they so transform the whole economic geography as in Malaya. Rubber trees were planted on tens of thousands of acres, and Malaya became the world's leading exporter of this product. Undoubtedly Malaysia would not have developed so populous a core area without its plantation economy.

Where there are no alluvial soils, no volcanic soils, and no profitable plantations, as in the rain-forest areas and in the higher areas where there is more slope and not very much level land, far fewer people manage to make a living. Here the practice of shifting, subsistence cultivation prevails, augmented sometimes by hunting, fishing, and gathering wild nuts, berries, and the like. This practice is similar to that we observed in South America and Equatorial Africa, and it is capable of sustaining a rather sparse population; land that has once been cleared for cultivation must be left alone for years to regenerate. Here in the forests and uplands of Southeast Asia the nonsedentary cultivators still live in considerable isolation from the peoples in the core areas; the forests are dense and difficult to penetrate, distances are great, and the strong relief adds to the obstacles in the establishment of surface communications. Over a major part of their combined length, the political boundaries of Southeast Asia lie through these rather sparsely peopled, inland areas. Here are the roots of some of the region's political troubles; these interior, unstable zones never have been effectively integrated in the states of which they are part. The people who live in the forested hills may not be well disposed toward those who occupy the respective core areas, and so these frontierlike inner reaches of Southeast Asia, with their protective isolation, distance from the seats of power, and dense forest cover, are fertile ground for revolutionary activities if not for farm crops.

Indochina

The French colonialists called their Southeast Asian possessions *Indochina,* and the name is appropriate to the whole mainland region, suggesting the two main Asian influences that have affected the region for the past 2000 years. Overland immigration was mainly southward from Southern China, resulting from the expansion of the Chinese empire. The Indians came by way of the seas, as Indian trading ships plied the coasts and Indian settlers founded colonies on Southeast Asian coasts in the Malay Peninsula, the Mekong region, on Jawa, and on Borneo.

Malaya and Malaysia

The historical and political geography of Malaysia is complicated. The former British dependency of Malaya (the Malayan peninsula excluding Singapore) attained independence in 1957, and Singapore achieved this in stages culminating in 1959. When independence came to Malaya, the state was named the Federation of Malaya. Then Singapore joined Malaya in 1963, and the new name *Malaysia* came into use to designate the new state, which also included the former British dependencies on the Indonesian island of Borneo, Sarawak and Sabah.

In 1965 the union between Malaysia and Singapore was terminated, and Singapore became an independent city-state. The administrative structure of Malaysia was modified with the establishment of West Malaya (the peninsula) and East Malaysia (on Borneo). In 1972 still another change was introduced: West Malaysia now became Peninsular Malaysia, and East Malaysia once again was called Sarawak and Sabah.

Thus the term *Malaya* properly refers to the geographic area of the Malayan peninsula, including Singapore and other nearby islands. The term *Malaysia* identifies the politico-geographical unit of which Kuala Lumpur is the capital city.

With the Indians came their faiths: first Hinduism and Buddhism, later Islam. The Moslem religion was also promoted by the growing number of Arab traders who appeared on the scene, and Islam became the dominant religion in Indonesia, where nearly 90 percent of the population adheres to it today. But in Burma, Thailand, and Cambodia Buddhism remained supreme; in all three countries over 85 percent of the people are adherents of Buddhism. In Malaysia, the Malays are Moslems (to be a Malay is to be a Moslem), and almost all Chinese are Buddhists, but Hinduism retained its early strength and still today over 70 percent of the country's 1.3 million South Asians live in accordance with Hindu principles.

This, then, represents the *Indo* part of Indochina: the Buddhist and Hindu faiths (Ceylonese merchants played a big role in the introduction of the former), Indian architecture, art (especially sculpture), writing and literature, and social structure. But the Chinese role in Southeast Asia has been even greater. Apart from the fact that upheavals within China itself contributed to the intermittent southward push of Sincized peoples, Chinese traders, pilgrims, sailors, fishermen, and others sailed from southeast China to the coasts of Malaya and the islands and settled there. Invasions and immigrations continued into modern times, and the relationships between recent Chinese arrivals and the earlier settlers of Southeast Asia have often been strained.

The Chinese initially profited from the arrival of the Europeans, who stimulated the growth of agriculture, trade, and industries; here they found opportunities they did not have at home. They established rubber holdings, found jobs on the docks and in the mines, cleared the bush, and transported goods in their sampans. They brought with them skills that proved to be very useful, and as tailors, shoemakers, blacksmiths, and fishermen they did well. They proved to be astute in business, and soon they not only dominated the region's retail trade, but also held prominent positions in banking, industry, and shipping. Thus their importance has always been far out of proportion to their modest numbers in Southeast Asia.

Southeast Asia has been a recipient of external cultural influences, but the realm also has generated its own, local cultural expressions. Most of what remains in tangible form, however, has resulted from the infusion of foreign elements. The main temple at Angkor Wat, constructed in Cambodia (Kampuchea) during the twelfth century, remains a monument in Southeast Asia to Indian architecture of the time. (George Holton/Photo Researchers)

The Europeans used them for their own designs but found the Chinese to be stubborn competitors at times—so much so that, eventually, they tried to impose restrictions on Chinese immigration. Much earlier the United States, when it took control of the Philippines, also had sought to stop the influx of Chinese into those islands.

When the European colonial powers withdrew and Southeast Asia's independent states emerged, Chinese population sectors ranged from nearly 50 percent of the total in Malaysia (1963) to barely over 1 percent in Burma (Fig. 10-3). The separation of Singapore, where Chinese constitute over 75 percent of the population of 2.5 million (1980), reduced the Malaysian Chinese component to 36 percent—still a very large minority. In Indonesia the percentage of Chinese in the total population is not high (an estimated 3 percent) but the Indonesian population is so large that even this small percentage represents a Chinese sector of about 4.3 million. Indonesia's Chinese were suspected of complicity in a communist plot to overthrow the government (a plot that was almost certainly of indigenous origin and not fomented by Communist China),

Figure 10-3

and hundreds of thousands of Chinese were killed. In Thailand, on the other hand, many Chinese have married Thais, and the Chinese minority of about 17 percent has become a cornerstone of Thai society, dominant in trade and commerce. In recent years, however, the Bangkok government has begun to express concern over the number of Chinese immigrants crossing the northern boundary (entering via Laos or Burma). But possibly the changing situation in China of the 1980s will stem that tide from within.

In general in Southeast Asia the Chinese communities remained quite aloof and formed their own separate societies in the cities and towns; they kept their culture and language alive by maintaining social clubs, schools, and even residential suburbs that in practice if not by law were Chinese in character. There was a time when they were in the middle between Europeans and Southeast Asians, and when the hostility of the local people was directed to white people as well as to the Chinese. But with the withdrawal of the Europeans, the Chinese became the chief object of this antagonism, which was strong because of the Chinese position as money lenders, bankers, and trade monopolists. In addition, there is the specter of an imagined or real Chinese political imperialism along Southeast Asia's northern flanks.

The *china* in Indochina thus represents a wide range of conditions; the source of most of the old invasions was in southern China, and Chinese territorial consolidation provided the impetus for successive immigrations. The Mongoloid racial strain carried southward from East Asia mixed with the earlier Malay stock to produce a transition from Chinese-like people in the northern mainland to dark-skinned, Malay types in the distant Indonesian east. Although Indian cultural influences remained strong, Chinese modes of dress, plastic arts, types of boats, and other cultural attributes were adopted in Southeast Asia. During the past century, and especially during the last half century, renewed Chinese immigration brought Chinese skills and energies that propelled their minorities to positions of comparative wealth and power in this region.

European Colonial Frameworks

One shortcoming of the term *Indochina* might be its failure to reflect the third cultural impress that has molded modern Southeast Asia: Europe's. As in Africa, the colonial politico-geographical structure that emerged in Southeast Asia during the nineteenth century

Brunei

Brunei is an anomaly in Southeast Asia—an oil-exporting Islamic Sultanate far from the Middle East. Located on the north coast of Borneo, sandwiched between Malaysian Sarawak and Sabah, the Brunei sultanate is a British-protected remnant of a much larger Islamic kingdom that once controlled all of Borneo and areas beyond. Brunei was scheduled to achieve full independence in 1983.

With a mere 5800 square kilometers (2230 square miles) and fewer than 300,000 people, Brunei is dwarfed by the other political entities in Southeast Asia. But the discovery of oil in 1929 heralded a new age for this isolated, remote, and stagnant territory. Today Brunei is one of the largest oil producers in the British Commonwealth, and new offshore discoveries suggest that production will increase. As a result, the population is growing rapidly by immigration (about half the people are Malay, one-fourth Chinese), and the sultanate enjoys one of the highest standards of living in Southeast Asia. Most of the people live near the oil fields in the western corner of the country, at Kuala Belait and Seria, and in the capital in the east, Bandar Seri Begawan. The evidence of a boom period can be seen in modern apartment houses, shopping centers, and hotels—a sharp contrast to many other towns in Borneo. There are some sharp internal contrasts as well; Brunei's interior still remains an area of agricultural subsistence and rural isolation, virtually untouched by the modernization of the coastal zone.

had the effect of throwing diverse people together in larger political units while at the same time dividing people with strong ethnic and cultural unity. The Thais, Malays, and the Shans of northern Thailand and eastern Burma all experienced the latter at the hands of the British, as did the Khmers and the Annamese of Indochina at the hands of the French. The peoples unified by the Dutch in their East Indies are numerous and varied; at least 25 distinct languages are spoken in this insular realm now comprising the state of Indonesia. Again, as in Africa, the internal divisions within the respective colonial realms had lasting significance, for eventually the individual fragments were to emerge as separate states. Thus the French organized their Indochinese dependency into five units—four protectorates (Cambodia, Laos, Tonkin, and Annam) and the colony of Cochin China in the Mekong Delta. Eventually Cambodia and Laos emerged as independent states, but Vietnam (consisting of Tonkin, Annam, and Cochin China) remained fragmented into a North (capital Hanoi) and a South (capital Saigon, now Ho Chi Minh City) until 1976.

The British held two major entities in Southeast Asia—Burma and Malaya—in addition to a number of islands in the South China Sea and the northern sector of Borneo. In Burma they created a colony incorporating many different peoples, but from 1886 to 1937 the government was based in far-away New Delhi. Burma's lowland core area is the domain of the majority of Burmese (or Burmans), but the surrounding areas are occupied by other peoples (Fig. 10-4). The Karens live to the east in the Irrawaddy Delta and the eastern margins, the Shans inhabit the eastern plateau that adjoins Thailand, the Kachins live on the Chinese border in the north, and the Chins are in the highlands along the Indian border in the west. Other groups also form part of Burma's complex population, and this complexity was compounded still further during the period of India-connected administration, when more than a million Indians entered the country. These Indians came as shopkeepers, moneylenders, and commercial agents, and their presence intensified Burmese resentment against British policies. The national division of Burma was brought sharply into focus during World War II, when the lowland Burmese welcomed the Japanese intrusion, while the peoples of the surrounding hill country, who had seen less of British maladministration and who had little sympathy for their Burmese countrymen anyway, generally remained pro-British. The independence of Burma (1948) has led to the expulsion of perhaps more than half of the Indian population.

Figure 10-4

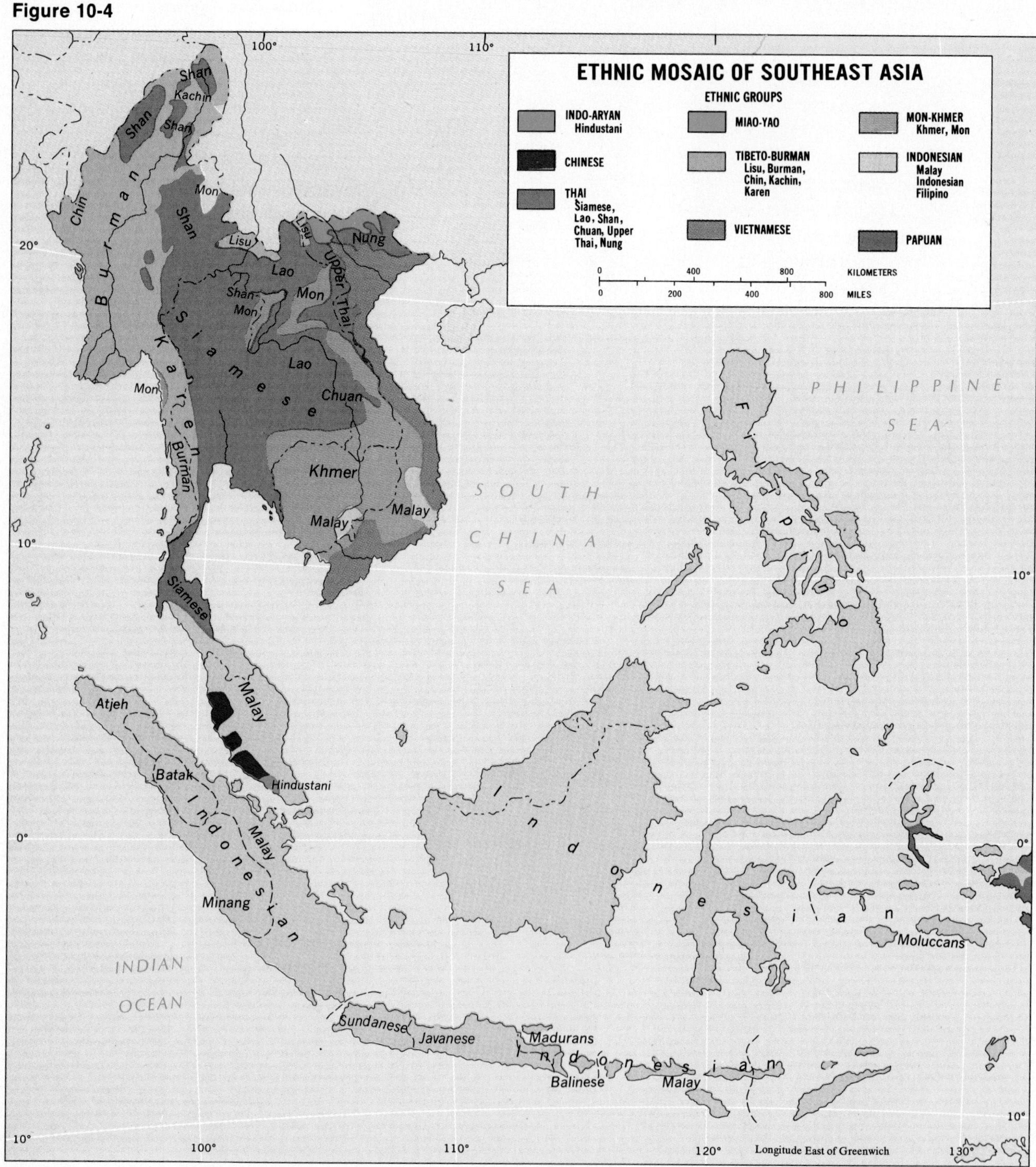

In Malaya, the British developed a complicated system of colonies and protectorates that eventually gave rise to the equally complex, far-flung Malaysian federation. Included were the former Straits Settlements (Singapore was one of these colonies), the nine protectorates on the Malayan Peninsula (born of sultanates of the Moslem era and developed from tiny river-mouth settlements into states covering entire drainage basins by the British), and the British dependencies of Sarawak and Sabah on the island of Borneo, plus numerous islands in the Strait of Malacca and in the South China Sea. The Federation of Malaysia survived in its original form for only a few years. Mainland Malaya achieved independence in 1957, but its structure was changed in 1963 by the accession of Singapore; then, in 1965, Singapore left Malaysia again and became a sovereign state (see page 522).

Malaysia's ethnic and cultural division is etched in its landscapes. The Malays are traditionally a rural people. They originated in this region and displaced the aboriginal peoples, now no longer significant numerically. The Malays, who constitute barely 50 percent of the Malaysian population of 14 million, have a strong cultural unity expressed in a common language, adherence to the Moslem faith, and a sense of territoriality that arises from their Malayan origins and their collective view of Chinese, South Asian, and other foreign intruders. Although they have held control over the government, the Malays often express a fear of the more aggressive, more commercially oriented, more strongly urbanized Chinese minority. Malay-Chinese differences were worsened by the course of the Second World War, when the Japanese (who occupied the area) elevated the Malays into positions of authority but ruthlessly persecuted the Chinese, driving many of them into the forested interior where they founded a communist-inspired resistance that long continued to destabilize the region. The British returned, then yielded after a system of interracial cooperation had been achieved, but racial tensions continued and resulted in 1969 in racial clashes in the capital, Kuala Lumpur (800,000). Kuala Lumpur's townscape reflects Malaysia's cultural mosaic; the minarets and domes of Islamic shrines rise above a city whose major structures (the railroad station, State House) have a Moorish architectural imprint, but trade and commerce are in the hands of the Chinese, around whose shops and businesses the city's life revolves.

The Hollanders took control of the "spice islands" through their Dutch East India Company, and the wealth that was extracted from what is today Indonesia brought the Netherlands its Golden Age. From the mid-seventeenth to the late eighteenth century, Holland could develop its East Indies sphere of influence almost without challenge, for the British and French were mainly occupied with the Indian subcontinent. By playing the princes of Indonesia's states off against each other in the search for economic concessions and political influence, by placing the Chinese in positions of responsibility, and by imposing systems of forced labor in areas directly under its control, the company had a ruinous effect on the Indonesian societies it subjugated. Java (Jawa), the most populous and most productive island, became the focus of Dutch administration, and from its capital at Batavia (now Djakarta) the company extended its sphere of influence into Sumatra (Sumatera), Borneo (Kalimantan), Celebes (Sulawesi), and the smaller islands of the East Indies. This was not accomplished overnight; the struggle for territorial control was carried on long after the company had yielded its administration to the Netherlands government. Northern Sumatera was not subdued until early in the present century.

Thus Dutch colonialism threw a girdle around Indonesia's more than 13,000 islands, creating the realm's largest political unit and one of its most complex states. In the broadest possible way, Indonesia's 143 million people can be divided into three groups: the aboriginal peoples (now a small minority), the peoples who grow rice on the islands' well-watered slopes, and the coastal communities. The aboriginal peoples have been pushed into the less productive, drier parts of the islands, where they practice shifting cultivation. The rice growers make up over half the total population and form the core of modern Indonesian society. Their cultural traditions have ancient roots and are expressed in architecture, dance, music, literature, as well as social structure and farming and land use. The coastal peoples developed from immigrations during and after the arrival of Islam, and Moslem communities exist on all the large islands.

Rice-growing Bali has been meticulously terraced by farmers who produce some of Southeast Asia's largest harvests per unit area. (Hans Hoefer/Woodfin Camp)

This differentiation barely begins to reveal Indonesia's cultural complexity, however. There are dozens of distinct aboriginal cultures; virtually every coastal community has its own roots and traditions. And the rice-growing Indonesians include not only the numerous Javanese, who are Moslems largely in name only and have their own cultural identity, but also the Sundanese (who constitute about 14 percent of Indonesia's population), the Madurese, Balinese, and others. Perhaps the best impression of the cultural mosaic comes from the string of islands that extends east from Jawa to Timur. The Balinese are mainly Hindu adherents; the population of Lombok is mainly Moslem, with some Hindu immigrants from Bali. Sumbawa is a Moslem community, but the next island, Flores, is mostly Roman Catholic. On Timur, Protestant groups remain, and Timur remains marked by its long-term division into a Dutch-controlled and a Portuguese-owned sector (Fig. 10-5). An independence movement on former Portuguese Timur was subdued by invading Indonesian forces; the Indonesian campaign was followed by severe dislocation and famine.

Dutch colonialism unified all this diversity and exploited Indonesia's considerable productive capacity under the so-called *Culture System,* involving forced-crop and forced-labor practices. The *Cultuur Stelsel* required the growing of stipulated crops on predetermined areas; Indonesians in the outer islands who had no land to cultivate were forced to make themselves available as laborers for the state. Undeniably, it stimulated production; sugar, coffee, and indigo were the leading exports, and they brought in increasing revenues as the system was made to work. Crops such as tea, tobacco, tapioca, and the oil palm (especially in Sumatera) were introduced or their strains were improved. But the price was high: the Javanese resented the principle, harsh and cruel methods were used to sustain the system—to such an extent that there was an outcry about it in the Netherlands. With

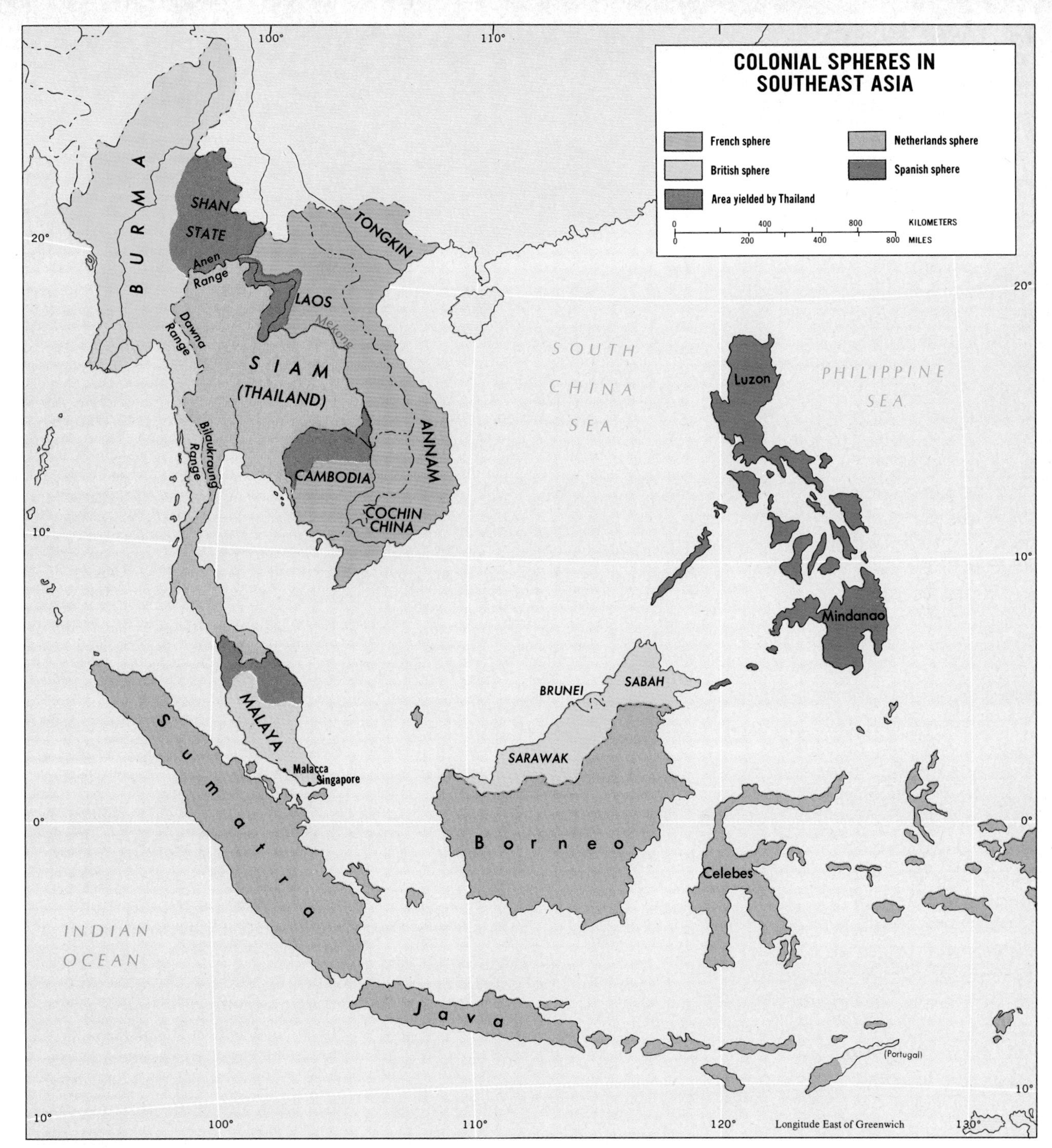

Figure 10-5

the Dutch and the Chinese in control of all phases of the trade based on the products of the forced croplands, including collection, treatment, packaging, and despatch, the local people got little or no experience in this aspect of their island's economy. There was little incentive, other than fear, to cultivate what was required. And lands that had been producing rice and other staples were periodically turned over to these cash crops, without (initially) satisfactory arrangements for adequate replacements of the food supplies thus lost.

The Culture System was officially abandoned in 1870 (although the forced growing of coffee continued until 1917), and a new and more liberal colonial policy evolved. But throughout the twentieth century the Indonesian drive for self-determination intensified, and following World War II, in 1945, independence was proclaimed. The Dutch fought a losing battle to regain their East Indies, finally yielding in 1949; the eastern territory of Irian Jaya or West Irian (on New Guinea) was awarded to Indonesia as recently as 1969, and formerly Portuguese Timur became Indonesia's 27th province in 1976 after Indonesian armed forces invaded the area.

The persistence of Indonesia as a unified state is another politico-geographical wonder on a par with India. Wide waters and high mountains have helped perpetuate cultural distinctions and differences; political centrifugal forces have been powerful and, in the late 1950s and 1960s, nearly pulled the country apart. But Indonesia's unity appears to have strengthened, just as the national motto, *Unity in Diversity,* underscores. With more than 300 discrete ethnic clusters and over 250 individual languages (and just about every religion practiced in the world) Indonesian nationalism has faced enormous odds—overcome to some extent by development based on the country's considerable resource base. This includes sizable petroleum reserves (Fig. 6-5), large rubber plantations (Sumatera shares Malaya's environments), extensive timber resources, major tin deposits offshore from eastern Sumatera, and soils that produce tea, coffee, and other cash crops. Copra and palm oil still rank among the exports as well. However, Indonesia's population continues to grow at the high rate of 2.7 percent, a long-term threat to the country's future. Rice and wheat already figure prominently among the annual imports.

Territorial Morphology

Having viewed territories long under British and Dutch colonial administration, we now consider French and Spanish imprints (and Thailand's singular independence) in another context. It is difficult to look at the political map the Europeans left behind in Southeast Asia (Fig. 10-1) without being struck by the several different spatial forms (that is, shapes) these states display. Indonesia is dispersed over a number of large and small islands, Vietnam extends narrowly along the east coast of Indochina, Burma and Thailand share the narrow, northern part of the Malay Peninsula. This aspect of the state—its shape—is a very important variable that can have a crucial effect on its cohesion and political viability. A state that consists of several separate parts, located far away from each other, obviously has to cope with problems of national cohesion not afflicting a state that has a contiguous territory. We saw what problems Pakistan's division caused; we know the advantages Uruguay derives from its compactness.

Political geographers identify five major shape categories on the world map, and four of them happen to be illustrated by the states of Southeast Asia; hence the region should be considered in this context. Consider for example, Cambodia, whose territory is quite nearly round in shape; in fact, it looks a great deal like Uruguay, and as it happens, it is almost the same size (181,000 square kilometers against Uruguay's 177,000). Cambodia is a *compact* state, which means that points on its boundary lie at about the same distance from the geometric center, near Kompong Thom on the Sen River. Theoretically, a compact state encloses a maximum of territory within a minimum of boundary, and this may hold advantages, for boundaries often still lie in sensitive areas. In the case of Cambodia, the common boundary with Vietnam is less than half as long as that of adjacent Laos; the Cambodians have sought to secure their border areas against intrusions, while the Laotian boundary is violated as though it did not exist. Compact countries such as Cambodia, moreover, do not face the problems confronting some other underdeveloped states, of integrating faraway islands, lengthy peninsulas, or other distant territorial extensions into the national framework. Effective national control is most easily established in a compact area, and for centuries the Cambodians came closer to national integration than perhaps any other Southeast Asian state. Cambodia did have the advantage of remarkable ethnic and cultural homogeneity in so complex a realm; perhaps as many as 90 percent of its people are Khmers, about 4 percent are Vietnamese, under 5 percent are Chinese.

Following the end of the Indochina War, a new communist government sought to reconstruct Cambodia as a rural society. The country's borders were sealed, thousands of opponents were eliminated, and the urban population of the capital, Phnom Penh, and other towns was forced to march into the countryside, there to engage in an unfamiliar subsistence lifestyle. Reports from the few who managed to cross into Thailand stated that hundreds of thousands may have died, especially the old, the very young, and the sick. In the aftermath of the Indochina War, Cambodia's compact manageability spelled an internal disaster.

In contrast, the states of Malaysia, Indonesia, and the Philippines are *fragmented* states, meaning exactly what the term implies—states whose national territory consists of two or more individual parts, separated by foreign territory or by international waters. Indonesia and the Philippines represent one of the three subtypes of this category, in that their areas lie entirely on islands. Malaysia is a second subtype, because it lies partly on a continental mainland and partly on islands. The third subtype is not present in Southeast Asia: in this case the major components of the fragmented state lie on a continental mainland. The United States of America is the best example, now that Pakistan has broken up.

Fragmented states must cope with problems of internal circulation and contact, and often with the friction of distance. Indonesia's government, based on Jawa, has had to put down secessionist uprisings in Sumatera, Sulawesi, and in the Moluccas (between Sulawesi and West Irian). The Luzon-centered government of the Philippines has had to combat rebels on Mindanao and Mindoro Island and elsewhere who took advantage of the insular character of the national territory. Far-flung Malaysia was forced to yield to the centrifugal forces inherent in its ethnic complexity and its spatial and functional structure, and expelled the microstate of Singapore. Even in the choice of a capital the fragmented state has difficulty; of all the separate parts of the country, one must be chosen to become the seat of government, the national headquarters. The choice may bring resentment elsewhere in the state, as it did in Pakistan. In Southeast Asia's fragmented states, there could not have been much doubt. Jawa contains almost two-thirds of Indonesia's nearly 143 million people, and the choice of Djakarta (the Hollanders' Batavia) as the capital could hardly be disputed. Nevertheless, the peoples of the Outer Islands have shown resentment against Javanese domination in Indonesia's affairs. In the Philippines, Luzon is the most populous island and Manila without doubt the country's primate city. In Malaysia, sparsely peopled Sarawak and Sabah can hardly compete with the mainland core area, and naturally Kuala Lumpur, the preindependence headquarters, continued as the new state's capital city.

Still another spatial form is represented by Burma and Thailand. The main territories of these two states, where their core areas are located, are essentially compact—but to the south they share sections of the Malay Peninsula. These peninsular portions are long and narrow, and states with extensions leading away from the main body of territory are referred to as *prorupt* states. Obviously Thailand is the best example: its proruption extends nearly 1000 kilometers (600 miles) southward from the vicinity of Bangkok. Where the Thailand-Burma boundary runs along the peninsula, the Thai proruption is in places less than 32 kilometers (20 miles) wide. Naturally such proruptions can be troublesome, especially when they are as lengthy as this. In the whole state of Thailand, no area lies as far from the core or from Bangkok as the southern extreme of its very tenuous proruption. But at least Thailand's railroad system extends all the way to the Malaysian border; in the case of Burma, not only does the railroad stop more than 480 kilometers (300 miles) short of the end of the proruption, but in 1978 there was not even a permanent road over its southernmost 240 kilometers (150 miles), reaching only as far as Mergui.

In Burma, the spatial structure of the state is compounded by a shift in the core area, a shift that has taken place during colonial times. Prior to the colonial period, the focus of the embryo state was in the so-called dry zone, between the Arakan Mountains and the Shan Plateau. Mandalay in Upper Burma, now with about a quarter of a million people, was the urban node. Then the British developed the rice potential of the Irawaddy Delta and Rangoon, occupied earlier by the British but long of lesser importance, became the new capital city. The old and new core areas are connected by the Irawaddy in its function as a water route, but the center of gravity in modern Burma lies in the south.

The fourth type of state shape represented in Southeast Asia is the *elongated* or attenuated territory. By this is meant that a state is at least six times

Singapore

Singapore in 1965 became Southeast Asia's smallest independent state in terms of territory (just under 600 square kilometers, 225 square miles) as well as in population (2.4 million in 1980). Situated at the southern tip of the Malay Peninsula where the Straits of Malacca open into the South China Sea and the waters of Indonesia, the port city had been a part of Britain's Southeast Asian empire and, briefly, a member state in the Malaysian federation. It had grown to become the world's fourth largest port by number of ships served, and it always was a distinct and individual entity—physically (the Johor Strait separates the island from Malaysia's mainland) and culturally (the population is 76 percent Chinese, 14 percent Malay, 7 percent South Asian). When Singapore seceded from Malaysia in 1965, a major conflict between city and country was averted.

The city-state has thrived since independence, capitalizing on its relative location and its function as an entrepot between the Malay Peninsula, Southeast Asia, and the industrialized nations, prominently including Japan. Crude oil from the Middle East is unloaded and refined at Singapore, then shipped to Asian destinations. Raw rubber from Malay and Sumatera is shipped to Japan, the United States, China, and other countries. Timber from Malaysia and rice and spices also are assembled and forwarded via Singapore; in turn, companies send machinery, automobiles, and equipment to Singapore for shipment to Malaysia and other countries in the realm.

Significantly, Singapore's manufacturing sector is expanding rapidly and in the late 1970s was in the process of overtaking the entrepôt function in terms of its contribution to the national income. Taking a page from Hong Kong's book, and aware of world price fluctuations and their effect on international trade, Singapore's planners have encouraged the diversification of industries. Foreign investment is attracted and given very advantageous terms. In addition to the refineries, Singapore now has shipbuilding and repair yards, food-processing plants, sawmills, and a growing number of small industries serving the local market.

as long as its average width. Familiar examples are Chile in South America, Norway and Italy in Europe, and Malawi in Africa. Elongation presents obvious and recurrent problems; even if the core area lies in the middle of the state, as in Chile, the distant areas in either direction may not be effectively connected and integrated in the state system. In Norway the core area lies near one end of the country, with the result that the opposite perimeter takes on the characteristics of a frontier. In Italy, the contrasts between north and south pervade all aspects of life, and they reflect the respective exposures of these two areas to different mainstreams of European-Mediterranean change. If a state is elongated and at the same time possesses more than one core area, there will be strong divisive stresses on it. This has been the case in Vietnam, one of the three political entities into which former French Indochina was divided.

French Indochina incorporated three major ethnic groups: the Laotians, the Cambodians, and the Vietnamese. Even before independence, the Laotians and the Cambodians possessed their own political areas (which later were to become the states of Laos and Cambodia), and this left a 2000-kilometer (1200-mile) belt of territory, averaging

Singapore, Southeast Asia's only developed country, is a mere city-state—but it owes its prosperity to relative location, effective planning, and superb spatial organization. (Georg Gerster/ Photo Researchers)

Singapore's population may be 76 percent Chinese, but its Chinese sector is by no means homogeneous. The Chinese immigration came from separate sources, Fujian Province and Guangdong Province among them. The Chinese communities speak different Chinese languages, but only one—Mandarin—is recognized as one of Singapore's official languages (the others are English, the lingua franca, Malay, and Tamil). Southeast Asia's largest port is a cultural crossroads as well as an economic hub.

under 240 kilometers (150 miles) in width, facing the South China Sea. This was the domain of the Annamites extending all the way from the Chinese border to the Mekong Delta. Administratively, the French divided this elongated stretch of land into three units: Cochin China in the south, Annam in the middle, and Tonkin in the north. The capitals of these areas became familiar to us during the Vietnam War; Saigon (now Ho Chi Minh City) was the focus of Cochin China and the headquarters of all Indochina; Hué, the ancient city, was the center of Annam; and Hanoi was the capital of Tonkin. Cochin China and Tonkin both were incipient core areas, for they lay astride the populous and productive deltas of the Mekong and Red Rivers, respectively.

In 1940 and 1941, the Japanese entered and occupied Indochina, and during their nearly four years of occupation, the concept of a united Vietnam emerged. The Japanese did not discourage the rising Vietnamese nationalism, especially when it became apparent that the tide of war was turning against them; they would rather have an Annamese government in Vietnam than a French one. So during the period of Japanese authority a strong coalition of proindependence movements arose in Indochina—the Viet

The Indochina War devastated much of Vietnam. These were the remains of the small town of Dong-Ha following an attack by insurgents in 1972. (Yves-Guy Burgès/Gamma Liason)

Minh League. In 1945, when the Japanese surrendered, this organization seized control and proclaimed the Republic of Vietnam a sovereign state under the leadership of Ho Chi Minh. For some time after the Japanese defeat, the Chinese actually occupied Tonkin and northern Annam, but theirs was a sympathetic involvement as well, and the Viet Minh grew in strength during that time. Meanwhile, the French had proposed that Vietnam, Laos, and Cambodia should join as associated states in the fourth-republic concept of the French union, but this plan was rejected everywhere in Indochina. As the French reestablished themselves where they had always been strongest—in Saigon and the Mekong Delta of Cochin China—they started negotiations with the Viet Minh. These talks soon broke down, and in 1946 a full-scale war broke out between the French and the Viet Minh. France's base lay in Cochin China, and the Viet Minh had their greatest strength in Tonkin, a thousand miles away. Vietnam's pronounced elongation favored the Viet Minh and their numerous nationalist, communist, and revolutionary sympathizers. After an enormously costly eight-year war, the crucial battle was fought in the north (at Dien Bien Phu), not far from the Chinese border, and lost by the French.

The Insurgent State

The sequence of events in Vietnam in the 1960s and 1970s still is fresh in memory. Opposition to the Saigon government in South Vietnam turned into armed insurrection; North Vietnam provided arms and personnel to aid the anti-Saigon insurgents; United States advisors and armed forces in support of the Saigon regime at one time exceeded a half million in number. Notwithstanding severe bombing of North Vietnam and its capital, Hanoi, and extensive defoliation efforts in the protective forests of the South, what had begun as a scattered rebellion ended in the defeat of a national government and the ouster of its ally, the United States. Political scientists recognize stages in the sequence of events; the first is a period of *contention,* when armed rebellions start to occur in remote locales, followed by *equilibrium,* when the insurgents manage to withstand government forces in critical areas of the country, and finally *counteroffensive,* a stage of regular ground and air war. The result may be the defeat of the insurgents, as happened in Malaya in the 1950s, or their success and takeover of the national government, as in Cuba in 1960.

In an article published in 1969, R. McColl translated these stages in territorial terms, suggesting that they chronicle the emergence of an *insurgent state* within the boundaries of a national state. During the stage of equilibrium, the insurgent state attains a degree of stability. It may be fragmented and consist of several cores, but it has headquarters, boundaries, an increasingly complex administrative structure, schools, hospitals, and even air traffic with neighboring countries that may sympathize with the insurgents' cause. The map of (then) South Vietnam's insurgent state in about mid-1967 (Fig. 10-6) shows areas still under Saigon control actually surrounded by Vietcong-held territory.

In 1978, the 10-year insurgent state that had existed in Moçambique from the mid-1960s had taken control of the government. In Ethiopia, an insurgent state still existed in the province of Eritrea. The Turkish segment of Cyprus, too, had qualities of an insurgent state, and in Pakistan there were signs that an insurgent state was emerging in the northwest

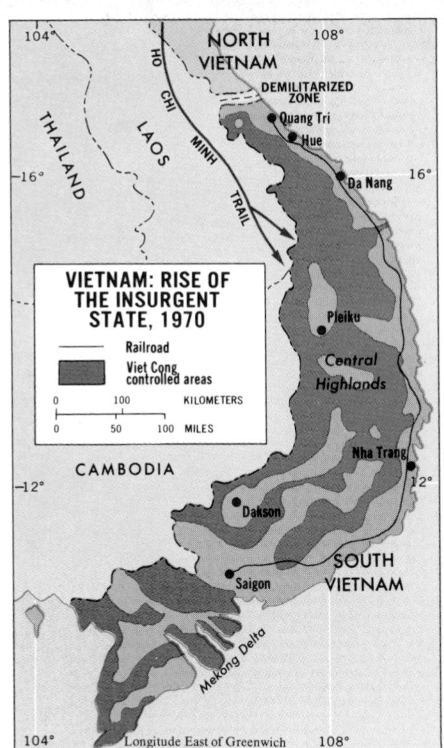

Figure 10-6

(see page 461). For years, the Philippines has been combating efforts by Moslem separatists in the southern islands, especially Mindanao, to create an insurgent state (and by Maoists throughout the fragmented country to accomplish the same). In 1976 the Philippine government promised to grant greater autonomy for the country's Moslem regions, thereby reducing the level of contention. The insurgent state is a product of current ideological contests, the proliferation and availability of modern weapons, and the effectiveness of modern communications.

The unification of North and South Vietnam in 1977 produced one of the world's most pronouncedly elongated states and, with a population of over 50 million, Southeast Asia's second largest. And the country's attenuation continues to dominate its regional geography. A central government notwithstanding, the old divisions between north, center, and south still prevail. Saigon may now be Ho Chi Minh City, but it remains a world apart from Hanoi and the distant north. Since unification Vietnam has been involved in an armed conflict with China along its northern boundary; it has also embarked on a campaign of invasion and domination in Cambodia. These events reflect the continuing specter of ancient animosities; in Cambodia, the Khmer people, old adversaries of the Vietnamese, in 1980 faced decimation if not extinction following a decade of war, organized genocide, foreign subjugation, and famine.

Domino Theory and Thailand

During the 1960s, as one country after another became engulfed by the Vietnam War (later to be called the Indochina War to signify its wider arena), Southeast Asian specialists began to refer to the "domino theory"—the idea that the fall of South Vietnam would inevitably lead to communist takeovers in Cambodia, Laos, Thailand, Burma, Malaysia, and ultimately Indonesia and the Philippines as well. This was no mere speculation, for the effect of adjacent sanctuary on the progress of the Vietnamese insurgency was crucial. The Ho Chi Minh Trail brought supplies from North Vietnam to the South via Laos; Vietcong insurgents enjoyed safety beyond South Vietnam's borders in Cambodia. When Cambodia and Laos were drawn into the conflict and the war took its course, those who supported the domino idea appeared to have been correct.

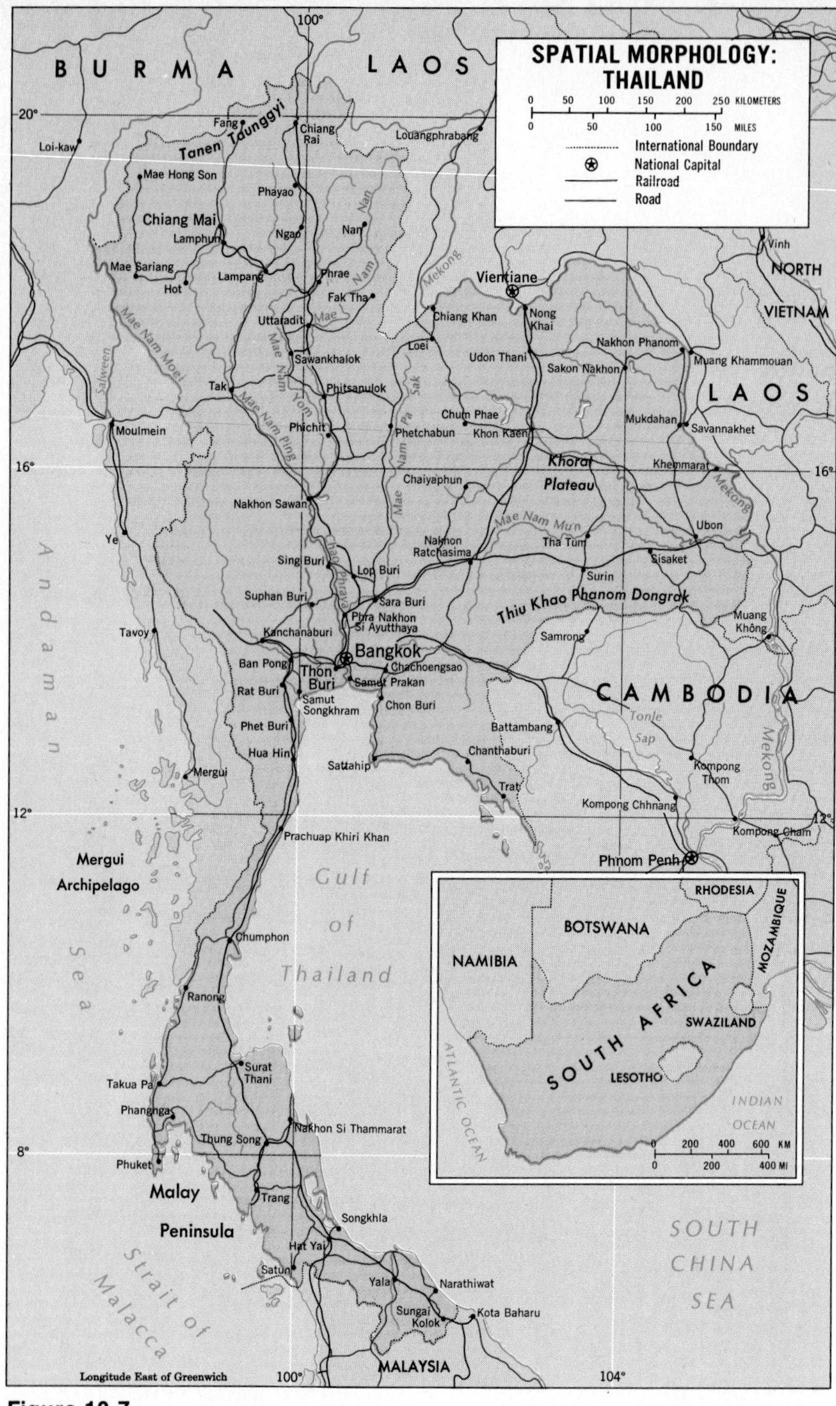

Figure 10-7

Further support for the domino effect can be derived from Africa, where Moçambique's FRELIMO insurgents could not mount their ultimately successful effort without staging areas in neighboring Tanzania, and where Rhodesia and Namibia became areas of contention only after adjacent Moçambique and Angola had seen revolutionary movements succeed. Opponents of the idea pointed to Malaysia, where a large-scale communist insurgency was defeated more than 20 years ago and where, political conflicts notwithstanding, no renewed threat of similar proportions arose during the Indochina War.

Both supporters and opponents of the domino concept point to Thailand, from where American warplanes took off on their bombing missions to North Vietnam and to where those Cambodians who could, fled when a communist regime took power in Cambodia. Thailand's strongly prorupt territory adjoins Laos as well as Cambodia; its northern areas lie within 150 kilometers (slightly under 100 miles) from China's borders. Those northern areas also are among Thailand's most mountainous, remote, and frontier-like, but there was colonial design in the creation of a British-French buffer between Thailand and China. Thailand retained its independence during the colonial period (when it was called Siam, until 1939 when it was renamed *Prathet Thai,* "Land of the Free"), but it was purposely separated from South China. Today, Thailand's core area lies in the lower basin of the Mae Nam (Chao Phraya), centered on the capital, Bangkok (the Bankok Metropolitan Area, including Thonburi, has a population of about 4 million, see Fig. 10-7).[2] This is the focal region for the 75 percent of the population who are Thais, but in the country's marginal areas live peoples with other roots. Khmers in the east, Shan and Yao in the north (with affinities to China), Karen in the west (movement continues across the

[2] The inset map on Fig. 10-7 shows the fifth of the spatial forms of states, the only case not represented in Southeast Asia—the *perforated* state. South Africa is shown perforated by the state of Lesotho.

border to and from Burma), Malays on the southern peninsula—these are but some of the peripheral minorities. Add to these some significant differences, ethnic as well as cultural, between the lowland Thais and the Thais in the mountains, and Thailand's regional vulnerability is clear.

Already, communist-inspired insurgents have established an area of contention in the far north, and separatist problems have emerged in the deep peninsular south, where the dominantly Buddhist country's Moslem minority is concentrated. The divisions between left and right, mountain peasant and central government, student and soldier have intensified in recent years. There can be no doubt about it: external ideological pressures and actual guerilla support have begun to destabilize a country that survived the colonial era but became heavily involved with the United States after the Second World War. As the United States withdrew from Vietnam and began to establish new relations with China, Thailand was forced to modify its long-term commitment to the West—a process that has already contributed to unprecedented governmental instability, political clashes, and regional unrest. When the insurgent state emerges in Thailand, it will be difficult to argue against a domino effect in Southeast Asia.

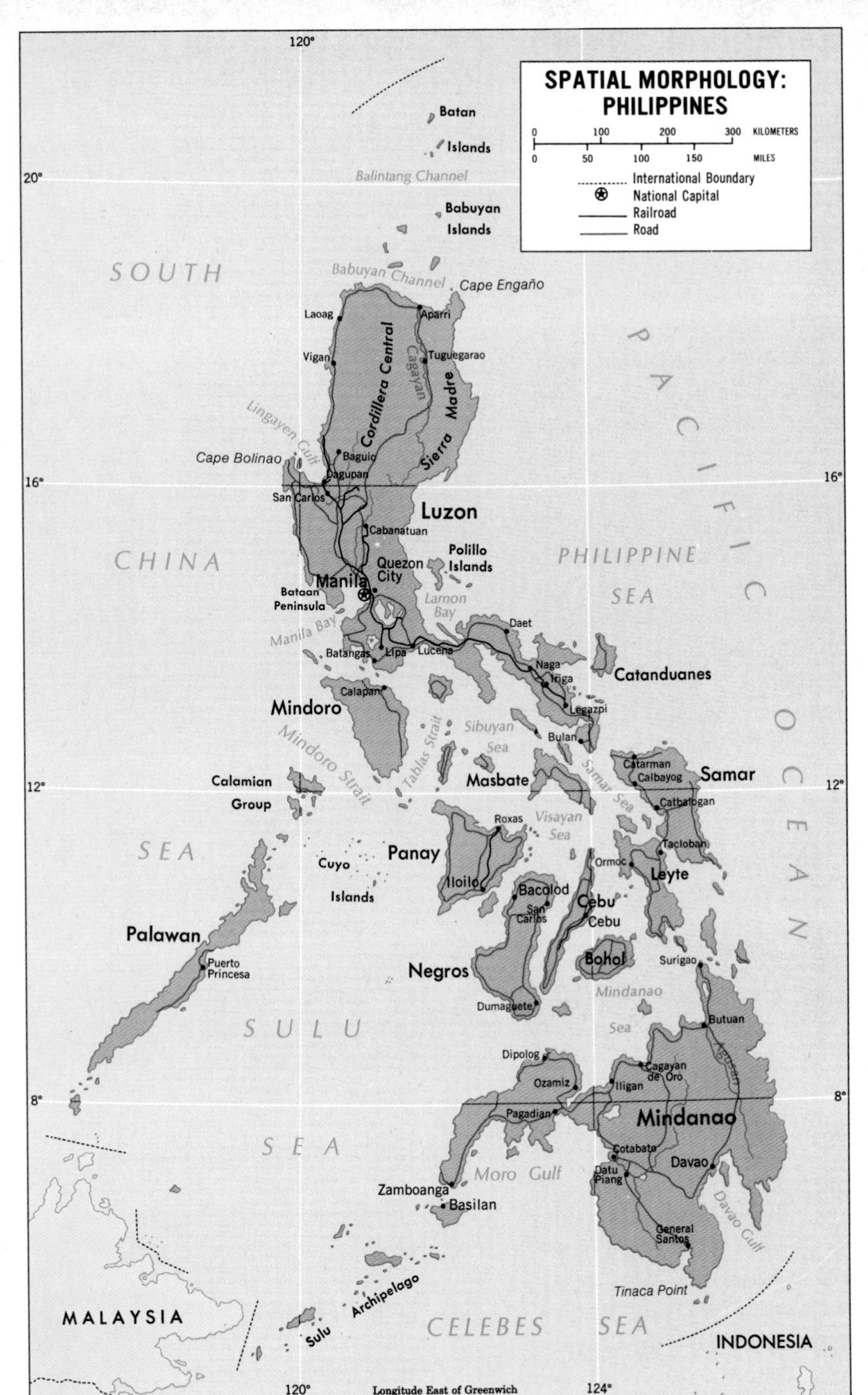

Figure 10-8

Philippine Fragmentation

After Indonesia and Vietnam, the Philippines, with over 49 million people, is Southeast Asia's next most populous state (Fig. 10-8). Few of the generalizations that can be made about this realm would apply without qualification to this country, however, and the Philippines' location relative to the mainstreams of change in this part of the world has had much to do with this. The islands, inhabited by peoples of Malay ancestry with Indonesian strains, shared with much of the rest of Southeast Asia an early period of Indian cultural influence, strongest in the south and southwest and diminishing northward. Next came a

Chinese invasion, felt most strongly in Luzon in the northern part of the archipelago. Islam's arrival was delayed somewhat by the position of the Philippines well to the east of the mainland and to the north side of the Indonesian islands. The southern Moslem beachheads were soon overwhelmed by the Spanish invasion of the sixteenth century, and today the Philippines, adjacent to the world's largest Moslem state, is about 85 percent Roman Catholic and only 5 percent Moslem.

Out of the Philippine melting pot, where Mongoloid-Malay, Arab, Persian, Chinese, Japanese, Spanish, and American elements have met and mixed, has emerged the distinctive culture of the Filipino. It is not a homogeneous or a unified culture, but in Southeast Asia it is in many ways unique. One example of its absorptive qualities lies in the way the Chinese infusion has been accommodated: the "pure" Chinese minority numbers a mere 1 percent of the total population (far less than normal for Southeast Asian countries), but in fact a much larger portion of the Philippine population carries a marked Chinese ethnic imprint. What has happened is that the Chinese have intermarried, producing a sort of Chinese-mestizo element—constituting more than 10 percent of the entire population. In another cultural sphere, the country's ethnic mixture and variety is paralleled by its great linguistic diversity. Nearly 90 Malay languages, large and small, are spoken by the 49 million people in the Philippines; about 1 percent still uses Spanish. Visayan is the language most commonly spoken; approximately 40 percent of the population is able to use English. At independence the largest of the Malay languages, Pilipino, based on Tagalog, was adopted as the country's official language, and its general use is being promoted through the educational system. English is learned as a second language and remains the chief *lingua franca*.

The widespread use of English in the Philippines, of course, results from a half century of American rule and influence, beginning in 1898, when the islands were ceded to the United States by Spain under the terms of the Treaty of Paris, following the Spanish-American War. The United States took over a country in open revolt against its former colonial master (the Philippines had declared their independence from Spain on June 12, 1898) and proceeded to destroy the Filipino independence struggle, now directed against the new foreign rulers. It is a measure of the subsequent success of United States administration in the Philippines that this was the only dependency in Southeast Asia that during World War II sided against the Japanese in favor of the colonial power (anti-Japanese movements did develop in Malaya and Burma as well). United States rule had its good and bad features, but the Americans did initiate reforms that were long overdue, and they were already in the process of negotiating a future independence for the Philippines when the war intervened.

The reforms begun by the United States in the Philippines, and continued by the Filipinos themselves after independence in 1946, were designed to eliminate the worst aspects of Spanish rule. Spanish provincial control was facilitated through allocations of good farmland as rewards to loyal representatives of church and state. The quick acceptance of Catholicism and its diffusion through the islands helped consolidate this system; exploitation of land and labor (often by force) were the joint objectives of priests and political rulers alike.

When, after three centuries of such exploitation, Spanish colonial policy showed signs of change during the nineteenth century, it was too late. Crops from the Americas were introduced (tobacco became a lucrative product), and a belated effort was made to integrate the Philippine economy with that of Spanish-influenced America. But what the Philippines needed most the Spaniards could not provide: land reform. As everywhere else in the colonial world, the main issue between the colonizers and the colonized in the Philippines was land, agricultural land. And, as elsewhere, the Spanish colonizers found that what had been easy to give away was almost invariably impossible to retrieve. Long after the Spaniards had lost their Philippine dependency the Americans, with a much freer hand, found the same still to be true—and even today the Filipino government, after more than three decades of sovereignty, still faces the same issue.

The Philippines' population, concentrated where the good farmlands lie in the plains, is densest in three general areas: the south-central and northwestern part of Luzon (Manila lies at the southern end of this zone), the southeastern proruption of Luzon, and the islands between Luzon and Mindanao,

Manila is the leading city and largest port of the Republic of the Philippines. Located in southwestern Luzon on the eastern shore of Manila Bay, the city and environs in 1980 contained about 15 percent of the country's population. Shown here is part of the townscape along Roxas Boulevard, one of the city's main thoroughfares. (David Burnett/Leo DeWys)

the Visayan Islands. The Philippine archipelago is reputed to consist of over 7000 mostly mountainous islands, of which Luzon and Mindanao are the two largest, accounting for nearly two-thirds of the total area. In Luzon the farmlands, producing rice and sugar cane, lie on alluvial soils, but in extreme southeastern Luzon and in the Visayan Islands there are good volcanic soils. When world market prices are high, sugar is the most valuable export of the agriculture-dominated Philippines; the forests yield valuable timber, while copra and coconut oil rank next—but most Filipino farmers are busy growing the subsistence crops, rice (in which the country temporarily achieved self-sufficiency in 1968) and corn. As in the other Southeast Asian countries, there is considerable range of supporting food crops. Along with other countries in the region, the Philippine state does not yet face a major overpopulation problem (although in parts of Luzon and the Visayan Islands the threshold is being reached). And when it comes to more intensive farming, few peoples in the world could provide a better example than northern Luzon's Igorots, who have transformed the hillsides in their domain into impeccably terraced, irrigated paddies.

Thus the Philippines provides both rule and exception in Southeast Asia. A fragmented island country, it must cope with internal revolutionary forces and intermittent encroachment from outside (Indonesia's former leader, Sukarno, while envisaging a Greater Indonesia sphere, at times expressed irredentist support for Mindanao's Moslems). Culturally it is in several ways unique; economically it shares with the rest of Southeast Asia the liability of imbalanced and painfully slow development. The Philippines' lumber and agricultural exports are supplemented by copper and iron ores, while revenues must be spent on the purchase of machinery, various kinds of equipment, fuels, and even some foods. All this exemplifies the country's persistent underdevelopment, and regional frustrations are in the process of deepening a political crisis that has been building for many years. As in the case of Thailand, the Philippine government is reconsidering its long-term alignment with Western powers in this part of the world while the politico-geographical map of Southeast Asia changes.

Land and Sea

In a region of peninsulas and islands such as Southeast Asia, the surrounding and intervening waters are of extraordinary significance. They may afford more effective means of contact than the land itself; as trade and migration routes, they sustain internal as well as external circulation. On the other hand, the waters between the individual islands of a fragmented state also function as a divisive force. There are literally tens of thousands of islands in Southeast Asia, and while some are productive and in effective maritime contact with other parts of the region and the world, many of these islands and islets are comparatively isolated. In this respect, of course, Southeast Asia is not unique in the world. In previous chapters we discussed the peninsular character of Western Europe and the insular nature of Caribbean America. In all these areas, peoples and governments have an awareness of the historic role of the seas. States with coasts have extended their sovereignty over some of their adjacent waters. This is not a new principle; in Europe, the concept that a state should own some coastal waters is centuries old. Initially these coastal claims—on a world scale—were quite modest. But as time went on, some states began to extend their territorial waters farther and farther out, and other countries followed suit. The widening maritime claims involved not only the ocean itself, but also the seabed below, the floor of the ocean and the rocks beneath. Technological advances after World War II made it possible to explore and mine the bottom of the ocean to ever greater depths, and soon only the deepest trenches may lie beyond the grasp of modern machines. Already the continental shelves (submerged zones adjacent to the continents to a depth averaging 200 meters) are intensively exploited, and over 20 percent of all the oil brought to the surface each year now comes from offshore wells. Minerals are being mined even from the deep ocean floor: manganese nodules, potato-sized concentrations of valuable minerals, are being drawn to the surface through a pipe lowered more than 3500 meters (11,000 feet) to the ocean floor.

Like the desert in the Middle East, the ocean floor near Southeast Asia's islands and peninsulas is yielding oil in increasing amounts. The rig shown stands in the Java Sea in Indonesia. (David Barnes/Photo Researchers)

These developments place the underdeveloped countries in a disadvantageous position, for they do not possess the technology to mine the resources off their own shores. They do, however, have collective influence in the United Nations, and their response has been to extend their claims to territorial waters far beyond prevailing limits. This process has become known as the "scramble for the oceans," and it is still in progress today. When World War II ended, as many as 40 countries claimed a mere 3 nautical miles of territorial waters, and only nine countries had territorial waters wider than 6 nautical miles. By the late 1970s only 24 countries still adhered to their 3-mile claim, 60 countries claimed a 12-mile territorial sea, and 10 countries demanded sovereignty over 100 to 200 miles of adjacent ocean and seabed.

In Southeast Asia, Indonesia announced in 1957 that it would claim as national territory all waters within 12 nautical miles of its outer islands *and* all waters between all the islands

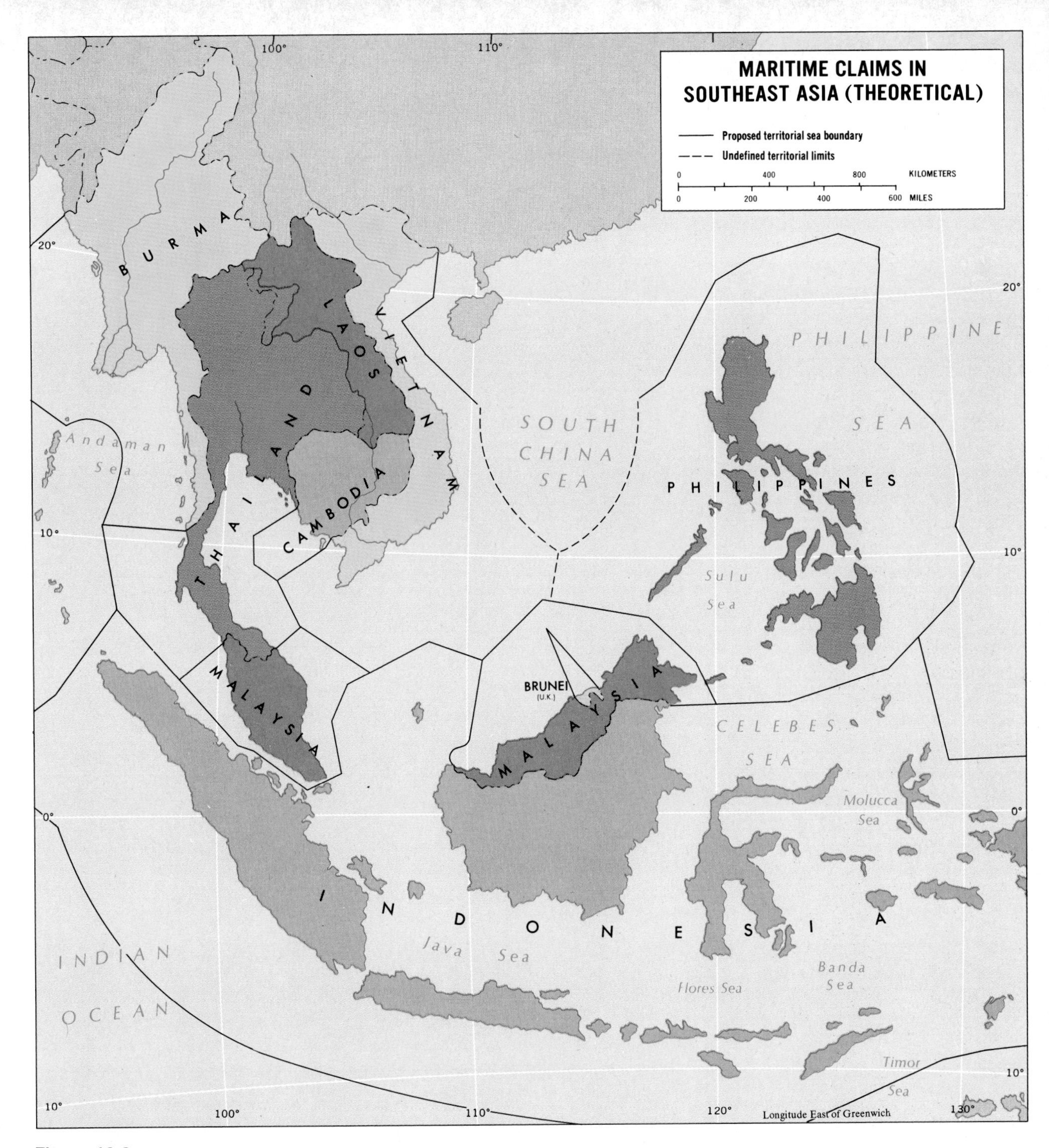

Figure 10-9

of its far-flung archipelago (Fig. 10-9). This had the effect of making the entire Java Sea, Banda Sea, and most of the Celebes (Sulawesi) Sea territorial waters. A similar claim by Malaysia, following establishment of its federation with Sarawak and Sabah, appropriated much of the South China Sea. These claims to territorial waters in Southeast Asia, as elsewhere, also involve the underlying continental shelf—known to contain petroleum reserves but only partially explored.

In the late 1970s the scramble for the oceans continued unabated. International agreement was reached in 1977 on a 200-mile fishing zone, and countries within 200 miles of each other (such as the United States and Cuba) began the difficult task of defining and delimiting their maritime boundaries. The effect of a 200-mile zone of national sovereignty on the world's oceans is shown on Fig. 10-10. The open oceans, or "high seas," are much diminished by the 200-mile limit. But the first steps *beyond* this 200-mile principle already have been taken in some areas. In Southeast Asia, the Caribbean region, and the North Sea, the 200-mile zone is irrelevant because the waters between countries are often narrower. In such cases other techniques for boundary definition are employed, and the result is a map such as Fig. 1-8, allocating geometrically delimited regions of the sea to coastal countries. Already this process has been extended to include all the world's oceans, and Fig. 10-10 reveals the boundary framework of the world if the "world-lake concept" were to be adopted. It seems inconceivable, at present, that this could actually come about; but it seemed equally inconceivable, 30 years ago, that a 200-mile limit would ever receive international sanction. The oceans are this planet's last frontier.

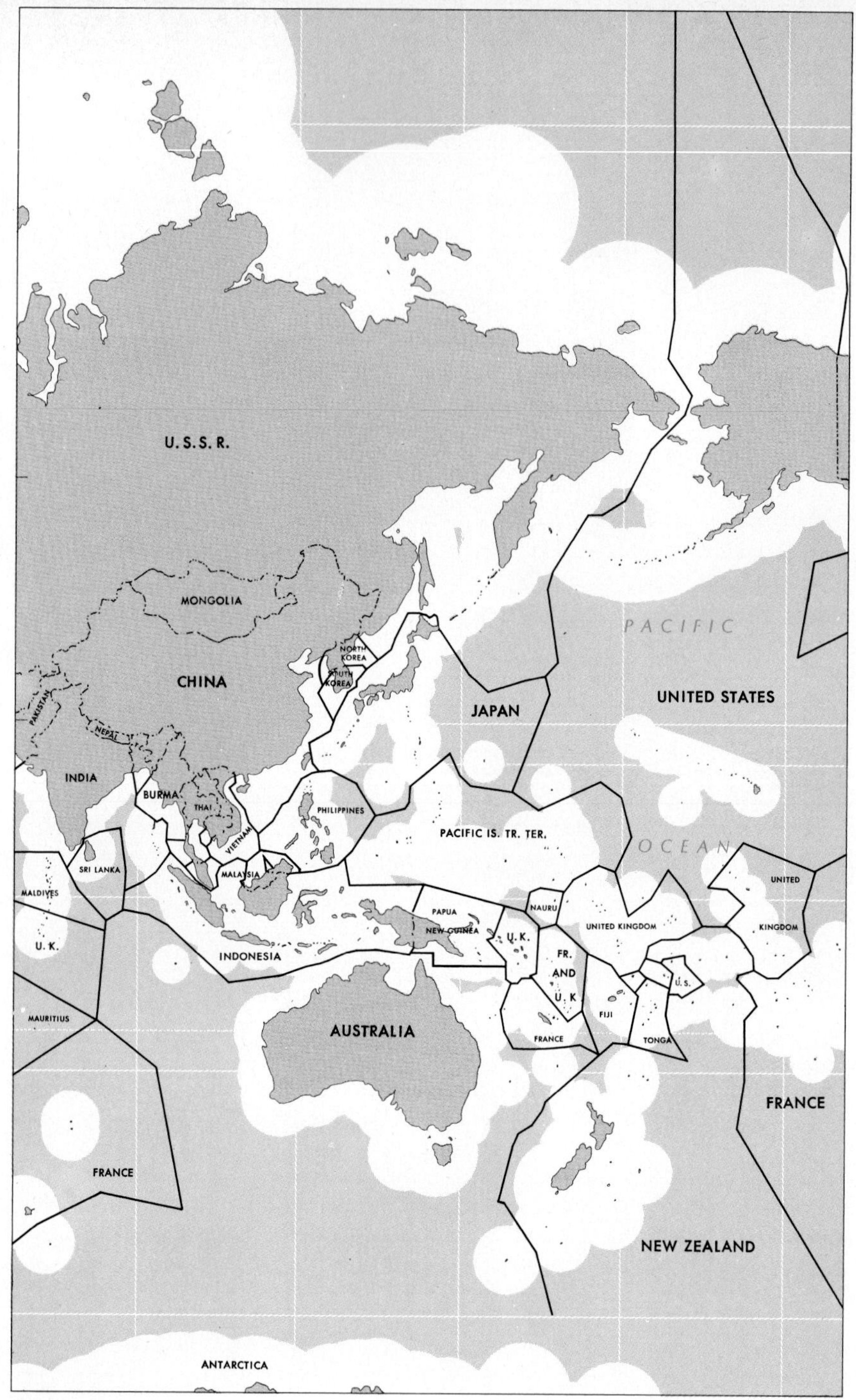

Figure 10-10

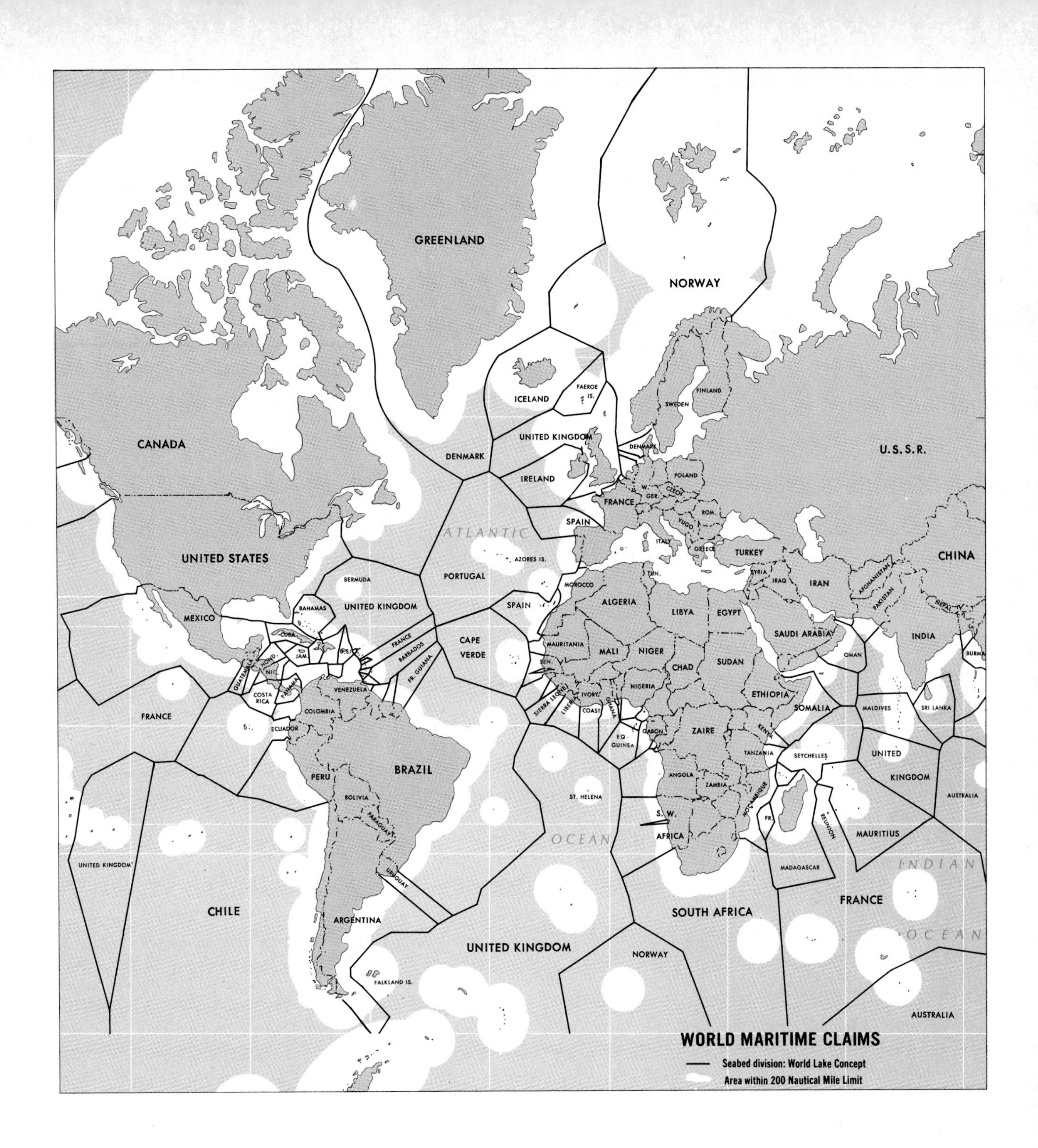

WORLD MARITIME CLAIMS
Seabed division: World Lake Concept
Area within 200 Nautical Mile Limit
GREENLAND
NORWAY
CANADA
U.S.S.R.
UNITED STATES
ATLANTIC
OCEAN
INDIAN
OCEAN
CHINA
BRAZIL
CHILE
SOUTH AFRICA
UNITED KINGDOM
FRANCE
AUSTRALIA

Additional Reading

Although geographers sometimes disagree on exactly what constitutes Southeast Asia, a number have written textbooks about the region as they conceive it. The most readable and up-to-date volume presently available is D. W. Fryer, *Southeast Asia: Problems of Development* published in the United States by Halsted, New York, 1978. E. H. G. Dobby's *Southeast Asia* is now in a seventh edition, published by London University Press in 1969. Another major work is C. A. Fisher's *Southeast Asia: a Social, Economic and Political Geography,* published in a second edition in New York by Dutton in 1966, and one of the standards is the book written by C. Robequain and translated from the French by E. D. Laborde, *Malaya, Indonesia, Borneo and the Philippines; a Geographical, Economic, and Political Description of Malaya, the East Indies, and the Philippines,* a second edition of which was published by Longmans in London, in 1958. Other basic works are G. Hunter's *South-east Asia: Race, Culture, and Nation,* printed by Oxford University Press in New York in 1966, and T. G. McGee, *The Southeast Asian City: A Social Geography,* a Praeger, New York, publication of 1967. Also see the useful *Atlas of South-East Asia,* published by St. Martin's, New York, in 1964. The volume by A. Kolb, *East Asia* (Methuen, London, 1971), contains a detailed discussion of Vietnam. An extremely useful bibliographic source is K. G. Tregonning's *South-East Asia: A Critical Bibliography,* published by the University of Arizona in 1969 and containing over 2000 items. On the Chinese in Southeast Asia, see V. Purcell, *The Chinese in Southeast Asia,* a second edition of which was published by Oxford University Press in New York in 1965, and L. E. Williams, *The Future of the Overseas Chinese in Southeast Asia,* published by McGraw-Hill in New York in 1966.

Individual countries also have been discussed. On Burma, see J. L. Christian, *Modern Burma, a Survey of Political and Economic Development,* an older book (it was published by the University of California Press, Berkeley, in 1942) but still useful. On Thailand, see one of the Country Survey Series of the Human Relations Area Files, W. Blanchard's *Thailand: Its People, Its Society, Its Culture,* published in 1958. On Vietnam see C. H. Schaaf and R. H. Fifield, *The Lower Mekong,* No. 12 in the Van Nostrand Searchlight Series. On Cambodia, another Human Relations Area Files volume, by D. J. Steinberg, *Cambodia: Its People, Its Society, Its Culture,* was published in 1959 and is still full of interest. From the same publisher and in the same series is a volume edited by F. M. LeBar and A. Suddard on *Laos,* brought out in 1960. On Malaysia, see F. C. Cole, *Peoples of Malaysia,* again an older book (1945), but still valuable. It was published by Van Nostrand, Princeton. Also on Malaysia, there is a good volume by Jin-Bee Ooi, *Land, People, and Economy in Malaya,* published by Longmans in London in 1963. On Indonesia, see the volume by B. and J. Higgins, *Indonesia: The Crisis of the Millstones,* No. 10 in the Van Nostrand Searchlight Series. And on the Philippines there are several good works. One of the best is an older volume, by J. E. Spencer, *Land and People in the Philippines; Geographic Problems in Rural Economy,* published by the University of California Press in Berkeley, in 1952. Later, Spencer collaborated with F. L. Wernstedt in *The Philippine Island World: a Physical, Cultural, and Regional Geography,* published by the University of California Press at Berkeley and Los Angeles, in 1967. Also see R. E. Huke, *Shadows on the Land: An Economic Geography of the Philippines,* published in Manila by Bookmark in 1963.

Maritime boundaries and associated problems are discussed by J. V. R. Prescott in *The Political Geography of the Oceans,* a Halsted/Wiley volume published in New York in 1975. Also see L. M. Alexander, *The Law of the Sea: Offshore Boundaries and Zones,* published by Ohio State University, Columbus, in 1967. R. W. McColl's article entitled "The Insurgent State: Territorial Bases of Revolution" was published in the *Annals of the Association of American Geographers,* Vol. 59, No. 4, December 1969.

Pacific Regions

IDEAS AND CONCEPTS

Maritime environments
High-low insular cultures
Acculturation

Between the Americas to the east and Asia and Australia to the west lies the vast Pacific Ocean, larger than all the land areas of the world combined. In this great ocean lie tens of thousands of islands, some large (New Guinea is by far the largest), others small (many are uninhabited). This fragmented, culturally complex realm, the preponderance of water notwithstanding, does possess regional identities. It includes the Hawaiian Islands, Tahiti, Fiji, Samoa—fabled names in a world apart.

Indonesia and the Philippines are not part of the Pacific realm, and neither are Australia and New Zealand. Before the European invasion, Australia and New Zealand would have been included: Australia as a discrete Pacific realm on the basis of its indigenous black population, and New Zealand because its Maori population (page 139) has Polynesian affinities. But black Australians and Maori New Zealanders have been engulfed by the Europeanization of their countries, and the regional geography of Australia and New Zealand today is dominantly Western, not Pacific. Only on the island of New Guinea do Pacific peoples remain the dominant cultural element.

Ten Major Geographic Qualities of the Pacific Realm

1. The Pacific realm's total area is the largest of all world realms. Its land area, however, is among the smallest.
2. The bulk of the land area of the Pacific culture realm lies in New Guinea.
3. New Guinea, with an estimated population of 6 million, alone contains over three-quarters of the Pacific realm's population.
4. The Pacific realm consists of three regions: Melanesia (including New Guinea), Micronesia, and Polynesia.
5. The Pacific realm is the most strongly fragmented of all world realms.
6. The Pacific realm's islands and cultures may be divided into volcanic "high-island" cultures and coral "low-island" cultures.
7. The Hawaiian Islands, the United States fiftieth state, lie in the northern sector of Polynesia. As in New Zealand, indigenous culture has been submerged under Westernization.
8. Polynesian culture nearly everywhere in Polynesia is under severe strain in the face of external influences.
9. Indigenous Polynesian culture has a remarkable consistency and uniformity throughout the Polynesian region, its enormous dimensions and dispersal notwithstanding.
10. The Pacific realm is in politico-geographical transition as islands attain independence or revise their political associations.

The Pacific realm is fragmented, scattered, remote—but it does possess regions. New Guinea lies at the western end of a Pacific region that extends eastward to Fiji and includes the Solomon Islands, New Hebrides, and New Caledonia (Fig. P-1). These islands are inhabited by Melanesian peoples, who have very dark skins and dark hair (*melas* means black); the region as a whole is called Melanesia. Some cultural geographers include the Papuan peoples of New Guinea in the Melanesian race, but others suggest that the Papuans are more closely related to the aboriginal Australians. In any case, Melanesia is the most populous Pacific region. New Guinea alone has a population of about 6 million (although statistics are unreliable); the island is divided into two halves by a geometric boundary that separates West Irian, now an Indonesian province, from newly independent Papua New Guinea. With about 3.4 million inhabitants, Papua New Guinea became a sovereign state in 1975 after nearly a century of British and Australian administration. It is one of the world's

A representative village and landscape near Mount Hagen, Papua New Guinea. (Jack Fields/Photo Researchers)

A village market near Mount Hagen, Papua New Guinea. (George Holton/Ocelot)

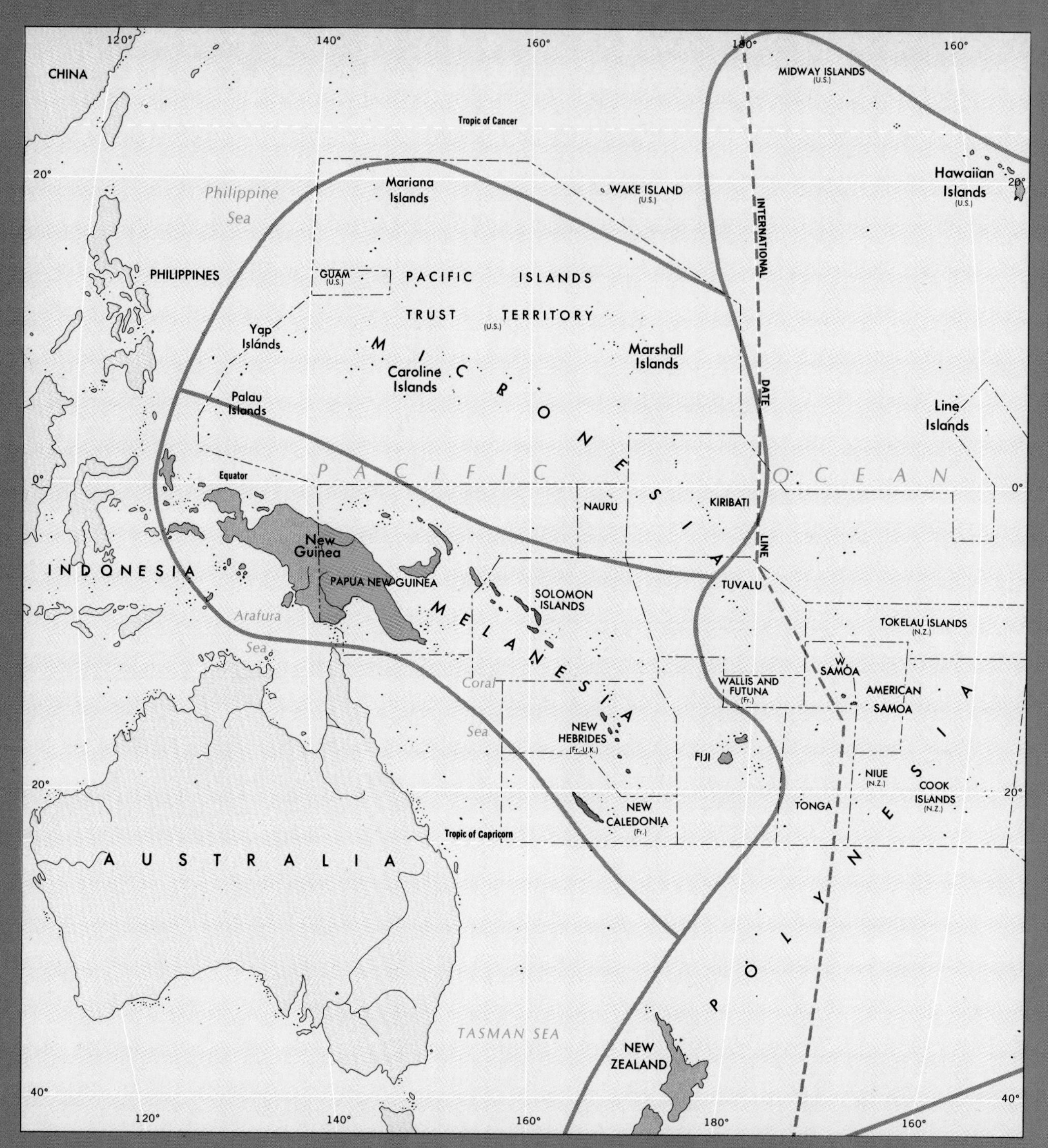

Figure P-1

poorest and least developed countries, with much of the mountainous interior (where the Papuan population is concentrated) hardly touched by the changes that transformed neighboring Australia. The largest town and capital, Port Moresby, has slightly over 100,000 inhabitants, but only about 12 percent of the people live in urbanized areas. Although English is used by the educated minority, about 85 percent of the population remains illiterate and over 700 languages are spoken by the Papuan and Melanesian communities. The Melanesians are concentrated in the north and east coastal areas of the country, and here as in the other islands of this region they grow root crops and bananas for subsistence.

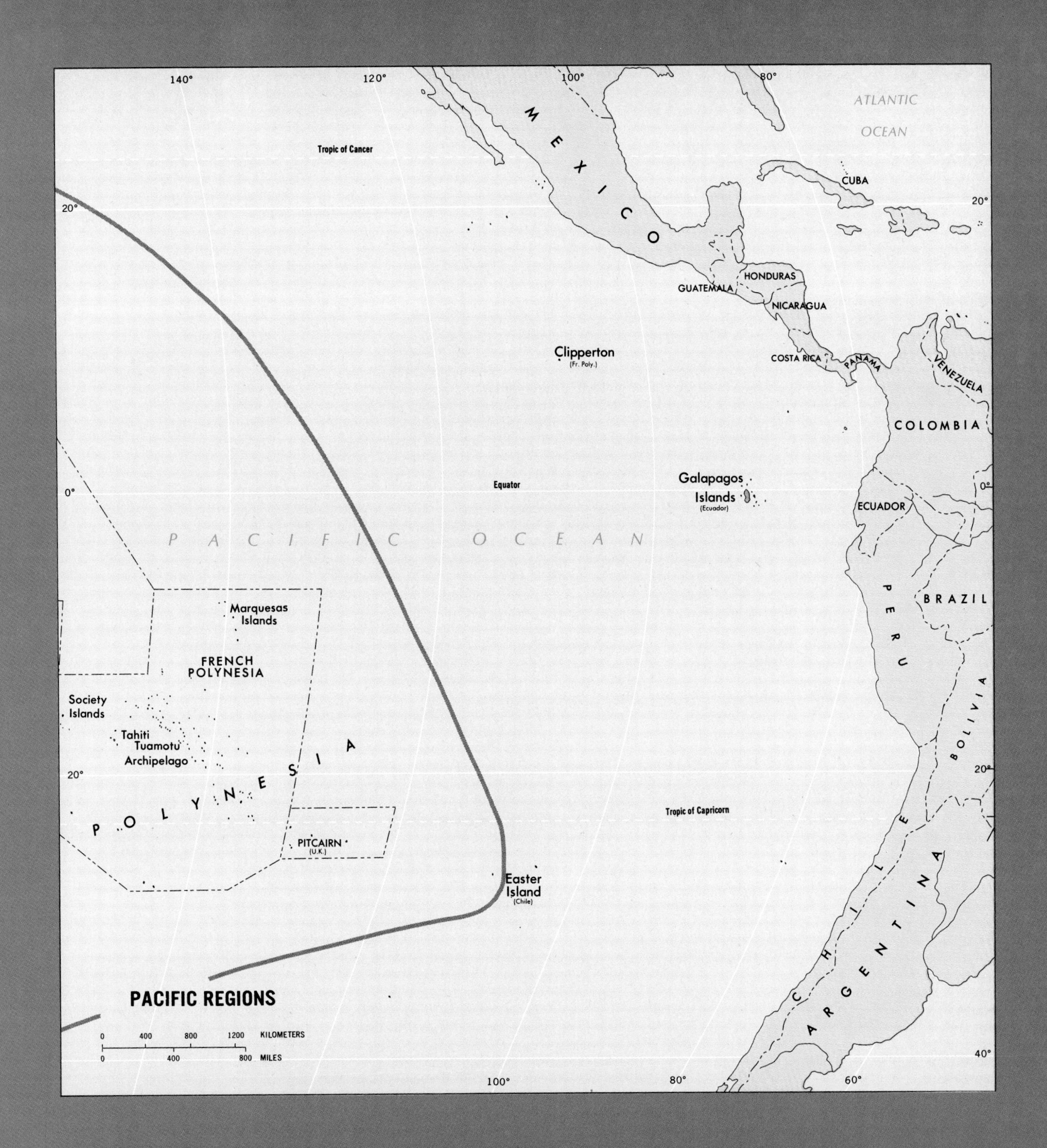

North of Melanesia (and east of the Philippines) lie the islands that comprise the region known as Micronesia (*micro* means small). In this case the name refers to the size of the islands, not the physical appearance of the population. The islands of Micronesia are not only small (many of them no larger than one square kilometer), but they are also much lower, on an average, than those of Melanesia. There are volcanic islands ("high islands," as the people call them), but they are outnumbered by islands made of coral, the "low islands" that barely rise above sea level. Guam, with 550 square kilometers (210 square miles) is Micronesia's largest island, and no island elevation any-

Movement in the Pacific realm always has been by water. Even today, the Polynesians are expert boaters, often choosing a water route when a road is also available. This canoe, loaded with people and goods, sets off from Suva (Fiji) to a small town about 12 kilometers away. The trip could also be made by vehicle around the bay, but these Fijians perfer the familiar alternative. (Harm J. de Blij)

where in Micronesia reaches 1000 meters. The region is largely a United States Trust Territory and includes the Mariana, Marshall, and Caroline Islands and Truk (Fig. P-1). The Micronesians, not nearly as numerous as the Melanesians, nevertheless form a distinct racial group in the Pacific realm. Culturally it is useful to distinguish their communities as "high-island" cultures based on the better-watered volcanic islands where agriculture is the mainstay, or "low-island" cultures on the sometimes drought-plagued coral islands where fishing is the chief mode of subsistence. These numerous Micronesian communities have developed a large number of locally spoken languages, many of them mutually unintelligible. But the islands do not exist in isolation. There is a certain complementarity between the islands of farmers and the islands of fishing people, and Micronesians—especially the low islanders—are skilled boaters. The trade for food and basic needs encourages circulation; sometimes the threat of devastating typhoon compels the "low islanders" to seek the safety of higher ground elsewhere.

To the east of Micronesia and Melanesia lies the heart of the Pacific, a great triangle from the Hawaiian Islands to Chile's Easter Island to New Zealand. This region is Polynesia, and New Zealand's Maori are Polynesians, but unlike New Guinea or Fiji or Hawaii, Europeans today dominate New Zealand's cultural landscapes. Polynesia is the region of numerous islands (*poly* means many), ranging from great volcanic mountains rising above the Pacific's waters (Mauna Kea on Hawaii reaches over 4200 meters or nearly 13,800 feet), clothed by luxuriant tropical vegetation and drenched by hundreds of centimeters of rainfall each year, to low coral atolls where a few palm trees form the only vegetation and where drought is a persistent and recurrent problem. The Polynesians have somewhat lighter skin and wavier hair than do other peoples in the Pacific realm; they often are described as having an excellent physique. Anthropologists differentiate between these original Polynesians and a second group, the Neo-Hawaiians, who are a blend of Polynesian, European, and Asian ancestries. In the state of Hawaii, Polynesian culture has not only been Europeanized but also Orientalized.

Notwithstanding its vastness and the diversity of its natural environments, Polynesia constitutes a geographic region in the Pa-

Superimposition of cultures: the skyscrapers of Honolulu on Oahu, Hawaii. Familiar Diamond Head, an extinct volcano, is visible in the left background. (Harm J. de Blij)

cific realm. Polynesian culture, spatially fragmented though it is, has a remarkable consistency and uniformity from one island to the next, from one end of the scattered region to the other. In vocabularies, technologies, housing, and artforms, this consistency is expressed. The Polynesians are uniquely adapted to their maritime environment, and long before European sailing ships began to arrive in Polynesian waters, the Polynesian seafarers had learned to navigate their wide expanses of ocean in huge double canoes as much as 45 meters (150 feet) in length. They traveled hundreds of kilometers to favorite fishing zones and engaged in interisland barter trade, using maps constructed from bamboo sticks and cowrie shells and reading the stars. Modern descriptions of a Pacific Polynesian paradise of emerald seas, lush landscapes, and gentle people distort harsh realities; Polynesian society was forced to accommodate much loss of life at sea when storms claimed their boats, families were ripped apart by accident as well as migration, hunger and starvation afflicted the local communities on smaller islands, and the island communities often were embroiled in violent conflicts and cruel retributions.

The political geography of Polynesia is complex. The Hawaiian Islands in 1959 became the fiftieth state to join the United States. The state's population is approaching 900,000, over 80 percent resident on the island of Oahu (Hawaii is the largest of the state's more than 130 islands). The Kingdom of Tonga, with a population of just over 100,000, became an independent country in 1970 after 70 years as a British protectorate. Western Samoa, a German possession until occupied by New Zealand during World War I, also achieved independence (1962) as one of the world's smallest countries with a mere 2934 square kilometers (1133 square miles) and under 200,000 inhabitants. The British-administered Ellice Islands were renamed Tuvalu; along with the Gilbert Islands to the north, they achieved independence from Britain in 1978. Other islands continued under French control (including the Marquesas Islands and Tahiti), under New Zealand administration (Rarotonga), and under British, United States, and Chilean flags. In the process of politico-geographical fragmentation, Polynesian culture has been dealt severe blows. Land developers, hotel builders, and tourist dollars have set Tahiti on a course along which Hawaii has already traveled far. The Americanization of Eastern Samoa has created a new society quite different from the old. Polynesia has lost much of its ancient cultural consistency; today the region is a patchwork of new and old—the new often bleak and barren, the old under pressure.

Additional Reading

An older standard work on the Pacific realm was edited by O. W. Freeman, *Geography of the Pacific,* published in 1951 in an edition that has since been reprinted a number of times (Wiley, New York). A more recent work is by H. Brookfield, editor, *The Pacific in Transition: Geographical Perspectives on Adaptation and Change,* published by St. Martin's, New York, in 1973. Also see a volume edited by R. G. Ward, *Man in the Pacific Islands,* a Clarendon Press, Oxford, publication of 1972. On Melanesia, a detailed discussion was written by H. C. Brookfield with D. Hart, *Melanesia: A Geographical Interpretation of an Island World,* published by Methuen in London in 1971. An unusual work that attempts to simulate Polynesian migration patterns was written by M. Levison, R. G. Ward, and J. W. Webb, *The Settlement of Polynesia.* The introductory discussion constitutes a superb overview of Polynesian environments and societies. This book was published by the University of Minnesota Press, Minneapolis, in 1977.

Glossary

Absolute location
The position or place of a certain item on the surface of the earth as expressed in degrees, minutes, and seconds of **latitude**[1] north or south of the equator and **longitude** east or west of Greenwich, England.

Accessibility
The degree of ease with which it is possible to reach a certain location from other locations. Some places can be reached by road, rail, and air; others are served by a single road that may become impassable during the wet season. Accessibility varies from location to location and can be measured.

Acculturation
Cultural modification resulting from intercultural borrowing. In cultural geography the term is used to designate the change that occurs in the culture of indigenous peoples when contact is made with a technologically more advanced society.

Action space
The area within which an individual or members of a community moves in the course of daily, seasonal, and lifetime activity. Action space of individuals is marked by locations about which they have information and which are viewed preferentially. Action space changes as accessibility changes. In an urban area, a new road or new shopping center can cause such change. In a wider context, a stronger and more aggressive community may encroach on a weaker and declining one, modifying individual and collective action spaces.

Aesthetics
Relating to things perceived to be beautiful.

Agglomerated (nucleated) settlement
Compact, closely packed settlement (small hamlet or larger village) sharply demarcated from adjoining farmlands.

Agglomeration
Process involving the clustering of manufacturing plants and businesses that benefit from close proximity because they share skilled labor pools and technological and financial amenities.

[1] Words in boldface type are defined elsewhere in this glossary.

Agrarian
Relating to the allocation and use of land, to rural communities, and to agricultural societies.

Alloy
Mineral such as chromium, cobalt, manganese, molybdenum, nickel, or tungsten (among others) that is smelted with iron to produce a steel of particular quality.

Alluvial
Refers to the deposits of mud and sand (*alluvium*) deposited by rivers and streams. Alluvial *plains,* consisting of such deposits laid down during floods, adjoin many larger rivers, creating fertile and productive soils. Alluvial **deltas** mark the mouths of rivers such as the Mississippi and the Nile.

Altiplano
High-elevation, elongated valley between even higher mountain ranges. In the Andes Mountains of South America altiplanos are at 3000 meters and even higher.

Antecedent
Antecedent, like *subsequent* and *superimposed,* is a term used in human as well as physical geography. Antecedent is something that goes before; from Latin *ante,* before, and

cedere, to go. In physical geography, a river that is antecedent is one that can be seen to be older than the landscape through which it flows; that is, it is older than the *structures* that underlie the landscape. While the rocks were being folded and the surface tilted by geologic processes, the stream was already there and just continued to cut down through the rocks. An antecedent boundary was there before the cultural landscape emerged—it "went before," and stayed put while all around it people moved in to occupy the surrounding area.

Anthracite
Highest-quality coal, formed under conditions of high pressure and temperature that eliminated most of the impurities (hydrocarbons). Anthracite burns almost without smoke and produces high heat.

Anthropogeographic boundaries
Political boundaries that coincide substantially with cultural discontinuities in the human landscape such as religious or linguistic transitions.

Apartheid
Literally, "apartness," the Afrikaans term given to South Africa's policies of racial separation. The term no longer has official sanction and has been replaced by "separate development."

Aquaculture
The use of a pond or other artificial body of water for the growing of food products, including fish, shellfish, and even seaweed. Aquaculture also may be carried out in controlled natural water bodies such as estuaries or river segments. Japan is among the world's leaders in aquaculture. When the activity is confined to the raising and harvesting of fish, it is usually referred to as *fish farming.*

Arable
Literally, "cultivable"—land that is fit for cultivation by one method or another.

Archipelago
A set of islands grouped closely together.

Area
A term that refers to a part of the earth's surface with less specificity than *region. Urban area* alludes very generally to a place where urban development has taken place while *urban region* requires certain specific criteria upon which a delimitation is based.

Areal interdependence
A term related to *functional specialization.* When one area produces certain goods or has certain raw materials or resources and another area has a different set of resources and produces different goods, their needs may be *complementary* and by exchanging raw materials and products they can satisfy each other's requirements. The concepts of areal interdependence and complementarity are related; both have to do with exchange opportunities between regions.

Arithmetic density
A country's population, expressed as an average per unit area (square kilometer or square mile), without regard for its distribution or the limits of arable land (see **physiologic density**).

Artesian
Water rising under pressure from an underground source, flowing from wells drilled through the overlying rock layers without any need for pumping. The source is replenished and the pressure kept up because the water-bearing rock stratum (*aquifer*) rises, buried under impervious layers, until it becomes exposed at a comparatively high elevation to precipitation.

Aryan
From the Sanskrit *Arya,* noble, a name applied to people who speak an Indo-European language and who moved into northern India. Although properly a language-related term, Aryan has various additional meanings, especially racial ones.

Autocratic
An autocratic government holds absolute power; rule is often by one person or a small group of persons who control the country by despotic means.

Atmosphere
The earth's envelope of gases that rests on the oceans and lands and penetrates the open spaces within soils. This layer of gases is densest at the earth's surface and thins with altitude. It is held against the planet by the force of gravity.

Balkanization
The fragmentation of a region into smaller political units; the breakup of countries.

Band
A territorially based group of people (usually numbering no more than several dozen) living together. This was the smallest community of early human society, smaller and much less complex than what has come to be known as a *tribe.*

Bantustan
Term formerly used to denote one of South Africa's African territories designated for "independence" under the terms of

"separate development" (see **apartheid**). The official designation today is Bantu Homeland.

Barrio
Slum development on the outskirts of a Middle American city.

Bauxite
Ore of aluminum. An earthy, reddish-colored material that usually contains some iron as well. Soil-forming processes such as leaching and redeposition of aluminum and iron compounds contribute to bauxite formation, and many bauxite deposits exist at shallow depths in tropical areas of high precipitation.

Bituminous coal
Coal of lesser quality than **anthracite** (more hydrocarbons remain) but of higher grade than **lignite.** Usually found in extensive horizontal layers, relatively undisturbed, between rock strata. When heated and converted to coking coal or *coke,* it is used in the great blast furnaces to make iron and steel.

Bolshevik
From the Russian, *bol'she* (large, majority). The name was taken by the radical left wing of the Russian Social Democratic Labor Party in Russia before the 1917 Revolution. Under the leadership of V. I. Lenin the Bolsheviks ousted the more moderate **Mensheviks** and took control of Russia in 1917.

Break-in-bulk
A location along a transport route where goods must be transferred from one carrier to another. In a port, ships' cargoes are unloaded and put on trains, trucks, and perhaps smaller river boats for distribution.

Buffer zone
A set of countries separating ideological or political adversaries. In Asia, Afghanistan, Nepal, Sikkim, and Bhutan formed a buffer zone between British and Russian-Chinese imperial spheres. Thailand was a *buffer state* between British and French colonial domains.

Caliente
The *tierra caliente* forms the lowest and most tropical of four vertical zones into which Middle and South American topography is divided, lying below 750 meters (2500 feet).

Cartel
An international syndicate formed to promote common interests in some economic area through the formulation of joint pricing policies and the limitation of market options for consumers.

Caste system
The strict social segregation of people (specifically in Hindu society) on the basis of ancestry and occupation.

Central business district
The heart of a city, the CBD is marked by high land values, a concentration of business and commerce, and a clustering of the tallest buildings.

Centrality
The strength of an urban center in its capacity to attract producers and consumers to its manufacturing and marketing facilities, the city's "reach" into the surrounding region. A *central place* possesses a certain measure of centrality and forms the urban focus for a particular region.

Central place theory
Geographic theory in search of explanations for the spatial distribution of urban places.

Centrifugal forces
A term employed to designate forces that tend to divide a country, such as religious, linguistic, ethnic, or ideological differences.

Centripetal forces
Forces that tend to unite and bind a country together: a strong national culture, a single faith (Islam forms a centripetal force in Arab countries), shared ideological objectives.

Charismatic
Personal qualities of certain leaders that enable them to capture and hold the popular imagination, to secure the allegiance and even the devotion of the masses.

Chernozem
Also known as the *Black Earth,* this is a Russian term for the fertile, dark-colored, **humus**-rich soil that prevails in the Soviet Union's Ukraine and in North America's Great Plains.

Chinampa cultivation
A system of cultivation developed by Indian peoples in indigenous Middle America that involved the creation of "floating gardens," a form of land reclamation in shallow lakes and marshy areas. Strips of aquatic vegetation, cut from the lake margins, were sunk in layers until the top one just reached the water surface. On top of the uppermost layer the Indian farmers spread fertile mud from the lake bottom to create a planting surface. This resulted in long artificial farm fields, 2 to 4 meters wide and more than 30 meters long, separated from each other by narrow canals.

City-state
An independent political entity consisting of a single city with (and sometimes without) an immediate **hinterland.** The ancient city-states of Greece have their

modern equivalent in Singapore.

Climate
A term used to convey a generalization of all the recorded weather observations at a certain place or in a given area. It represents an "average" of all the weather that has occurred there. In general, a tropical location such as the Amazon Basin has a much less variable climate than areas located, say, midway between the equator and the pole. In low-lying tropical areas the *weather* changes little, and so the climate is rather like the weather on any given day. But in the middle latitudes there may be summer days to rival those in the tropics and winter days so cold that they resemble polar conditions. Here, when we refer to climate, we must combine the warm summers and the cold winters in our generalization.

Climax vegetation
The final, stable vegetative community that has developed at the end of a succession under a particular set of environmental conditions. This vegetation is in dynamic equilibrium with its environment, including other elements of the ecosystem such as its animal life.

Collectivization
The reorganization of a country's agriculture involving the expropriation of private holdings and their incorporation into relatively large-scale units farmed and administered cooperatively by those who live there. The system has transformed Soviet agriculture, has been resisted in communist Poland, and has gone beyond the Soviet model in China's program of communization.

Common Market
Name given to a group of European countries that formed an association to promote their economic interests. Official name is European Community (EC).

Compact
A politico-geographical term to describe a state that possesses a roughly circular, oval, or rectangular territory in which the distance from the geometric center to any point on the boundary does not vary very much. Poland, Belgium, and the U.S. state of Colorado are examples of this shape category.

Complementarity
Regional complementarity exists when two regions, through an exchange of raw materials and finished products, can substantially satisfy each other's requirements.

Concentric
Having a common center.

Concentric zone theory
A geographical model of the city that suggests the existence of five concentric regions.

Condominium
In political geography, the term *condominium* denotes the shared administration of a territory by two governments. The African state of Sudan long was administered jointly by the United Kingdom and Egypt as a condominium.

Coniferous forest
As the word suggests, a forest of cone-bearing trees—needleleaf, evergreen trees with straight trunks and short branches, including spruce, fir, and pine.

Contagious diffusion
The spread of an idea, innovation, or some other item through a population by contact from person to person, without regard to social status or position in the society.

Contiguous
A word of some importance to geographers, for whom it has spatial implications. It means, literally, to be in contact with, adjoining, or adjacent. Sometimes we hear the continental United States without Alaska referred to as contiguous. Alaska is not contiguous to these states, for Canada lies between; neither is Hawaii, separated by thousands of miles of ocean. The ancient Hellenic civilization was not contiguous either, for its settlements were separated by sea, peninsulas, and rugged mountains.

Continental shelf
Beyond the coastlines of the continents the surface beneath the water, in many areas, declines very gently until a depth of about 100 fathoms (600 feet; somewhat under 200 meters). Thus the edges of the landmasses today are inundated, for only at the 100-fathom line does the surface drop off sharply to the ocean bottom along the *continental slope.* The submerged continental margin is called the continental shelf, and technically it extends from the coastline to the continental slope.

Contour plowing
Plowing along the contour line (thus along horizontal rows) rather than up and down the slope.

Conurbation
Imprecise term used to identify an urban area marked by some degree of coalescence of two or more urban centers.

Copra
Meat of the coconut, fruit of the coconut palm.

Cordillera
Mountain chain consisting of sets of parallel ranges.

Core area
In geography, a term with several connotations. *Core* refers to the center, heart, focus; in physical geography the core area of a continent identifies the ancient shield zone. In human geography the core area of a nation-state is constituted by the national heartland, the largest population cluster, most productive region, the area with greatest centrality and accessibility, probably containing the capital city as well.

Corridor
A land extension that connects an otherwise landlocked state to the ocean. History has seen several such corridors come and go. Poland once had a corridor (it now has a lengthy coastline); Bolivia lost a corridor to the Pacific Ocean between Peru and Chile; Finland possessed a corridor to the Arctic Ocean. *Secondary* corridors may lead to another means of egress, for example a navigable river. Colombia, although it has coasts on the Caribbean Sea and the Pacific Ocean, has a secondary corridor to the Amazon River.

Cultural diffusion
The spreading and dissemination of a cultural element from its place of origin over a wider area.

Cultural ecology
The interactions between cultural development and prevailing conditions of the natural environment.

Cultural landscape
The "forms superimposed on the physical landscape by the activities of man," to quote Carl O. Sauer.

Culture area
A distinct, culturally discrete spatial unit; a region within which certain cultural norms prevail. Culture *region* might be more appropriate.

Culture complex
A related set of culture traits, such as prevailing dress codes, cooking and eating utensils.

Culture hearth
Heartland, source area, place of gestation of one of the world's major cultures.

Culture realm
A cluster of regions in which related culture systems prevail. In North America, the United States and Canada form a culture realm, but Mexico belongs to a different culture realm.

Culture trait
A single element of normal practice in a culture, such as the wearing of a turban.

Cyclic movement
Movement (migratory, for example) that has a closed route repeated annually or seasonally.

Death rate
The *crude death rate* is expressed as the number of deaths per thousand individuals in the population during a given year.

Deciduous
A deciduous tree loses its leaves at the beginning of winter or the start of the dry season.

Definition
In political geography, the description (in legal terms) of a boundary between two countries or territories.

Delimitation
In political geography, the translation of the written terms of a boundary treaty (the *definition)* into cartographic representation.

Delta
Alluvial lowland at the mouth of a river, formed when the river deposits its alluvial load upon reaching the sea.

Demarcation
In political geography, the marking of an international political boundary on the ground by means of barriers, fences, posts, or other markers.

Demographic variables
Births (fertility), deaths (mortality), and migration are the three basic demographic variables.

Demography
Study of population: birth and death rates, growth patterns, longevity, and related patterns.

Density of population
The number of people per unit area. For specific forms, *see* Arithmetic, Physiologic, and other types of density measures.

Desert
An arid area supporting very sparse vegetation, usually exhibiting extremes of heat and cold because the moderating influence of moisture is absent.

Desertification
The encroachment of desert conditions upon moister zones along the desert margins, where plant cover and soils are threatened by desiccation, in part through overuse by humans and their domestic animals and, possibly, also because of inexorable shifts in the earth's environmental zones.

Despotic rule
Absolute, often tyrannical rule by an individual or a small group of individuals, negating human rights.

Determinism
See Environmental determinism

Devolution
In political geography, the disintegration of the nation-state as a result of emerging or reviving regionalism.

Dhows
Wooden boats with characteristic triangular sail, plying the seas between Arabian and East African coasts.

Diffusion
The dissemination of a culture element (such as a technological innovation) or some other phenomenon (a disease outbreak, for example). Geographers recognize *expansion* diffusion and *relocation* diffusion.

Dispersed settlement
In contrast to **agglomerated** or **nucleated** settlement, dispersed settlement is characterized by the wide spacing of individual homesteads along rural roads and waterways. The pattern is characteristic of rural North America.

Distance decay
The various degenerative effects of distance on spatial processes. The degree of spatial interaction diminishes as distance increases, but there is no simple way of saying exactly how this happens. In migration, for example, the information flow from the new homelands of the emigrants proved less accurate the farther they had to travel.

Diurnal
Daily.

Divided capital
In political geography, a country whose administrative functions are carried on in more than one city is said to have divided capitals.

Doldrums
Equatorial belt of variable winds and frequent extended calms. The name is associated with the problems of sailing vessels that were often becalmed, sometimes with disastrous results, in this zone of undependable winds.

Domestication
The transformation of a wild animal or wild plant into a domesticated animal or a cultivated crop. The process involves changes in the physiology and tameness of the animal and modifications in the growing habits of plants.

Double cropping
The planting, cultivation, and harvesting of two crops, successively within a single year, on the same piece of farmland.

Doubling time
The time required for a population to double in size.

Ecology
Strictly speaking, *ecology* refers to the study of the many interrelationships between all forms of life and the natural environments under which they have evolved and continue to develop; the study of ecosystems. The term is also used in more specific ways, as in *cultural ecology* and *urban ecology,* involving the study of the relationships between human cultural manifestations and the natural environment in which they are occurring.

Ecomienda
The reward, in terms of land and sometimes people to work it, received by early Spanish invaders of South America, notably those in the Inca-Peru sphere.

Elite
A social class whose power and privilege give it control over a society's political, economic, and cultural life. The members of an elite usually are small in number but extremely influential in their country's affairs.

Emigrant
A person migrating away from a country or area; an outmigrant.

Enclave
A piece of territory that is surrounded by another political unit of which it is not a part.

Entrepôt
A place, usually a port city, where goods are imported, stored, and transshipped. Thus an entrepôt is a **break-in bulk** location, but it has special facilities for temporary storage prior to redistribution.

Environmental determinism
The view that the natural environment has a controlling influence over various aspects of human life including cultural development and even physical attributes. Also referred to as *environmentalism.*

Erosion
A combination of processes resulting in the modification of landscapes. Running water, wind action, and the force of moving ice combine to wear away soil and rock. Human occupation often speeds erosional processes through the destruction of natural vegetation, careless farming practices, and overgrazing by livestock.

Escarpment
The cliff or steep slope that forms the edge of a plateau.

Estuary
The widening mouth of a river as it reaches the sea. An estuary forms when the margin of the land has subsided somewhat and seawater has invaded the river's lowest portion. Islands often mark the estuary of a major river.

Evapotranspiration
The loss of moisture to the atmosphere through the combined processes of evaporation from the soil and plants and from transpiration from plants.

Exclave
A bounded piece of territory that is part of a particular state but lies separated from it by the

territory of another state. Note: an island does not qualify as an exclave under this definition.

Expansion diffusion
The spread of an innovation or an idea through a population in an area in such a way that the number of those influenced grows continuously larger, resulting in an expanding area of dissemination.

Extraterritoriality
Politico-geographical concept that suggests that the property of one state lying within the boundaries of another state actually forms an extension of such a state. The principle was used by colonial powers in China to carve sections of cities out of the Chinese national state.

Favela
Shantytown on the outskirts or even well within an urban area in Brazil.

Fazenda
Coffee plantation in Brazil.

Federation
The association and cooperation of two or more nation-states or territories to promote common interests and objectives.

Fertile crescent
Semicircular zone of productive lands extending from near the southeast Mediterranean coast through Lebanon and Syria to the **alluvial** lowlands of Mesopotamia. Formerly better-watered and more fertile than today, this is one of the world's source areas of agricultural innovation.

Feudalism
Prevailing politico-geographical system in Europe during the Middle Ages, when land was owned by lords, counts, barons, and dukes and worked by peasants and serfs. Feudal lords held absolute power over their domains and inhabitants, and Europe was a mosaic of private estates. Feudalism existed also in other parts of the world and the system persisted into the twentieth century in Ethiopia and Iran, among other areas.

Fjord
Narrow, steep-sided, elongated, inundated coastal valley deepened by glacier ice that has since melted away, leaving the sea to penetrate. Norway's coasts are marked by fjords, as are parts of the coasts of Canada, Greenland, and Southern Chile.

Floodplain
Low-lying area adjacent to a river, often covered by **alluvium** and subject to the river's floods.

Forced migration
Human migration flows in which the migrants have no choice whether or not to move to a new abode.

Formal region
A type of region marked by a certain degree of homogeneity in one or more phenomena; also called *uniform* region or *homogeneous* region.

Forward capital
Capital city positioned in actually or potentially contested territory, constituting an advance bastion confirming the state's determination to maintain its presence in the region in contention.

Fragmented state
A state whose territory consists of several separated parts, not a contiguous whole. The individual parts may be separated from each other by the land area of other states or by international waters.

Francophone
Describes country or region where other languages are also spoken but where French is the **lingua franca** or the language of the **elite.** Quebec is Francophone Canada.

Fria
In Middle and South America, *tierra fria* denotes the highest zone of human occupation below the snow line, from an average of 1700 meters (5500 feet) to around 3000 meters (10,000 feet) or more. There are pastures in the cold zone; wheat and potatoes also provide sustenance in this often difficult environment.

Frontier
Area of penetration, of contention; an area not yet fully integrated in a national state.

Functional region
A region marked less by any internal consistence or sameness than by its integrity as a functioning system; because of its focus upon a central node, also called *nodal* region.

Functional specialization
The production of particular goods (or services) as a dominant activity in a particular location. Certain cities produce automobiles or iron and steel; others serve tourists. All these places also perform other functions, but their specialization lies in the production of certain items.

Geomancy
A Chinese belief in the power and influence of geomancers, philosopher-teachers who were able to communicate with the spirits thought to reside in the landscape, in hills and mountains, soil and plants, water and air.

Geometric boundaries
Political boundaries defined and delimited as straight lines or arcs.

Geopolitik (Geopolitics)
A school of political geography that involved the use of quasi-academic research to encourage a national policy of expansionism and imperialism. Its origins are attributed to a Swede by the name of Kjellen but its most famous practitioner was a German named Haushofer. The misuse of *Geopolitik* does not necessarily invalidate its position as a field worthy of academic pursuit.

Ghetto
An urban region marked by particular ethnic, racial, religious, and economic properties, usually (but not always) a low-income area.

Green revolution
The successful development of higher-yield, fast-growing varieties of rice and other cereals, leading to increased production per unit area and a temporary narrowing of the gap between population growth and food needs.

Gross national product
The total value of all goods and services produced in a country during a certain year. In some underdeveloped countries, where a substantial number of people practice subsistence and where the collection of information is difficult, GNP figures may be unreliable.

Growing season
The number of days between the last frost in the spring and the first frost of the fall.

Growth pole
An urban center with certain minimal attributes that, if augmented by a measure of support, will stimulate regional economic development in its **hinterland.**

Hacienda
Literally, a large estate in a Spanish-speaking country. Sometimes equated with *plantation,* but there are important differences between these two types of agricultural enterprise.

Heartland
National or regional focus, the center of activity and productivity, the source of culture and ideology and the foundation of power.

Hierarchical diffusion
A form of diffusion in which an idea or innovation spreads from one individual to another in a particular class or group rather than (as in the case of expansion diffusion) through the population as a whole.

Hierarchy
An order or gradation of phenomena; each grade or level being subordinate to the one above it and superior to the one below. Simple words can designate a hierarchy: hamlet, village, town, city, metropolis.

High seas
Areas of the oceans still beyond national jurisdiction, open and free for all to use.

Highveld
A term used in southern Africa to identify the grass-covered, high-elevation plateau that dominates much of the region. *Veld,* sometimes misspelled veldt, means grassland in Hollands and in Afrikaans. The lowest areas in South Africa are called lowveld; areas that lie at intermediate elevations are the middleveld.

Hinterland
Literally, "country behind," a term that applies to an area lying aroung a city, where that particular city is the chief service center, the focus of commercial activity, the major urban influence.

Humus
Dark-colored upper layer of a soil, consisting of decomposed and decaying organic matter such as leaves and branches, nutrient-rich and giving the soil a high fertility.

Iconography
The identity of a region; its particular cultural landscape and atmosphere.

Immigrant
A person migrating into a particular country or area; an immigrant.

Imperialism
The drive toward the creation and expansion of an empire and, once established, its perpetuation.

Indenture
Contract labor; the sale of one's labors for a certain period of time. Tens of thousands of South Asians came to East Africa as indentured workers at British behest; working conditions often did not match contract stipulations, but indentured workers did not have the option to cancel their agreements.

Industrial revolution
A series of related innovations and inventions that had the effect of intensifying industrial production in England during the eighteenth century through the introduction of machines and inanimate (nonhuman, nonanimal) power.

Infrastructure
The foundations of a society; urban centers and their amenities, communications, existing farms, factories and mines, and facilities such as schools, hospitals, postal services, police, and armed forces.

Innovation
Something new: a new method, a new device, a new idea.

Insular
Having the qualities and properties of an island. Real islands are not alone in possessing such properties: an **oasis** in the middle of a large desert also has qualities of insularity.

Insurgent state
Territorial embodiment of guerrilla activity, the antigovernment insurgents establishing a territorial base in which they exercise permanent control; thus, a state within a state.

Intermontane
Literally, "between mountains." The location can bestow certain qualities of natural protection or isolation to a community; it also affects climate and weather.

Internal migration
Migration flow within a nation-state, such as westward and northward movements in the United States and eastward movement in the Soviet Union.

International migration
Migration flow involving movement across international boundaries.

Intertropical convergence
(Also ITC or ITCZ.) Belt of surface convergence or air masses in the Northeast and Southeast Trades along the axis of the equatorial trough.

Intervening opportunity
The presence of a nearer and perceived-to-be better opportunity that diminishes the attractiveness of sites even slightly farther away.

Introduced capital city
A capital city chosen to further certain national objectives such as seaward orientation, territorial redirection, or centralization; a disputed term.

Irredentism
A policy of cultural extension and potential political expansion aimed at a national group living in a neighboring country.

Irrigation
The artificial watering of croplands. *Basin* irrigation is an ancient method involving the use of floodwaters, trapped in basins on the floodplain and released in stages to augment rainfall. *Perennial* irrigation requires the construction of dams, barrages, and irrigation canals for year-round water supply.

Isobar
Line connecting points of equal atmospheric pressure. An *isotherm* connects points of equal temperature; an *isohyet* links points of equal rainfall.

Isolation
The condition of being cut off or far removed from a mainstream (or streams) of thought and action. It also denotes a lack of receptivity to outside influences and stimuli. Distance and inaccessibility are among its causes.

Isthmus
A land bridge, a comparatively narrow link between larger bodies of land. Central America forms an isthmian link between North and South America.

Jihad
The Moslem holy war that carried Islam into areas of Africa and Eurasia.

Juxtaposition
When places or areas are located very near each other, they are said to be juxtaposed or in close juxtaposition.

Land alienation
One society or culture group taking land from another. In Africa, white Europeans took land from black Africans and put it to new uses, fencing it off and restricting settlement. Land alienation also occurred in the Americas and in Asia.

Land reform
The spatial reorganization of agriculture through the allocation of farmland (perhaps expropriated from landlords) to peasants and tenants who never owned land; also, the consolidation of excessively fragmented farmland into more productive, perhaps cooperatively run farm units.

Landlocked
A country or state that is surrounded everywhere by land is identified as landlocked. Without coasts, a landlocked state is at a disadvantage in a number of ways—in terms of access to international trade routes and in the scramble for possession of areas of the **continental shelf.**

Latitude
Latitude lines are those that are aligned east-west across the globe, from 0 degrees latitude at the equator to 90 degrees north and south latitude at the poles. Areas of low latitude therefore lie near the equator, in the tropics; high latitudes are those near the poles and in the Artic and Antarctic. The United States and Europe lie in what is commonly called the middle latitudes.

Lignite
Also called brown coal, this is a woody variety of coal, just a little better than *peat* but not nearly as good as the next higher grade, **bituminous coal.** Lignite cannot be used in the large blast furnaces, but it is important as a fuel nevertheless.

Lingua franca
The term derives from "Frankish language" and applies to a tongue spoken in ancient Mediterranean ports and consisting

of a mixture of Italian, French, Greek, Spanish, and even some Arabic. Today it refers to a "common language," a language that can be spoken and understood by many peoples although they speak other languages at home.

Littoral
Coastal, along the shore.

Llanos
Name given to the savanna-like grasslands of the Orinoco River's wide basin in parts of Colombia, Venezuela, and Guyana. Elsewhere these grassy plains have different names—for example, *chaco* in Northern Argentina.

Location theory
Comprehensive term for the theories and models developed by geographers to explain the relative location of various human establishments, including towns and cities, factories, shopping centers, and so on.

Loess
Deposit for very fine silt or dust, laid down after having been wind-borne over long distances, perhaps hundreds of kilometers. Loess is characterized by its fertility under irrigation and its ability to stand in steep vertical walls when eroded by a river or (as in China) excavated for human use.

Longitude
Angular distance east or west measured from the prime **meridian** (standard meridian).

Maghreb
Western lands of the North African-Southwest Asian realm, consisting of Morocco, Algeria, and Tunisia.

Malthusian
Designates the viewpoints of Malthus; Neo-Malthusian is used to describe those who subscribe to Malthusian positions in modern contexts.

Marchland
A frontier or area of uncertain boundaries subject to various national claims and an unstable political history. The term refers to the movement of armies from various nationalities across this zone.

Matrilineal
In a matrilineal society, descent is traced through the female line; so are inheritance, succession, and other functions. In a *patrilineal* society, descent is traced through the male line.

Megalopolis
Term used to designate large, coalescing supercities that are forming in diverse parts of the world—in Japan, Western Europe, and eastern North America. In the American example, a megalopolis is emerging along the eastern seaboard from Boston to Washington.

Mensheviks
The moderate faction in the Russian Social Democratic Labor Party of prerevolutionary Russia, whose representatives actually constituted a majority but who lost the ideological struggle with the **Bolsheviks** led by V. I. Lenin. Menshevik leaders shrunk into oblivion and exile after the Revolution of 1917.

Mercantilism
Protectionist policy of European states during the sixteenth to the eighteenth century, promoting a state's economic position in the contest with other countries. The acquisition of gold and silver and the maintenance of a favorable trade balance (more exports than imports) were central to the policy.

Meridian
Line of **longitude,** aligned north-south across the globe, part of the grid system of which **parallels** also form parts. The 0 degree meridian (the prime or *standard* meridian) lies through Greenwich, England; the line marking 180 degrees lies on the opposite side of the globe in the Pacific Ocean. Measurements are given in degrees east or west of Greenwich to 180 degrees.

Mesoamerica
Meso means middle or in-between; Mesoamerica is another way of saying Middle America. Since it would be confusing to identify the old culture hearth as well as the geographical region as Middle America, Mesoamerica is used to denote a historically significant part of Middle America.

Mestizo
The root of this word is the Latin for *mixed,* and it means a person of mixed ancestry. Specifically, it is used to identify a person whose ancestry is both white and American Indian.

Mexica Empire
The name the ancient Aztecs gave to the domain over which they held hegemony in Middle America.

Migration
See Internal, International, Forced, and Voluntary migration.

Migratory movement
Human movement from source to destination without a return journey, as opposed to *cyclic* movement.

Monotheism
The belief in and worship of a single god.

Mulatto
A person of mixed ancestry with African (black) and European (white) antecedents.

Multinationals
The multinational (or global) corporations that strongly influ-

ence the economic and political affairs of many nation-states.

Multiple-nuclei model
A model that shows a city to consist of several nuclear growth points.

Nation
Legally, a term encompassing all the citizens of a state, it has also taken on other connotations, and dictionary definitions tend to refer to bonds of language, ethnicity, religion, and other shared cultural attributes. Such homogeneity prevails within very few states.

Nation-state
A country whose population possesses a substantial degree of cultural homogeneity and unity.

Natural increase
Population growth measured as the excess of live births over deaths per thousand individuals per year. Natural increase of a population does not reflect either immigration or emigration.

Natural resource
As used in this book, any valued element of the environment, including minerals, water, vegetation, and soil.

Nautical mile
By international agreement, the nautical mile—standard measure at sea—is 6076.12 feet in length, equivalent to approximately 1.15 statute miles.

Negritude
Philosophy promoting black cultural pride, with roots in Caribbean America, French universities, and Francophone West Africa.

Network, transport
The entire physical system of connections and nodes through which movement can occur.

Nomadism
Movement from place to place. Nomadic people are mobile, wandering, mostly pastoral people.

Norden
A regional concept that includes the three Scandinavian countries, Finland, and Iceland.

Nucleated settlement
A clustered pattern of population distribution.

Nucleation
Clustering, agglomeration.

Oasis
An area, small or large, where the supply of water transforms the desert into a green cropland. An oasis may consist of just a few palm trees and a mud-baked building or two, but it will nevertheless be the most important focus for human activity for miles around. It may encompass a large, densely populated zone along a major river where irrigation projects stabilize the water supply—as along the Nile River in Egypt, really an elongated set of oases.

Occidental
Occidental and *Oriental* are words that derive from the Latin for *fall* and *rise,* and thus with the direction in which the sun is seen to rise and fall. *Oriental* (rise) therefore refers to the east, and Oriental peoples today are peoples of the Far East; *Occidental* (fall) relates to the west, and it means Western.

Oligarchy
Political system involving rule by a small minority, an often corrupt **elite.**

Oral history
Historical detail as preserved in orally transmitted records (that is, by mouth, through legends, myths, folk tales, children's stories, plays, and so forth).

Organic theory
Concept that suggests that the state is in some ways analogous to a biological organism with a life cycle that can be sustained through cultural and territorial expansion.

Oriental, Occidental
The root of the word *Oriental* is the Latin for *rise,* and thus it has to do with the direction in which one sees the sun "rise"—the east; *oriental* therefore means eastern. *Occidental* originates from the Latin for *fall,* or the "setting" of the sun in the west; *occidental,* then, means Western.

Pacific ring of fire
Zone of crustal instability, marked by earthquakes and volcanic activity, surrounding the Pacific Ocean.

Pangaea
A vast landmass consisting of most of the areas of the present continents, including the Americas, Eurasia, Africa, Australia, and Antarctica, that existed until near the end of the Mesozoic Era when plate divergence and continental drift broke it apart. The "northern" segment of Pangaea is called *Laurasia,* the "southern" part *Gondwana.*

Parallel
Line that connects points of the same latitude on the earth's grid. A parallel, as a line of **latitude,** is intersected at right angles by a **meridian,** a line of **longitude.**

Paramós
Highest zone in Latin American vertical zonation, above the tree line in the Andes.

Pastoralism
A form of economic pursuit: the practice of raising livestock, although many peoples described as pastoralists actually have a mixed economy—they may fish, or hunt, or even grow a few

crops. Few peoples are "pure" herders. But pastoral peoples' lives chiefly revolve around their animals.

Peasants
In a stratified society, peasants are the lowest class of people who depend on agriculture for a living, but who often own no land at all and must survive as tenants or day workers.

Peninsula
Comparatively narrow stretch of land extending from the main body into the sea. Italy constitutes a peninsula.

Peon
Term used in Latin America to identify people who often live in serfdom to a wealthy landowner, landless peasants in continuous indebtedness.

Per capita
Capita in this context means "individual"; income, production, or some other measure is often given per individual.

Periodic market
Village market that opens every third or fourth day (or at some other regular interval), part of a regional network of similar markets in a preindustrial, rural setting where goods are brought to market on foot (or perhaps by bicycle) and barter remains a major mode of exchange.

Periodic movement
A form of migration involving intermittent but recurrent movement, such as temporary relocation for college attendance or service in the armed forces.

Permafrost
Permanently frozen water in the soil, regolith, and bedrock, as much as 300 meters in depth, producing the effect of completely frozen ground. Permafrost may begin a few centimeters below seasonally warmed and "soft" surface soil or deeper.

Permanent capital
Capital city that has historically been the headquarters of the state, such as Rome and Athens.

Physiographic-political boundaries
International political boundaries that coincide with prominent physiographic breaks in the natural landscape, such as rivers or the crests of mountain ranges.

Physiographic region (province)
A region within which there prevails a substantial degree of landscape homogeneity.

Physiologic density
The number of people per unit area of arable land.

Pidgin
A language that consists of words borrowed and adapted from other languages, originally developed from commerce among peoples speaking different languages.

Pilgrimage
A journey to a place of great religious significance by an individual or by a group of people. When Islam entered West Africa, pilgrimages to Mecca attracted tens of thousands of Moslems annually. Many took up permanent residence along the way and the human map of savanna Africa was transformed.

Plantation
A large estate owned by an individual, family, or corporation and organized to produce a cash crop. Almost all plantations were established within the tropics; since 1950 many have been divided into smallholdings or reorganized as cooperatives.

Pleistocene epoch
Recent period of geologic time covering the rise of humanity, marked by repeated advances of glacial ice and milder *interglacials.* Although the most recent 10,000 years are known as the *Recent* Epoch, Pleistocene conditions seem to be continuing and the present is likely to be another Pleistocene *interglacial;* the glaciers will return.

Population explosion
The rapid growth of the world's human population during the past century, attended by ever-shorter doubling times and increasing *rates* of increase.

Population structure
Representation of a population by sex and age, usually in the form of a population *pyramid.*

Primacy
At the top of a hierarchy, in first place, the highest ranking. A *primate* city is the dominant urban place in a country.

Primary industries
Industries engaged in the direct exploitation of natural resources, such as mining, fishing, lumbering, and farming.

Primate city
A country's largest city, most expressive of the national culture, its focus, usually (but not always) also the capital city.

Proletariat
Lowest class in a community or society; people who own no capital or means of production and who live by selling their labor.

Prorupt
A type of state territorial morphology that exhibits a narrow, elongated land extension.

Protectorate
In Britain's system of colonial administration, the protectorate was a designation that involved

the assurance of certain rights and guarantees to peoples who had been placed under the control of the Crown. One of these guarantees related to the restriction of European (white) settlement and land alienation. In some protectorate areas, whites were not allowed to own land at all. It did not always work out that way, but such were the intentions.

Public architecture
The architecture of buildings such as government headquarters, states offices, and other official structures.

Push-pull concept
The idea that migrations from rural to urban areas (and other migration streams as well) are stimulated by conditions in the source area that tend to drive people away and also by the perceived attractiveness of the destination.

Region
A commonly used term and a geographic concept of central importance. An area on the earth's surface marked by certain properties.

Relative location
The position or situation of a point or place relative to the position of other places. Distance, accessibility, connectivity, and other factors affect relative location.

Relic boundary
An international political boundary that has ceased to function but the imprint of which on the cultural landscape can still be detected.

Relief
Vertical difference between the highest and lowest points within a particular area.

Relocation diffusion
Sequential diffusion process in which the things being diffused evacuate the old areas as they move (relocate) to new areas. A disease can move from population cluster to population cluster in this way, running its course in one area before fully invading the next. A population movement can carry the process, for example the movement of the black population from the South to northern cities in the United States.

Renaissance
Literally, "rebirth." Specifically, the revival of Europe from the fourteenth into the seventeenth century with a resurgence of the arts, literature, and science.

Rural density
A measure that indicates the number of persons per unit area actually living in rural parts of the country, excluding the urban concentrations; provides an index of rural population pressure.

Sahel
Zone across West Africa between the southern margins of the Sahara and the moister coastal forest and savanna zone.

Satellite state
A state whose sovereignty has been compromised by the dominating influence of a larger power. Such a situation can arise when the ruling political party in such a country represents the ideology of the larger power, or when the economic influence of the larger power is so great that it is in virtual control of production in the satellite state.

Scale
Representation of a real-world phenomenon at a certain level of reduction.

Scandinavia
Denmark, Sweden, and Norway.

Secondary industries
Industries that process raw materials and transform them into finished products; the *manufacturing* industries.

Secular
Worldly, nonreligious, nonspiritual. Secularism holds that ethical and moral standards should be formulated and adhered to for life on earth, and not to accommodate the prescriptions of a deity and promises of a comfortable afterlife. The secular state is the opposite of theocracy.

Sedentary
Permanently attached to a particular area, fixed in some locale. The opposite of **nomadic.**

Sequent occupance
The notion that successive societies leave their cultural imprints on a place, each contributing to the total cultural landscape.

Shantytown
Slum development on the margins of major cities, umplanned and mostly without any amenities; dwellings and shelters made of scrap wood, iron, even pieces of cardboard. Brazil's *favelas,* North Africa's *bidonvilles,* India's *bustees,* Mexico's **barrios.**

Sharecropping
Relationship between a large landowner and farmers on the land in which the farmers pay rent for the land they farm by giving the landlord a share of the annual harvest.

Shatter belt
Zone caught between stronger cultural-political forces, under stress and frequently broken up by aggressive rivals. Eastern Europe is the type region.

Shifting agriculture
Cultivation of crops in forest clearings, soon to be abandoned in favor of more recently cleared forest land. Also known as slash-and-burn agriculture and by various regional names.

Sinicization
Giving a Chinese cultural imprint; Chinese **acculturation.**

Site
The physical setting of a place, its local relief and other physiographic attributes.

Situation
The relative location of a place; its position with reference to other places and to resources and areas of productive capacity.

Slash-and-burn agriculture
The system of "patch" agriculture that involves the cutting down and burning of the vegetation in a part of the rain forest and the temporary cultivation of crops on the clearing. Also see *milpa* and **shifting agriculture.**

Stratification
Layering. In a stratified society, the population is divided into a **hierarchy** of social classes. In industrialized society, the **proletariat** is at the lower end; those who possess capital and control the means of production are at the upper level. In traditional India, the "untouchables" form the lowest caste and the still-wealthy remnants of the princely class are at the top.

Subsequent
A word that has specific meaning in physical geography—just the opposite of antecedent: the river formed *after* the geologic forces had created the main regional structure.

Subsequent boundary
An international political boundary that developed contemporaneously with the evolution of the major spatial elements of the cultural landscape in an area.

Subsistence
Existing on the minimum necessities to sustain life.

Supranational
Involving three or more national states: political, economic, or cultural cooperation to promote shared objectives. The European Common Market is a supranational organization, and in another area, so is OPEC.

Take-off
Economic concept to identify a stage in a country's development when conditions are set for a domestic industrial revolution, as happened in the United Kingdom in the late eighteenth century and in Japan after the Meiji Restoration; it is happening in China today and appears to be approaching in Brazil.

Tell
The lower slopes and narrow coastal plains along the Atlas Mountains in Northwest Africa, where the majority of the region's population is clustered.

Templada
Altitudinal zone in Latin America immediately above the tierra **caliente** and below **fria.** The elevation of this zone varies somewhat with **latitude** and exposure, but generally lies between about 750 meters (2500 feet) and 1700 meters (5500 feet).

Terracing
The transformation of a hillside or mountain slope into a step-like sequence of horizontal fields for cultivation.

Territoriality
A community's sense of property and attachment toward its territory as expressed by its determination to keep it inviolable and strongly defended.

Territorial sea
Zone of water adjacent to a country's coast, held to be part of the national territory and treated as a segment of the sovereign state.

Tertiary industries
Industries and activities that engage in services such as transportation, banking, and education.

Theocracy
State whose government is under the control of a ruler who is deemed to be divinely guided and inspired, or a group of religious leaders (monks, priests).

Tierra caliente
The lowest of three vertical zones into which the topography of Middle and South America is divided according to elevation. The *tierra caliente* is the hot, humid coastal plain and adjacent slopes up to between 750 and 900 meters (2500 and 3000 feet) above sea level. The natural vegetation is the dense and luxuriant tropical forest; the crops are bananas, sugar, cacao, and rice in the lower areas and coffee, tobacco, and corn along the somewhat higher slopes.

Tierra fria
The highest zone below the snow line in Middle and South America, over about 2100 meters (7000 feet) and up to around 3000 meters (10,000 feet). Coniferous trees stand here, and upward they change into scrub and grassland; there are important pastures on the *tierra fria.* Wheat can also be cultivated. Several major population clusters in the region lie at these altitudes. Above the *tierra fria* lies the *paramós,* the highest and coldest zone.

Tierra templada
The intermediate altitude zone in Middle and South America, between about 750 to 900 me-

ters (2500 and 3000 feet) and 1800 to 2100 meters (6000 to 7000 feet). This is the "temperate" zone, with moderate temperatures compared to the *tierra caliente.* Crops include tobacco, coffee, corn, and some wheat.

Totalitarian
A government whose leader or leaders rules by absolute control, tolerating no differences of political opinion.

Transculturation
Cultural borrowing that occurs when different cultures of approximately equal complexity and technological level come into close contact. In **acculturation,** by contrast, an indigenous society's culture is modified by contact with a technologically more developed society.

Transhumance
Seasonal movement of people and their livestock in search of pastures. This movement may be vertical—that is, into highlands during the summer and back to lowlands in winter—or horizontal, in pursuit of seasonal rainfall.

Underdeveloped countries (UDCs)
Countries that, by various measures, suffer seriously from negative economic and social conditions, including low per capita incomes, poor nutrition, inadequate health, and related circumstances.

Unitary state
A nation-state that has a centralized government and administration.

Urbanization
A term with several connotations. The proportion of a country's population living in urban areas is its level of urbanization. The process of urbanization involves the movement and clustering of people in towns and cities (it continues to affect most countries today). Another kind of urbanization occurs when a sprawling city absorbs rural countryside and transforms it into suburbs, developments, and squatters' settlements.

Urbanized area
The entire built-up, nonrural area and its population, including the most recently constructed suburban developments. Gives a better picture of the dimensions and population of such an area than the delimited municipality or municipalities that form its heart. The population of a "city proper" is always less than the entire urban area within which it lies.

Veld (also see *Highveld*)
Open grassland on the South African plateau, becoming mixed with scrub at lower elevations where it is called *bushveld.* As in Latin America, there is an altitudinal zonation into *highveld, middleveld,* and *lowveld* ranging from nearly 3000 meters in Lesotho to under 500 meters in the Swaziland "Lowveld."

Völkerwanderung
Literally, a "wandering of peoples." The word is German, and it has come into general use to identify the mass migrations in Europe following the disintegration of the Roman Empire.

Voluntary migration
Population movement in which people relocate as a result of perceived opportunity and its attractions rather than compulsion.

Water table
When precipitation falls on the soil, some of the water is drawn downward through the pores in soil and rock under the force of gravity. Below the surface it reaches a level where it can go no farther; there it joins water that already saturates the rock completely. This water that "stands" underground is *ground water,* and the upper level of the zone of saturation is the *water table.*

Author Index

Subject Index